Judi Margulis

4 00

ELEMENTS OF BIOLOGY

second edition

Charles K. Levy
Boston University

ELEMENTS OF BIOLOGY

second edition

ADDISON-WESLEY PUBLISHING COMPANY

Reading, Massachusetts
Menlo Park, California
London Amsterdam Don Mills, Ontario Sydney

This book is in the
Addison-Wesley Series in
The Life Sciences

Second printing, June 1978

Copyright © 1978, 1973 by Addison-Wesley Publishing Company, Inc.
Philippines copyright 1978, 1973 by Addison-Wesley Publishing Company, Inc.

All rights reserved. No part of this publication may be reproduced, stored in a retrieval system, or transmitted, in any form or by any means, electronic, mechanical, photocopying, recording, or otherwise, without the prior written permission of the publisher. Printed in the United States of America. Published simultaneously in Canada. Library of Congress Catalog Card No. 77–77746.

ISBN 0-201-04335-1
ABCDEFGHIJ-RN-798

Cover photograph of owl: Earl Kubis/
Tom Stack & Associates

IN MEMORY OF
LT. CMDR. MAURICE E. LEVY U.S.N.R.
AND
FIRST LT. MATTHEW J. LEVY U.S.A.F.R.

PREFACE

This text was written to provide a brief, systematic introduction to living phenomena for college nonscience students. Because of the diversity of living phenomena to be discussed and the variety among nonscience students, this task is a challenging one. Within the limitations not only of space but of time, as dictated by the one-semester course for which it is intended, this text introduces students to the full range of phenomena in the living world and our knowledge of it.

I have tried to achieve a balance between descriptive phenomena of organisms and mechanistic explanations of those phenomena. I have sought to balance the discussions of procaryotic and eucaryotic cells, of plants, animals, and man. Perhaps most important is balance in seeing the big picture above the small details. I have made a constant effort to ensure that the concepts are not obscured by details. However, enough details are presented to support each general concept.

In my view, too many textbooks for the nonscience student have been simply watered-down versions of larger texts. Often difficult to read, they have tended to offer what is familiar to the instructor rather than what is interesting and useful to the student. Some books are superficial in their coverage and consequently condescending and patronizing in their tone. Others have been very selective in the phenomena they discuss. In this book I have tried to provide a balanced presentation, directed specifically at the nonscience student who comes to biology with uncertain interests and weak background preparation, especially in chemistry.

An early elementary discussion of chemistry is an attempt to overcome the most important problems that introductory students have with this material. I try to make it clear that without this understanding, it is really impossible to explore and appreciate the complexities of life processes as science presently understands them. One simply cannot participate in the excitement of recent discoveries in biology without understanding some chemistry.

There are several keys to creating a text that is accurate, comprehensive, balanced, readable, and appealing to the imagination. The first of these keys is organization. The book has been organized around functions common to all living systems. Within each chapter, unity of process or function and diversity of forms are given equal emphasis. Again I hope I have achieved a balance that is realistic for student and instructor.

A second key is the language used to inform and involve students. The introductory course of any discipline inevitably opens up an entirely new vocabulary, and biology is no exception. I have been careful to avoid unnecessary technical terms, bearing in mind that the familiar word that does the job is not to be scorned. An equally important aspect of language is its ability to excite. I have tried to instill through the language used in the text some of the flavor of probing the unknown.

Third, having taught biology on varying levels to students of varying interests has given me an appreciation for areas that need patient explanation and clear, simplified illustrative material. Consequently, the coverage of some subjects has been expanded, some illustrations have been added, and others have been altered.

CHANGES AND NEW FEATURES

The second edition of *Elements of Biology*, in addition to being thoroughly updated in its contents, has a number of new features.

In accordance with a perceptible classroom trend in recent years toward emphasizing descriptive biology, the discussion of taxonomy and diversity of the living world has been expanded to three chapters in the second edition. A full-color portfolio stressing the diversity of life is entirely new, as are three additional photographic inserts, each focusing on an important unifying theme.

In the interests of bringing the course closer to student experience, a number of essays on human biology have been added in appropriate places. Included are such topics as diabetes, recombinant DNA, and the effects of exercise.

At the end of each chapter is a summary of the chapter's contents. This is a valuable study aid and can be used before as well as after reading the chapter. There is also a list of references and suggested supplementary readings. New review and discussion questions are designed to help the student understand the principal focal points in the chapter and apply them to both everyday and imaginary situations.

New terms are printed in boldface and defined when first encountered in the text. Full definitions appear in the extensive glossary at the end of the book. For each entry the pronunciation in simplified form is also given.

To the extent possible, legends accompanying illustrations are complete and self-contained. That is, the student examining the illustration has direct access to its full explanation in the legend.

SUPPLEMENTS

Accompanying the text are a study guide, an instructor's manual, and test items. The study guide is designed to help students interrelate concepts from various sections and chapters and to review the important points of the text. For each chapter of the text, it includes questions for study, discussion, and application.

THE LANGUAGE OF BIOLOGY

On a matter that today calls for increasing sensitivity, I have made a considerable effort to avoid sexist language, though sometimes with only limited success because of the special problem biologists face. Nobel prizewinner S. E. Luria has so lucidly expressed the problem that I can do no better than incorporate his statement here.

> In deference to the well-justified concern for equality between sexes even in the relatively trivial matter of the use of words, I tried to avoid the use of "man" and "his" as referents to the human species as a whole but found no way to do so without being bogged down in the clumsiness of alternative expressions. So "man" it is, in the deplorable sexist tradition of the English language, a tradition that proves hard to overcome. "Man" stands for the species—that is, about equal numbers of women and men. An understanding of biological facts and an awareness of the depth of biological interdependence

> among all members of a species may help foster the recognition of the justness of social equality between members of the two sexes and among all groups of mankind.*

ACKNOWLEDGMENTS

No one author can have complete grasp of such a broad field as biology, and four of my colleagues at Boston University were kind enough to help in their areas of special competence. I am particularly grateful to Javier Penalosa for his contribution in expanding and modifying the chapters on taxonomy; to William Feder for his help in botany; to Don Patt for guidance in covering genetics; and to the late Frank Belamarich for his suggestions on cell chemistry, cell biology, and metabolism.

I would also like to acknowledge the fine work of a number of reviewers. Some went over the entire first edition and made many valuable suggestions and corrections. Others reviewed parts of the revision and gave me the benefit of their expertise. My sincere thanks to Albert Baccari, Montgomery County Community College; Walter C. Biggs, University of North Carolina; Richard K. Boohar, University of Nebraska—Lincoln; Wayland L. Ezell, St. Cloud State College; John R. Jungck, Clarkson College; Laurence A. Larson, Ohio University; William A. McEowen, Mesa Community College; Barbara Raisbeck, formerly of Northeastern University; LeRoy J. Scott, St. Petersburg Junior College; William Stockton, California State University—Long Beach; and Tommy E. Wynn, North Carolina State University.

Thanks also go to several persons whose special talents helped to bring this book and its supplements together: Meredith Nightingale for her enthusiastic efforts to provide superior photographic materials; Linda Fritz, Mary Pratt, and Ronald Waife for their help in translating biology into English prose; Candace Kasper for a researcher's thoroughness; Alice Holtz for the care and imagination she brought to the assembling of materials for the study guide; and a sizable group of able and cooperative editorial and production staff members at Addison-Wesley.

Boston, Massachusetts
August 1977

C. K. L.

* S. E. Luria, *Life, The Unfinished Experiment* (New York: Scribner's, 1973), pp. 4–5.

CONTENTS

part one THE FUNDAMENTALS OF LIFE 1

THE STUDY OF LIFE 3

ATOMS, MOLECULES, AND CHEMICAL BONDS 13

THE CHEMISTRY OF LIFE 31

part three

DIVERSITY OF LIFE: INTRODUCTION 411

DIVERSITY OF LIFE: PLANTS AND FUNGI 431

DIVERSITY OF LIFE: ANIMALS 457

PRINCIPLES OF ECOLOGY 477

ZONES OF LIFE 503

THE FUNDAMENTALS OF LIFE

part one

THE STUDY OF LIFE
chapter one

Four and a half billion years ago the Earth, a molten rock spinning through space, began to cool. Water vapor condensed to rain, which by steady pounding washed minerals from the rocks. The rain-water carrying minerals ran into rivers and into great depressions that formed the oceans, where living things gradually originated.

Four and a half billion years ago the Earth, a molten rock spinning through space, began to cool. Water vapor condensed to rain, which by steady pounding washed minerals from the rocks. The rainwater carrying minerals ran into rivers and into great depressions that formed the oceans.

In the ocean, living things gradually originated—untold billions of substances accumulated, a few of which came together into living assemblies. They changed, slowly at first, then more rapidly as these simple organisms themselves changed their surroundings. From the odd molecules and their emerging assemblies, which some have termed the "primeval slime," there arose a vast number of diverse plants and animals. Most of these were unable to adapt to the constantly changing environment and became extinct. But some of their offspring had changed, or mutated, just enough to survive within the existing environment. This slow process of extinction of the unfit, chance mutation, and survival of the fittest constitutes what we term "evolution."

Through evolution, there eventually arose an erect, manipulative, vocal, and very curious creature—*Homo sapiens*, conventionally called man. To date, man has been the only organism to exhibit secondary evolution—that is, the ability to accumulate knowledge, to create complex tools, and to harness and use energy. In time, man even developed an ability to travel to another world in outer space in search of other living things. But attaining this goal took years of thought and research—research that began with close and thorough study of the obvious.

THE WORLD WE SHARE

One area of human research began with a close examination of the obvious fact that man shares the planet with many other living plant and animal species. From this observation man was rather easily able to see that if many other things appear to be as alive as he is, then those things must share some characteristics with the human organism. Second, he assumed that the study of other living things, given their degree of commonality with the human species, can yield highly valuable information on matters concerning man himself.

Third, in view of the human fondness for organizing and systematizing whatever initially appears to be without order, the wide diversity of living things stimulates man's intellect, almost compels it, to somehow gain control through understanding of this huge assortment of life. Indeed, having carefully observed the shades of similarity and differences among plant and animal life, man has succeeded in establishing for himself an extraordinarily developed system of classification in which most of the plant and animal species on this planet have been described. Man has thus come to the conclusion that there is an underlying order and even unity among living organisms.

Finally, a respect for living things comes from observing that among the living there are several million species of plants and animals in every available habitat. And just as remarkable, each species is the product of a continuous and cruel competition for survival in an ever-changing environment.

Characteristics of life

It is not surprising that signs of life were the first concerns of moon-bound scientists. Indeed, biologists have traditionally been concerned with identifying the common characteristics of living things as a means for distinguishing them from nonliving matter. However, the search for such unifying factors has not been easy, because there is great diversity among living things and, furthermore, because some nonliving things possess some characteristics comparable to those of certain organisms.

Living things always display at least five characteristics that are essentially unique to them. Briefly, the recognizable characteristics of living matter are the continual maintenance of some structure, the performance of metabolic functions, irritability or response to stimuli, reproduction, and the potential ability to evolve.

STRUCTURE

The structural diversity of living organisms is enormous. Nevertheless, all living things are composed of basic structural units called cells, and each cell is composed of molecules that are further subdivided into atoms. Although the same molecules and atoms exist in both living and nonliving matter, in most nonliving matter they tend to form random combinations or simple repeating patterns. In living cells, however, the atoms and molecules combine in very specific and complex arrangements. This complexity appears even in unicellular organisms, in which the entire anatomy is limited to a single cell. The microscopic examination of such an organism shows a highly complex structure in pronounced contrast to the structure revealed by a similar examination of a drop of pure water or a speck of dust.

Multicellular organisms are even more complex. They are composed of many cells, some of which are highly specialized to perform particular functions. In the higher organisms, the specialized cells combine to form tissues and organs. Thus the complexity of an organism's overall structure may be readily understood in terms of the number and arrangement of its cells. Although most cells contain internal structures specialized for the efficient and rapid performance of given tasks, it is the cell itself that is the basic functional unit common to all living things.

Interestingly enough, this generalization has led to some controversy. There is a category of things, the viruses, which lack most of the components found in even the most primitive cells, but which appear in many ways to be alive. Thus it is probably safer to say that the cell shows the minimum degree of structure that constitutes an organism, but there are also viruses, and they may possibly be an exception to this generalization.

METABOLIC ACTIVITY

The maintenance of structure requires that an organism obtain substance and energy from its environment. It accomplishes this goal by means of two biological processes. The first is nutrition, the acquisition by the organism of nutrients from the environment. In the second process, **metabolism,** the organism breaks down the nutrients through respiration or fermentation. It is from the metabolism of nutrients that energy is released to be used by the organism to perform its many functions, all of which have evolved under selection pressures and which work to maintain organization.

IRRITABILITY

All organisms must respond to alterations in their environment if they are to remain alive. In short, organisms must be able to maintain, within optimum ranges, their special internal environments, despite variations of external conditions. In order to survive, organisms must be capable of responding to changes of such environmental stimuli as light, temperature, and the concentration of chemicals in the environment. Unicellular organisms perform these necessary responses within their single cells. In the most advanced multicellular organisms, this characteristic of irritability is the function of special cells and organs. Examples are the cells that bend a leaf toward the light, or receptors that detect signs of environmental change and activate nerve cells to send the signals and to assemble them into some meaningful piece of work.

REPRODUCTION

No living thing continues to exist forever, nor does time affect all organisms equally. Some, like the bacterium, have a life span of hours; some, like the sequoia, live for thousands of years. Although all individual organisms must eventually die, the extinction of a species is usually avoided by the production of offspring by individuals prior to death.

Reproduction can occur at different levels, in accordance with the complexity of the organism. However, the process always depends on some form of cellular division, accompanied by the passing on of traits of the original cell to both of the new cells that result. Unicellular organisms grow to some size and divide to form two daughter cells. These can grow to form two adult unicellular organisms identical to the parent. Some multicellular organisms are dependent on two parents and are said to reproduce sexually, but here, too, cell division is required to produce the special sex cells, or gametes, from which the new individual arises. Each new generation of organisms thus reproduced is a continuation of the particular species of its parents.

ABILITY TO EVOLVE

Even though a species continues to survive through reproduction, no species—not even man—is unchanging. Through evolution, each is continually adapting to its environment. This process occurs because organisms within a species differ in form, structure, and physiology. An example of this variety, known as *phenotypic variation,* is the size of human limbs in different parts of the world.

Within an environment, each variation of a species has a different survival rate, known as *differential fitness.* For example, people living in icy areas were best able to survive the cold weather if they had short limbs, which reduced skin area and therefore the amount of heat that could escape from the body. People in hot, dry areas were better off with longer limbs, which increased skin area and thus heat loss. Since heat escapes from the body through sweat, which cools the body as it evaporates, exposed skin is an advantage in East Africa, a disadvantage in Alaska. In each area, those with the limbs best adapted to their particular environment have had a better chance of survival and thus have been more likely to reproduce. Because offspring inherit characteristics relating to body size, the trait is passed down to future generations. Those with modifications that are best adapted to their existing environment are the most successful at survival and most likely to reproduce. Through heredity, the trait is passed on, and the entire species gradually evolves.

Vitalism versus mechanism

Characteristics of life—structure, metabolic activity, irritability, reproduction, the ability to evolve—merely describe life; they do not define it. To define it brings one into the philosophical debate of vitalism versus mechanism.

Throughout history the vitalist has contended that some special life-determining substance, or *vital force,* differentiates the living from the nonliving. That force drives all forms of life to survive and to adapt to new conditions within the environment. Unable to accept the hypothesis that life began from lifeless molecules, the vitalist points to science's inability to create life in a test tube as proof of a differentiating force between the living and the nonliving.

At one time, the vitalist claimed that man could not duplicate in a laboratory any of the chemical operations performed by living tissues. Indeed, the vitalist felt that the creation of any and all organic matter was beyond the capacity of man. However, he was forced to abandon this position when Friedrich Wöhler in 1828 converted inorganic matter into organic urea. Subsequently, even the chemistry performed by living tissues proved a laboratory possibility, when Eduard Büchner late in the nineteenth century fermented fruit juice into alcohol without the presence of living yeast cells.

Vitalism has not disappeared completely, but today's vitalist sees the special "force" not as a mystical substance but as a property—the organization and complexity of chemical matter in living things—that makes the whole organism more than the sum of its parts. Thus the modern-day vitalist, more commonly referred to as an organicist, considers the entire organism as a unit of life.

Whereas vitalism in one form or another has always been a part of the history of ideas, the opposing philosophy of mechanism dates back only a few centuries. Some scientists in the seventeenth century began to suggest that living things operate like machines and are, in essence, no more than machines. Arms and legs, for example, function like levers and pulleys. This notion was gradually modified into the nineteenth-century position that all living things, like all nonliving matter, can be reduced to chemistry and physics. The discoveries of Wöhler, Büchner, and others confirmed that chemical and physical principles of inorganic chemistry do indeed apply to living organisms. And yet neither side can be said to have the whole truth.

The argument that remains between the vitalist and the mechanist is an argument grounded less in scientific evidence than in philosophy. The vitalist sees the living system as a complete entity, uniquely different from nonliving matter. The mechanist sees life as a chain of physical and chemical reactions occurring in the basic molecules. This split is the basis for the controversy between—to use today's terminology—the organicist and the molecular biologist.

The scientific method

The aim of any science, whether biology, physics, sociology, or chemistry, is primarily to establish principles and thereby to acquire knowledge. This task can be difficult since we often cannot understand events occurring in the world around us. So many variants are operating at one time that it is usually impossible to determine what is causing an event. Indeed, variability might be included as a characteristic of life. These varying factors are so interrelated that changes in any one of them are likely to affect the nature of the others. Moreover, the human mind is itself highly variable and capable of self-delusion and unconscious bias, and thus the facts that are perceived can be distorted.

For these reasons, it is not enough for science merely to gather information; the information must be reliable.

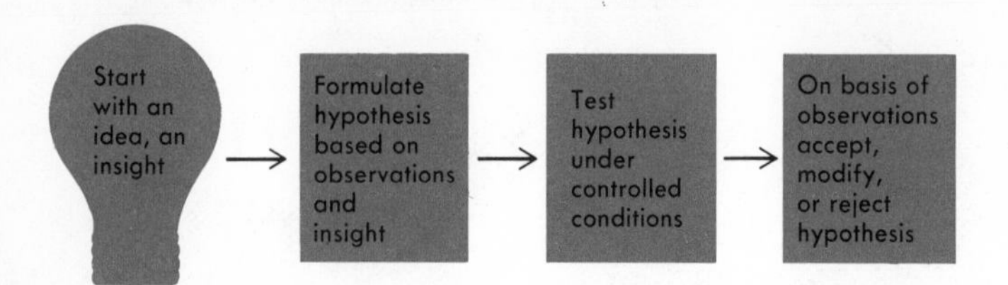

FIGURE 1.1
The scientific method.

Hence science must systematize a way of amassing knowledge. It is this systematization—known as the scientific method—that marks the essential difference between a science and any other discipline (Fig. 1.1).

A scientist may begin a study with an idea or question. It may be a brillant insight made following a casual observation directly out of daily life. Or it may develop from observations made in previous laboratory work. On the other hand, the scientist may not observe anything that triggers an idea until after beginning to investigate a field of study. Typically, before beginning to study a question of interest, the scientist examines the research already done and the data obtained. To accomplish this task, he or she may read the appropriate journals, attend meetings at which scientists present research papers, and perhaps even acquire the skills and techniques of computerized research. At the very least, this procedure lessens the possibility of duplicating efforts of others, thereby avoiding the wasteful rediscovery of previously understood phenomena. Since negative results are also reported, it additionally prevents the scientist from repeating others' mistakes.

After becoming familiar with previous research in the intended field of study, the scientist is prepared to make fresh observations of conditions and phenomena within the field. It is on the basis of these observations, as well as the information obtained from the studies of others, that the scientist builds a generalization, or a **hypothesis.** In the hypothesis, an attempt is made to explain the observations or offer tentative answers to questions.

To test the hypothesis, the scientist designs an experimental framework in which one of the variables (or factors) can be independently altered, so that the effect of the changes in the variable can be examined. For example, he may measure the effect of the variable temperature on the rate of reproduction of a bacterium. In order to be valid, the biologist's research must meet at least two requirements: (1) there must be evidence that only the one experimental variable is responsible for any observed changes, and (2) the results of the experimentation must be reproducible.

The scientist meets the first requirement by using what is known as a **control group** (see Fig. 1.2). The control group consists generally of a group or setup similar to the experimental group, but one in which the experimental variable is never manipulated. For example, a biologist investigating the effect of heat on the reproductive rate of bacteria would vary the temperature in the experimental group but maintain a constant temperature in the control group. If similar observations are made in both the experimental and control groups, the observed changes in both groups can be considered as due to random changes in the organisms. But if the experiment produces data significantly different from the control data, there is basis for assuming that they are the result of changes in the experimental variable.

How can a scientist be reasonably certain that the results obtained are reliable? The safest way is to repeat the experiment several times and, in addition, make certain that the work is reproducible by other scientists. In order to ensure reproducibility, the scientist must precisely record the techniques employed in the experiment, as well as observations made, so that if the experiment is conducted exactly the same way again and again, statistically similar data will be obtained.

Finally, the scientist draws conclusions from the data obtained. Do the experimentally derived data support the hypothesis? To what degree of statistical probability? That is, in what percentages of cases can we expect the hypothesis to hold true? If the hypothesis is supported, the scientist reports the results of the experiments to colleagues, who critically examine the hypothesis and the data. If they feel the hypothesis is supported, they may base their own work in the field on it. The hypothesis may then attain the status of or contribute to a **theory,** or broadly accepted generalization. If the data do not support the hypothesis, the scientist modifies it or abandons it and seeks another.

This procedure of observing, formulating a hypothesis, and testing it experimentally, in order to confirm, modify, or reject the hypothesis, is only a generalized method, which science uses to develop and systematize reliable information. It does not describe how an individual scientist approaches a given problem. It also does not describe how any scientist achieves the creative insight that makes possible the formulation of a hypoth-

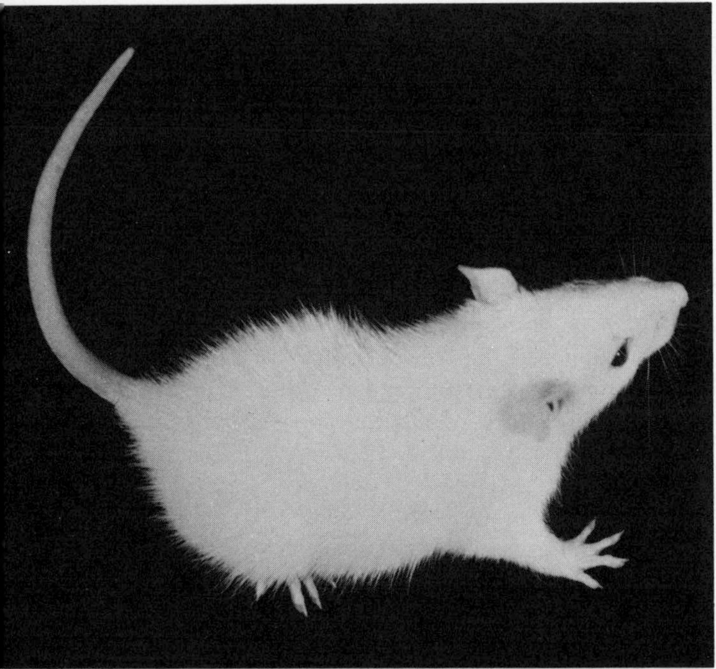

FIGURE 1.2

Dietary experiment on trace-element deprivation in rats, illustrating the use of a control (top) for comparison with the experimental subject (bottom). Both animals received essentially the same diet, except that tin, vanadium, and fluorides were excluded from the food of the experimental rat for 20 days. The later addition of tin and vanadium alone did not restore normal growth, but when potassium fluoride was added to the diet as well, the experimental rat returned to normal health.

esis. In a general sense, however, the scientific method does describe how science proceeds and how any given hypothesis will be judged over time.

It is through use of the scientific method that scientists have arrived at the theory explaining the emergence of life and the diversity of organisms. This theory contends that living things did not acquire their various adaptive qualities spontaneously but that these arose gradually, over extended periods of time. The process of slow, irreversible adaptation to new quantities or qualities in the environment has been termed **evolution.**

Evolution: the key to diversity

The property of irritability enables the individual organism to respond and adjust to short-term environmental changes. But considering the harsh realities of the Earth's environments, where life is always potentially threatened, one must necessarily assume that organisms possess a certain long-time flexibility of adaptation, and that species respond to overall changes in their environment with changes in their structure and function. How else could organisms accustomed to living in very cold climates produce generations of offspring that were increasingly able to tolerate warmth? Surely this is what must have happened when the great glacial blocks of ice receded from parts of northeastern America. Such changes in the overall characteristics of groups of organisms are the result of evolution. For evolution produces *heritable* changes, not in the parents of a given generation, but in their offspring and their offspring's offspring.

Some would say that the age of modern biology began with the theory of evolution by natural selection, and it is therefore little more than a century old. In 1859, Charles Darwin, a British naturalist, published *On the Origin of Species*—an extremely important work on his observations of species differences and the changes that, over incredibly long periods of time, produced these different species. He argued that the changes were the result of random, heritable alterations of some organisms in a population.

FIGURE 1.3

Evolution sometimes results in unusual associations between organisms. As the clownfish swims among the poisonous tentacles of the sea anemone, it remains safe from predators. When feeding, it may drop scraps of food the anemone can eat. Such associations arise through the process of natural selection. They are perpetuated because of the survival value to both species.

To account for the fact that many features of organisms appear uniquely suited to meet specific needs, Darwin proposed the concept of **natural selection.** According to this concept, random differences that exist among members of a population improve their prospects for survival and reproduction. These differences, being heritable, may be passed on genetically to future generations. The alterations in appearance or function may be slight with each generation, but over many thousands of generations, slight differences gradually accumulated to bring about substantial changes in the organisms bearing them—and of course, eventually in the species.

Although species have evolved and are presently evolving, it must be emphasized that this evolution of traits and characteristics did not and probably will not involve any abandonment of the basic properties of life. Thus, after millions of years of evolution, all living things still possess the basic qualities of distinct structure, metabolic activity, irritability or response to stimuli, reproduction—and the continuing ability to evolve. In short, although organisms may have evolved differences in these qualities—differences in structure, metabolic processes, and the manner in which they reproduce and respond to stimuli—they have not lost any of these basic characteristics of life. Indeed, the potential for evolutionary change itself is considered an additional property of life, the very property responsible for its vast panorama.

Summary

1. Through science man gazes on his origins and his place in the universe, taking the "obvious" as a starting point and discerning the underlying order and unity. Biology is the science of life.

2. Life is characterized by the continual maintenance of structure, with the cell as the basic unit. The integrity of the structure is maintained through the processes of metabolism. Irritability permits living things to preserve optimum metabolic conditions by responding to changes in the environment. Reproduction allows the structure to be duplicated, thereby perpetuating the species. The ability to adapt to the environment results in the gradual change of organisms over long periods of time into new species by the process of evolution.

3. The vitalist—today's organicist—believes that a special force or property distinguishes the entire living organism from nonliving matter. The mechanist, whose philosophical descendant is today's molecular biologist, believes that all matter, living as well as nonliving, is reducible to common principles of chemistry and physics. Basically, the positions of the vitalist and the mechanist differ in philosophy rather than in supporting scientific evidence.

4. Systematic knowledge about the living world is acquired through the scientific method of observation, hypothesizing, experimental testing, and confirmation, modification, or rejection of the hypothesis.

5. Through the use of the scientific method biologists have arrived at a general theory to explain the diversity of life: the theory of evolution. This theory contends that living things gradually diversified over extended periods of time through random heritable changes that enhanced their chances for survival and procreation. The principal mechanism of evolution is natural selection.

REVIEW AND STUDY QUESTIONS

1. What characteristics of life are cited in this chapter? Are there others?

2. Does a rock show any of the characteristics of life? A river? A candle flame?

3. Imagine that you are a diehard vitalist in 1828, learning that Friedrich Wöhler has converted inorganic matter into organic urea. What defense of your position might you offer?

4. Devise a simple hypothesis and explain how you would test it scientifically.

5. Give an example of a question that cannot be answered by the scientific method, and explain what makes it unanswerable.

6. What does natural selection mean?

SUGGESTIONS FOR FURTHER READING

McElroy, W. D. 1971. "The Role of Fundamental Research in an Advanced Society." *American Scientist* 59: 294–297.

The public attitude to science is changing because of increased awareness of health hazards, pollution, and technological advances related to research. Dr. McElroy suggests a "take five" period to assess the influence of these attitudes on the future of science.

Snow, C. P. 1958. *The Search.* Scribner's, New York.

Both a scientist and a novelist, Lord Snow vividly portrays the atmosphere of the laboratory in this exciting story of a young English researcher on the verge of a major scientific breakthrough.

Stent, G. S. 1972 (December). "Prematurity and Uniqueness in Scientific Discovery." *Scientific American* Offprint no. 1261. Freeman, San Francisco.

A molecular geneticist reflects on two general historical questions: (1) What does it mean to say a discovery is "ahead of its time"? (2) Are scientific creations any less unique than artistic creations?

Watson, J. D. 1968. *The Double Helix.* Atheneum, New York (paperback, Signet, 1969).

An exciting and even suspenseful narrative by the Nobel prize-winning participant, about the fundamental research breakthrough in molecular genetics, the discovery of the structure of DNA.

Weisskopf, V. F. 1972. "The Significance of Science." *Science* 176: 138–146.

The author interprets the attitudes to modern science from three positions: the need for basic science; the limitations in science; and the intrinsic value of science.

ATOMS, MOLECULES, AND CHEMICAL BONDS

chapter two

No one will ever know with certainty what the surface of the primitive Earth looked like, but this artist's conception includes the components about which there is common agreement.

How did life begin? For centuries man has comtemplated, theorized, and philosophized about this question. From ancient to modern times, the belief that all forms of life were divinely created has been a part of many cultures. On the other hand, for most of the same period, strong support came from many sources for the conflicting idea of **spontaneous generation**—the idea that living things arise directly out of nonliving matter. Despite efforts of imaginative pioneers of science over the centuries, it was long impossible to dislodge the belief in spontaneous generation. The popular notion was strengthened by such common observations as maggots appearing in spoiled meat and lizards crawling out of mud. In fact, spontaneous generation was not decisively refuted until the middle of the last century, notably by the eminent French scientist Louis Pasteur in a series of remarkable experiments. Yet, ironically, a variation of the concept of spontaneous generation not only has reappeared in this century, but has come to dominate scientific thinking about the origin of life. The very important difference in the current version is the hypothesis that life arose from nonliving matter only at its very beginning, billions of years ago, under atmospheric conditions that were very different from those of the present.

THE CURRENT HYPOTHESIS

The theory that most scientists today accept was first set forth in 1924 by Alexander Oparin, a Russian biochemist. Oparin hypothesized that many millions of years ago, living things chemically evolved from inorganic gases, including water vapor, in the primitive Earth's atmosphere. He proposed that energy from sunlight, lightning, and heat activated very simple molecules in the atmosphere, and in strict accordance with the laws of physics and chemistry, more complex organic compounds resulted. Condensing and dissolving into the Earth's primitive oceans, these organic compounds provided the building blocks for the evolution of simple living organisms.

A few years later, the English biochemist J. B. S. Haldane independently developed and published a similar theory. For years the theoretical position taken by Oparin and Haldane was highly controversial, and it seemed likely to remain so, since there was little prospect of either proving or disproving it experimentally.

In 1953, however, important support for their theory came when Stanley L. Miller, then a graduate student at the University of Chicago working under Nobel prizewinner Harold Urey, conducted a now classic experiment. Working from Urey's proposition that the early atmosphere contained gaseous ammonia, methane, water, and hydrogen, Miller attempted to recreate that primitive environment in his laboratory. He used a simple apparatus (Fig. 2.1) equipped with electrodes for discharging electric sparks, which simulated lightning.

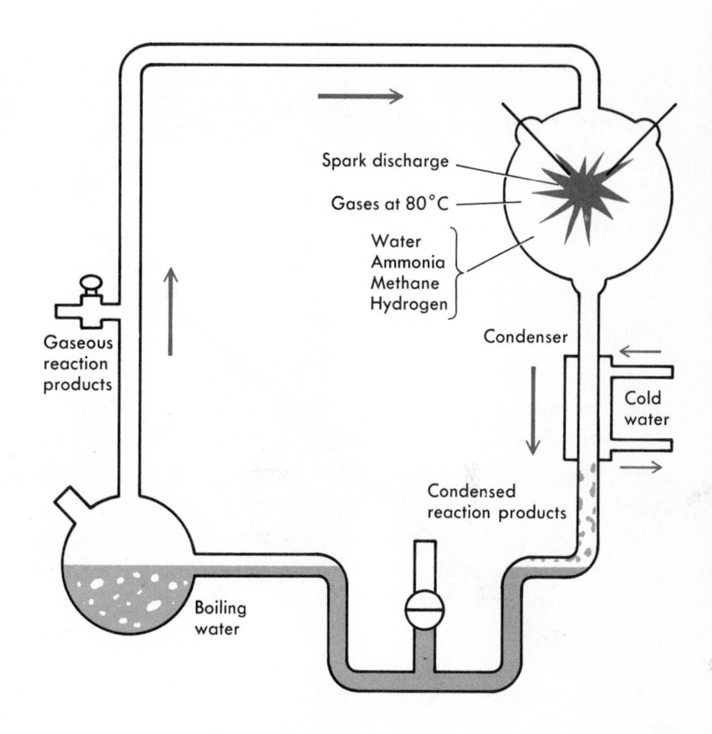

FIGURE 2.1

Laboratory apparatus simulating the primitive, lifeless environment on the Earth. The components were high temperature, electrical storms, and a gaseous atmosphere containing ammonia, water vapor, methane, and hydrogen. The apparatus was designed for the purpose of discovering whether the basic building blocks of living matter could form under such conditions.

In the apparatus, Miller circulated a mixture of ammonia, methane, hydrogen, and water vapor in the proportions in which they are thought to have existed in the early atmosphere. After exposing the artificial atmosphere to the electric sparks for several days, he analyzed the products. The results were astounding; included, among other things, were several amino acids—the building blocks of proteins, which are essential ingredients of all living systems.

The relative ease with which the amino acids were formed is of great significance. For if the conditions simulated experimentally did in fact prevail on the primitive Earth at a time before life appeared, life may then have arisen through processes similar to those carried out in this experiment. In many other experiments that have followed Miller's breakthrough effort, other forms of energy have been successfully used in the laboratory to create not only amino acids but other critical biological molecules. Thus it appears that no special obstacles would have interfered with the construction of the essential building blocks of life on the primitive Earth, given the amount of time now believed to have passed

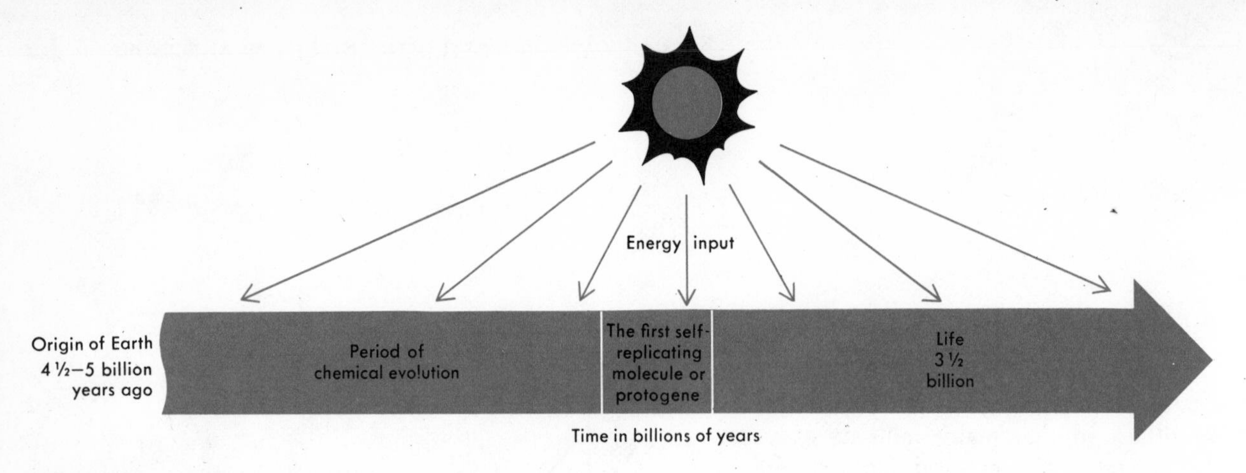

FIGURE 2.2

A possible time scale showing the origin of the Earth. Although there is no certain way of determining the age of the Earth or the origin of life, geological evidence permits scientists to make reasonable guesses.

since the formation of the Earth (see Fig. 2.2). The current hypothesis holds that these building blocks, which were still subjected to enormous inputs of energy, interacted to form larger and more complex molecules, including the essential proteins. Further experimentation has shown that proteins under certain conditions form membrane-bound structures that have many properties in common with living cells. Although no scientist has yet created life in the laboratory, these experiments have led many biologists to accept the idea that once air and ground conditions on Earth were generally suitable to support life, life was inevitable. Therefore, if the origin of life is a highly probable event when certain conditions prevail, there is also a strong possibility that on other planets in the universe similar events occurred (or will occur) leading to life. It is perhaps the search for other life that has led man toward the beginnings of his exploration of outer space.

Why did the chemical reactions occur in the mixture of gases in the laboratory or in the primitive atmosphere? How could electricity flashed through an inorganic chemical mixture produce a group of organic chemicals that were far more complex than the original parts of the mixture?

Essentially two major events occurred. First, the energy of the electric sparks was converted into chemical energy and heat. Second, the heat and chemical energy sped up the motion of the ammonia, methane, hydrogen, and water, causing their molecules to collide, break apart, and recombine into more complex substances—for example, amino acids. The original inorganic gases had within them a quantity of potential energy. Some other form of energy was necessary to bring about its release and transformation into **kinetic energy,** the energy of motion.

The process that occurred in the series of experiments performed by Miller and later investigators is typical of what goes on in living things. All organisms, whether they consist of one cell or many, are continuously converting energy. In essence, all life depends on the stepwise transfer of energy from one form to another.

If life processes were unable to direct energy into various forms, they would be considerably less efficient. No sooner would an organism finish one meal—using part of the energy content and wasting the rest—than it would have to start another. Instead, as we shall see, the cells can store energy in immediately usable forms.

In speaking of living things, then, we are speaking necessarily of chemical reactions. All things are made up of chemicals, but life has its own kinds of chemical reactions. Thus a basic understanding of chemistry in general—with particular emphasis on the chemistry involved in the life processes—is an invaluable asset for the understanding of modern biology.

The states of matter

All things, living or nonliving, are forms of matter, and all matter consists of chemicals. Matter is anything that has mass and occupies space. (Mass is best understood in terms of the amount of force needed to move an object.)

Portfolio 1: Variety in Cellular Structure and Function

1.1
A great many different mechanisms allow for cell movement. In this scanning electron micrograph, hamster kidney cells, which have been grown in tissue culture, are moving by means of the ruffling process.

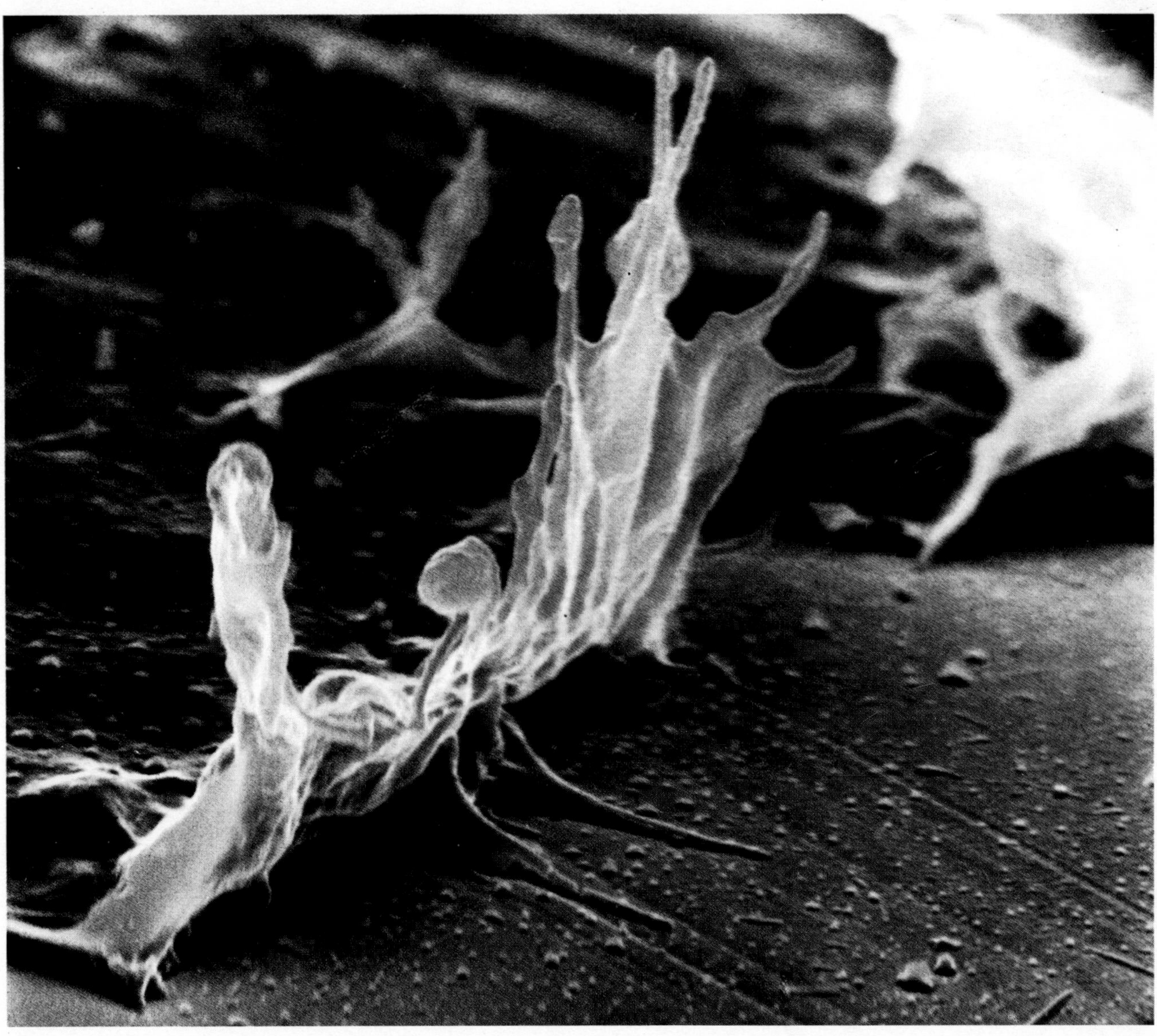

1.1 12,960X

1.2
This SEM of a vascular bundle from the leaf of the pitcher plant shows the variety of cell structures used to move water and photosynthesized carbohydrates through the plant.

1.3
The structure of a cell can be revealed by the scanning electron microscope. This cancer cell, grown in a laboratory, has many long thin projections, whose function may be related to the invasive capability of such cells.

1.4
In many ways this sunflower cell, seen in cross section, is a typical plant cell. Note the chloroplasts, the heavy cell wall, and the vacuole.

1.5
The internal architecture of a typical animal cell, shown in this TEM of a rat liver cell, reveals the complexity of the organelles and their numbers.

1.6
Starches—large polymers of sugar molecules—are stored forms of energy. The stores are packaged as large granules within membrane-bound sacs or vacuoles in the cytoplasm of a cell. Starch granules are clearly visible in this transmission electron micrograph.

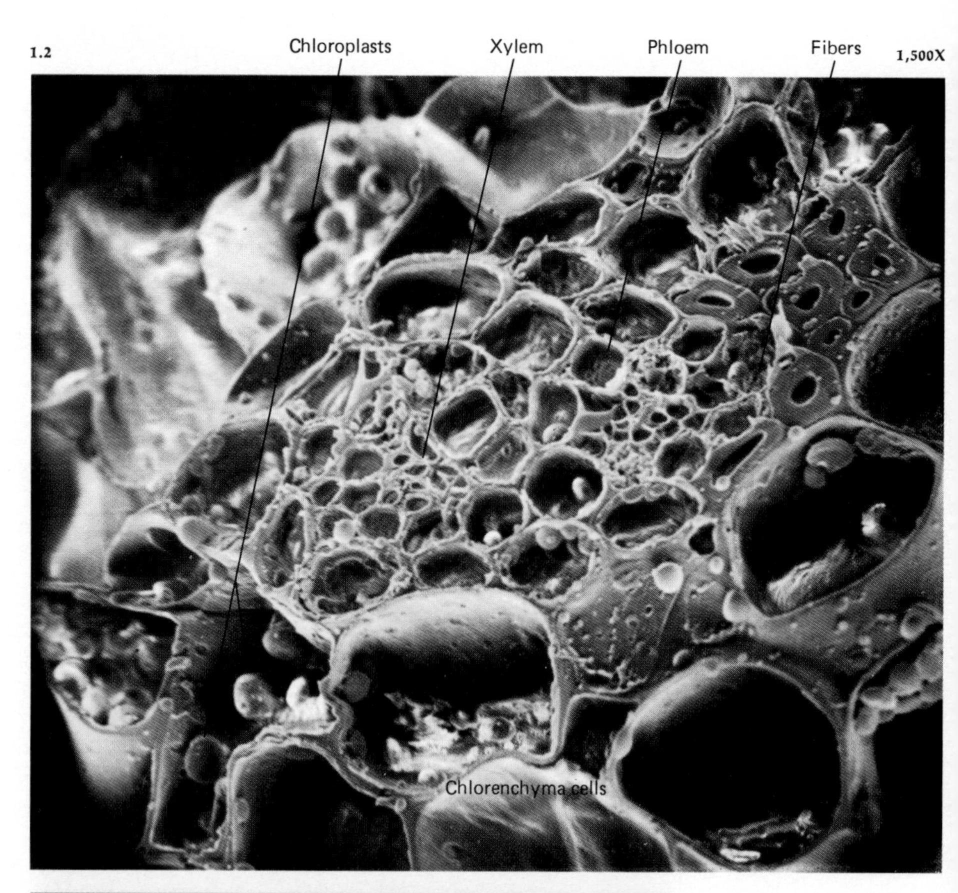

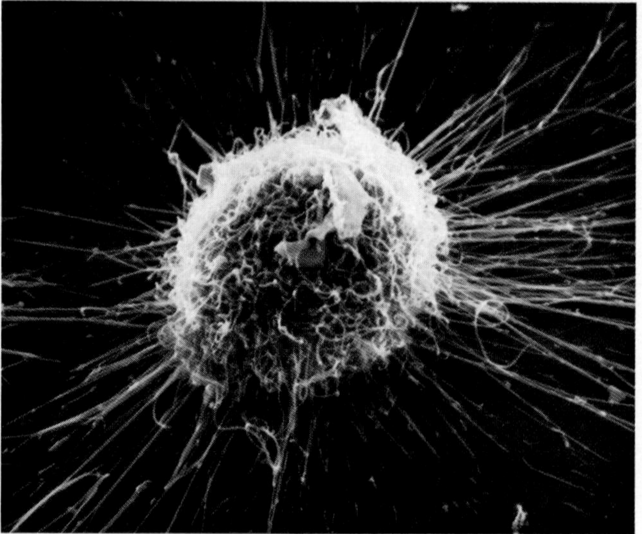

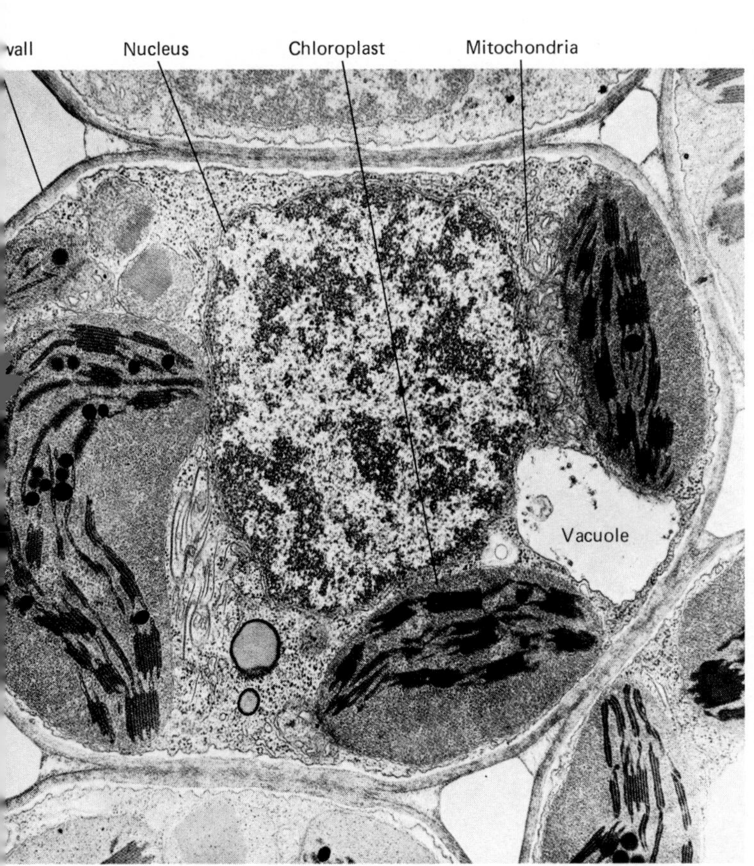

1.5

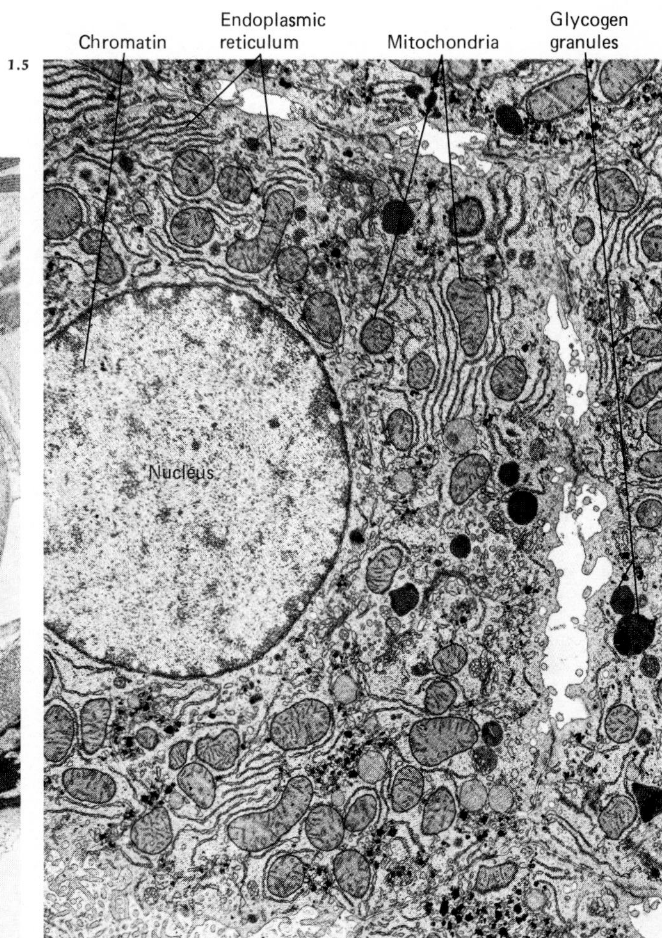

11,900X

Starch granules

1.6

11,000X

1.7

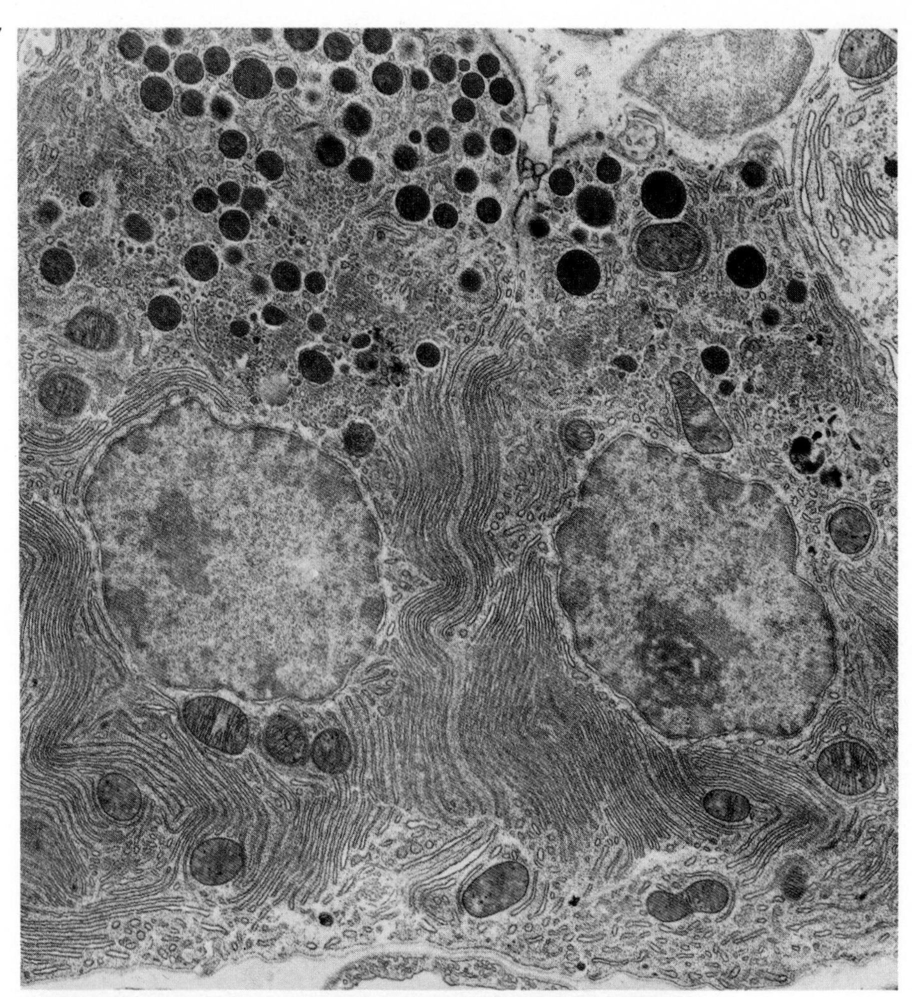

10,800X

1.8

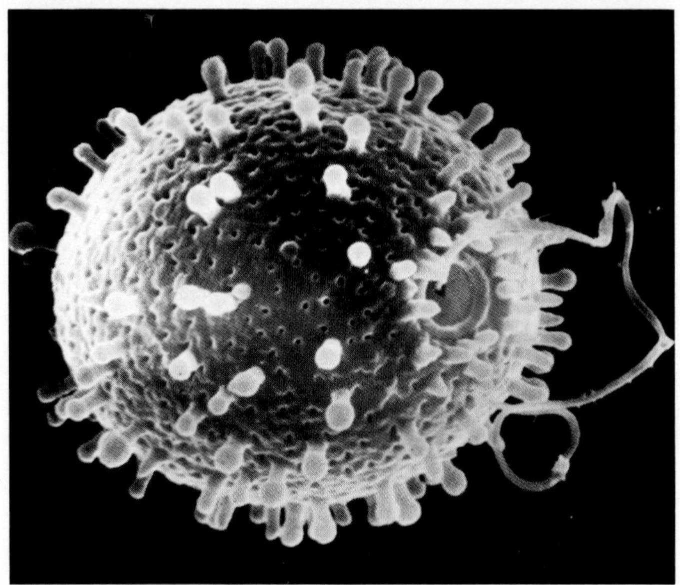

2,000X

1.7
Secretory cells typically have elaborate membranous synthetic machinery. This pancreatic cell has an extensive endoplasmic reticulum and dark-staining storage vacuoles crammed with zymogen granules (precursors of enzymes), which can be released as needed.

1.8
One-cell organisms show great variety in form and structure. This SEM shows the exterior of *Trachelomonas*, **a free-living euglenoid flagellate found in ponds and lakes, especially in the springtime.**

Despite its almost infinite diversity of form and shape, matter consists of chemicals in any one of three different physical states—solid, liquid, or gas. Water is an example of a form of matter readily existing in all three states, depending on its temperature. On a dinner table, for example, one may find ice cubes (water in its solid state) floating in a glass of water (the liquid state) while steam (water as a gas) rises from a nearby cup of hot coffee.

When two or more units of matter are physically combined, they form a mixture. One form of mixture, a **solution,** consists of very small particles, called the **solute,** scattered evenly throughout a fluid carrier, called the **solvent,** to form a homogeneous mixture (Fig. 2.3). When the fluid carrier is water, which is the commonest of all solvents, the mixture is called an **aqueous solution.**

Looking back at that steaming cup of coffee on the dinner table, we note that the cup holds an aqueous solution in which minute particles of roasted coffee bean (the solute) are distributed evenly in water to form a mixture that will remain stable. Sugar may of course be a second solute.

In some mixtures—for example, sand and water—the particles cling together in tiny clumps, larger than the individual molecules. Called a **suspension,** such a mixture requires continuous stirring if the clumps are to be prevented from settling to the bottom of the container.

Intermediate between the solution and the suspension is the third type of mixture, the **colloid.** In the colloid, the particles are too large to dissolve completely to form a solution but are not so large that they sink to the bottom. Instead, they tend to bounce off one another because of their inherent heat energy and because their identical electric charges cause the molecules to repel one another. The net effect is that the particles remain suspended permanently in a fluid state. When a colloid is in a liquid state, it is called a **sol,** whereas in a semisolid state it is a **gel.** Sols and gels change back and forth with changes in the concentration of the suspended particles or changes in the colloid's temperature.

The effect of temperature on matter illustrates how one form of energy can be transformed into another.

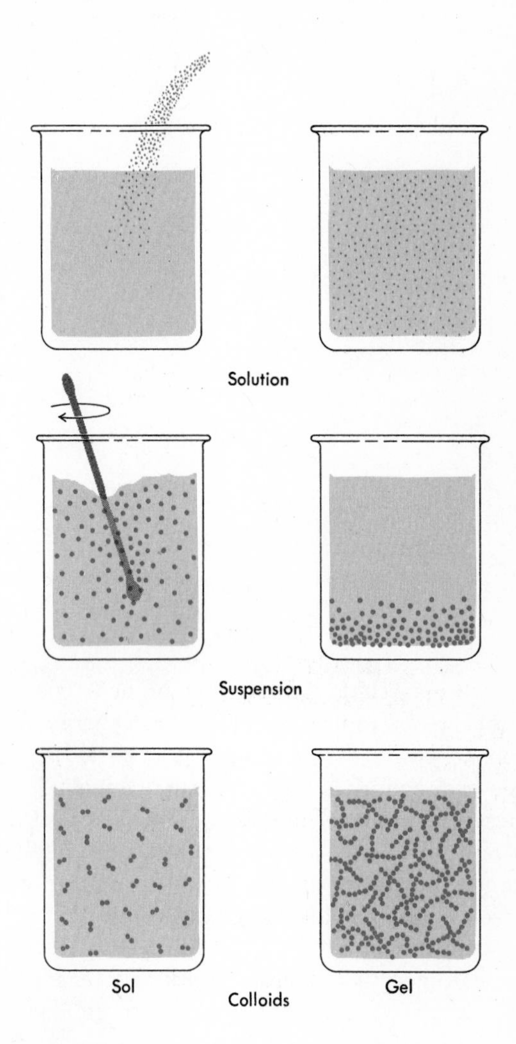

FIGURE 2.3

The three types of mixtures. In a solution, the small particles (solute) are scattered evenly throughout the carrier fluid (solvent) and will not settle out. In a suspension, larger particles will eventually settle out unless the mixture is stirred continuously. The particles of a colloid are intermediate in size and do not settle out readily. When the particles in a colloid are suspended throughout a continuous liquid medium, the colloid is in a sol state. However, under proper conditions the particles can form a more or less rigid suspension with the continuous medium; the colloid is then in the gel state.

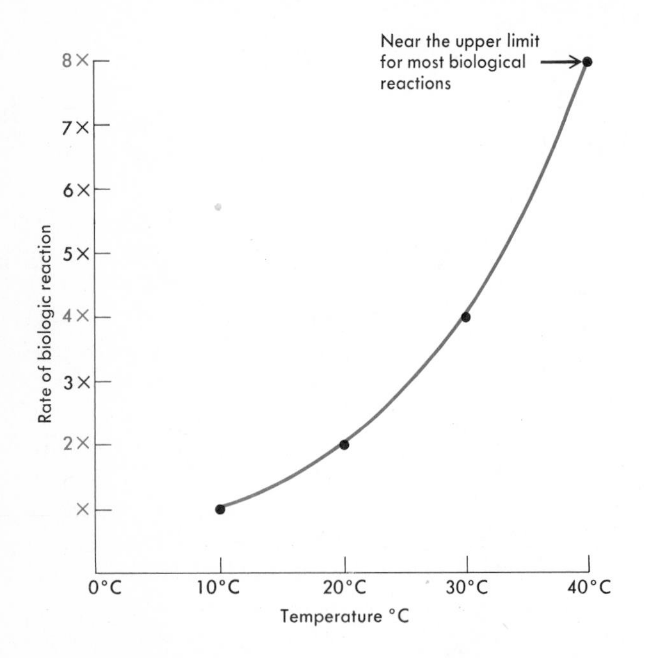

FIGURE 2.4

The influence of temperature on biological reaction rates. As temperature increases, energy is imparted to reacting molecules, and reaction rates typically double, as indicated here, or triple with every rise of 10°C. Most biological reactions occur within the narrow temperature range shown, 0° to 40°C. Above 40°C, molecules that are critical to life, particularly proteins and nucleic acids, may be destroyed.

Temperature is simply a measure of heat energy; the higher the temperature, the more heat energy there is, and consequently, the more of it is imparted to matter. Because heat energy is really a form of **kinetic energy**—the energy of motion—increasing it makes the particles of matter move faster and farther apart. At higher temperatures, therefore, many forms of matter exist as gases, for it is in the diffuse gaseous state that particles can have maximum motion. At lower temperatures, there is less heat energy. Therefore less kinetic energy is imparted, and the particles of matter are more likely to remain bound together as a liquid or solid.

The effects of temperature and pressure on matter are critical for all chemical reactions. Most life processes are carried on within highly specific temperature ranges, for biological reaction rates (symbolized as Q) double or triple with every 10°C rise in temperature, as shown in Fig. 2.4. This temperature effect occurs because higher temperatures increase the motion of matter, thereby causing a higher probability of collisions between reacting chemicals. For all organisms there is an upper temperature limit, since many critical protein molecules are very temperature-sensitive and may be irreversibly altered.

About 150 million years ago, certain animals evolved the ability to stabilize their body temperatures within a very narrow range. This event was crucial in the development of birds and mammals, since their success is largely attributable to the fact that they can maintain their body temperatures physiologically. Thus they are freed from having to spend as much time as reptiles and amphibians do in regulating their temperatures through behavioral traits.

The composition of matter

The components of the mixture from which Stanley Miller synthesized amino acids were relatively simple. They were atoms of oxygen, carbon, hydrogen, and nitrogen combined in the various proportions that constitute the molecules of the gas mixture—ammonia, methane, hydrogen, and water vapor. However, oxygen, carbon, hydrogen, and nitrogen are only four of the 105 known fundamental types of matter, called **elements.** An element is a substance that cannot be decomposed by chemical reaction into substances of simpler composition. The atoms of each element, which are the smallest particles that retain the chemical properties of that element, may combine with each other or with atoms of other elements to fashion the myriad forms of matter present in the universe. Note, however, that only a rela-

TABLE 2.1
Names and symbols of elements commonly found in organisms

NAME	SYMBOL
Carbon	C
Hydrogen	H
Oxygen	O
Nitrogen	N
Calcium	Ca
Copper	Cu
Iron	Fe
Magnesium	Mg
Potassium	K
Sodium	Na
Phosphorus	P
Sulfur	S
Chlorine	Cl

tively small number of the 105 elements are commonly found in organisms (see Table 2.1).

INSIDE THE ATOM

Surprisingly, the extremely small atom consists mostly of empty space. It resembles a miniature solar system where, like planets orbiting the sun, particles called **electrons** orbit a central nucleus. Electrons are so named because they bear an electric charge; that is, they attract some particles and repel others. Opposite charges always attract one another, whereas like charges repel. Thus the electron, which is usually negatively charged, is repelled by other negatively charged electrons and other negatively charged particles. What then keeps the electrons in any one atom from repelling each other and flying off in all directions? They are held at certain distances from the atomic nucleus by positively charged particles, or **protons,** located within the nucleus. The proton has about 1836 times the mass of an electron. An atom in its elemental state contains an equal number of electrons and protons, and therefore it is electrically neutral. The nucleus may also contain electrically neutral particles, or **neutrons.**

In essence, the number of protons in an atom (termed the **atomic number**) distinguishes the chemical properties of one element from the chemical properties of another. An atom whose nucleus has one proton is an atom of the element hydrogen; it reacts in very specific ways with other elements. But if an atom has two protons in its nucleus, as does helium, its chemical properties are quite different: whereas hydrogen combines explosively with oxygen in the atmosphere—much to the chagrin of the early dirigible designers—helium is virtually inert and will not combine readily with oxygen. This important difference in the reactivities of hydrogen and helium is correlated with their atomic numbers.

Although the number of protons in the atoms of a given element is constant, the number of neutrons may vary. Since a neutron is equivalent to a proton fused with an electron, and since an electron is of negligible mass (only about 1/1836 of a proton), the neutron is said to weigh the same as the proton. If we arbitrarily give each proton and neutron the mass of 1 and then total them, we can determine the atomic mass for each element. The atomic mass of an element is indicated by a superscript before the symbol of that element, which is an abbreviation of the element's Latin or English name. For example, regular oxygen is designated ^{16}O, since there are eight protons and eight neutrons in the nucleus of every oxygen atom.

Atoms with the same number of protons but a different number of neutrons are called **isotopes.** One isotope quite useful in scientific study is carbon 14, abbreviated as ^{14}C. The most commonly found carbon isotope, ^{12}C, has six protons and six neutrons. However, ^{14}C has an atomic mass of 14 and thus has two more neutrons than the ^{12}C isotope. The additional neutrons make some isotopes unstable; many emit radiation in the form of a stream of electrons released from neutrons breaking apart in their nuclei. The radioactive decay of carbon 14 proceeds constantly, although the instant at which any particular atom decays is perfectly random. Therefore, in order to get an accurate estimate of the rate of radioactive decay, one measures radioactivity over a few minutes', hours', or even days' time and then averages the results.

Although the various isotopes of an element differ in atomic mass, they have almost identical chemical

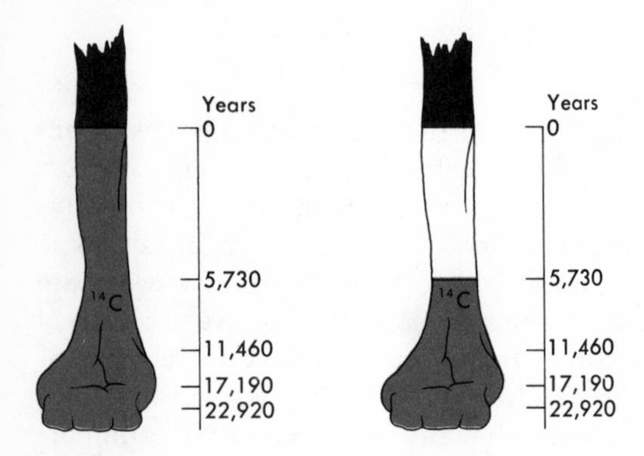

FIGURE 2.5

Carbon dating. Very small amounts of the radioactive isotope of carbon, ^{14}C, become incorporated into living tissue and then decay at a determinable rate. Because the incorporation of carbon stops at death, scientists can determine how long ago some part of an organism was alive by measuring the radioactive carbon, which gives off detectable beta rays. Half of the remaining radioactivity is lost each 5730 years (known as half-life), and by this method it is possible to determine age accurately up to approximately 70,000 years. (The diagrammatic representation here of the ^{14}C proportion of a bone is exaggerated for demonstration purposes.)

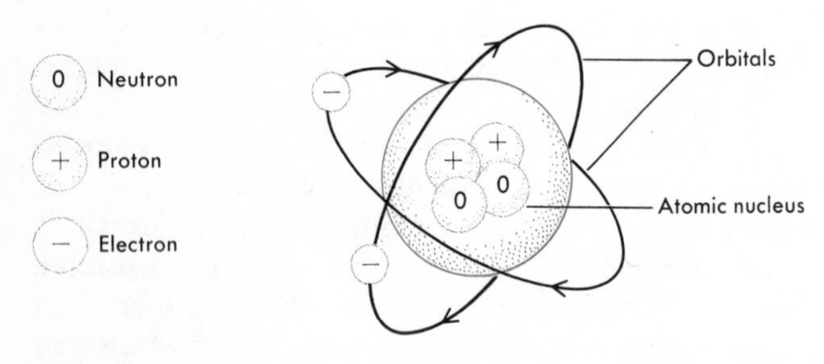

FIGURE 2.6

Three-dimensional view of helium, a relatively simple atom. Whatever the size of a given atom, each of its electrons moves around the nucleus in a separate path, or orbital, giving the atom its characteristic shape. The orbitals of the innermost electrons are spherical, as shown here, but additional electrons follow orbitals of various other shapes, creating a pattern of considerable complexity.

properties. Thus atmospheric carbon dioxide may contain either ^{12}C or ^{14}C, both of which plants may use in the production of sugars and starches.

During their lifetimes, living organisms acquire carbon atoms, of which a fixed proportion is radioactive. Plants acquire them in the course of using carbon dioxide, and animals acquire them by eating plants or other animals. When the organisms die, they no longer take carbon atoms from their environment, but the radioactive ^{14}C atoms they have taken while alive continue to decay. By ascertaining the ratio of the amount of ^{14}C to the amount of ^{12}C in a material that was once alive, one can determine the age of the material (see Fig. 2.5).

The electrons The atoms of each element have a characteristic pattern of electrons surrounding the nucleus. An atom can have as few as one or as many as seven "shells," or areas, each larger than the one below it, within which the electrons may move around the nucleus (see Fig. 2.6). The first shell can contain only two electrons, the second shell eight electrons, and the third shell as many as eighteen electrons. Although the remaining shells can hold more than eighteen electrons, only eight are required to complete a shell. With few exceptions, only the outermost shell will ever be incomplete, because before the outermost shell can begin to fill with electrons, all those beneath it must be filled with the number of electrons appropriate for those shells.

Atoms tend to complete their shells, or become stable. Their ability to do so depends on the number of electrons in their outermost shell. An atom with a complete outer shell is already in a stable state; it will ordinarily not react with other substances. However, an atom whose loss or gain of even a single electron will result in a completed outer shell is not stable. Energy must be expended in the gaining or giving up of the electron. The instability of an atom, then, accounts for its potential energy.

Similarly, the instability of electrons accounts for their potential energy. The outermost electrons are least stable since they are farthest from the attractive force of the nucleus; they also have the greatest energy. Each shell represents an energy level, with the outer level being the highest. The farther the electron from the nucleus, the less stable or secure it is, and consequently the more potential energy it has.

One way in which atoms can achieve stability is through reacting with other substances, forming chemical compounds. The chemical properties of the atoms commonly found in organisms—for example, what elements these atoms can combine with, and what kinds of compounds they can form—depend on the number of electrons in their outer shells.

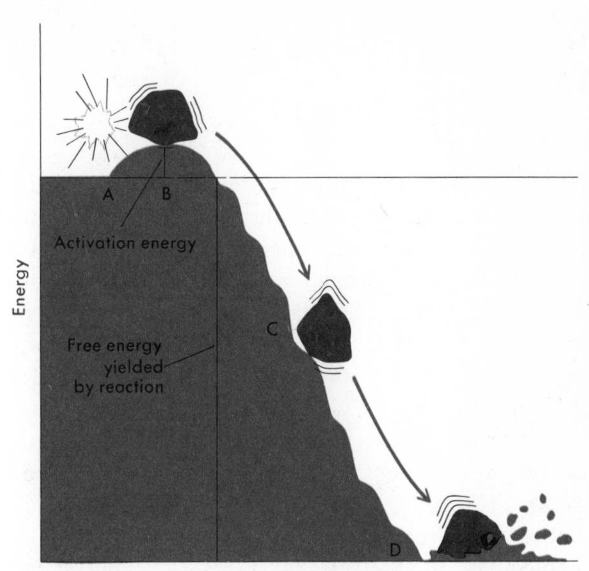

FIGURE 2.7
Graphic and pictorial representation of activation energy and free energy. When an atom is energized, electrons move to new orbitals, where they are unstable. The process is analogous to pushing a boulder up a small incline to the edge of a cliff (from A to B), from which it then drops, giving up some of the stored energy at each bounce (C). At the bottom (D), it has released all the stored energy and has come to rest. In biological reactions, energy is released gradually in small amounts until the atom again reaches its stable state.

Compounds and molecules

A chemical compound consists of two or more atoms joined together in specific proportions by virtue of a mutual rearrangement of the electrons in their outermost shells. For example, water is a chemical compound composed of atoms of hydrogen and oxygen in a ratio of two to one. Hydrogen has an atomic number of 1; oxygen's atomic number is 8. This means that there is one electron in the hydrogen and eight in the oxygen. Hydrogen, therefore, has only one shell with one electron; but oxygen has two shells, the first complete with two electrons and the second incomplete with six. Hydrogen needs an additional electron to achieve a stable configuration, whereas oxygen needs two. When hydrogen and oxygen atoms are mixed and a small amount of heat energy is applied, the atoms collide and come together intimately enough to *share* each other's electrons. Thus it takes two hydrogens and one oxygen for both to be mutually satisfied, and as noted above, that is the ratio of the atoms in water. In the process of combining, **molecules** are formed.

A molecule is the smallest particle of a chemical compound; it consists of two or more atoms joined together and may have chemical and physical properties that are different from those of the atoms composing it. In the formation of a molecule of water, each of the two hydrogen atoms shares its one electron with an atom of oxygen. At the same time, the oxygen atom shares two of its six outermost electrons with the hydrogen atoms. The two hydrogens then each have their one additional outer electron for a stable shell configuration; the oxygen, with two additional electrons, changes from an incomplete outer shell of six to a stable outer shell of eight. The water molecule is designated by the chemical formula H_2O, indicating that two atoms of hydrogen are combined with an atom of oxygen.

How and why does this happen? The outer shell of the oxygen atom is not filled; it has space for two more electrons. When the hydrogen atom approaches (collides with) the oxygen atom, the proximity of the positive charge on the oxygen nucleus, which is greater than that of the hydrogen nucleus, tends to pull the hydrogen's lone electron away. When another collision occurs, the oxygen nucleus again draws away the hydrogen electron, thereby obtaining a second hydrogen atom.

The electrons here may be likened to a boulder tottering on the edge of a cliff, as suggested in Fig. 2.7. A nudge (a form of mechanical energy input) might be sufficient to make the boulder plummet, converting its potential energy into kinetic energy. The same is true of the electrons in hydrogen and oxygen. When the gases are mixed, the brief nudge of a lit match (a form of heat energy input) is enough to start the electrons rearranging themselves with force, losing their potential energy in the process of achieving more stable states. The energy provided by the match is the **activation energy,** or the minimum energy required to initiate a chemical reaction.

Once the hydrogen and oxygen are combined as water molecules, their electrons are more stable. Consequently, the electrons are less likely to be separated from their shell, and therefore they have less potential energy than they had in the atoms prior to the reaction. A reaction like the formation of water, in which energy is given off in the form of heat, light, and noise, is termed an **exergonic** reaction. One can split the water molecule by using electrical energy, a process that involves shoving the electrons back into less stable arrangements (that is, increasing their potential energy). A reaction like this, in which energy is absorbed, is termed an **endergonic** reaction. All chemical reactions are either exergonic or endergonic.

CHEMICAL BONDS: THE LINKS OF MOLECULES

Water is a chemical compound because each molecule of it consists of three atoms (two hydrogens and one oxygen) joined together through a mutual rearrangement of the electrons in their outermost shells. This rearrangement of electrons into a more stable configuration constitutes the chemical bond that holds the atoms together.

Because the electrons in the water molecule are shared between the hydrogen and oxygen atoms, the bond they form is called a **covalent bond.** A covalent bond is a chemical bond that results from the sharing (as opposed to the giving and taking) of a pair of electrons. Covalent bonds are not the only kinds of bonds, however; there is another important type of bond, the ionic bond, as well as various weaker bonds. We shall consider each of these types of bonds in the following paragraphs.

Covalent bonds Life depends on the ability of carbon atoms to form covalent bonds. Carbon is the principal element of **organic** compounds, which include the proteins, carbohydrates, and fats that are associated mainly with living things. The atomic number of carbon is 6; each atom therefore has two shells, the inner with two electrons and the outer with four.

Carbon *shares* its four electrons with four electrons from a different atom or from several atoms, and it may also share electrons with other carbons (see Fig. 2.8). Thus a carbon atom can bond with as many as four other atoms. The bond that results is, of course, a covalent bond; the prefix "co" itself suggests sharing. Because the electrons in a covalent bond are shared between atoms, the atoms do not carry an electrical charge, nor will they become charged very easily even when placed in a polarized medium.

As a result of carbon's propensity to form covalent bonds, it can form molecules with many different shapes and sizes, from the long carbon-to-nitrogen chains of proteins to the compact rings of sugars and the pyramid-shaped molecule of the gas methane—the compound mentioned earlier that is believed to have been present in the Earth's primitive atmosphere.

As with water, whose molecules are polarized, covalent molecules are not always strictly covalent. In these covalent bonds the shared electrons are located closer to one atom than to the other. Such bonds are called **polar covalent.** The particular configuration of bonded electrons in a molecule is one of the determinants of the molecule's shape. The polarity of electrons in the water molecule, for example, is one factor that causes the hydrogen and oxygen atoms to be joined at an angle of 104.5°. Thus the molecule is shaped like a *V* with the oxygen apex bearing a negative charge and the two hydrogen ends bearing a positive charge. As we shall soon see, this particular shape of the water molecule enables water to perform certain critical biological functions.

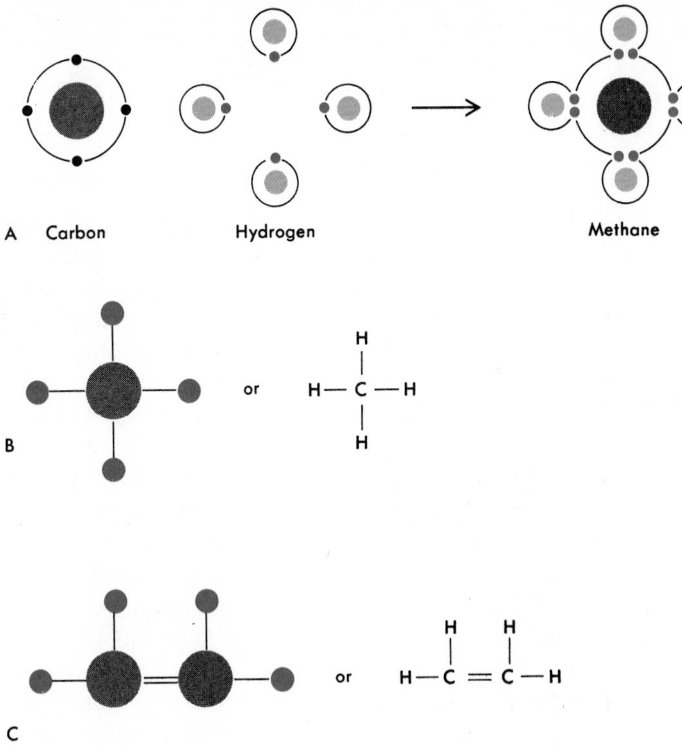

FIGURE 2.8

Schematic and symbolic representations of covalent bonding, or electron sharing. (A) Covalent bond formation between an atom of carbon has four electrons that it may share with other atoms. Each atom of hydrogen has one electron to share. When the four hydrogen atoms bond to the carbon atom, four electron pairs form; this completes the outer shell of the carbon atom of each hydrogen atom. (B) Each electron pair is commonly represented by a short line that connects the symbols for the covalently bonded atoms. (C) Carbon may also bond covalently to other carbon atoms. Here two carbon atoms have bonded with each other, sharing four electrons, or two pairs. The other two electrons of each carbon atom are shown bonded to hydrogen atoms. The bond between the two carbon atoms is called a double bond.

Ionic bonds An **ionic bond** occurs between atoms when an atom of one element donates one or more electrons to, or accepts one or more electrons from, an atom of another element. The atoms become electrically charged (i.e., they are converted from atoms into **ions**), and they are held together by the electrostatic attraction between them. The ionic bond in sodium fluoride, for example, can be understood in terms of the electron configurations of atomic sodium and fluorine.

Sodium has an atomic number of 11; fluorine has an atomic number of 9. This means that sodium has one electron in its outer shell, and fluorine has seven. Sodium needs to lose its outermost electron to achieve a stable configuration, and fluorine needs to gain one electron to fill its outermost shell with eight. Each sodium atom donates its outer electron to a fluorine; the result is that both atoms have eight electrons in their outer shells. The sodium atom now has 11 protons in its nucleus and 10 electrons around it; it is therefore no longer electrically neutral but, because of the loss of one electron, carries a net charge of +1. The fluorine atom has nine protons in its nucleus, and 10 electrons in its shells; the additional electron gives it a net charge of −1. It is the attraction of the two unlike charges that holds the sodium fluoride molecule together, and the molecule is electrically neutral. The ionic bonding of sodium and fluorine to form sodium fluoride is illustrated in Fig. 2.9.

Valence The sodium ion is designated as Na^+, and the fluoride ion as F^-, the superscripts indicating the kind of charge present and its force. Each superscript also indicates the element's **valence.** The valence of an element is its capacity to lose or gain electrons when it combines chemically with other elements. Sodium has a valence of +1. This means that a sodium atom has the capacity to lose one electron and combine chemically with an atom of another element that will accept one electron.

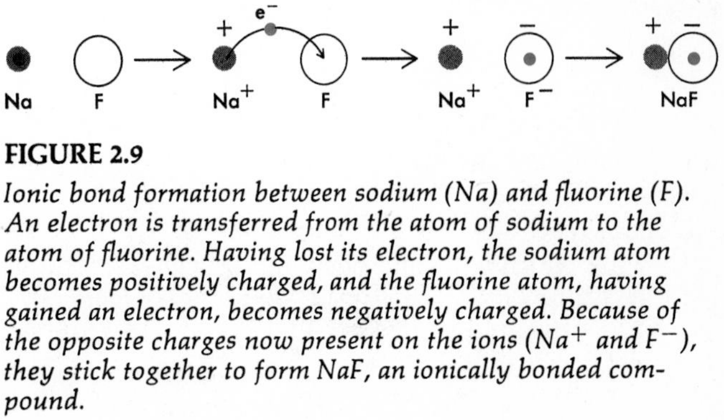

FIGURE 2.9

Ionic bond formation between sodium (Na) and fluorine (F). An electron is transferred from the atom of sodium to the atom of fluorine. Having lost its electron, the sodium atom becomes positively charged, and the fluorine atom, having gained an electron, becomes negatively charged. Because of the opposite charges now present on the ions (Na^+ and F^-), they stick together to form NaF, an ionically bonded compound.

Similarly, oxygen has a valence of −2, because each atom of oxygen will gain two electrons when it combines with one or more atoms.

Oxidation-reduction The type of reaction in which electrons are donated by one atom and received by another is called an **oxidation-reduction reaction. Oxidation** is the loss of one or more electrons; **reduction** is the gain by an atom of one or more electrons. (The term "reduction" refers to the reduced positive charge on an atom as a result of its having gained negatively charged electrons.) Oxidation and reduction always take place simultaneously, since an atom that loses electrons donates them to an atom that will accept them. Generally, atoms with fewer than four electrons in their outer shell will be oxidized (lose their outer electrons), and atoms with more than four electrons in their outer shell will be reduced (gain electrons from another atom).

A major characteristic of ionic compounds is that, in contact with water molecules, which have polar charges on them, the ionic compound is pulled apart by the attraction of water's charges for the charges on its own atoms. With the ionic compound sodium fluoride, for example, the sodium and fluoride tend to dissociate (separate) in water with their charges intact. The fluoride takes with it the electron it received from the sodium, and the sodium goes into solution minus an electron. Since they are not really sodium and fluorine atoms but have configurations different from the original atoms and are thus electrically charged, they are now ions instead of atoms. Hence an ion is an atom that has gained or lost electrons and that consequently carries an electrical charge.

The capacity of ionic compounds to dissociate in water into charged particles is of biological importance. Many ionic compounds required by the body can be used only in their ionized form; the ions function in biological reactions, not the compound itself. The readiness with which an ionic compound, such as sodium chloride (table salt), dissociates in water into sodium and chloride ions is vital in furnishing the body with its needed supply of these ions. For example, the sodium ion (Na^+) is essential in the conduction of nerve impulses. Some other inorganic ions essential to life are potassium, calcium, iron, and iodine. Some ions are needed in large amounts, but only minuscule amounts of other ions are essential. For example, there is only a small amount of iodine ion in the body, and almost all of it is concentrated in the thyroid gland. Without that small amount, however, an infant human fails to grow properly, becoming a dwarf whose nervous system fails to mature. In both the infant and the adult, a lack of iodine leads to a lowering of the rate at which food is metabolized, and the individual becomes lethargic and shows other effects.

Because the oxygen nucleus in the water molecule has a greater positive charge than the hydrogen nucleus, the negative electrons tend to be pulled closer to the oxygen nucleus. This polarizes the water molecule, giving the oxygen end a negative charge and the hydrogen end a positive charge, and thereby makes water somewhat similar to an ionic compound. The charges on the surrounding water molecules may attract the charges on a particular water molecule sufficiently to dissociate it into hydrogen ions (H^+) and hydroxyl ions (OH^-). Thus pure water is continually dissociating into a relatively small quantity of H^+ and OH^- while, simultaneously, the latter are recombining to form H_2O. However, the number of H^+ and OH^- ions formed by water is very small: there is only one H^+ and one OH^- present for every 5.55×10^8 (555 million!) molecules of water.

The quantity of free H^+ in the body is critical to many of its functions. Large concentrations of H^+ are essential to the proper functioning of the stomach, but similar concentrations in the brain would be lethal. The proper concentrations of H^+ in various parts of the body are so crucial that the human body has evolved a number of mechanisms for precisely regulating H^+. When these mechanisms go awry, acidosis, too much H^+ in the blood, or alkalosis, too little H^+ in the blood, may occur and can result in death. The degree of acidity is a measure of the concentration of H^+ in an aqueous solution.

A high concentration of H^+ is called a high degree of acidity; a low concentration results in low acidity. In pure water, the concentration of H^+ is equal to the con-

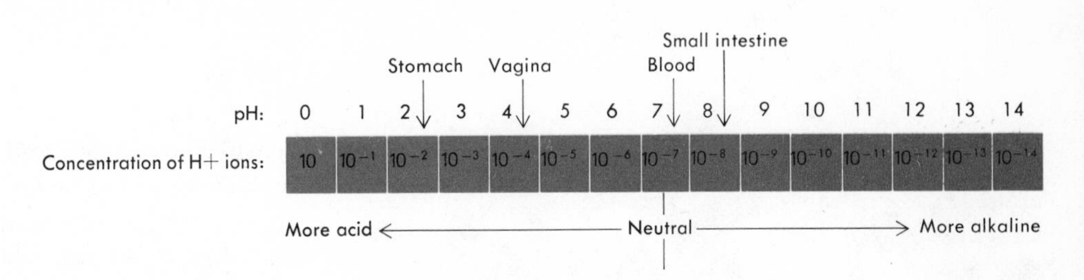

FIGURE 2.10

The pH scale on which is represented logarithmically the concentration of hydrogen ions in moles per liter of solution. An increase of one pH unit indicates a tenfold decrease in hydrogen ion concentration and a corresponding tenfold increase in OH ion concentration. Although most biological reactions occur in the slightly alkaline range (a little higher than pH 7), the range of pHs encountered in organisms is wide. Some approximate pHs found in different parts of the human body are shown. At pH 7, or neutral, the concentrations of H ions and OH ions are equal.

centration of OH^-, and the water is neutral. When certain compounds are added to water, however, they dissociate and the concentration of H^+ increases; other compounds decrease the concentration of H^+ but increase the concentration of OH^-. Compounds that increase H^+ concentrations are called **acids.** Compounds that increase OH^- concentrations are called **bases.**

The hydrogen nucleus has only one proton (the atomic mass of hydrogen is 1), and so the hydrogen ion (the atom less the electron) is really only a proton. When an acid is mixed with a base that contains OH^-, the hydrogen ions (that is, protons) combine readily with hydroxyl ions (neutralizing them) to form water. Because of this property of acids to combine with bases, acids may be defined as proton donors. Similarly, a base may be defined as any substance, such as OH^-, that accepts protons. Actually, the neutralization reaction is an oxidation-reduction process. By combining with the OH^-, the H^+ accepts the extra electron on the OH^- and is thereby reduced; by giving up the electron, the OH^- is oxidized. In living organisms neutralization is usually accomplished by **buffers,** which are capable of controlling the amount of the H^+ present. (We will consider the action of buffers in Chapter 8.)

The degree of acidity (the concentration of H^+) is measured by a logarithmic system known as the pH scale. By definition pH is the negative logarithm of the hydrogen ion concentration of a solution. Values for pH normally run from 0 to 14. The lower the value, the more acid a solution is—that is, the more hydrogen ions it contains. A pH value below 7 is considered acid, and a pH value above 7 is considered alkaline. At neutrality, or pH 7, the concentration of hydrogen ions equals the concentration of hydroxyl ions. Most chemical processes in living things take place within a pH range of from 6 to 8, with the average pH of most body fluids being slightly on the alkaline side, pH 7.4 to 7.6. Exceptions, of course, are biological activities such as digestion that require high levels of acidity (the human stomach has an internal pH of about 2). Figure 2.10 is a graphic representation of the pH scale, with identification of some approximate pHs found in different parts of the human body.

Weak chemical bonds The ability of hydrogen to enter into a polar covalent bond (as previously discussed in the example of oxygen-hydrogen bonds in water) and consequently to carry a weak positive charge allows hydrogen to form a weak bond, called a **hydrogen bond,** with other atoms (see Fig. 2.11). Because of the weakness of the bond, comparatively little energy is required to break it.

Hydrogen bonds are the key structural links in many compounds, including all protein chains. They play a role in the function of enzymes—the proteins that **catalyze** or accelerate biochemical reactions—but they are not unique to complex organic compounds. Water molecules are themselves held together weakly by hydrogen bonds. These bonds occur between the hydrogen of one molecule and the oxygen of another. They serve to arrange the water molecules into a latticework that is responsible for a characteristic quality known as **surface tension**—a cohesion responsible for the formation of a molecular film on the surface of water. Surface tension enables certain insects to stand on water without sinking

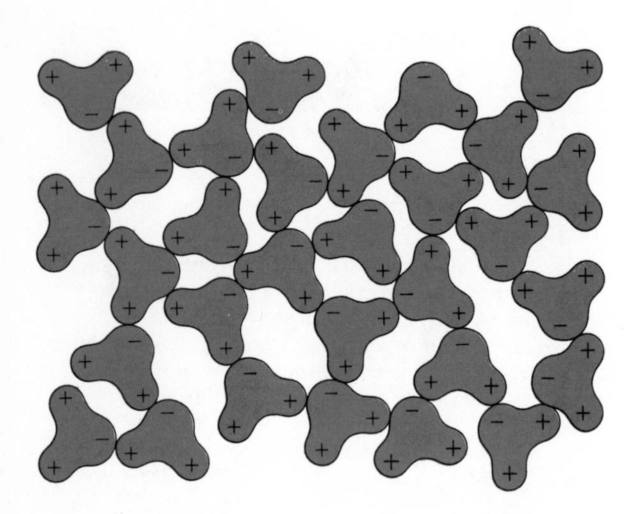

FIGURE 2.11

Hydrogen bonding occurs in water as well as in other molecules. The slight positive charges on hydrogen atoms and slight negative charges on oxygen atoms in other water molecules attract each other, holding the molecules loosely together. One water molecule can form such hydrogen bonds with four other water molecules. Because of the distribution of electric charges, the water molecule is polar, and it is this polarity that allows water to act as a solvent for other polar molecules, such as sugar.

and makes it possible for long columns of water to be drawn up the lengths of trees. The hydrogen bonds, forming a latticework, act as a kind of chain, with one molecule pulling another along.

Chemical equilibrium

All life processes are based on the fact that chemical reactions are reversible. For example, although hydrogen and oxygen can combine to form water, the water is not necessarily a permanent substance. It can be transformed back into hydrogen and oxygen under the appropriate conditions. To be reversed, however, many reactions require very little energy. For example, during the reaction in which hydrogen and acetate combine to form acetic acid, some of the newly formed molecules of acetic acid are already decomposing back into acetate and hydrogen, according to the formula

$$H^+ + Ac^- \rightleftarrows HAc.$$

At the beginning of the reaction, when acetate and hydrogen are first brought into contact, the reaction proceeds primarily in the direction of producing acetic acid. As the quantity of acetic acid increases, however, the rate of the reverse reaction increases, but more slowly. Acetic acid in solution is relatively stable; that is, it has a tendency to stay together. However, some of it does come apart, or dissociate, to produce free hydrogen (H^+) and acetate (Ac^-) ions. Each compound has a characteristic dissociation constant, which reflects the direction of equilibrium. Some totally dissociate, and others only partially dissociate. In the acetic acid reaction formula, the arrow pointing to the right would be much larger than that pointing to the left, and the equilibrium is such that for each free ion there are about 10,000 molecules of stable acetic acid.

These two possibilities—reversibility and equilibrium—have important implications for biological systems. The fact that reactions are reversible enables living things to switch reactions along two paths—from reactants to products, or products to reactants, recycling the same materials over and over. This ability helps the living system to conserve much energy that would otherwise have to be expended in reacquiring materials already used. The second possibility—that many reactions within a living system will eventually reach equilibrium—compels living things to employ certain mechanisms to avoid equilibrium. (Equilibrium in a living system means that what it worked so hard to produce is lost by being transformed back into the raw materials.) These mechanisms involve the removal of the products of a reaction immediately as they are produced, so that they do not revert to the original material. Another mechanism involves the use of stepwise series of reactions to circumvent the possibility of equilibrium inherent in the use of a single one-step reaction.

RATES OF REACTION

Although all chemical reactions are reversible, the rates at which reactions take place often depend on the chemical and physical conditions in their surroundings. Factors that affect the rate of reaction are: (1) the concentration of the reactants; (2) the temperature of the reactants; (3) the pressure of the reactants; and (4) the presence or absence of a catalyst.

The concentration of reactants affects the reaction rate because when there are more atoms and molecules available, as in a concentrated solution, they collide more. There is thus an increase in chemical interactions, and the reaction proceeds more rapidly. Conversely, in a less concentrated solution there are fewer atoms and molecules to collide, and there are fewer chemical interactions and a slower rate of reaction. This dependence on concentration is known as the **law of mass action.** It states that the rate of a chemical reaction is proportional to the product of the concentrations of the reacting substances (see Fig. 2.12).

Temperature, too, affects reaction rate. The higher the temperature, the more kinetic energy is possessed by the reacting atoms or molecules; they move about at greater speeds and over greater distances, colliding more frequently and therefore increasing the reaction rate. On the other hand, the lower the temperature, the less the kinetic energy, and the slower the reaction. Since biological systems depend on reactions that proceed at precise rates, temperature is clearly a factor that is critical to the life of any organism (see Fig. 2.12).

Pressure—the degree to which the atoms or molecules are packed together—influences reaction rates in that the closer reactants are packed, the more collisions there will be, and the faster the reaction is likely to proceed. Conversely, the more spread out they are, the fewer the collisions, and the slower the reaction is likely to be.

Finally, the reaction rate can be affected by the presence of a **catalyst**—a substance that increases the rate at which other substances react but is not itself altered by the reaction. Recalling Fig. 2.7, one can think

A Concentration

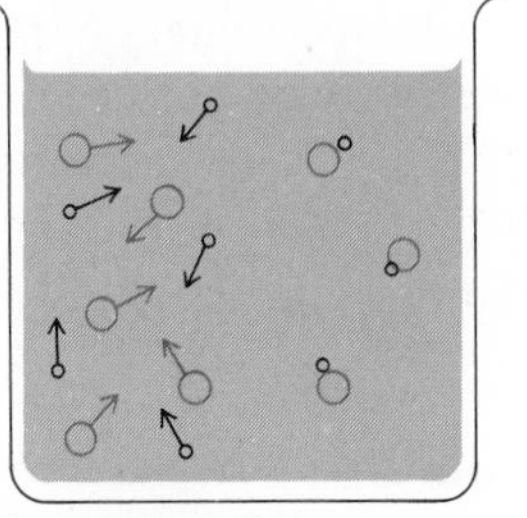

Low concentration
Few collisions
Slow reaction

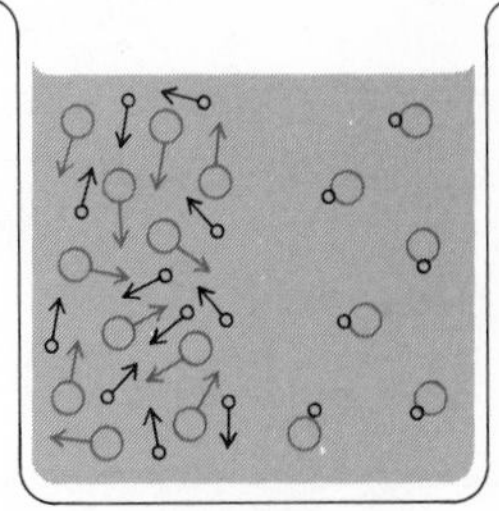

High concentration
Many collisions
Fast reaction

B Temperature

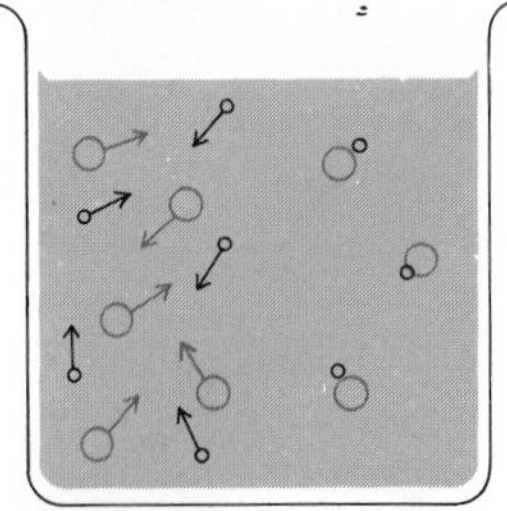

Low temperature
Slow-moving particles
Slow reaction

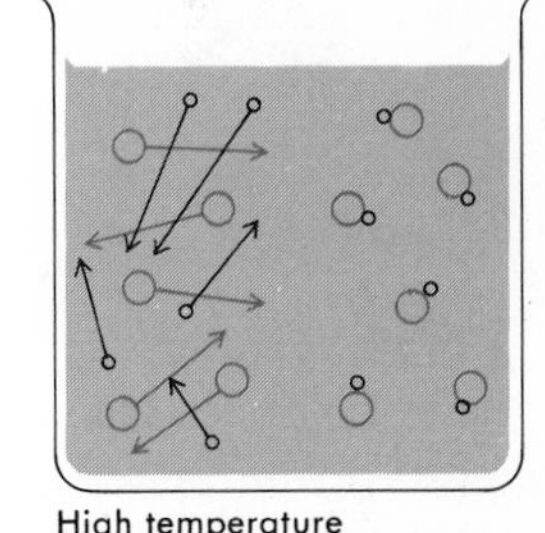

High temperature
Fast-moving particles
Fast reaction

FIGURE 2.12

Reaction rates. Two of the most important factors in determining the rates at which reactions take place are the concentration and the temperature of the reactants. (A) In low concentration there are few molecules of the reacting substances and thus few collisions between them to form the product. In high concentration the reacting compounds are crowded together, offering many opportunities for collisions between them. The reaction rate, or number of molecules of product formed each second, is higher. In chemical reactions the concentration of reactants drops as they are used to form products, and as the concentration drops, the formation of the product occurs more slowly. (B) At low temperatures the reacting molecules move around slowly and do not collide with each other as often as at high temperatures. The result is that the number of product molecules formed in each second is smaller at low temperature than at high temperature.

of a catalyst as reducing the **activation energy** required to push the boulder over the edge, thus starting the reaction sooner. Stated another way, it is easier to go from one valley to another via a tunnel than it is to go over the mountain that separates them.

In living systems, special proteins called **enzymes** are the catalysts. Enzymes are critical to life processes. Reactions that are completed by enzymes in a matter of seconds in a cell would require hours if carried out in a beaker in the biology laboratory without enzymes.

Summary

1. The key to life is the organization of chemical compounds. Simple organic compounds have been synthesized under laboratory conditions simulating the Earth's primordial atmosphere and seas.

2. All matter consists of atoms and molecules of various elements and compounds existing in one of three different physical states—solid, liquid, or gas.

3. The temperature of matter is a measure of the kinetic energy of its atoms or molecules. The greater the temperature the faster the atoms and molecules move, increasing the average distance and the collision frequency between them.

4. The effects of temperature are particularly critical in biological systems because organic molecules are large, complicated, and fragile, and they are therefore particularly vulnerable to the destructive effects of increased kinetic energy.

5. The atom is composed of a central nucleus of positively charged protons, neutrons carrying no charge, and other particles around which move negatively charged electrons.

6. Electrons move about the atomic nucleus within certain fixed regions called shells. Each shell can hold only a certain number of electrons, and an atom achieves greatest stability when each shell is completed by sharing, losing, or acquiring electrons through interaction with other atoms.

7. Atoms combine with other atoms to fill their shells with electrons and thereby become more stable. An atom's capacity to gain or lose electrons in a chemical combination is expressed as its valence.

8. Activation energy is the energy required to start a chemical reaction.

9. Atoms may be chemically joined together primarily by one of two kinds of bonds: the covalent bond, where electrons are shared; and the ionic bond, where one bonding partner takes electron(s) from the other and an electrostatic attraction binds the partners together.

10. Molecules which are ionically bonded come apart or ionize when dissolved in water.

11. When dissolved in water, acids release hydrogen ions while bases release hydroxyl ions or have the capacity to bind hydrogen ions. When acids and bases are mixed together, neutralization occurs as hydrogen ions combine with hydroxyl ions to form water.

12. Covalently bonded molecules become polarized if the electrostatic charges in the molecule are not symmetrically distributed. Polarization affects the shape and the chemical and physical properties of the molecule.

13. Hydrogen tends to form weak electrostatic bonds with negatively charged portions of other molecules. These hydrogen bonds are frequently involved in the three-dimensional folding interactions of large organic molecules.

14. The rate at which a chemical reaction occurs is dependent on four factors; 1) the concentration, 2) the temperature, 3) the pressure of the reactants, and 4) the presence or absence of a catalyst.

REVIEW AND STUDY QUESTIONS

1. Why was it difficult to dislodge the idea of spontaneous generation?

2. Name as many different forms of energy as you can, and indicate what conversions are possible from one form of energy to another.

3. What would happen to you and to other forms of life if normal atmospheric temperatures all over the world suddenly increased or decreased by an appreciable amount? Consider both short- and long-range effects.

4. What are the advantages of carbon dating? What are the limitations?

5. Discuss the statement "Life depends on the ability of carbon atoms to form covalent bonds."

6. Why do oxidation and reduction always take place simultaneously?

7. What is the difference between an atom and an ion?

REFERENCES

Dyson, F. J. 1971 (September). "Energy in the Universe." *Scientific American* Offprint no. 662. Freeman, San Francisco.

Goldsby, R. A. 1967. *Cells and Energy*. Macmillan, New York.

Lehninger, A. L. 1971. *Bioenergetics*, 2nd edition. Benjamin, New York.

SUGGESTIONS FOR FURTHER READING

Asimov, I. 1960. *The Wellsprings of Life*. Signet (Science Library), New York.

Asimov is at his best in this exciting description of the evolution of living matter. Within this context, he also gives the reader a concise history of biochemistry from the eighteenth century up until the Crick-Watson discovery of the structure of DNA.

Baker, J. J. W., and G. E. Allen. 1974. *Matter, Energy, and Life*, 3rd edition. Addison-Wesley, Reading, Mass.

Excellent introductory discussion of the chemical structures of organic compounds, including a separate chapter on matter and energy. Many clear illustrations.

Bush, G. L., and A. A. Silvidi. 1961. *The Atom: A Simplified Description*. Barnes & Noble, New York.

Clear, basic description of atomic structure, including energy sublevels. Good for the beginning student.

Hein, H. S. 1971. *On the Nature and Origin of Life*. McGraw-Hill, New York.

Excellent treatment of the philosophical positions underlying the scientific approaches of various scholars on the origin of life.

Keosian, John. 1968. *The Origins of Life*, 2nd edition. Reinhold, New York.

Excellent short survey.

Margulis, L., ed. 1969. *Origins of Life*. Proceedings, Vol. 1, First Interdisciplinary Conference on the Origins of Life. Gordon and Breach Science Publishers, New York.

Full recording of the debate among participants at the conference.

Oparin, A. I. 1953. *The Origins of Life*, revised edition. Dover, New York.

Translation of Oparin's 1936 book. This classic should be included in the reading of anyone interested in the subject.

Orgel, L. E. 1973. *The Origins of Life*. Wiley, New York.

An outline of the theory and experiments related to the development of nucleic acids in the origin of life.

THE CHEMISTRY OF LIFE

chapter three

Of the many different protein compounds in the body, collagen is the most abundant. Collagen fibers, each a chain of protein molecules joined by hydrogen bonds, constitute a tough connective tissue that resists stretching and gives the body considerable strength.

Chemistry is conventionally divided into two major categories: organic and inorganic. Organic chemistry is the study of chemical compounds held together by carbon-to-carbon links and primarily produced in nature by living things. Inorganic chemistry is the study of the elements in general and of compounds not necessarily associated with life.

Inorganic compounds of life

Inorganic compounds are the units from which are made all of the organic compounds found in living things. And a few of the waste products are inorganic compounds. Water, carbon dioxide, ammonia, and oxygen are the inorganic substances from which organisms synthesize organic compounds—the carbohydrates, lipids, and proteins that are part of all living matter. The organic compounds are made of these inorganic substances (as well as many others) in varying proportions and amounts.

WATER

Almost without exception, when any living thing loses even a relatively small percentage of its water, it dies. Water is essential to all life, as both an internal medium and part of the environment. We have already noted that the water molecule has polar covalent bonds and thus causes ionic compounds to dissociate and dissolve in it. Many other polar substances (substances containing polar covalent or ionic bonds) dissolve in water, too. Thus water is an excellent solvent. It serves as an internal fluid within a cell or a body, in which dissolved particles may come together to react chemically. The property of dissolving ionic substances makes aqueous solutions excellent conductors of electricity; a necessary part of nervous transmission relies heavily on this property. In the cell fluid, particles may also be transported from place to place. Roughly 90 percent of the composition of human blood plasma, for example, is water. Besides serving as a solvent, water plays an essential and active role in certain chemical reactions, such as the digestion of proteins and sugars.

Another function of water in living systems is derived from its thermal properties. Water can absorb relatively large amounts of heat without greatly changing its temperature. This effect is most obviously reflected in weather: the climate near coastal regions is much more moderate than it is over large land masses. Water's capacity to absorb heat enables living things to use water in various ways to keep their own temperatures from varying greatly. Many aquatic organisms, such as fish, depend on this temperature-buffering (resistance to change) property of water to prevent thermal shock—rapid temperature variations in their environment that might prove fatal. In addition, water absorbs a great deal of heat when it evaporates from a surface. A panting dog, from whose saliva water evaporates as he breathes hard on it, a laboring man who bares his sweating back to cooling breezes, and a hippopotamus who takes a dip in a river and then comes out to dry, all use the cooling effect of the evaporation of water to keep their body temperatures within very narrow limits despite rises in the external temperature.

At least four other properties contribute to the special importance of water for living systems:

1. Water is densest at 4°C, and it expands when it drops below that level and freezes. Hence lakes freeze from the top down instead of from the bottom up. In deep lakes, fish and other organisms can live in a fairly constant thermal environment all winter long with no danger of freezing.

2. Water is very cohesive. Plants are therefore able to pull long columns of water through thin tubal elements to great heights. The cohesiveness of water is also crucial in keeping the contents of a cell together and providing structure for the cell.

3. Water has a significant surface tension. Some insects can walk on water without sinking through the surface film.

4. Water gives up energy as heat before it freezes and absorbs heat from the sun when it melts. As a result, fluctuations in temperature are kept within a range that is tolerable for living things.

CARBON DIOXIDE

Life on earth is dependent on atmospheric carbon dioxide (CO_2) as well as water and oxygen. The fact that carbon dioxide constitutes only a fraction of a percent of the atmosphere, however, makes the margin for existence a slim one indeed. Carbon dioxide is vital to life because it is the major source of carbon for organic compounds—the principal components of living substances. Green plants use atmospheric carbon dioxide together with water and light energy to manufacture sugars in the process of photosynthesis: animals eat plants and obtain for food the compounds produced by the plants. These plant materials are then digested by the animals and later undergo respiration—the process by which complex compounds are broken down into simpler ones and energy is released. (Plants respire also and release carbon dioxide as well as oxygen into the atmosphere.) The end product of respiration is carbon dioxide, which is returned to the atmosphere.

Besides its important structural role as the building block of organic chemicals, carbon dioxide has a vital chemical role. In many cells it dissolves in cellular water to form a buffer which regulates pH within a narrow range. We will discuss this vital role in detail in Chapter 8.

Carbon dioxide is a covalent compound whose molecules consist of one atom of carbon bonded doubly to each of two atoms of oxygen. The double bond consists of two pairs of electrons, one pair donated by each atom. In the carbon dioxide molecule the carbon shares two of its electrons with each of the two oxygen atoms; thus all three atoms then have eight electrons in their outer shells. The molecular formula of carbon dioxide is designated as CO_2; the **structural formula**—that formula indicating where the double bonds are and which atoms are bonded to which—is designated

$$O{=}C{=}O$$

where each line in the structural formula represents one pair of electrons.

AMMONIA

Ammonia (NH_3) is a gas that plays a paradoxical role in living things. It is a principal component of amino acids, of which proteins are built, yet it is toxic when released into the living system from protein decomposition. The amino groups ($-NH_2$) of amino acids were originally derived from ammonia. In the amino acid, the amino group consists of an atom of nitrogen bonded covalently to two atoms of hydrogen; the third hydrogen, which would be present in ammonia, is replaced in amino acids by

```
    H  O
    |  ||
  —C—C—OH
    |
    H
```

or a more complicated chain of carbons and other atoms. (Nitrogen has an atomic number of seven, so that there are five electrons in its outer shell; thus it has a capacity for acquiring three electrons. It can do so by sharing three of its own electrons either with three from another atom, or with one of each of three different atoms.)

The structural formula for the amino group is

```
H
 \
  N—
 /
H
```

When proteins are decomposed, amino groups are broken off from their carbon atoms, and they combine with a hydrogen atom to form ammonia gas. In the human body the ammonia is converted by the liver into urea and is excreted in the urine.

OXYGEN

Oxygen plays a twofold role in living things. First, it assumes the critical function in respiration of oxidizing ("burning") complex compounds and thereby releasing energy. Second, it serves as a vital component of carbohydrates, lipids, and amino acids—the basic foodstuffs that are manufactured by plants and animals and that are nutrients vital to all organisms.

What happens in oxidation—a process that may involve elements other than oxygen, where electrons are lost by one atom and gained by another—is similar to what happens when a match is put to a piece of newspaper; heat from the match is converted into kinetic energy of electrons. This enables oxygen in the air to remove electrons from the molecules in the paper, breaking these down into carbon, carbon dioxide, and water, and releasing heat energy. If no oxygen were present, the paper would not burn.

The oxygen level in the atmosphere today remains constant at about 21 percent (although it seems to have varied greatly early in our planet's history), largely because it is released by green plants as a by-product of the process of sugar manufacture (photosynthesis). Plants require water and carbon dioxide to manufacture sugar; some of the oxygen contained in the water molecules is released as a gas into the atmosphere, from which animals obtain it for use in respiration. The profusion of plant life assures the release—through photosynthesis—of a sizable quantity of oxygen, replacing that used in animal respiration.

Organic compounds of life

CARBOHYDRATES

A quick and easy way to obtain a chunk of carbon is to pour some sucrose, which is ordinary table sugar, into a test tube and hold a flame under it; in a few minutes the crystals of sugar will have been transformed into a clump of black carbon, and the sides of the test tube will be coated with droplets of water. What has happened is that the sucrose molecules have been oxidized and broken down into the substances—carbon and water—out of which they were composed. Sucrose and the class of organic compounds—consisting of carbon, hydrogen, and oxygen—in which it is included are called **carbohydrates.**

If any one organic compound can be thought to play a pivotal role in the chemical reactions of life, it is most certainly the sugar glucose; every kind of compound found in living matter—whether protein, amino acid, fat, or starch—can be broken down into substances that can ultimately be converted into this one sugar. And it is from the oxidation of this one sugar that energy is obtained for most life processes, yielding carbon dioxide and water as the end products.

Glucose abounds in grapes and so is known as grape sugar; it is also found in the blood, where its concentration is precisely regulated. Glucose is also the building block of cellulose, the fibrous constituent of wood. Because cellulose is both strong and resilient, it can be used to make furniture and paper. Glucose consists of a series of six carbon atoms linked together, usually, as a ring configuration and sometimes as a chain; to these carbon atoms are attached atoms of hydrogen and oxygen. The **empirical formula**—the chemical formula that indicates the elements present in each molecule of a compound and the numbers of atoms of each—for this sugar is $C_6H_{12}O_6$. Figure 3.1 shows the two forms in which glucose is found.

Glucose is a monosaccharide—a carbohydrate monomer often referred to as a simple sugar. A **monomer** is a basic unit—whether a carbohydrate or an amino acid—that, when coupled one to another, results in longer compounds called **polymers.** There may be several monosaccharide monomers that have the same empirical formula. These monomers differ from each other in their physical structures. Compounds of this sort are called

Straight chain form

Ring form

FIGURE 3.1
The structural formulae in chain and ring form of a single carbohydrate, glucose. All carbohydrates are composed entirely of carbon, hydrogen, and oxygen.

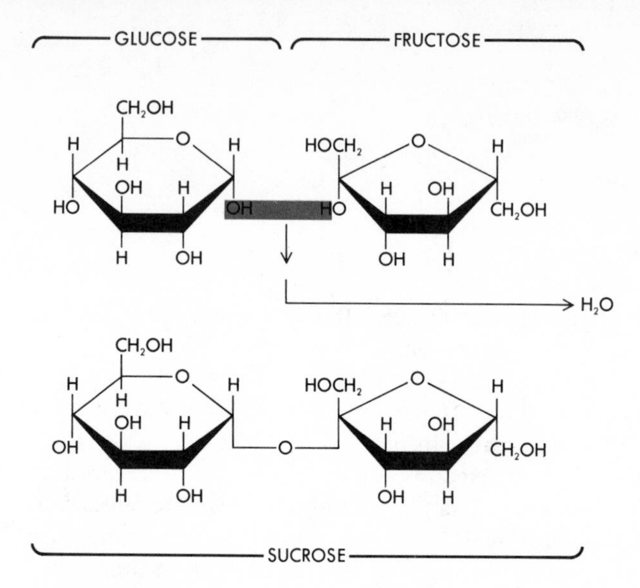

FIGURE 3.2

A condensation reaction between molecules of glucose and fructose, linking these two monosaccharides to form the disaccharide sucrose (common table sugar). A molecule of water is removed in the reaction. This is a dehydration reaction. In the converse reaction, called hydrolysis, sucrose is split into its component parts by the addition of water.

isomers. Two isomers of glucose are fructose, the sugar found in fruits, and galactose, the sugar found in milk.

Glucose is a member of a class of six-carbon **monosaccharides** (the name given to monomer sugars in a polymeric sugar molecule), all of which have the same $C_6H_{12}O_6$ formula. These sugars, as well as others that consist of as few as three carbon atoms, are the simplest of carbohydrates and are the units out of which the larger carbohydrates are made, being composed of one monosaccharide ring attached to another. Two monosaccharides joined together are called a **disaccharide;** three are called a **trisaccharide;** many linked together make up a **polysaccharide.**

Monosaccharides combine chemically to form larger compounds through a link formed between a carbon atom of one and an oxygen of another. (The OH, or hydroxyl, group that was on the carbon of one monomer pulls off the hydrogen of the OH group on the second monomer to form a molecule of water, which splits off. This leaves a carbon that is free to form a new bond. Thus the carbon attaches itself to the oxygen that lost its hydrogen, so that the two monomers are linked by an oxygen atom shared between them.) This reaction—where a molecule of water splits off so that two monomers combine—is called a **condensation** (or **dehydration) reaction.** The disaccharide sucrose, for example, is formed by a condensation reaction between a molecule of glucose and a molecule of its isomer, fructose (Fig. 3.2). Another disaccharide, lactose, is formed by a condensation of glucose and another of its isomers, galactose.

Just as polymers are formed by the removal of water through a condensation reaction, they are broken apart by a reaction called **hydrolysis**—the reverse of condensation, in which two monomers are separated by the insertion of a molecule of water. (A hydrogen from the water combines with the oxygen that binds the monomers; the OH combines with the carbon that now has a free bonding capacity by virtue of the separation from the oxygen.)

The three polysaccharides of major importance to living things are the starches, glycogen, and cellulose. Starches are composed of many glucose monomers. They are used by plants as storage compounds for glucose molecules, the oxidation of which yields energy for other chemical reactions. Just as plants use starches for energy storage, animals use glycogen—a polysaccharide slightly different from plant starch. (Glycogen is also known as animal starch.) Cellulose differs markedly from starch and glycogen. One difference of major importance is that the bonds in cellulose are not easily hydrolyzed, and cellulose is therefore insoluble in water, a factor that enables it to serve as the supporting material for plants. In fact, the resistance of cellulose to hydrolysis is the reason that man and most other animals are not able to subsist on grass; the human body does not contain the enzymes necessary to hydrolyze cellulose. The reason that cows can feed on grass and that termites can reap sustenance from pieces of wood (wood is made largely of cellulose) is that cows and termites have in their intestines microorganisms that supply the cellulose-digesting enzyme. (Such microorganisms are called symbiotic because they live in a close and dependent relationship—**symbiosis**—with another organism.) If the microorganisms were not present in a termite's gut, the insect could chew away an entire lumberyard but die of starvation.

LIPIDS

The oxidation of a gram of glucose will yield about four calories of heat energy; the oxidation of a gram of fat—a member of one of the classes of water-insoluble compounds called **lipids**—yields approximately nine calories. (A **calorie** is the amount of heat required to change the temperature of one gram of water 1°C, or centigrade.) Indeed, because of their chemical structure, fats yield more energy than any of the other organic compounds in the body. The carbons and hydrogens in a lipid are assembled in such a way that more hydrogens are accessible to oxidation than in a carbohydrate. The more hydrogens capable of being oxidized, the greater the energy yield. Thus fats yield much more energy than glucose. Unfortunately for the weight-conscious, carbohydrates are *easier* to oxidize than fats; thus in a diet containing both fats and carbohydrates, the carbohydrates are used preferentially for energy, and only a part of the fat may be oxidized. The rest is deposited in fat "depots" which are located in various sightly and unsightly places in the body.

Fats are composed of three molecules of fatty acids bonded to one molecule of glycerol. Glycerol is an alcohol. Alcohols are distinguished by having one hydroxyl (OH) group bonded to a chain of carbon atoms. Glycerol differs from ethyl ("drinking") alcohol; whereas ethyl alcohol contains two carbons with one hydroxyl group, glycerol has three carbons to which are bonded **three** hydroxyl groups:

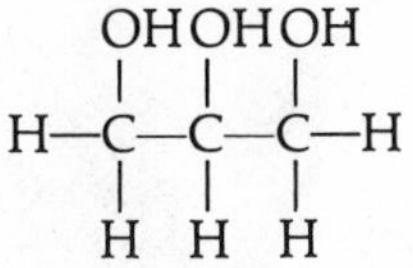

$$H-\underset{H}{\overset{OH}{C}}-\underset{H}{\overset{OH}{C}}-\underset{H}{\overset{OH}{C}}-H$$

Fatty acids combine with glycerol at the sites of the hydroxyl groups, and the glycerol molecule is thus attached to three fatty acids. **Fatty acids,** the second component of fats, obtain their acid property (that of donating protons) from a carboxyl group

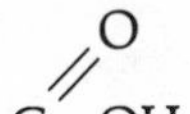

$$C\overset{\displaystyle O}{\diagup\!\!\diagup}\!\!-OH$$

bonded at some point to the molecule. A fatty acid combines with glycerol through a condensation reaction. The acid donates the hydrogen of its carboxyl group to an OH group on the glycerol. Then a molecule of water splits off from between the two molecules. In the reaction, the oxygen that lost the hydrogen combines with the glycerol carbon that lost the hydroxyl group. Like polysaccharides, fats are digested by hydrolysis—the reverse of the condensation (dehydration) reaction. The basic structure of a lipid is indicated in Fig. 3.3.

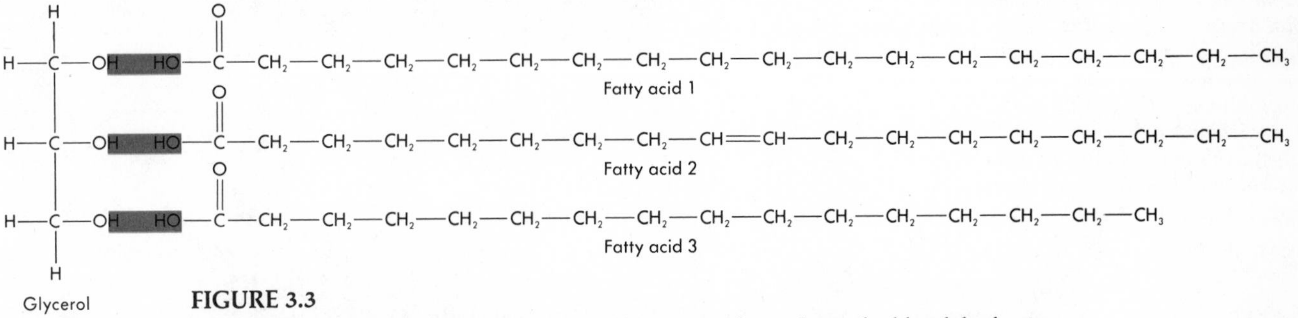

$$H-\overset{H}{C}-OH\quad HO-\overset{O}{\overset{\|}{C}}-CH_2-CH_2-CH_2-CH_2-CH_2-CH_2-CH_2-CH_2-CH_2-CH_2-CH_2-CH_2-CH_2-CH_2-CH_2-CH_2-CH_3$$

Fatty acid 1

$$H-C-OH\quad HO-\overset{O}{\overset{\|}{C}}-CH_2-CH_2-CH_2-CH_2-CH_2-CH_2-CH_2-CH=CH-CH_2-CH_2-CH_2-CH_2-CH_2-CH_2-CH_2-CH_3$$

Fatty acid 2

$$H-\underset{H}{C}-OH\quad HO-\overset{O}{\overset{\|}{C}}-CH_2-CH_2-CH_2-CH_2-CH_2-CH_2-CH_2-CH_2-CH_2-CH_2-CH_2-CH_2-CH_2-CH_2-CH_3$$

Fatty acid 3

Glycerol

FIGURE 3.3

The structural formula of a lipid, which consists of glycerol attached by dehydration synthesis to the carboxyl groups of three fatty acid molecules. This lipid has three different fatty acids, but they can also be all the same. Fatty acids without double bonds are saturated whereas those with double bonds are unsaturated. The nature of the fatty acids is responsible for the nature of the lipid.

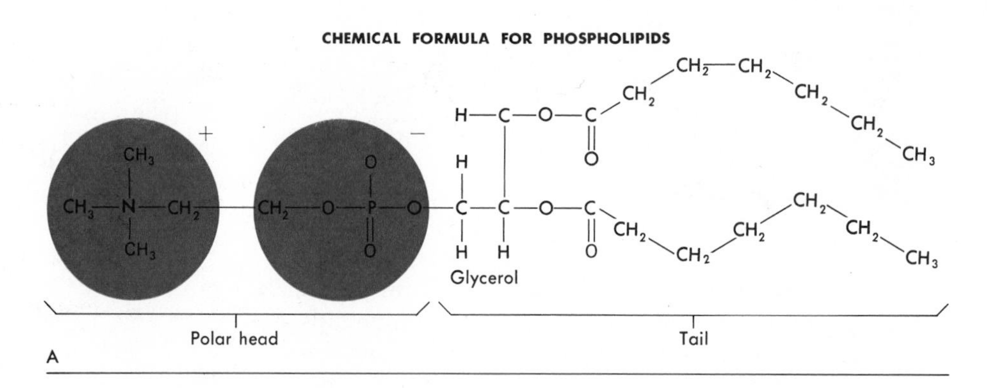

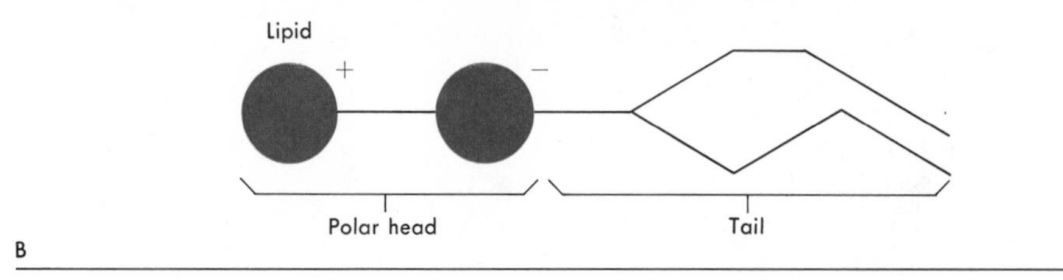

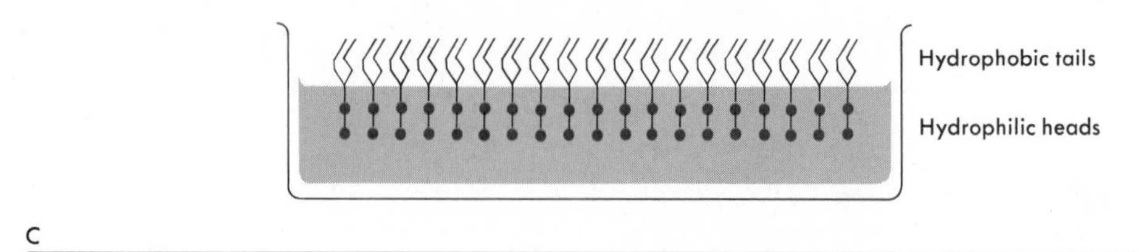

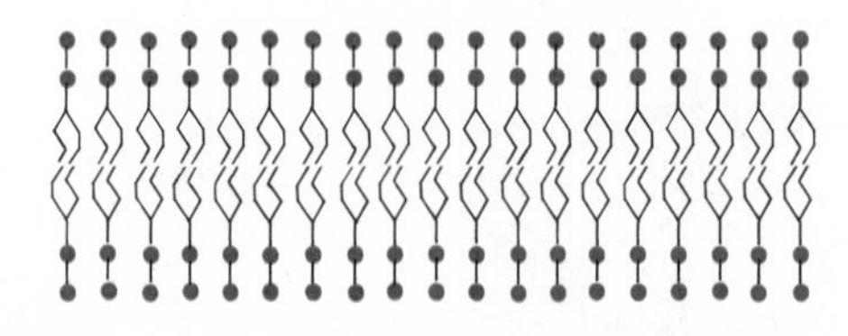

FIGURE 3.4

Cell membrane. The molecular structure of cell membranes is still not precisely known, but it can be shown that molecules of phospholipid and protein will form membranes very much like the supposed structure of cell membranes. The organization of these molecules in laboratory experiments is solely the result of the physical properties of the molecules, and no living organism is necessary to form an artificial membrane. (A) A phospholipid molecule consists of glycerol and two molecules of fatty acid, which compose the tail, and phosphoric acid, which forms the head. Because of the distribution of electrical charges, the phospholipid is polar. (B) The symbolic representation of phospholipid. (C) When placed in water, phospholipid molecules spread out to form a thin skin or film on top of the water. The molecules are arranged so that the heads, which are hydrophilic (water-loving), immerse themselves, and the tails, which are hydrophobic (water-hating), usually stick out. Two such phospholipid films are thought to compose much of the cell membrane (D).

A second class of lipids includes the **phospholipids.** Phospholipids are closely related to the fats. They differ only in that one of the three glycerol carbons is bonded to a carbon chain, parts of which have in it a phosphate group,

```
      O
      ||
—O—P—O—
```

and a nitrogen atom. (The phosphate is negatively charged and the nitrogen is positively charged.)

In water, a phospholipid molecule will usually have its electrically neutral "tail" sticking up through the surface; its "head"—with a positive and a negative charge on it—orients itself downward because of the attraction between its charges and those on the water molecules. This tendency of phospholipid molecules to orient themselves in definite patterns enables them to serve as a basic structural component of cellular membranes (see Fig. 3.4). A phospholipid called lecithin is extremely important for this reason.

Another group of lipids consists of the waxes. Waxes like cutin make the glossy layer or cuticle on leaves that serves as a protective coating, and they also coat the skin of some animals.

A fourth group of lipids consists of the **steroids.** Testosterone, the male sex hormone, and the estrogens, the female sex hormones, are steroids, as are the bile salts, the emulsifiers in fat digestion, and one of the vitamins (vitamin D). Steroids also play a role in cell metabolism, acting with the phospholipids in the cell membrane to regulate what comes into the cell and what leaves. Steroids differ structurally from the other lipids; they do not even contain a molecule of glycerol but are made of four rings of carbon with a carbon chain attached. Steroids are classified as lipids because, like other lipids, they are insoluble in water but dissolve in oil. In humans, all steroids are formed from a substance called cholesterol, which is carried by the blood. The sex glands obtain the cholesterol from the blood and transform it into the sex hormones. Cholesterol is an essential precursor to many key molecules, but in excess it can be detrimental since high blood cholesterol seems to play a role in coronary blood vessel disease and heart disease. But cholesterol is not all derived from the diet; the body can also synthesize it from acetic acid (CH_3COOH).

PROTEINS

Proteins (meaning first things) are large informational molecules that contain carbon, hydrogen, oxygen, about 16 percent nitrogen, and often phosphorus and sulfur. Some are structural molecules in tendons, ligaments, and bones; some are working contractile molecules in muscle; some are instruction-carrying hormones; and some of the most important—the enzymes—do chemical work by **catalyzing** all essential metabolic reactions.

Proteins are made up of elongated chains of amino acids. There are 20 different biologically important amino acids that make up the alphabet of proteins. With such a large working alphabet, the variety of proteins that are possible is virtually limitless, and each species —indeed, each individual—has literally thousands of highly distinctive proteins.

All amino acids subunits (monomers) of protein have the same basic structural components—a carboxyl group (COOH) and an amino group (NH_2) attached to a central carbon—as well as some other organic entity, which for the sake of convenience is labeled simply R in the following structural formula:

```
     R  O
     |  ||
  H—C—C—OH
     |
     N
    / \
   H   H
```

Amino acids bond together by a condensation reaction, in which the hydrogen from an amino group of one amino acid combines with the OH (hydroxyl) group from the carboxyl end of another amino acid to form a molecule of water; the resulting bond vacancy is filled by the nitrogen of one amino acid bonding to the carbon of the carboxyl on the second.

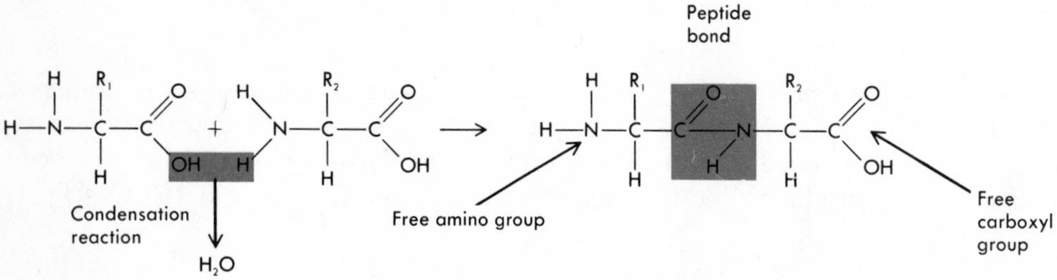

FIGURE 3.5

Peptide bond. When two amino acids bond together by a condensation reaction, the hydrogen of the amino group of one combines with the hydroxyl from the carboxyl group of the other to form a molecule of water. The carbon of one amino acid is thus linked to the nitrogen of another in a peptide bond. The symbols R_1 and R_2 stand for the different side groups, whose structures may be simple or complex.

FIGURE 3.6

Polymerization, the chemical reaction in which a polymer is made from smaller molecules. Many molecules of biological importance are synthesized by polymerization. (A) Two identical sugar units (monomers) may join in pairs to form a dimer (such as maltose) or in longer chains to form a homopolymer. (B) Many different units, such as the five amino acids shown, may join to form a heteropolymer.

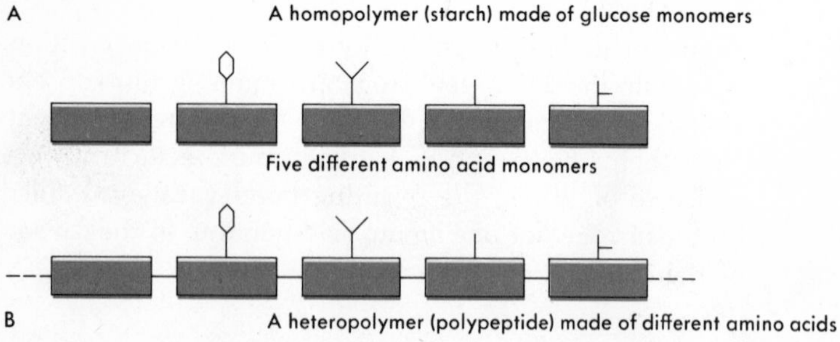

This carbon-to-nitrogen bond linking two amino acids is called a peptide bond (Fig. 3.5); many amino acid molecules bonded together by peptide bonds form a **polypeptide.** Polypeptides are really polymers composed of amino acid monomers; the formation of a polypeptide is a **polymerization** reaction (see Fig. 3.6). A single polypeptide or a group of them bound together into an integral unit of structure or function is called a protein.

The sequence of amino acids in a protein is called the primary structure, and even a modest protein will contain about 200 amino acids. However, the secret of protein's success is not to be found in its linear sequence but in the way the molecule occupies three-dimensional space; that is, shape is crucial.

Although the research that has determined the structures of proteins began only recently, considerable progress has been made. The structures of a number of proteins are now known in full detail. Proteins rarely exist as straight chains, and many assume pretzel-like configurations. One chain may be twisted spirally

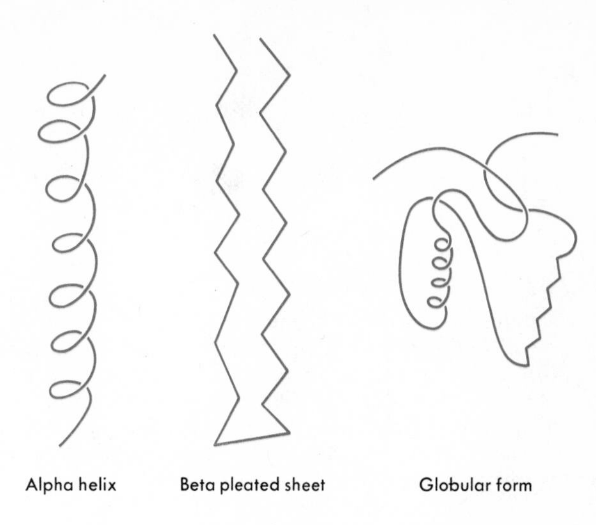

FIGURE 3.7

Protein configurations. Proteins may assume many different shapes, only some of which are suggested in the sketches included here. Note that the globular form contains within it segments that reflect some of the simpler shapes. Coils and pleats are held together by weak hydrogen bonds, whereas the stronger disulfide bonds are commonly found creating connections between polypeptide chains and bends in the coils or pleats.

around another, or a single chain may be folded back on itself in a variety of ways (Fig. 3.7). These various shapes are made possible by two kinds of bonds—the hydrogen bonds and the disulfide (S–S) bonds of the amino acid cysteine. For example, one spiral configuration common among proteins is the **alpha helix.** The curves in the alpha helix are formed by hydrogen bonds spaced at regular intervals along the chain. These bonds occur because the hydrogen attached to the nitrogen of one amino acid is attracted to the carboxyl oxygen (C=O) of another.

Although spirals are formed by hydrogen bonds, it is through the disulfide (S–S) bond that a protein chain can fold on itself. The sulfur atom of one cysteine molecule can break its bond with the adjacent sulfur atom to form a bond with the sulfur atom in a molecule of cysteine somewhere else in the chain. One part of the chain may thus be linked to, and folded in on, another.

As a result of the hydrogen and sulfur bonds and the consequent shapes assumed by protein chains, proteins are three-dimensional structures. Because of the limitless variety possible in a protein's amino acid composition, each protein has a highly specific structure. This structure restricts the protein in its chemical reactivity, so that most proteins can react with only a limited number of other substances.

If a protein is exposed to harsh conditions, such as excessive heat and acidity, the hydrogen bonds may be broken, and the protein will become misshapen. This loss of shape is called **denaturation.** When a protein is denatured, it will have lost its normal chemical reactivity. Although it is possible for a denatured protein to be restored to its former structure, denaturation may sometimes proceed so far that the protein loses its solubility and coagulates. For example, an egg, which consists largely of protein, cannot be unfried. Cooking heat denatures it irreversibly.

Proteins may consist of other substances as well as amino acids. Proteins composed only of amino acids are called **simple proteins.** However, proteins may be attached to groups—called prosthetic groups—that are not amino acids or polypeptides. Proteins associated with such prosthetic groups are called **conjugated proteins.** Prosthetic groups may be carbohydrates (the protein is then called a glycoprotein), or nucleic acid (the protein is a nucleoprotein), or a lipid (the protein is a lipoprotein).

Enzymes As mentioned earlier, enzymes are the catalysts for the chemical reactions taking place in living things. They speed up the reactions without themselves being altered in the process. It is because of the enzymes that these reactions require little activation energy (see Fig. 2.7) and so can proceed without raising the temperature of the organism. For example, each time we eat a pancake or other starchy food, enzymes digest the starch into sugars at the normal temperature of the body. The same digestion could take place without an enzyme, but only at a temperature which would irreversibly damage the structural proteins of the body.

The manner by which enzymes accomplish their roles as biological catalysts derives from the fact that enzymes are proteins. As a protein, each enzyme has a very specific three-dimensional structure that is determined by hydrogen bonds, disulfide bonds, and the particular sequence of its amino acid composition. Structural specificity is so particular that each enzyme has its own distinctive peaks and valleys, notches and protru-

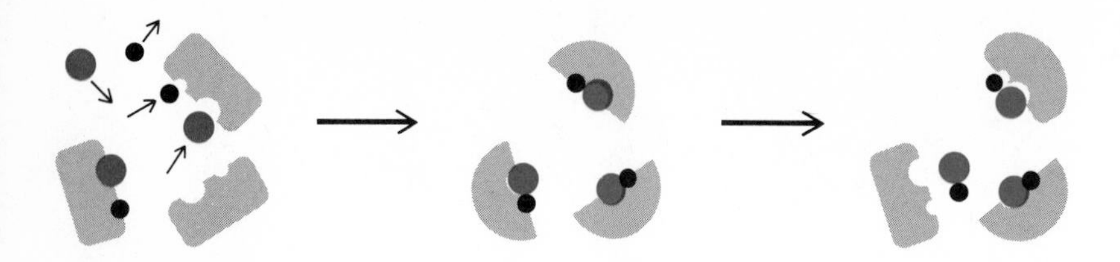

FIGURE 3.8

Enzymatic catalysis. At present the most widely accepted model for the activity of an enzyme is that in which the enzyme, whose specific surface configuration enables it to bond specifically with particular substrates, deforms once it has attached to the substrates. The deformation is thought to bring the reacting substrate molecules close enough to bond chemically. Once they have bonded, they shoot away from the enzyme molecule, which resumes its native shape and is free to engage in another reaction. A catalyzed reaction requires less activation energy than does an uncatalyzed reaction.

sions. Obviously, these distinctive features can match those of only very few other molecules; for every protrusion on an enzyme molecule, there must be a corresponding notch or groove in the molecule (or substrate) whose reaction the enzyme catalyzes, and vice versa. This is the reason that enzymes are able to act on only a few other substances, and many can do so with only one. Figure 3.8 presents in visual form one of the hypothetical mechanisms proposed for the action of enzymes.

It is thought that the temporary attachment of the substrate to the enzyme serves to hold the substrate in place. In this way it can thoroughly and quickly react with other nearby substances. It is also possible that the linkup puts a strain on the chemical bonds of the substrate so that they are more easily broken. Thus the reaction can proceed rapidly and without much activation energy.

Since enzymes are proteins, they are sensitive to variation in such conditions as heat and acidity. Too much of either may change the disulfide and hydrogen bonds and denature the enzyme (see Fig. 3.9). Once the enzyme loses its particular shape, it can no longer attach itself to a substrate. Thus the starch-digesting enzyme amylase, found in the mouth, is denatured when it enters the extremely acid stomach. In the stomach, however, the protein-digesting enzyme pepsin works quite well in its hydrochloric-acid surroundings. Each enzyme has an optimal pH at which it works best.

The specificity of enzymes for particular substrates enables the body to hold its many chemical reactions in check. If an enzyme could catalyze a great many chemical reactions, it would be dangerous to have; these reactions would soon get out of hand, and probably kill the organism. In addition, enzymes provide organisms with a means for controlling the reaction rate. More enzyme will accelerate a particular reaction; less will slow it. Thus when there is too much product of an enzyme reaction, the supply of the particular enzyme may be cut

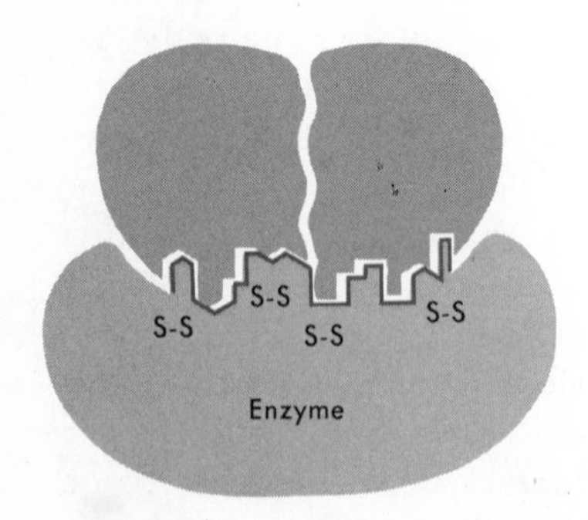

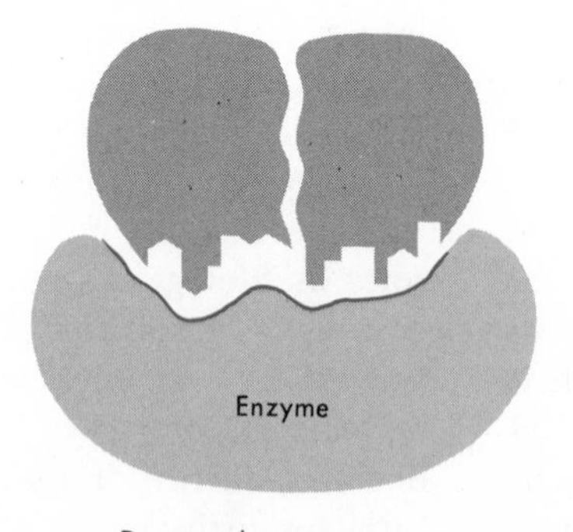

FIGURE 3.9

The catalytic properties of an enzyme molecule are the result of its specific surface shape, which enables it to fit together with two reacting molecules. Its surface shape is stabilized by S–S (disulfide) bonds. Heating the protein enzyme molecule denatures it by breaking the S–S bonds, allowing the surface shape to change so that the molecule no longer fits and reacts with the reacting molecules.

back so that no more of the product is manufactured. This mechanism, as we shall see, plays an exceedingly important role in a cell's regulation of its nutrient levels.

THE NUCLEOTIDES

A nucleotide consists of a nitrogen-containing base (a purine or a pyrimidine), a sugar (ribose or deoxyribase), and an inorganic molecule (a phosphate). There are three major groups of nucleotides: (1) the adenosine phosphates; (2) the nucleotide coenzymes; and (3) the nucleic acids.

There are three **adenosine phosphates.** Adenosine monophosphate (AMP) plays a major role as an internal messenger within cells. Adenosine diphosphate (ADP), which has been likened to an uncharged battery, has the capacity to pick up a third phosphate and become the energy source for all biological activities. This crucial molecule—the "charged battery"—is adenosine triphosphate (ATP), and as we shall see later, the energy from all the fuels we burn is eventually stored in usable form as ATP molecules.

The **nucleotide coenzymes,** which are closely related to the B vitamins, work with enzymes to play a major role in metabolism. The two most important nucleotide coenzymes are flavin adenine dinucleotide (FAD) and nicotinamide adenine dinucleotide (NAD). Both act as hydrogen carriers, and they are recycled over and over again.

Nucleic acids, such as deoxyribonucleic acid and ribonucleic acid, are long polymers of ribo- or deoxyribonucleotides. They are found especially in the nuclei of cells, where they play a role in the transmission of hereditary characteristics and in the control of cellular activity through the synthesis of proteins (see Table 3.1).

TABLE 3.1
Nucleotides and related chemicals: Their role in metabolism

ABBREVIATION	FULL NAME	ROLE
cAMP	cyclic adenosine monophosphate	intermediate in many hormone reactions
ADP	adenosine diphosphate	can be transformed into the energy-carrier ATP
ATP	adenosine triphosphate	energy carrier
NAD	nicotinamide adenine dinucleotide	electron carrier in many metabolic reactions
NADP	nicotinamide adenine dinucleotide phosphate	electron carrier in many metabolic reactions
FAD	flavin adenine dinucleotide	an electron carrier in metabolic reactions
Co A	coenzyme A	important enzyme in metabolism of carbohydrates and fatty acids
DNA	deoxyribonucleic acid	stores and transmits genetic information
RNA	ribonucleic acid	intermediate between DNA and protein

As we have seen in this chapter, the chemistry of life is the chemistry of individual molecules, atoms, and even electrons. We can now enumerate five reactions that are basic to all life processes and that, aside from the relatively more complex molecules being acted upon, really involve as the key performers only the simplest of compounds and particles. These reactions are (1) hydrolysis, in which molecules are broken down into smaller units through the interposition of a molecule of water; (2) dehydration-condensation, in which molecules are linked together by the removal of a molecule of water; (3) transfer, in which a part of a molecule, such as a methyl group, an amine, or a phosphate, is separated and attached to another molecule; (4) oxidation, which is the loss of electrons by one atom to another, resulting in the formation of a bond between the atoms; and (5) reduction, which is the gain of electrons by one atom from another.

Summary

1. There are two major categories of chemistry. Organic chemistry concerns compounds structured by carbon linkages and primarily produced in nature by living things. Inorganic chemistry concerns those compounds not necessarily associated with life.

2. Water is indispensable for life. Because it is an excellent solvent, it serves as a medium for cellular reactions, and performs support and transport functions as well. Because it retains heat and does not change temperature easily, it helps organisms to maintain constant internal environment. It also plays an essential role in chemical reactions such as the breakdown and synthesis of organic compounds.

3. The structure of organic compounds and the existence of life itself are dependent on the bonding properties of carbon, an atom which can form four different covalent bonds at one time, leading to the synthesis of complex molecules.

4. Many living organisms carry on respiration, the oxygen-requiring process in which carbohydrates are broken down, releasing the energy necessary for life processes.

5. The four major classes of biologically important organic compounds are carbohydrates, lipids, proteins, and nucleotides.

6. Enzymes are organic catalysts made of protein. Each enzyme acts on a particular substrate, lowering the activation energy necessary for a specific reaction.

7. Life processes occur by virtue of five types of chemical reactions: hydrolysis, the breakdown of molecules through the interposition of a molecule of water; dehydration, the linkage of molecules through the removal of water; transfer, the movement of parts of molecules to other molecules; oxidation, the creation of a bond through the loss of electrons by one atom to another; and reduction, the creation of a bond through the gain of electrons by one atom from another.

REVIEW AND STUDY QUESTIONS

1. What are the important properties of water? How do those properties affect your general well-being?

2. In what ways is carbon dioxide essential to life in general and to your life in particular?

3. What "paradoxical role" does ammonia play in the scheme of life? How does the human system adapt to that role?

4. What are "organic" compounds? Why are they so named?

5. What are the component parts of a protein? A carbohydrate?

6. What is an unsaturated fat? What tells you that a fat is unsaturated?

7. What does an enzyme do?

REFERENCES

Kendrew, J. C. 1961 (December). "Three-Dimensional Structure of a Protein Molecule." *Scientific American* Offprint no. 121. Freeman, San Francisco.

Koshland, D. E., Jr. 1973 (October). "Protein Shape and Biological Control." *Scientific American* Offprint no. 1280. Freeman, San Francisco.

Lambert, J. B. 1970 (January). "The Shapes of Organic Molecules." *Scientific American* Offprint no. 131. Freeman, San Francisco.

Lehninger, A. L. 1971. *Bioenergetics*, 2nd edition. Benjamin, New York.

White, A., P. Handler, and E. L. Smith. 1973. *Principles of Biochemistry*, revised edition. McGraw-Hill, New York.

SUGGESTIONS FOR FURTHER READING

Bailar, J. C., Jr. 1971. "Some Coordination Compounds in Biochemistry." *American Scientist* 59: 586–592.

Reviews the role of metals in life processes. Focuses on their importance in cytochromes, coenzymes, enzymes, and metaloproteins.

Baker, J. J. W., and G. E. Allen. 1974. *Matter, Energy, and Life*, 3rd edition. Addison-Wesley, Reading, Mass.

Excellent introductory discussion of the chemical structures of organic compounds, including a separate chapter on matter and energy. Many clear illustrations.

Blum, H. 1969. *Time's Arrow and Evolution*, revised edition. Harper Torchbooks, New York.

Thoughtful examination of the laws of thermodynamics, the limitations placed on living systems by these laws, and the mechanisms that have developed in order to compensate for these limitations.

Rose, S. 1968. *The Chemistry of Life*. Penguin Books (Pelican), Balitimore.

Analysis of basic chemical systems and control systems, both intracellular and intercellular. Well documented and highly readable.

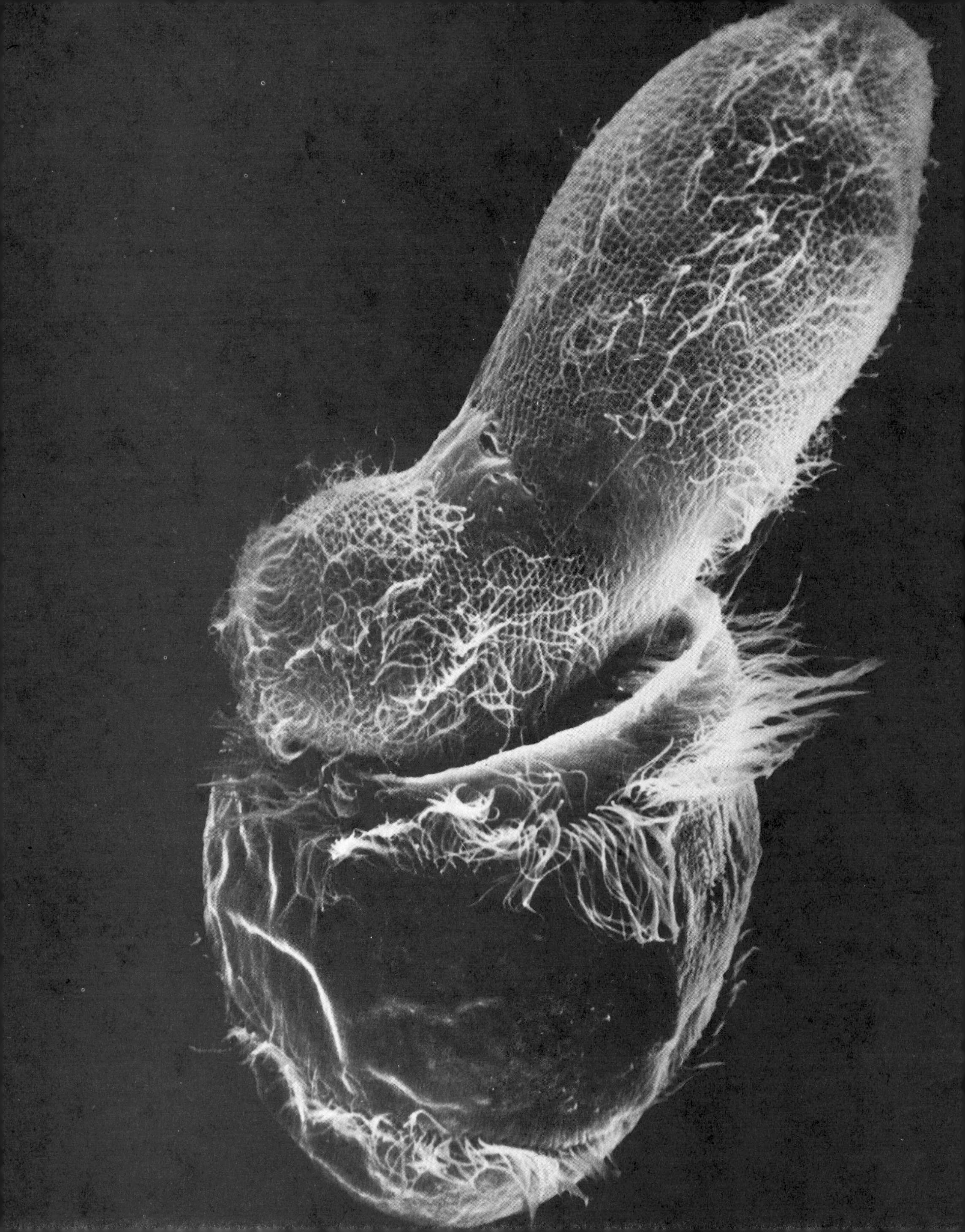

THE CELL
chapter four

In this scanning electron micrograph, the barrel-shaped didinium is making a meal of the paramecium by engulfing it. Both organisms are one-celled ciliates, specialized protozoa that ingest their food and propel themselves with beats of their hairlike cilia.

Eager to define life and tempted to create it, we have turned to science for an explanation of the chemistry of living things. We can, in fact, list the chemical compounds that organisms are made of. For many of these chemical compounds, we can also specify the quantities that are generally present, in relation to other substances, in a living thing. Yet we would no more expect a living being to come out of a test tube containing such a mixture than we would expect to drive away in an automobile whose parts were laid out in their proper relationships on a garage floor. Just as an automobile will operate only when correctly assembled, an organism will function only when its chemical components are properly arranged.

Life, then, depends on the precise *organization* of the various chemical compounds that compose and sustain it—compounds such as structural proteins, lipids, and carbohydrates, as well as many functional molecules, such as the enzymes needed to catalyze necessary chemical reactions.

If life depends on the precise organization of components, then a study of this organization might be expected to yield profound insights into living things. Initially, however, we face the problem of deciding where to begin this study of biological organization. Varieties of life abound; there are complex organisms, such as man or an oak tree, with their many organs, tissues, and specialized cell types, as well as simpler systems, such as the sponge, the bacterium, or the lichen.

Faced with this diversity, the biologist has searched for a basic unit of organization that is common to all forms of life. That common denominator is the cell, the smallest structural unit of an organism that is unquestionably alive.

All organisms—from the simplest of the bacteria to man—are either individual cells or massive collections of cells. A man, though he develops from a single cell, consists of billions of cells; each of the bacteria is a single cell, complete in itself. Regardless of their number in an organism, all cells perform the functions that are essential to life. All cells receive and transform energy; all synthesize large molecules from simpler substances; all regulate their internal environment and are capable of growth, replication, or division; and all react to changes in the outside environment.

Cell theory

Only since the early part of the nineteenth century has it been known that all organisms are composed of cells. Prior to that time, scientists had applied reasoning similar to ours in their approach to the study of living organisms. That is, they attempted to find some structural unit that was common to all living things. Unlike the atom, whose existence was postulated from observations of chemical reactions, the cell could not be inferred merely from observations of organisms, or even from studies of organs. The discovery of cells had to await the advent of the microscope.

In 1665 Robert Hooke, an Englishman, published a treatise in which he reported some of the observations he had made using an early microscope (with a 42X magnification). Examining a thin slice of cork—reputedly from a recently drained wine bottle—he found, to his amazement, that it consisted of row upon row of tiny chambers of the same size and shape. These chambers he called **cells,** though they were in fact cell walls, devoid of living substance (see Fig. 4.1).

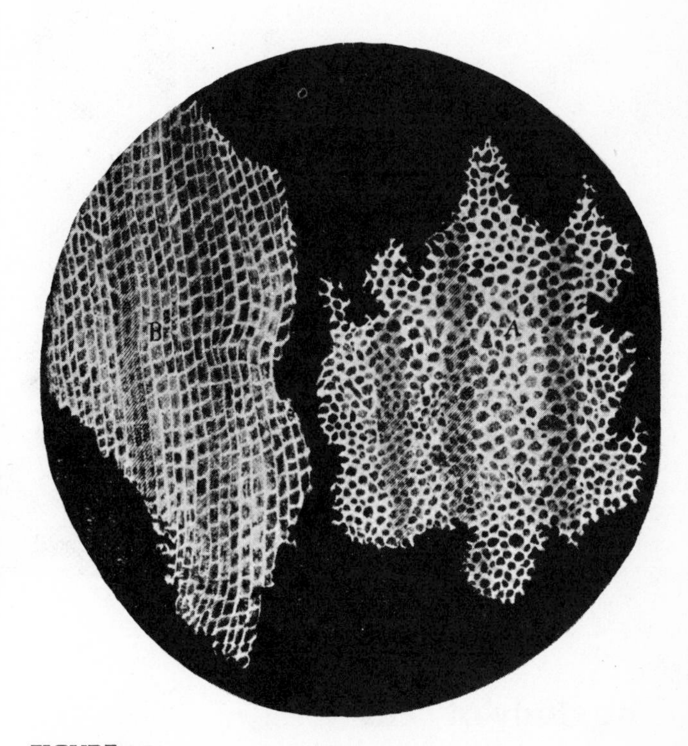

FIGURE 4.1
A reproduction of a drawing made by Robert Hooke and published in 1665. The drawing shows the cellular nature of a piece of cork, which Hooke examined under a primitive microscope. The living contents of the chambers have died, leaving only the walls that surrounded the individual cells.

It remained for later scientists to formulate the theory that plants and animals were cellular in nature. As early as 1824, a French physiologist, R. J. H. Dutrochet, stated the cell theory in a precise manner, but his pioneer effort is often overlooked. The origin of the cell theory is usually set at 1838, when the German scientist M. J. Schleiden summarized experimental findings on cells in a wide variety of plants. The next year Theodor Schwann, a German zoologist working independently, announced his discovery of cells in animals. These two discoveries served as the basis for the cell theory, which states that all living things are composed of cells, that there is no life that is not characterized by and organized into cells.

It was not until 1858, however, that the idea was advanced by Rudolph Virchow, another German, that cells begat cells—that no cell could be generated spontaneously from nonliving matter. Virchow's thesis was supported by experiments conducted by the French scientist Louis Pasteur, among others. Pasteur demonstrated that a flask of boiled broth could be kept free from microorganisms so long as it was protected from the air, whereas a similar flask, exposed to the air, was soon teeming with organisms. Pasteur's experiments indicated that the microorganisms that appear in broth, bread, or once-fresh meat do not arise from these nonliving substances, but that they come from the air to which these substances are exposed. The experiment supported Virchow's theory that cells arise only from preexisting cells—that living things are generated by other living things of their own kind. However, these experiments did not disprove that spontaneous generation had ever occurred—for example, in the origin of life on earth.

FIGURE 4.2
The microscope used by Robert Hooke (A) and a typical light microscope of today (B). The primitive microscope could magnify an object about 40 times, as compared with magnification of 1000 by the modern instrument. Much greater magnifications are possible with modern electron microscopy.

The study of cells

Much has been learned about the cell since the days of Hooke, Schleiden, and Schwann. Infinitely better microscopes and improved techniques of preparing cells for study have enabled us to observe not only cells, but the minute specialized structures (called **organelles**) within the cell, and the complex processes in which they take part.

One instrument that has enabled biologists to make great progress in cell biology is the electron microscope. Whereas ordinary microscopes use light, which is bent by lenses, to magnify the object being studied, the electron microscope uses beams of electrons with wavelengths that are thousands of times shorter than those of visible light. The physical properties of the electron beam make it possible for the electron microscope to resolve structures as small as 0.0005 micron. (A micron, commonly abbreviated by the symbol μ, is 1/1000th of a millimeter, or 4/100,000ths of an inch.) Two kinds of electron microscope are widely used in biology. The transmission electron microscope allows very high resolution, but only on very thin sections (see Fig. 4.3A). The more recently developed scanning electron microscope reveals features of the surface of a specimen by sweeping a narrow beam back and forth across it (see Fig. 4.3B). There is less likelihood of distortion, because specimens do not have to be sliced thin.

Other types of microscope, notably phase-contrast, interference, fluorescence, and polarizing microscopes, are able to show various cellular structures in greater detail, through delicate manipulations of normal light conditions.

A major shortcoming of the use of some types of microscope, including the electron microscope, is that most of the various techniques employed to prepare cells for study—fixing the organelles in place and staining them—kill the cell in the process. Thus cell structure can be viewed under the microscope, but how the organelles actually function in a live cell taken from, say, a kidney, cannot be observed. Even in the observation of cell structure, problems sometimes arise. Techniques such as fixation may result in our seeing "artifacts" or structures which either are not really parts of the actual

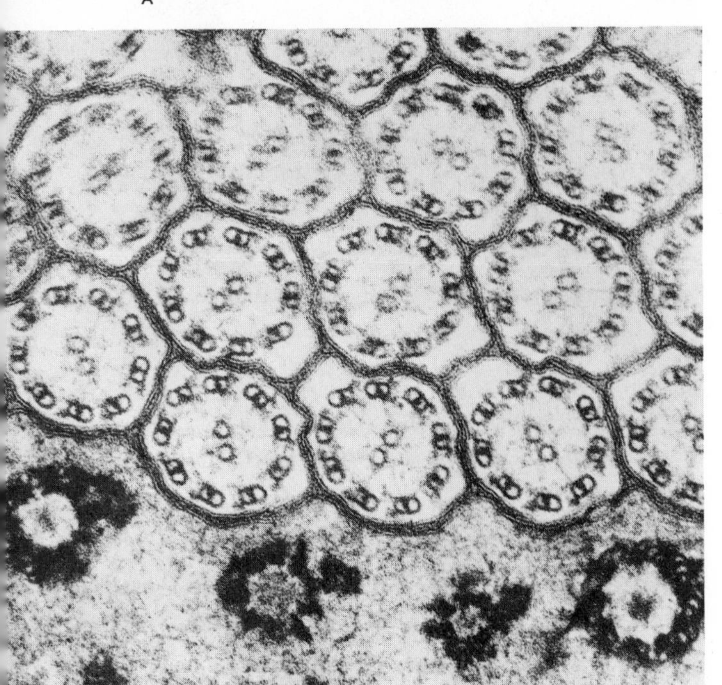

FIGURE 4.3

Contrasting views of cilia produced by the two kinds of electron microscopy. (A) The transmission electron micrograph (TEM) reveals the orderly structure of each cilium, with its two central fibrils, or microtubules, and its nine pairs around the outer edge (see also Fig. 4.13). (B) The scanning electron micrograph (SEM) shows in typical three-dimensional form a group of cilia on the outer surface of a planarian.

cell or represent serious distortions of the actual structure.

In addition to using microscopes, biologists explore cell structure by means of centrifuges, machines that use a spinning mechanism to separate substances of different densities, and thus make possible the isolation of different cellular components for study. Centrifuges of different speeds are used, the fastest of which is the ultracentrifuge. The centrifuge technique is another example of the compromise sometimes made in the quest for biological information. Preparing the various organelles for study necessitates killing the cell itself. Nevertheless, the techniques for separating cell organelles and keeping them active in an artificial but cell-like medium have proved to be invaluable aids for the study of their functions. Figure 4.4 illustrates some of the techniques of centrifugation and chromatography, another method employed in the study of cells.

To study cells in their living state, biologists resort to more delicate and imaginative methods. For example, chemical reactions may be followed by injecting radioactive isotopes into the cell. Even the organelles may be removed from a cell by a process known as microsurgery; in this way, studies can be made of how a cell gets along without, perhaps, its nucleus, or how it functions with the nucleus from some other cell. These and other methods have contributed much information—and inspired much speculation—on the processes of the cell.

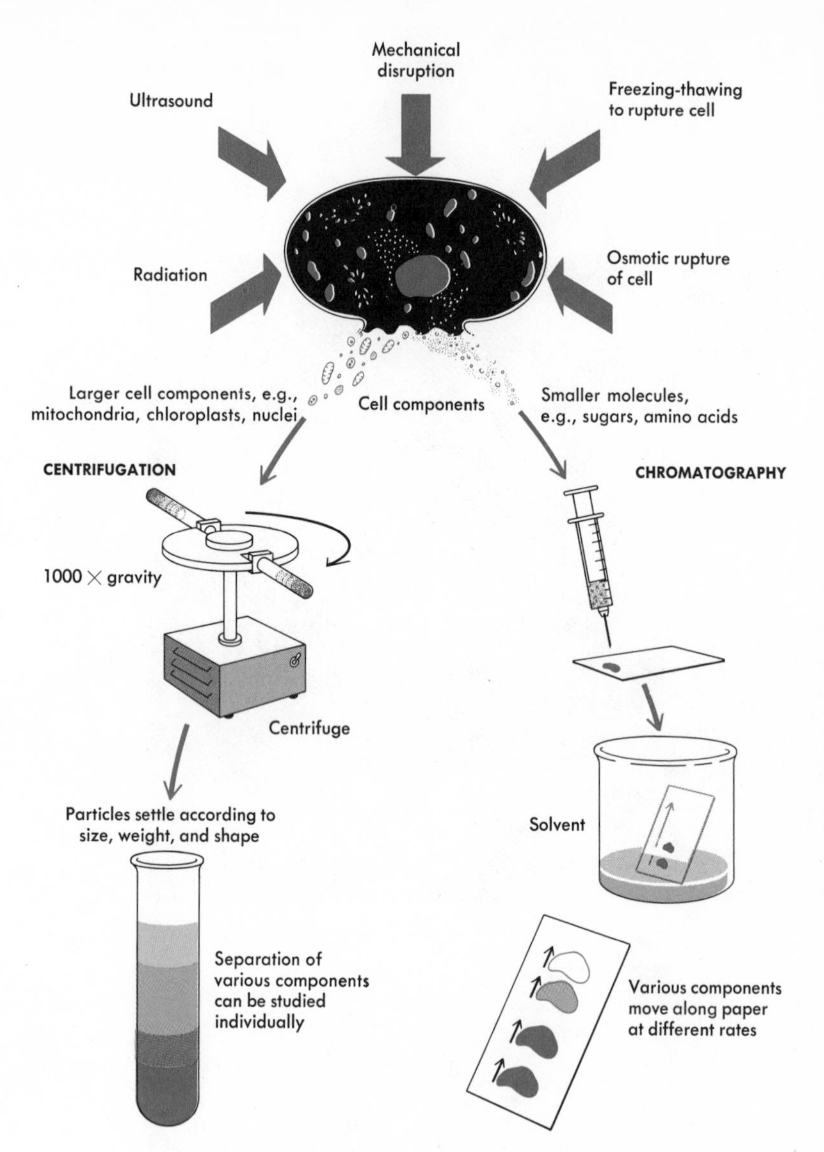

FIGURE 4.4
Study of cells. Cellular components may be studied in a number of ways. Intact and functioning organelles may be obtained by breaking many cells open by mechanical means or other disruption of the cell membrane. Centrifugation at very high speeds in a special medium separates the organelles on the basis of their specific gravity, and layers form in the centrifuge tube. Each layer contains only one kind of organelle. The intact organelles may then be carefully withdrawn from each layer and their functions studied in the test tube. The smaller molecular components of a cell may be studied by chromatography. For example, the pigments in a leaf may be extracted from broken cells and spotted onto a piece of filter paper dipped in a solvent. As the solvent moves up the paper, it carries the pigments along with it at different rates. Spots form of each pigment in the mixture, and each pigment can thus be studied separately.

Cell form and function

The cell is the common denominator of life; yet there is no such thing as a common or typical cell. Cells vary enormously in their shapes, sizes, and components. In multicellular organisms the variation is particularly noticeable because most mature cells have undergone a process of **differentiation,** or change, in order to perform a specific function in an organism. Such differentiation enables the organism to house cells specializing in all the operations necessary to its survival—in energy transfer, the synthesis of complex molecules, or reproduction, for example.

In biology it is often observed that function and structure account for the same fact. For example, the fact that the tiger's claw is sharp and strong (structure) and the fact that this claw is used to tear the flesh of the tiger's prey (function) are, in a sense, identical. Similarly, in cell biology, the structure and function are virtually inseparable.

The fish eggs in caviar are really only single cells, though unusually large ones. Most of each egg's bulk consists of the yolk, the function of which is to store food and water that would be needed by a growing embryo should the egg be fertilized. On the other hand, the red cells in blood are quite small and can be seen

Microscopes—Revealing More than Meets the Eye

In 1665, Robert Hooke viewed a slice of cork through a primitive microscope that magnified his specimen 42 times. Today, electron microscopes can achieve useful magnifications of a million times on their viewing screens. The word "useful" is important—a specimen can be magnified a million times, but it is useful only if it's in focus. The ability to clearly reveal detail in a magnified image is called the *resolution,* or *resolving power,* of a microscope. For example, if a microscope has a resolution of 75 angstroms [1 angstrom (Å) = 1/250 millionth of an inch], it can give a clear image of a cell membrane, which is 75–100 Å thick. Objects smaller than 75 Å will appear blurred or distorted.

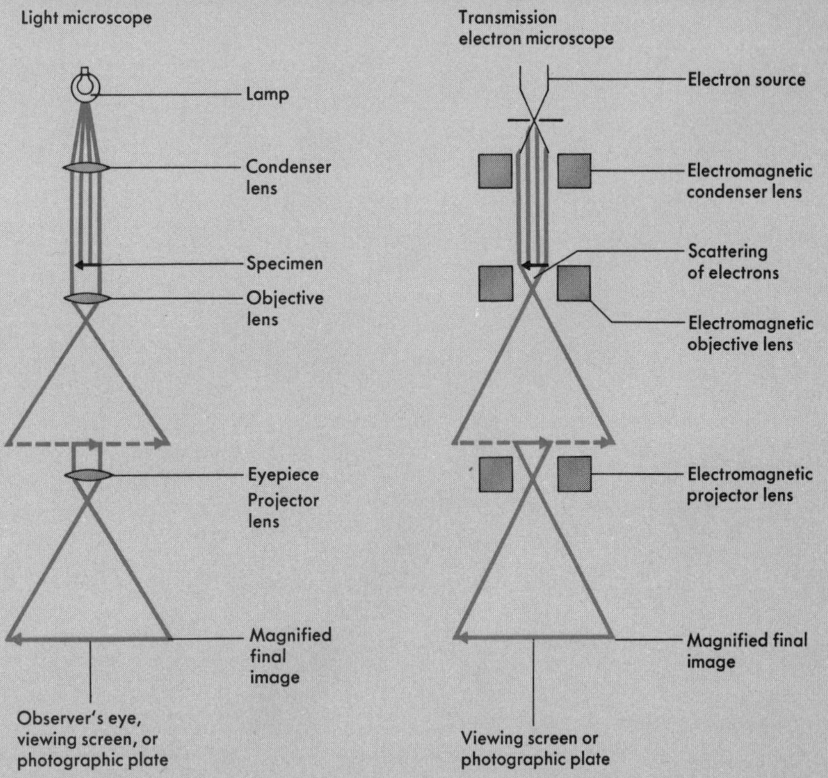

Light microscope

Lamp
Condenser lens
Specimen
Objective lens
Eyepiece Projector lens
Magnified final image
Observer's eye, viewing screen, or photographic plate

Transmission electron microscope

Electron source
Electromagnetic condenser lens
Scattering of electrons
Electromagnetic objective lens
Electromagnetic projector lens
Magnified final image
Viewing screen or photographic plate

Light versus electrons. A light microscope (LM) uses waves of light to produce an image. The beam of light is refracted (bent) through glass lenses and focused on the whole surface of the specimen. As it passes through the specimen, the light beam is scattered. The scattered rays are then refocused by glass lenses to produce a magnified image on a viewing screen, a photographic plate, or a person's eye. Because light can be focused on only one level of the specimen at a time, the image projected is two-dimensional.

Visible light travels in wavelengths of approximately 3500–7500 Å, and with special types of lenses, it is possible to see specimens as small as 2000 Å. The highest useful magnification of a light microscope is about 1000.

In an electron microscope (EM) a stream of electrons—with properties similar to a beam of light—is emitted from an electron gun, which is somewhat like a filament in a light bulb. This stream of negatively charged electrons is focused on the specimen by a series of electromagnetic "lenses." Since the electron beam is charged, it can be bent by electromagnets, just as glass lenses refract light.

The *transmission electron microscope* (TEM) is similar to the light microscope in operation. The electron beam focuses on the surface of the thinly sliced specimen (approximately 100 Å thick). After passing through the specimen, the scattered electrons are focused by more electromagnets onto a viewing screen or photographic plate, which converts the electrons into a visible two-dimensional image (see Fig. 4.3A for a transmission electron micrograph). Since electrons in motion are guided by waves 1–2 Å long, electron microscopes have greater resolution than light microscopes. A TEM has a resolution of 1–5 Å and a maximum useful magnification of more than a million.

The *scanning electron microscope* (SEM) works more like a television than a microscope. It can have a resolution of about 100 Å, depending on magnification, which can range to 100,000. But more important, the SEM can give depth of focus to an image and produce a three-dimensional impression of a specimen (see Fig. 4.3B). Neither the

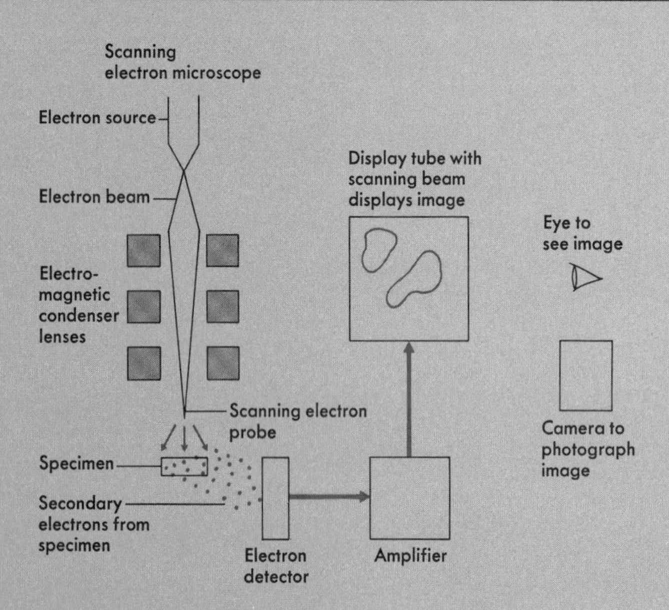

TEM nor the LM can produce such a spectacular image.

As in the TEM, there is a stream of electrons, but electromagnets in the SEM reduce the stream to an intense beam of electrons, an *electron probe*. This probe slowly moves across, or *scans*, the specimen point by point. Since a specimen may have a million or more points to scan, the process takes several minutes. The probe follows and focuses on *all* levels of the specimen's surface—bumps, depressions, and wrinkles.

Because SEM's focus on multiple levels of the specimen, the specimen doesn't have to be as thinly sliced as for an LM or TEM. Therefore the SEM can depict a three-dimensional image, and you can see a more realistic view of the specimen.

As the probe hits each point, it causes the specimen to give off electrons of its own, called *secondary electrons*. The secondary electrons are gathered by an electron detector—equivalent to a TV camera—and passed through an amplifier, which changes the electrons into electronic signals of varying strength. To complete the formation of the image, an electron beam scans the face of a display tube—just like a TV picture tube. The two scanning beams, one on the specimen and one on the tube, move at the exact same time in the exact same pattern. The varying strengths of the electronic signal from the specimen affect the strength of the display tube's electron beam. The variations in strength produce light and dark areas on the picture tube, which together form the image of the specimen presented on the screen. Photographs taken of the image are called scanning electron micrographs.

We have pointed out the similarities in the operation of these microscopes, but physically and financially, they are not at all alike. A light microscope can be carried from table to table, requires no special installation, and costs about $1000, depending on how complex it is. However, an electron microscope requires a great deal of space. It also requires special electrical wiring as well as technicians to complete a complicated and precise installation. An electron microscope is not as easy to operate or as portable as an LM, and the cost can range from $15,000 to $250,000, depending on how elaborate it is.

Importance of electron microscopes. All these microscopes have played an important role in many scientific disciplines. They have enabled scientists to confirm and test hypotheses in a way that was not available to them as recently as a half century ago. Biologists in particular have furthered their knowledge of cell structures, functions, and chemical activity with EMs. Before the discovery of the EM, disease-causing viruses could not be seen. Now scientists know their structure and chemical composition, as well as how they get into cells and reproduce themselves.

New applications for electron microscopes are still being explored in hospitals and laboratories. Although first developed in the 1930s, they were not generally available until the 1960s, and so they are relatively new. Perhaps one application in the near future will be the analysis of DNA—the location and identification of separate genes and the furthering of our understanding of how we become what we are.

only through a microscope. Since red blood cells must course throughout the body, traveling in the narrowest of blood vessels, it follows that microscopic size is of tremendous selective value. Cell size is therefore clearly related to the function of the cell in the organism and to the type of environment in which it exists.

A striking example of the structure–function relationship can be seen in the extraordinary shape of human nerve cells, which have extensions as long as 4 feet and are often less than 1 μ thick. These dimensions make the nerve cells particularly well suited to conducting electrical impulses between the brain and the innumerable structures of the body's internal environment, as well as those that interface with the external environment. Another structural adaptation is evident in the cells of certain invertebrates' skin. These cells often take the form of polyhedrons, solid figures with many faces. Their shape allows many cells to fit together snugly, so that not much open space is left between them. The skin can thus function, in one sense, as a protection against entering organisms and consequent infection.

Although cells vary in size from large-diameter egg cells down to the 0.02 μ width of certain bacteria, most cells range in diameter from 0.5 μ to 40 μ. This apparent limitation in diameter suggests that cell size and shape may be determined by factors other than the general functions of the cell.

One factor limiting the size of a cell is the **surface-to-volume ratio.** The volume of a cell cannot be too large relative to the surface area. The larger the volume compared to the surface, the less will be the capacity of the surface to allow materials such as oxygen, water, and sugar to pass into the cell and waste products to leave. Although it is quite tempting to think that all the cell needs to do is to increase its surface area right along with any increases in volume, the fact is that as the cell grows in size, its volume increases disproportionately to the increase in its surface area. The surface area increases as the square of the radius of a sphere, but the volume increases as the cube.

It follows, therefore, that the more active the cell—that is, the more materials its surface area must allow passage for within a given time—the smaller the cell volume will be. Many unicellular organisms have overcome the problem of limited surface area by the development of cytoplasmic projections, which increase the surface area available for the exchange of materials.

Another factor limiting cellular size is information control; the larger the cell, the farther away are the various parts from the cell's information control center, the nucleus. This principle applies to any kind of organization: for example, the more governmental agencies and departments there are, the less efficient the government is likely to be in administering its own vast bureaucracy.

Structures of the cell

Because a cell's overall structure varies in accordance with that cell's function in the organism, we find numerous structural differences between cells. However, cells by definition have certain functions in common, and they have evolved an array of specialized internal structures that enable them to carry out these common functions. (In the sections that follow, we will examine many of these structures, but bear in mind that there is no such thing as a typical cell. Some cells have only a few of the structures to be discussed, and a few primitive ones, in fact, have none of them.) The electron microscope reveals that cells do vary in their possession of these specialized structures, and quite possibly there is variation in their use as well (see Fig. 4.5). All higher cells by definition share at least one feature—membranous partitions that divide the cell into the highly specialized structural and functional units known collectively as organelles. Both animal and plant cells generally have organelles which serve to keep physically and chemically separate such chemical processes as protein synthesis, respiration, and photosynthesis. This separation prevents one kind of process from interfering with another and improves the overall efficiency of the cell.

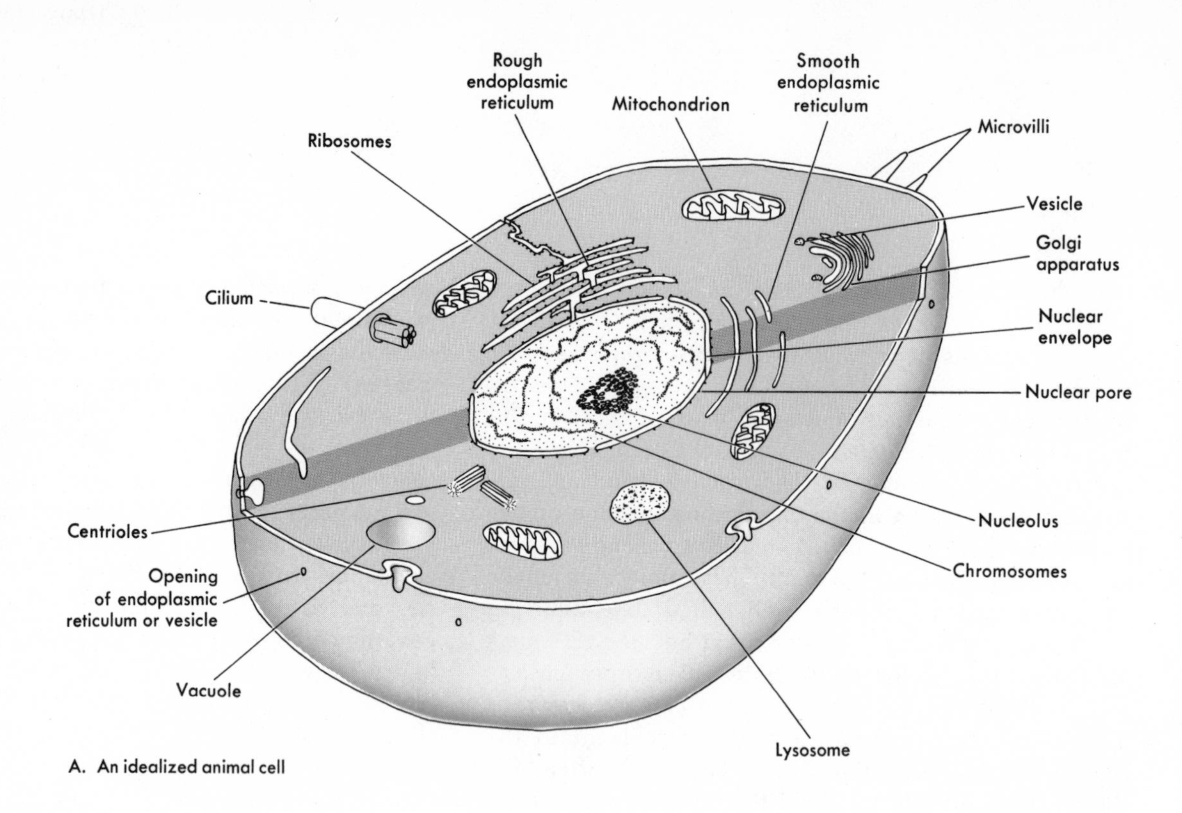

A. An idealized animal cell

CELLULAR MEMBRANES

Under the microscope, the most distinctive feature of a cell is its system of membranes—the outer cell membrane that defines the shape of the cell, the nuclear membrane that encircles the nucleus, and the membrane configurations that criss-cross its interior. (A review of Fig. 3.4 is suggested at this time.)

The chemistry of cellular life is, in fact, controlled by membranes. As the cell's immediate link to its environment, the outer **cell membrane** acts as a selective barrier to molecules entering and leaving the cell. This membrane is in no sense a purely physical or mechanical barrier; it is capable of constantly responding to the physiological "needs" of the cell, admitting and expelling various materials. (Even when the contents of a cell are removed, the membrane may remain differentially responsive, as has been shown experimentally in a variety of cells.) Internal membrane structures are similarly important. They may provide a surface along which protein synthesis occurs, and, in addition, form boundaries between areas in which different chemical reactions are taking place.

Early observations on the permeability of the outer cell membrane showed that particles soluble in lipids appeared to dissolve in the membranes, passing easily through. Thus it was hypothesized that the membrane consisted of a layer of lipids. However, water-soluble particles also passed through, and that could not occur if the membrane was composed solely of lipids. These and other observations led to the hypothesis of a lipid-protein configuration.

Such a configuration was the basis for a structural model of the cell membrane proposed in the 1930s by two biologists, J. H. Danielli and H. Davson. They hypothesized that the cell membrane is a double layer of lipids sandwiched between two continuous layers of globular proteins. To account for the accepted fact that water-soluble molecules pass through the cell membrane, the model included "pores" lined with protein and extending through the entire thickness of the membrane.

Danielli and Davson were working under the handicap of studying the cell membrane only indirectly. With the development of the electron microscope in the early 1950s, it became possible to view the membrane itself. J. D. Robertson's measurements of the thickness of cell membranes supported the Danielli-Davson hypothesis in general. Robertson saw a three-layer structure that was about 75 angstroms thick. (An angstrom is 1/10,000 of a micron, and the unit is usually abbreviated as Å.)

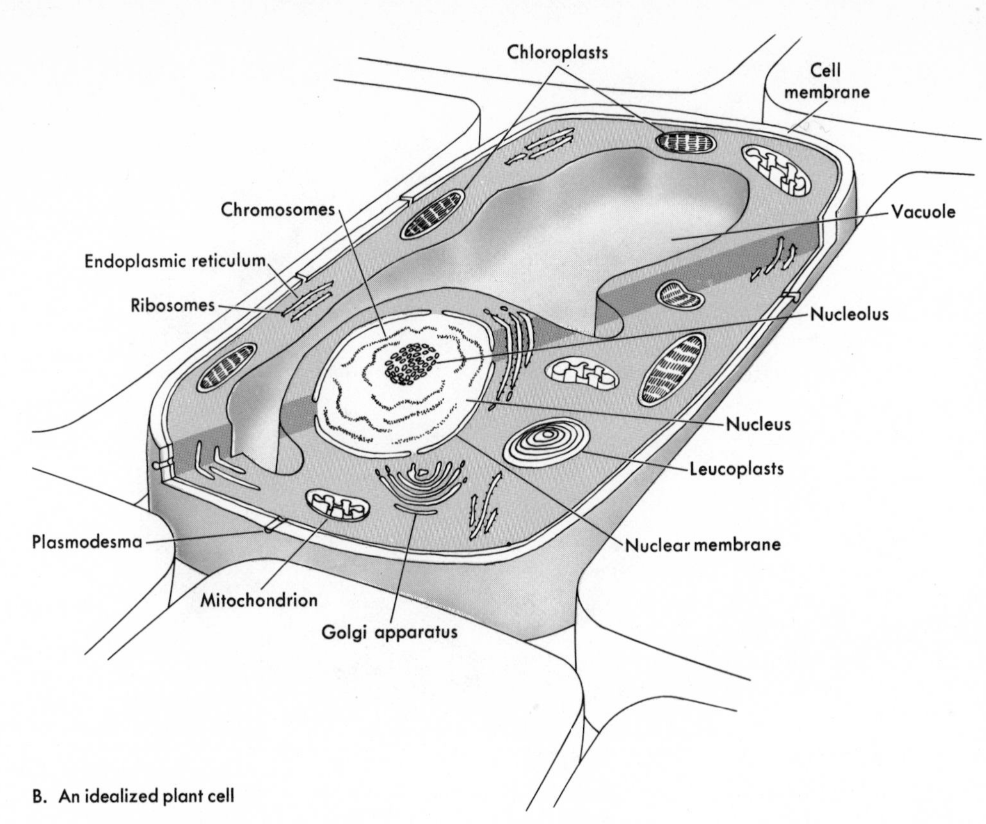

B. An idealized plant cell

FIGURE 4.5

Cellular form. Although the cell performs the basic functions of life in both plants and animals, its form differs in the two. Because the animal cell (A) has no rigid wall, its shape is often less regular than that of the plant cell. The plant cell (B) typically has a large central vacuole for storage of wastes, so its organelles (mini-organs) are distributed in the cytoplasm around the outer edge of the cell. The cells pictured here are composites that include representative organelles.

Robertson suggested that the proteins were spread in flattened sheets over the double layer of lipids, rather than being globular. It was the arrangement of the proteins in this three-layer structure on which scientists next began to focus their attention. Did the proteins really form continuous layers, as Danielli, Davson, and Robertson had postulated? Improved methods of preparing specimens for detailed examination led to the discovery of globular structures *within* the membrane, that is, in the portion previously thought to be composed exclusively of lipids. Although the chemistry of these globules is still under study, evidence that they are proteins warranted a major revision of the Danielli-Davson model pertaining to the arrangement of protein molecules in the cell membrane.

In the 1960s, S. J. Singer and D. Wallach, working at separate universities in the United States, independently arrived at a structural model—now known as the fluid-mosaic model—in which globular proteins are located within the cell membrane, extending through it rather than forming its outer layers (Fig. 4.6). These globules are visualized as floating in the double lipid layer like icebergs floating in a sea. Some are buried in the lipids with only their tips projecting on either side of the membrane. Others extend all the way through

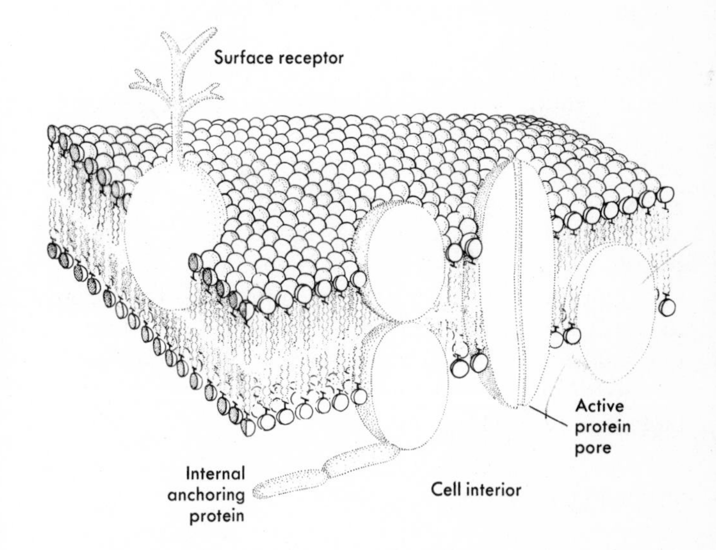

FIGURE 4.6

The fluid-mosaic model of a membrane. Globular proteins of various sizes and shapes are embedded in the bilayer of lipids. Some protrude on only one side; others extend through the entire bilayer and transport materials into and out of the cell. On the surface are specialized receptor proteins. Cell shape is in part determined by anchorlike proteins attached to the membrane interior.

the double layer of lipids. Since the interior of the lipid layer repels water (or is **hydrophobic**), the part of the protein molecule embedded in it must also repel water. On the other hand, since the projecting tips are continuously bathed in water, they must attract water—that is, they are **hydrophilic.** A protein molecule can both attract and repel water only if it folds over, becoming globular. But how does a molecule that is essential to a cell's existence pass through both a water-attracting tip and a water-repelling center? In other words, how do molecules of nutrients and waste products pass through the cell membrane? Present thinking suggests that there is more than one answer to this question.

Polar molecules (e.g., water), sodium, potassium, and chloride ions, and other small molecules can pass by simple diffusion through the pores in the membrane. Physiological evidence supports the existence of such pores even though they have not as yet been seen. For transporting larger substances, Singer postulated carriers in the form of water-filled channels in the protein globules. A molecule entering or leaving the cell floats into such a channel, which expands and contracts, thus squeezing the molecule through. Oil-soluble molecules can pass directly through the lipid layer of the membrane.

It is not only possible but even probable that scientists will modify and refine this currently accepted working model of the cell membrane as laboratory work proceeds in one of the most active and exciting areas of biological research.

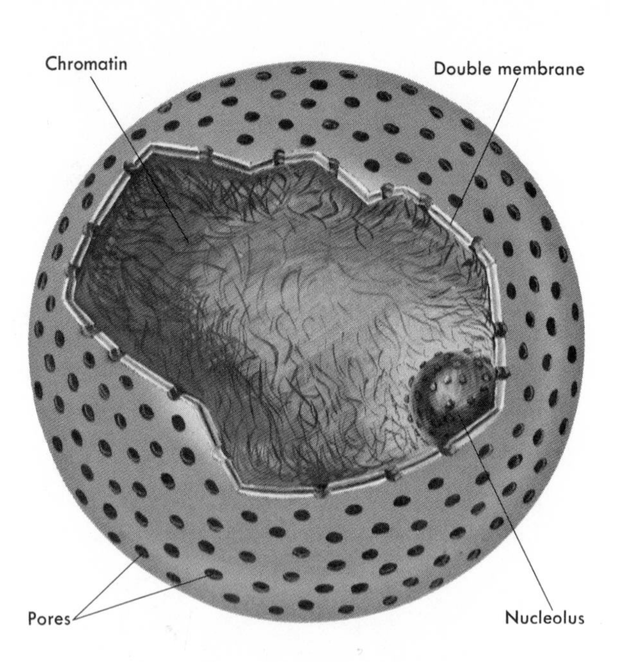

FIGURE 4.7
Diagrammatic drawing and micrograph of a cell nucleus. The nucleus is usually spherical. The nuclear envelope is composed of a highly perforated double layer of membrane, which permits materials to pass between the nucleus and the cytoplasm. Contained within the nucleus are masses of tangled genetic material, chromatin, and one or more nucleoli, which direct the synthesis of ribosomal RNA.

THE CELL NUCLEUS

An organelle located at the center of many cells, the nucleus is actually the cell's control center, bearing roughly the same relationship to the cell as does man's brain to his body. Some cells have no nuclei (an important example is the red blood cell). When a nucleus is present, it is usually spherical although in some cells it is oval or lobed like beads on a thread.

Every nucleus is surrounded by a membrane with properties and composition similar to those of the cell membrane. One important difference, however, is that the nuclear membrane is doubled, consisting of the equivalent of two cell membranes, one inside the other. The nuclear membrane is punctured by distinct pores, as suggested in Fig. 4.7. These pores are thought to facilitate the passage of the relatively large nucleic acids or the movement of other complex molecules.

The nucleus contains the nucleic acids, RNA and DNA, molecules that code for the manufacture of cell components and regulate most activities of the cell. It is by means of a certain type of RNA, discharged into the cell cytoplasm, that the nucleus directs the cell's activities—altering the rate at which chemicals are built and destroyed, controlling the cell's reproductive processes, and transmitting the hereditary instructions. Important to these functions are the genes, portions of DNA molecules carrying chemical codes of genetic information. DNA is the protein-bound nucleic acid (the DNA-protein complex is sometimes called chromatin) which contracts to form rod-shaped chromosomes. It

may be observed under the microscope with the aid of dyes.

In addition to the nucleic acids and chromosomes, the nucleus contains small oval bodies called the nucleoli. Nucleoli synthesize ribosomal RNA, which in turn manufactures protein.

THE CYTOPLASM

Lying outside the nucleus and inside the cell membrane is the **cytoplasm,** which may be considered either as a descriptive term for the region or as the contents of the region.

Cytoplasm is essentially a complex mixture of chemicals. It includes water, proteins, fat globules, simple sugars, amino acids, and various inorganic elements and compounds. How is such a watery mixture able to hold itself together to assume even a remote semblance of structure, let alone the complicated structural integrity that it has? The ability of cytoplasm to serve as the basic structural medium of the cell is largely due to its capacity to exist in varying stages from a liquid to a semisolid gel. These stages occur simultaneously; the deeper portion of the cell is usually more fluid than the outer regions, which are often in the semisolid gel state.

Cytoplasm can exist in several stages of fluidity at the same time because it is a watery mixture, and substances mixed in water can exist in three forms, depending on the size of the particles mixed into the water. A solution consists of very small particles scattered evenly throughout. In a suspension, somewhat larger particles cling together in tiny clumps and tend to settle to the bottom. Intermediate between these two forms is the colloid, which can also be seen in two states: a sol, or liquid state, and a gel, or semisolid state. In the cell, part of the cytoplasm is in the form of a solution, part is in the sol state, and part is in a gel. However, most of the cytoplasm is a colloid consisting of large protein molecules and tiny droplets of fat that are scattered in aqueous solution (the cytosol).

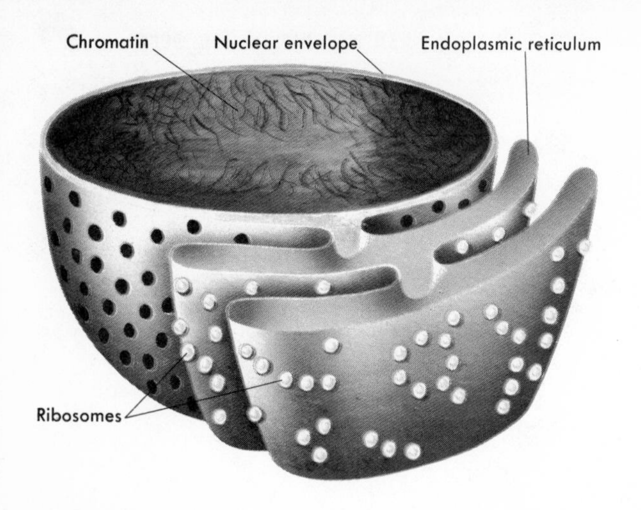

FIGURE 4.8

The relationship between the endoplasmic reticulum (ER) and the nuclear envelope. The internal spaces of ER often appear to be connected to the space between the inner and outer membrane of the nuclear envelope, perhaps permitting exchange of materials between the nucleus and the ER.

The endoplasmic reticulum and ribosomes An organelle called the **endoplasmic reticulum** (usually referred to as ER) is a vast network of membrane that has surrounded portions of the cytoplasm to form what appear to be flattened spaces or sacs, called **cisternae.** When seen through the electron microscope, the membranes look like wavy lines and the enclosed cisternae like long channels or canals (see Fig. 4.8). In many cells the ER connects with the nucleus and other organelles.

The ER can be a site of protein synthesis because sometimes, but not always, it appears studded with tiny granules, structures called **ribosomes,** which convert

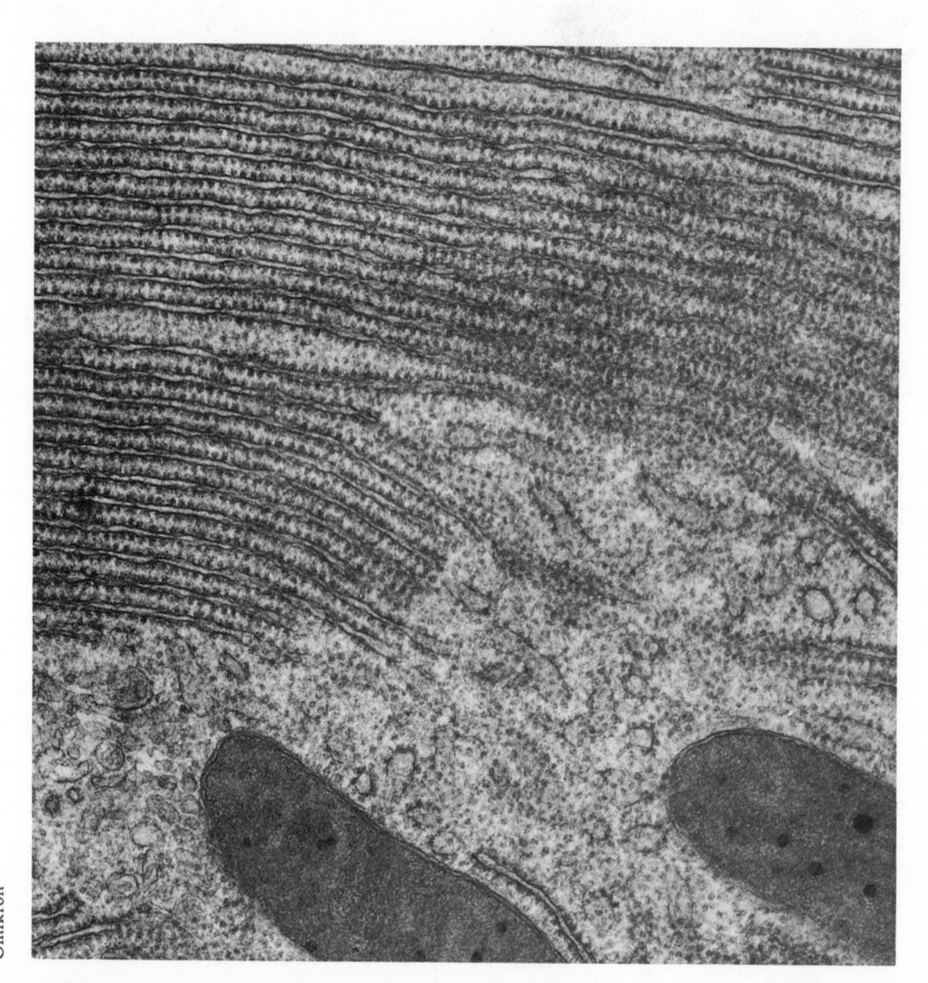

FIGURE 4.9

Endoplasmic reticulum in a bat pancreas. Groups of ribosomes cluster on the surface of the membranous endoplasmic reticulum (ER), which fills large sections of cytoplasm in eucaryotic cells. Ribosomes are the organelles of protein synthesis.

amino acids into complex protein molecules. An ER dotted with ribosomes is termed **rough ER;** a reticulum without ribosomes is called **smooth ER** (see Fig. 4.9).

Studies of ER using the ultracentrifuge have revealed that its ribosomes contain a large proportion of ribonucleic acid. When the ribosomes form chains, called polyribosomes, they contain most of the nucleic acids outside the nucleus. To determine the function of these cytoplasmic nucleic acids, researchers introduced radioactive amino acids, the building material of proteins, into a cell. The ribosomes became radioactively labeled within 10 to 15 minutes, indicating ribosomal function in protein synthesis.

Ribosomes occur either in association with the ER or in the cytoplasm by themselves. Those situated on the ER enjoy certain advantages; the wavy nature of the ER creates little chambers in which the ribosomes can carry out protein synthesis undisturbed by the chemical activity of the surrounding cytoplasm.

The ribosomes associated with the ER are situated in such a way that they can import more efficiently those raw materials they use, manufacture materials without chemical interference, and so release the finished products to destinations that may include the nucleus, other parts of the cytoplasm, or areas completely outside the cell after being "packaged" by the Golgi apparatus.

The Golgi apparatus Discovered by Camillo Golgi in the late nineteenth century, the **Golgi apparatus** is similar in structure to the ER, in that it consists of vesicles and vacuoles bound by a unit membrane. Biologists have exposed the Golgi apparatus by a tracing technique, using radioactive amino acids which expose photographic plates as the amino acids move through the cell (see Fig. 4.10). The apparatus has been shown to be an extension of the endoplasmic reticulum. It is not involved in protein synthesis and thus has no ribosomes. Rather, it acts as a collection and processing center, receiving specific substances (notably lipids and proteins) secreted by the ribosomes, and organizing and altering them for use by the cell. If we think of the cell as a manufacturing plant, the nucleus represents the board of directors and upper-level management, the ER and ribosomes are the as-

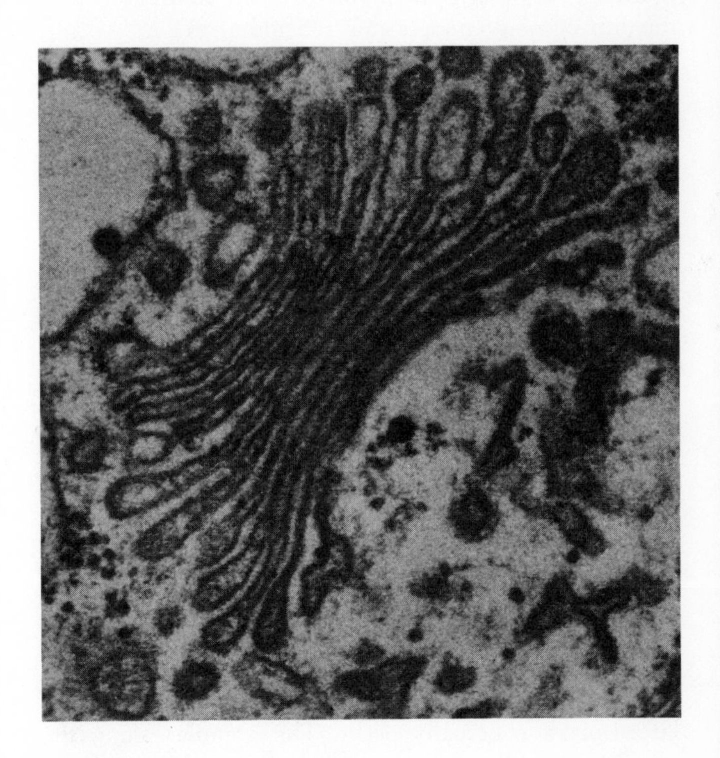

FIGURE 4.10

The Golgi apparatus. The three-dimensional drawing and the micrograph show the sacs, stacked like perforated pancakes. The Golgi apparatus is active in the synthesis of complex polysaccharides and in adding carbohydrates and lipids to proteins. It is also involved with packaging various substances manufactured elsewhere in the cell into membrane-bound vesicles.

sembly line for the company line of products, and the Golgi apparatus is the storage and shipping department. The exact mechanism involved in the alteration of these secreted substances is still a subject of controversy (see Fig. 4.10).

Golgi apparatus are especially apparent in cells with secretory functions (such as those in a secreting gland), and for this reason they are generally assumed to be involved in cellular secretion. However, their continuity with the endoplasmic reticulum may indicate participation in intercellular transport. It has also been suggested that Golgi apparatus are involved in the continuous production of new endoplasmic reticulum.

Mitochondria Mitochondria function as "generators" for the cell, in that they are involved in the transformation of simple sugars into energy, a process known as **respiration.** A mitochondrion is more complex in structure than either the ER or the Golgi apparatus. Like the nucleus, it is enclosed by two membranes, but a mitochondrion is barely large enough to be detected under the regular light microscope. Unlike the double membrane of the nucleus, however, the two membranes of the mitochondrion are not similar in structure. The outer membrane is smooth, relatively free of folds, and slightly different in composition from the inner one, which is folded in a serpentine fashion to form projections called **cristae.** A necklace of beadlike granules coats both the outermost side of the smooth outer membrane and the surface of the cristae (see Fig. 4.11).

The cristae, like the vesicles in the ER and the Golgi apparatus, provide extensive surfaces where the energy-producing work of this organelle can take place. It is in the cristae that the chemical reaction sequences of cellular respiration actually occur. The granules along the membrane surfaces are probably tiny packages of enzymes that are used in breaking down the sugar molecules. Cells that metabolize at very high rates for sustained periods and require greater quantities of energy, such as the muscle cells of the heart, contain substantially more and larger mitochondria than do less vigorous cells.

Some biologists maintain that the mitochondria were originally snipped off from portions of the cell membrane. A different hypothesis suggests that the mitochondria may once have been independent organisms that invaded larger cells and established a permanent host-parasite symbiotic relationship. (Presumably most cells did not carry on oxidative respiration before that invasion.) According to this theory, the outer membrane of the mitochondrion is formed by the larger cell, as a kind of barrier against the invader, and the inner folded membrane is actually the cell membrane of the small parasite. Support for this hypothesis comes from the fact that mitochondria contain nucleic acids that differ in certain marked ways from the nucleic acids found in the cell nucleus, and from the additional fact that they can reproduce themselves. The theory of symbiotic organelle evolution has attracted considerable controversy, and new data, both pro and con, are constantly being reported.

Plastids The yellows and oranges of autumn leaves; the luscious yellow of a ripe pear; the green of a blade of grass—these colors are caused by the pigments contained in **plastids.** Like the mitochondria, plastids are large enough to be seen under the light microscope, are able to reproduce themselves, and contain nucleic acids. Therefore, they are also thought to have evolved by symbiosis. Unlike mitochondria, however, which occur in cells of animals as well as plants, plastids occur in plant cells only. Whereas mitochondria perform the function of decomposition, converting simple sugars into energy, certain plastids perform the function of synthesis, using energy from the sun to manufacture sugars and starch for food.

The two categories of plastids are the **chromoplasts,** those with color, and the **leucoplasts,** those that are colorless. Of the chromoplasts, the most important are the **chloroplasts.** They contain light-sensitive chlorophyll, a green pigment which is essential to the chemical reactions of photosynthesis—the manufacture of sugar. Other pigments contained in the chloroplasts are the yellow and orange **carotenoids.**

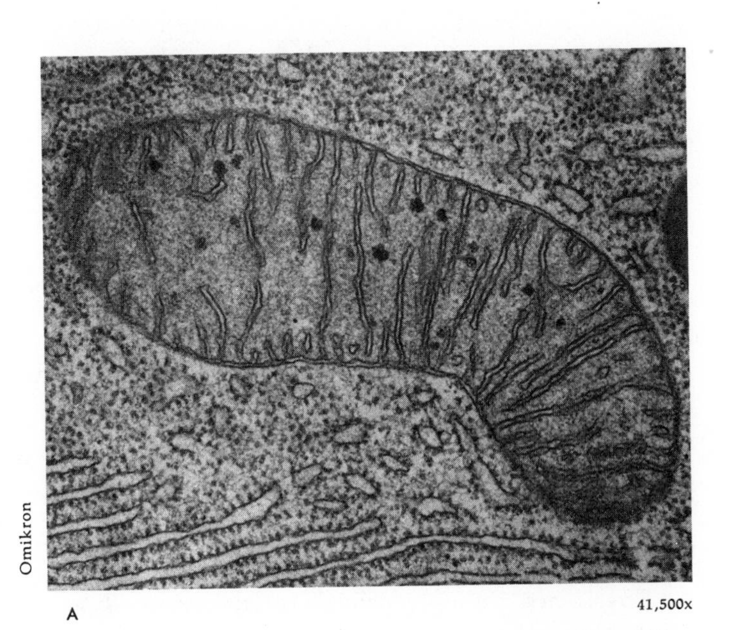
Omikron

A 41,500x

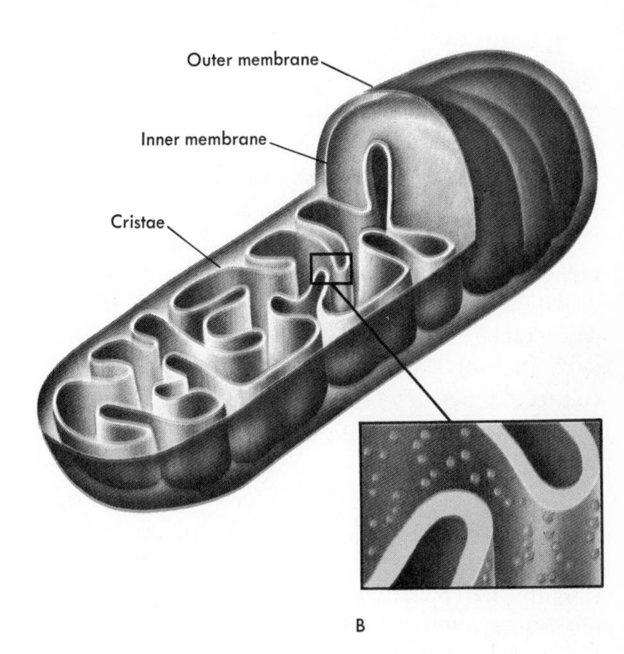

B

FIGURE 4.11

The mitochondrion, powerhouse of the cell. (A) Micrograph of a mitochondrion in a bat pancreas. (B) The drawing shows the smooth outer membrane cut away to reveal the folds, or cristae, of the inner membrane. The cristae are dotted with small spheres, which are clusters of enzymes involved in cellular respiration. (C) A sperm cell of a bat shows mitochondria in orderly rows along the flagellum, or tail, where they supply the energy for its lashing motion.

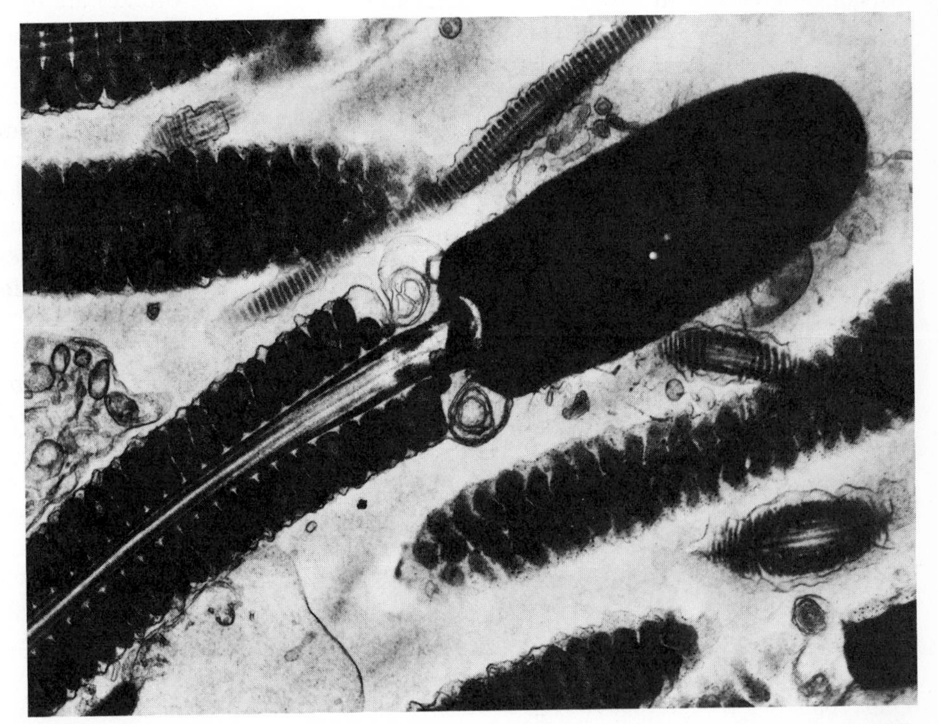

C 16,000x

As might be expected from the fact that plastids and mitochondria both employ enzymes—plastids to manufacture sugars and mitochondria to metabolize them—the two organelles somewhat resemble one another structurally. Like mitochondria, plastids have an outer membrane and a system of inner membranes. The inner membranes are arranged to form many disc-like sacs, called **thylakoids,** which are stacked like rolls of coins. The stacks, called **grana,** are embedded in the general substance of the plastid, called the **stroma,** and interconnected by an extensive system of tubules (see Fig. 4.12). The grana membranes, as seen under the electron microscope, appear to be composed of tiny granules or particles, which are arranged in an orderly fashion. The particles, called **quantasomes,** are composed of the pigments and other molecules that perform the important function of carbohydrate synthesis, half of the process known as **photosynthesis.** The other half of photosynthesis is known as **photophosphorylation,** a process in which light energy is converted into chemical energy. The orderliness of the quantasome arrangement in the grana is thought to play a vital role in making photosynthesis more efficient. Leucoplasts—the colorless plastids—are also engaged in the manufacture of food. However, their role is to make starch from simpler sugars, the same sugars that were originally synthesized by the chloroplasts, and to store it.

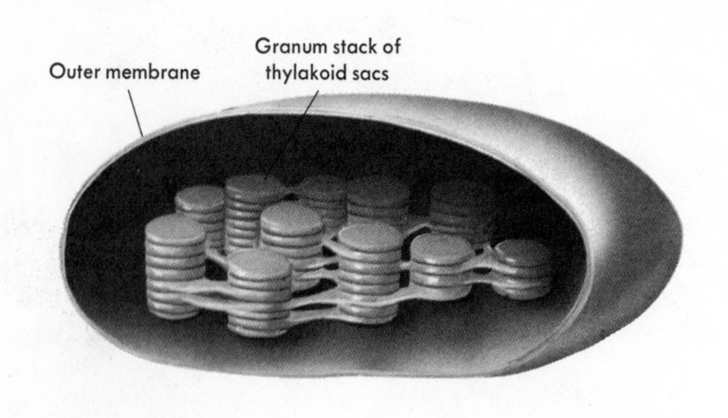

FIGURE 4.12

Chloroplast cut open to reveal the interior. Shown are the chloroplast envelope, its membranous covering, and stacks or grana composed of thylakoid sacs that are interconnected by internal membranes.

Lysosomes Lysosomes are small membrane-enclosed sacs containing enzymes that, if released, can decompose the contents of the cell. For this reason their discoverer referred to them as "suicide bags." The enzymes cannot damage the cell so long as they are sealed inside the membrane of the lysosome, but under certain chemical conditions, the lysosomal membrane ruptures and the enzymes flow out, killing the cell. For example, shrimp quickly spoil when they are not refrigerated properly because, under prolonged exposure to room temperature, their lysosomal membranes rupture, and the released contents of the lysosomes destroy the meat, causing it to have a foul odor.

Lysosomes are aptly named, for the prefix "lyso" comes from a Greek word meaning to loosen or untie. Lysosomes are organelles that untie or tear down. They are usually associated with the process of digestion. For example, some one-celled organisms, such as the amoeba and the paramecium, absorb food into food vacuoles (phagosomes), which then fuse with lysosomes, and the digestive enzymes released into the vacuolar space break down the food. The products of the breakdown are then released into the cytosol.

Lysosomes are found in most animal cells, and similar enzyme-containing structures are also found in plant cells.

Vacuoles A vacuole is a fluid-filled space—a droplet of water containing certain chemicals in solution or suspension—surrounded by a membrane. Hence, like the lysosomes, vacuoles are appropriately named; the word vacuole has the same prefix as the words vacuum and vacant.

Vacuoles assume various functions, depending on the type of cell within which they are situated. In some cells, the vacuoles digest food. Or they may serve as receptacles into which lysosomes pour their enzymes instead of emptying them indiscriminately into the open cell. Many plant cells are characterized by the presence of large vacuoles that occupy nearly the entire volume of the cell, pushing the cytoplasm to the edge.

Microtubules and microtubule-containing bodies Microtubules are extremely small, hollow tubes that are found in the cytoplasm of both plant and animal cells. Visible only with a powerful electron microscope, microtubules are smaller than endoplasmic reticulum and consist of proteins arranged in a circular pattern. The function of these tiny tubules is still largely a matter of speculation. Since they are most densely clustered near the inside of the cell membrane, some researchers believe they may act as a "skeleton" for the cells of both animals and plants. Microtubules may also serve as canals for the flow of cytoplasm within the cell and, in certain animals, as pathways through which digestive enzymes reach food trapped in the animals' tentacles. In plants they may act as passageways for materials used in building the cell wall. Furthermore, microtubules may function with the centriole, since they are more numerous during cell division.

Centrioles are small rod-shaped organelles made up of nine bundles of three microtubules each positioned around their outer surface. They usually appear in pairs, one perpendicular to the other, and some investigators believe they are involved in the formation of the spindle fibers that play a role in cell reproduction. However, centrioles are not essential to all cell reproduction, since they have not been found in the cells of higher plants.

Closely resembling the centriole, the **basal body,** or kinetosome, also has nine bundles of microtubule triplets and is self-reproducing. The two organelles are so similar that scientists believe the basal body evolved from the centriole.

Also microtubule-containing are the **cilia** and **flagella,** which are hairlike structures extending outside the cell. Their principal function is to move the cells to which they are attached. A cell that has cilia usually has a large number of them, whereas a cell usually has only a single flagellum. Cilia are shorter than flagella and move like oars in a rowboat; flagella move in waves, somewhat like a snapped whip. As indicated in Fig. 4.13, both flagella and cilia have two central tubules (missing in the centrioles) surrounded by nine bundles, but each bundle consists of two tubules rather than the three of the centriole and basal body.

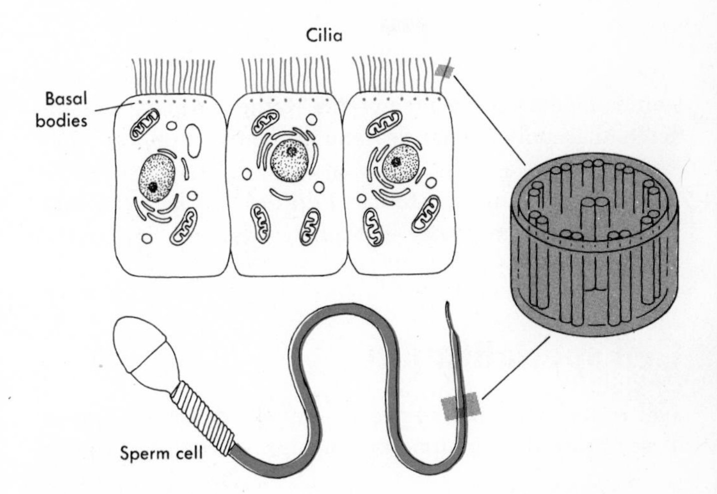

FIGURE 4.13

Cilia and flagella have the same basic structure. Enclosed by a double membrane is an outer ring of nine pairs of fibrils surrounding two inner central fibrils. There is a similar ring structure in the centrioles, which have three rather than two fibrils in the outer sets and no central fibrils. The basal bodies, from which the cilia and flagella grow, have this same outer ring of triplets and are believed to be derived from the centriole. (See Fig. 4.3 for cilia structure as shown by electron microscopy.).

Cell coatings The last structures that we shall consider are the diverse coatings found on the outsides of bacterial, plant, and animal cells. In bacteria and plants the coatings—called **cell walls**—are made chiefly of long polymeric chains of sugar monomers. In higher plants, for example, the cell wall is composed mainly of cellulose, the fibers of which lend rigidity to the lining part of the cell. It is the stiff cell wall which prevents the cell from bursting and permits instead the development of hydrostatic or **turgor** pressure when the cells take in water.

Animal cells may also have coats of material external to their membranes. Amoebae, for example, have a fuzz composed of numerous fine filaments which coat the membrane. In animal tissues, cells are rarely stuck together membrane-to-membrane. Rather, a matrix of intercellular polysaccharides and proteins glues the cells together into a mass. Only in special tissues (such as the mucosa lining the stomach) do the cells lack coatings and fit closely together instead.

Cell specialization

Not every cell contains every organelle described above. A structure that is numerous or highly developed in one cell may simply not appear in another. Differences in cell structure, of course, indicate differences in function. It is, in fact, the differentiation of cells that enables them to specialize in the numerous operations necessary to life.

On the unicellular level, certain organisms are hardly specialized at all. These cells, termed **procaryotic cells,** contain no organelles whatever—not even a nucleus. However, some procaryotic cells do have a region consisting of chromosomes and nucleic acids, and they include regions that perform specific metabolic functions. Examples of procaryotic cells are the bacteria and the blue-green algae.

Much more complex are the **eucaryotic cells**—those cells, whether they are unicellular organisms or the structural units of multicellular creatures, that do contain distinct organelles. The paramecium is a well-studied eucaryotic cell; it contains not only many of the organelles described earlier, but a mouthlike oral groove as well. Table 4.1 summarizes some of the basic differences between procaryotic and eucaryotic cells.

TABLE 4.1
Comparison of procaryotic and eucaryotic cell

	PROCARYOTIC	EUCARYOTIC
Cell membrane	Yes	Yes
Nucleus	No	Yes
Chromosomes and nucleic acids	Sometimes	Yes
Cytoplasm	Yes	Yes
Endoplasmic reticulum	No	Sometimes
Ribosomes	Yes	Yes
Golgi apparatus	No	Sometimes
Mitochondria	No	Sometimes
Plastids	No	Only in plants
Chlorophyll	Sometimes (blue-green algae)	Only in plants
Lysosomes	No	Sometimes
Vacuoles	Yes	Only in plants and lower forms of animals
Microtubules	No	Sometimes
Centrioles	No	Only in animals and lower plants
Basal bodies	No	Sometimes
Cilia and flagella	Only simple flagella	Sometimes
Cell coatings	Yes	Sometimes

Contemporary evolutionary theorists have suggested that modern procaryotic cells resemble the first cell to appear on earth, and that eucaryotic cells are those that have evolved greater degrees of specialization. It is believed that both types of organisms evolved from a common ancestor, a procaryote-like cell.

On the multicellular level, specialization varies considerably. There is the *Pandorina*—a colony of one-celled green algae whose specialization lies only in the fact that the algae live together, obtaining food more efficiently and sharing it among themselves. And there is the *Volvox,* also a colonial alga, in which some of the cells specialize in the role of reproduction. This in turn enables other cells to specialize in the function of obtaining food, so that the colony is able to increase its food supply.

Among the more complex multicellular organisms, cells are combined to form **tissues**—groups of specialized cells that are similar in structure and function. In the higher forms of life, tissues serve as the material of organs. First we will examine the various specialized cells that go into the formation of plant tissues. Then we will consider the cells that make up the tissues of animals.

SPECIALIZATION IN PLANT CELLS

Plant cells vary greatly in their degree of specialization. Certain plant cells are relatively unspecialized, to the extent that they are able to become modified into other more specialized types. Other plant cells are so specialized that they are restricted to very specific roles in the life of the plant. We call these **differentiated cells.** The relatively unspecialized types of cells are considered **undifferentiated.** Types of plant cells are illustrated in Fig. 4.14 and discussed in the following paragraphs.

Meristematic cells Meristematic cells are quite active each spring, when they produce the new leaves and buds of plants that bloom annually. Their sole function

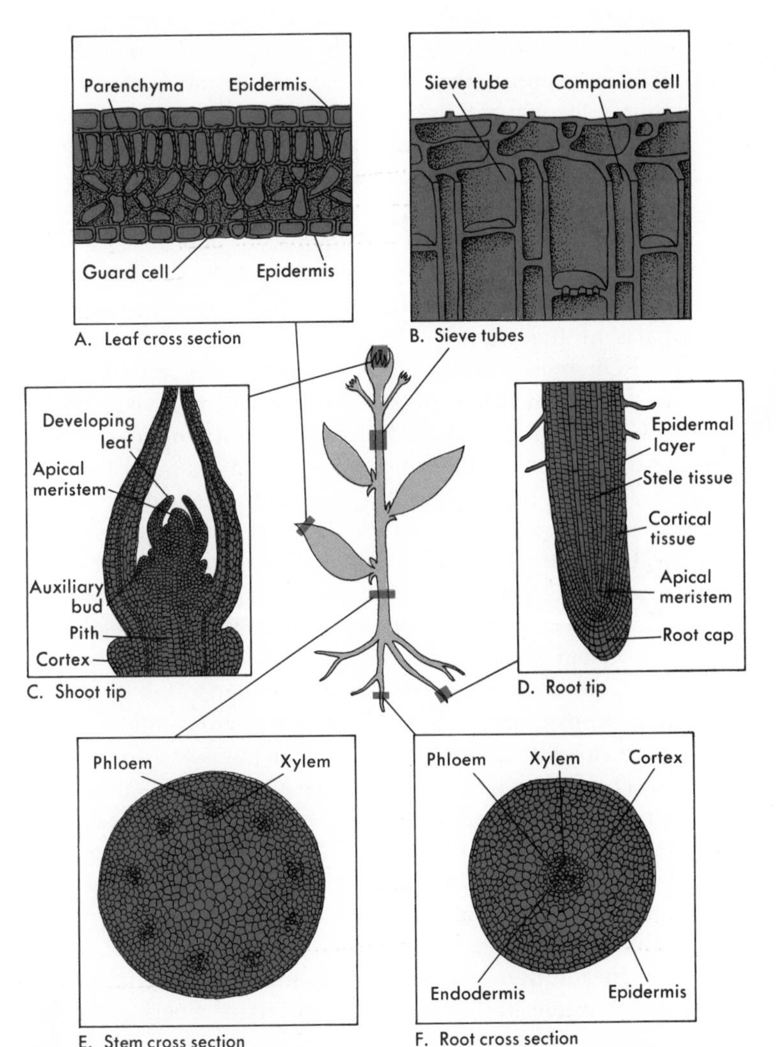

FIGURE 4.14

Diagrammatic representations of important cells and tissues in higher plants. (A) The leaf, main site of photosynthesis. (B) Sieve elements and their companion cells. The sieve elements, connected end-to-end to form long strands, are responsible for transport of nutrients. (C) The apical meristem of the shoot, producing leaves and stem tissues. (D) A longitudinal section of a root tip. The root cap cells protect the tip of the root as it grows downward into the soil, and the meristem cells divide to produce new cells for the growing root. (E) A cross section of the stem of a dicot. The vascular bundles, composed of xylem, phloem, and fibers, are in a ring near the perimeter. (F) A cross section of the root of the dicot. In the center are the cells of the xylem, which carry water and dissolved minerals from the root up into the stem and leaves. Near the xylem are bundles of phloem cells.

is to divide and produce new cells so that the plant, or its parts, may grow. Meristematic cells are found in regions of the plant body called **meristems.** These regions are located, for example, at the tip of the root, at the apex of the stem, and along the edges of young, growing leaves. Through the specialized processes of cell division, the meristematic cells multiply to produce the cells which differentiate and later mature into the various parts of the plant body.

In view of their reproductive functions, meristematic cells would be expected to have the very large nuclei that they in fact do have, for the nucleus is so important to cellular reproduction. Similarly, it is not surprising to find that their cell walls are thin, since a thicker cell wall would hinder them in their reproductive activities.

Epidermal cells Epidermal cells make up the outer covering of a plant and may be considered analogous to the skin of animals. They constitute the plant's protective tissue, known as the **epidermis.** One need look no further than the stiff leathery covering of the leaves of a philodendron for an example of a protective structure formed by the epidermis.

Epidermal cells are more differentiated than meristematic cells. They often secrete a waxy coating, called the **cuticle,** which helps to defend the more vulnerable internal plant tissues from invasion by bacteria and viruses. The cuticle also helps to prevent excessive water loss through evaporation.

Parenchyma cells The green color of a summer leaf is a result of the chloroplasts contained in special **parenchyma** tissues called **mesophyll.** Photosynthesis occurs in the mesophyll cells. Similar cells, lacking chloroplasts, are also found in the roots and stems, where instead of building sugars, they store water and plant nutrients. They contain large vacuoles which may be used to store aqueous solutions of carbon dioxide or sugar. The thin-walled parenchyma cells are relatively undifferentiated and so are able to divide and to develop into more specialized cell types. Perhaps their reproductive functions is one reason their cell walls are rather thin.

Collenchyma cells Elongated parenchyma cells with thickened corners and thick cell walls are called **collenchyma** cells. Whereas epidermal cuticle provides a measure of support from outside the plant, collenchyma cells lend support from the inside. They are important for the support and flexibility they give to both leaves and stems.

Sclerenchyma cells The reason walnuts are hard to crack and sunflower seeds do not readily yield their tasty kernels is that their shells and coatings are made up of **sclerenchyma** cells. Although sclerenchyma cells function in support, they nevertheless are more specialized than collenchyma tissue. They give themselves over to the task of support so completely that they die after they have laid down their very thick cell walls.

Xylem and phloem cells Xylem and phloem cells—the most specialized of all the plant tissues—provide a means of transporting water and nutrient material to all parts of the plant.

One needs only to look at the leg of a wooden table to see xylem tissue; wood is really xylem. Xylem tissue consists of several kinds of cells. In softwood—gymnosperms, such as the pines—the xylem contains specialized water-conducting tubes called **tracheids.** Similar fiber-tracheids, whose internal spaces are too narrow for water conduction and whose walls are hard and stiff, provide the woody tissue with support. In angiosperms, which include most varieties of common garden plants, the xylem tissue is more complex. Angiosperm xylem is composed of tracheids that are similar to the tracheids in gymnosperms, as well as **vessel elements**—open-ended wide tubes that are connected end-to-end to form long water-conducting pipelines. These pipelines permit water and nutrients to pass from the ground to the farthest reaches of the plant body. The thick cell walls of the tracheid and vessel element provide support for the plant, but they are so thick in the mature cell that the cell dies.

In addition to the tracheid and vessel element, the xylem tissue in angiosperms consists of fibers (sclerenchyma), which provide support, and parenchyma cells, which conduct water horizontally and store various substances.

An important difference between xylem and phloem tissue is that the cell walls of phloem tissue are not so thick that they kill the cytoplasm within. The somewhat thinner walls make it possible for the phloem not only to provide some support but also to serve as a complete transportation network for nutrients. That is, the phloem cell walls are thin enough to allow carbohydrates and amino acids to pass in and out. Another difference is that materials move downward as well as upward in traveling through the phloem network. Thus it is the phloem tissue that carries the sugars manufactured in the leaves down through the plant.

Phloem tissue in angiosperms consists of parenchyma, sclerenchyma, and two types of specialized cells: **sieve elements** and **companion cells.** A vertical row of elongated sieve elements forms a **sieve tube,** which contains **sieve plates**—groups of pores in the end walls and occasionally in the side walls of the sieve elements. (Nutrient conduction through the phloem tissue will be discussed in more detail in Chapter 7.)

The companion cells are so named because they lie alongside the sieve elements. Note, however, that they are found only in angiosperms, not in gymnosperms. A companion cell is recognizable for its very thin walls and numerous organelles, including a nucleus (the sieve cell lacks a nucleus). Consequently companion cells are thought to play a role in the metabolic activities of the sieve elements.

SPECIALIZATION IN ANIMAL CELLS

Animal cells vary in their degree of specialization, as do plant cells. Unlike plant cells, however, the major types of animal cells are too differentiated to be transformed into other types of cells. In fact, differentiation of animal cells is so pronounced that some of the cell categories we will discuss pertain only to higher animal classes, such as the mammals.

Epithelial cells In animals, as in plants, the outer layer of tissue is called the **epidermis.** It is composed of epithelial cells and is protective and supportive in function. The protective function in animals, however, is far more extensive than it is in plants. For animals are generally more complex in their structure than plants; they have internal organs and systems that require protection and insulation from one another almost as much as they do from the environment outside. Consequently, epithelial cells are located on every surface area of the animal body, even seemingly internal areas such as the inner wall of a blood vessel (where they are called endothelium) or of the stomach. Because the environment at the outer end of the cell is different from that at the inner end, these cells are polar in structure, with special coverings or cilia on the free end and cell organelles concentrated in the attached end of the cell. The cells fit tightly together and thus play a significant role in regulating what enters and leaves the body and its various organs and systems.

In the human body, epithelial cells come in three shapes: squamous—flat cells with several sides; cuboidal—cells that resemble ice cubes; and columnar—tall cells that look like rows of index fingers (see Fig. 4.15). Squamous cells sometimes constitute the outer layer of epithelium, such as the epidermis of the skin, and beneath them lie several layers of cuboidal and columnar cells. Occasionally, however, squamous cells are absent altogether.

Connective cells Connective cells constitute such tissues as bone, cartilage, and blood. It is the function of these tissues to hold together many specialized areas of the animal body so that the animal can function as a distinct entity.

It seems logical to assume that cells whose function is to hold the body together would have to be long and tough—capable of withstanding great stress. However, cells of that nature probably could not survive. They would be likely to have a destiny similar to the cells in plants whose thick cell walls seem to constrict and kill the cytoplasmic contents. Moreover, such cells would

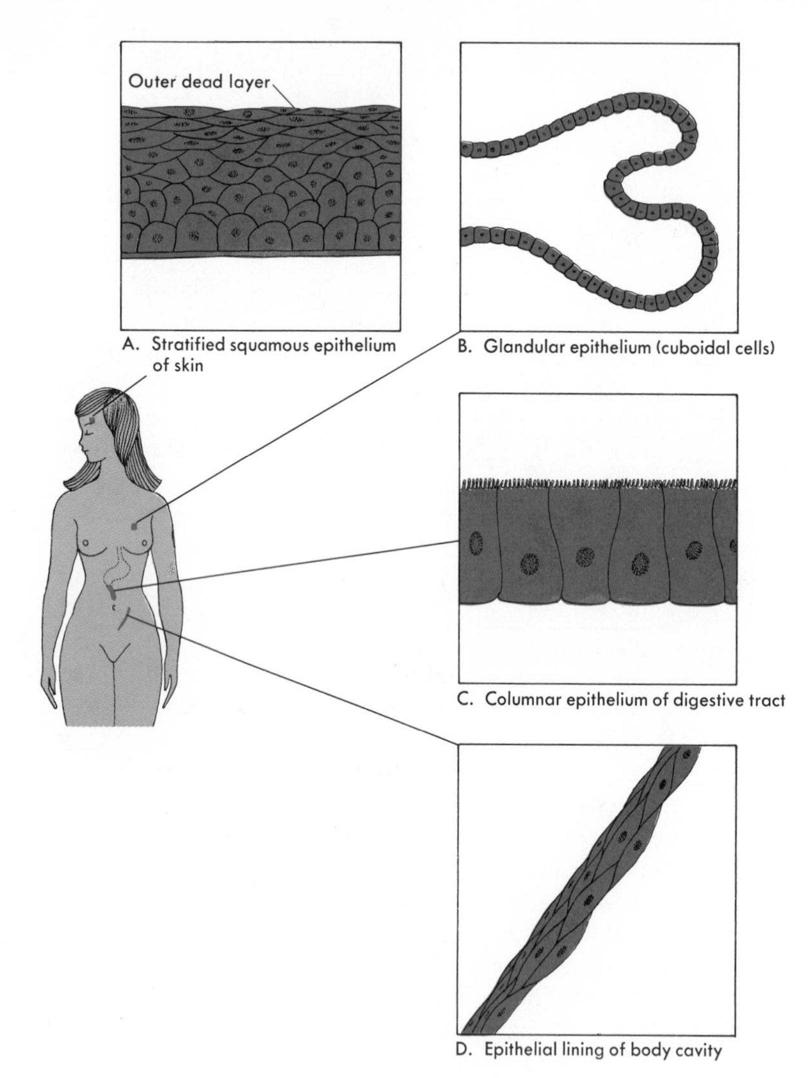

FIGURE 4.15

Diagrammatic representations of human epithelial cells, which take a variety of shapes and are found performing special functions in various locations. (A) Squamous cells are flattened and thin, and they are sometimes stratified into layers, like the stratified squamous epithelium of the skin. Because of their roof-shingle shape, the cells water-proof the body, and their hard texture protects the underlying tissues from mechanical damage. The cells are continuously rubbed off the surface by mechanical contact with clothing or other objects, and they are replaced by the underlying layers. (B) Cuboidal cells, found in only a few regions of the body, sometimes have a secretory function. The skin glands, for example, produce sweat, skin oil, milk, and earwax. (C) Columnar epithelial cells have a nearly cylindrical shape. Those found in the digestive tract may have many small projections (called microvilli) of the cell membrane, which serve to increase the surface area of the membrane available for absorption. (D) Squamous epithelial cells form thin flattened sheets, only one cell thick, which protect various parts of the body. They may be found, for example, in the peritoneal lining of the abdominal cavity.

have to be so numerous and located in so many places that their demand for nutrients might be too great for the animal to supply.

This dilemma is solved by a structure called **matrix,** the medium in which connective cells are embedded. In most of the connective tissues, the matrix is secreted by the cells themselves. It can be hard, as it is in bone; or it can be flexible, as it is in cartilage. Since the matrix is not alive, it does not require nutrients and can be distributed throughout the body with a minimum of energy required for its maintenance. The cells remain alive and widely dispersed within the matrix, which binds the cells together. Types of connective cells are shown in Fig. 4.16 and discussed in the following paragraphs.

Bone It is somewhat difficult to think of bone as a tissue because of its rocklike texture. The reason that bone is so hard is that its matrix is composed principally of calcium salts. In bone, the connective cells are called **osteoblasts.** They secrete fibers of a protein called collagen—the basic structural material of the matrix. After secreting the fibers, the osteoblasts mature into **osteocytes.** The osteocytes receive their nutrients through tiny tubes called **canaliculi,** which are connected to larger tubes

FIGURE 4.16

Diagrammatic representations of connective tissues in the body. Human connective tissues consist of cells embedded in a nonliving matrix, which they may secrete. The matrices are of different types and give the connective tissue types their distinctive properties. The connective tissues hold various parts of the body together and provide support. (A) Fibrous cartilage (fibrocartilage) consists of cells embedded in a dense, stretch-resistant, stiff matrix composed largely of fibers of collagen. It is found, for example, in the discs between vertebrae in the spinal column. (B) Hyaline (clear) cartilage is a translucent tissue composed of cartilage cells embedded in a matrix of collagen. Much of the skeleton is formed of hyaline cartilage in the embryo and later becomes ossified, or hardened into bone. Hyaline cartilage remains as a slick, low-friction covering of the ends of movable bones. (C) Blood is an unusual connective tissue consisting of various types of cells (erythrocytes, leucocytes, and platelets) in a fluid matrix. Blood carries gases, nutrients, hormones, and wastes around the body. (D) White fibrous cartilage is a tough, nonelastic tissue that is formed into cords called tendons, which hold muscles to bones. (E) Bone is a connective tissue in which the bone cells secrete a very hard rigid matrix consisting largely of calcium salts. Although the tissue is very hard and thus capable of supporting and protecting the body, it is nevertheless alive and is constantly being remodeled in response to individual patterns of use and breakage.

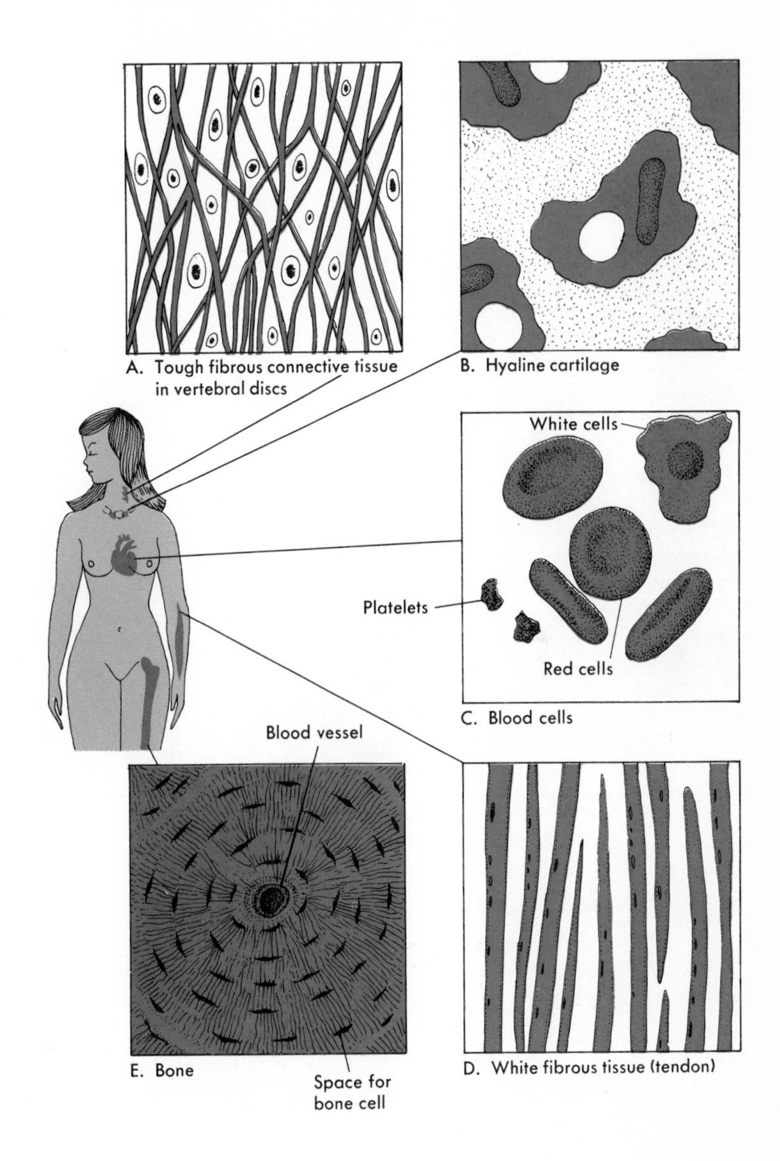

called **Haversian canals,** containing blood vessels and nerves. The cells and canals form subunits of bone called **Haversian systems.**

Cartilage Cartilage is a tissue having both stiffness and a measure of flexibility. The skeleton of the human embryo is formed first of cartilage, but subsequently most of the cartilage is replaced by bone. However, some cartilage is retained in many parts of the adult body. The discs of cartilage between the hard, bony vertebrae of the vertebral column, for example, act as a kind of shock absorber and provide flexibility to this rather rigid column. In cartilage, the connective cells are called **chondrocytes.** Like the osteocytes of bone, chondrocytes secrete the very thin fibers of their matrix, in which the cells themselves are widely spaced.

Fibrous connective tissue This type of connective tissue may be thought of as the body's wrapping tape. It is found in tendons and ligaments, tough fibrous structures that connect muscles to bones and bones to each other, respectively. One kind of cell in this tissue is the **fibroblast.** Like the cells in bone and cartilage, fibroblasts secrete the fibers of their matrix—fibers that are made

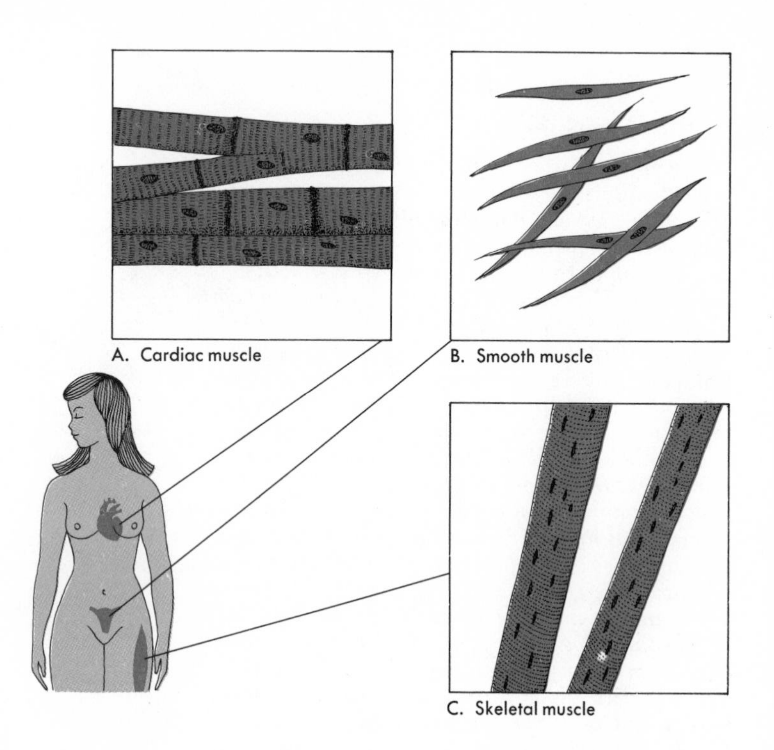

FIGURE 4.17
Diagrammatic representations of the three types of muscle found in the human body. (A) Cardiac muscle, or heart muscle, is found only in the heart. It is capable of powerful, rhythmic contractions that last for the lifetime of the organism. The muscle has striations similar to those of skeletal muscle, but it seems, like smooth muscle, to be divided into individual cells. The boundaries between cells take the form of dark, intercalary discs. (B) Smooth muscle, also called involuntary muscle, is found in many organs, such as the uterus, the intestines, and the walls of blood vessels, where it produces sustained contractions. The muscles are composed of spindle-shaped cells, each containing only one nucleus. (C) Striated muscle, also referred to as voluntary muscle, is attached to the bones of the skeleton and moves them. It is composed of long multinucleate filaments marked by prominent bands, or striations. The striations are a reflection of the muscle's underlying chemical organization, which is essential for contractility.

of collagen and another protein, called **elastin.** Fibrous connective tissue is characterized by abundant fibers that give it the toughness and flexibility it requires to hold structures together.

Blood Just as it is difficult to think of hard, stony bone as a tissue, it is equally difficult to consider fluid blood as such. But no one would deny that blood, which courses through every part of the body, has a connective function. The fact is that connective tissue matrix need not be a solid. With blood, the surrounding medium is a fluid called the **plasma.** Plasma contains dissolved foods and waste products, hormones, and antibodies, and it serves as the medium in which blood cells travel.

The major types of blood cells are the red blood cell, or **erythrocyte,** and the white blood cell, or **leucocyte.** Erythrocytes are filled with the protein hemoglobin, a chemical that has an extraordinary capacity to combine with oxygen, and that can also combine—though less readily—with carbon dioxide, one of the waste products of cellular metabolism. Human erythrocytes are so specialized for their function—delivering oxygen to all the cells in the body and removing their carbon dioxide waste—that they contain no organelles whatever, not even a nucleus. Consequently, they cannot reproduce and continuous replacement is necessary.

Leucocytes represent one contingent of the body's defense forces. They can pass through the walls of blood vessels easily and are quickly mobilized to destroy infecting agents, such as bacteria. A third type of blood cell, the **thrombocyte** or platelet, is essential in the clotting mechanism.

Muscle cells Muscle cells are responsible for the movement of parts of the body as well as movement of the body as a whole. Their power is enormous, as demonstrated, for example, by the muscular legs of a jaguar or the strong arms of an ape.

Muscle cells are long, and they band together to form elongated bundles of tissue, known commonly as muscles. It is their extraordinary capability to contract that enables muscles to perform their work. Just as the cell membrane is able to stretch and fold because of the protein chains within it, muscle cells, too, are able to contract and relax largely because they contain long strands of protein fibers.

As shown in Fig. 4.17, there are three basic types of muscle cell: **striated, smooth,** and **cardiac.** Striated, or voluntary, muscle is the kind of muscle attached to some component of the skeleton, as in the leg; it can be moved at will. Striated muscle cells are multinucleated because they result from the fusion of many single cells. Smooth muscle and cardiac muscle are not generally subject to conscious will; rather, they perform automatically. Smooth, or involuntary, muscle is found in the walls of various organs—for example, in the stomach, where it is responsible for that organ's churning of its food content without the conscious direction of the animal. Cardiac muscle is unique in that it occurs in only one organ—the heart. This extraordinarily hard-working muscle normally goes on contracting rhythmically, second after second, for 75 years or more. Cardiac muscle cells—in accordance with their huge energy requirements—have a preponderance of very large mitochondria.

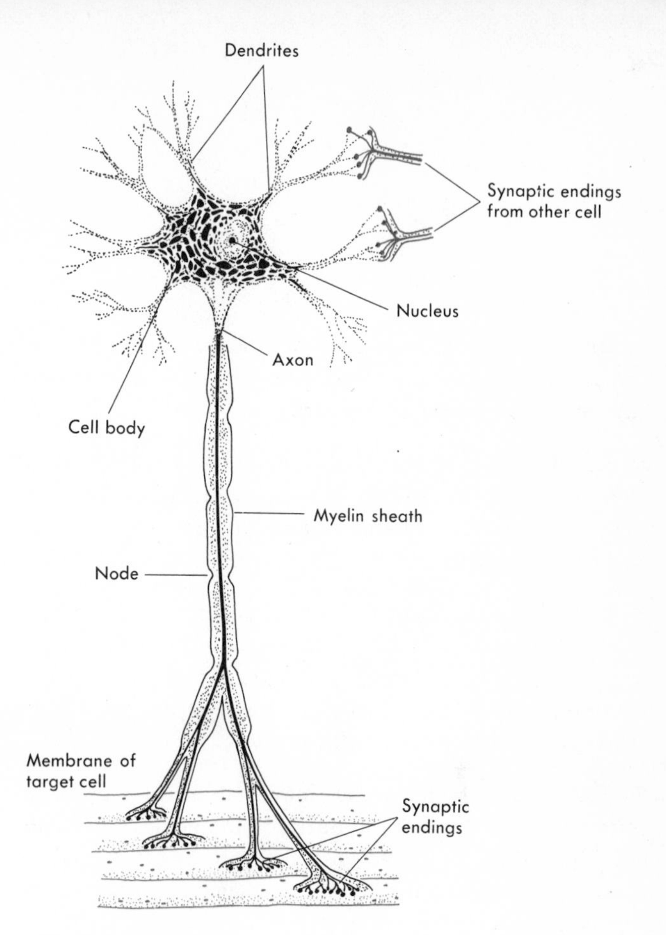

FIGURE 4.18

One of several types of nerve cell. The dendrites receive signals from other nerve cells and pass them to the cell body. Electrical charges pass through the axon and, via its synaptic endings, to other cells. The nerve cell is part of a communication system, in which the dendrites receive information and the axon carries it toward its destination.

Nerve cells Some nerve cells, or **neurons,** are longer by far than any other type of cell, measuring in some instances as much as 3 or 4 feet in length. This extraordinary feature particularly suits their function—that of conducting electrical impulses between all parts of the body (see Fig. 4.18).

Nerve cells consist of three major sections: the **cell body**—a small oval structure containing the nucleus and other organelles, of which the Golgi apparatus is particularly large; the **dendrites**—short and thin filaments of cytoplasm that branch out from the cell body; the **axon**—a long wire-like extension of cytoplasm that accounts for the unusual length of certain neurons.

Dendrites receive impulses and convey them to the cell body. The cell body in turn sends them on to the axon. Impulses are transmitted along the axon either to the dendrites of another neuron, to a gland, or to a muscle cell, whichever the axon is connected with. The axon is often wrapped in a "jelly roll" composed of Schwann cells—cells whose membranes contain the lipid myelin. Such axons are said to be myelinated, a condition that speeds up the transmission of impulses.

Summary

1. Life functions depend on the organization and interaction of organic chemical compounds.

2. The smallest independent structural unit common to all living things is the cell.

3. Cells perform all the functions essential to life: they receive and utilize energy; synthesize large molecules from simpler substances; regulate their internal environment; are capable of growth and reproduction; and react to changes in their external environment.

4. Cells were first discovered in 1665 by Robert Hooke, who used an early microscope to observe tiny walled chambers in a thin slice of cork.

5. Knowledge of cell structure has grown with the development of increasingly powerful microscopes and new techniques for chemical and physical analysis of cell components.

6. Cell structure is intimately related to cell function, and most mature cells have undergone a process of differentiation in which their physical characteristics have developed in a way that enables them to perform their specialized function more efficiently.

7. The size to which a cell may successfully grow is limited by the number of nutrient molecules that can enter the cell as compared with the volume of the cell to be nourished. This limiting factor may be expressed as the surface-to-volume ratio.

8. Cells contain a number of specialized internal structures called organelles that perform certain life functions.

9. Cellular membranes composed of lipids and proteins act as selectively permeable barriers between different parts of the cell and between the cell and the external environment.

10. The cell nucleus controls the heredity and life functions of the cell through the nucleic acids DNA and RNA. DNA is the major component of the chromosomes, which contain the cell's genetic material. RNA is responsible for directing cell activities through controlling the synthesis and breakdown of the proteins and enzymes necessary for cellular structures and chemical reactions.

11. The cytoplasm, the protoplasm outside the nucleus and inside the cell membrane, contains a mixture of organic and inorganic chemicals including water, amino acids, proteins, fat globules, simple sugars, and salts. It serves as an intracellular circulatory system and a support medium for numerous organelles.

12. Most cellular organelles are found in both plant and animal cells. Exceptions are plastids and nonliving external cell walls, which are seen only in plants; lysosomes, which are found only in animals; and centrioles, which are found only in animals and lower plants.

13. Cells vary in their degree of specialization. Procaryotic cells are very unspecialized and contain no organelles; eucaryotic cells are more complex and contain a variety of organelles.

14. In more complex multicellular organisms, groups of closely associated cells specialized to perform the same function are called tissues.

15. In plants, meristem tissue is responsible for growth; xylem and phloem, for intraplant transport; parenchyma tissue, for photosynthesis and storage; collenchyma and sclerenchyma tissue, for support; and epidermal tissue, for external protection.

16. In higher animals epithelial tissue provides external protection and support; connective tissue (bone, cartilage, fibrous connective tissue, and blood) holds together separate body parts in a unified whole; muscle tissues provide voluntary and involuntary motion; and nerve tissue provides electrical communication and integration.

REVIEW AND STUDY QUESTIONS

1. What contributions to cell theory were made by Hooke, by Schleiden and Schwann, by Virchow, and by Pasteur?

2. Explain the importance of size to the functioning of a cell. In what ways do cells overcome factors that tend to limit cell size?

3. Give an example of a cell type that illustrates the structure-function relationship.

4. Explain the functions of cellular membranes, ribosomes, the Golgi apparatus, lysosomes, and mitochondria.

5. Name and explain the functions of organelles that are found in plant cells but not in animal cells, and vice versa.

6. What is the principal difference between procaryotic and eucaryotic cells?

7. Explain what is meant by cell specialization. What are its advantages?

8. Name three types of animal tissue and give an example of each. Explain the function of each type.

REFERENCES

Burke, J. D. 1970. *Cell Biology.* William and Wilkins, Baltimore.

De Duve, C. 1963 (May). "The Lysosome." *Scientific American* Offprint no. 156. Freeman, San Francisco.

Fox, C. F. 1972 (February). "The Structure of Cell Membranes." *Scientific American* Offprint no. 1241. Freeman, San Francisco.

Green, D. E., and R. F. Bruckner. 1972. "The Molecular Principles of Biological Membrane Construction and Function." *BioScience* 23(1): 13–19.

Jensen, W. A., and R. B. Park. 1967. *Cell Ultrastructure.* Wadsworth, Belmont, Calif.

Lehninger, A. L. 1965. *The Mitochondrion.* Benjamin, New York.

Loewy, A. G., and P. Siekevitz. 1970. *Cell Structure and Function,* 2nd edition. Holt, Rinehart and Winston, New York.

Novikoff, A. B., and E. Holtzman. 1970. *Cells and Organelles.* Holt, Rinehart and Winston, New York.

Porter, K. R., and M. A. Bonneville. 1973. *Fine Structure of Cells and Tissues,* 4th edition. Lea and Febiger, Philadelphia.

SUGGESTIONS FOR FURTHER READING

Crichton, M. 1969. *The Andromeda Strain.* Knopf, New York (paperback, Dell, 1971).

This national best seller follows the Wildfire team as it seeks to locate and destroy the outer-space virion that threatens humanity. The story calls on physiology, microbiology, pathology, and other disciplines of biology.

Du Praw, E. J. 1968. *Cell and Molecular Biology.* Academic Press, New York.

The early chapters describe the morphology and analyze the functions of cell structures. The importance of the relationship between structure and function is emphasized.

Gillie, O. 1971. *The Living Cell.* Funk & Wagnall (World of Science Library), New York.

Handsomely illustrated, popularly written, and far-ranging introduction to general cell structure and function.

Luria, S. E. 1973. *Life: The Unfinished Experiment.* Scribner's, New York.

Highly readable presentation by a distinguished biologist of the basic concepts of modern human biology. Intended for the layman who wants to gain understanding of the material basis of life.

Swanson, C. P. 1969. *The Cell,* 3rd edition. Prentice-Hall, Englewood Cliffs, N.J.

For the introductory student, this book provides a well-integrated description of cell structures and differentiation. Useful charts and drawings accompany the text.

Whaley, W. G., M. Dauwalder, and J. E. Kephart. 1972. "Golgi Apparatus and Influence on Cell Structures." *Science* 175(4022): 596–599.

A review article on the Golgi apparatus, highlighting its function in the synthesis of carbohydrate-protein macromolecules through selected enzymes. This complements the catabolic function of the organelle.

LIFE PROCESSES WITHIN THE CELL

chapter five

All living organisms depend on the radiant light energy of the sun, which is converted by plants into energy-storing glucose during the process of photosynthesis.

Life on Earth depends on energy—energy whose only original source is the sun. The challenge for green plants is to convert this light energy from the sun into the chemical energy that is every organism's fuel for life.

Fundamental to our understanding of how energy is converted are the **laws of thermodynamics.** The first law states that energy can be neither created nor destroyed. The total amount of energy in the universe is constant and can change only its form—from chemical energy to mechanical energy, for example. On a molecular level, every compound holds in its bonds the amount of chemical energy necessary to link its atoms together.

The second law of thermodynamics says that the amount of *usable* energy—that is, the amount of energy available to do work—is always decreasing. In every chemical reaction, some of the potential energy stored in the bonds of the original compound is dissipated as heat, and this energy is no longer available for use by the organism. This constant loss of *usable energy* means that all living organisms must depend on a nonliving source for energy. That source is the sun.

A vast amount of the sun's light energy reaches the Earth. A third of this energy is reflected back into space. Much of the rest is absorbed by the Earth's surface and atmosphere and converted into heat, which sets in motion the weather cycles of the planet. The small remainder of the light energy from the sun—less than one percent—reaches the leaves of plants, is absorbed, and is converted into chemical energy.

ENERGY CONVERSION

The name given to the conversion process is **photosynthesis.** Every organism, from a tiny bacterium to a 400-foot-tall giant redwood tree, relies on one class of molecules—the chlorophylls—to trap solar energy in molecules that can be used by other living forms that lack chlorophyll. Chlorophyll molecules act as **transducers,** substances that convert one form of energy into another—in this case, light energy into chemical energy. The chlorophyll found in many bacteria, all blue-green algae, and all green plants uses radiant light energy to produce oxygen and energy-storing molecules called glucose—a simple six-carbon sugar—from carbon dioxide and water.

For solar energy to have an effect, it must first be absorbed by the chlorophyll molecules. This absorbed light energy excites, or energizes, the electrons in the chlorophyll molecule very briefly, pushing them to a higher energy level. Such excited electrons have a specific amount of energy—comparable to a boulder pushed to the edge of a cliff (see Fig. 2.7). When the unstable excited electron returns to its stable state, the potential energy is transduced to the kinetic energy needed to manufacture new chemical bonds.

Once locked up in the chemical bonds of the sugar glucose, the energy can be used to create ATP molecules, which are the suppliers of energy for all cellular work within an organism. The energy stored in glucose may also be passed from organism to organism, as one eats another. Thus the life-sustaining energy produced in plants reaches organisms that lack chlorophyll of their own by which to obtain supplies of glucose.

Having been brought to cells and to organelles within cells, glucose is split in a series of energy-releasing steps called cellular respiration. The chemical energy trapped in the chemical bonds of glucose is transferred and stored in many molecules of adenosine triphosphate (ATP). From the high-energy bonds of ATP, energy may be quickly obtained and used for many vital cellular processes, such as synthesis of materials for growth, synthesis of enzymes, active transport, secretion, movement, generation of electrical impulses, and cellular division. The by-products of respiration—carbon dioxide and water—are used again by the plant in the photosynthesis cycle.

The vital processes of photosynthesis and respiration depend on the transportation of nutrients from place to place within cells, and from cell to cell within organisms. This movement is often dependent on very simple physical properties of cell membranes, chemical molecules, and the environment, and it is usually accomplished through processes known as diffusion and osmosis.

Mechanisms of transport

DIFFUSION

Many essential nutrients, secretory products, wastes, and other chemical substances *move* between cells, and from organelle to organelle within the cytoplasm of a single cell, by the simple physical process of **diffusion.** This is a process which is *passive;* that is, it is dependent solely on physical properties, such as the sizes of molecules, the temperature, and the **concentration** of molecules—the degree to which they are crowded together. All molecules can diffuse; they all possesss a quantity of heat energy (as indicated by their temperature) that causes them to move around incessantly but randomly (see Fig. 5.1). This random movement causes them to bump into each other; naturally they bump into each other more often when they are crowded together—when their concentration is high—than when they have space around them—when their concentration is low. Thus molecules move more readily in low concentration than in high concentration. This is the basic principle of diffusion: because of their freer movement, *molecules move from areas in which their concentration is high to areas in which their concentration is low.* This difference in concentration between any two areas is called a **concentration gradient.**

The rate (distance moved per second) at which diffusion, the movement of molecules from a region of high concentration to a region of low concentration, occurs is

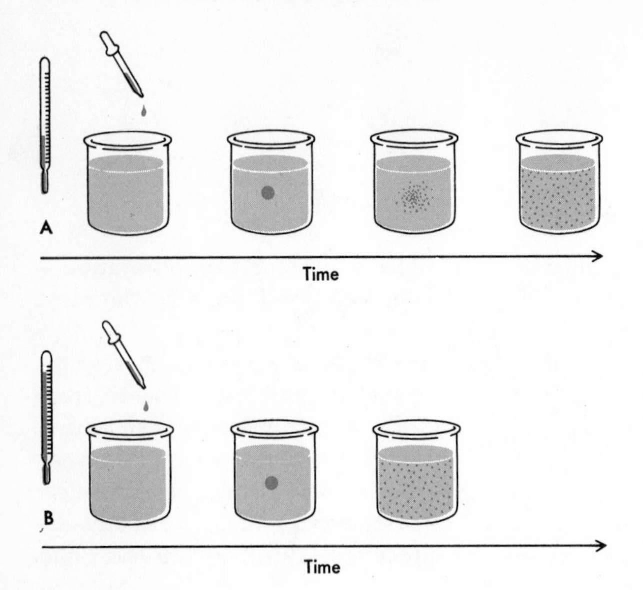

FIGURE 5.1

The influence of temperature on the rate of diffusion. A drop of ink placed in water spreads out to color the entire vessel because of diffusion. The different vessels represent the passage of time; the rate is equal in the two examples. At the lower temperature in (A), molecular movement is relatively slow, and a longer time is required for the colored molecules to move away from each other to color the contents of the container evenly. In (B) the temperature is higher, and diffusion of the colored molecules requires less time.

dependent on any physical factors which affect their rate of motion. The mass of the diffusing molecules, for example, affects diffusion rate: large molecules, such as proteins, diffuse much more slowly than small molecules like water. Here a simple physical principle operates: molecular movement decreases as molecular weight increases, and hence diffusion (movement) rate decreases as molecular weight increases. Similarly, increasing temperature, which increases the rate of molecular movement, also increases the diffusion rate, as does increasing the concentration of diffusing molecules.

The practical consequences of this simple process are widespread. Diffusion of gas molecules, such as oxygen and carbon dioxide, ensures that an organism can continue to respire—oxygen concentration is always higher in the air around an organism than inside cells in which it is used up. Hence oxygen always diffuses *into* respiring cells. Similarly, carbon dioxide is always present in higher concentration inside respiring cells than outside. Hence the waste gas departs from the interior of a cell, by diffusion, and enters the air around the cell. Many other molecules needed by cells or produced by cells move from place to place, at no cost of energy to the cell or organism, by diffusion. This process, in which substances other than water diffuse through a selectively permeable membrane, is called **dialysis.**

OSMOSIS

Osmosis is a special kind of diffusion process: the diffusion of a solvent (usually water) through a selectively permeable membrane. The selective properties of cell membranes are a result of their physical and chemical characteristics. These selectivities may change with the death of a cell; however, many of them can be duplicated in artificial membranes. Experiments with such membranes have led to the hypothesis that much of the selectivity of cell membranes is due to pore size and solubility in the lipid portion of the membranes. Thus a cell may determine which substances may leave and enter through its membrane, or any of the membranes of its organelles, at *no cost of energy to the cell.* Osmosis is a *passive* process.

Because osmosis is a passive process, one can study it in the laboratory by means of simple nonliving membranes, such as the cellophane "membrane" or casing of sausages. Cellophane sausage casing is surprisingly porous. Filled with water and tied at both ends, a short segment of casing quickly loses its puffy, sausage-like appearance as water passes out through the pores of the casing and evaporates from the surface. Given enough time, all of the water inside the casing will pass through the membrane and evaporate, leaving a completely collapsed casing, still sealed at either end, but dry inside. Diffusion ensures that the water molecules move from the region of their higher concentration, the liquid inside the casing, to a region where water molecules are less prevalent, the air around the casing. Because of the diameter of its pores, water molecules (which have a small diameter) easily diffuse through the seemingly solid cellophane membrane. Indeed, the diameter of the pores in cellophane is the very key to its effectiveness as an osmotic membrane.

Instead of filling a sausage casing with pure water, one might fill it with a solution containing water and some larger molecules, such as soluble starch. If the casing is again sealed at the ends and left hanging in the air, water molecules will again diffuse out through the cellophane, forming tiny droplets of water which evaporate into the air. But starch molecules are much larger than water molecules; they are, in fact, larger than the pores or holes in the cellophane. Thus the forces of diffusion cause the starch molecules to move toward the surface of the cellophane, but these molecules cannot pass through. Consequently, a nonliving membrane is able to precisely select, on the basis of molecular size, which molecules will pass through by diffusion.

We have seen that eventually the starch contents of the sausage casing will dry up as all of the water departs from the interior. But this situation, useful as a laboratory demonstration, is not strictly analogous to the situation in which most living cells and their membrane-bound organelles find themselves. Instead of being surrounded by air, most cells are surrounded by solutions of many solutes (chemical substances) dissolved in water. Similarly, membrane-bound organelles are immersed in a cytoplasmic "solution" in which many different solutes, nutrients and wastes, are dissolved in water. Thus we must alter our sausage casing analogy slightly to account for the physical situation in which living cells most often find themselves.

Let us imagine again our sausage casing, filled with a starch-water solution in which 1 g of starch is dissolved in 100 g of water. Instead of hanging it in air, we can immerse it in a beaker, in which it may be surrounded by a starch-water solution containing only 0.1 g of starch dissolved in 100 g of water (see Fig. 5.2). In

FIGURE 5.2

Osmosis. Osmosis may be demonstrated with sausage casings filled with 1 percent starch solution suspended in beakers of starch solution of varying concentrations. (A) In an isotonic solution (1) concentrations of water are equal inside and outside, so net movement in both directions is equal, and the casing remains unchanged. When the starch solution outside is hypotonic (2), water molecules diffuse into the casing, swelling it. When the outside solution is hypertonic (3), water diffuses out of the casing, where its concentration is higher. (B) A diagrammatic representation of osmosis. Pores in the selectively permeable membrane permit the free movement of the solvent, water, but not the larger molecules of the solute, starch. Because of the higher concentration of water molecules on the left, more of them will randomly strike the pores and pass through, and the net movement of water will be from left to right.

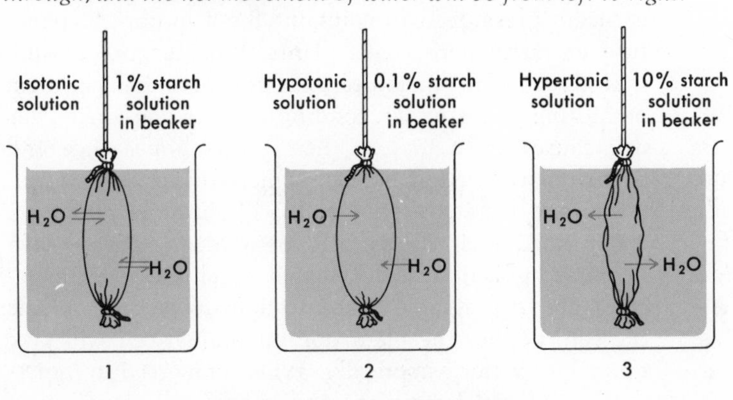

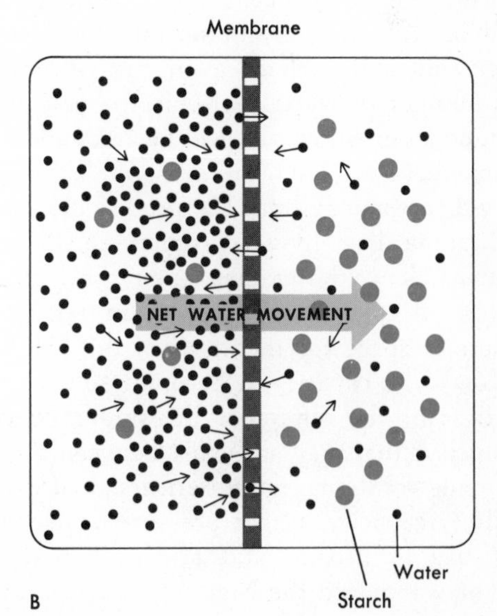

these solutions, the starch concentration (the number of molecules of starch in each milliliter volume of water) is higher inside the sausage casing than outside, and the concentration of water, conversely, is higher outside the casing than inside. If free diffusion were permitted by the membrane, one would expect the starch molecules to diffuse out of the bag and water molecules to diffuse in, until there was no longer a difference in concentration of either substance, inside and outside the bag. However, bump against the membrane as they might, the starch molecules cannot depart from the interior; nor can they enter. The water molecules, however, can and will travel through the membrane. As we expect from the law of diffusion, they will move from the region of their higher concentration—outside the membrane—to the region of their lower concentration—inside the membrane. Movement of water molecules will continue into the bag until one of two things makes them stop. First, as they leave the solution outside the bag, the concentration of the outside solution changes, as does the concentration of the inside solution as they enter. Eventually **osmotic equilibrium** may be reached, in which the concentration of starch and water, called the **osmotic concentration,** on both sides of the membrane is equal (at this point, the solutions are isotonic, or of the same osmotic concentration). Although water molecules will still randomly bump into the membrane from either side and perhaps even pass through, the total (net) number of water molecules passing through the membrane from either side will be zero, and further osmotic movement of water will cease.

A second factor may limit the process. The interior of a membranous bag, living or nonliving, has only a given volume. Hence water may enter the bag (under the conditions of concentration we began with) only as long as there is space for them; when the bag fills up with solution to its capacity, it becomes stiff, or **turgid,** and regardless of the difference in osmotic concentration (or **tonicity**) that may still exist between the inside and the outside solutions, *no more water will enter.* As the bag fills to capacity, **turgor pressure** develops. This causes the bag to become stiff and prohibits further movement of water into the bag.

The concentration of solutes is not always higher inside cells than outside. Hence, if the osmotic concentrations are reversed, a cell may lose water instead of gaining it. Such a loss occurs when a person drinks large amounts of sea water—that is, salt water. Because of the high concentration of osmotically active solutes (the salt) in sea water, these solutes get into the blood plasma. The plasma then shows an increase in osmotically active solutes (the salt) and is **hypertonic** to the cells it bathes. Because water tends to move more readily across cell membranes than other substances do, water diffuses out of the cells into the blood plasma, instead of the solutes moving into the cytoplasm. This diffusion of water will continue until the water concentration becomes equal in the cells and the blood plasma or, as is more usual, all the man's cells die of dehydration.

Plant cells as well as animal cells are greatly affected by osmosis. Plant cells immersed in a **hypotonic** solution (in which the water concentration outside the cells is higher than the water concentration inside the cells) will gain water, until the entry of more water molecules is impeded for lack of space within the cells. The space within a plant cell is relatively limited by the presence of the stiff cellulose cell wall, which, we recall, gives a fixed shape and volume to a cell. As more and more water molecules enter by osmosis, they become packed tighter and tighter together, thereby raising the water pressure within the cell. The water pressure, which presses outward against the inside of each cell wall much like the air in an inflated balloon, is turgor pressure. It is turgor pressure that maintains the firm upright structure of many herbaceous plants. It is turgor pressure that gives fresh vegetables their crispness; salting them or storing them under conditions in which their cells dry causes them to lose their turgid appearance and become flaccid. Animal cells, immersed in *hypotonic* solution and lacking a stiff wall, will continue taking up water until their flimsy cell membranes burst. When immersed in **isotonic** solutions, in which the water molecules are in the same concentration inside and outside the cell, neither the plant nor the animal cell will take up or lose water osmotically. When immersed in *hypertonic* solutions, both plant and animal cells *lose* water;

plant cells **plasmolyze** (their cytoplasmic contents shrink as water is lost) and animal cells **crenate** (they also shrink as water is lost). Similarly, any organelle within any cell, because of the presence of its selective membrane and the simple physical laws of diffusion, is able to retain large molecules, such as enzymes and structural molecules, while gaining or losing water and other small nutrient molecules at no cost of energy to effect the movement.

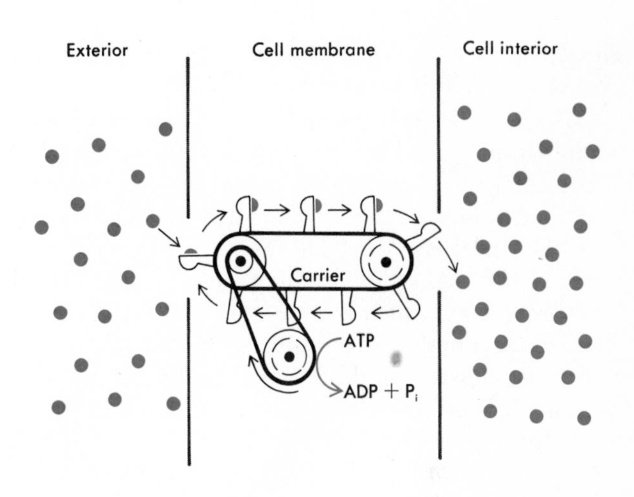

FIGURE 5.3

A mechanical analogy of active transport. The transport mechanism can accumulate particles, such as ions, inside the cell in higher concentration than outside. The carrier located within the membrane is moved by the energy released as ATP is broken down into ADP.

PASSIVE AND ACTIVE TRANSPORT

The life processes within the cell depend on movement of materials across membranes by dialysis and osmosis. They are the means by which many nutrients required for survival pass into the cell. Small molecules and ions dissolved in the surrounding fluids will enter the cell along a concentration gradient, but larger molecules cannot get through the pores of the membrane. This automatic, selective passage of materials into the cell is a form of **passive transport.** However, some ions apparently defy the laws of diffusion. They may be concentrated either inside or outside the cell membrane. Because these substances will not pass automatically through the membrane, they must be transported across it. This process, called **active transport,** involves cellular work, the expenditure of energy, to move substances against a concentration gradient. As shown in Fig. 5.3, this method of transport is accomplished by means of energy-driven carrier molecules. Still other molecules, some that are too large to pass through the membrane pores, would passively diffuse too slowly into the cell for it to survive. Such substances are also carried along their concentration gradient into the cell with specific carriers. However, this so-called **facilitated transport** does not require energy as active transport does.

The carrier molecules are located in the cell membrane, and there is some experimental evidence to suggest that certain lipoproteins of the membrane function as the carriers. Present thinking is that the carrier molecules form a temporary chemical bond with the molecule that is about to enter the cell. The combined molecules then diffuse across the membrane along a concentration gradient (from the cell periphery—an area where the molecules are more concentrated—to areas inside the cell where the molecules are less concentrated). Once the carrier has crossed the membrane, the bond is broken, probably by enzymes, and the transported molecule is released. As the carriers accumulate on the inner surface of the membrane, their concentration builds so that the gradient is reversed. The carriers then diffuse along the gradient back across the membrane, where they can form bonds with other molecules that are then transported into the cell.

Besides active and facilitated transport, two other mechanisms are used to move materials into and out of cells. These processes are **pinocytosis** (cell-drinking) and **phagocytosis** (cell-eating). In both of these processes, the cell membrane envelops the substance and, in essence, swallows it. When the particles are relatively small or the substance is a liquid, the process is called pinocytosis. When the engulfed material is larger, the process is called phagocytosis (see Fig. 5.4).

All these mechanisms of transport, whether passive, facilitated, or active, can provide the cell with vital nutrients and energy-containing foods. The cell digests the foods and thereby releases the nutrients they brought to it. The nutrients fuel all the varied processes and functions of the cell—including active transport, the same process by which some of the fuel was originally obtained. A combination of processes, known overall as **exocytosis,** constitutes the reverse mechanism by which the cell excretes waste materials or secretes substances useful for a given function.

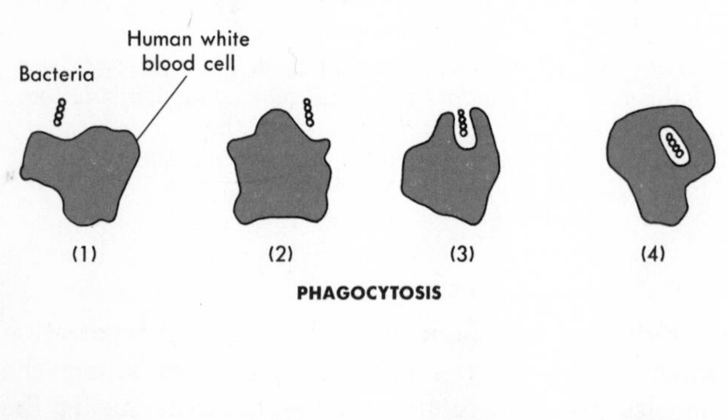

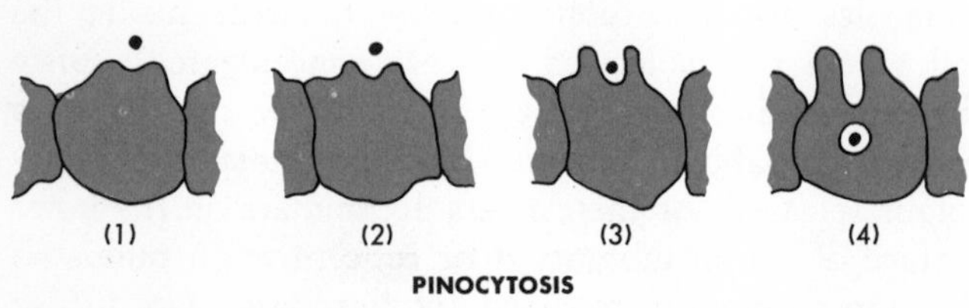

FIGURE 5.4

Supplementary mechanisms of transport. In phagocytosis, particles (e.g., bacteria) are engulfed by fingerlike extensions of the cell and enclosed within a membrane-bound vacuole. In pinocytosis, a small particle and fluid flow into a deep, narrow channel. At the end of the channel, the particle and fluid are enclosed in a vesicle or small vacuole, which moves to the interior of the cell. Pinocytosis takes place in cells whose movement from place to place is restricted because they are multicellular tissue.

Components of energy reactions

Energy is obtained mainly from the oxidation of carbohydrates—organic compounds (such as sugars and starches) that may be oxidized readily and that yield a comparatively large amount of energy—but other nutrients, fats (lipids) and protein, may also be oxidized. The cell either obtains its supply of carbohydrates, lipids, and proteins from the environment or manufactures them itself from simple precursor molecules. Unicellular or multicellular organisms that manufacture their own foodstuffs by using the sun's light energy as a power source are called **autotrophs.** Organisms that obtain their supply of energy-containing compounds by eating other organisms and digesting the compounds in them are called **heterotrophs.** Higher plants, algae, and certain pigmented bacteria are autotrophs. Heterotrophs include all animals, many bacteria, and fungi.

All living things—whether autotrophs or heterotrophs—depend on the manufacture of carbohydrates from simpler inorganic substances. Animal life could not exist without the green plants that are the ultimate producers of sugars, lipids, and proteins. Thus the process of photosynthesis, briefly described earlier, is fundamental to our understanding of life. In our discussion, we will concern ourselves mainly with photosynthesis as it occurs in leafy plants.

LEAF STRUCTURE

Although photosynthesis may occur in any green part of the plant, such as the flat stem of a cactus or the roots of orchids, in most higher plants the process usually occurs in the leaf. The flat shape of most leaves enables the plant to make the best use of the factors needed for photosynthesis. More sunlight can be absorbed; carbon dioxide can be taken in rapidly; and more sugar can be produced because with increased surface area, many more cells that manufacture sugar can be accommodated near the leaf's surface, exposed to light and air.

Most leaves consist of a stemlike structure (termed a petiole) and a thin sheetlike structure called the **blade.**

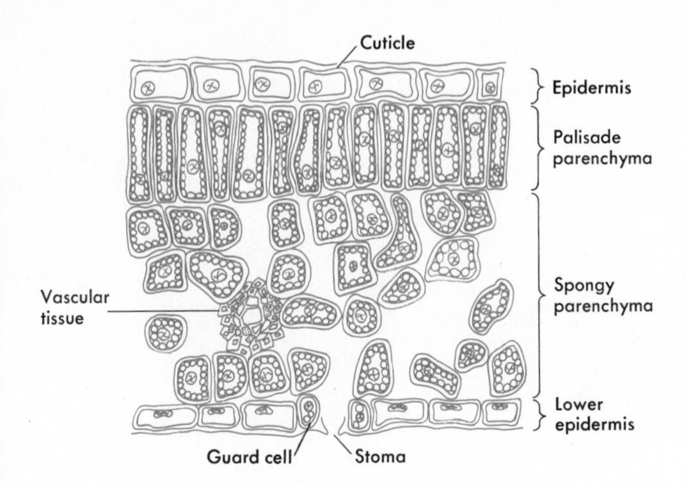

FIGURE 5.5
A cross section of a leaf of a higher green plant. Chloroplasts are located in the parenchyma tissue. Stomata openings in the lower surface permit the exchange of gases. The vascular tissue transports water from the roots to the leaf and photosynthesized material from the leaf to other parts of the plant, including the roots.

The blade consists of two parts characterized by two different types of cells. The upper and lower surfaces of the leaf (the epidermis) are made up of a layer of **epidermal** cells. This layer is usually only one cell thick, and it serves to limit water loss from the interior of the blade and offers structural support to the interior cells. In some plants the outer surface of the epidermal cells secretes a waxy **cuticle** that further helps to reduce water loss. (Waxes offer great resistance to the passage of water, as is evident from anyone who has used waxed-paper or paper cups coated with wax.) The interior of the blade is designated the mesophyll tissue and made up of parenchyma (palisade parenchyma and spongy parenchyma) and vascular tissue (Fig. 5.5).

Photosynthesis takes place in the parenchymal cells of the leaf, which are divided into two layers. The upper layer is composed of column-shaped cells that are close to each other and give the appearance of hanging from the upper epidermis like stalactites from a cave roof. This layer is called the **palisade mesophyll.** The lower layer, called the **spongy mesophyll,** consists of loosely packed cells interspersed with numerous spaces. The spaces permit the free circulation of both water and the gases (carbon dioxide and oxygen) involved in photosynthesis.

Openings in the epidermis—the stomata—allow the passage of oxygen and carbon dioxide between the mesophyll and the external environment. On each side of a stoma, special kidney-shaped guard cells regulate the stomatal diameter to prevent the vital mesophyll cells within from drying out. The guard cells are the only epidermal cells equipped with chloroplasts.

Because of the large amount of intercellular space, the leaf has a large internal surface area. Almost every mesophyll cell has direct contact with leaf "veins" (vascular tissue) composed of xylem and phloem cells. The veins are considerably branched, and their strong, thick-walled cells help support the blade, keeping it out flat and preventing it from curling. In the vascular tissue xylem cells transport water and other nutrients to the leaf from the roots. The phloem cells transport carbohydrates synthesized in the leaf to other parts of the plant.

CHLOROPLASTS

The chloroplast is the highly structured cellular organelle in which the reactions of photosynthesis are accomplished. Chloroplasts can vary in size and shape, but they are always surrounded by two membranes. The interior of the chloroplast also contains membranes. Many of these membranes are gathered in stacks called grana (see Fig. 5.6).

The chlorophyll and enzymes necessary for photosynthesis are believed to be bound to these internal membranes and to be arranged in a precise order. The proper arrangement of chlorophyll and enzymes is necessary for the efficiency of photosynthesis. The raw materials for photosynthesis are soluble in the water-filled regions that surround the grana membranes. The car-

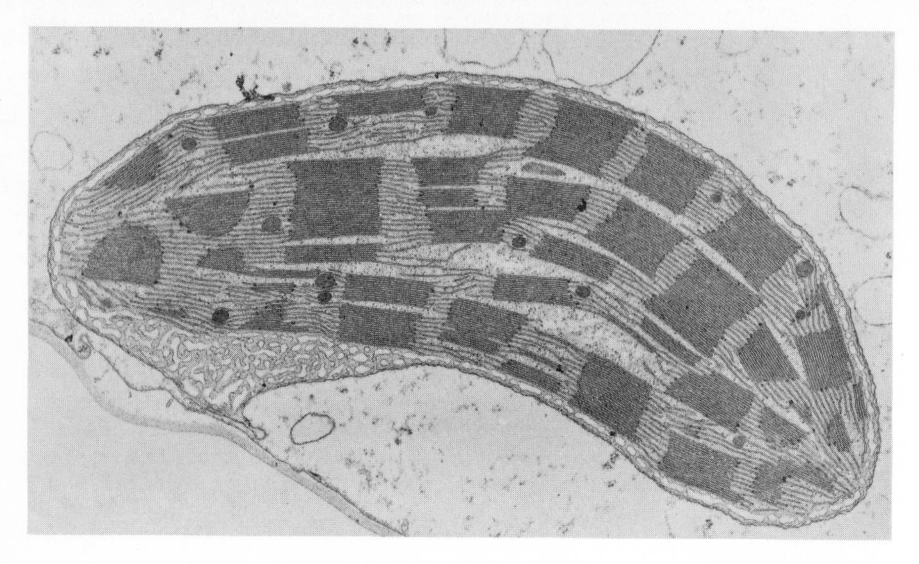

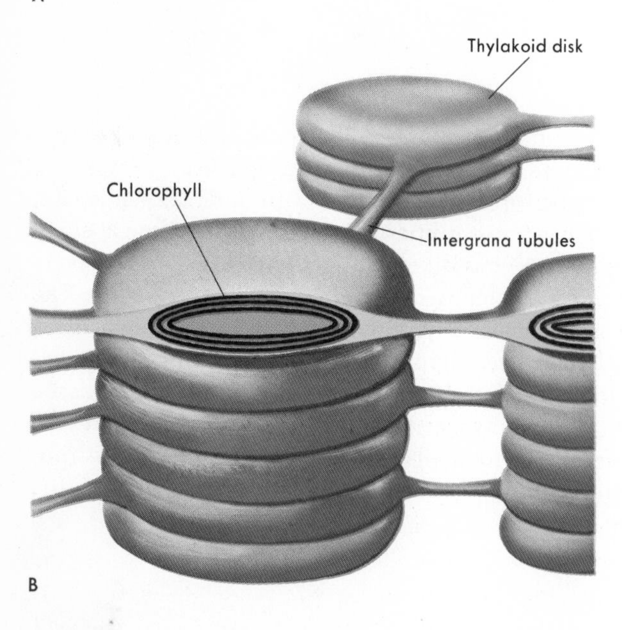

FIGURE 5.6

Chloroplast of a higher plant and its internal structural details. (A) Electron micrograph of a cross section, showing the stacking of membrane sacs (thylakoid disks) to form the grana, which are shaped like rolls of pennies. The grana are interconnected by tubules that stretch from thylakoid to thylakoid in adjacent grana. (B) Three-dimensional representation of the thylakoid disks forming grana and the tubules that connect adjacent grana. The membranes composing the thylakoids contain chlorophyll and the enzymes necessary for the light reactions in photosynthesis.

bohydrate end-products of photosynthesis are also soluble and are thus carried away from the membrane sites of synthesis.

Photosynthesis involves a series of at least 30 chemical reactions that must take place in a precise sequence. It is therefore understandable that the structural relationship of chlorophyll and the enzymes is very important. Membranes serve to maintain this relationship.

Chlorophyll Chlorophyll molecules play the vital role in photosynthesis of absorbing light energy from the blue and red ends of the visible light spectrum.

The molecular configuration of chlorophyll enables the compound to absorb light energy very efficiently. When quantities of light energy, called **photons,** are absorbed by the chlorophyll molecule, they boost at one or more of its electrons to a higher position in its shell. These are termed excited electrons. During the fraction of a second that an electron is in this position, it has greater potential energy, and the chemical machinery of the chloroplast is able to trap the energy from the excited electron for use in the synthesis of carbohydrate. If the energy of the excited electron is not used, the electron will fall back to its original level, releasing its potential energy as light—a process called **fluorescence** (Fig. 5.7).

Photosynthesis

Enormous quantities of radiant energy from the sun fall on the surface of the earth. If only a greater portion of this energy, wasted mainly as heat, could be trapped

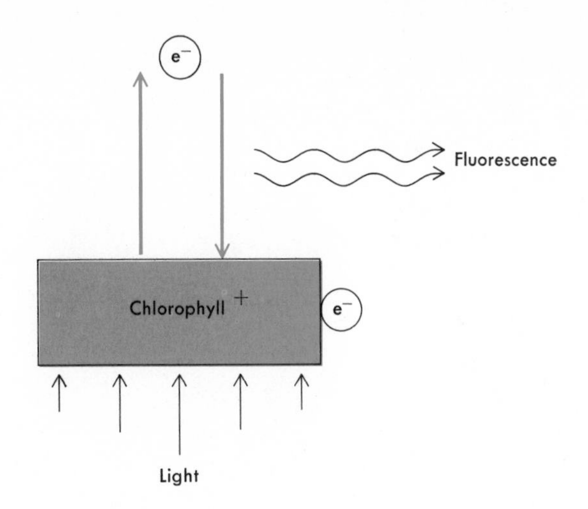

FIGURE 5.7

Fluorescence occurs when chlorophyll is isolated from the rest of the photosynthetic machinery and the energy of the excited electron goes unused. The electrons rejoining the chlorophyll molecule emit a red light (fluorescence).

and channeled in useful ways, many of the energy-hungry world's problems might be solved. For millennia, however, the energy *has* been harvested to some extent, and the impact of the presence of the harvesting mechanism on the development of life has been vast. The conversion of the sun's radiant energy into chemical forms of energy is the special province of photosynthetic plants, and it has been the evolution of the photosynthetic-chemical process that is largely responsible for the presence of oxygen in our atmosphere and the evolution of forms of animal life as we recognize them today.

The enormous importance of photosynthesis as a process, and green plants as the processors, in the energy and food economy of the entire world must not be understated. A single acre of corn plants can produce about 100 bushels of the yellow edible fruit; in addition to the fruit, about four tons of plant parts, consisting in the main of the walls of the cells in the leaves, stalks, and stems, may be produced. When we consider the enormous tracts of land planted in corn in this country, we can only begin to imagine the vast amount of plant matter produced by corn alone. From the seemingly useless portions of plants, substances as diverse as food for cattle and building materials may be produced. From the fodder may be produced the flesh of beef animals, by the special metabolic processes that cows use to convert cellulose into meat. Corn is only a familiar example of a food-producing plant; there are even vaster acres of water, with their own harvest of algae and other phytoplankton which serve as food for all the aquatic animals.

Surprisingly, the uncounted tons of solid matter produced by plants is derived by chemical conversions from omnipresent water and the ordinary waste gas, carbon dioxide, which is produced by living animals (and noisy factories!). And the energy source for this conversion is sunlight. Thus green plants are able to *trap* radiant energy and use it to build chemical bonds between carbon atoms, thereby producing food and other useful materials. In this section we shall examine the process, photosynthesis, by which the trapping and conversions are done.

PHOTOSYNTHESIS: AN OVERVIEW

A green plant may be likened to a small chemical factory. From carbon dioxide and water, using the radiant energy of sunlight, it produces carbohydrates. These carbohydrates can be transported to other parts of the plant (roots, for example) to fuel growth, or they may be stored for later use. The plant is essentially self-sufficient, and from the carbohydrates and other inorganic (mineral) nutrients available, it can synthesize all the other chemicals of which it is composed: proteins, carbohydrates, lipids, hormones, and others. In the photosynthetic process only a relatively few chemical reactions require light energy. Other chemical reactions lead to the production of a simple carbohydrate, called **phosphoglyceraldehyde (PGAL).** From this substance, by a great number of chemical reactions, are produced the wood of a tree, the succulent flesh of a pear, the sticky sweetness of maple syrup. The chemical reactions can be simplistically summarized. We must bear in mind, however, that although the following reaction describes the overall process of photosynthesis, it merely hints at what actually occurs.

$$CO_2 + H_2O + \text{light energy} \rightarrow (CH_2O)_n + O_2$$

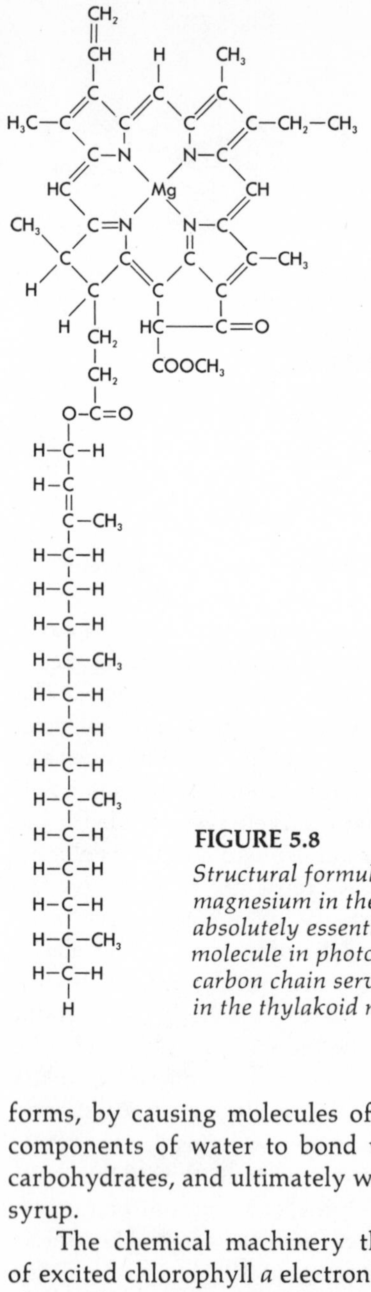

The reactions alluded to above occur in many diverse parts of plants. In all these parts, we associate a particular color, green, with the process. This green color is due to the presence of chlorophyll. Let us examine now the relationship of chlorophyll to the feeding of the world.

Energy trapping: chlorophyll In order to use the energy contained in sunlight, a photosynthetic plant must first absorb or trap the light. Absorption is done by colored molecules, called **pigments.** In plants, the most prominent pigment is **chlorophyll,** a large complex molecule with the metal magnesium in its center (Fig. 5.8). Chlorophyll is actually the name of a whole family of pigments, very closely related to each other in chemical structure and light absorption capabilities. We will discuss the activity of the one found in higher plants, chlorophyll *a*, recognizing that the basic light absorption and energy conversion processes occur in all green plants.

Chlorophyll, as we have noted, absorbs blue and red light and reflects green. This reflection of the green component of white sunlight gives leaves their green color. But such reflection is wasteful, and higher green plants have evolved ways of using other pigments, chiefly chlorophyll *b* and the **carotenes** (responsible for the oranges and yellows of fall leaves), to improve the efficiency of light trapping. These molecules absorb light energy and channel it to the chlorophyll molecule.

Energy, as we may recall from Chapter 2, has the ability to do work, or move objects. In photosynthesis, the objects moved are relatively loosely bound electrons in the molecules of chlorophyll *a*. On impact of a quantum, or package, of absorbed light energy, freely movable electrons in molecules of chlorophyll are actually displaced from their normal positions. During the extremely brief period of their displacement (10^{-9} second), the electrons, now farther from the nucleus to which they are attracted, have extra energy, potential energy. If the energy is not "harvested" by the chemical machinery, the excited electron's energy is wasted, producing only heat and red fluorescence. Fortunately, however, chemical mechanisms *have* evolved which may convert the energy of the excited electron to useful forms, by causing molecules of carbon dioxide and the components of water to bond with each other to form carbohydrates, and ultimately wood and pears and maple syrup.

FIGURE 5.8

Structural formula for chlorophyll a. *The magnesium in the center of the molecule is absolutely essential for the function of the molecule in photosynthesis. The long hydrocarbon chain serves to anchor the molecule in the thylakoid membranes.*

The chemical machinery that harvests the energy of excited chlorophyll *a* electrons is housed in the chloroplast. The initial steps of photosynthesis are known as the **light reactions** because they are obliged to operate when light is present. Later steps, which use chemical products of the light reactions, may proceed in darkness or light, since no further absorption of light energy is necessary. Usually called the dark reactions, they are also referred to as the Calvin cycle, after Nobel prizewinner Melvin Calvin, who first discovered the sequence of reactions for the fixation of carbon dioxide. (A second process for CO_2 fixation will be considered later in this chapter.)

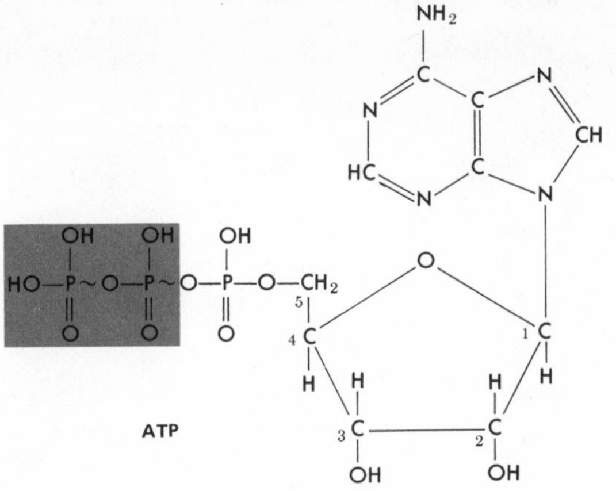

FIGURE 5.9

Structural formula of ATP, consisting of the purine adenine, the sugar ribose, and three phosphate groups.

The light reactions accomplish the production of two substances: the so-called "universal energy currency of the cell," **ATP** (adenosine triphosphate), and "reduced" **NADP** (nicotinamide adenine dinucleotide phosphate). The hydrogen atoms of reduced NADP are used to reduce molecules of carbon dioxide to form carbohydrates. ATP and $NADP_{reduced}$ are molecules that *temporarily* store energy in chemical form, and that move easily from the chemical reactions in which they participate. These are reduction and oxidation reactions, or "redox" reactions for short. Oxidation, as we have seen, is the loss of electrons, and reduction is the gain of electrons. However, the electrons gained and lost do not exist free in the chemical environment; instead, they are attached to other molecules, called electron carriers.

One of the most important of these electron carriers in photosynthesis is NADP, which may exist in two forms: $NADP_{oxidized}$ when it lacks the two electrons it has the capacity to carry, and $NADP_{reduced}$ when it has its two electrons. In its oxidized state, NADP may participate in photosynthetic reactions that reduce it by giving it two electrons; with its burden of two electrons, $NADP_{reduced}$ also has a negative charge. Because of the negative charge, $NADP_{reduced}$ attracts positively charged H^+ (hydrogen) ions to itself, and carrying its electrons and hydrogens, it diffuses to the sites of carbohydrate synthesis. There the electrons and hydrogens are given up so that they may participate in the formation of new bonds necessary for the creation of energy-rich carbohydrates from CO_2. In giving up its electrons, the carrier molecule is oxidized. Having performed its carrying task, $NADP_{oxidized}$ diffuses back to the light-reaction sites in photosynthesis to pick up more electrons and H^+ ions.

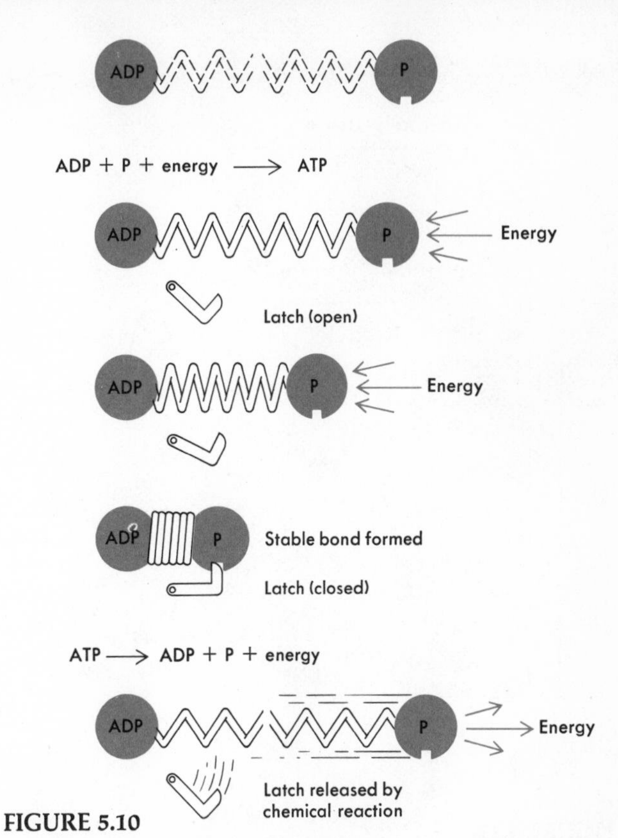

FIGURE 5.10

The storage of energy in ATP and its release, shown in the mechanical analogy of a spring. Energy input is required to form a bond between ADP and phosphate. The bond, symbolized by a spring, must be compressed in order to bring ADP and P close enough to form a stable compound (ATP). Once formed, ATP may be transported to any site in the cell in which energy is required. There the bond may be released, and the energy stored in the bond is liberated as ADP and P spring apart.

Energy currency: ATP ATP is a rather unusual molecule, with properties of such extreme value to all living things that it is found in every living cell, plant or animal. It is a relatively small molecule. Being soluble in the water of the cell, it diffuses rapidly. It is composed of an organic component, adenosine, to which three inorganic phosphate groups are linked in sequence (Fig. 5.9). The bond linking adenosine to the first phosphate group is ordinary. However, the bonds linking the first phosphate to the second and the second to the third are high-energy bonds. These bonds, which require more energy for their formation than most other bonds, serve as energy stores. They are, in fact, analogous to the compression of a spring, as shown in Fig. 5.10. Energy required to move the spring to its compressed position or (by

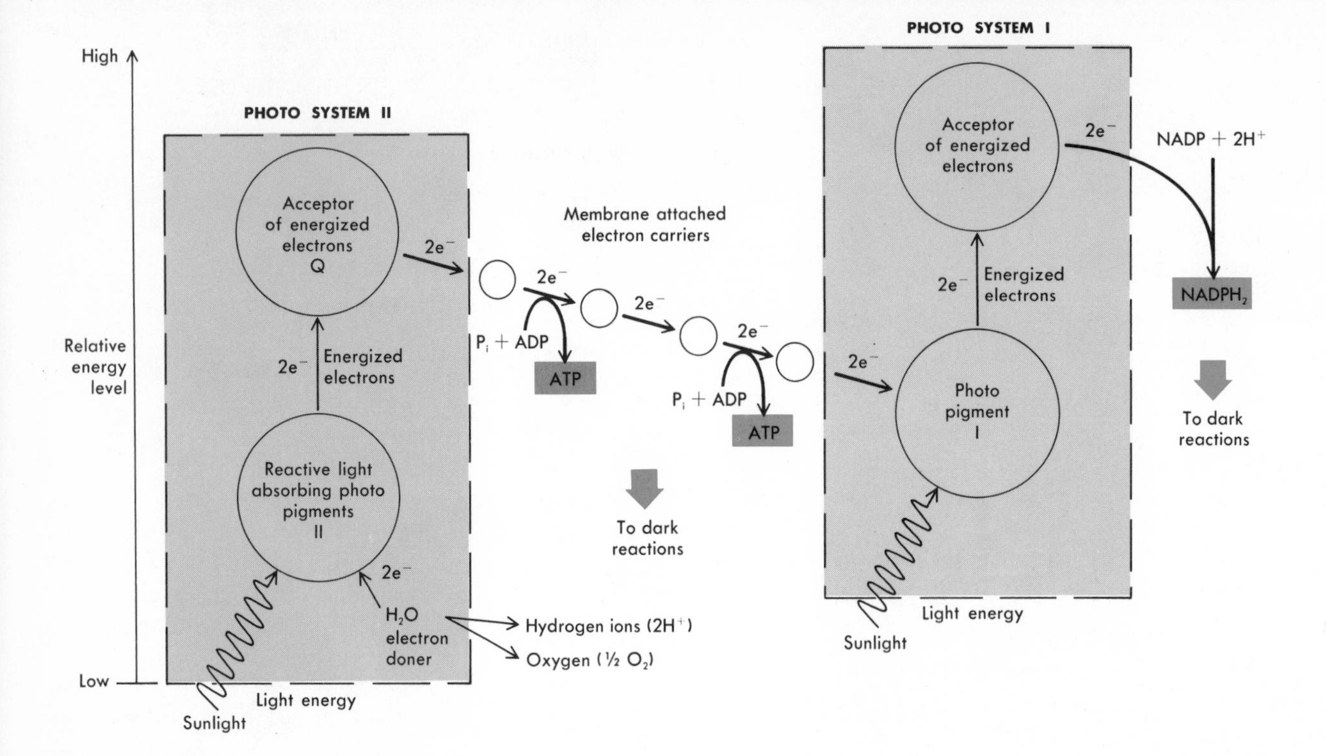

FIGURE 5.11

Simplified diagram of the photosynthetic light reactions. Light energy absorbed in photosystem II energizes reactive photopigment and thus can pull electrons away from water molecules. Hydrogen ions and oxygen are produced as by-products. The energized electrons are boosted to a higher energy level and picked up by an acceptor, Q. The excited electrons on Q go downhill, giving up enough energy in the process to recharge two ADP molecules into two ATP molecules. These electrons then become part of photosystem I. They are picked up by photopigment I and with light energy are boosted uphill again, picked up by another electron acceptor, and as they lose some of their excess energy, convert NADP to $NADPH_2$, picking up free hydrogen ions (H^+) in the process. NADP will carry electrons from photosystem 1 to the dark reactions (the Calvin cycle).

analogy) to push molecules of phosphate close enough to react becomes locked into the molecule of ATP when the bonds form. Although the bonds are stable, as is the locked compressed spring in the model shown, they may be broken, and the energy is then released to do useful work. In our analogy of the high-energy bonds of ATP to compressed springs (Fig. 5.10), we included a "latch" to give stability. In the cell some trigger is required to release the energy of ATP for a chemical reaction. This trigger is an enzyme.

This convenient storage molecule, ATP, is formed in many different energy-conserving reactions in cells from adenosine diphosphate (**ADP**) and phosphate (P). It diffuses to places in the cell where energy is required and is broken on demand into ADP and P, releasing the stored energy. Thus, energy-releasing reactions are separated in an organism from energy-using reactions, and the small energy-carrier molecule ATP is shunted back and forth from sites of energy production to sites of energy use. As we shall see, ATP can be used in almost every overall synthetic process or chemical process that accomplishes biological work in organisms, from making sugar to moving muscles. All the details of the chemical steps in which ATP is formed are not fully known. In fact, the precise mechanism for ATP formation in chloroplasts or mitochondria is the subject of some of the most difficult and exciting current research. Nonetheless a great deal is known, and we will now describe ATP formation (**photophosphorylation**) during photosynthesis by green plants.

TWO LIGHT REACTIONS

Within the chloroplasts of higher plants, the enzymes that perform the light reactions are thought to be arranged in two separate but interconnected systems. Each system called a **photosystem,** is built around a molecule of chlorophyll *a* and contains **electron carriers,** attached to the grana membranes, that are capable of actually passing electrons from one molecule to the next. The details of the chemical processes involved in the light reactions are not yet fully understood. Therefore we shall present a composite scheme derived from the hypotheses of several researchers (see Fig. 5.11). We do know that light energy is trapped and converted into "energy of excitation" of electrons. Such excited electrons can have one of two fates: (1) they can be transferred from carrier to carrier with a stepwise loss of energy coupled to the synthesis of the high-energy bonds of ATP; or (2) they may be removed entirely from the sites of light reactions by NADP and used later to reduce CO_2 and make sugars.

The two photosystems are distinguishable from each other by the fate of the electrons they contain, and by the nature and environment of the chlorophyll molecule at the heart of each photosystem. The first system we shall discuss, photosystem II, contains electrons that are mainly transferred to photosystem I. During this energy-requiring transfer process, ATP is made from ADP and P. The second system, termed photosystem I, transfers its electrons to an electron carrier, NADP, which becomes reduced NADP and carries the electrons away from photosystem I for use in the dark reactions.

The light reactions in photosystem II begin when a quantum of light energy strikes a molecule of chlorophyll *a,* or another accessory pigment (such as carotene), which passes the energy to a molecule of chlorophyll *a.* The absorbed energy excites one or more electrons in the chlorophyll molecule by moving it out of its normal position. The excited electrons are trapped, actually removed from the chlorophyll molecule by some chemical whose identity is not established but which is called Q. Once trapped, the electrons are passed to a series of electron carriers (called the cytochrome series) for conversion of their energy of excitation into chemical energy of ATP bonds.

The **cytochromes** are proteins that contain iron atoms attached to them. Several different kinds of cytochromes are present in photosystem II. They are attached to the surfaces of the grana membranes and arranged physically in a series so that they may remove excited electrons from chlorophyll *a* and transfer electrons from one cytochrome to the next (Fig. 5.12). The cytochromes differ slightly in their molecular composition; each has a slightly different affinity for an electron. The electron can thus be kept moving from one cytochrome to the next, since each cytochrome attracts an electron slightly more avidly than its neighbor. As an electron passes from one cytochrome protein to the next, the cytochromes are alternately reduced and oxidized.

To clarify the concepts involved in electron excitation, transit through cytochromes, and coupled ATP synthesis, let us consider an analogy of a waterfall. A volume of water at the top of the waterfall, which has **potential energy,** is comparable to the electron excited to a higher energy level. As this water tumbles over the edge, its potential energy is converted to **kinetic energy** —the energy of motion. The falling water can be used to turn a paddle wheel, which can be considered comparable to the cytochromes. The paddle wheel can be coupled to a generator to produce electrical energy to do other useful work. This sequence is analogous to the transformation of the potential energy of an electron into the chemical energy of ATP. Once at ground level, the water has no more energy to impart, and it must be raised up

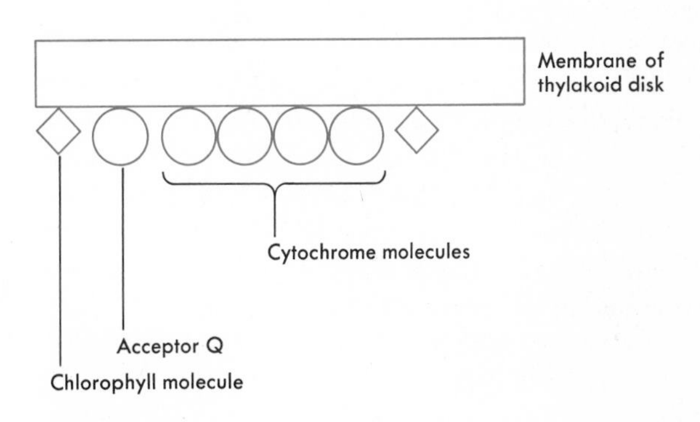

FIGURE 5.12
Alignment of chlorophyll and electron acceptor molecules along a thylakoid membrane.

to great height again by the evaporative power of sunlight. The paddle wheel, however, can be used to drive other machines since it has, in its motion, the energy formerly stored in water. The paddle wheel in this analogy is a form of transducer, converting or transducing the mechanical energy of falling water to the electrical energy of a generating plant.

The cytochromes are *coupled* or chemically tied to enzymes that use part of the falling electrons' energy to create the high-energy bonds of ATP.

Photosystem I shares with photosystem II the same principles. Light absorbed by a molecule of chlorophyll *a* in photosystem I excites its electrons. However, as we can see in Fig. 5.11, the electrons never follow a path back to a chlorophyll molecule. Instead, they are taken to the electron carrier, NADP, which is free to move physically away from the site of photosystem I. As molecules of NADP remove electrons from photosystem I, they become $NADP_{reduced}$. And as electrons are removed by the impact of light quanta from chlorophyll *a* in photosystem I, the chlorophyll molecules lose electrons or become oxidized.

Now a chemical dilemma seems to appear. If all the molecules of chlorophyll *a* in all the type I photosystems in a chloroplast became oxidized, photosynthesis would stop. However, a mechanism evolved that continuously supplies electrons to photosystem I and in the process creates molecular oxygen, the gas needed to sustain animal life.

Because of the physical proximity of photosystem II to photosystem I, electron transfers probably occur between them. Electrons from photosystem II, instead of being returned to the molecules from which they came, are transferred to chlorophyll molecules in photosystem I. But this process would leave photosystem II molecules oxidized and useless, lacking electrons and unable to engage in ATP production. However, present in the neighborhood of both photosystems is always the universal solvent, water. Although water ordinarily dissociates to only a slight degree,

$$H_2O \xrightarrow{\text{dissociation}} H^+ + OH^-$$

the light reactions of photosynthesis seem to be able to drive or force this dissociation, thereby causing the photolysis, or splitting, of water.

$$H_2O \xrightarrow{\text{photolysis}} H^+ + OH + e^-$$

The photolysis (literally, splitting by light) seems to take place in association with the absorption of light by photosystem II. Two of the products of photolysis, electrons and H^+ ions, have related fates in photosystem I. The electrons are accepted by oxidized chlorophyll *a*, permitting it to continue absorbing light and exciting electrons. The remainder of water's O and H atoms present after photolysis recombine in some little-understood way to form oxygen and a small amount of water.

$$4\ OH \rightarrow O_2 + 2\ H_2O$$

The oxygen is released from the chloroplast as a gas and may be used by the cell for the oxygen-requiring reactions of respiration, or it may be released into the surrounding atmosphere or water.

Let us restate the light reaction processes. There are two kinds of *light reactions*, which occur in two different, but linked, *photosystems*. Light energy absorbed by photosystem II is used to phosphorylate ADP, creating ATP, and to split H_2O molecules into free oxygen, hydrogen ions, and electrons. Light absorbed by photosystem I is used to reduce $NADP_{oxidized}$, forming $NADP_{reduced}$. Electrons are passed from photosystem II to photosystem I because of their close proximity. The electrons removed from chlorophyll molecules in photosystem II are restored by the ionization and photolysis of water. Thus the activation of photosystem II leads to a photooxidation step that generates reducing power that will be required in the photoreduction step needed in the activation of photosystem I. The molecules of ATP and $NADP_{reduced}$ formed during the two light reactions may migrate away from the photosystems and be used during the dark reactions to make the energy-rich bonds found in carbohydrates. Hydrogen ions formed in photolysis will end up on the carbohydrate.

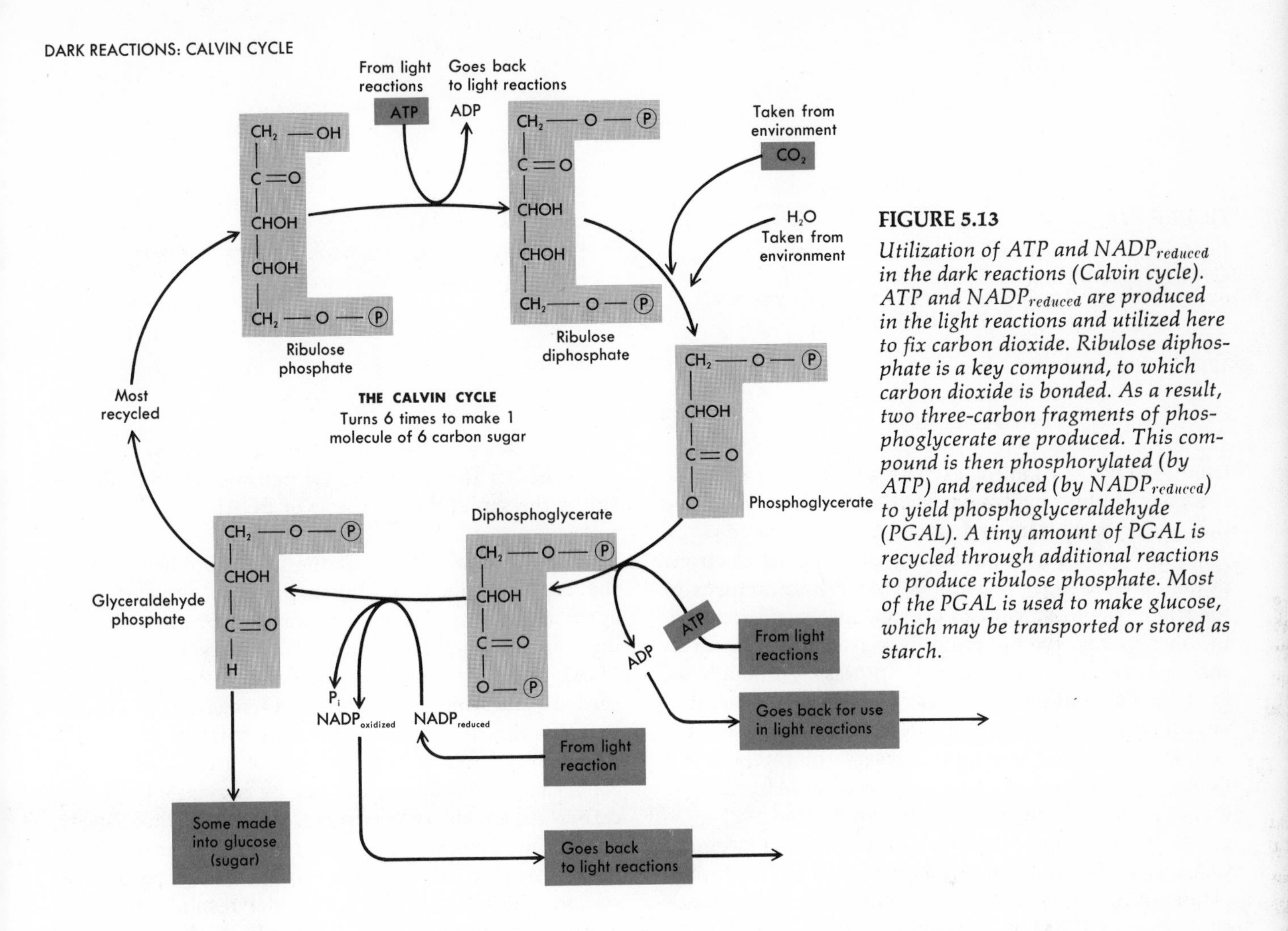

FIGURE 5.13

Utilization of ATP and $NADP_{reduced}$ in the dark reactions (Calvin cycle). ATP and $NADP_{reduced}$ are produced in the light reactions and utilized here to fix carbon dioxide. Ribulose diphosphate is a key compound, to which carbon dioxide is bonded. As a result, two three-carbon fragments of phosphoglycerate are produced. This compound is then phosphorylated (by ATP) and reduced (by $NADP_{reduced}$) to yield phosphoglyceraldehyde (PGAL). A tiny amount of PGAL is recycled through additional reactions to produce ribulose phosphate. Most of the PGAL is used to make glucose, which may be transported or stored as starch.

DARK REACTIONS: THE CALVIN CYCLE

The enzymatic machinery of the chloroplast uses the energy from the ATP molecules and the electrons from the $NADP_{reduced}$ molecules to make carbohydrates. The entire sequence of reactions, in which carbon dioxide is fixed into longer molecules containing *energy-rich bonds,* can occur any time that ATP and $NADP_{reduced}$ are present. Carbohydrates are usually produced during periods of illumination, and their production will continue until the reserves of ATP and $NADP_{reduced}$ are used up. Because the subsequent reactions of carbon dioxide fixation do not *require* light, they are often referred to as the **dark reactions.** However, these reactions do not necessarily occur in the dark; in fact, they progress rapidly in the light.

The entire sequence of chemical reactions leading to the production of carbohydrates from the input CO_2 molecule is quite complicated. We shall therefore mention only the major reactions of the two principal pathways for CO_2 fixation that occur in higher plants. One pathway uses only three-carbon molecules ("C3"). The second uses four-carbon molecules first ("C4") and has one more step than the C3 pathway.

In the C3 pathway (the Calvin cycle), carbon dioxide combines in an enzymatic reaction with a five-carbon sugar, ribulose diphosphate (see Fig. 5.13). This reaction forms a temporary and very unstable six-carbon "intermediate" compound that breaks apart, yielding two three-carbon fragments. Phosphates from ATP are added to the three-carbon fragments, and they are then reduced by the addition of hydrogen ions and electrons

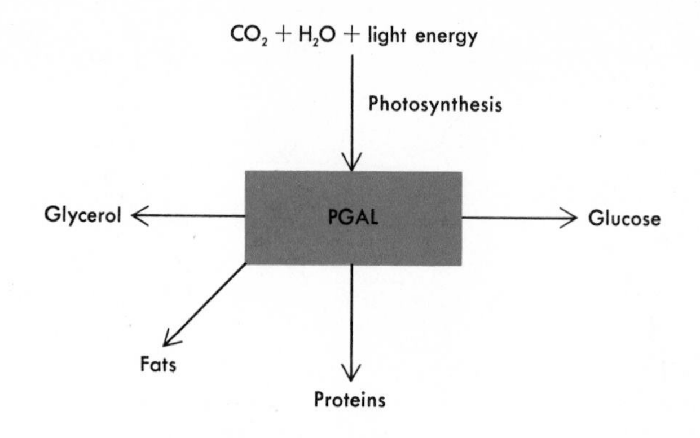

FIGURE 5.14
Phosphoglyceraldehyde (PGAL), a crucial molecule in the synthesis of many different substances vital to life. PGAL derived from photosynthesis may be converted by several different enzymatic reactions to the substances shown.

from $NADP_{reduced}$ molecules. The end product of these reactions is phosphoglyceraldehyde (PGAL), a molecule that is very important and should be remembered.

The NADP molecule, having given up its electrons in the dark reactions to become $NADP_{oxidized}$, returns to the site of photosystem I to become once again reduced. In other words, the light and dark reactions interact in such a way as to create a cyclic process. Although we have here simplified the process of photosynthesis, it is important to remember that the continual input of carbon atoms from CO_2 creates a reserve of three-carbon PGAL molecules. From that reserve of PGAL, ribulose diphosphate and other sugars are synthesized.

In the mid-1960s, another pathway for transforming CO_2 into carbohydrates was discovered in certain higher plants. In this process, the CO_2 combines with phosphoenolpyruvate (PEP) in a reaction that is considerably more efficient than the reaction of CO_2 with ribulose diphosphate in the Calvin cycle. The reaction with PEP produces two four-carbon compounds (hence the name "C4"). These C4 compounds are rapidly converted within the plant to carbon dioxide and the usual three-carbon compounds that eventually yield PGAL as in the C3 pathway. The result of using the C4 pathway is a much more efficient plant that can conduct photosynthesis with smaller concentrations of carbon dioxide than the C3 plants require. This difference is particularly important in very hot and dry climates, where any opening of the stomata to obtain more CO_2 also causes severe water loss. The C4 pathway has great agricultural significance: several important hot-weather crops, including corn, sorghum, and sugar cane, are C4 plants. A less desirable result of the efficiency of C4 plants can be seen on your lawn: hardy, fast-growing crabgrass.

Whether the C3 or the C4 pathway is used, the result is the same: the synthesis of PGAL.

The role of PGAL Most of the PGAL produced during the dark reactions in photosythesis is recycled, used again to reform the carbon dioxide acceptor, ribulose diphosphate, in the Calvin cycle. However, excesses of PGAL do accumulate after the dark reactions have proceeded for a while, and as Fig. 5.14 suggests, PGAL can be converted to various substances vital to life. PGAL has a crucial role in the nutrition of a cell and may be converted, in reactions catalyzed by enzymes, to the six-carbon sugars glucose or fructose. Either of these sugars may diffuse from the chloroplast to other parts of the cell to serve as food energy sources. In these other parts of the cell, in which the process of cellular respiration occurs, glucose is again broken down into PGAL before further reactions occur. Thus glucose appears to be a sort of nutrient-storage molecule, in which units of PGAL may be safely carried from the site of production to the site of use. But often nutrients must be carried much farther than from one part of a photosynthetic cell to another part of the same cell; indeed, in very tall trees nutrients must be transported for perhaps hundreds of feet from the photosynthetic leaves to the nonphotosynthetic—but nutrient-requiring—cells of the roots. Thus a larger molecule is needed so as to transfer more at a time. For this purpose, part of the glucose may be enzymatically transformed to fructose, and the two six-carbon sugars polymerized to form the disaccharide sucrose. Sucrose is the sugar chiefly transported through the plant.

At the site of production of carbohydrates, large excesses may accumulate during periods of very vigorous photosynthesis. These may be temporarily stored as granular deposits of the very large polymer starch, a molecule which is far too large to transport. Such granules may also form in nonphotosynthetic tissues that function in long-term storage of carbohydrates, when the amount of sucrose transported is far in excess of the immediate needs of the nonphotosynthetic cells of the storage organ. Such an organ of storage is the tuber of the common white potato. The starch stored there may be hydrolyzed to form transportable sugars, upon the demand of newly growing shoots, and brought to sites of use. There glucose may be enzymatically re-formed and transformed into PGAL for cellular nutrition.

In addition to serving a nutrient function, PGAL may be diverted to an equally crucial series of enzymatic reactions, in which structural chemicals are formed (for example, cellulose for cell walls, proteins, lipids, nucleic acids, etc.) and add to the substance of the cells of an organism, enabling it to grow.

Thus PGAL, the output of the dark reactions of photosynthesis, serves a pivotal role for all living organisms on our planet. From it may be formed nutrients, energy for all the organisms on the earth, and structural components from which the substance of plants, and ultimately all organisms, is made.

Respiration

The reactions of photosynthesis, as we have seen, produce carbohydrates, molecules in which the carbon atoms are highly reduced (compared with CO_2). Such molecules, which require energy to synthesize, can be thought of as energy stores, for in the enzymatic processes of **cellular respiration** they may be oxidized, and their bond energy may thus be liberated for use in the respiring cell. The processes of cellular respiration, however, are relatively cumbersome and slow to yield energy. As a consequence, an indirect but very effective method of supplying energy to the diverse chemical reactions in organisms evolved. During respiratory reactions, glucose is oxidized in a series of many enzymatic steps; the oxidation yields energy, bit by bit, which is utilized to create molecules of ATP from ADP and phosphate. ATP is a relatively small molecule, and it is very soluble in the water environment of the cell. Consequently, it may be transported very rapidly to any cellular site in which there is work to be done: ATP stores in muscle, built up during respiration, may be quickly tapped to supply energy for rapid movement; within nerve cells it may drive the ion-separation machinery which is responsible for the production of nerve impulses; in any cell that is growing or building proteins, ATP may serve as the source of energy.

The term **cellular respiration** is used to describe a sequence of many chemical reactions in which oxygen is used in the breakdown of nutrient molecules (glucose) to produce ATP, CO_2, and water. These reactions and the enzymes that catalyze them are localized within the cell. There are three major groups.

1. Glycolysis, or glucose-splitting, in which six-carbon molecules of glucose are oxidized—electrons and hydrogen ions are removed—and broken down into smaller three-carbon molecules and finally into **pyruvic acid,** with the attendant formation of small amounts of ATP. These reactions take place in the cytoplasm.

2. The Krebs cycle, in which pyruvic acid produced by glycolysis is split and oxidized again, liberating CO_2 and electrons, which may be used in the production of ATP. The reactions of the Krebs cycle are localized within the mitochondria.

3. The electron transport system, a series of reactions in which cytochrome-proteins are alternatively reduced and oxidized by electrons produced in the reactions of glycolysis and the Krebs cycle. The release of energy in the stepwise oxidation and reduction of cytochromes is coupled to the synthesis of ATP from ADP and phosphate. In other words, ATP captures some of this energy released in oxidation. To operate the electron transport system requires the presence of O_2. Oxygen is the final acceptor of electrons and hydrogen ions to yield water, H_2O.

The several different cytochrome-proteins of the electron transport system have a specific fixed relationship to one another. These molecules are not free to

FIGURE 5.15

Glycolysis. Anaerobic glycolysis occurs in the cytoplasm in a series of enzymatically catalyzed steps that will chip away, bit by bit, the energy stored in the glucose chemical bonds. Initially energy is put in to drive the reactions. This activation energy is provided by splitting two ATPs (investments). Eventually two molecules of PGAL are formed for each glucose split. In the subsequent reactions, two ATPs are formed for each PGAL molecule converted to pyruvate. These are energy dividends, and since there are two PGALs per glucose, glycolysis produces four ATPs. But because two ATPs were used to start the reaction, the net gain is two. Note that NAD is reduced along the way, and that will provide more interest on the investment later.

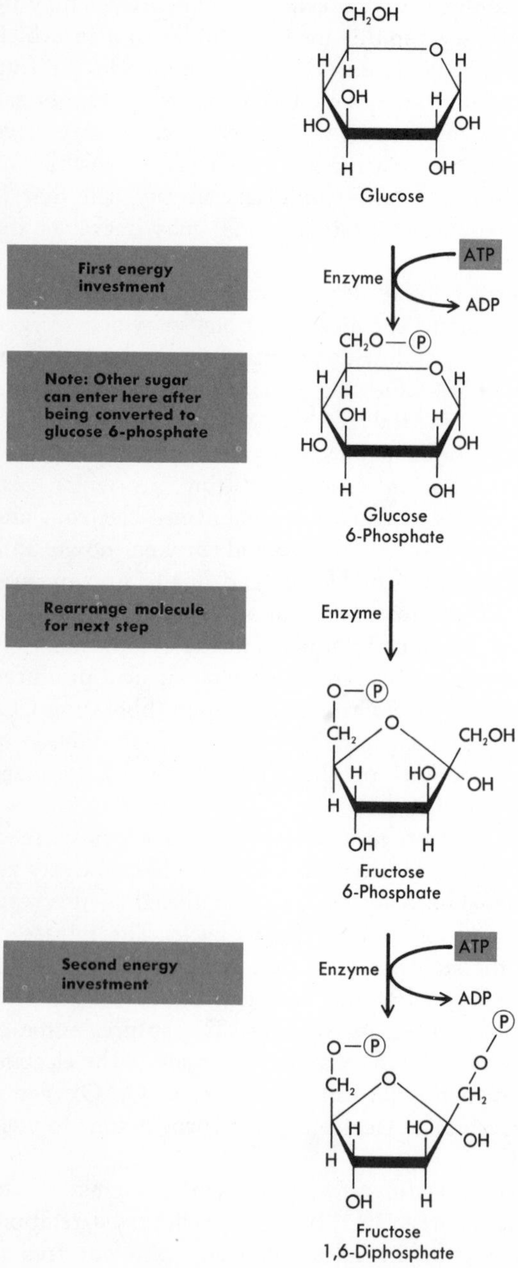

Glucose

First energy investment

Enzyme ATP ADP

Note: Other sugar can enter here after being converted to glucose 6-phosphate

Glucose 6-Phosphate

Rearrange molecule for next step

Enzyme

Fructose 6-Phosphate

Second energy investment

Enzyme ATP ADP

Fructose 1,6-Diphosphate

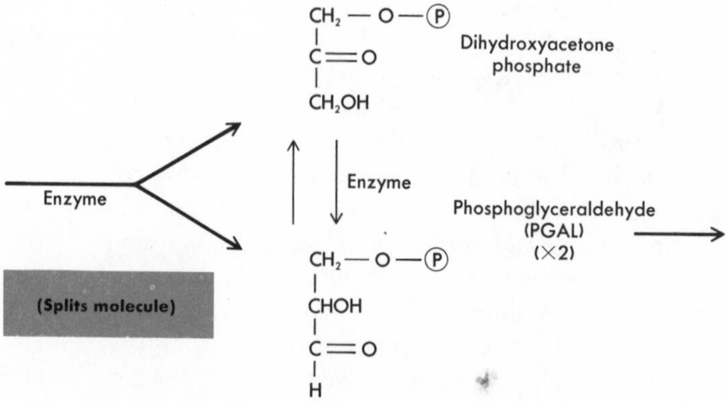

Enzyme

(Splits molecule)

Dihydroxyacetone phosphate

Enzyme

Phosphoglyceraldehyde (PGAL) (×2)

move around. They are localized on the inner membranes of the mitochondria.

These three components, though not always present in all cells, are widely distributed throughout plants, animals, and bacteria.

Although it is not actually a part of respiration, we will also discuss an **anaerobic** (requiring no oxygen) process—fermentation—in which energy stored in glucose is liberated in glycolysis and used to make a few ATP molecules, as in respiration. However, because of the lack of oxygen, means other than the Krebs cycle and the electron transport system are used to continuously remove electrons from glucose. Pyruvic acid from glycolysis is transformed by fermentation into either ethyl alcohol (the drinkable kind) and carbon dioxide (the bubbles in beer), or into lactic acid. Like the carrier system, this process removes electrons from glucose, permitting glycolysis and its ATP production to continue. Thus, in many cells glycolysis is a common pathway for either aerobic respiration or anaerobic fermentation.

GLYCOLYSIS

Glycolysis is performed by enzymes that are soluble in the cytoplasm of the cell. By this sequence of chemical reactions, each six-carbon sugar (glucose) is broken down into two three-carbon compounds (pyruvate) with the release of a small amount of stored energy. This energy is captured in ATP.

The process of glycolysis is initiated by the expenditure of two molecules of ATP (see Fig. 5.15) to produce a six-carbon sugar bearing two phosphate groups. This sugar compound is broken down to produce two three-carbon molecules—phosphoglyceraldehyde (PGAL). Each of these molecules then undergoes a very special and complicated reaction. PGAL is oxidized by NAD to produce $NADH_2$. During this reaction, a second molecule of inorganic phosphate, P_i, is added to the three-carbon compound to yield 1,3-diphosphoglyceric acid. There then follows a series of reactions that transfer the phosphates from this compound to ADP, yielding ATP and a three-carbon compound, pyruvate (see Fig. 5.15). This kind of synthesis of ATP is called substrate-level phosphorylation.

By studying Fig. 5.15, you will see that there is only one oxidation step in the process of glycolysis. The $NADH_2$ produced there may be oxidized by other reactions in the cytoplasm (in fermentation, for example), or it may make its way to the mitochondrion. In either case, oxidation of $NADH_2$ yields NAD, so the process of glycolysis can continue.

Because glycolysis involves only one oxidation step, only a small part of the energy stored in the glucose molecule during photosynthesis is released. The many steps of the process are necessary to ensure that energy

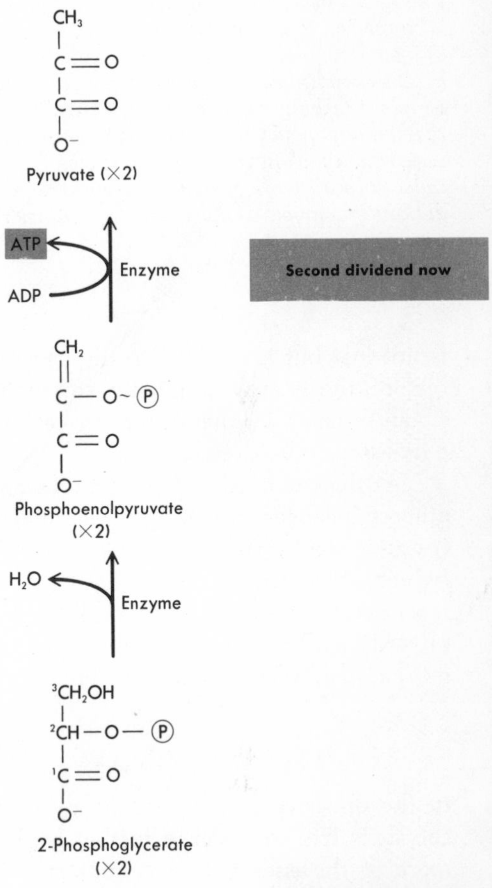

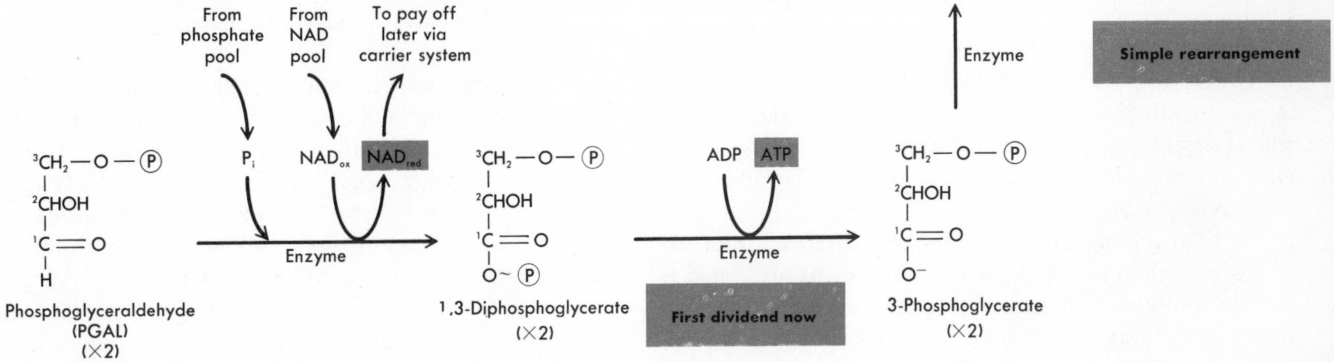

Phosphoglyceraldehyde (PGAL) (×2)
From phosphate pool
From NAD pool
To pay off later via carrier system
P_i
NAD_{ox}
NAD_{red}
Enzyme
1,3-Diphosphoglycerate (×2)
ADP
ATP
Enzyme
First dividend now
3-Phosphoglycerate (×2)
Enzyme
Simple rearrangement
2-Phosphoglycerate (×2)
H_2O
Enzyme
Phosphoenolpyruvate (×2)
ADP
ATP
Enzyme
Second dividend now
Pyruvate (×2)

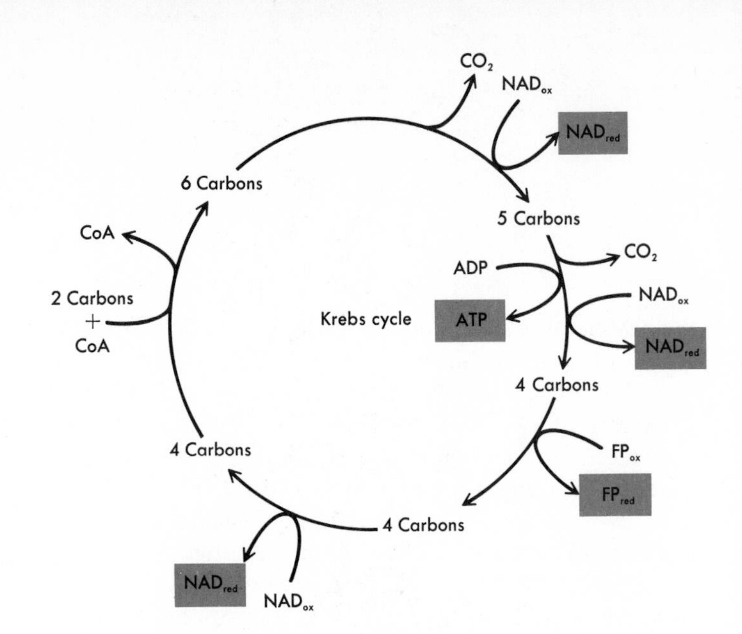

FIGURE 5.16

The Krebs cycle. Pyruvic acid (three carbons) from glycolysis is converted to acetic acid (two carbons) and the production of a single molecule of carbon dioxide (one carbon). Acetic acid is then linked to a four-carbon compound to form citric acid (six carbons). Citric acid will then be split in a series of reactions to yield the four-carbon compound that initially reacted with acetic acid. In this process, carbon dioxide is twice split off from citric acid, and there are repeated oxidations (removals of electrons) to produce reduced electron carriers.

is not lost but is coupled to the synthesis of ATP. The rest of the energy originally stored in the six-carbon sugar remains in the two three-carbon compounds of pyruvate (pyruvic acid).

At the end of glycolysis, therefore, each molecule of glucose has been broken into *two* molecules of pyruvate (pyruvic acid), four molecules of ATP are produced (which, if we count the two *used* to start the glycolytic reactions, leaves a *net* of *two* ATPs), and two molecules of $NAD_{reduced}$. These reactions occur *anaerobically*, without oxygen, anywhere in the cytoplasm of the cell.

KREBS (CITRIC ACID) CYCLE

In the presence of oxygen, the pyruvic acid from glycolysis is free to undergo further oxidation and to have more of the energy that is contained in its carbon-to-hydrogen bonds utilized in the manufacture of more ATP. Although there is some substrate-level oxidation (in which ATP is created directly) in the Krebs cycle, this is only a minor source of ATP. The major outcome of this series of reactions is the reduction of $NAD_{oxidized}$ by electrons removed during the complete oxidation of pyruvic acid to CO_2 (Fig. 5.16).

It is in the ordered and compartmentalized interiors of the mitochondria that the oxidation of pyruvic acid is carried to its completion. Along the precisely arranged membrane folds (cristae) of the mitochondria lie enzymes in a highly organized array—the enzyme necessary for one reaction situated in close proximity to the enzyme necessary for the succeeding reaction—so that the reactions that produce the molecules of ATP can proceed as quickly and efficiently as possible.

Before the reactions of the Krebs cycle take place, pyruvate is broken down in an oxidation that uses NAD and yields $NADH_2$, CO_2, and acetic acid (a two-carbon compound). This is the first reaction in which one of the common end products of respiration, CO_2, is liberated. During that oxidation of pyruvate, two electrons and two H^+ ions are removed. They combine with $NAD_{oxidized}$ to become $NAD_{reduced}$ ($NADH_2$). As we shall see shortly, $NAD_{reduced}$ donates its electrons to cytochromes, and their energy is used to phosphorylate six molecules of ADP, yielding six ATP.

Acetic acid enters the Krebs cycle by combining with a four-carbon compound to form a six-carbon compound, citric acid (Fig. 5.16). (The cycle is sometimes called the citric acid cycle, although it is usually referred to as the Krebs cycle after its discoverer, Nobel prizewinner Sir Hans Krebs.)

The citric acid is now oxidized through a series of steps, producing a five-carbon compound and several four-carbon compounds. Two molecules of CO_2 are given off in the process and released into the environment. During the Krebs cycle oxidations, eight electrons and H^+ ions are removed from intermediate molecules in the cycle (see Fig. 5.16). As we have seen, every oxidation of a molecule implies a coincident reduction of some other molecule. In this case the molecules reduced are $NAD_{oxidized}$ and flavin adenine dinucleotide ($FAD_{oxidized}$), which become $NAD_{reduced}$ and $FAD_{reduced}$, respectively. These two electron carriers diffuse to the

The Economic Value of Fermentations

Over the centuries fermentation has become a valuable process in the preparation and preservation of foods, the manufacture of such alcoholic beverages as wine, beer, and whiskey, and the manufacture of certain chemicals. The different end products of various fermentation processes depend on the type of substrate (sugar, starch, etc.) and the type of organism used to carry out the fermentation.

The most familiar fermentation process is that employed in making alcoholic beverages. For making wine, the substrate is fruit juice. The most common fruit is grapes, though any fruit juice with high sugar content will work. The fermenting organism is yeast. The fruit juice is first placed in a very large vat or crock and seeded with a desirable strain of yeast. Fermentation is allowed to proceed for a few days, during which the vat bubbles with carbon dioxide. During this period the sugar is consumed and ethyl alcohol is produced. When the alcohol content reaches 14%, the yeast cells can no longer survive, and the process eventually comes to a halt. The contents of the vat are then poured into storage containers that allow limited access to air. The raw wine sits in these containers for a month or more, during which time many impurities (including tannins and the dead yeast cells) settle out. After several transfers the wine is bottled.

The sweetness of a wine is largely determined by the amount of sugar added at the beginning—or after the initial fermentation has been completed. The color of the wine depends on whether red or white grapes are used. The flavors of various wines, on the other hand, are a result of many factors: the strain of yeast used, the type of grape, and environmental factors such as the soil in which the grapes were grown and the climate that particular year. This accounts for the interest shown by wine connoisseurs in the national and/or regional origin of the wine as well as the year it was produced. In order to produce wine of consistent taste and quality, winemakers must learn to control these variables as accurately as possible.

One of the most serious problems winemakers face is contamination of the fermentation vats with bacteria. Since bacteria ferment sugars to lactic acid rather than alcohol, their presence gives the wine a very sour taste. When contamination of wine by bacteria precipitated the great wine calamity in France during the mid-nineteenth century, the wine industry called in a young chemist, Louis Pasteur. Pasteur demonstrated that the contaminating substance was lactic acid. He went on to show that living organisms, bacteria, produced the lactic acid. His solution involved the introduction of sterile procedures for handling the grape juice and inoculating it with yeast. These procedures, later applied to milk as the well-known "pasteurization" process, made it possible to avoid bacterial contamination while still allowing yeast to ferment the grape juice. Pasteur's discoveries about the role of bacteria in infections grew out of his early work with the French wine industry.

Beer is produced much like wine, except that the starting substrate is malt rather than fruit juice. Malt is derived from the seeds of barley plants, which

---

From Jeffrey J. W. Baker and Garland E. Allen, *The Study of Biology*, 3rd edition (Reading, Mass.: Addison-Wesley, 1977), pp. 210–211.

are softened by soaking in water to stimulate germination. The malt is fermented by a special strain of yeast known as "brewer's yeast." The flavor of beer is produced by adding hops (the tops of certain flowers of the mulberry family, which produce a bitter substance) to the fermentation vats. After a few days the fermented liquor is drawn off and, like wine, stored for a period of time in closed containers. It is important in making both wine and beer to be sure that too much air does not become available during the fermentation process. This would stimulate the yeast cells to carry out aerobic respiration, consuming the sugar and producing only carbon dioxide and water rather than alcohol.

Hard liquor, such as bourbon, scotch, or gin, is produced by a more complex process. In each case, a different substrate is used for the initial fermentation: for bourbon, corn mash, and for gin, potatoes. Fermentation proceeds in large open vats for a few days in a manner similar to that for wine and beer. The liquor is then poured off and allowed to settle for a period of time to remove the least tasty impurities (mostly tannic acid). The liquid is then poured into distillation vats, where it is heated until it evaporates. Since alcohol vaporizes at a much lower temperature than water, it is driven off first and collected in condensing coils (distillation). By distillation the alcoholic content of the final product can be made much greater than is possible for either wine or beer. Alcoholic content of hard liquors is reckoned as "proof," 100-proof being 50% alcohol. Further distillation of the fermented liquid drives other substances into the condensing apparatus as well, adding to the pure alcohol flavors derived from the original fermented liquid.

After distillation the liquor is stored in wooden vats or barrels. The kinds of wood used and the way it has been processed determine the final flavor of the liquor. Bourbon, for example, is usually stored in maple barrels whose insides have been charred. The liquor leaches out various substances from the charred wood, giving bourbon a characteristic taste. The materials leached from the wood also give bourbon and scotch their color. Since gin is not stored in wooden vats, it remains clear.

The preparation of commercial alcohol involves complete distillation of a fermented mixture, giving relatively pure ethyl alcohol. Commercial alcohol can be 200-proof, or 100% alcohol. "Denatured" alcohol is merely ethyl alcohol with an additive to make it undrinkable.

Lactic acid fermentation, so injurious to the wine and beer industries, has been used beneficially in the preparation and preservation of certain foods. For instance, the bacterium *Lactobacillus casei* is employed in the production of sour cream from milk. Other kinds of bacteria are used to produce cottage cheese, yogurt, and cheeses. *Lactobacillus plantarum* is used as a preservative in the preparation of pickles and sauerkraut.

Controlled fermentations are also used to produce chemicals on an industrial scale. The yeast *Citromyces* ferments sugars to citric acid, and *Aspergillus niger* ferments sugars to oxalic acid. Citric acid is used commercially in producing carbonated soft drinks and in blueprinting. Oxalic acid is used in the leather and textile industries to condition raw materials, as well as in the manufacture of dyes.

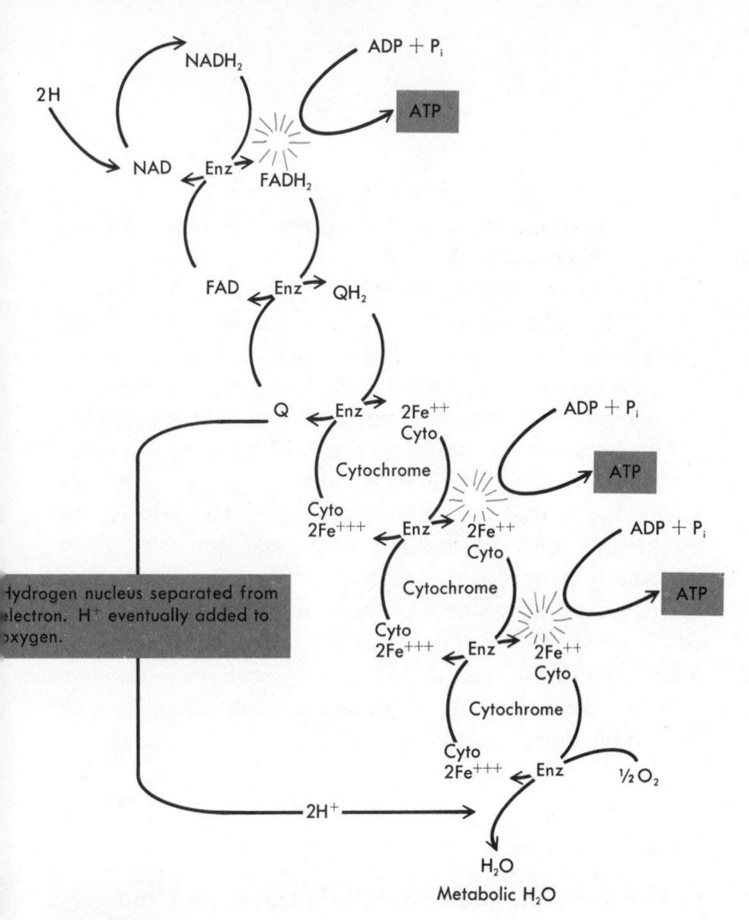

FIGURE 5.17

The electron carrier system. Hydrogens split off from the breakdown of sugars, fats, and amino acids are usually picked up by electron carriers, such as NAD and FAD. In the mitochondria these carriers and their associated enzymes are lined up along the cristae. The greatest potential energy is in $NADH_2$. As $NADH_2$ passes its hydrogen on to FAD, enough energy is released to make one new ATP. At other steps down this cascade of energy, there is enough energy released to make two additional ATPs. For each reduced NAD entering the carrier system, three new ATPs are made.

electron transport chain, where they deposit their electrons, become oxidized ($NAD_{oxidized}$ and $FAD_{oxidized}$), and diffuse back to the Krebs cycle to carry more electrons. In the electron transport chain, as we shall see, the energy of these electrons is tapped to make many molecules of ATP. In addition to electrons being yielded at several oxidative steps, a molecule of ATP is synthesized directly during one of the steps, as in the substrate-level phosphorylations of glycolysis.

The Krebs cycle intermediate compound remaining after one turn of the cycle is the four-carbon compound with which the cycle began. This molecule is again ready to combine with another molecule of acetic acid for another passage of the cycle.

To completely oxidize the two molecules of pyruvate derived from a single molecule of glucose, two "turns" of the Krebs cycle are necessary.

ELECTRON TRANSPORT CHAIN

As you will recall, at several steps during glycolysis and the Krebs cycle, oxidations took place, with the removal of electrons (and H^+). The electrons were accepted by $NAD_{oxidized}$ to produce $NAD_{reduced}$ (also written $NADH_2$). The numerous molecules of $NAD_{reduced}$ produced in the complete oxidation of glucose to CO_2 donate to an "electron transport system." This system is a series of electron carriers bound to the inner membrane of the mitochondrion. $NAD_{reduced}$ reacts only with the first protein in the electron transport series, passing its electrons to that molecule, thereby reducing it. $NAD_{oxidized}$, the immediate product, can diffuse back for use in the Krebs cycle or glycolysis. In the passing of electrons from $NAD_{reduced}$ to the electron transport protein (FAD), energy is released, and the release is coupled to the synthesis of ATP from ADP and P_i. This process is known as oxidative phosphorylation.

The reduced electron transport protein ($FADH_2$) is then oxidized by the adjacent electron transport protein, Q, which becomes QH_2. As Fig. 5.17 makes clear, this process is repeated several times as a pair of electrons is passed from one electron transport protein to the next. The final electron acceptor is oxygen (O_2), and the product is water, but before that point is reached, a total of three molecules of ATP have been synthesized. For every pair of electrons transported down the electron transport chain, three molecules of ATP are synthesized and one molecule of H_2O is produced.

By this process we have the complete oxidation of one molecule of glucose to yield six molecules of CO_2, six of H_2O, and energy. The energy is captured in the form of ATP. For each molecule of glucose oxidized, a total of 36 molecules of ATP are produced. By far the greatest number of them are produced in oxidative phosphorylation coupled to the electron transport system. Very few are produced by substrate-level phosphorylation.

This compact energy system is comparable in efficiency to the best engines man has been able to make. Its efficiency depends on the electron transport system, which uses oxygen as the final electron acceptor and yields water as the nontoxic waste product. Of the 36 ATP molecules produced from each glucose molecule, only two are produced in glycolysis (substrate-level phosphorylations, with no oxygen directly involved). Of the remaining 34 ATP, 32 are produced by oxidative phosphorylation (using oxygen as the final electron acceptor). The remaining two molecules of ATP are produced by substrate-level phosphorylations in the Krebs cycle.

FERMENTATION

Oxygen is not always available for respiration. Indeed, in the primitive stages of life, free oxygen was probably not present at all in the atmosphere. Hence the first respiratory mechanism that evolved worked anaerobically, or without oxygen.

Under anaerobic conditions, the hydrogen ions removed by NAD during glycolysis must have somewhere to go. The $NADH_2$ must be oxidized to NAD to permit the contained functioning of glycolysis. This necessary step is accomplished by a reaction in which $NADH_2$ reduces pyruvate to form lactic acid. Many animals (including man) produce lactic acid under certain conditions. It is also produced by many one-celled microorganisms—for example, the common *Lactobacillus* found in yogurt. In some microorganisms, such as yeasts, pyruvate can be transformed into ethyl alcohol and carbon dioxide. Whether the end product is lactic acid or ethyl alcohol, anaerobic (partial) oxidation of glucose is always called **fermentation.**

Fermentation is not a particularly efficient way to make molecules of ATP. The reason for the small yield in energy molecules is that the pyruvic acid—by being transformed into lactic acid—is chemically blocked from being further oxidized.

The fermentation process is used in manufacturing wine from glucose (grape sugar) and beer from malt (malt sugar). The process is also used by the muscles of the four-minute miler during his finishing kick because the diffusion of oxygen from the blood into the cells is not fast enough to supply all the ATP the muscles need. Therefore ATP is produced by glycolysis. The only disadvantage is that the accumulation of lactic acid in the trackman's muscles produces pain and may limit activity. As the pace of activity decreases, lactic acid is reconverted to pyruvate, which then enters the Krebs cycle and is *completely* metabolized, producing a great deal of ATP. This abundance of ATP after a brief rest—familiar to us as "second wind"—will support the resumption of strenuous muscle activity.

Energy yield

In summary, the processes of photosynthesis and respiration are oxidation-reduction reactions. In photosynthesis, molecules of carbon dioxide are reduced by hydrogen (the resulting C—H bonds contain energy) and built into the more complex molecules of glucose. In respiration—glycolysis, the Krebs cycle, and the electron transport system—molecules of glucose are oxidized (the carbon-to-hydrogen bonds are broken and the electrons involved are removed and transported down an energy gradient, where their energy is at some point utilized in the coupled phosphorylation of ADP to ATP). ATPs can be used as a quick source of energy anywhere in an organism where work is to be done.

How much energy is captured from the respiration of a molecule of glucose? The answer may be expressed in terms of the calories of heat produced by burning the various compounds. A given amount of glucose yields roughly 700,000 calories. The amount of ATP that can be produced by this quantity of glucose yields roughly 300,000 calories. Thus the energy harnessed from the oxidation of a molecule of glucose is a little over 40 percent.

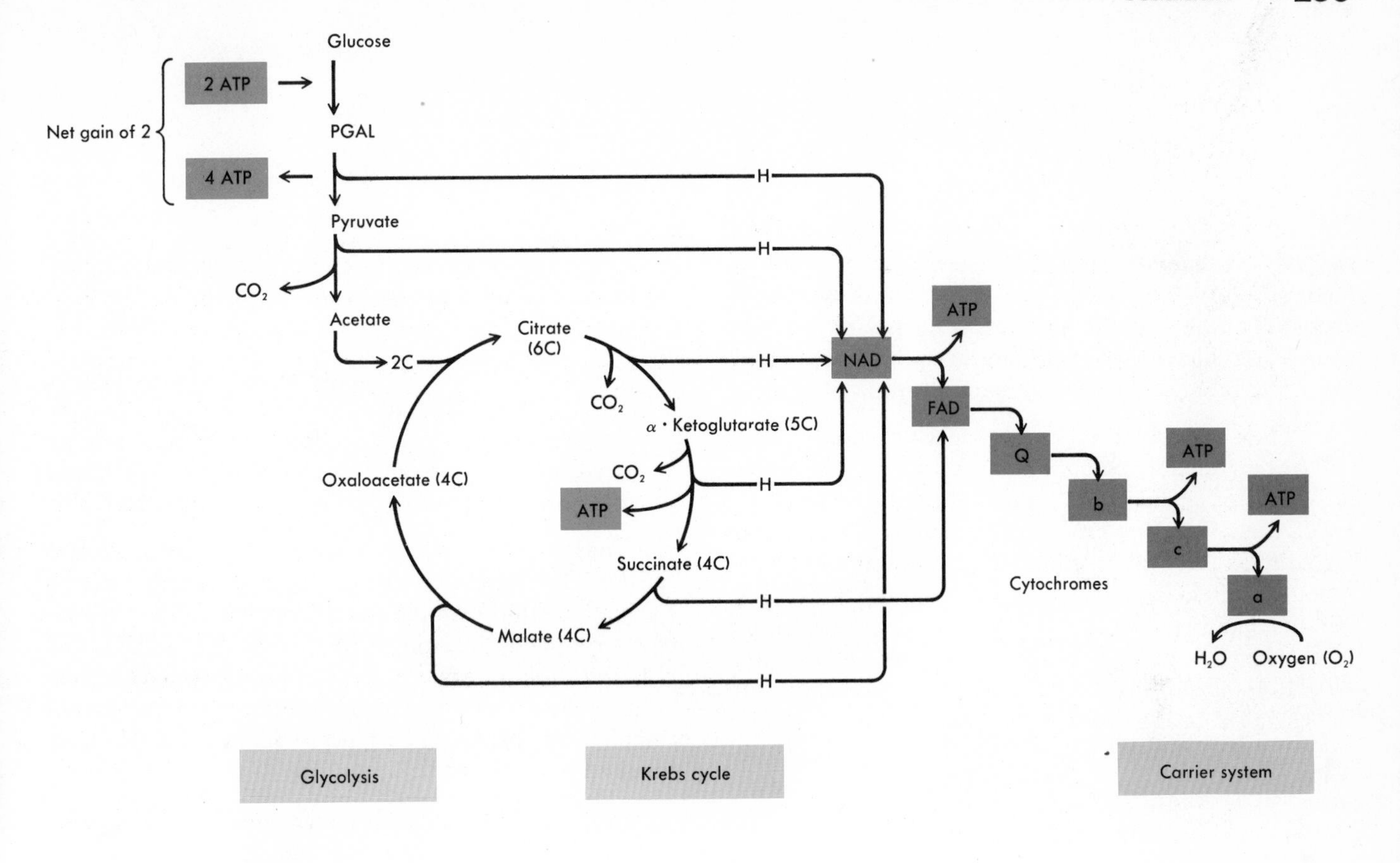

FIGURE 5.18

Summary of energy-yielding reactions. As glucose is broken down into two molecules of pyruvate, there is a net gain of 2 ATP. Another 6 ATP will result from the two $NADH_2$ produced in glycolysis. In the conversion of two pyruvate to two acetate, another 6 ATP will be produced from $NADH_2$. Two turns of the Krebs cycle yield three molecules of $NADH_2$ per turn which produce 18 ATP, and one molecule of $FADH_2$ per turn, which yields 4 ATP; in addition, 2 ATP are produced directly by substrate-level phosphorylation. The total number of ATP produced for each glucose can therefore be as high as 38.

Summary

1. The laws of thermodynamics are fundamental to an understanding of energy conversion processes in living organisms. The first law says that the total amount of energy in the universe is constant. The second law states that the amount of *usable* energy is always decreasing.

2. Substances move between cells and within cells by the process called diffusion. The rate of diffusion is determined by physical properties of the diffusing molecules: their size, concentration, lipid solubility, and temperature. Osmosis is a special kind of diffusion process, the diffusion of a solvent, usually water, through a selectively permeable membrane.

3. In addition to the passive processes of diffusion and osmosis, there are various energy-requiring processes for moving substances into and out of cells. These forms of active transport require cellular work to move substances *against* a concentration gradient, and they involve carrier molecules located in the cell membrane.

4. Energy for cellular work is obtained mainly from carbohydrates, and every organism either obtains its supply of carbohydrates from the environment or manufactures them. Organisms that manufacture their own foodstuffs by using the sun's light energy as a power source are called autotrophs, and those that obtain them by eating other organisms are heterotrophs.

5. Photosynthesis, the process by which green plants and certain other autotrophs manufacture carbohydrates, is a series of chemical reactions that can be divided into two groups. The first group require sunlight and are referred to as the light reactions. The second group do not require sunlight and are called the dark reactions. During the light reactions, radiant energy is transformed into chemical energy with the formation of ATP and $NADP_{reduced}$. During the dark reactions, the energy in the chemical bonds of these compounds is used to form various carbohydrates needed for cellular work. Photosynthesis may be summarized in the chemical formula:

$$CO_2 + H_2O + \text{light energy} \rightarrow (CH_2O)_n + O_2$$

6. In higher plants the light reactions of photosynthesis occur in two types of enzyme-chlorophyll groups called photosystems II and I. Light-excited electrons from the chlorophyll in photosystem II pass by means of electron carriers to photosystem I with the production of ATP. Excited electrons in photosystem I—received from outside sources or from the photosystem's own chlorophyll molecule—are transferred by electron carriers to molecules of NADP in noncyclic electron flow. Finally, through the photolysis of water, electrons are restored to the oxidized chlorophyll via photosystem I, hydrogen ions are released and bonded to $NADP_{reduced}$, and molecular oxygen and a small amount of water are formed.

7. In the Calvin cycle, the dark reactions of photosynthesis, ATP and $NADP_{reduced}$ drive a series of reactions through the donation of energy, electrons, and hydrogen ions. By means of these reactions, atmospheric CO_2 is used to form a three-carbon carbohydrate called phosphoglyceraldehyde (PGAL). Most PGAL is recycled back into the dark reactions during periods of sunlight, but excesses of the compound are combined enzymatically to form glucose or fructose, both six-carbon sugars. PGAL is also used in the formation of lipids and proteins.

8. Glucose serves as a short-term food energy storage molecule. It is transported easily to all cells and may be used to form larger molecules for long-term storage or broken down for energy. In plants large excesses of photosynthates are stored as the long polymer starch.

9. There are four series of reactions in which ATP is formed. The Krebs cycle and hydrogen and electron transport require aerobic conditions and collectively effect a complete breakdown of pyruvic acid to CO_2 and H_2O with a large yield of ATP. Glycolysis and fermentation, which occur under anaerobic conditions, do not complete the oxidation of glucose and yield a relatively small amount of ATP.

10. In glycolysis, glucose is enzymatically broken down into two molecules of pyruvic acid with the attendant net yield of two ATP and two $NAD_{reduced}$. From glycolysis, metabolism may take one of two pathways: aerobic respiration or fermentation.

11. In fermentation, glucose is broken down by a series of enzymatic reactions into either lactic acid in animals and some bacteria or ethyl alcohol and CO_2 gas in plants. The primary effect of fermentation is to regenerate $NAD_{oxidized}$ and thus allow oxidation of glucose in the absence of oxygen.

12. The aerobic pathway uses the Krebs cycle and the electron-hydrogen transport chain. These reactions occur in the mitochondria along the membrane folds, where the necessary enzymes and cytochromes are situated.

13. In the Krebs cycle, pyruvic acid is oxidized to CO_2 gas with an attendant production of a small amount of ATP and removal of electrons and hydrogen ions by NAD and FAD.

14. The electron and hydrogen transport process is coupled to the Krebs cycle, for the NAD and FAD reduced during the Krebs cycle donate their electrons and hydrogens to the transport chain. As the electrons and hydrogens pass along the chain, much ATP is formed in conjunction with the oxidation of several electron carriers. At the end of the chain the electrons and hydrogens are donated to oxygen, thereby forming water.

15. The aerobic oxidation of glucose by glycolysis, the Krebs cycle, and the electron transport chain is very efficient, producing up to 36 molecules of ATP for each molecule of glucose.

16. In animal muscle during strenuous activity, the available ATP supply may become depleted, and then anaerobic fermentation results in the formation and accumulation of lactic acid (oxygen debt). When energy expenditure returns to normal, lactic acid is reconverted to pyruvic acid, and the organism returns to the usual metabolic pathway.

17. Cells can harness approximately 40 percent of the energy of glucose through the formation of ATP.

REVIEW AND STUDY QUESTIONS

1. How do the two laws of thermodynamics apply to the energy that drives all forms of life?

2. Osmosis is defined as a special kind of diffusion process. How is it different from simple diffusion?

3. When a sausage casing containing a starch solution is suspended in a container of plain water, what happens? Explain.

4. How are phagocytosis and pinocytosis similar, and how are they different?

5. A popular bumper sticker reads "Have you thanked a green plant today?" For what?

6. Using the analogy of a waterfall suggested in this chapter, explain how solar energy trapped during photosynthesis is transformed into chemical energy stored in the bonds of ATP molecules.

7. Recent studies have led to the discovery of the C4 pathway used by certain hot-weather crops for carbon fixation. What is the significance of this discovery for a world threatened with food shortages?

8. What does cellular respiration mean? Why is that term used for a process that seems to have little relationship to "breathing"–the more familiar meaning of respiration?

REFERENCES

Brachet, J. 1961 (September). "The Living Cell." *Scientific American* Offprint no. 90. Freeman, San Francisco.

Calvin, M. 1962. "The Path of Carbon in Photosynthesis." *Science* 135(3): 879–889.

Capaldi, R. A. 1974 (March). "A Dynamic Model of Cell Membranes." *Scientific American* Offprint no. 1292. Freeman, San Francisco.

Holter, J. 1961. "How Things Get into Cells." *Scientific American* Offprint no. 96. Freeman, San Francisco.

Loewy, A. G., and P. Siekevitz. 1970. *Cell Structure and Function*, 2nd edition. Holt, Rinehart and Winston, New York.

McvElroy, W. D. 1971. *Cell Physiology and Biochemistry*, 3rd edition. Prentice-Hall, Englewood Cliffs, N.J.

Robertson, J. D. 1962 (April). "The Membrane of the Living Cell." *Scientific American* Offprint no. 151. Freeman, San Francisco.

Satir, B. 1975 (October). "The Final Steps in Secretion." *Scientific American* Offprint no. 1328. Freeman, San Francisco.

SUGGESTIONS FOR FURTHER READING

Giese, A. C. 1973. *Cell Physiology*, 4th edition. Saunders, Philadelphia (paperback, Boxwood, 1975).

Classic and authoritative text, which is simultaneously comprehensive and understandable to the beginning student.

Jensen, W. A. 1970. *The Plant Cell,* 2nd edition. Wadsworth, Belmont, Calif.

Current analysis of the structure and function of plant cells.

———. 1966. *Plant Biology Today,* 2nd edition. Wadsworth, Belmont, Calif.

Excellent standard reference. See especially L. Bogarad's material on photosynthesis.

Lehninger, A. L. 1971. *Bioenergetics,* 2nd edition. Benjamin, New York.

For those who want to delve into more detail on the energy reactions in living systems, this book provides accounts of the variety of means for transforming and transferring energy.

Went, F. 1963. *The Plants.* Time Inc., New York.

A Life *magazine special publication written in nontechnical terms with illustrations and explanations of plant anatomy and physiology.*

STRUCTURE AND FUNCTION OF THE ORGANISM

part two

NUTRITION: OBTAINING AND PROCESSING FOOD

chapter six

An adult elephant normally consumes more than 300 pounds of vegetable matter every day. Driven by population pressures and shrinking food supplies, elephants have turned to the baobab tree, every part of which is edible. These gigantic herbivores may rip apart whole trees to obtain the moisture and nutrients in the fibrous wood.

Every organism is part of a food chain—a sequence of food-getting activities in which all living things remain alive either by utilizing materials found in the environment or by consuming other organisms. In a food chain there are primary producers—plants which make organic nutrients for themselves from carbon dioxide and minerals (by photosynthesis and other synthetic processes), and consumers—organisms which eat plants **(herbivores)**, organisms which eat herbivores **(carnivores)**, or organisms which eat plants, other carnivores, and herbivores **(omnivores)**.

The autotrophs, as we have seen, are generally photosynthetic plants and chemosynthetic bacteria, organisms that use simple inorganic substances, such as carbon dioxide, water, and minerals (inorganic nutrients dissolved in water), to make the complex organic molecules necessary for their growth and functioning. Heterotrophs, which are unable to make many of the compounds needed for maintaining life, must obtain those compounds preformed. They must then transform them into the nutrients suitable for their own use, because the proteins, lipids, and carbohydrates in a plant autotroph are not the same as the proteins, lipids, and carbohydrates in an herbivore, such as a cow. However, the cow cannot manufacture these substances from inorganic molecules and must therefore obtain them from the vegetation it eats. To use the nutrients contained in the vegetation, the cow must then **digest** it, or break down the compounds of the vegetation into simpler compounds, absorb them, and convert the simple compounds to more complex molecules of its own.

All living things convert nutrient substances into simple chemicals and then transform them into molecules unique to the organism. However, it is probably safe to generalize that the more complex the life form, the more complicated this conversion process becomes.

FIGURE 6.1

Food chain. In a community of organisms, nutrients may be passed from organism to organism in a food chain. For example, the plant behind the rodent uses simple nutrients and sunlight to make the complex nutrients in the berries. The rodent eating the berries digests these nutrients and from them synthesizes proteins, carbohydrates, lipids, and other substances that constitute its body. The owl that eats the rodent digests the tissues of the rodent and from them synthesizes the proteins, lipids, and carbohydrates found in the owl's body. Thus, the nutrients originating in the berries are passed along the food chain, from one organism to another.

The basic problem of any organism is that of obtaining nutrients from the environment and transporting them to its cells. For many unicellular organisms, the problem of obtaining nutrients is considerably diminished because the organisms are in direct contact with their environment. Gases and other materials needed by the organisms are obtained from the environment either through a passive process, diffusion, or through an active mechanism such as active ion transport or phagocytosis. Other unicellular organisms, such as the ciliate protozoa, have surprisingly complex structural and behavioral mechanisms of trapping, ingesting, and digesting prey. In complex, multicellular organisms many cells are insulated from the environment by other cells. Somehow, materials must be brought from the environment to the insulated cells. The actual transport processes whereby nutrients are distributed among the cells of higher organisms will be the subject of Chapter 7; here we will consider **nutrition,** the ways that organisms acquire their nutrients.

Autotrophs and heterotrophs

Organisms may generally be classified according to their mechanism for obtaining nutrients. An autotroph is an organism, whether unicellular or multicellular, that absorbs relatively simple raw materials and uses them to manufacture the nutrients it requires, usually by the process of photosynthesis. A heterotroph is an organism that cannot manufacture its own food and must obtain it from other organisms. Heterotrophs require more com-

TABLE 6.1
Autotrophic and heterotrophic organisms

AUTOTROPHS	HETEROTROPHS
Chemosynthetic bacteria	Most bacteria
Photosynthetic flagellate	Most protozoa
Protozoa	Fungi (molds, yeasts, rusts, mushrooms)
Algae	
Mosses and liverworts	Parasitic plants (mistletoe)
Higher green plants	Animals

FIGURE 6.2
Algae are autotrophs; that is, they can meet their nutritional needs by using the energy of the sun to synthesize food from inorganic substances in the environment. A pair of green algae, Micrasterias radiata, *are shown immediately after each of the one-celled organisms has reproduced itself by cell division.*

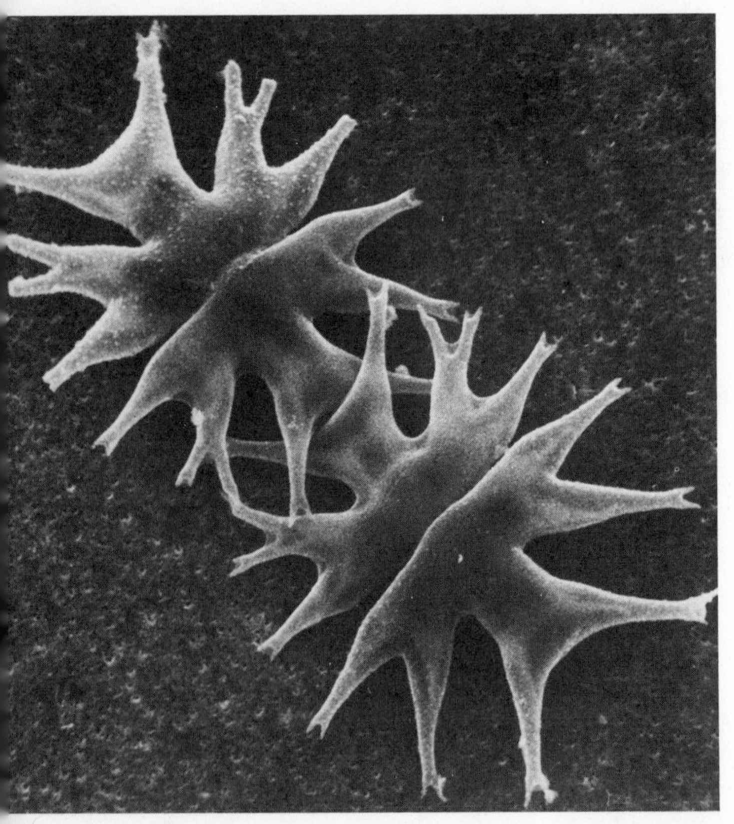

plicated mechanisms for obtaining nutrients, since they must break down and absorb the relatively complex substances of which other organisms are composed. In heterotrophs food is usually first ingested and then broken down (digested) into simpler substances before it can be absorbed—through cell membranes—into cells and then reassembled into the specific kinds of materials that a heterotroph's cells can use.

Table 6.1 indicates some generalizations, but it is important to realize that the classification of organisms as autotrophs or heterotrophs is often not clear-cut. Green plants are generally autotrophs and animals are generally heterotrophs, but there are exceptions. The flagellated alga *Euglena,* like other green plants, can manufacture its food photosynthetically, but it can also eat preformed organic nutrients. It is therefore necessary to study how organisms satisfy their nutritional requirements by focusing on the nutritional mechanisms they employ, rather than simply considering whether the organisms are plants or animals.

Autotrophs are the primary energy producers, making vital food substances out of materials derived from the air and soil. Heterotrophs acquire their vital substances from the organisms (such as autotrophs or other heterotrophs) or organic matter (the often nutrient-rich wastes or remains of other organisms) that they eat. These substances are ultimately returned to the air and soil either as waste products or as the products of the decay of heterotrophic organisms after they have ceased to live. The raw materials are "recycled" so that the process continues over and over again. Certain substances are basic to *all* forms of life, both autotrophic and heterotrophic. These, and their nutritional roles, are summarized in Table 6.2.

Autotrophs: the primary energy producers

Some of the first living forms that evolved were autotrophs. These primitive organisms probably resembled the present-day procaryotic blue-green algae one can

TABLE 6.2
Major elements necessary for life

ELEMENT	FUNCTION
Carbon	Forms the "skeleton" of organic compounds (carbohydrates, lipids, proteins, nucleic acids, vitamins, and other compounds of which living organisms are composed).
Hydrogen	Essential component of all organic compounds and of water. Its bonding properties are in large part responsible for shape of complex organic molecules. Its removal from sugar during respiration is coupled to the release of energy for cell processes.
Oxygen	Essential component of water and of most organic compounds. Molecular oxygen is required by all organisms that uses aerobic respiration to release stored food energy for cell processes.
Nitrogen	Essential component of the amino group of amino acids (therefore of proteins), of DNA, RNA, ATP, and other nucleotides.
Phosphorus	Center of phosphate group, which is a component of nucleic acids, enzymes, energy-storing molecules such as ATP and NADP, and phospholipids, which are a major component of cell membranes.
Sodium	In ionic form, active in maintenance of osmotic equilibrium, acid-base balance within cells. In animals, active in transmission of nerve impulses. In plants, not essential as far as is known.
Potassium	In animals, in ionic form it is active in maintenance of osmotic equilibrium, acid-base balance within cells, nerve impulses, and muscle contraction. Essential for plant growth, but it appears to be effective only at relatively high concentrations.
Calcium	In plants, deposited as an intercellular cement in cell walls. In animals with skeletons, deposited in bone matrix.
Sulfur	Component of amino acids (cystine, cysteine, and methionine), plant and animal hormones, thiamine, coenzyme A. In some chemotrophs, it is used in inorganic respiration instead of oxygen.
Magnesium	Essential for phosphate metabolism. In green plants, the metal ion component of chlorophyll. In animal skeletons, deposited in bone matrix.
Chlorine	In plants, it takes part in the reactions of photosynthesis leading to oxygen evolution. In animals, in ionic form (chloride ion) it is necessary for water balance and osmotic pressure regulation.
Micronutrients (trace elements)	Minerals required by living things in very small amounts. *Examples in plants:* boron, manganese, copper, zinc, iron, molybdenum. *In animals,* they include manganese, copper, zinc, iodine, cobalt, fluorine. Function often unknown. Iron provides example of known use.
Iron	Essential component of cytochromes. In green plants, necessary for proper formation of chlorophyll. In animals, vital component of hemoglobin.

see floating on the surface of stagnant ponds. The needs of these organisms were simple—energy from the sun and a relatively small quantity of gases and minerals from their watery environments. They took whatever simple materials they required and produced their own more complex nutrients, just the way the simplest autotrophs do today. Autotrophs may be classified according to their degree of complexity. The simplest appear to be the blue-green algae, procaryotes that resemble bacteria. Next in complexity appear to be the unicellular and multicellular algae—whose cells are eucaryotic—such as *Valonia* and *Ulva*. The most complex of autotrophs are undoubtedly the vascular plants, those with special cells that conduct water and nutrients to every part of the plant. The most familiar examples of the latter are trees and grasses.

All photosynthetic autotrophs have similar needs: they must all obtain gases—chiefly carbon dioxide and oxygen—water, and mineral ions from their environment. Thus their nutrient intake processes consist of gas absorption, water absorption, and mineral absorption. We shall discuss each separately.

GAS ABSORPTION

Gas absorption in autotrophs involves the intake of carbon from the air as CO_2 and oxygen as O_2. Such absorption may occur in the leaves of higher plants, or directly from the soil or water in which simpler plants dwell. The processes of gas absorption will be discussed more fully in Chapter 7.

The carbon from absorbed carbon dioxide enters into the formation of most of the solid components of a plant; about 90 percent of the dry weight of a tree, for example, is synthesized from carbon dioxide. Carbon dioxide is similarly employed by other photosynthetic autotrophs to make structural materials and metabolic nutrients.

WATER ABSORPTION

Water absorption occurs by diffusion in autotrophic cells. No matter what kind of environment an autotroph lives in or how complex the autotroph is, water absorption by the organism will always be the result of osmosis.

The first and most primitive autotrophs obtained water directly from the seas and lakes in which they lived. However, the transition from an aquatic environment to a terrestrial habitat was made possible by the evolution of specialized structures. The first such structures were probably rhizoids (simple rootlike extensions), but root systems appeared later. Once autotrophs had evolved rhizoids and roots, they were no longer dependent on the sea for the water and other materials they needed but could absorb them directly from the air and soil.

Roots in water absorption Roots are long and intricate structures that serve to anchor a plant in the ground and to increase that part of the plant's total surface area that is beneath the ground. The more surface area that is exposed, the more water the plant can absorb. There are two patterns of root development. A **taproot** system consists of one main root from which many tiny rootlets grow. A **fibrous root** system has no main root, but instead many roots of roughly similar size that branch off into side roots. Although the methods differ, the function of both systems is the same, that is, to increase surface area. Most **dicots** (so named because their seeds have two leaves, or **cotyledons**) have taproots. Examples are the oak tree and the dandelion. Many garden plants, such as the grasses and lilies, have fibrous systems, and these plants are all **monocots** (that is, plants that have one cotyledon in the seed).

A longitudinal section of a common root (see Fig. 6.3A) shows how specialized the organ is. At the tip of the root is the root cap, a group of thick-walled cells that protect the nearby internal meristem cells that are responsible for growth. Above the meristem cells are the cells they produce: the xylem and phloem, or **vascular** tissue, consisting of a network of pipes or vessels that transport nutrients throughout the plant. Adjacent to the vascular tissue is a thick layer of parenchyma cells called the **cortex.** The outermost layer of the root consists of small cells that constitute the epidermis.

The root as it appears in cross section (Fig. 6.3B) shows a central core of vascular tissues (the **stele**) that contains both vascular and parenchyma cells. The outer parenchyma cells of the stele, known collectively as the **pericycle,** initiate lateral roots. Between the central stele and the cortex is the **endodermis,** a single layer of cells whose walls are thickened by a layer of waxy substance, the **Casparian strip,** which prevents the passage of water through the cell walls (Fig. 6.4). Numerous intercellular spaces exist between the cells of the cortex.

To help increase the total surface area of the root system, some cells in the epidermis develop **root hairs.** Root hairs are tiny extensions of the epidermal cell walls that grow in number and size according to the availability of water. In locales where water is scarce, root hairs tend to be longer and more numerous. The relationship between the appearance of root hairs and the availability of water may be readily seen by comparing the root hairs of a radish root grown in moist air with the hairs of a root grown in water.

Water absorption begins in the epidermal cells of the root hair. The concentration of osmotically active solutes is higher inside an epidermal cell than in the soil outside. This means, conversely, that the concentration

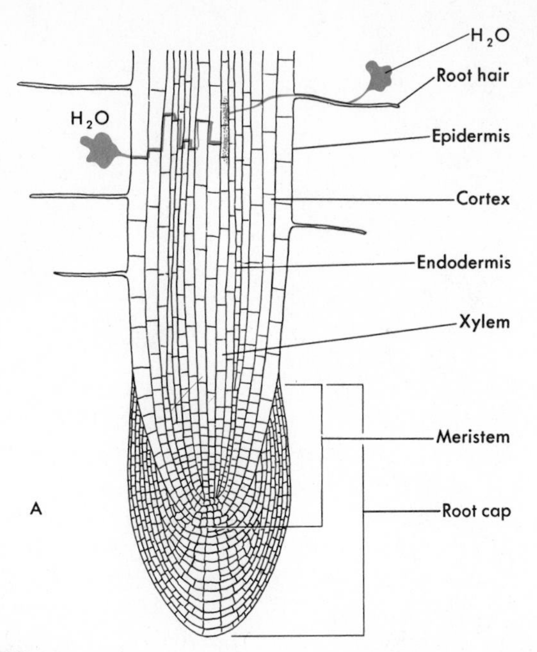

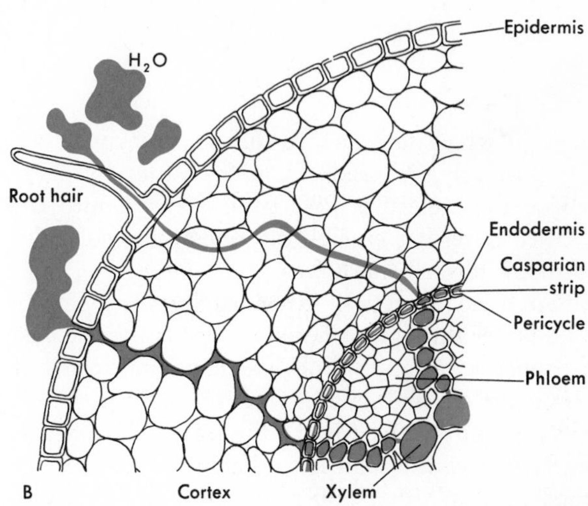

FIGURE 6.3

Longitudinal section (A) and cross section (B) of a root. In both sections the path of absorbed water and dissolved minerals is traced in color. In (A) the outer layer of cells above the root tip forms the epidermis, a tissue that provides protection and an absorptive surface for the root. Water enters the root through extensions of epidermal cells called root hairs or through spaces between epidermal cells. It then passes through the cortex and endoderm into the stele, which contains the vascular tissues (xylem and phloem). After entering the water-carrying xylem, it is pulled upward toward the leaves. In (B) one can see at the center the circular stele surrounded by the endoderm. In the stele the large water-carrying xylem tubes are arranged in the shape of a star, with bundles of the nutrient-carrying phloem between the arms of the star.

of water molecules is greater in the soil than it is in the root's epidermal cells. Thus water tends to move into the epidermal cells—where the concentration of water molecules is lower. Since the membranes of epidermal cells are quite permeable to water molecules, they diffuse into the cell easily. Once inside the epidermis, water continues to move along a concentration gradient—always moving to an area where the concentration of water is lower. Hence water molecules tend to move inward toward the xylem in the stele, which can bring water from the roots to all other parts of the plant. They may either pass directly into the stele *through* the cortex cells or pass through the very porous intercellular spaces between the walls of the cells. Passage into the root is comparatively easy for water molecules until they confront the thickened, waxy walls of the endodermal cells. Here resistance to passage may be encountered by the water molecules as they reach the hard, waxy layer that constitutes the Casparian strip. The water is thus obliged to pass through the **protoplast** (the living part of a plant cell contained within the cell membrane) of each endodermal cell. This phase of the absorption process operates as a

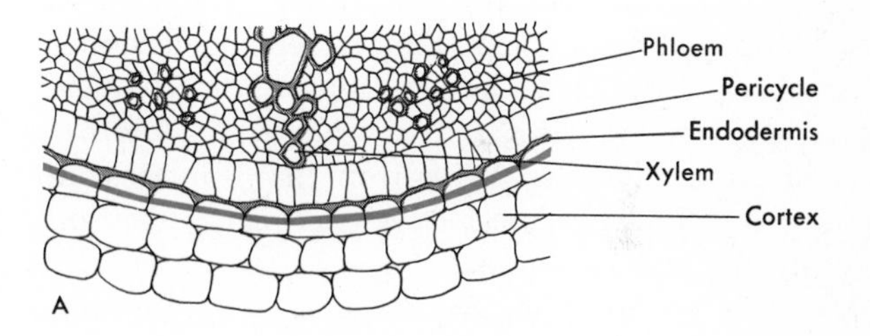

FIGURE 6.4

(A) Part of a root cross section, showing the position of the endodermis in relation to the water-transporting cells of the stele (the xylem). (B) The endodermal cells have an obvious waxy layer, which coats the upper, lower, and radial walls of each cell, waterproofing them. Water must thus pass through the living part of the cell, from its outer face to its inner face.

6.5x

FIGURE 6.5
Nodules on the roots of clover containing nitrogen-fixing bacteria. The bacteria, living within the nodules, convert molecular nitrogen from the atmosphere into soluble nitrates that can be used by the plant. Farmers often grow crops of such plants on fallow fields, eventually plowing them into the soil, where the soluble nitrogen compounds may be used by other crop plants.

final stage in which undesirable substances may be actively or passively "filtered" from the water by these cells before the molecules pass from the root into the xylem. Once water has entered the xylem, the molecules move directly toward the leaves.

Although no energy is expended by the plant in absorbing water through osmosis—a mechanism of passive transport—it is quite probable that some forms of active transport occur in addition. It has been observed that roots "pump" water toward the leaves. This action is called **root pressure,** and is probably responsible for the drops of water that appear at the ends of leaves occasionally (called **guttation**). However, it is not yet known precisely how root pressure operates.

MINERAL ABSORPTION

Mineral absorption is accomplished readily by autotrophs because all the necessary minerals appear dissolved in their water supply. The minerals dissolve in water as it flows over rocks or soil en route to the plant (this is called leaching of the soil) and then can enter cells either by diffusion or by an active ion transport mechanism such as the one we discussed in Chapter 5. Another source of plant minerals derives from man's interest in keeping his autotrophs productive—fertilizing the soil to increase its mineral content.

Minerals always enter the plant as ions in water, but only sometimes through the passive mechanisms of osmosis. Plants often have to obtain their minerals through active transport—expending energy themselves to remove ions from the water of soil. This is because minerals are often present in a higher concentration inside the cell than outside, and they are not, in such cases, subject to osmosis. To get new mineral ions inside, cells must force them to travel against a concentration gradient and must expend energy to accomplish that end.

METHODS OF OBTAINING NITROGEN

In addition to water and minerals, plants require nitrogen to make amino acids, the building blocks of proteins. Even though nitrogen is in plentiful supply—it constitutes about 78 percent of the atmosphere—that abundance is deceptive, because most plants cannot use nitrogen in its atmospheric form, molecular nitrogen (N_2). Plants must obtain their nitrogen from soluble nitrate compounds, and some have developed remarkable adaptations for doing so.

Certain microorganisms can convert atmospheric nitrogen to soluble nitrates by a process known as **nitrogen fixation.** They include the blue-green algae, a few fungi, and a few bacteria. Some of the nitrogen-fixing bacteria live in nodules on the roots of legumes, such as peas, beans, and clover (Fig. 6.5). They live with the plants in a symbiotic relationship that is mutually beneficial: the bacteria get the nutrients they need for their existence, and the plants get nitrates in return. Because of the nitrogen-fixing bacteria, legumes not only grow

FIGURE 6.6
Most flowering plants are photosynthetic autotrophs, but a few are heterotrophs. One of the most interesting is the delicate, frosty-white Indian pipe (Monotropa uniflora), *which contains no chlorophyll whatever. The Indian pipe was long classified as a saprophyte (living on decaying organic matter), but recent findings suggest that it obtains nutrients through a complex relationship with a fungus and another flowering plant.*

successfully in nitrogen-poor soil but even improve the soil itself. The symbiotic bacteria tend to synthesize more nitrates than the plants can use, and the excess enriches the growing medium. Because most other types of plants draw nitrogen from the soil, farmers often plant legumes in rotation with wheat, corn, and other staples.

A limited number of plants obtain preconverted nitrogen by a method that can be called "nitrogen trapping." For example, *Dionaea muscipula*, better known as Venus's-flytrap, inhabits the nitrogen-poor bogs of the Carolinas. The plant evolved the ability to trap insects rich in protein and to utilize some of the insects' nitrogen. Plants such as *Dionaea* are called insectivorous; some 400 different species use some method to trap insects and obtain nitrogen. None of them eat humans!

Heterotrophs: the consumers

Heterotrophs are plants and animals that are unable to manufacture their own food. They require their nutrients preformed, a factor that makes them *totally* dependent on other organisms for sustenance. Consequently, all heterotrophs are somehow adapted to consume either autotrophs and/or other heterotrophs that have ingested autotrophs.

Some heterotrophs, such as fungi, may be classified as either **saprophytes,** which sustain themselves on nonliving organic matter, or **parasites,** which live at the expense of other organisms known as their "hosts." The fungus we recognize as mold on bread is a typical saprophyte, whereas the fungus that causes the disease known as "athlete's foot" is parasitic. Christmas mistletoe, which is a flowering plant, or angiosperm, is also parasitic. Mistletoe hangs from the bark of trees and removes nutrients from the phloem of these hosts.

However, most heterotrophs are animals—from the simplest unicellular amoeba to man—and they fall into four different categories:

1. Herbivores—such as the hippopotamus or elephant—are strictly plant eaters. A hippopotamus eats about 600 pounds of plant material each day to nourish its two-ton bulk.
2. Carnivores eat only flesh. Examples are the dog and the wolf, but the tiny shrew is also a carnivore that eats eight times its own weight of flesh a day.
3. Omnivores—notably man—eat a mixed diet of plants and flesh.
4. Parasites, such as the tapeworm and the leech, live by absorbing either nutrients that have already been digested by their hosts or blood and other tissue fluids.

HETEROTROPHIC NUTRITION

All heterotrophs, plant and animal, need certain essential nutrients to survive—carbohydrates, proteins, lipids, and water—many of which they cannot make for themselves and must obtain preformed from other organisms that they eat.

Carbohydrates, or sugars and starches in one's food, are partly what gives one a feeling of satiety after a meal; the high level of glucose in the bloodstream helps to diminish the appetite. Since most heterotrophic activities are supported by the energy released by glucose metabolism, heterotrophs have extensive requirements for carbohydrates.

Heterotrophs obtain carbohydrates in bulk foods. The glycogen (animal starch), plant starches, and sugar components of the foods are then converted to glucose, which is in turn systematically broken down to release the energy needed for all life processes.

Proteins are essential to the formation of all cellular membrane structures and the formation of enzymes. One need only observe a child with protein deficiency —which produces the symptoms of the increasingly familiar disease, kwashiorkor—to realize the effects. Without an adequate supply of proteins to maintain the integrity of cell structures and enzymes, an animal heterotroph's muscles, in which proteins play a vital structural and functional role, wither; similarly, the brain's functions, as well as other vital functions, slow down.

Every heterotrophic organism must obtain amino acids from the proteins it eats in order to make its own proteins. Heterotrophs have a number of mechanisms for converting amino acids from one type of amino acid to another, but certain conversions cannot be made. Man, especially, depends on certain essential preformed amino acids that his body cannot synthesize.

Much research has been conducted on the problem of protein deficiency and how to solve it. For example, the United States government has for some time been providing protein-rich foods to undernourished people in underdeveloped countries, but there is some question as to what kinds of protein are needed. Plant-derived foods alone will supply some proteins, but hardly enough to supplement a poverty diet that is deficient in meat, fish, or fowl. Ground fish meal—a food that is both rich in protein and inexpensive—seemed to be one source of protein which could be easily shipped to areas lacking adequate dietary protein. But dietary habits and taboos often outweigh the logic of dietary change, and the introduction of this valuable dietary supplement has faced some difficulties. Similarly, the change in diet from relatively protein-poor corn to wheat is difficult for people who have either never seen the new food or have no idea how to prepare and eat it. For these reasons scientists are currently experimenting with new varieties of corn that are rich in essential amino acids. Such a grain, relatively rich in the amino acid lysine, as well as other important amino acids, may, if widely enough planted and consumed, help to alleviate debilitating protein starvation.

Lipids are oily substances that are used to store energy, as well as manufacture cellular membranes. One category of lipids consists of fats—molecules composed of glycerol and fatty acids—deposited as fatty droplets in the cells of adipose tissue, which are generally located around the abdomen and hips and in other places under the skin. In addition to storing energy, fats help to keep organisms warm in cold weather, and they play a vital role in insulating nerve cells. Heterotrophs may obtain lipids from synthetic reactions carried out on carbohydrates, but in some animals, including man, certain lipids cannot be synthesized.

Vitamins are complicated molecules that function as coenzymes in many different cellular reactions. Heterotrophs require vitamins in order to live but cannot ordinarily synthesize them. Thus animals must rely on plants and bacteria to supply their vitamins. For example, carotene in plants can be converted into vitamin A, and vitamin K can be manufactured by intestinal bacteria (intestinal "flora"). There are two kinds of vitamins —those that are soluble in water and those that dissolve in lipids. Most heterotrophs must ingest water-soluble vitamins daily because these vitamins cannot be stored. The familiar warning against overcooking vegetables is based directly on the water-solubility of vitamins. Lipid-soluble vitamins are stored in fatty droplets inside cells, so they need not be ingested on a regular basis. Most of the vitamins that people require must be obtained from their diets, but some are available from other sources. For example, vitamin D is one substance that the human body can synthesize from normal exposure to sunlight,

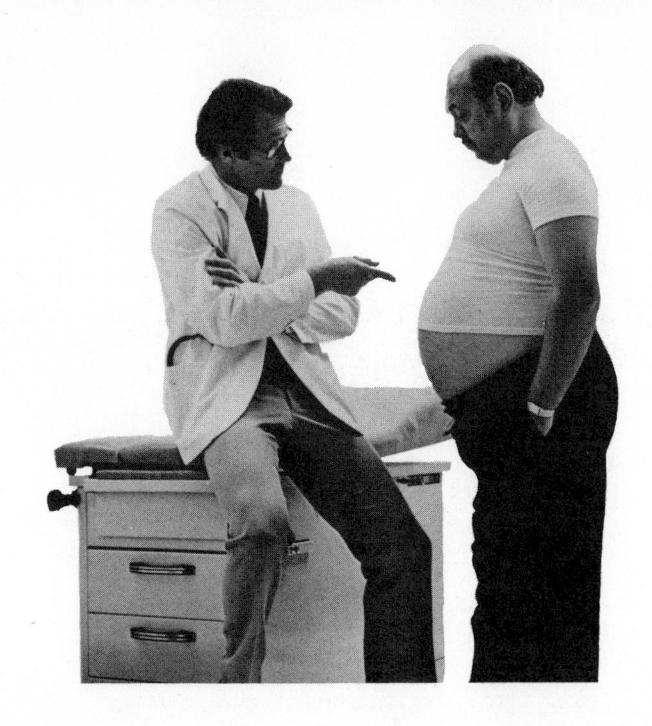

FIGURE 6.7

An excess of 3500 calories in the food ingested produces one pound of fat. Fat is deposited in adipose tissue, where one mile of blood vessels are required to maintain it. Obesity is correlated with a number of life-shortening diseases.

and most people do not need supplementary vitamin D. (Prison populations, however, do require vitamin D supplements. And babies require special doses of vitamin D because they are especially prone to vitamin D deficiency if they are kept indoors for long periods. This is why milk contains an additive called **ergosterol**—a "provitamin" which the body can convert to vitamin D.) Man's intestinal flora, which produce vitamin K, also provide some of the B vitamins, so that people usually get all the B vitamins they need from food and production by flora, without resorting to supplements.

Specific vitamin deficiencies sometimes occur in man. Vitamin C deficiency (caused by lack of fresh fruit and vegetables) results in scurvy, a disease that was common on sailing ships before refrigeration was invented, when there were no methods for keeping fruits and vegetables fresh on board. Vitamin D deficiency causes rickets, a rubbery condition of the bones resulting from inadequate calcium salts in the diet. Vitamin A deficiency produces night blindness, a deterioration of the eye's sensitivity to light. Vitamin A is a component of the eye's light-sensitive pigment (retinene). Except during an illness in which the body is prevented from absorbing vitamins normally, or when one is following a fad diet of foods that are nutritionally incomplete, there is probably no good reason to use commercial vitamin supplements.

Water is essential to heterotrophs. Unlike autotrophs, they do not manufacture their foods from carbon diox-

FIGURE 6.8

Zebra come to the waterhole to replace vital fluids lost from their bodies during the hot African day. They drink hastily, not lingering in such an exposed position, where they are especially vulnerable to predators.

Nutrition Labeling and the Recommended Daily Allowances

The Food and Drug Administration requires U.S. Recommended Daily Allowances to be listed on packages and in advertisements for any food that claims to have nutritional value. The U.S. RDAs inform consumers just what they can expect to get, in terms of nutrition, from the food they eat.

U.S. RDAs are the Recommended Dietary Allowances established by the Food and Nutrition Board of the National Research Council. These Dietary Allowances represent the *average nutrient requirement* for protein and several vitamins and minerals, plus an extra 10–50 percent "margin of safety" to cover variations in individual needs, nutrient loss due to cooking and storage, etc. These amounts are *not* minimum or even ideal intakes. Dietary Allowances are purposely set high to meet the needs of practically all healthy persons in the United States. They provide a guide for nutrition projects (school-lunch programs and hospital-food programs) and, ideally, encourage the kind of dietary practices that will lead to a healthy population.

Dietary Allowances are figured for both sexes in many age categories. To make this mass of information practical and usable on labels as U.S. Recommended *Daily* Allowances, the figures for the various age and sex categories are condensed. For example, the Dietary Allowances for a teenage boy are the basis for the U.S. RDAs for "adults and children 4 or more years of age," the category used for most nutrition labeling. Since a growing teenage boy needs a large supply of proteins, vitamins, and minerals, the RDA will be higher than the needs of some people included in this group. For example, many adults may need only 75 percent of the RDA, children only 50 percent.

U.S. Recommended Daily Allowances (U.S. RDA)

VITAMINS, MINERALS, AND PROTEIN	UNIT OF MEASUREMENT[a]	INFANTS	ADULTS AND CHILDREN 4 OR MORE YEARS OF AGE	CHILDREN UNDER 4 YEARS OF AGE	PREGNANT OR LACTATING WOMEN
Vitamin A	IU	1,500	5,000	2,500	8,000
Vitamin D	IU	400	400[b]	400	400
Vitamin E	IU	5.0	30	10	30
Vitamin C	mg	35	60	40	60
Folic Acid	mg	0.1	0.4	0.2	0.8
Thiamine	mg	0.5	1.5	0.7	1.7
Riboflavin	mg	0.6	1.7	0.8	2.0
Niacin	mg	8.0	20	9.0	20
Vitamin B_6	mg	0.4	2.0	0.7	2.5
Vitamin B_{12}	μg	2.0	6.0	3.0	8.0
Biotin	mg	0.5	0.3	0.15	0.3
Pantothenic Acid	mg	3.0	10	5.0	10
Calcium	g	0.6	1.0	0.8	1.3
Phosphorus	g	0.5	1.0	0.8	1.3
Iodine		45	150	70	150
Iron	mg	15	18	10	18
Magnesium	mg	70	400	200	450
Copper	mg	0.6	2.0	1.0	2.0
Zinc	mg	5.0	15	8.0	15
Protein	g	18[c]	45[c]	20[c]	

[a]Units of measurement: IU = International Units; μg = micrograms; mg = milligrams; g = grams.
[b]Presence optional for adults and children 4 or more years of age in vitamin and mineral supplements.
[c]If protein efficiency ratio of protein is equal to or better than that of casein, U.S. RDA is 45 g for adults, 18 g for infants, and 20 g for children under 4.

Source: U.S. DEPARTMENT OF HEALTH, EDUCATION, AND WELFARE, Food and Drug Administration, Revised January 1976.

ide and water, but heterotrophs require water to transport materials throughout the body and to meet other needs. About 90 percent of the weight of a heterotrophic cell is water. Some cells may absorb water through osmosis or pinocytosis. Animals generally supply their cells with water by the indirect method of drinking the fluid through a mouth. The tall cool drink that helps one cope with a hot summer day provides fluids that will pass easily through the blood vessels, eventually cooling the body by being evaporated, and working to distribute nutrients efficiently among all cells. Some desert animals are known not to drink any fluids; they survive by using the water that is produced during metabolic reactions as well as by using a variety of water-conserving mechanisms. As we shall see in Chapter 8, some of the major control mechanisms in organisms are strategies for using and retaining water. Man's water requirements are great; the average person must consume *several* glasses of fluid (1500–2000 cc/day) each day—in some form or another—to maintain normal body functions.

Digestion

Once a heterotroph has procured the various foodstuffs required for its survival, these substances must be broken down, or digested, into the simpler constituents that the organism uses for its various life processes.

Among the heterotrophs, **intracellular digestion** is primarily employed by the unicellular protozoa, which form food vacuoles around ingested prey and then digest them. The more complex, multicellular organisms contain specialized cells to digest food. These cells secrete digestive enzymes to accomplish **extracellular digestion.**

INTRACELLULAR DIGESTION

The first problem facing any hungry heterotroph is to get food into the body. Invagination (phagocytosis), a process the amoeba performs to *engulf* or surround its food, is the simplest way of ingesting substances. White blood cells use this method also to trap bacteria, protecting the human body from infection and feeding themselves. The amoeba navigates to its prey with the aid of **pseudopodia** (false feet), which it projects from various parts of the cell. These projections also enable the amoeba to enclose a food particle completely, after which the particle is wrapped in a membrane. This process results in the formation of a food vacuole. The vacuole is then fused with a lysosome, whose enzymes aid in the intracellular digestion of the food particles. The vacuole membrane is permeable to the nutrients so that, after digestion has occurred, the nutrients are able to pass through the membrane into the cytoplasm.

The paramecium is slightly more specialized than the amoeba. The paramecium has a permanent structure —called the **oral groove**—that is lined with cilia, and the rhythmic motion of the cilia creates currents that help to sweep the organism's prey directly into the oral groove. Once food has reached the base of the oral groove, it becomes wrapped in membranes and is digested intracellularly as in the amoeba.

EXTRACELLULAR DIGESTION

Extracellular digestion may occur entirely outside the organism. For example, beer yeasts digest sugars and absorb the small fragments, helping to turn malt to beer; and bread molds survive by breaking down starches in bread and absorbing the nutrients produced. Intestinal parasites rely on the digestive fluids inside another animal for their food. For example, the tapeworm has suckers and hooks to anchor it to the host's intestinal tract, but it has no other digestive organs, since its whole body surface is used to absorb nutrients from the host.

Digestion in the simple multicellular hydra consists of a more sophisticated mechanism than that employed by either the amoeba or the paramecium; the hydra is one of the simplest heterotrophs to engage in extracellular digestion within its body cavity. The hydra's body is hollow and consists of only two layers of cells that are held together by an acellular middle layer, called the **mesoglea.** Despite the simplicity of its body structure, the hydra possesses a definite mouth surrounded by ten-

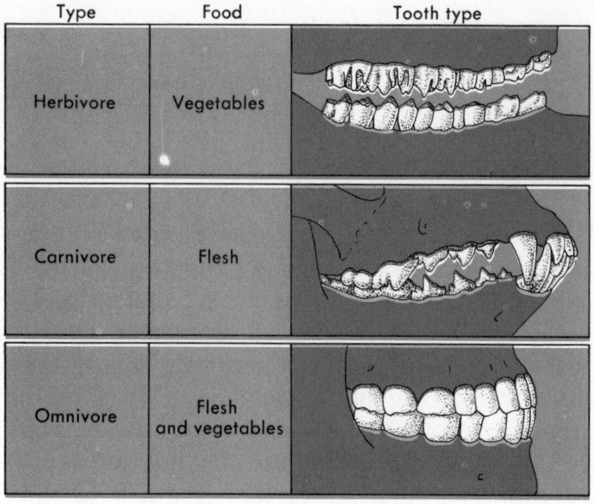

Type	Food	Tooth type
Herbivore	Vegetables	
Carnivore	Flesh	
Omnivore	Flesh and vegetables	

FIGURE 6.9

Stylized drawings demonstrating the relationship of tooth structure to the function and nutritional habits of organisms. An herbivore has flattened teeth that crush plant materials and shred the tough cellulose cell walls. A carnivore has sharp, pointed teeth, well suited for tearing chunks of flesh that can be readily swallowed. Omnivores may have a combination of both types—relatively sharp front teeth that can tear food, and back teeth that are flattened to form grinding surfaces to reduce foods to a soft, pulpy mass.

tacles, which have harpoon-like **nematocyst** cells that can penetrate and stun prey. When the prey is shoved into the body cavity by the tentacles, enzymes are secreted by the cells lining the cavity, and extracellular digestion occurs. Absorption of the digested nutrients is no problem, since none of its cells is more than a cell's width away from the digestive cavity. When digestion has been completed, wastes are eliminated along the route that was originally used to ingest the food; they are emptied through the mouth. This dual-purpose opening is characteristic of an **incomplete digestive system.**

In animals whose cells are further removed from the digestive cavity, greater absorption surface is necessary. For example, the planaria, a free-living flatworm, has a branching gut cavity to solve this problem. Planaria feeds through the pharynx, a tube projected through the mouth, and uses its much-branched cavity to bring the digested food into all regions of the body.

The transition among organisms from intracellular to extracellular digestion involves the elaboration of a more complex and efficient feeding apparatus. Heterotrophs with extracellular digestion depend on only a fraction of the total number of their cells to secrete digestive enzymes. In higher organisms, extracellular digestion involves a number of organs that constitute the digestive system; foods are adequately separated from wastes, and enzymes can digest the foods directly without the wastes interfering. Most important, in higher heterotrophic organisms, the food processing system is a one-way tube, in which food enters at one end and wastes exit at the other.

In a **complete digestive system,** food first enters the mouth, a process which may be aided by the action of claws (as in the lobster) or by a set of teeth. Teeth may be highly specialized structures in which form tends to follow function according to the kinds of food that the organism eats (see Fig. 6.9). The diet of an elephant (or of any herbivore) consists of plant life, so the elephant's teeth are flat like grindstones, shaped to enable them to pulverize vegetable matter. Dogs and wolves (carnivores), whose diet consists of meat, need sharper, stronger teeth to kill their prey and to tear apart the flesh. Man (an omnivore) has two types of teeth: sharp anterior (front) teeth adapted to tearing off parcels of flesh, and flat posterior (back) teeth designed to grind food into a soft mass. Despite such differences, there are two basic functions of teeth. They break food down into smaller particles and thus make swallowing easier; and they increase the surface area of the food, exposing more of it to oral enzymes, which in turn increase digestion in the oral cavity.

Another structure involved in the digestive process is the tongue. The tongue moves the food around while it is being chewed and helps to push the food down the throat toward the stomach. Moreover, the tongue is involved in the rather pleasant function that we call taste. Chemoreceptors in the tongue and olfactory receptors in the nose enable a person to identify the food he is eating and to decide whether or not the substance is to his liking. The action of the receptors also makes it possible for an animal to follow a scent so that prey can be tracked down.

Once food passes through the mouth and enters the tube system, it may be stored temporarily in specialized organs. Thus the heterotroph does not need to be constantly foraging for food because normally there is a residual quantity available in its digestive system. This freedom from the constraints of an incessant search for food has high adaptive value for the survival of the species. Only when the animal has consumed a sufficient amount of nutrients can it devote time to its reproductive and other functions. A grazing animal such as the deer has a many-chambered stomach, so that it can spend as little time as possible eating in a pasture exposed to predators. Later, in a safer place, the food is returned to the mouth as a cud for thorough chewing.

The earthworm has a **crop,** or thin-walled pouch, where bits of organic matter and soil are held until digestion begins. Because the worm has no teeth, the **gizzard,** a structure just behind the crop, churns and grinds the food, often with the help of small stones lodged there. Then the food moves on to the intestine, where enzymes free the nutrients for absorption and where wastes are passed along to be excreted through the anus.

From the example of the earthworm, we can see that a complete digestive system contains a mouth or anterior opening, an internal storage space, an internal digestive apparatus, a mechanism for absorbing nutrients and water, and a specialized system for eliminating undigested wastes.

FIGURE 6.10

Life-sized reconstruction of the jaws of an extinct shark that must have been the most prodigious carnivore on Earth in its day. But don't worry, it has been among the missing for more than 30 million years!

Humankind as ultimate consumers

As the most advanced heterotroph, man occupies a singular position among organisms. Although his diet may consist of plants as well as certain animals, man is unique among organisms in that his feeding behavior is likely to have a significant impact on his environment by disrupting natural food chains. Man seldom returns his waste to the soil from which his food has been derived, and he is rarely the prey of other organisms.

Complex and adaptive as man may be, he is like other organisms in requiring the distribution of nutrients to each of the billions of cells in his body. This process of preparing food for **cellular absorption** by the human organism constitutes one of the most intricate digestive systems of any life form. Its structure is basically similar to the system in any higher heterotroph, consisting of a one-way tube into which food is admitted at one end and waste is expelled from the other. The digestive organs are separated from one another by a number of valves. Digestive enzymes break down the materials that are taken in, and internal storage is provided for. It is only the degree of specialization of man's digestive system that distinguishes it from similar systems in other heterotrophs.

TABLE 6.3
Human digestive enzymes

LOCATION	NAME	SUBSTRATE	PRODUCT
Mouth	Ptyalin (amylase)	Starches: amylose, amylopectin, glycogen	Soluble starch, erythrodextrins
Stomach	Pepsin	Proteins	Proteoses, peptones
	Rennin	Casein of milk	(Paracasein) milk curdle
	Gastric lipase	Fats (esp. butter, egg yolk)	Fat fragments
Duodenum	Aminopeptidase	Complex protein fragments	Amino acids
	Amylase	Starches	Maltose
	Maltase	Maltose	Glucose
	Lactase	Lactose	Galactose, glucose
	Sucrase	Sucrose (cane sugar)	Fructose, glucose
	Lipase	Fat, fat fragments	Fatty acids, glycerol, neutral fats
	Nucleases	Nucleic acids	Nucleotides
	Nucleotidases	Nucleotides	Phosphoric acid, nucleotides
	Enterokinase	Trypsinogen (inactive precursor in pancreatic juice)	Trypsin
Pancreas	Trypsin	Proteins	Proteoses, peptones, polypeptides, amino acids
	Chymotrypsin	Proteins	Amino acids
	Carboxypeptidase	Polypeptides	Amino acids
	Pancreatic amylase	Starch	Maltose
	Pancreatic maltase	Maltose	Glucose
	Pancreatic lipase	Neutral fats	Fatty acids, glycerol
	Nucleases	Nucleic acids	Nucleotides

Man's enzymes, like those of other higher vertebrates, are highly specialized; each is designed to decompose a specific substance, in a process called **enzymatic hydrolysis.** In this process, chains of food molecules are decomposed when the enzymes cause water molecules to enter at points of linkage, so that the molecules break apart.

Carbohydrases are enzymes that specifically break down carbohydrates by enzymatic hydrolysis. Carbohydrases enable the human system to break down starches quickly at normal body temperature, so that energy may be extracted in respiration. Without carbohydrases, hydrolyzing starch to sugar would require very high temperatures.

Lipases are enzymes that hydrolyze lipids. Since lipids do not dissolve in water (for the same reason that water does not mix with oil), lipases alone are not enough to render lipids water-soluble. Some other factor is needed, and it is provided by bile salts.

Bile salts, which are not enzymes, are secreted by the liver to help lipases break down fats. These salts have the same emulsifying action as a good detergent, causing fat drops or globs to break up in water, thereby increasing their total surface area. Thus the lipids have more of their surface exposed and are more accessible to attack by lipases.

Proteases are enzymes that hydrolyze proteins. All protein molecules are built up of combinations of amino acids. Proteases break apart the protein chains by hydrolyzing the bonds that link the amino acids. A summary of the human digestive enzymes and their functions is given in Table 6.3.

Thus the digestive enzymes have two important features. First, they all perform their functions by hydrolysis. Second, the enzymes begin to act on foods from the time the nutrients enter the body, and they continue until nutrient particles are small enough to be absorbed by the individual cells of the small intestine.

THE HUMAN MOUTH

The process of digestion begins in the mouth, where food is torn apart by the anterior teeth and ground into a pulp by the posterior teeth. The tongue serves to manipulate the food so that it is masticated (chewed), and finally helps to propel the food into the digestive tract. The tongue, through its chemoreceptors, also receives chemical sensations of taste, which the brain interprets. **Salivary glands** near the floor of the mouth (the submaxillary and sublingual glands) and near the cheeks (the parotid glands) produce saliva, which performs three different functions. First, the saliva moistens the food with water so that enzymatic reactions will be facilitated. Second, it lubricates the food with mucus—a viscous, slippery substance—so that it can pass down the tube smoothly. Finally, the saliva secretes *salivary amylase* (ptyalin), a carbohydrase that begins the digestion of carbohydrates while they are still in the mouth. The amylase attacks molecules of starch, breaking alternately every second bond in the chain, thus producing molecules of maltose.

THE HUMAN ESOPHAGUS

After each morsel has been sufficiently masticated in the mouth and has been formed into a lubricated ball, called the **bolus,** it is propelled into the esophagus in the act of deglutition, or swallowing. To prevent the bolus from accidentally entering the trachea (windpipe), which is located just ventral to (in front of) the esophagus, the **epiglottis**—a structure that is positioned anterior to the esophagus and trachea and functions somewhat like a trap door—closes over the tracheal opening as food is swallowed, thus blocking the food from entering and preventing the onset of serious complications, such as choking.

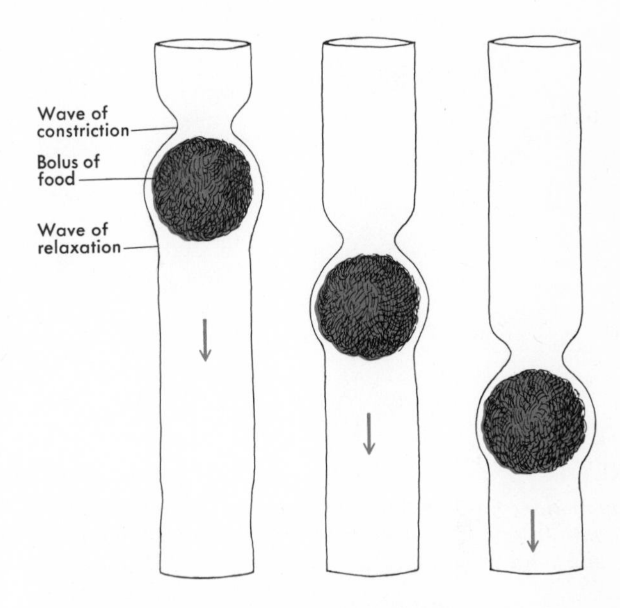

FIGURE 6.11

A wave of peristaltic constrictions pushes a bolus of food downward in the esophagus on its way through the digestive system. These constrictions are the result of coordinated contraction of smooth muscles.

With each swallow, a small volume of air enters the system and forms an "air ball" in the stomach. Ordinarily the presence of air is not noted. However, the gulped air that is taken into the digestive tract with hastily swallowed food can cause a "nervous stomach" and **eructation,** or belching of some of the acid contents back into the lower end of the esophagus. Because the esophagus is ill equipped to cope with acids, pain and burning (heartburn) may result.

Once in the esophagus, the food bolus is transported down the tube by a series of muscle contractions known as **peristalsis** (Fig. 6.11). When the moist food reaches the end of the esophagus, it stimulates a valve-like ring of muscle called the **cardiac sphincter** to open and permit the bolus to enter the stomach.

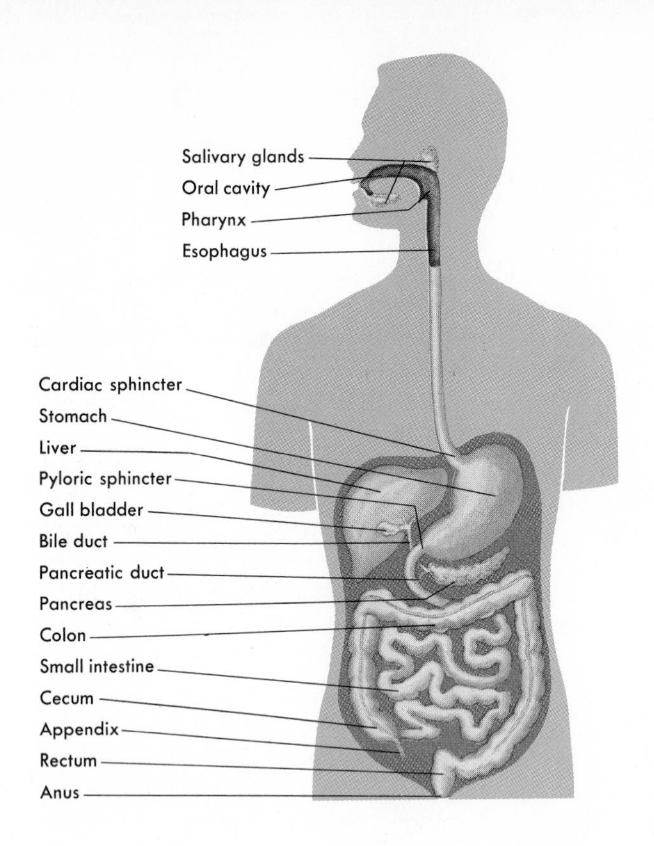

FIGURE 6.12

Diagrammatic representation of the human digestive system from mouth to anus. The liver, pancreas, and salivary glands secrete bile and various digestive enzymes into the digestive tract.

THE HUMAN STOMACH

The stomach is involved in several important functions. One is the storage of food; digestion is interrupted while food lies in the main body of the stomach. The stomach's function as a storage site enables man to get along without active food intake for several hours at a time. Periodically, the muscles in the stomach's wall contract, thus forcing portions of the bolus toward the pyloric end of the stomach from which they eventually pass through the **pyloric sphincter** into the small intestine.

Cells in the stomach's inner walls secrete hydrochloric acid (HCl) and **pepsin**—an enzyme that hydrolyzes proteins. Hydrochloric acid macerates food by dissolving the cement that holds together the cells composing it until it forms a loose, watery mass called **chyme.** Water secreted along with HCl serves as a solvent for enzymatic reactions and as a transport medium for moving food easily from one part of the alimentary canal to the other. Pepsin, another enzyme, can hydrolyze proteins most efficiently at the highly acidic levels normally found in the stomach, that is, a pH range of from 1.5 to 2.5. This high level of acidity, together with the maceration caused by HCl, also destroys several kinds of harmful bacteria that may have been ingested with the food. The latter action is fortunate because our food is usually not sterile, and the bacteria, some of which are pathogenic, or disease-causing, could thrive and multiply if they were ever to reach the small intestine.

To protect the delicate cellular and muscle components of the stomach's walls from the highly acidic environment, tightly packed mucosal cells—which stand so close together that acid cannot usually penetrate between them—line the stomach walls. Moreover, the cells lining the stomach constantly **desquamate,** or flake off, leaving young, new, undamaged cells to interface with the acid and pepsin. The human stomach sheds about 5.0×10^5 cells each *minute*—indeed, about 25 percent of the dry weight of each bowel movement is composed of dead stomach and intestinal cells.

THE SMALL INTESTINE

After the food bolus has been thoroughly transformed into liquid chyme by the action of HCl and pepsin, it is squirted a little at a time through the pyloric sphincter into the duodenum. The **duodenum** is the initial portion of the small intestine, in which enzymes from the pancreas and liver converge on the chyme to extract the

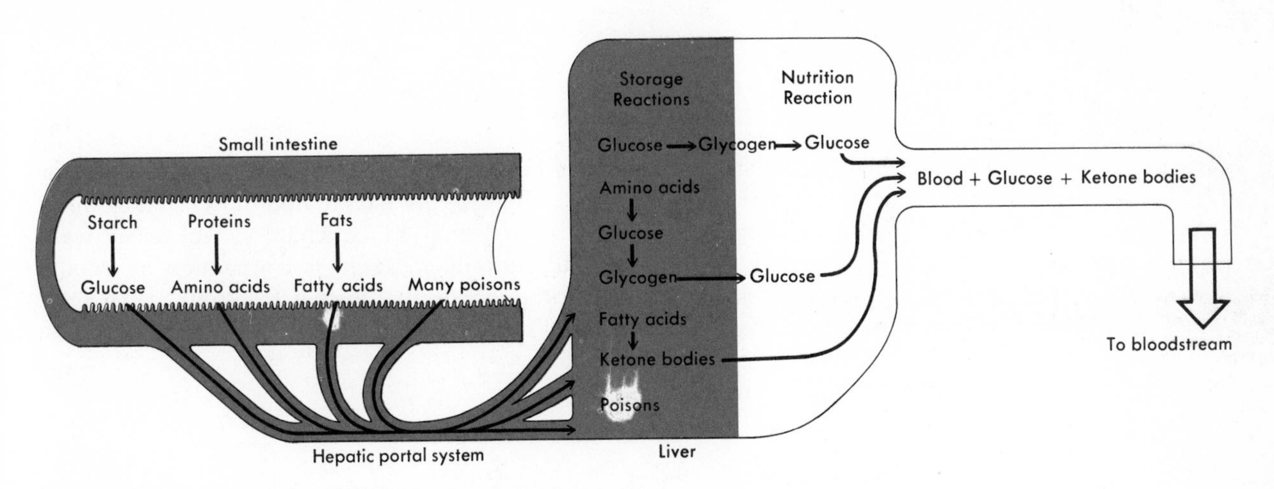

FIGURE 6.13

In mammals the small intestine and liver are connected by the hepatic portal system of veins. Digested nutrients absorbed by the intestinal villi are brought to the liver for storage and detoxification. When nutrient levels in the blood drop (for example, during fasting periods), the stored nutrients are mobilized and sent from the liver into general circulation.

useful nutrients still remaining after the digestion that occurred in the stomach. The pancreas secretes three enzymes: pancreatic lipase, pancreatic protease, and pancreatic carbohydrase. These enzymes hydrolyze lipids, proteins, and carbohydrates, respectively. Bile is also secreted (by the liver) to help emulsify fats, and it is stored temporarily in the gall bladder until needed.

Aside from providing an optimal climate for enzymatic reactions, the small intestine also is actively involved in absorption. Its surface area is increased by finger-like projections called **villi.** Each villus is long and well equipped with blood vessels (blood capillaries and lymph capillaries called **lacteals**), so that digested nutrients can pass directly into the circulatory system.

THE HUMAN APPENDIX

Between the small intestine and colon lies a distended pouch known as the **cecum,** which has a finger-like extension in humans called the **appendix.** The appendix is a remnant of our evolutionary history, and in man it serves no significant function except to become infected. Usually antibiotics will be sufficient, but sometimes it must be removed by a surgical procedure, an appendectomy.

THE HUMAN LIVER

From the small intestine, absorbed nutrients pass into the liver through a special branch of the circulatory system called the **hepatic portal system** (see Fig. 6.13). Unlike most blood vessels, the hepatic portal system terminates at each end in capillaries that originate from the wall of the elongate intestine and terminate within the liver. The liver is basically a metabolic factory, detoxification center, and storage organ with one major function —to maintain at the appropriate levels the quantity of nutrients in the body. It secretes bile salts, carbonates, and pigments. In terms of function, the liver is the most diverse tissue in our bodies. It also has a fantastic repair capacity. Removal of as much as 90 percent of the liver can be tolerated because the organ regenerates so rapidly. It performs its regulatory process in three different ways. A sizable amount of glucose is usually deposited in the liver after a meal. Here the glucose is transformed into glycogen (animal starch), which can later be hydrolyzed for energy. Glycogen is thus stored and is readily available to meet the body's energy needs as they occur. Second, the liver removes nitrogen from excess quantities of amino acids, producing from them carbohydrates that can be used for energy. The liver also processes fatty acids and stores the products as **ketone bodies,** which may be later released as nutrients for active muscles. Finally, the liver assumes an important role in preventing certain poisons from harming the body. Poisonous substances, such as DDT and some of the compounds that are related to it, can actually be rendered harmless in the liver by being either broken down or bound to another molecule that renders them incapable of doing damage. This process is called **detoxification.**

THE HUMAN COLON

The primary function of the colon (large intestine) is to reabsorb the large quantities of water used in digestion and then return the fluid to the circulatory system. Water is crucial throughout the digestive process for lubrication, enzymatic hydrolysis, and transportation. However, the water must be returned to the circulation to prevent the body from dehydrating. After the water has been reabsorbed, the residue, known as **feces,** consists of indigestible vegetable matter, a lot of dead bacteria from the intestine, many desquamated cells from the intestinal lining, a small amount of water, and brownish bile pigments. Various gases, probably produced by bacterial action upon these wastes, may also be present. The feces are excreted from the intestine by a wave of peristaltic muscle contractions which squeeze material down the tube toward the rectum.

THE HUMAN RECTUM

When feces enter the rectum, a feeling of fullness is usual and serves as a stimulus for elimination. If for some reason elimination cannot occur, the rectum can store feces for a considerable period of time. However, prolonged storage may result in the formation of a dehydrated, hard stool that is difficult to expel. This condition, known as constipation, is a major affliction of older people. Elimination may also be postponed through conscious inhibition, a cultural phenomenon that exists in certain societies in which such behavior is encouraged. The process of elimination involves the exercise of voluntary muscles, increasing the possibility of *learning* eliminative behavior. Some infants are able to learn the muscular techniques of elimination faster than others. The natural reflex is to eliminate when rectal walls stretch under strain, initiating a contraction of smooth muscles in the rectum that pushes feces out of the anus.

Human nutrition

During the twelve-hour journey from the mouth to the large intestine, the food we consume is broken down sufficiently to be absorbed by our cells and put to use for one or more of the following purposes.

Energy. The energy stored in food is converted into the energy that maintains life, supports growth and movement, and powers the many chemical reactions within the cell.

Construction (growth and repair). From conception to the completion of adult growth, humans experience a tremendous increase in body size in terms of both linear growth (height) and cell mass (body weight). Much of the food intake during this period is retained as body structure. In addition, since the human body is never a "finished" product, much of the food provides material for replacing, repairing, and renewing the body's cells and tissues. The most visible examples of this continuous renewal are hair and fingernails, but the process also occurs in muscles, blood, intestines, and other locations.

Regulation. Small quantities of vitamins and minerals found in foods regulate the release of energy for body activities, the acidity level and water balance of our systems, and other biological reactions.

NUTRIENT CATEGORIES

As noted earlier, there are five basic nutrient categories: carbohydrates, lipids (fats), proteins, minerals, and vitamins. Table 6.4 shows the primary role or roles that each category plays in general nutrition, as well as its major

TABLE 6.4
Nutrient categories: sources, roles, and functions

NUTRIENT CATEGORY	MAJOR FOOD SOURCES	PRIMARY ROLES			FUNCTIONS
		ENERGY	CONSTRUCTION	REGULATION	
Carbohydrates	Sugars, cornstarch, raisins, flour, bread, potatoes	++	+	0	Primary source of energy; provides bulk to diet (prevents constipation); promotes development of intestinal bacteria
Lipids	Meat, fish, poultry, fats and oils (butter, vegetable oils), milk, cheese	+++	+	0	Source of energy; energy reserve; satiety value; important in blood clotting; carries fat-soluble vitamins; component of cell membrane; insulation; prevents excessive water loss
Proteins	Meat, fish, poultry, eggs, dairy and cereal products, fruits, vegetables	+	+++	0	Source of eight "essential" amino acids, necessary for growth of new tissue; necessary to formation of hormones, enzymes, coenzymes, hemoglobin
Minerals (e.g. calcium, phosphorous, sodium, iron, potassium)	Dairy products, vegetables, eggs, poultry, fish, meat, table salt (depending on mineral)	0	++	+	In general: maintain acidity level of internal environment; catalysts for biological reactions; components of hormones, enzymes, vitamins; maintain water balance; play part in transmission of nerve impuses; growth of tissue; regulation of muscle contraction
Vitamins	See Table 6.5	0	0	+++	See Table 6.5
Water	Beverages, foods				Solvent; transports nutrients and waste products; medium for reactions; lubricant; regulates body temperature

food sources and its functions. A balanced diet should contain adequate amounts of each category.

Carbohydrates Carbohydrates (sugars and starches) are the cell's first choice for an energy source, even though lipids have a higher energy content. As long as there are enough carbohydrates in the diet, the fats that are ingested will not be used to any appreciable extent.

In the United States, it is estimated that 50 percent or less of the total energy intake comes from carbohydrates. In some of the poorer countries, carbohydrates may account for as much as 80 percent of the energy intake because they are a relatively cheap source of energy compared with lipids or proteins.

One carbohydrate is important in a healthy diet even though it has no nutritional or energy value to humans. **Cellulose,** a polysaccharide of glucose, is the major component of cell walls and of the structural and fibrous parts of plants (leaves, stems, roots, and seed and fruit coverings). Cellulose (commonly referred to as "fiber," "bulk," or "roughage") cannot be broken down by the digestive enzymes of humans and other carnivores. Therefore much of the plant material we eat remains undigested. This indigestible fiber helps move the food mass through the gastrointestinal tract by stimulating peristalsis, and it adds bulk (volume) to the food and digestive residues in the intestines by absorbing water.

The daily requirement for fiber is estimated to be between 0.14 and 0.25 oz. The best sources of fiber are (in the order given) dried fruits, whole-grain cereals, nuts, fresh fruits, and vegetables. The needed amounts of fiber may be provided, for example, by one serving of whole-grain cereal or bread, two servings of vegetables, and two servings of fruits.

Proteins Proteins are the most essential structural components of the body. They are involved not only in the growth and repair of tissues and cells, but also in the structure of hormones, hemoglobin, and enzymes, all of which are necessary in the biological reactions that keep us going.

Many people in the United States and throughout the world do not get sufficient protein, and protein deficiency is one of the most widespread types of malnutrition. People suffering from a protein deficiency may be getting some proteins in their diet but not the right kinds. Of the twenty amino acids that humans need and get from proteins, eight are labeled "essential." Nutritionists use this term to distinguish the amino acids that our bodies cannot manufacture and that must be obtained through diet. However, not all essential amino acids are present in all protein sources. Plant proteins, for example, are low in lysine, and therefore vegetarians must be certain to supplement their diets with eggs and cheese, both of which include lysine, or with commercial protein preparations. Animal proteins, on the other hand, supply a more complete line of essential amino acids.

Lipids Lipids, which have the highest energy content of all nutrients, function as energy reserves and work in a variety of other capacities. In the United States, 40–45 percent of our energy intake comes from foods with high fat content. The following three types of lipids are important.

The *triglycerides* are the high-energy lipids that release their stored energy when the carbohydrate supply is too low to meet the body's energy requirements.

The *phospholipids* are important in the structure of cell membranes, in the nervous system, and in the formation of thromboplastin, which is essential to the blood-clotting process.

Cholesterol is also an important lipid in cell membranes. Found in eggs, butter, and animal fat, cholesterol is also manufactured by body cells, especially those in the liver. A considerable amount of current research is directed toward the possible connection between the cholesterol level in the blood and heart disease.

The "average American" consumes approximately 4 oz. of fats a day. The amount may be too high because fats play an aesthetic role in our diet. They make food more palatable (anyone on a low-fat diet can attest to that), and they satisfy the appetite. Since lipids leave the stomach more slowly than other nutrients, their presence there tends to stall the need for the next meal. A final but important feature of lipids is that they form the medium in which fat-soluble vitamins (A, D, E, and K) dissolve prior to being absorbed by the cells.

Vitamins The human body needs vitamins in small amounts for many purposes, mainly regulatory. In this capacity, vitamins play a role similar to that of motor oil in a car. Oil is not a part of the car's structure, nor is it fuel for the engine, but it is necessary to keep the machinery operating smoothly and efficiently. Most vitamins (A and D excepted) are not stored in the body; any excess over what can be used immediately is excreted. Table 6.5 summarizes information about the principal vitamins.

Recently there has been publicity about the therapeutic effects of vitamins in amounts exceeding daily requirements. Examples are Vitamin C (ascorbic acid) for preventing colds and Vitamin E for healing and for retarding the aging process. As yet no satisfactory solution to this controversy has been reached because there is

TABLE 6.5
The principal vitamins

VITAMIN	DEFICIENCY DISEASE	SOURCES	OTHER INFORMATON
A	Night blindness	Milk, butter, fish liver oils, carrots, other vegetables	Precursor in the synthesis of the light-absorbing pigments of the eye Stored in the liver Toxic in large doses
Thiamine (B_1)	Beriberi Damage to nerves and heart	Yeast, meat, unpolished cereal grains	Coenzyme in cellular respiration
Riboflavin (B_2)	Inflammation of the tongue Damage to the eyes General weakness	Liver, eggs, cheese, milk	Is an active component of FAD in cellular respiration
Nicotinic acid (niacin)	Pellagra (damage to skin, lining of intestine, and perhaps nerves)	Meat, yeast, milk	In an altered form, this is a component of NAD and NADP—two important coenzymes for redox reactions in the cell
Folic acid	Anemia	Green leafy vegetables Synthesized by intestinal bacteria	Used in synthesis of coenzymes of nucleic acid metabolism
B_{12}	Pernicious anemia	Liver	Each molecule contains one atom of cobalt
Ascorbic acid (C)	Scurvy	Citrus fruits, tomatoes, green peppers	Coenzyme in synthesis of collagen
D	Rickets (abnormal bone and tooth development)	Fish liver oils, butter, steroid-containing foods irradiated with ultraviolet light	Synthesized in the human skin on exposure to ultraviolet light Toxic in large doses
E	No deficiency disease known in humans	Egg yolk, salad greens, vegetable oils	
K	Slow clotting of the blood	Spinach and other green leafy vegetables Synthesized by intestinal bacteria	Necessary for the synthesis of prothrombin, an essential agent in the clotting of blood

evidence leading to both negative and positive conclusions.

Minerals Humans require minute amounts of about 17 different minerals in their diets. These minerals contribute to the structure of hormones, enzymes, and vitamins; help to maintain the acidity level and water balance in our systems; and are active as catalysts for biological reactions. Some of the more important minerals are covered briefly in the following paragraphs.

Iron is an essential mineral in our diets, even though we need extremely small quantities of it. We are frequently admonished by television advertisers to avoid "iron-poor blood." Because iron is part of the hemoglobin molecule (the major oxygen-carrying component of red blood cells), iron deficiency will decrease the ability of these red blood cells to carry oxygen to body cells, where it is needed for cellular respiration.

The decrease in quality and/or quantity of red blood cells that results is called **anemia.** Its symptoms include pallor, easy fatigue, and decreased resistance to infections. One can reverse all these effects, however, by including foods high in iron content in the diet, such as liver, apricots, and raisins.

Calcium and *phosphorus* (found in dairy products, meat, and fish) are both important in the formation of teeth and bones. Consequently, children need large amounts of calcium and phosphorus in their diets, and so do pregnant women and nursing mothers.

Sodium (a component of table salt) and *potassium* (found in a great many foods) help to maintain a balance of body fluids and also take part in the transmission of nerve impulses.

Water Water is probably the most essential of the nutrients. A human being, who can live for weeks or years without some of the components of a balanced diet, usually cannot survive for more than a few days without water. In our bodies, water acts as a solvent for transporting nutrients and waste products, as a medium for cellular reactions, as a lubricant, and as a regulator of body temperature.

Although about two-thirds of our water intake comes from beverages (40 percent tap water, 60 percent milk and other beverages), we can and do get water from other sources. Most "solid" foods contain some water. Refined (white) sugar contains only 0.5 percent of its weight as water, but 96 percent of the weight of lettuce is water. As you will recall from Chapter 5, water is one of the end products of the oxidation of carbohydrates, fats, and proteins.

To make up for water excreted from the body in urine and feces and lost through the skin in perspiration, fluid intake should be around two quarts per day. This amount varies with each individual, depending on age and on environmental temperature and physical activity (both of which affect the amount of water evaporating from the skin). The young, people involved in strenuous activity, and people living in tropical climates all require more fluid per unit of body weight than adults, the nonactive, and those living in temperate climates (see Table 6.6).

TABLE 6.6
Fluid requirements per unit of body weight

	FLUID OZ. (PER POUND)
Infants	1.69
10-year-old children	0.61
Adults	
72°F	0.75
100°F	1.28

MALNUTRITION

Too little food Evidence is mounting that malnutrition during fetal development and the first years of growth can adversely and irreversibly affect the development of the brain and, therefore, affect intellectual abilities. Throughout the remaining years of life, as we have pointed out, man must supply his body with the nutrients it needs to keep functioning properly.

Since a large part of the world's population cannot obtain all the ingredients of a balanced diet, malnutrition with its ensuing illnesses constitutes the greatest health hazard in the world today. At least one billion people are underfed, 500 million suffer from chronic hunger, and hundreds of thousands face starvation.

Although the number of deaths from starvation is estimated to be in the hundreds of thousands each year, no figure can be exact since the cause of death is not always listed as "starvation." Many deaths that are attributed to infections or parasites are actually due to malnutrition—the *cause* of the cause of death since it weakens the individual's system until it can no longer fight the infection effectively. Paul Ehrlich, leading spokesman for population control and environmental pollution, defines death by starvation as "any death that would not have occurred if the individual had been properly nourished regardless of the ultimate agent."

Most malnourished people live in poor and underdeveloped countries, but the problem is hardly unknown in developed nations. In the United States, Public Health studies have uncovered the fact that between 10 and 15 million Americans are chronically undernourished.

Because of high nutritional demands during the growth period, children constitute a majority of the victims of malnutrition in the United States and worldwide. The protein deficiency disease kwashiorkor is estimated to affect as many as 50 percent of the children in poorer countries, where limited economic and natural resources result in a diet consisting primarily of starches and sugars. Another disease that primarily affects children, marasmus, is caused by poor nutrition in general, not just by protein deficiency. The effects of both kwashiorkor and marasmus are reversible up to a certain time if the proper foods are eaten. However, if not treated, the child grows weaker, and such infections as measles and pneumonia, which a healthy child usually survives, often lead to death.

The best solution to the problems of kwashiorkor and marasmus—and any other form of deficiency malnutrition, of course—is prevention in the form of a diet of high-quality animal and plant proteins that provides all eight essential amino acids and an adequate number of calories. But for the poorer populations of the world such a solution is difficult. They may not have the land, money, or agricultural technology (fertilizers, irrigation methods) to develop the foods for such a diet.

Efforts have been made by various institutions throughout the world to process foods rich in vitamins and minerals and especially in animal protein (such as fish, flour, and coconut seed cakes) into palatable cereal or beverage form. Plant geneticists have also studied and developed grains with high nutritive value. The problems in developing such foods are many. Not only must they be "super" nutritional, but

1. they must be available or capable of being produced in the areas where they are needed;
2. they must be inexpensive enough so that the people who need them can afford them;
3. they must keep well for long periods of time in the tropical climates of many underdeveloped nations;
4. they must have acceptable taste, odor, and consistency; and
5. they must not have any toxic effects.

Too much food Not all malnutrition results from a lack of food and is characterized by thinness. A person can eat as much as he wants every day of the week, be overweight, and still be undernourished. It is the *quality* of the food that counts, not the quantity. So, while citizens of the underdeveloped countries are suffering from a lack of quality and quantity, many citizens of the more affluent nations are suffering from an abundance. In the United States, it is estimated that there are at least 25 million people fighting the "battle of the bulge"—and indeed they should be. Not only is excess weight unsightly, but the mortality rate for overweight people is at least 50 percent higher than for those of normal weight, and a number of diseases can accompany a spare tire, among them heart and circulatory disease, kidney disease, diabetes, and cirrhosis of the liver.

Calories. When treating this kind of malnutrition, we have to consider another dimension of our food: **calories.** Food intake is calculated in calories, which measure the energy available from food. Whether you gain, lose, or stay at the same weight depends on how much food you take in and how much of it you use for energy. If you are involved in a highly active profession (dancer, construction worker), you can and should eat more to fill the energy requirements of your body. Growing children and teenagers also need more calories and nutrients than an adult, who no longer needs large amounts of energy or nutrients for growth. However, if you eat more food than your body requires, the excess energy is stored in fat (adipose) cells and tissue, and you gain weight.

To lose this weight you must consume fewer calories. Remember, however, that reducing your food intake should not involve decreasing your supply of nutrients to an inadequate level. Diets to gain, lose, or maintain weight should always be balanced. Therefore doctors often object to "fad" diets, such as the low-carbohydrate diet. Such unbalanced diets may result in rapid weight loss, but they fail to correct the poor eating habits that caused the weight gain to begin with.

Although obesity is caused by eating too much, there may be many reasons behind the urge to overindulge. In rare cases, the brain's appetite control center may be malfunctioning, or some people may have more fat cells (adipose tissue) than others. Some studies show (though not conclusively) that there may even be a genetic tendency toward obesity. And in recent years, there has been some acceptance of the idea that psychological problems (loneliness, depression, frustration) may cause some people to overeat.

Summary

1. From the simplest aquatic plant to man, all living cells must absorb the same basic materials (carbon, hydrogen, oxygen, and various other chemicals combined in various ways) to survive and grow.

2. Autotrophic cells can produce their own food with a few raw materials—gases, water, and minerals extracted from the environment. Autotrophs are the primary source of the energy trapped in foods. Without autotrophs, there would be no other organisms.

3. Heterotrophic cells cannot manufacture their own foods and must rely on autotrophs to supply them with nutrients that are preformed. Since most heterotrophic cells are part of multicellular organisms and thus cannot absorb food directly from the external environment, the organisms require digestive systems to hydrolyze the preformed nutrients into their simpler components.

4. Generally, the more complex an organism is, the more complicated are its food-getting, digestive, and absorption mechanisms. To obtain the simple inorganic compounds from which they synthesize food, autotrophic plants must carry on three processes: gas absorption, water absorption, and mineral absorption.

5. Plant roots are specialized organs that anchor the plant and absorb water, minerals, and gases. Roots absorb water by the passive process of osmosis. To obtain minerals, however, the root cells often must expend energy and use active transport to draw mineral ions into the cells against a concentration gradient.

6. Heterotrophs are divided into two groups: nongreen plants and animals. These subgroups can further subdivide according to the way the organisms derive their nourishment. Thus, nongreen plants may be either saprophytes or parasites, and animals may be herbivores, carnivores, omnivores, or parasites.

7. The basic nutritional needs of a heterotroph are water, minerals, and four types of organic compounds: carbohydrates, proteins, lipids, and vitamins.

8. To absorb and use food, a heterotroph must break it down into its component molecules by either intracellular or extracellular digestion.

9. Most unicellular organisms use intracellular digestion. Their food is encased in a food vacuole and then broken down by enzymes secreted by lysosomes.

10. Most multicellular organisms use extracellular digestion. It occurs in an enclosed cavity lined with cells and is accomplished by enzymes secreted in a specific sequence by specialized tissues. Such a system may be complete or incomplete.

11. In the human organism, the system is highly specialized and intricately coordinated and can transform a quantity of bulk food into individual nutrients to feed billions of cells.

12. Regardless of how organisms obtain their nutrients, all organisms use nutrients for the same purposes—survival, growth, and reproduction. Nutrients constitute the source of energy that maintains the order and intricacy of life, and nutrients are the source of the substances from which an organism is made. Their source is limited, so they are continually being recycled.

REVIEW AND STUDY QUESTIONS

1. Name the three categories of consumers and explain how they differ from one another. What category are you in?

2. Explain the importance of vascular tissue in plants.

3. Discuss the significance of the process called nitrogen fixation.

4. Nutritionists sometimes disagree as to whether water can be called a "nutrient." What are the arguments on both sides of this question? Which position do you favor?

5. Why is it incorrect to associate bacteria only with disease? What are some of the benefits we derive from bacteria?

6. Trace a grape through the human body from ingestion to excretion.

7. Explain the statement "A person can eat as much as he wants, be overweight, and still be suffering from malnutrition."

REFERENCES

Barrington, E. J. W. 1962. "Digestive Enzymes." *Biochemistry* 7: 1–65.

Beaton, G. H., and E. W. McHenry, eds. 1964–1966. *Nutrition: A Comprehensive Treatise.* 3 vols. Academic Press, New York.

Davenport, H. W. 1972 (January). "Why the Stomach Does Not Digest Itself." *Scientific American* Offprint no. 1240. Freeman, San Francisco.

SUGGESTIONS FOR FURTHER READING

Guyton, A. C. 1969. *Functions of the Human Body,* 3rd edition. Saunders, Philadelphia.

Detailed examination of the organs of the digestive system. Includes many excellent illustrations.

Jennings, J. B. 1973. *Feeding, Digestion and Assimilation in Animals,* 2nd edition. Pergamon Press, New York.

Comprehensive discussion of the subject, including many interesting examples of food-getting devices.

Lappe, F. M. 1971. *Diet for a Small Planet.* Ballantine, New York.

A comprehensive guide, written in familiar language, to the economics of protein nutrition. Includes chapters on protein theory, recipes, and tables of food value.

Larimer, J. 1973. *Introduction to Animal Physiology,* 2nd edition. Brown, Dubuque, Iowa.

Standard physiology text, with comprehensive discussion of digestion in ruminants.

Nyotti, S., and W. Dutty. 1971. *You Are All Sanpaku.* Award Books, New York.

A witty, easy-to-read text on macrobiotics, which also includes a useful summary at the end of each chapter.

Ray, P. M. 1972. *The Living Plant,* 2nd edition. Holt, Rinehart and Winston, New York.

Readable, concise description of plant functions.

Shepro, D. S., F. A. Belamarich, and C. K. Levy. 1978 (in press). *Human Anatomy and Physiology,* 2nd edition. Holt, Rinehart and Winston, New York.

Vander, A. 1975. *Human Physiology,* 2nd edition. McGraw-Hill, New York.

Standard reference work for physiology. Excellent illustrations.

INTERNAL TRANSPORT AND GAS EXCHANGE

chapter seven

Every organism must bring substances vital to life from the external environment to each cell in its body. In most higher forms of life, blood is the transport medium, carrying nutrients and oxygen to the cells. The lung of terrestrial animals and the gill of many aquatic forms provide the large surface across which oxygen diffuses into the blood and carbon dioxide diffuses out. The axolotl, a salamander of the southwestern United States, retains its particularly prominent gills, even as an adult.

Regardless of size, all organisms—from procaryote to man—face certain common problems. All must carry on active exchange of materials with their environment, taking from it nutrients—solid, liquid, and gas—and returning to it wastes and heat. These exchanges occur at the surface of the organism. The materials taken in through the surface must be distributed internally, and wastes produced in the interior must be brought to the surface for removal. Specialized systems evolved for exchanging materials at the surface and for transporting materials internally. In the last chapter we discussed the intake of solid and liquid nutrients from the environment and the production of solid wastes. In this chapter, we shall examine the structures that evolved for the intake and disposal of gases, as well as the mechanisms by which gases and digested nutrients are distributed or transported within the organism.

Exchange of materials is not a great problem for a unicellular organism, with its entire surface exposed to the environment. Neither is internal distribution of materials, for within the confines of a single cell, materials may be transported from organelle to organelle either by movements of the cytoplasm (cyclosis) or by diffusion. But as multicellular organisms evolved, the problems of transport of nutrients and gases to the interior of the organism became more complex. As the number of cells in evolving organisms increased, the number of cells which might face the external environment decreased. (For a graphic demonstration of this fact, see Fig. 7.1.) Many cells within a multicellular organism have no exposure whatever to the external environment; for them the environment consists solely of neighboring cells.

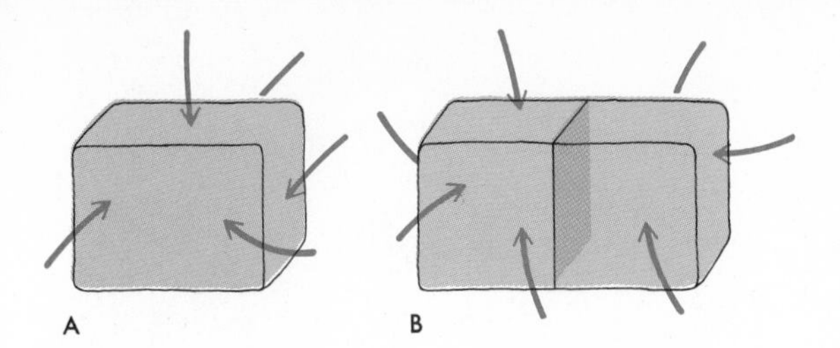

FIGURE 7.1

Surface-volume relationships. As multicellular organisms evolved, less and less of the surface of any cell was exposed to the outside environment. In a single cell (A), all sides of the cell may engage in exchanges of materials with the environment. In a two-celled organism (B), the common surface between the cells does not face the environment, reducing the surface area available to each cell for exchange. In a multicellular organism, many cell surfaces are necessarily internal and cannot carry on material exchange with the environment. In such organisms, systems of internal transport evolved.

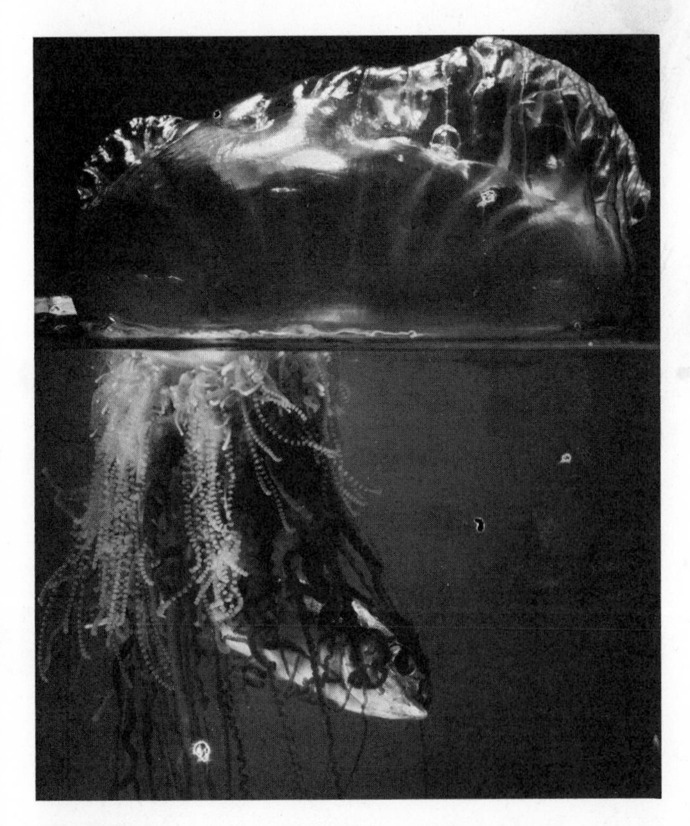

FIGURE 7.2

Portuguese man-o-war (Physalia). *Commonly known as a jellyfish, this organism is a colonial coelenterate made up of different types of cells hanging on long tentacles. Some cells have specialized to make the colorful gas-filled float, the pneumatophore, that buoys the remainder of the colony. Caught in the tentacles is a fish that has been killed by the stinging cells.*

INTERNAL TRANSPORT

When no direct exchange with the environment is possible, indirect exchange becomes essential; the external environment, or a facsimile of it, must be brought to every cell. And so, as evolution favored multicellularity, organisms developed internal transport systems capable of providing the proper environment for internal cells.

The complexity of internal transport systems that evolved is related to the complexity of the organisms they serve. The internal transport systems of higher terrestrial plants, systems that distribute effectively the water, dissolved nutrients, and gases required for life and growth, are highly developed. More complex yet are the transport systems in higher terrestrial animals, whose energy requirements are generally more demanding.

In the more advanced and complex animals, including the mammals, there evolved a unique distribution system—a system in which blood, the transport medium, travels through more than 60,000 miles of blood vessels to reach within 0.005 inch (or 0.12 mm) of every one of the billions of body cells.

GAS EXCHANGE

As part of its ongoing life processes, a living cell is continuously engaged in the chemical breakdown of stored food to obtain energy. In the cells of animals and most plants, this breakdown involves the chemical combination of the food with oxygen and produces, as one of its waste products, the gas carbon dioxide. Many living cells are thus constantly taking in oxygen and giving up carbon dioxide, as they perform the process of respiration.

We have seen that green plants manufacture their food through a process called photosynthesis. This process utilizes the energy of sunlight to promote the chemical buildup of nutrients from water and carbon dioxide, and oxygen is the waste product. Since all life depends on this continual process of photosynthesis, it is clear that oxygen and carbon dioxide play exceedingly important roles in the energy economy of our ecosphere. The means by which these two important gases enter and leave organisms is known as **gas exchange.**

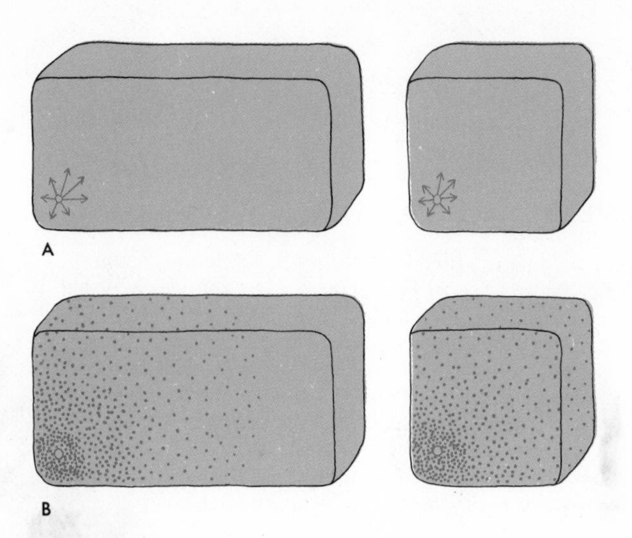

FIGURE 7.3

Cell size and diffusion. The ability of a cell to use diffusion alone for the distribution of molecules is partly dependent on the size of the cell. (A) A drop of a diffusible substance is shown in the corner of each of two cells, one larger than the other. (B) After a period of time, the diffusing substance in the larger cell has not yet spread to the end of the cell. In the smaller cell the diffusing molecules have reached the edges of the cell.

The basic mechanism of gas exchange in all organisms is the diffusion of gases across moist membranes. In individual cells, gases can be easily transported across the cell membrane, so long as the cell membrane is moist and the gases are in solution. This diffusion of gases into and out of a cell follows simple physical laws. For example, in cellular respiration carbon dioxide is produced inside the cell. When the internal concentration of the gas exceeds the external concentration, carbon dioxide diffuses out. Similarly, as oxygen is used up, the internal concentration of the gas falls below the external concentration, and oxygen diffuses inward. In those green plant cells where photosynthesis is also taking place, more carbon dioxide than oxygen is used, so that the net concentration of gases is the opposite of that in cells where respiration only is taking place. In photosynthesizing cells, it is therefore carbon dioxide that diffuses into the cell and oxygen that diffuses out.

In the atmosphere, availability of the two gases poses no significant problems for gas exchange. The atmosphere contains approximately 21 percent oxygen, enough to satisfy the respiratory needs of any organism that breathes air directly. And although air is only 0.03 percent carbon dioxide, the absolute amount of carbon dioxide is still sufficiently large to satisfy the photosynthetic needs of all land plants.

In water, the conditions for gas exchange are slightly different. Carbon dioxide is highly soluble; it is abundantly available to meet the photosynthetic needs of aquatic plants. Oxygen, however, is poorly soluble in water, and the oxygen balance of a small or medium-sized body of water is easily upset. The outflow of heated water from a power plant (thermal pollution) kills fish because heated water cannot hold as much oxygen as cold water. A second way in which the oxygen balance in water can be upset involves **eutrophication**—from a Greek word meaning "well nourished." If high-phosphate detergents or agricultural fertilizers, both of which contain chemicals that promote plant growth, are washed into a lake, a rapid overgrowth of algae will result. The algae die at the end of each growing season, sink to the bottom, and are decomposed by bacteria that use up most of the dissolved oxygen in the lake to satisfy their respiratory needs. Without sufficient oxygen, the other living things in the lake die.

Simple organisms—with simple structure, low volume, and ample surface area for gas exchange—handle their gas exchange problems easily and efficiently. But large, complex organisms have gas exchange mechanisms that are coupled to those of internal transport of nutrients and wastes. Animals have gills and lungs, both of which rely heavily on the blood circulation system for their proper functioning. In plants, insects, and certain simple organisms, other mechanisms have evolved.

Unicellular organisms

For the simplest of organisms, material interchange with the environment, as we have seen in Chapter 6, is no problem. Such organisms may employ any of several methods of exchange that involve a minimum of complexity. Gases are obtained through diffusion.

To accomplish internal transport of gases and other materials, unicellular organisms may rely on diffusion. The time required for diffusion depends, of course, on distance, as suggested by Fig. 7.3. Over the small distances within a single cell, diffusion time is brief and the process is practical for distributing nutrients to all areas.

Another method of internal transport that is utilized by unicellular organisms is **cyclosis**—an internal streaming of the cell cytoplasm. Cyclosis can be readily observed in the locomotion of the amoeba. The amoeba moves by pushing out a pseudopodium—a kind of temporary "false foot" into which the cytoplasm flows. The currents of cytoplasm contain droplets and other nutrient particles so that as the amoeba moves, nutrients are circulated. Cyclosis also can be observed in plant and animal cells that do not alter their form and that do not move about (Fig. 7.4). In these cells, the streaming of cytoplasm serves, perhaps exclusively, to distribute nutrients.

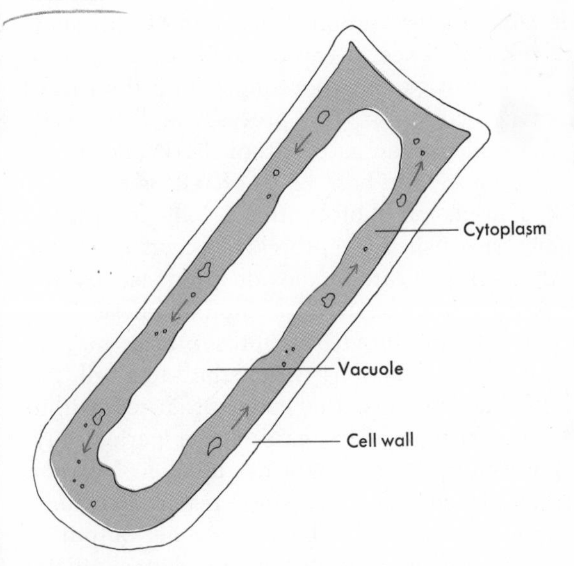

FIGURE 7.4

Cyclosis. In many plant cells cyclosis, or cytoplasmic streaming, is quite easily seen. In the very large cylindrical cells of the alga Nitella, *cyclosis carries many particles swiftly around the cell.*

Lower multicellular plants

The internal transport and gas exchange problems of simple aqueous forms of plants, such as algae, are almost as easily solved as those of unicellular organisms.

INTERNAL TRANSPORT

The cell walls of the simpler plants have many capillary spaces that facilitate diffusion of gases and nutrients into and out of the organism. The algae, moreover, consist of many cells that are photosynthetic. These produce carbohydrates from carbon dioxide and water under the influence of light. Since the cells that do not photosynthesize are located near those that do, they obtain their nutrients by diffusion from the photosynthetic cells. The process of diffusion probably provides nutrients even to the huge number of cells in *Ulva*—the large flat green sheets of sea lettuce often washed onto marine beaches. Despite the size of the sheets, *Ulva* is very thin—only two cells thick—and diffusion into its cells through its huge surface is quite easy.

As multicellular terrestrial forms evolved, internal transport of materials inevitably became more difficult. Water is the medium of transport used by aqueous plants, yet water is not so readily available on land. Terrestrial forms, then, needed specialized structures to search out water. Yet structures with special functions would obviate the possibility that transport could be accomplished by the simple process of diffusion; specialized transport facilities were needed.

Nature can and often does advance in small steps. For plants, the first type of terrestrial environment was very wet soil of the sort found along estuaries and rivers. In fact, such lower terrestrial plants as the bryophytes (mosses and liverworts) are to be found only in moist environments, usually growing flat against the damp soil. Moisture and dissolved mineral nutrients are absorbed by cells that are in direct contact with the soil, and both diffuse to the internal cells. This enables the bryophytes to live in a terrestrial environment without requiring a highly developed system for internal trans-

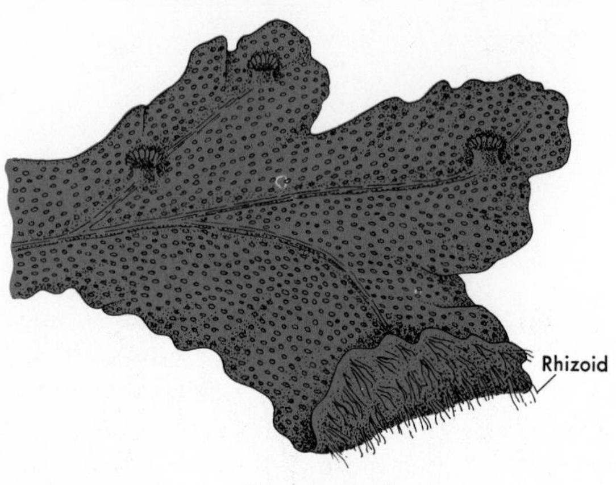

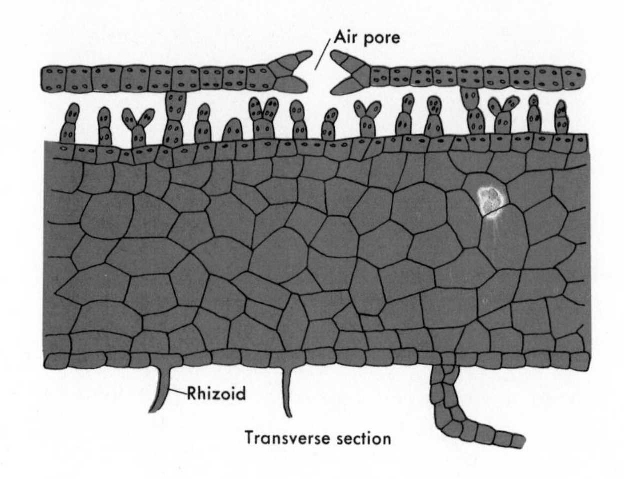

FIGURE 7.5

The bryophytes are simple, flat plants, up to 30 cells thick, that live in moist soil. Although they lack true roots and internal transport systems, they do have rhizoids, rootlike structures that anchor the plants in soil and probably help in water absorption by increasing the surface area. Gas exchange is also simple. The top surface of the plant has pores that permit access of air to the spongy inner cells.

port (see Fig. 7.5). Although some bryophytes possess **rhizoids** (simple rootlike structures) that take water from deeper soil layers, bryophytes do not possess a special internal transport system. Fungi generally handle transport in an equally simple fashion.

GAS EXCHANGE

Naturally, land plants also evolved solutions to the problem of gas exchange. One such solution is illustrated by the liverwort *Marchantia*. Although its plant body is thin, the liverwort is nevertheless more than two cells thick, and so diffusion of gases into the plant's outermost cell layers is not adequate to satisfy the plant's gas exchange needs. The problem of increasing available surface area has been solved for *Marchantia* by the process of **invagination;** that is, numerous small internal air chambers are connected with the environment by means of tiny pores. Inside the air chambers, the gas exchange needs of each individual plant cell can be satisfied.

Gas exchange in higher plants

The invagination principle reaches the height of complexity in the vascular plants, or **tracheophytes**—the club mosses, ferns, conifers, and flowering plants that are the most highly evolved plants on earth. Each part of the plant—each leaf, stem, and root—has its own set of internal passageways and openings to the environment. There is no single, interconnecting *gas* transport system or any special organ of gas exchange, although as we shall see, there are systems of transport of other nutrients.

In vascular plants, both gas exchange and photosynthesis are carried on primarily in the leaf, which consists of both photosynthetic and nonphotosynthetic cells. During periods of illumination, when photosynthesis is occurring, the leaf takes in carbon dioxide, the photosynthetic cells fix it by incorporating it into the carbohydrates manufactured in the process, and the cells release water and oxygen as by-products. While this is happening, respiration in both kinds of cells is going on. Some of the needs of photosynthetic cells for carbon dioxide may be satisfied by processes occurring inside the cells themselves—carbon dioxide produced by the metabolism of food can be used for photosynthesis. Similarly, any oxygen produced by photosynthesis may be used for respiratory functions within the same cell. Although the cells may use the gases produced within them, the net exchange of the gases varies periodically. During photosynthesis there is a net uptake of carbon dioxide from the environment, since respiration alone cannot suffice as a source; and there is a net output of oxygen, since the oxygen production in photosynthesis is greater than the amount used in respiration. During nonphotosynthetic periods, such as nighttime, the net exchange of gases is reversed. Oxygen must be taken from the environment to meet respiratory needs, and the carbon dioxide that is produced is released.

Leaves have evolved specialized openings, or **stomata** (singular: **stoma**), through which gases pass back and forth between the outside air and the intracellular spaces within the leaf. Stomata are found scattered within the epidermal layers of the leaves of all vascular plants (Fig. 7.6). To conserve water (which is lost by evaporation), stomata are usually more numerous on the underside of the leaf than they are on the side directly exposed to the sun and wind. The number of stomata on a leaf varies enormously from one plant to another. A begonia, for example, may have an average of only 4,000 stomata per square centimeter of leaf surface, whereas a scarlet oak has 100,000 in the corresponding area. Plants that have adapted to survive in hot, dry climates may have as few as 1,000 stomata per square centimeter of epidermis, yet some plants that live in tropical rain forests have almost 130,000.

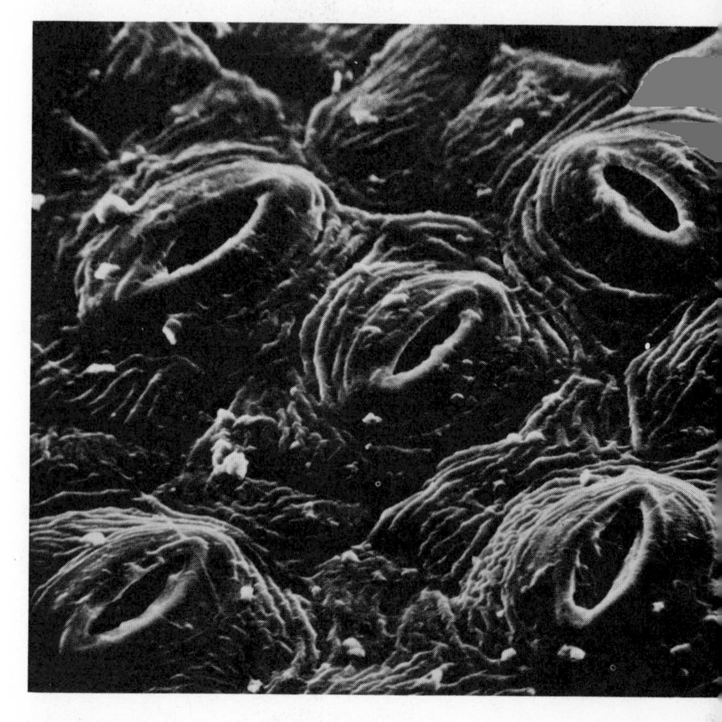

FIGURE 7.6

Scanning electron micrograph of stomata on the lower surface of a leaf. The opening and closing of a stoma is the result of changes in the turgor of its guard cells. When turgor is low, the cells become flaccid and sag, causing the lips of the stoma to close. When turgor is high, the guard cells fill, and the pore between them opens. It is quite easy to manipulate experimentally the opening and closing of stomata, but we still do not completely understand how changes in turgor in response to water loss occur naturally.

The stomata are the primary mechanisms for the regulation of water loss. Of the great quantities of water lost from the leaves and stems of plants, more than 90 percent is lost through the stomata, and less than 10 percent from the cells of the leaf epidermis. This small water loss from the epidermis is largely a result of the presence of **cutin,** a waxy substance that coats the cells of both the upper and lower epidermis of the leaf. The thickness of the epidermis may also prevent water loss; it may be several layers thick in plants like *Ficus elastica,* the India rubber tree that can be successfully cultivated in dry, air-conditioned buildings (because there will be no appreciable water loss).

The stomata of vascular plants are somewhat more complicated structures than the simple pores found in such plants as liverworts. Each stoma is bordered by two special epidermal cells, called **guard cells,** shaped roughly like a set of parentheses. Unlike other epidermal cells, guard cells contain chloroplasts and are thus sites of photosynthesis. Nutrient production by photosynthesis within guard cells results in an altered osmotic relationship with neighboring cells, bringing about changes that trigger the behavior of the guard cells. When photosynthesis stops, as at night, the turgor of the guard cells decreases, the cells assume a flaccid position, and the stoma closes. When photosynthesis resumes, the guard cells become turgid again, and the stoma opens. However, heavy water loss by evaporation, due to excessive heat during the daytime, will cause the stoma to close again.

The opening and closing of the stomata have the apparent effect of conserving as much water as possible. They open during the day to permit the gas exchange necessary for photosynthesis. When daytime heat and dryness become excessive, the stomata close, preventing too much water loss from the vital photosynthetic cells

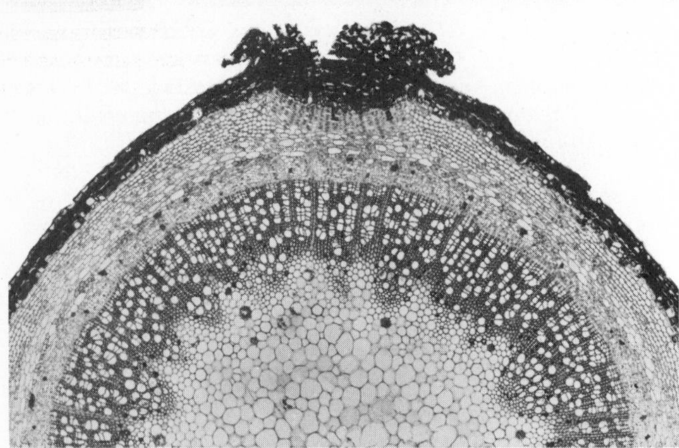

Courtesy of Carolina Supply Company

FIGURE 7.7

Gas exchange in mature trees. (A) Lenticels in the bark of trees like this birch take the form of slashes through which cork cells protrude. (B) In cross section, the lenticel appears as a pore in the otherwise impermeable bark. The loosely packed cork cells spilling through the opening permit an exchange of gases for the living tissues.

of the leaf. The stomatal mechanism thus balances the gas exchange needs of the plant against the possibility of permanent damage from excessive loss of water.

In the stem of a grassy or herbaceous plant, or even in the stem of a very young tree, gas exchange takes place as it does in the leaf—through stomata. But in older, woody plants, the thin epidermis of the stem is replaced by cork cells that form bark. As the bark forms, other tissue changes also take place. Beneath each stoma, the cork tissue forms masses of loosely packed cells through which gases can easily pass. Finally, these cells break through the outer layer of bark to form the special structures known as **lenticels.** In bark-covered stems, lenticels take the place of stomata as organs through which gases can pass. Since their openings are packed with cells, albeit loosely, lenticels probably do not permit gas exchange as readily as stomata. But since the living tissues in a woody stem are confined to a very thin outer layer, the distances through which diffusion must occur are relatively short, and lenticels suffice as pores for gas exchange. Lenticels may form very prominent markings on the bark of certain trees, such as the long horizontal slashes that are so characteristic of the birch (Fig. 7.7), or the raised horizontal dots of the cherry.

Transport in higher plants

Although the gas exchange mechanisms of higher plants evolved without any direct relationship to a system of internal transport, movements of dissolved nutrients within the plant imposed special problems. Taller plants had the advantage of obtaining more sunlight, but this particular form of vertical evolution introduced new transport problems. Cells at the top of a tall plant would get most of the sunlight and do most of the work of photosynthesis, but cells nearer to the ground, distant from the site of sugar production, had to be supplied with nutrients. Moreover, an increase in plant height required greater anchorage in the soil, and the resulting extension of the root system increased the distance over which nutrients needed to be transported. Fortunately the increased surface area provided by these roots may have enhanced their ability to absorb water, so that tall plants could inhabit relatively dry land.

OBTAINING WATER: ROOTS

Lower tracheophytes (club mosses and horsetails) are among the earliest of the higher plants. They grow to several feet in height but maintain a relatively simple architecture. They are commonly found in moist, shady places where water is relatively abundant and water losses by evaporation are minimal. Thus lower tracheophytes require only a simple root system. The roots of these plants grow out from horizontal stems called **rhizomes.**

Higher tracheophytes are more complex plant forms and require root systems that are more elaborate. Higher tracheophytes are the dominant land plants and include gymnosperms (cone-bearing plants such as pines and firs), and angiosperms (flowering plants such as hickory, walnut, and other dicot trees; soft species such as tomato and carrot; monocot "trees" such as palms and bananas; and herbaceous species such as lawn grass, corn, and onion).

The roots of these higher tracheophytes, often extensive, are of two types. Most of the dicots have a taproot system. The first root formed, called the taproot, is dominant, and the remaining root structure branches out from it. The monocots characteristically have a fibrous root system. This system has no dominant root but is rather a highly branched structure arising **adventitiously,** or from the stem. Both root systems are illustrated in Fig. 7.8. In both dicots and monocots the multibranched root system provides much absorptive surface. Yet an increase in surface area is not altogether without liability, for the distances water must travel to other parts of the plant are increased proportionately.

In order to perform their vital function of transporting water and nutrients over great distances, the cells of a root must absorb oxygen so that they can metabolize foods and stay alive. Oxygen is also required for their growth, which provides new root surface to accommodate the needs of a growing plant. Thus roots typically absorb oxygen and produce carbon dioxide as a waste product. Despite the location of roots beneath the earth, gas exchange is not necessarily a problem, for good soil, in which most plants can grow, contains subsurface air held in pores in the soil. The principal barrier to proper root aeration is, paradoxically, an excess of water—another substance vital to plant growth. Too much water in the soil will reduce the amount of air available by occupying needed air spaces. Although many plants can survive a short time without absorbing oxygen through their roots, the eventual result is the death of the root.

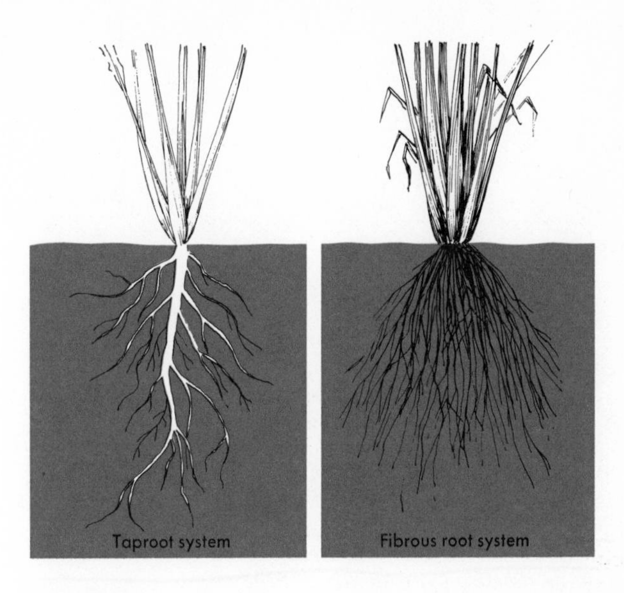

FIGURE 7.8

Root systems. In most dicots there is a prominent central taproot from which all other roots in the system branch. The monocot root system is typically diffuse, with no taproot.

Certain plants do have special mechanisms that enable them to survive with their roots under water. In some water plants, air spaces extend from the leaves down into the roots; in effect, the roots breathe through the leaf stomata. Other plants—the black mangrove, for example—have special breathing tubes called **pneumatophores** that extend above ground from their roots and presumably absorb oxygen directly from the air. However, the process of root respiration, particularly in water-saturated soil, is not yet thoroughly understood.

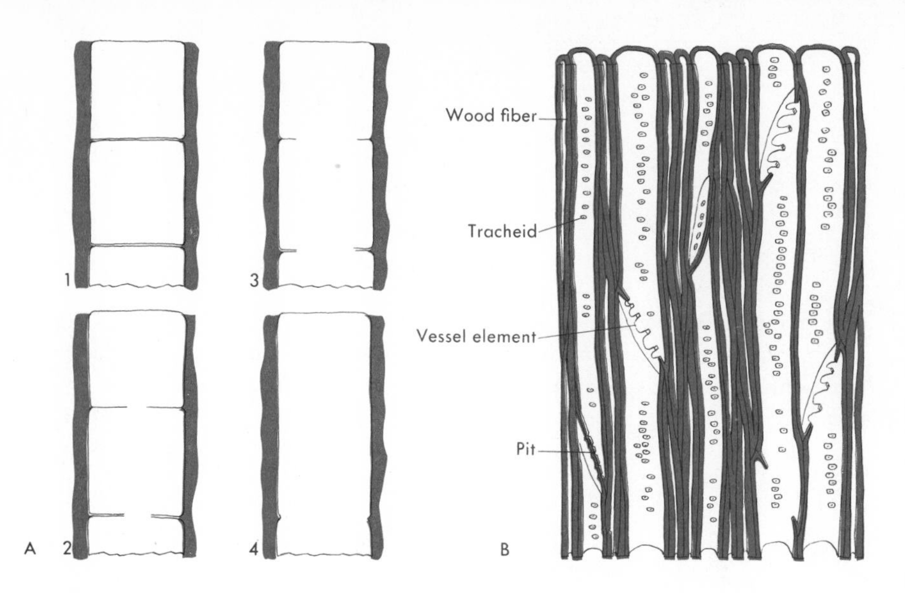

FIGURE 7.9

Water and nutrient transport. Tracheids and vessels, which transport water and dissolved minerals from roots to leaves, are formed in meristematic tissues located in roots, stems, and leaves. (A) A vessel, which is a tube composed of individual xylem cells called vessel elements, is formed when the living part of each cell in the tube digests its end walls, as shown in the sequential drawings. After this maturational step, the living part of each cell dies, leaving only the pipeline (4) consisting of the tubular cell walls. (B) The relationships of different cells in xylem tisue of angiosperms. The widest cells, with no end walls, are vessels. Narrower tracheids, with pitted end walls, carry water also. Fibers, which have virtually no internal space to transport water, seem to act as support for xylem tissue.

CONDUCTIVE TISSUE: XYLEM

Once inside the root system, water (and the minerals dissolved in it) must be transported upward to the leaves, which in tall trees may be almost 400 feet above the ground. Water must also be routed to all cells along the way, for 95 percent of the composition of all cells is water, and cells therefore require water replenishment. The solution to the problem of water transport was the evolution of xylem conductive tissue—familiar to all of us as wood.

Xylem may consist of two types of specialized conductive cells: tracheids and vessel elements. A tracheid, the more primitive type of conductive cell, is an elongated structure that may reach 5 mm in length. This is several hundred times as long as the meristematic cell, the original undifferentiated embryonic cell from which the tracheid develops. A vessel element, though also elongated, is shorter and wider than a tracheid. In both types of cells, a very stiff secondary wall is deposited. The secondary wall consists largely of cellulose microfibrils that brace the wall against collapse. As the secondary wall is formed, lignin and other waterproofing materials are deposited.

In tracheids, thin circular spots or depressions, called **pits,** form on the inner face of the protoplast. Pits, however, are not holes. The primary wall of the tracheid remains intact and continues to be porous to water. On the other hand, the end walls in vessel elements are either perforated or dissolved completely. Both tracheids and vessel elements are the remains of dead cells; in each type of cell the protoplast dies after laying down the primary and secondary walls, leaving a waterproof tube. The space within the tube, which was formerly occupied by the protoplast, is now free for water conduction.

Tracheids and vessel elements are arranged in long strands, as shown in Fig. 7.9. Although water passage is easier through vessel elements because of the greater porosity (or even absence) of their end walls, the conifers —some of which, the California coast redwoods, are the tallest of all trees—nevertheless depend solely on tracheids for upward transport of water. In more familiar angiosperm trees such as the maple, and herbs such as common lawn grass, xylem contains both tracheids and vessels (strands of connected vessel elements) for water conduction.

Support Plants obtain the structural support they require largely from the thickened, lignified conducting cells in xylem. Xylem tissue also may contain fibers—tracheid-like cells with lumens, or inner channels, which are too narrow to transport significant quantities of water, but which serve the plant as additional supporting rods. Moreover, the role of xylem as a supporting tissue is critical to the success of photosynthesis. Xylem cells form a network throughout the blades of leaves; in a

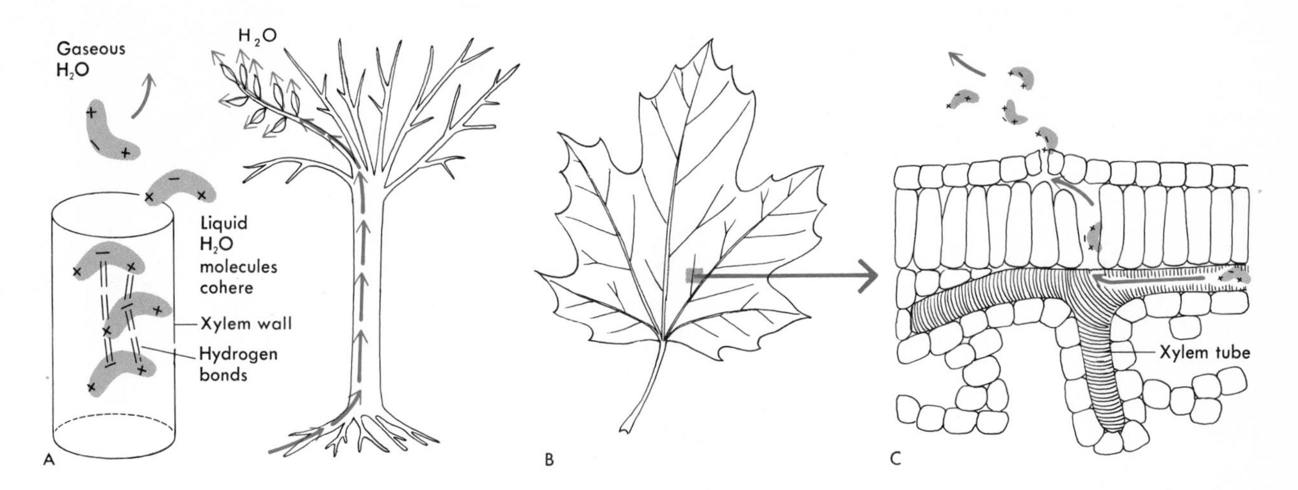

FIGURE 7.10

Transpiration. Water moves from the roots of a plant toward the leaves because of the forces of transpirational pull. (A) The unbroken column of water molecules that reaches from the roots to the leaves seems to behave like a long, cohesive thread because of hydrogen bonding between hydrogen and oxygen in nearby water molecules that hold the molecules together. (B) In the leaf, xylem tubes branch extensively and ramify the tissues of the leaf. (C) Water molecules leave the ends of xylem cells by evaporation, enter the air spaces between cells, and then depart through the stomata.

single square centimeter of leaf area, thousands of xylem tubes may terminate. These tubes serve as conduits of water to the leaf. They also support the blade so that its flat shape is maintained and it remains properly oriented to capture the sunlight necessary to maximize the efficiency of photosynthesis.

MOVEMENT OF WATER

The xylem system begins just above the root tips and extends, with much branching along the way, to leaves. Water with its dissolved minerals is able to flow freely (by the mechanisms described in Chapter 6) through the open lower ends of xylem cells. But how does water ascend? How do the huge volumes of water required by large trees move many feet—in some trees, several hundred feet—upward?

Water molecules have polarity, which allows them to cohere or to form weak bonds with each other. Indeed, in a very narrow tube, such as that formed by xylem cells, cohesive forces can give a column of water more tensile strength than that of a steel wire having the same cross-sectional area. As water molecules evaporate from the end of each xylem water column in the leaf and pass out through the stomata, the column's water level rises. The water molecules—linked together by cohesion—are apparently pulled upward from the roots, and they then diffuse into the area of the xylem tube vacated by the molecules that evaporated. This evaporation process in leaves, shown in Fig. 7.10, is called **transpiration.** In most plants, 98 percent of all water entering the plant is lost by transpiration; only 2 percent is actually used.

At one time scientists thought that roots might absorb water from the soil and push it upward, developing so-called root pressure. Most of us are familiar with the effects of root pressure: walking barefoot in the grass early in the morning we become aware of a pleasant wetness. This is partly the result of guttation; droplets of water are forced out of special openings along the leaves by root pressure. But laboratory measurements show that root pressure is small, only strong enough to push water about one-quarter of the way up the trunk of a tall tree. And scientists have accumulated other evidence which minimizes the importance of root pressure in water transport during the day. The trunk of a tree actually gets thinner during the day, showing that water is being drawn up under suction, a process that collapses the xylem cells slightly. Since water forced into the xylem tubes under pressure from the roots would tend to make them thicker (just as air forced into a long narrow balloon makes it thicker), it must be transpiration that *pulls* water up the stem.

Water can move through xylem at speeds up to 75 cm or more per minute. The rate at which water

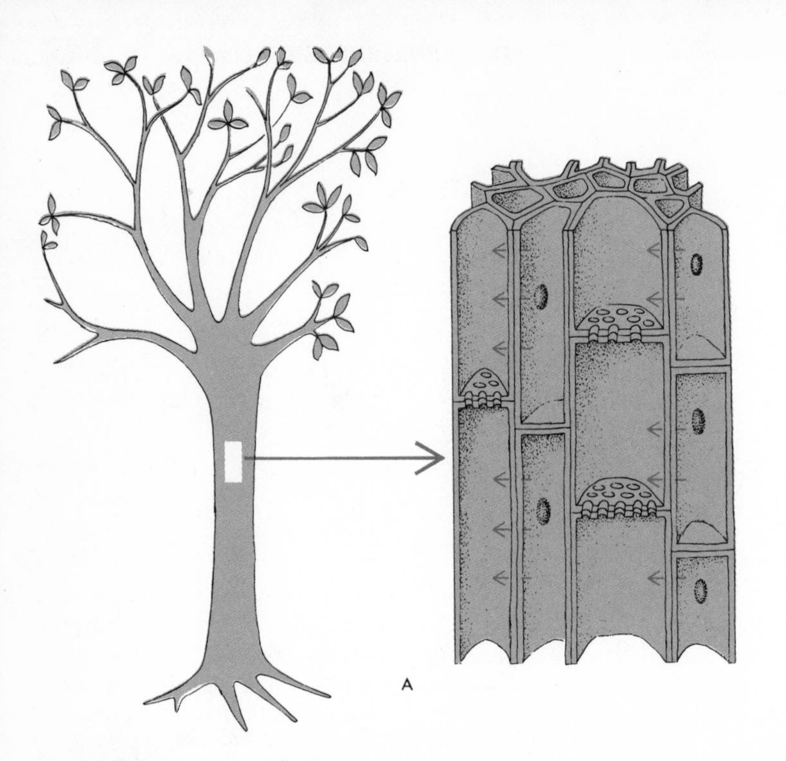

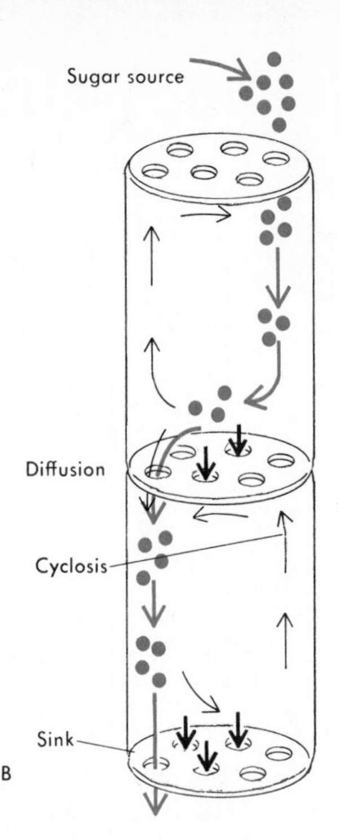

FIGURE 7.11

Nutrient transport. (A) Many organic nutrients are transported in the phloem, a system of cylindrical sieve cells (called sieve elements in angiosperms) connected end to end to form a living pipeline. At the ends of each cell in the tube are sieve plates, peculiar perforated cell walls that seem to permit movement of materials from cell to cell. In angiosperms each sieve element has a companion cell, which appears to supply the enucleate sieve element with vital materials (colored arrows) necessary for transport activities. (B) In the two sieve elements shown, colored arrows indicate the presumed mechanisms of transport of nutrients such as sucrose from cell to cell in a tube. At the source end of a tube, cells such as leaf mesophyll produce sugar, which enters the cytoplasm of the phloem tube. Other cells serve as a sink. Nutrients move toward a sink, probably by the diffusion and cyclosis that occur within the phloem sieve elements.

moves through the plant is linked to the rate of water loss through transpiration. Water transport, however, may lag behind the rate of transpiration. It may be several hours after transpiration has peaked before the volume of water that evaporated is replaced. This accounts for the phenomenon known to many gardeners in which, on a hot summer afternoon, many plant leaves wilt temporarily. The reason the leaves wilt is that the rate of transpiration exceeds the rate of transport. Thus water is transpired from the leaves more quickly than it is replaced. At night, however, even if the plants are not watered, the leaves fill out again as the rate of transport exceeds the rate of transpiration.

NUTRIENT CONDUCTION: PHLOEM

In addition to the need for internal transport of water and minerals, plants require the transport of many other nutrient materials, including those produced by photosynthesis in the leaves. These other materials are translocated (transported) by phloem, another conductive tissue that like xylem, is distributed vertically and laterally throughout the plant. Unlike xylem, phloem cells function only while they are alive.

There are two types of phloem cells in an angiosperm, or flowering plant: sieve elements, cells which function in transporting materials, and companion cells, which appear to carry on vital metabolic processes for which the sieve elements are not equipped. A vertical row of the elongated sieve elements forms a **sieve tube.** Characteristically, sieve tubes contain **sieve plates,** definite groupings of pores in the end walls of the sieve elements. The pores also are occasionally found in the side walls.

Although the actual mechanism by which carbohydrates and other nutrients are transported in phloem is still not clear, Fig. 7.11 suggests current thinking on the subject. In some fashion—possibly by diffusion, possibly by cyclosis—bulk movement is accomplished. Indeed, given the small size of the pores in the phloem conducting elements, the quantity of material that is moved into fruits, tubers, or fleshy roots is awesome. For example, material weighing more than a dozen grams is moved into a pumpkin during each day of a 33-day growing period. This material is moved through the phloem in one slender stem that connects the fruit with the vine. Using materials that were radioactively tagged, investigators have measured movement rates greater than 21 cm per hour in the phloem of cotton and up to 100 cm per hour in the phloem of some forest trees.

Essentially, the conduction system in plants is not a circulating system, as it is in animals and man. Conduction in plants is, rather, a process of *source* and *sink:* materials are dumped into one end (the source) and are removed at the other (the sink). This type of system, however, meets plant needs efficiently.

Gas exchange in animals

The gas exchange and transport mechanisms in animals involve many of the same basic principles as these processes do in plants. In both, gas exchange consists of an exchange of oxygen and carbon dioxide by diffusion across a moist membrane, while transport within cells may be effected by diffusion. However, as in plants, the evolution of multicellularity favored the evolution of complex transport systems capable of bringing gases and nutrients to cells, and carrying metabolic wastes away for disposal. The gas exchanges of animals are different from those of plants, in that animals take in only oxygen, whereas plants also absorb carbon dioxide. For an animal, carbon dioxide is a metabolic waste product to be expelled from the body.

The gas exchange mechanisms of an animal may depend on whether it lives in air or water. Although air is rich in oxygen (about 21 percent at sea level), it tends to dry out the membrane across which gases must diffuse. And water, though obviously a source of abundant moisture, is a poor solvent of oxygen. Even oxygen-rich water contains only about one part of oxygen to 100 parts of water.

Many water-dwelling animals are able to exchange gases across a simple body-surface membrane. Simpler invertebrates, such as the ribbon-shaped flatworms, rely in part on surface diffusion, which is in principle similar to the method used by the sea lettuce *Ulva*. Other aquatic invertebrates utilize the principle of invagination. Some simple animals, like hydra, jellyfish, and planaria, have a well-defined system of internal spaces constituting a digestive cavity that, in addition to its digestive function, provides another surface through which gases can be exchanged. Moreover, the starfish and their relatives have a complex system of branching tubes through which oxygen-laden water is conducted into their body interiors, thus exposing a maximum of surface area to the sparsely oxygenated water. The examples cited indicate rough parallels between many gas exchange mechanisms in plants and animals. Higher animals have developed more complex systems of transport and gas exchange—often intimately linked—for rapid conduction of necessary materials to the sites where they are used.

Internal transport in animals

Animals have nutrient requirements that are more varied than those of plants. Moreover, the locomotion of animals consumes much energy and makes rapid transport a necessity. Thus, there were evolutionary selection pressures for more and more complex transport mechanisms, such as a reusable, circulating fluid from which cells could withdraw nutrients and into which they could dump wastes, a vascular pump to keep the fluid circulating, and channels to bring the fluid to areas where its constituents are needed.

The fluid that evolved and met the transport requirements of animals is blood; the pump is the heart; the channels that circulate the blood are called blood vessels (although in many forms indistinct channels called sinuses carry blood). Among genera of animal organisms, the structure and composition of the blood, heart, and vessels are extremely variable.

BLOOD

Blood is a complex mixture of water, dissolved chemicals, and cells. The exact proportions of water, dissolved proteins and inorganic ions, cells of various types, and other substances, is unique for a given species. For most animals, the concentration of inorganic ions in the blood is similar to that found in ancient oceans—a fact that suggests the marine origins of virtually all forms of life. Protein concentrations are fairly constant—up to 10 percent (weight per volume)—in all species. The blood composition of most species includes several kinds of proteins, but the blood of cephalopods (such as the octopus or squid) is an exception. Cephalopod blood consists of only one protein, hemocyanin, which carries oxygen to tissues. Insect blood similarly differs from the blood of most other animal species in that it contains protein enzymes that promote the digestion of tissues during metamorphosis but does not have the ability to bring oxygen to tissues.

As organisms evolved and their physiological processes became more complex, the oxygen and waste-carrying ability of blood became more efficient. In such animals, gas exchange is a multistep process. First, oxygen is absorbed from the environment by the bloodstream; then it is transported to the cells, where it passes from the bloodstream into the cells. Simultaneously, carbon dioxide passes from the cells into the bloodstream, which transports the gas to where it can be exchanged with oxygen.

Thus evolved several different gas-carrying pigments—colored conjugated-protein molecules containing a variety of metal ions, such as copper and iron. These components of the blood, in organisms as diverse as arthropods (such as crabs) and mammals, vastly improved their gas-carrying efficiency and permitted better aeration of their tissues. These conditions were vital for the development of rapid movement, better coordination, and a complex brain and nervous system. Eventually, special blood cells evolved that carry these pigments.

In many higher organisms, such as vetebrates, special cells called erythrocytes contain the gas-carrying pigment—hemoglobin. But in lower forms, such as insects, cells of another type—called **haemocytes**—found in the blood have a different function. Haemocytes collect at the sites of injuries, where they form plugs to help seal wounds, and where they proliferate to remove dead and discarded cells by phagocytosis. Haemocytes also function to defend the insect from invading organisms. They surround and engulf parasites that enter the insect, much as bacteria that infect a man are attacked by the white blood cells in his bloodstream. A variety of different kinds of cells, with different functions, evolved as organisms, and their needs became more complex. Vertebrate blood, as we shall see in the discussion of human blood, is relatively complex in structure and function, and it contains several different types of cells, as well as many noncellular components.

SPECIALIZED STRUCTURES FOR GAS EXCHANGE

Many of the higher animals possess special structures through which gases are exchanged between the environment and the bloodstream. In water-living animals, these structures are called **gills.** In fish, gills are **evaginated;** they are branched-out portions of the circulatory system that form a very fine, highly complex network of blood vessels sandwiched between membranes. It is across the membranes that the diffusion of gas between the water and the fish's bloodstream occurs (see Fig. 7.12).

A particularly efficient mechanism, known as **countercurrent flow,** enables fish gills to extract a maximum amount of oxygen from the water. In simple terms, countercurrent flow refers to the fact that water flows across the gills in a direction opposite to the flow of blood within the gills. Thus blood entering the gills encounters water that has already given up most of its oxygen. Because this blood is so poorly oxygenated, oxygen diffuses into it quite readily, despite the low oxygen concentration of the water.

Gills are sufficient to supply the oxygen needs of fish and even larval amphibians, such as tadpoles and baby salamanders. But out of water, gills cannot function because their unprotected, evaginated membranes

A

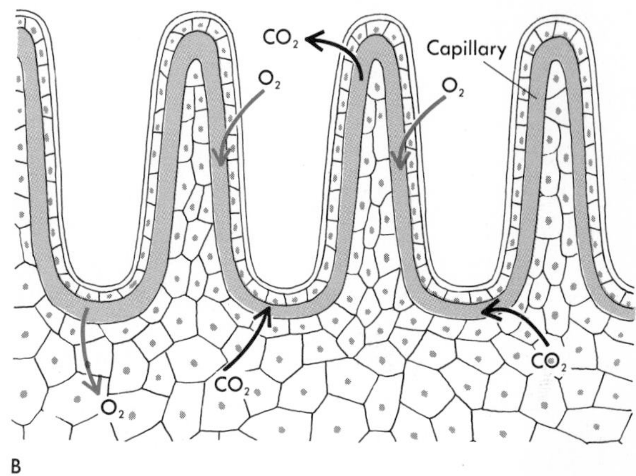

B

dry up. In addition, a terrestrial habitat does not provide the buoyancy of water, and the gills collapse.

Land-dwelling animals have developed gas exchange mechanisms other than gills. The various mechanisms that evolved have met the metabolic needs of specific organisms. The simple diffusion of gases across a moist body-surface membrane—the mechanism used by earthworms—is not adequate for larger animals. Frogs get some of their oxygen through their skin, but few other large land animals are able to do so.

A structurally and functionally simple gas exchange mechanism is used by the insects—the commonest of all land animals. The body of an insect is permeated with branching air tubes, called **tracheae**, which open to the exterior through pores called **spiracles.** The tracheae conduct gases deeply enough into the body to ensure intimate contact of all body cells with the required amount of oxygen. Simple muscular exertions by the insect are sufficient to pump air into and out of its body. The insect's blood has no gas transport function; it transports only nutrients and wastes. The principal disadvantage of the system is that, because air enters the finest inner tracheae by diffusion rather than by muscular action, and diffusion cannot take place at a useful rate over more than a few millimeters, the size of insects in quite limited. An insect the size of a dog would be an impossibility, since the cells of the body would receive inadequate supplies of oxygen (see Fig. 7.13).

Some spiders and scorpions, distant relatives of the insects, handle their gas exchange needs slightly differ-

FIGURE 7.12

Gas exchange. (A) A marine tubeworm extends a beautiful fan of delicate gills that filter food and exchange gases between its bloodstream and the surrounding water. (B) Generalized structure of a gill. The gas exchange surface, in contact with the water, is a very thin and sometimes flattened filament or protrusion. Blood-carrying capillaries (color) reach into each filament. Oxygen diffuses into the blood from the water and then may be transported to all cells in the organism. Carbon dioxide produced by cells is transported to the gills by the blood and diffuses into the water.

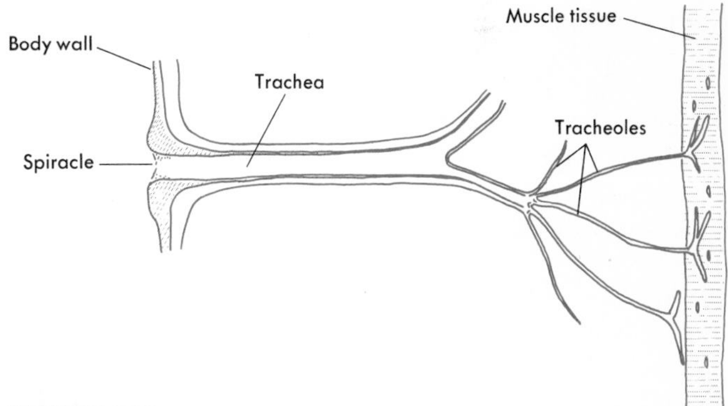

FIGURE 7.13

Trachea of an insect. An insect trachea is an invaginated tube that branches to form very small tracheoles, whose walls are permeable to gases. Oxygen enters the external opening, the spiracle, and diffuses through the tracheoles into the internal cells. Carbon dioxide leaves via the same route. The gas exchange is direct between the tracheal tube system and the tissues, since no gases are carried by the blood.

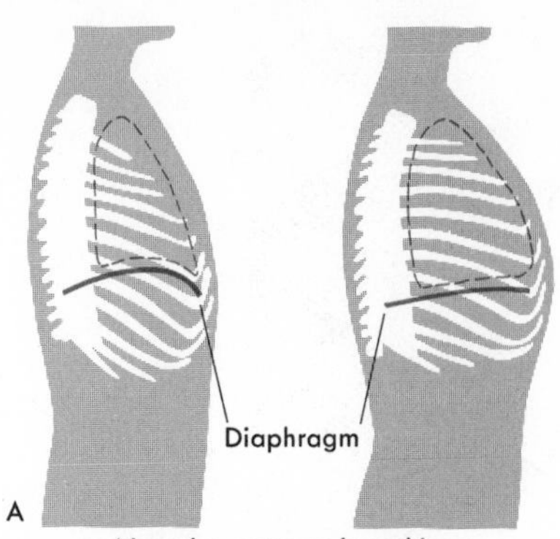

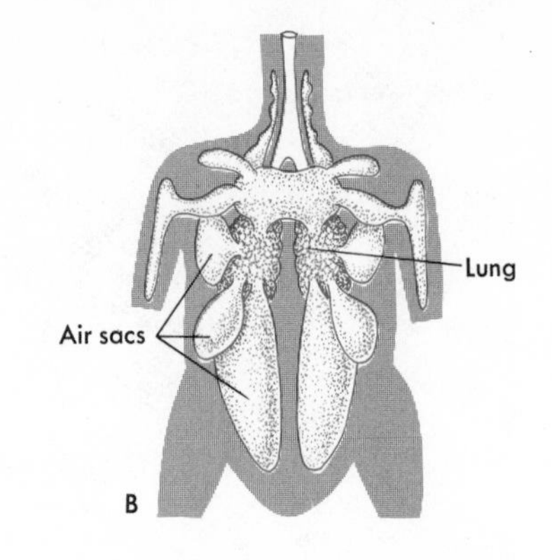

Negative pressure breathing

FIGURE 7.14

Lungs and breathing methods. Mammals use negative-pressure breathing. Expansion of the chest cavity by contraction of the diaphragm and rib muscles creates a partial vacuum. Air rushes in through the trachea, expanding the lungs to fill the cavity. (B) Birds also use negative-pressure breathing. However, attached to the lungs are many air sacs, which ramify many hollow spaces in the bird's body and evidently provide additional air-holding capacity for the gas exchange surfaces. Air passes over the gas exchange surfaces of the lungs twice—once on the way into the air sacs and again on the way out.

ently. Like insects, the bodies of these organisms contain spiracle openings, but they open into internal sacs called **booklungs.** Each booklung consists of delicate membranes that lie parallel like the leaves of a book. The membranes contain blood that absorbs oxygen through the membrane surfaces. Thus these simple terrestrial organisms have evolved gas exchange surfaces, lungs, which work in conjunction with a medium of internal transport, blood.

Only in the vertebrates do true lungs appear. There is considerable evidence that the earliest bony fish possessed lungs as well as gills and could breathe both in water and in air. In most descendants of the bony fish, however, the lungs evolved into a different structure—the swim bladder—whose principal function is to permit the fish to regulate its buoyancy by the intake or expulsion of air. Some descendants of the earliest bony fish kept their lungs, lost their gills (at least in their adult form), and moved onto the land. These were the first amphibians—probable ancestors of all later air-breathing vertebrates, the reptiles, birds, and mammals.

Lungs are structurally quite different from gills. Whereas gills are evaginations of the circulatory system, lungs are **invaginations.** This invagination of the lung surface enables the cell membranes utilized in gas exchange to be hidden inside the body cavity, where they are adequately supported by body tissues and where body fluids can keep the membrane surfaces moist. In this way, the two principal drawbacks of gills in terrestrial life were met by the vertebrate lung. Except in birds, however, the lung is not a particularly efficient structure for gas absorption. Mammalian lungs, for example, absorb only about 25 percent of inhaled oxygen. But this inefficiency of the lung is compensated for by the relatively high oxygen content of air (about 21 percent oxygen).

Since lungs are internal, a specialized method is required for getting air into and out of them. Amphibians, for example, inhale a mouthful of air, then squeeze it into their lungs by pressure of the mouth and throat—a mechanism that uses **positive pressure** (equivalent to filling the lungs with a bellows).

Reptiles, birds, and mammals, on the other hand, employ **negative pressure** (equivalent to sucking air into the lungs by means of a vacuum cleaner). They use their chest muscles to expand and contract the volume of their chest cavity, causing internal changes in air pressure. Upon expansion, air pressure in the lungs drops and air rushes in from outside, filling the elastic lung sacs. In expiration, the process is reversed. The chest cavity gets smaller, air pressure in the lungs increases, and the air is simply pushed out of the elastic lung sacs. The principal muscles of respiration in birds and reptiles are those that expand and contract the ribs. In mammals, there is an additional arched muscle—the **diaphragm**—separating the chest and abdominal cavities. The contraction of this muscle results in inspiration (air intake). Two examples of negative-pressure breathing are illustrated in Fig. 7.14.

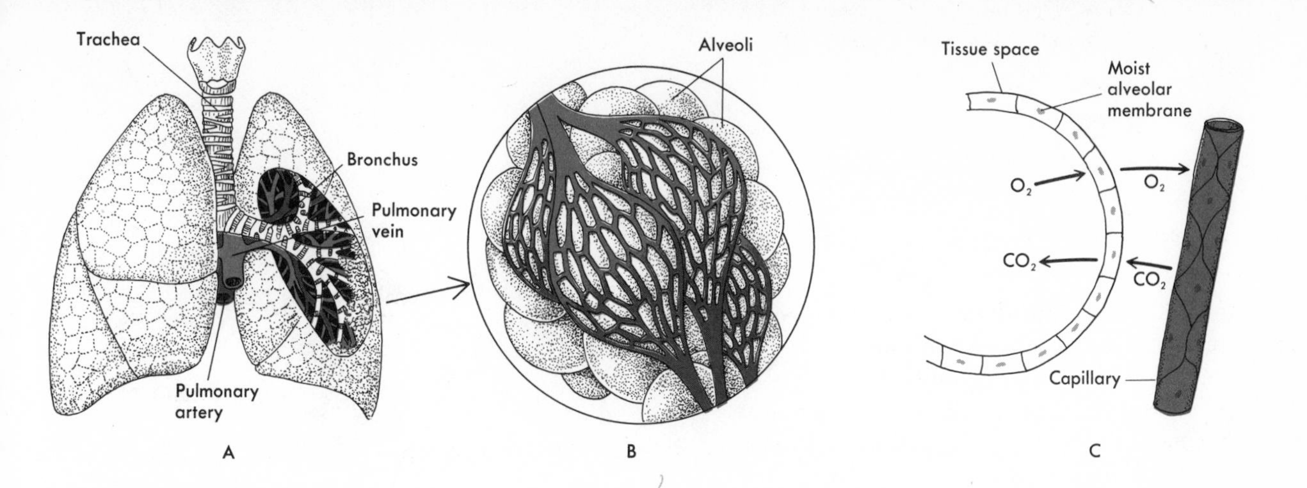

FIGURE 7.15

(A) The gross appearance of the lungs of a human, showing the connection of the trachea to the bronchi and the successively finer branches of air passages, which terminate in alveoli. (B) The alveoli as clusters of thin, membranous, air-filled sacs, surrounded by a complex network of capillary blood vessels. (C) An alveolus in relation to the capillary blood vessels. The alveolar and capillary membranes are very thin, permitting easy gas exchange across the membranes. Blood flowing in the capillaries around each alveolus brings carbon dioxide to the alveolus, where it diffuses into the air and oxygen diffuses into the blood for transport to internal tissues.

In amphibians, the lungs are fairly simple structures that only partly fill the need for supplying adequate surface for gas exchange. In the frog, for example, oxygen diffuses into the bloodstream through the lining of the mouth cavity as well as through the lungs. On the other hand, the reptile lung is complex and efficient, since it must serve all the respiratory needs of the animal. The more highly complex lungs of birds and mammals evolved from that of reptiles.

Unlike reptiles, birds are warm-blooded; their heat-generating metabolism requires a greater consumption of oxygen to maintain their body temperature than reptiles do. The gas exchange mechanism of a bird must therefore be much more efficient than that of a reptile. In fact, the respiratory system of a bird is a marvel of complexity. Extending out from a bird's lungs is a system of air storage sacs, so that when a bird inhales, air passes through its lungs into these sacs; when the bird exhales, air passes from the sacs back through the lungs. In this way a bird's blood is oxygenated during inspiration as the air passes through the lungs on its way to the sacs, and during expiration as the air passes back out through the lungs. The bird lung differs from the lungs of all other vertebrates in that it consists of a network of interconnecting passageways, rather than simply a collection of dead-end sacs.

The mammalian lung—and therefore the human lung—is invaginated deeply into the chest, or thoracic cavity. An opening, the **glottis,** in the floor of the throat leads into the **trachea,** a tube supported by cartilage rings. The trachea divides into two **bronchi,** each of which leads into a lobe of the lungs. Each bronchus in turn divides and subdivides into additional branched tubes, the **bronchioles,** which finally open into the **alveoli**—tiny sacs through which gases are exchanged. Within the alveoli the gases are exchanged across their membranous walls between the air in the air sacs and the blood cells in the **capillaries**—the tiniest of blood vessels (see Fig. 7.15). There are about 150 million alveoli in the human lung, and in surface area the alveoli exceed 600 square feet. This would be equivalent to the surface of a cube 10 feet on a side, or a doubles tennis court.

With each breath, an average young adult in a relaxed state takes into his chest about 500cc—about a pint—of air, and he takes an average of 17 breaths a minute. Exercise may accelerate the breathing rate to as much as 75 breaths a minute, and a deep breath followed by the most forceful expiration possible moves an average of 4,500 cc of air—nine times the amount exhaled in the relaxed state. Even after forceful exhalation, however, approximately 1,500 cc of residual air still remain in the lungs.

The lungs provide the body with precisely the amount of oxygen needed under conditions ranging from complete rest to vigorous activity. The master control center for breathing is situated in the lower brain, in the regions known as the pons and the medulla. From this breathing center, nerve impulses are transmitted that stimulate the various coordinated muscular activities that are involved in the act of breathing.

The respiratory center itself responds to a variety of bodily stimuli. The center is sensitive to nerve impulses that relate information concerning the degree of stretch of the lungs and the breathing muscles. The center is also sensitive to the carbon dioxide content of the blood flowing through it. In addition, two of the larger arteries —the **carotid** and the **aorta**—contain areas that are sensitive to blood pressure, to a lowered oxygen content of the blood, and to the acidity of the blood. Stimulation of these areas causes reflexes that affect breathing. Scientists are only beginning to understand how the various nerve and vascular mechanisms maintain a breathing rate that meets perfectly the need of the human body. Once gas exchange between the outside environment has been effected, refreshed, oxygenated blood must be brought to internal cells, where oxygen is released for their use and carbon dioxide is removed for transport to the gas exchange surface. This movement of blood, the internal transport fluid, is effected by the heart.

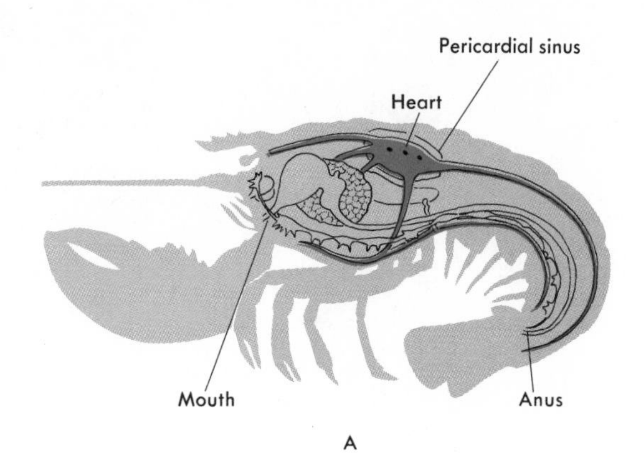

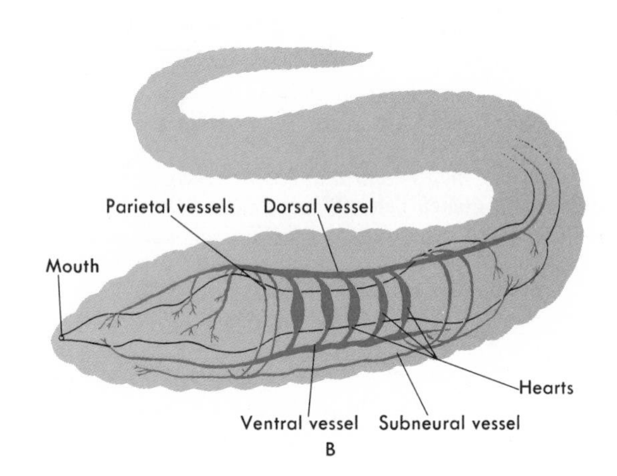

FIGURE 7.16

Circulatory systems in animals. With the evolution of increasing structural complexity came more efficient movement of blood and less mixing of oxygenated and deoxygenated blood. (A) In the lobster, an arthropod, blood is carried from the heart in an artery and returns to the heart in a sinus, or blood pool. Blood flow is sluggish and poorly channeled. (B) In the earthworm, an annelid, blood is carried both to and from the heart in distinct channels. Movement of blood in the channels is affected by five pairs of pulsating vessels (hearts). A system of valves in the vessels keeps the blood flowing in only one direction in a given vessel.

The heart and circulation

Only in animals possessing contractile cells, called muscle, could hearts have evolved. Although there are many variations in heart structure, all hearts have a similar function: they are essentially pulsating devices or pumps. An effective pump requires muscular walls capable of rhythmic contraction and one-way valves of various types to prevent blood, once pumped out, from flowing backward.

The evolution of a variety of tubes—arteries, capillaries, and veins (collectively called the **vasculature**)—made possible the delivery of blood pumped by the heart to all areas of the organism and its return for recirculation. The **arteries** are thick-walled, elastic tubes with well-developed muscular layers that conduct blood away from the heart. The **veins** are thin-walled, slightly muscular tubes that return the blood to the heart. In most species, networks of tiny capillaries are situated between the arteries and the veins. It is through the capillaries that blood moves to reach individual cells; exchange of nutrients and wastes occurs through the thin capillary walls. Also present in some vascular systems are **sinuses** —structurally indistinct channels in which blood may pool. **Lymph vessels,** structures that return some of the animal's tissue fluid to the blood, may be present in some animals.

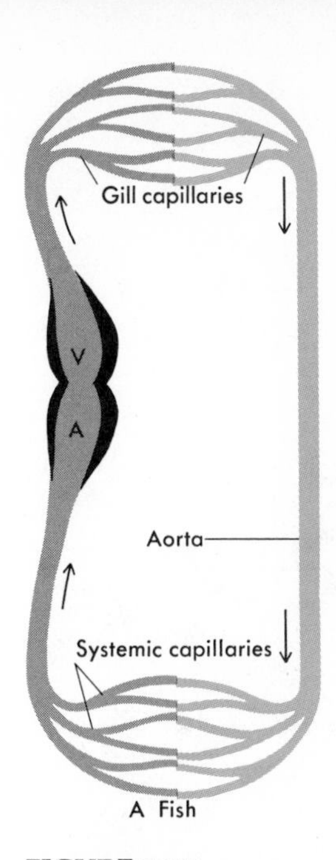

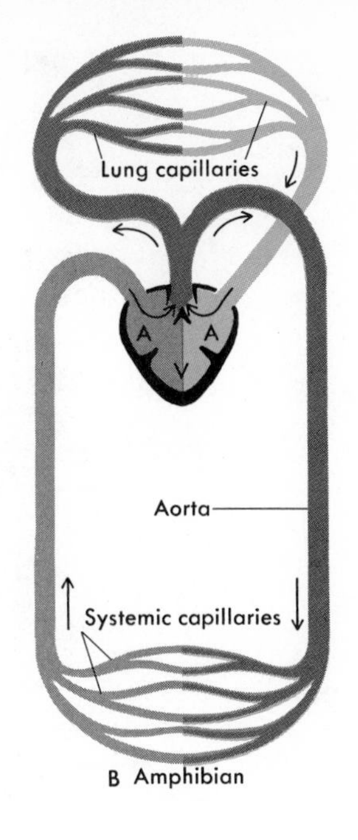

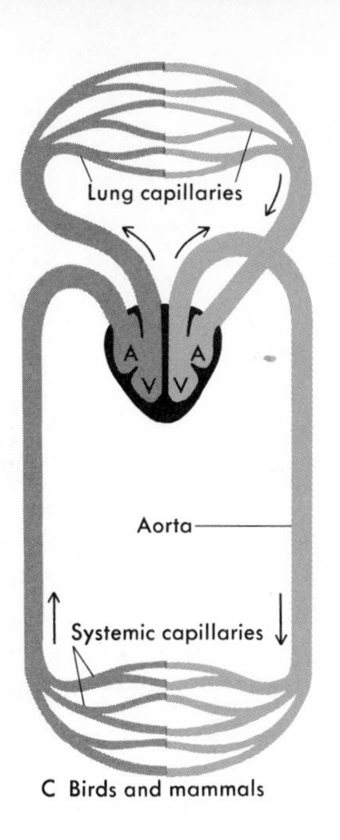

FIGURE 7.17

The circulatory systems of vertebrates, such as fish, amphibians, birds, and mammals, are similar in many respects. Blood is carried in completely closed channels that direct it to gills or lungs for oxygenation and to tissues. The efficiency of pumping and channeling varies as shown. (A) The fish heart has only one ventricle and one atrium. Blood is channeled first to the gills for aeration and then to the tissues. (B) The amphibian differs from the fish in that it has lungs instead of gills. The heart is three-chambered, and some mixing of oxygenated and deoxygenated blood occurs. Some of the higher reptiles (e.g., crocodiles and alligators) have partially divided ventricles—an evolutionary step between three- and four-chambered hearts. (C) Birds and mammals have completely separated circulation to the lungs and other tissues. Used blood from cells is pumped through the lung capillaries, where it is oxygenated, and the recharged blood is pumped to the tissues.

The range of structural and functional complexity found in the circulatory systems of various organisms is considerable. At one extreme is the open circulatory system—the system found in arthropods (such as insects and lobsters) and molluscs. The only distinct vessels in the open system are the main channels leading from the heart; the remainder of the system consists only of sinus spaces that are rather ill defined. Thus, over part of its course, blood flow is poorly directed and sluggish. The relative inefficiency of the open system is not critical because in the species in which it is found, the open system is often not involved in gas (oxygen, carbon dioxide) transport. Of greater complexity than the open circulatory system is the type of circulatory system that is found in such species as the earthworm or in its annelid relatives, such as *Nereis*, the seashore worm found on the New England coast. These species of worms have a closed, though relatively simple, circulatory system in which multiple little pulsating vessels, or "hearts," force blood through the body. For a comparison of the circulatory systems of a typical arthropod and a typical annelid, see Fig. 7.16.

The circulatory systems found in many lower vertebrates are closed and rather elaborate, with fairly complex hearts. Though complex and more efficient than invertebrate hearts, those of amphibia (for example, frogs) and reptiles (for example, snakes) permit fresh blood to be mixed with used blood. Such mixing reduces the efficiency of the circulatory system. The fresh blood is contaminated by wastes from the used blood, so that the capacity of fresh blood for delivering nutrients and also for returning wastes is reduced.

The most complex circulatory systems are those of mammals. The human circulatory system consists of a heart and a network of blood vessels whose total length may exceed 60,000 miles. Moreover, the composition of human blood, together with the other elements in the system, ideally suits the human circulatory system to meet complex human needs. Vertebrate circulatory systems of varying complexity are compared in Fig. 7.17.

The human circulatory system

HUMAN BLOOD

Although the ancients did not understand the function of blood, they conceived of blood as the seat of the soul. It is hardly any wonder that today, although we know better than to assign to blood any mystical functions, we nevertheless consider it the most vital component of the body. In the average adult there are 13 pints of blood, which serve to provide the body's tissues with food and oxygen for life, and bring to them other materials that are needed for growth and repair. Blood removes wastes, dissipates the body's heat, and transports the hormones that control bodily functions. Blood also serves as the fluid medium for antibodies and cells that fight invading organisms and foreign matter, and it transports the drugs that are administered to combat disease.

It is vital that the flow of blood be constant. For example, if blood flow to the brain is interrupted for only five seconds, the human organism loses consciousness. Within 20 seconds, the body begins to twitch convulsively. After prolonged periods—four minutes of blood deprivation—many of the brain's mental powers are irreversibly damaged. Death follows several minutes of deprivation.

COMPOSITION OF BLOOD

Blood is slightly heavier than water and several times as viscous, or resistant to flow. About 55 percent of blood is composed of plasma, an amber fluid, in which are suspended the solid components of blood: red cells, white cells, and platelets. The erythrocytes (red blood cells) are biconcave discs, a shape which improves the efficiency of their gas transport by providing more surface area than would a flat disc. They are formed in the marrow of flat bones, such as the sternum or breastbone, and lose their nuclei as they mature. Red blood cells contain the red pigment hemoglobin, the compound that enables the cells to perform their function of oxygen transport. Each red blood cell contains about 280 million molecules of hemoglobin that are dissolved in water.

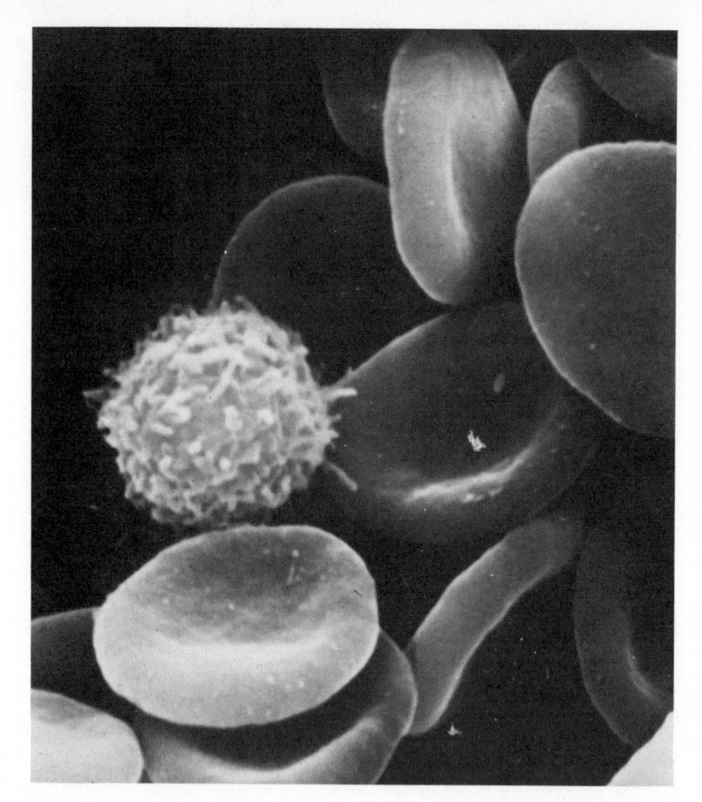

5,250X

FIGURE 7.18

Human blood cells as shown by the scanning electron microscope. The white cell (leucocyte) is surrounded by red cells (erythrocytes), each of which is about 7 microns in diameter. A cubic millimeter of blood contains about 5 million erythrocytes and about 7500 leucocytes.

There are about 30 trillion erythrocytes in the average adult male and 27.5 trillion in the average adult female. The life span of each red cell is about 120 days, and new cells are constantly being produced as old cells die and are destroyed. Production of red cells is carefully regulated by a variety of internal signals, such as the loss of blood or lowered levels of oxygen. The red cells are manufactured in the red marrow of the bone by a process called **hematopoiesis,** which involves a series of cell divisions and progressive differentiation of blood-forming cells.

The white blood cells (leucocytes) are much fewer in number than erythrocytes. (The ratio is approximately one white for every 600 red blood cells.) There are several different types of white cells, all of which, along with the red blood cells and platelets, are formed in the marrow of flat bones. White blood cells may be distinguished by the type of nucleus they contain (the nuclei vary in size and shape) and by the amount of granular material that is visible in the cytoplasm. The

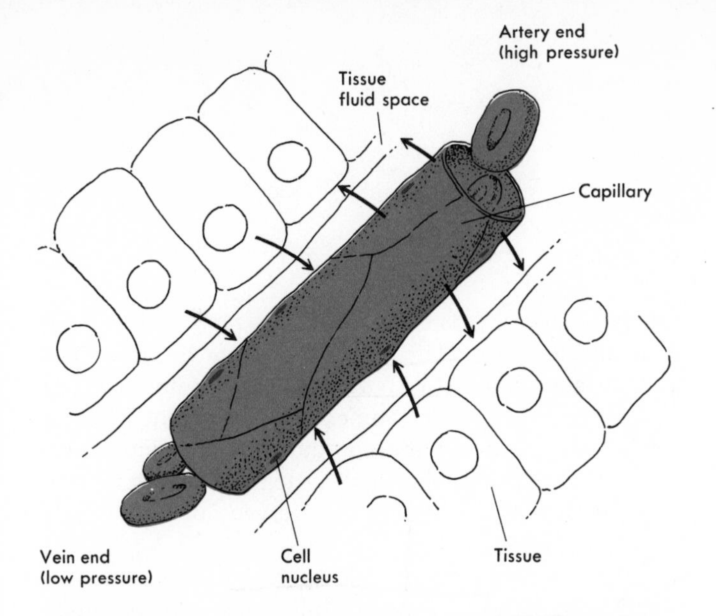

FIGURE 7.19

Nutrient and waste transport to and from cells through capillaries. At the end of a capillary nearest the arteriole, hydrostatic pressure forces nutrients out of the capillary into the tissue. Near the venous end of the capillary, hydrostatic pressure is relatively low, and wastes and other materials move back into the capillary from tissue fluids. The change in pressure results from osmotic pressure produced by blood proteins.

neutrophil—a type of **granulocyte** (with cytoplasmic granules), and the **monocyte**—a type of **agranulocyte** (without cytoplasmic granules), are phagocytic. They can engulf bacteria and other invading foreign matter and thus help to protect the body against disease. Other agranulocytes produce specific protein **antibodies** in response to exposure to foreign substances, or **antigens.** These antibodies are an essential component of the body's defenses against viruses and bacteria.

The **platelets,** about 1.5 trillion in the average adult, are non-nucleated, disc-shaped cells that are formed in the bone marrow. Platelets are instrumental in the formation of clots, which plug any breaks in the blood vessels and thus prevent profuse bleeding, or hemorrhaging. About 91 to 92 percent of plasma is water. (**Serum** is the watery portion of the plasma that remains after clotting has removed its content of solid material.)

In addition to the blood cells and platelets, plasma contains three valuable kinds of proteins—**albumins, globulins,** and **fibrinogens.** These proteins serve a number of important functions. They help maintain the osmotic concentration that is needed for the exchange of materials between blood and tissue spaces. They also help to give viscosity—thickness—required for normal blood pressure, and to maintain a proper pH (balance of acids and bases). Fibrinogen is involved in the clotting mechanism, and globulins produce specific antibodies to react with and to combat invading organisms.

Plasma also contains nutrients, including the sugars derived from carbohydrate and protein foods, as well as hormones, chemical messengers which it transports to all parts of the body. There are about 70 to 120 mg of nutrient material in every 100 ml of plasma. Three other components of plasma are fats—triglycerides of fatty acids, which are transported to various parts of the body for use or storage, phospholipids, and cholesterol. Phospholipids help in the transport and use of fats. Cholesterol is derived from foods and also synthesized in the body. Cholesterol is thought to prevent blood cells from precipitating and also to prevent phospholipids from causing **hemolysis**—the splitting of red blood cells. When present in overabundance, cholesterol can form plaques which clog vital arteries leading to the heart.

Other plasma components are the salts of sodium, potassium, calcium, and magnesium, which help to maintain a suitable osmotic pressure, maintain the proper pH, and provide a physiological balance between blood and tissues. Finally, plasma contains wastes from the tissues: **urea**—the end product of protein catabolism; **uric acid**—the end product of nucleic acid metabolism; and **creatinine**—the end product of muscle metabolism.

FUNCTIONS OF BLOOD

As fresh blood first enters the thin-walled capillaries, passing to them from the arteries, the force of blood pressure is so much greater than the osmotic forces which tend to bring water into the capillaries that it can move the nutrient cargo through the capillary walls into the tissue fluids. By the time blood nears the venous side of the capillaries, the force of the blood pressure has been reduced. At this point, the combined force of osmosis and tissue fluid hydrostatic pressure exceeds the blood pressure. As a result, the wastes from the tissue fluids are forced into the capillaries (Fig. 7.19).

Part of the precious cargo transported by fresh blood is oxygen. This vital gas is carried by the **hemoglobin** molecules in the red blood cells. (Hemoglobin is remarkable because, without it, a liter of blood could carry no more than 3 ml of oxygen. It is only because of the red pigment's great affinity for oxygen that the blood is able to carry a volume of oxygen 70 times greater than it could without the hemoglobin present.) At the gas exchange surfaces in the lungs, hemoglobin bonds with oxygen to form oxyhemoglobin (HbO). As the red cells pass through the capillaries, in single file, hemoglobin gives up its oxygen, which passes into the tissues. The tissues in turn release their carbon dioxide wastes, which dissolve in the deoxygenated blood. The hemoglobin and carbon dioxide are transported to the lungs, where the carbon dioxide is excreted and where the hemoglobin again picks up its cargo of oxygen.

The hemoglobin molecule is so vital to life that even minor impairments of its structure may lead to serious disability or even death. For example, the serious inherited disease, sickle-cell anemia, is due to a minute change in hemoglobin structure, which greatly reduces its oxygen-carrying capacity and deforms erythrocytes so that they get stuck as they pass through capillaries. Pain and death may follow. Also, the toxicity of carbon monoxide, the gas formed as the result of incomplete combustion in gasoline engines, is due to the much greater affinity of hemoglobin for carbon monoxide than for oxygen. When carbon monoxide is present, it combines readily with the oxygen binding site on the hemoglobin molecule, thus diminishing the amount of oxygen that can be carried and eventually causing death.

In addition to oxygen transport, another major function of blood is communication. Blood carries the hormones produced by the various endocrine glands—the pituitary, thyroid, adrenals, and others (all of which will be discussed in Chapter 9). The hormones are messenger agents which control the functions of various parts of the body. When an individual faces a particularly dangerous situation, for example, and must fight or flee, the adrenals pour epinephrine into the blood. The epinephrine is carried to the heart and the muscles and mobilizes them for action. The effectiveness of blood as a medium for communication is enhanced by its rapid rate of flow. In man, blood completes its circuit from the heart and back in approximately 20 seconds. Nerve impulses are transmitted in much less time, and consequently nerves are better suited for the coordination of rapid movement. Nevertheless, the vital system of chemical communication through hormones is served quite effectively by the circulatory system.

Still another property of the blood is its ability to repair leaks in blood vessels. This ability was of evolutionary advantage, in that reliance on blood as a transport fluid might have been fatal to organisms without the clotting or repair property of blood. Even a small injury could lead to a loss of most of the blood and to death for the organism. Therefore organisms that evolved blood-clotting mechanisms to repair such damage were more likely to survive. The clotting mechanism is remarkable in several respects. One of the components necessary to form a clot is kept separated from the others until a break occurs. This separation ensures that they do not ordinarily combine to form a clot that might clog a vital blood vessel (although in certain diseases this does occur, leading to heart attacks or strokes). The components are **thromboplastin,** a lipid-protein complex found in platelets in the blood; calcium ions, metal ions dissolved in the water of blood; **prothrombin,** made by the liver and secreted into the blood plasma; and fibrinogen, a protein found dissolved in the blood plasma. When a break occurs in a blood vessel a sequence of reactions is initiated; the break damages the platelets, causing them to release their contents, thromboplastin, into the blood near the injury. This triggers the sequence of reactions shown in Fig. 7.20. The ordinarily soluble fibrinogen is converted to insoluble **fibrin,** a stringy fibrous protein that forms a strong meshwork around the site of injury. The meshwork traps blood cells as they begin to escape, and eventually causes the formation of a leakproof plug. In some people one of the components of a clot is missing, either through dietary deficiency or inherited disease. Vitamin K, normally ingested in adequate amounts in a diet containing leafy green vegetables, is necessary for the production of prothrombin in the liver. A vitamin K deficiency can thus lead to a

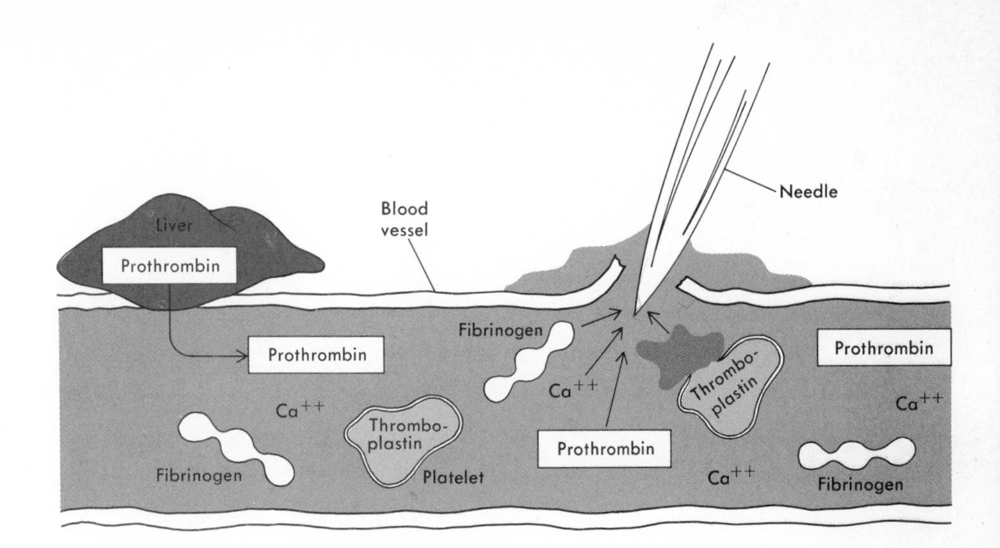

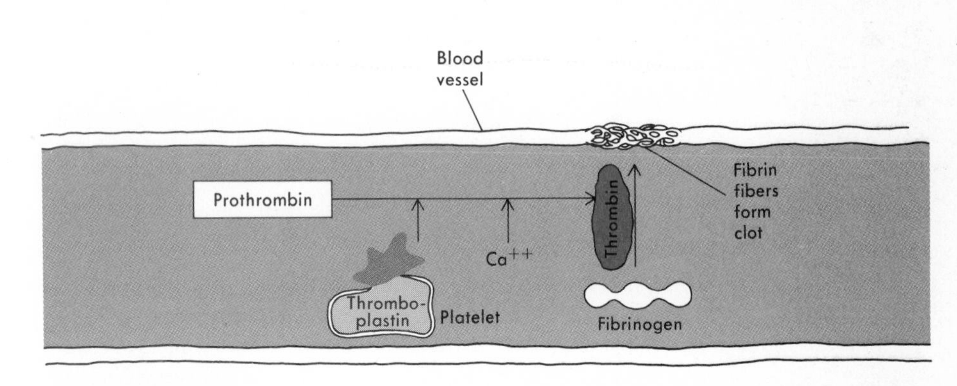

FIGURE 7.20

A simplification of the human clotting mechanism. Until a break in a vessel occurs, thromboplastin is kept inside special blood cells called platelets. When platelets encounter a wound or break in a blood vessel, they release thromboplastin into the blood around the break. A series of reactions forms a strong meshwork of stringy protein, fibrin, and these fibers, together with trapped blood cells, form the clot. As the clot contracts, serum oozes out.

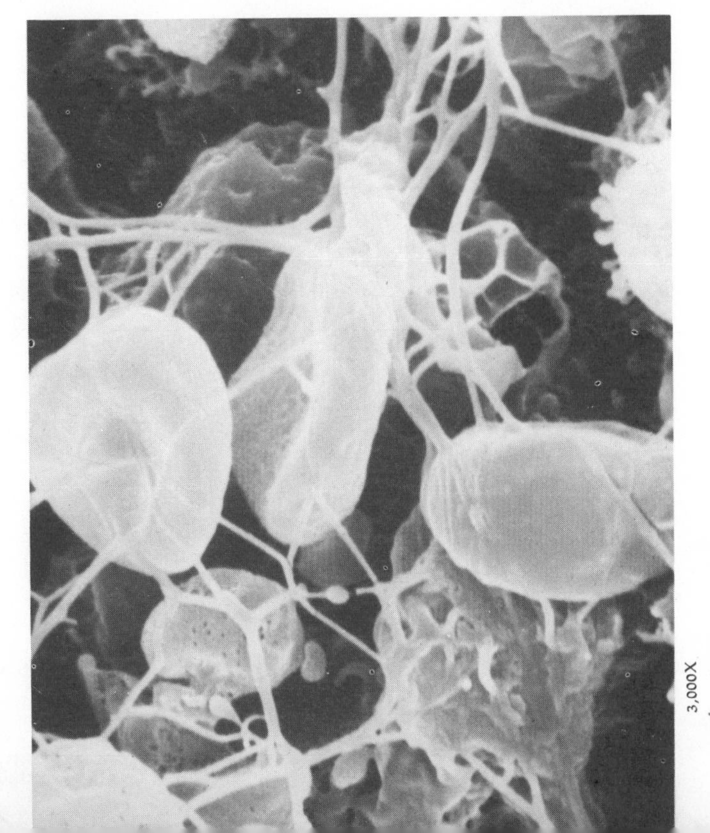

3,000X

bleeding disease. Hemophilia, one such disease, is an inherited inability to make one of the vital components of a clot. A person with hemophilia bleeds freely in response to the slightest injury and may die from a minor wound.

THE HUMAN HEART

The human heart, when fully grown, weighs less than a pound and is only slightly larger than a clenched fist. Yet, in a lifetime of 70 years, this organ pumps nearly 18 million barrels of blood.

It is a muscle that under normal conditions never tires of beating. It begins to beat by the fourth week of fetal life, and it continues to beat until death. Provided it is supplied with nutrients, the heart will continue to beat even when removed, for example, from the body of an accident victim.

The heart is actually a double pump. Surrounded by its muscular wall—the **myocardium**—are four hollow chambers, two on the right side and two on the left. The upper chambers are the thin-walled **atria,** or receiving

Hypertension: An Internal Time Bomb

Nearly 20 percent of the American population carry within themselves a silent time bomb called hypertension (high blood pressure). Since hypertension in its early stages gives few noticeable hints of its presence, these people may not be aware of their condition. With the passage of time, however, internal changes are taking place that make the victims of hypertension likely to experience various types of serious and even fatal damage to the blood vessels to the brain, lungs, and heart. Such cardiovascular accidents, the major killer diseases in the United States, claim more than 600,000 American lives a year. Fortunately, hypertension is easily detected, and it can usually be controlled with changes in diet and life style and with a variety of antihypertension drugs.

In healthy people, the "normal" blood pressure within the circulatory system is constantly regulated by a variety of control mechanisms (both chemical and neural). Short-lived increases in blood pressure during heavy work represent a normal response, ensuring that the active organs (e.g., the muscles) get enough blood to supply their needs. When the work is done, blood pressure drops back to its normal range. Blood pressure may also rise temporarily during moments of emotional stress (fear, excitement), but again it always readjusts to its normal level when the stress-provoking stimuli cease. Hypertensive people, however, lack the capacity to lower their pressure, which is always higher than normal. Sometimes the hypertension is slight, but blood pressure can remain dramatically elevated over long periods of time.

"Normal" blood pressure actually consists of a range of pressures that is considered safe—a high normal blood pressure is around 130/85 millimeters of mercury. These figures mean that the pressure exerted by contraction of the ventricles of the heart, when transmitted through the arteries, will support a column of mercury 140 millimeters high. This is the systolic pressure. When the heart relaxes between contractions, there is still some pressure in the artery, the diastolic pressure, which is the second number. An elevated diastolic pressure is the best indicator of hypertensive changes. With a pressure of 140/90, the person is considered a borderline hypertensive.

Even a moderate rise from this level increases the probability of complications and increases the risk of heart disease or stroke. For example, insurance company statistics suggest that a 35-year-old man who maintains a blood pressure of 150/100 mm of mercury has a life expectancy seventeen years shorter than that of a 35-year-old man with normal blood pressure (120/70).

WHAT ARE THE CAUSES OF HYPERTENSION?

Some forms of hypertension are well understood. For example, damage to the kidney produces a substance that ultimately causes arterioles to contract and blood pressure to rise. Some tumors of the adrenal gland can secrete hormones that elevate blood pressure. We also know that the deposition of a fatty substance called cholesterol in the arteries decreases the elasticity of the vessels and narrows their diameter, thus elevating blood pressure. However, there are many other variables that may be involved in the development of hypertension.

Although the reasons are not understood, diet, obesity, heredity, race, a previous history of diabetes, socioeconomic level, and stress all may play a part in increasing blood pressure. Some known relationships follow.

1. Diets high in sodium may be related to high blood pressure (possibly by increasing blood volume).
2. Hypertensive parents are more likely than normal parents to have hypertensive children (a hereditary implication).
3. In the United States, blacks are almost four times as susceptible as whites to hypertension (perhaps a reflection of diet or socioeconomic stress since black Africans are seldom hypertensive).
4. Lower socioeconomic classes of all races are more susceptible than the upper classes (again a possible reflection of both diet and socioeconomic stress).
5. Personality may also be involved. The successful, aggressive, go-go businessman—a product of the stress of our Western urban/industrial society—seems to run an increased risk of cardiovascular disease.
6. Recent research has implicated excessive saturated fat intake and lack of exercise as factors that increase risk.

From only the foregoing brief list of known relationships, it is apparent that high blood pressure and heart disease are not going to be readily cured, since the roles of so many different related variables remain to be clarified.

Although the exact causes of hypertension are not known, its effects—especially in the heart, kidneys, and brain—are. As blood pressure rises, the heart works harder to pump blood through narrowed blood vessels. In response to its increased workload, the heart may become enlarged, stretched, or weakened, thus losing pumping efficiency and starting a vicious cycle that can eventually cause heart failure. As time passes, if the blood pressure of an untreated patient continues to rise, the larger blood vessels become damaged, producing such symptoms as weakness, dizziness, pounding headache, and shortness of breath. Such warning signs of hypertensive crisis indicate a dramatically increased risk of strokes and heart attacks.

Most hypertensive crises are avoidable. They may often be successfully treated with a variety of drugs that lower blood pressure. Some of the drugs are diuretics, whose main effect is to decrease blood volume by increasing the excretion of sodium and water. Other drugs act in the nervous system to prevent vasoconstrictor substances from being released at nerve endings. There are also drugs that act directly on constricted blood vessels, causing them to dilate and thus lower pressure. Unfortunately, many of the drugs that relieve hypertension have such side effects as drowsiness, dizziness, and depression. Because of the side effects, many patients under treatment stop taking the medicine prescribed by their physicians and as a consequence continue to suffer from high blood pressure.

To combat the tendency of patients to be careless, clinics often have a vigorous follow-up program in which patients are reminded about check-ups. Detailed records provide a complete history of each patient's use of medication. Despite our advances in treatment of high blood pressure, many people with hypertension don't even suspect there is anything wrong with them. To overcome this hidden and treatable disease, health organizations have launched many early detection campaigns, all of which provide easy access to free blood pressure readings. If the blood pressure is high, the person is informed and referred to his or her physician for treatment.

High blood pressure remains one of our most significant national health problems. It is also one of the most expensive. But hypertension is primarily a problem of the individual. Only the individual can decide to have his blood pressure checked. Only the individual *hypertensive* can decide to take medication. So far, only about 30 percent of those with high blood pressure are doing something about it. As researchers continue to investigate the causes behind hypertension, the public is becoming more aware that this "silent killer" exists and that something positive can be done about it.

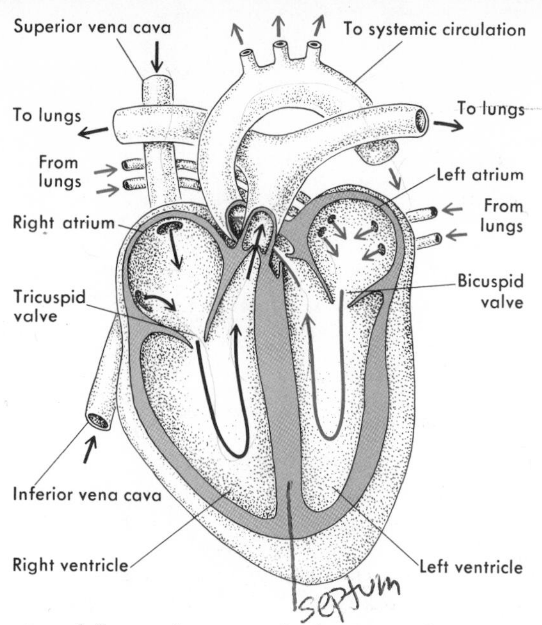

FIGURE 7.21

The human heart is a four-chambered double pump slightly larger than a man's clenched fist. The two atria, or upper chambers, receive blood—the right atrium receives used blood from the systemic circulation and the left atrium receives freshly oxygenated blood from the lungs. The right ventricle pumps used blood to the lungs, and the left ventricle pumps oxygenated blood to the systemic circulation. Black arrows represent deoxygenated blood, and colored arrows represent oxygenated blood.

chambers; the lower chambers are the thick-walled **ventricles,** or pumping chambers (Fig. 7.21). The atrium and ventricle on the right are separated from the two chambers on the left by a solid partition of muscle and conductive tissue, called the **septum.** Thus the heart functions as two separate pumps, one alongside the other. As a result, oxygen-rich (oxygenated) blood is kept separate from deoxygenated blood so that the two do not mix as they do in the hearts of amphibians and most reptiles.

The cardiac muscle, of which the heart is largely composed, has an inherent rhythmicity, or tendency to beat, even in the absence of external stimuli. But normally the beat is regulated by nerve impulses originating in the **pacemaker**—a regulatory tissue imbedded in the muscles of the right atrium which controls heart rate.

CIRCULATION: HEART AND VESSELS

The two sides of the heart function as two separate circulatory systems: pulmonary circulation, which carries blood to and from the lungs; and systemic circulation, which transports blood to and from the other tissues. Deoxygenated blood, having traveled through the body and having given up its oxygen and collected waste carbon dioxide, returns to the heart through either of the two great veins, the superior or the inferior **vena cava.** The venae cavae empty directly into the right atrium.

Shortly after the right atrium has filled with the returning venous blood, it contracts. This contraction forces the blood through a valve, the **auriculoventricular** or tricuspid **valve,** into the right ventricle. This valve, as well as the others in the heart and veins, ensures that blood flows in the proper direction. The valves all consist of flaps of tissue which open when pressure behind them is greater than in front, as it is when blood flows through in the right direction, but which close tightly when the pressure in front is higher than in back, as it is when blood tries to rush backward. Next, the right ventricle contracts and forces the blood past the **pulmonary semilunar valve** into the pulmonary artery. The pulmonary artery channels the blood to the lungs through its branches. In the lung capillaries, which have diameters of about 1/2500th of an inch, the red cells and plasma give up their carbon dioxide and receive oxygen. The blood then moves from the lungs into the pulmonary veins through which it enters the heart's left atrium.

The left side of the heart, having been furnished with oxygenated blood by the pulmonary system, now assumes the quite separate job of propelling this blood through the systemic circulation.

As the left atrium fills with oxygenated blood from the pulmonary veins, it contracts. This forces the blood through the mitral valve into the left ventricle. With a strong contraction, the left ventricle propels the blood through the aortic valve into the **aorta**—a great trunkline artery with a diameter of about an inch. From the aorta, many arteries branch off to carry blood to all body areas. After completing its circuit through the body, delivering oxygen and picking up carbon dioxide, the blood returns to the right atrium. The entire process is then repeated. In the course of its journey through the systemic circulation, blood passes through all the organs in the body. During one circuit the kidneys receive 25 percent of the heart's output for cleansing, and the liver helps regulate the nutrient level of the blood. The intestines and glands replenish the blood's nutrient and hormone content. The brain, which weighs only 2 percent of body weight, gets 10 to 15 percent of cardiac output to satisfy its great demand for oxygen.

Although the two sides of the heart pump blood through separate pathways, they actually contract in concert. As the right atrium receives blood from the venae cavae, the left atrium receives blood from the lungs. Signals originating in the pacemaker begin the contraction, or **systole,** of the heart; the pacemaker impulses begin at the upper corner of the right atrium and spread relatively slowly, in a wave, to all parts of the heart. As the wave passes, the cardiac muscle contracts, squeezing the blood in front of the contraction. The squeezing occurs in a pattern that "wrings" the heart, twisting it and contorting its chambers so as to ensure that the blood is emptied from every recess. First the atria contract, wringing their contents into their respective ventricles; then the ventricles wring their contents into their respective arteries. One-way flow, vital to efficient channeling of the blood in the right direction, is ensured both by the way the heart squeezes blood and by the heart valves. These normally slap shut after the blood has passed through them, giving the characteristic "lubb-dupp" sound audible at the surface of the chest wall. The "lubb" is the booming sound of the tricuspid and mitral valves closing after the right and left ventricles have emptied; the short, sharp "dupp" sound is made by the closing of the pulmonary and aortic valves after the blood has left the right and left ventricles. Following this wave of contraction is a rest or relaxation period called **diastole.**

HEART CONTROL MECHANISMS

The cardiac muscle of which the heart is mostly composed has, as we have said, its own inherent rhythm, or tendency to beat, and it will contract and relax without any outside influence. But its rate and strength of contraction can be modified to meet any emergency need in the body. The controls over rate and strength of beat are both hormonal and neural.

There are two neural control centers in the brain: an accelerator center and a cardioinhibitory center. The **accelerator center** can send rapid volleys of impulses to the pacemaker, via sympathetic nerve fibers (discussed in Chapter 10), which speed up the contraction rate of the heart muscle. The **cardioinhibitory center** can transmit volleys of slow-down impulses through the vagus nerve to the pacemaker, reducing its inherent rate and slowing the contraction rate of the heart muscles. Both the accelerator and inhibitory centers can respond to a variety of stimuli, such as chemicals (particularly lactic acid and carbon dioxide), blood pressure changes resulting from increases in muscular exertion, and emotional stimuli originating in higher centers of the brain. All these stimuli exert their effects in a **feedback loop,** a nerve circuit in which alterations in normal rate or pressure are sensed, information is transmitted to the brain, and corrective signals are generated in one of the two cardiac centers of the brain and sent via nerves to the heart's pacemaker. As the heart's rate and contraction strength return to normal, corrective signals cease. Thus the heart's rate is stabilized, and even minor deviations are corrected quickly. Of the large number of different feedback loops, we will consider here only one example.

The accelerator center can act in time of stress, excitement, or emergency to increase the heart's rate of contraction. For example, during muscular exertion, such as might occur as a result of fast running or heavy work, the skeletal muscles deplete the oxygen content of the blood, and release wastes, such as carbon dioxide and lactic acid. Chemoreceptors—special nerve endings— in the carotid artery sense the presence of these chemicals in the blood and send information to the cardiac accelerating center in the brain. This center sends speed-up signals to the heart's pacemaker, which increases the rate of heart contraction (and consequently the rate of flow of blood) and also increases the blood pressure. These increases serve to bring nutrients and oxygen more rapidly and in larger volume to the muscles under exertion. But as the heart rate and blood pressure increase, signals are again sent via the inhibitory fibers of the vagus to return the heart to its resting rate. This interplay between the pacemaker and the accelerator and cardioinhibitory centers (see Fig. 7.22) permits adjustment of the heart's rate to meet special demands, but it

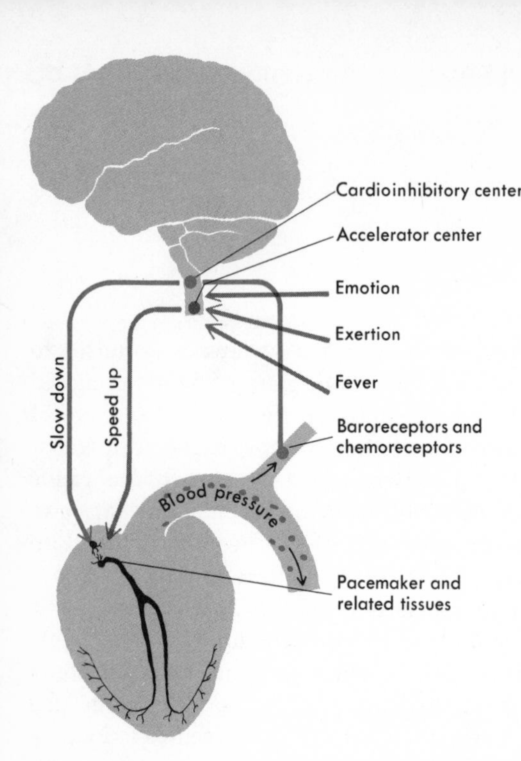

FIGURE 7.22

Control of the heartbeat rate is the result of an interplay between the heart's pacemaker signals and the cardiac acceleratory and inhibitory centers in the brain, which receive sensory inputs from receptors in the aorta, great veins, and carotid artery. Initiation of heart muscle contraction results from signals generated in the pacemaker. When the heart rate is too slow to meet the demands of the body, various signals may be sent to the cardiac accelerator center in the brain, which in turn sends impulses to the heart, causing an increase in the beat rate. But as the rate increases, receptors in the carotid artery send signals to the cardioinhibitory center. When the demands of the body have been met, volleys of signals from the inhibitory center effectively return the heart to its resting rate.

also ensures that when the special demands cease, the rate of heartbeat will return to and stabilize at its resting rate. Many other changes in the body, such as fear and anxiety, excitement, and the presence of hormones, can alter the normal rate.

NOURISHMENT OF HEART MUSCLE

The heart muscle, like any other muscle, requires nourishment. This nourishment, however, is not derived from the blood within the heart's chambers. Instead, the heart has a separate circulation of its own, called the coronary system. Two coronary arteries branch from the base of the aorta as it emerges from the left ventricle; these extend to the surface of the heart. It is through these arteries and their branches that the heart itself is fed. Sometimes, however, these arteries become blocked as a result of disease.

If a coronary vessel becomes obstructed, a portion of the heart muscle, the myocardium, dies for lack of nourishment (infarction) and a heart attack may ensue.

BLOOD PRESSURE

Blood pressure, a force exerted against artery walls, is produced primarily by the heart's pumping action. With each heart contraction (systole), blood pressure increases, and the pressure measured at the maximum point of contraction is called **systolic pressure.** With each relaxation between heartbeats (diastole), the pressure diminishes to a low point called the **diastolic pressure.** Blood pressure plays a key role in circulating the blood and also in overcoming osmotic forces, driving nutrients out of the blood and into the tissues.

Blood pressure is affected by thin vessels called **arterioles,** the very smallest of arterial branches. Arterioles have muscles called **precapillary sphincters** wrapped like fingers around their walls; these can expand and contract. When they expand, resistance to blood flow is reduced and blood pressure decreases. When they contract, resistance increases and pressure rises. The sphincters act much like the nozzle on a garden hose. Narrow the nozzle opening and pressure in the hose rises; enlarge the opening and pressure decreases.

It is by means of the nozzle-like sphincters that blood can be directed to where it is most needed. After a meal, for example, more blood is needed in the digestive tract for digestion. To accommodate this need, sphincters in other areas, such as the muscles, may contract, while those in the intestinal tract expand. This combined action diverts blood from the other areas of the body and increases the flow to the intestinal tract.

AUXILIARY CIRCULATION

In addition to the vascular elements already discussed, there is also an auxiliary system of vessels. The **lymphatic system** carries a clear tissue fluid, the lymph, which resembles blood plasma. Most of the tissues in the body contain lymphatic capillaries charged with collecting lymph—tissue fluids that normally escape osmotic reabsorption into the blood capillaries. The lymphatic capillaries transport the lymph through a fine network into larger vessels and finally into a large duct (thoracic duct). The thoracic duct empties into a tributary of the superior vena cava, where the lymph mixes with returning venous blood. The lymph thus becomes part of the plasma that enters the right side of the heart. The lymphatic system also consists of lymph nodes—oval or bean-shaped masses of tissue, varying in size from a pinhead to a lima bean. These act as filters, preventing any undesirable components of tissue fluid from entering the bloodstream.

In addition to the lymphatic system are the portal systems, special vascular pathways with capillaries at both ends. One of the most important, the **hepatic portal system,** carries nutrients in the blood directly from intestinal capillaries to the capillaries of the liver. This activity is of vital importance in regulating blood composition.

Summary

1. All organisms employ two forms of transport: they take in nutrients and expel wastes across the membranes that separate organism from environment and they circulate substances internally.

2. In all organisms gas exchange with the environment occurs across a moist membrane along a concentration gradient.

3. In most unicellular organisms diffusion and cyclosis ensure the movement of nutrients and wastes within the cells.

4. In simple water- and land-dwelling plants, such as algae and mosses, the movement of substances from cell to cell occurs by diffusion.

5. In higher plants the problem of acquiring adequate amounts of necessary gases for all cells was solved by invagination; the development of internal air spaces provided increased surface area where diffusion could occur and easier gas access to internal cells.

6. During nonphotosynthetic periods green plants carry on respiration, taking in oxygen and giving off carbon dioxide. During photosynthetic periods respiration continues, but photosynthesis usually occurs at such a rate that the net gas exchange is a release of oxygen and an uptake of carbon dioxide.

7. Most gas exchange and water loss in green plants occurs through stomata. To avoid excess water loss the highest concentration of stomata is usually on the under side of leaves, and in vascular plants the width of stomata openings is regulated by guard cells. Gas exchange in woody stems occurs through lenticels.

8. As plants moved from a water to a dry land environment they evolved roots, specialized structures for water absorption and stabilization. In higher plants there are two types of true root systems: the fibrous root system characteristic of monocots and the taproot system found in most dicots. In addition to water, most roots absorb gases from air pockets in the soil.

9. Highly specialized green plants contain xylem tissue composed of vessel elements and tracheids. Xylem cells conduct water throughout the plant; their thick cellulose walls also provide support.

10. Water moves through xylem tissue in one direction only, from the roots to the leaves. The pull of transpiration in the leaves is primarily responsible for the movement of water, although root pressure may provide a small push.

11. The transport of photosynthates and other nutrients throughout higher green plants is accomplished by the conductive tissue phloem. Phloem is composed of sieve cells and companion cells.

12. As in plants, gas exchange in animals must occur across a moist membrane. In aquatic animals gas exchange may be accomplished at the body surface, in the digestive cavity, in invaginated air sacs and passages, or in gills. In land animals it may occur through moist skin in tracheae, booklungs, or true lungs.

13. In animals internal transport of oxygen, nutrients, and wastes is accomplished by the blood and the circulatory system. Although the composition of blood and the structure of circulatory systems vary greatly between animal genera, blood typically is composed of water, dissolved chemicals, and cells, and circulatory systems of a heart, arteries, and veins. Other common circulatory system elements are capillaries, sinuses, and lymph vessels.

14. In humans gas exchange between the external environment and the blood occurs at the alveoli walls.

15. Breathing rate is controlled by a respiratory center in the pons and medulla of the brain. The center responds to nerve stimuli indicating the degree of stretch of lungs and breathing muscles and changes in the pressure, oxygen content, and acidity of the blood.

16. The blood-clotting mechanism enables blood to repair damage to blood vessels. In clot formation, the release of thromboplastin from damaged platelets triggers a series of reactions which results in the formation of a fibrin meshwork at the site of injury. The meshwork entraps blood cells, thereby plugging up the damaged vessel wall.

17. The human heart ventricles pump blood through one of the body's two circulatory systems: the right ventricle serves the pulmonary system while the left ventricle serves the systemic system.

18. Heart rate is regulated by a system of checks and balances exerted by the accelerator and cardioinhibitory centers in the brain. Acting on information about the changing chemistry and pressure of the blood received from their respective feedback loops, the accelerator center transmits speed-up impulses while the cardioinhibitory center transmits slow-down impulses according to the ever-changing needs of the body. The impulses are received by the heart's pacemaker, which regulates the actual contraction of the heart muscles.

19. The cardiac muscle is nourished by arteries and veins of the coronary system. A heart attack occurs when a coronary vessel is blocked, causing death of a portion of heart muscle.

20. Blood pressure is the force responsible for the circulation of the blood through the body. Pressure is created primarily by the pumping of the heart, but it is modified by the precapillary sphincters in arterioles.

21. The lymphatic system conducts lymph, a tissue fluid, from cells through a series of nodes which act as filters to the thoracic duct, where the lymph empties into the bloodstream.

REVIEW AND STUDY QUESTIONS

1. What does "gas exchange" mean? Why is it important to all living organisms?

2. Compare the process of gas exchange in your own body with gas exchange in a flowering plant. What are the important differences?

3. Describe the action of stomata. What effect does this action have?

4. Trace a drop of rain falling in a forest back to a rain cloud.

5. In what ways are xylem and phloem similar, and in what ways are they different?

6. Trace a drop of blood from one extremity of your body through the system and back to its starting point. Name the various parts of the system it passes through.

7. From what you know about the function of blood in the human body, what reasons can you give for the fact that any interruption of the blood supply to the brain is considered an emergency of the highest seriousness?

8. Why are there valves in veins but not in arteries?

REFERENCES

Comroe, J. H., Jr. 1966 (February). "The Lung." *Scientific American* Offprint no. 1034. Freeman, San Francisco.

Crafts, A S., and C. E. Crisp. 1971. *Phloem Transport in Plants.* Freeman, San Francisco.

Johansen, K., and A. W. Martin. 1965. "Comparative Aspects of Cardiovascular Function in Vertebrates." In W. F. Hamilton, ed., American Physiological Society's *Handbook of Physiology*, Vol. 3.

Mayerson, H .S. 1963 (June). "The Lymphatic System." *Scientific American* Offprint no. 158. Freeman, San Francisco.

Perutz, M. F. 1964 (May). "The Hemoglobin Molecule." *Scientific American* Offprint no. 196. Freeman, San Francisco.

Schmidt-Nielsen, K. 1971 (December). "How Birds Breathe." *Scientific American* Offprint no. 1238. Freeman, San Francisco.

Solomon, A. K. 1971 (February). "The State of Water in Red Cells." *Scientific American* Offprint no. 1213. Freeman, San Francisco.

Wiggers, C. J. 1957. "The Heart." *Scientific American* Offprint no. 62. Freeman, San Francisco.

Wood, J. E. 1968 (January). "The Venous System." *Scientific American* Offprint no. 1093. Freeman, San Francisco.

Zimmermann, M. H. 1963 (March). "How Sap Moves in Trees." *Scientific American* Offprint no. 154. Freeman, San Francisco.

Zweifach, B. 1959 (January). Microcirculation of the Blood." *Scientific American* Offprint no. 64. Freeman, San Francisco.

SUGGESTIONS FOR FURTHER READING

Florey, E. 1966. *An Introduction to General and Comparative Animal Physiology.* Saunders, Philadelphia.

A well-written, clear, well-illustrated text on comparative physiology. Chapter 9 ,"Respiration," is a good presentation of methods and results of comparative study of gas exchange.

Gordon, M. S., et al. 1968. *Animal Function: Principles and Adaptations.* Macmillan, New York.

Combining zoological, comparative, and evolutionary approaches, Gordon examines the adaptive functioning of animals in relation to the survival of whole organisms. Includes many good illustrations.

Meyers, B. S., D. B. Anderson, and R. H. Böhning. 1973. *Introduction to Plant Physiology*, 2nd edition. Van Nostrand, Princeton.

Deals extensively and in depth with absorption, diffusion, and translocation of gases, solutes, and liquids in plants. This is the revised short edition of a classic text.

Steward, F. C. 1964. *Plants at Work.* Addison-Wesley, Reading, Mass.

One of a series of treatises on plant physiology. Chapter 10 thoroughly covers the plant's respiratory activities.

Weichert, C. K. 1970. *Anatomy of the Chordates*, 4th edition. McGraw-Hill, New York.

Chapter 6 is a fifty-page discussion of the chordate respiratory system; it gives many clear examples of gills, swim bladders, lungs, and related or analogous structures.

Weier, T. E., C. R. Stocking, and M. G. Barbour. 1974. *Botany: An Introduction to Plant Biology*, 5th edition. Wiley, New York.

Chapter 12 is a brief but clear and well-illustrated account of transportation, conduction, and absorption in vascular plants.

Weisz, P. B. 1973. *The Science of Zoology*, 2nd edition. McGraw-Hill, New York.

Chapter 9, "Supply, Removal, Transport," discusses the comparative anatomy of circulatory systems, especially vertebrate, and draws them into a frame of reference that embraces the related processes of breathing, alimentation, and excretion. Good photographs and drawings.

Wilmoth, J. H. 1967. *Biology of Invertebrata.* Prentice-Hall, Englewood Cliffs, N.J.

Systems and processes concerned with locomotion, neural development, respiration, excretion, food-getting, and reproduction receive extensive coverage. Also included are the usual phylum classification with interesting side views of invertebrate adaptations.

Wolf, J. K. 1970. "Diagnosis and Acute Treatment of Stroke." *Health News* 47: 2–6.

A review of thrombic and embolic strokes and cerebral hemorrhages with diagnostic aids and preventive action. A glossary of terms for the nonprofessional is included.

MAINTAINING THE INTERNAL ENVIRONMENT
chapter eight

The adult emperor penguin is protected from the cold by an insulating layer of fat and a covering of stiff, dense feathers lying close to its body. Both parents take turns shielding eggs and young chicks from the harsh Antarctic climate by tucking them between the tops of their feet and a warm, bare brood patch under the belly. The fact that the brood patch is also an erogenous zone may explain why adults will brood an egg faithfully—without food and at temperatures of 40° below zero—for 62 to 64 days at a time.

A moment's thought reveals the astounding adaptability and variety of living things. Life is found in salty depths of the ocean and freshwater rivers, in the hot springs of Yosemite and the frigid reaches of Antarctica. But the chemical reactions that typify life can occur only under a narrow range of conditions. In order to survive for even its brief allotment of time on this planet, an organism must be able to maintain optimum internal conditions quite different from a hostile or changing environment.

The first living organisms probably developed in the sea, a relatively stable environment where many materials occurred readily in solution and where the physical and chemical properties of water cushioned the effects of temperature, light, and chemical changes. These primitive living systems were relatively simple and probably had very little capacity to alter their own chemical processes to adapt to changes in the external environment. But life became more functionally and structurally complex, and instead of altering its entire chemical makeup with each evolutionary change, the organism made small adjustments—adjustments that altered the organism in such a way that some of the benefits that the sea once conferred were internalized. Today even the most complex land animals still carry around the rudiments of their own ancient internal seas, where life processes can function under optimum conditions reminiscent of those in which their ancestors arose.

NEED FOR REGULATION

The internal environment is not entirely stable, however. Digested foods absorbed into the bloodstream may drastically alter its chemical composition: toxic (poisonous) wastes that are produced when foods are metabolized accumulate in cells and tissues; water is taken into the body by ingestion and lost through the skin or in the urine. Each of these changes in the chemical balance of the body, if not regulated and controlled, would pose a threat to life. As life forms became more and more complex, control mechanisms evolved. Tissues and organs became specialized for the task of conserving the relative composition of vital internal constituents while removing wastes and toxins. The liver, for example, has evolved as an organ that regulates the nutrient and toxin level of blood, and the kidneys have evolved as organs of water and solute control. The continuous adjustment and optimization of living conditions for the community of cells within a multicellular organism, or for a unicellular organism, is called **homeostasis.**

Homeostasis is a dynamic process. Although the overall quantity of any factor, such as water concentration, may appear to be constant, it actually varies from time to time. When the concentration of water rises or falls outside the optimum range, water is eliminated or absorbed to correct the imbalance. The regulation process, which is typical of all homeostatic mechanisms, involves **feedback control,** a process in which the level or concentration of a substance is monitored, the information about the level is transmitted to an integrating center, and corrective signals (feedback) are sent to an effector which can make the appropriate corrections. Sometimes the regulation is simple and passive, such as the control of pH by buffers, which requires no energy input by the cell; other mechanisms, such as the separation of valuable substances from wastes by the kidney, require energy expenditure by the cell and organism. A very simple model of a homeostatic mechanism is shown in Fig. 8.1.

Such dynamic regulation has limits, however. Even if homeostatic controls can return conditions to optimum, they are of no value if a sudden fluctuation has destroyed the cell or organism before optimal conditions are reached. So homeostasis also implies **stabilization** to avoid wide deviations from optimum levels. In unicellular organisms, for example, an enzyme for a particular product may be turned off as soon as a certain level of the product accumulates. In multicellular organisms, whole organs may be responsible for homeostasis. For instance, vertebrate kidneys regulate salt, water, and waste levels in the body. Nonetheless, many of the same homeostatic mechanisms used by free-living cells are found in the individual cells of multicellular organisms.

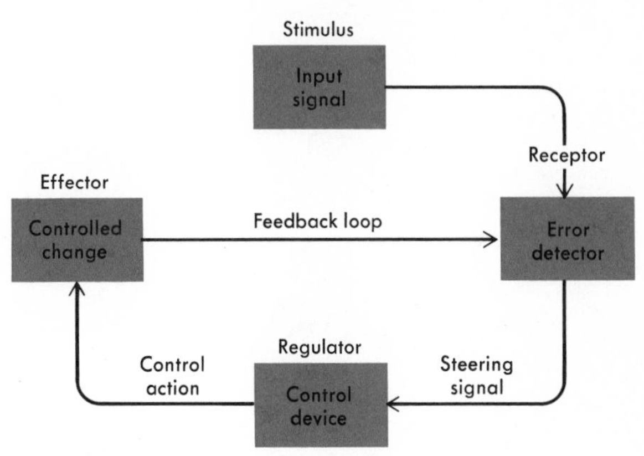

FIGURE 8.1

Feedback control. In a homeostatic mechanism there are continual adjustments as changes initiated by the input signal are detected and subjected to correction.

Cellular homeostasis

An eminent French physician-scientist of the nineteenth century, Claude Bernard, recognized that the ability to regulate the internal environment was essential for the maintenance of life. All aspects of growth, differentiation, genetic expression, and metabolism are carefully controlled by a variety of positive and negative regulatory mechanisms.

END-PRODUCT INHIBITION

Feedback mechanisms are the key to the cell's maintenance of homeostasis. The cell must manufacture or acquire several essential substances, but it must also be able to stop production or intake when the desired level

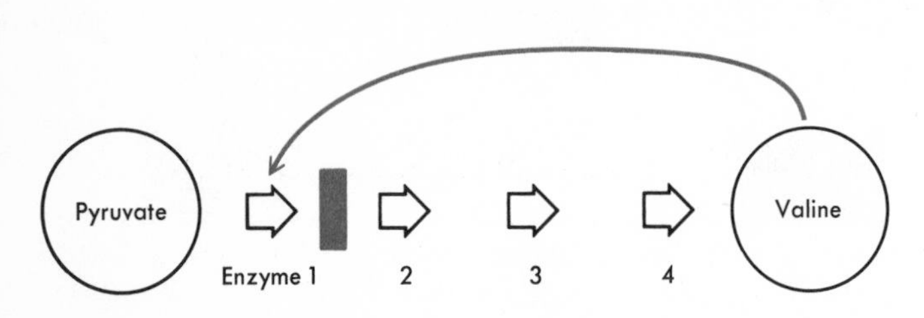

FIGURE 8.2

End-product inhibition. The presence of enough valine (an end product of this reaction) in the cell will stop the catalyzing action of the first enzyme in the reaction chain. End-product inhibition is a form of negative feedback.

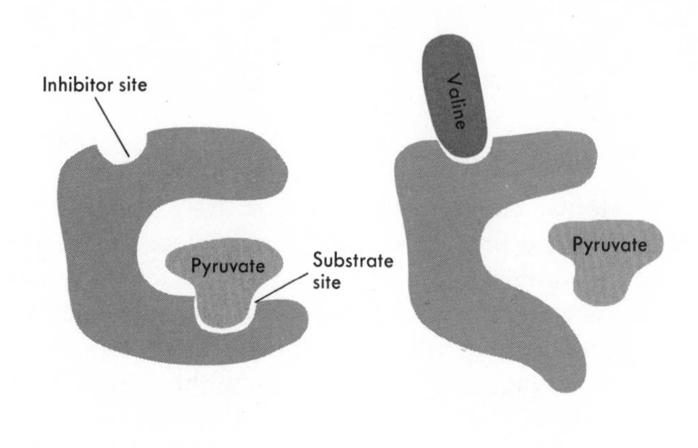

FIGURE 8.3

Allosteric regulation. When the end product binds with the enzyme at the inhibitor site, there is a resulting change in shape of the enzyme and consequently of the substrate site. This change in the substrate site prevents the substrate (pyruvate) from binding to the enzyme.

of those substances is reached. Knowing when to stop (and when to start) is accomplished by feedback control —the same biological mechanism used by organisms to monitor everything from heartbeats to hormone production to the regulation of genes. In cellular chemistry, feedback control is provided by **end-product inhibition.**

End-product inhibition is exactly what its name suggests. The cell uses a chain of enzymatic reactions to produce a needed substance. The end product of such a reaction sequence may be amino acid, a sugar, ATP, or any of various other substances the cell may need. If too much of this end product were to be produced, the cell's homeostatic balance would be upset. Instead, when enough of it has been manufactured (or obtained from outside the cell), the end product itself interrupts the catalyzing action of the first enzyme in the chain, stopping production early in the process.

An example of end-product inhibition is given in Fig. 8.2. The amino acid value is synthesized from pyruvate in a series of chemical reactions involving four different enzymes. When enough valine is present in the cell, the valine molecule itself inhibits the action of the first enzyme in the metabolism of pyruvate. The mechanism of end-product inhibition is a process called **allosteric regulation** (Fig. 8.3). The enzymes in the reaction chain have two separate binding sites—known as the substrate site and the inhibitor site. In the pyruvate-valine chain, the pyruvate combines with the enzyme at the substrate site. However, when valine—the end product—is present in sufficient quantity, it binds to the enzyme at the inhibitor site. This binding action changes the shape of the enzyme, altering the substrate site and making it unable to accept the pyruvate. The reaction chain is broken.

End-product inhibition thus provides the cell with an automatic mechanism for controlling the production of essential substances. It can also save the cell a tremendous amount of energy. In the pyruvate-valine example, when enough valine stops the reaction chain, there is a saving of 38 energy-rich ATP molecules that the cell would need to synthesize the valine from pyruvate. Energy is saved and cellular homeostasis is preserved.

REGULATION OF ACIDITY

Alterations of internal pH may jeopardize the chemistry of the cell. Too much variation can destroy the hydrogen bonds that stabilize enzymes' structure, and this in turn may alter the configurations of functioning enzymes, consequently affecting enzymatic reaction rates. Although this interference may occur at specific places in the organism in order to stop the action of enzymes whose work has been completed, *uncontrolled* changes in pH pose a threat to the cell. A majority of the chemical reactions in the cell are enzyme-mediated and so closely interrelated that alterations in a single reaction rate may interfere with the functioning of the rest of the organism (see Fig. 8.4).

Acids and bases that alter cytoplasmic pH may be used or created in a wide variety of reactions. Ammonia, for example, is a waste product of protein metabolism; it is a highly alkaline substance and would quickly elevate cytoplasmic pH if there were no pH-regulating agents inside the cell. Carbon dioxide, on the other hand, is a waste product of respiration and fermentation, which lowers the pH of the cell by forming carbonic acid when it dissolves in cytoplasmic water. Hydrolysis of fats can also lower cellular pH because the process liberates weak acids. Because of the number of reactions that alter pH, a wide variety of passive and active mechanisms have evolved to control these levels.

Buffers A number of compounds found in the body have the capacity to regulate changes in free hydrogen ions. They do so by literally sopping up free H^+ or OH^- ions and incorporating them into relatively stable compounds (Fig. 8.5). A buffer is the salt of a weak acid and a strong base and dissociates—that is, ionizes—readily in water. Strong acids also dissociate readily, yielding large quantities of free H^+ ions; weak acids dissociate less readily and thus bind free H^+ ions. Buffers increase the concentration of weak acids at the expense of strong acids. The body's capacity to buffer acids is enormous, and many proteins, including hemoglobin, have a considerable buffering capacity, but HCO_3 (bicarbonate), HPO_4, and other salts carry the main burden of buffering.

Regulating carbon dioxide levels Carbon dioxide is one of the major products of metabolism. When dissolved in cellular water, carbon dioxide forms carbonic acid (H_2CO_3). Though a weak acid, H_2CO_3 can appreciably lower cytoplasmic pH and thus adversely affect the cell. Carbon dioxide is also essential in the regulation of brain functions that control respiration, heart rate, and blood flow. It is not surprising that a number of homeostatic mechanisms regulating carbon dioxide concentration evolved over the course of time.

Elimination and regulation of the amount of carbon dioxide in the cell depends on the relatively simple physical process of diffusion. Cellular metabolism raises the concentration of intracellular carbon dioxide to a point higher than its concentration in the surrounding milieu (air or water), and the gas simply diffuses out along the

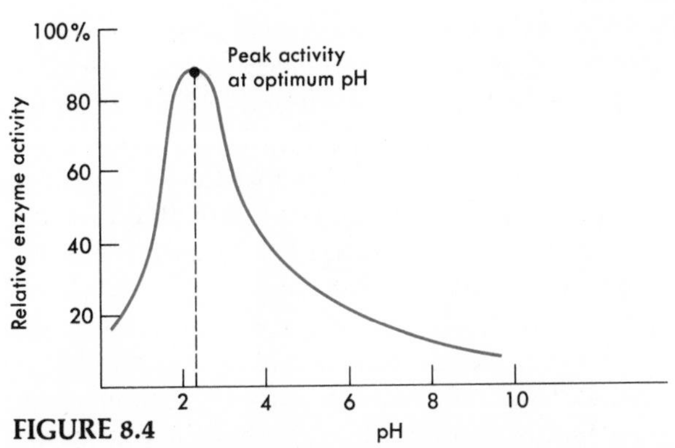

FIGURE 8.4

Optimum pH levels. The maintenance of pH is critical for many reactions in multicellular organisms. For example, the stomach enzyme pepsin works most effectively at a pH of slightly above 2. Pepsin's efficiency in breaking down proteins diminishes if the pH is either higher or lower than this optimum. Other enzymes have their own optimum pH ranges.

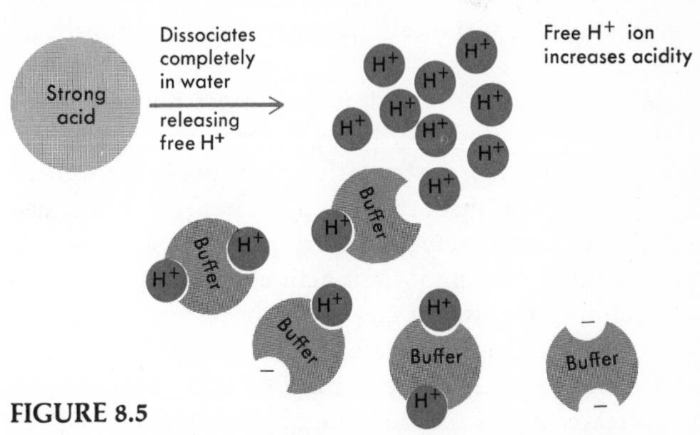

FIGURE 8.5

Buffering. A buffer is a substance that can control pH fluctuations without any expenditure of energy by the cell or organism. Strong acids dissociate almost completely in solution, liberating many H^+ ions. A buffer has an affinity for free H^+ ions, thus trapping them and preventing them from substantially altering the pH.

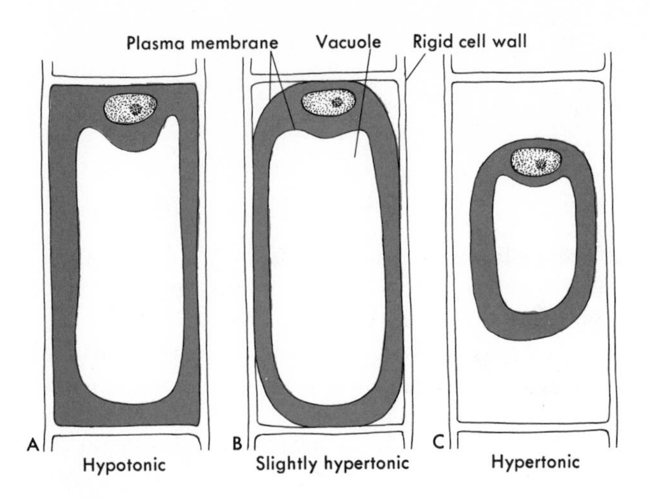

FIGURE 8.6

Plasmolysis. In the epidermal cell of a plant, plasmolysis occurs when the cell body loses additional amounts of water with each increase in the osmotic concentration of the surroundings. As more water leaves, the cell body shrinks away from the rigid cell wall. Plasmolysis is usually reversible if the osmotic concentration outside the cell decreases.

concentration gradient (from an area of high concentration to an area of low concentration). In multicellular animals, removal from the cell does not eliminate the problem; this will be dealt with in a later section.

Elimination of nitrogen-containing wastes Nitrogen-containing wastes may accumulate within a cell as the result of the metabolism of nitrogen-containing proteins or other substances. Although we do not fully understand how these substances are removed from a unicellular organism, simple diffusion may adequately account for its disposal. Since these nitrogen-containing wastes are in higher concentration inside the cell than outside, they would move along the concentration gradient and leave the cell.

REGULATION OF SALTS AND WATER

Regulation of salts Salts (or the metal ions of salts) are vital to a cell because they are components of many biochemical molecules. As explained in Chapter 5, however, salts create problems by contributing to the osmotic potential of cells. Excessive rise in salt concentration within an animal cell may lead to osmotic uptake of water, resulting in cell rupture. Conversely, a salt concentration that is low relative to its surroundings may cause osmotic loss of water from the cell, possibly resulting in cellular death.

Since salt concentration is of utmost importance to a cell, evolution gradually favored organisms with regulatory mechanisms, particularly mechanisms concerned with active ion transport, in which energy is expended to maintain adequately high levels of intracellular salt concentration. One important active transport system is the sodium-potassium pump, which brings potassium ions into the cell and excretes sodium ions. As we shall see, this mechanism also plays a vital role in multicellular animals in the production of electric potential in neurons. These mechanisms usually ensure salt concentrations that closely resemble those that probably existed in the primeval seas in which life is thought to have begun. Thus, even cells that have evolved to live apart from the sea still maintain internal environments with rather high salt concentrations.

Water balance Water is of critical importance to all cells for several reasons. First, almost everything needed by the cell will dissolve in this universal solvent; water is an excellent medium of transport for nutrients, gases, and wastes. Second, water is the major component of cellular matter. Hence great selective value was placed on the evolution of mechanisms to regulate water intake and disposal.

In plants, water regulation involves simple processes (see Fig. 8.6). Plant cells expend energy to acquire salts, but as intracellular salt concentration rises, water enters by osmosis. Therefore, an expenditure of energy to accumulate salt ions usually is accompanied by a passive uptake of water. Plant cells, however, are able to avoid excessive water uptake because of the rigidity imposed by their cellulose walls. As water enters the cell, turgor pressure develops inside the cell wall. Fortunately, the wall has a limited amount of give and resists any change in the shape of the cell. Because water is incompressible and only a limited amount can fit within the confines of the cell, water uptake is limited.

Animal cells have much the same requirement for water as do plant cells. However, mechanisms for acquisition and disposal of water tend to be more complicated in animal cells than in plant cells. As free-living animal cells evolved adaptations for living in fresh water, water regulation became a difficult problem because the hypotonic medium that bathes them can cause cell rupture.

Water may enter animal and plant cells by osmosis, or by pinocytosis (see Chapter 5). However, the passive control afforded by plant cell walls is not available to animal cells. Instead, some animal cells (notably the Protista) evolved the contractile vacuole, an active, energy-requiring mechanism for bailing out excess water (Fig. 8.8).

FIGURE 8.7

During the hot dry season in Africa, large animals such as the elephant avoid overheating and dehydration by resting in any available shade. The flapping of their great ears increases heat radiation from their bodies. The baobab tree under which they stand also regulates its internal condition; it will shed its leaves during severe drought to reduce water evaporation.

REGULATION OF NUTRIENTS

The level of nutrients in plant cells and animal cells must be carefully regulated to prevent the cell from starving for lack of the necessary material for metabolic reactions. In many kinds of plant cells excess sugars manufactured during photosynthesis are converted to starch granules—in the chloroplast or elsewhere—and are stored as food reserves. At night, or during times of need, the plant cell breaks down the starch to maintain necessary nutrient levels until dawn, when sugar production can begin again.

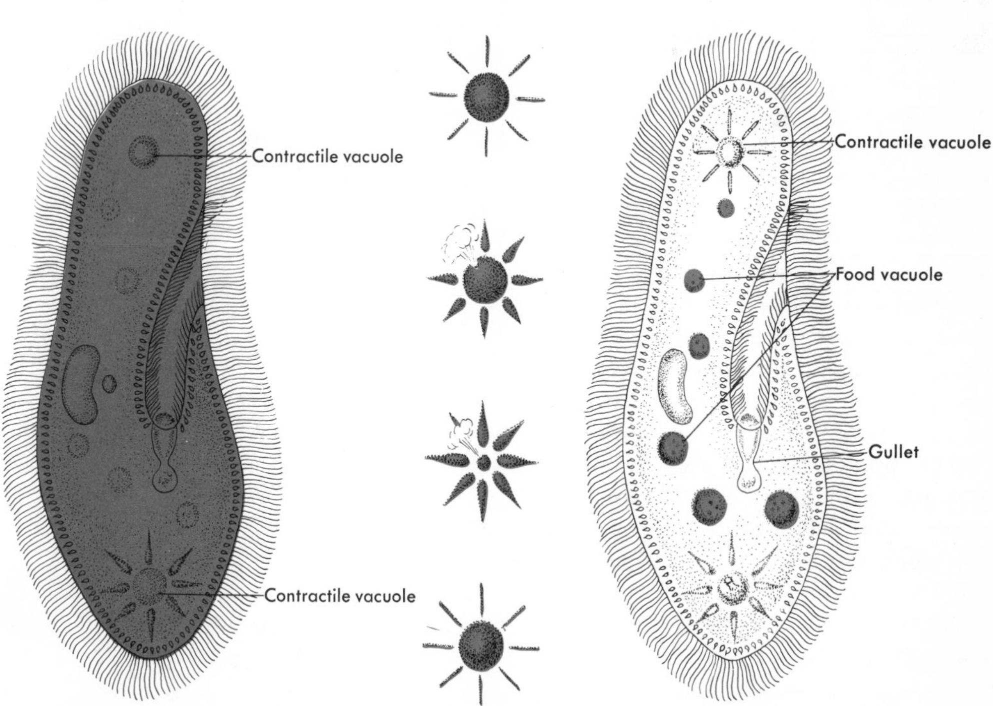

FIGURE 8.8

Water and nutrient regulation. Contractile vacuoles are a mechanism of water elimination in which osmosis causes water influx into the cell. The contractile vacuole actively pumps water back out to prevent cell-bursting. In paramecia the vacuole is surrounded by radiating tubules. As the tubules fill with water, they swell. Then they contract, emptying their contents into the vacuole, which, when full, contracts and squirts water out of the organism. This cycle repeats rhythmically.

Bacteria contained in the food vacuoles are digested there, and the nutrients are absorbed, so that each vacuole shrinks as it moves posteriorly from the gullet.

Since simple animals cannot create nutrients from simpler, inorganic substances by photosynthesis, they must forage almost continuously for food in order to acquire energy and substance for growth. The paramecium, for example, seems to constantly sweep food into its oral groove. When an excess of nutrients exists, feeding movements may continue until several food-digesting vacuoles have been created. Eventually, however, some form of self-regulation of the number of food vacuoles may occur, since the space for these organelles in a paramecium is somewhat limited (see Fig. 8.8).

Nutrient control at the molecular or genetic level may also be exercised to some extent. Many organisms can turn on and off their production of enzymes (such as digestive enzymes). In these organisms, controlling mechanisms operate on a chemical level to provide the enzymes that must be present when foods require processing. When digestion is completed, the production of enzymes ceases.

Homeostasis in multicellular animals

Multicellular organisms have more complex problems of homeostasis, for they must control and integrate many specialized areas or systems. We have seen, for example, the specialized role of the stomata in controlling water loss in multicellular plants. Many of the other regulating systems in plants are covered in other chapters.

Homeostatic specialization is greatest in the multicellular animals. Because of this specialization, some cells may lose their capacity for regulating a particular process, such as water balance. If there is too much water in the tissues, for example, erythrocytes (red blood cells) will continue to take in water until they burst. Only the presence of the other cells which regulate water balance ensures that red blood cells will not encounter this problem. These areas of specialized cells working in concert are called organs. Two of the most important homeostatic organs in the vertebrates are the liver and the kidney.

THE MAMMALIAN LIVER

The liver in mammals is the largest and most complex of all organs of the body. In evolutionary terms, the vertebrate liver probably began as a diverticulum (pouch) of the digestive tract and eventually evolved to become a secretory organ. Although the mammalian liver secretes bile, the liver in primitive organisms may have secreted digestive enzymes as well. In vertebrates, the liver's homeostatic function far outweighs this minor secretory role.

Liver tissue is composed of three types of cells. They are the cells of the biliary duct system, the star-shaped Kupffer cells, and the hepatic cells. Kupffer cells serve to eliminate foreign bodies and dead or injured erythrocytes from the body's circulation. The hepatic cells are the most important units of liver tissue because they are involved in the metabolism of all basic food elements, including vitamins. Moreover, through their metabolic function the hepatic cells are responsible for heat production and may detoxify certain poisons and drugs in the circulation. Some antibiotics may be broken down in the liver, so that they do not reach toxic levels in the blood.

The connection between the liver and the intestines is well suited to the liver's regulatory function. The main blood vessel of this connection, the hepatic portal vein, is unusual in that it has capillary terminations at both ends (Fig. 8.9). Most of the nutrients that are absorbed after digestion pass directly from capillaries in the intestinal villi through the portal vein into the liver. The vein breaks into a capillary bed at its termination in the liver to distribute substances throughout the organ. Nutrients that are present in excess, such as glucose, are withdrawn from the blood, and some are stored as glycogen in the liver. These substances can be released gradually to meet the body's needs. If the products of digestion were instead dumped directly into the general circulation from the intestines, the pH and osmotic concentration of the blood would be severely disturbed.

Glycemic regulation Glucose, the end product of carbohydrate digestion, circulates in the bloodstream. The

Portfolio 2: Cells That Perform Specialized Functions

2.1
The anatomical structure of a cell is not determined at random. Each aspect of shape is related to some function. Flexible disc-shaped red blood cells, shown surrounding a single white cell with many small fingerlike projections, have large surface areas for gas transport. The white cell has a considerably larger surface covered with antibodies that are invisible even under the electron microscope.

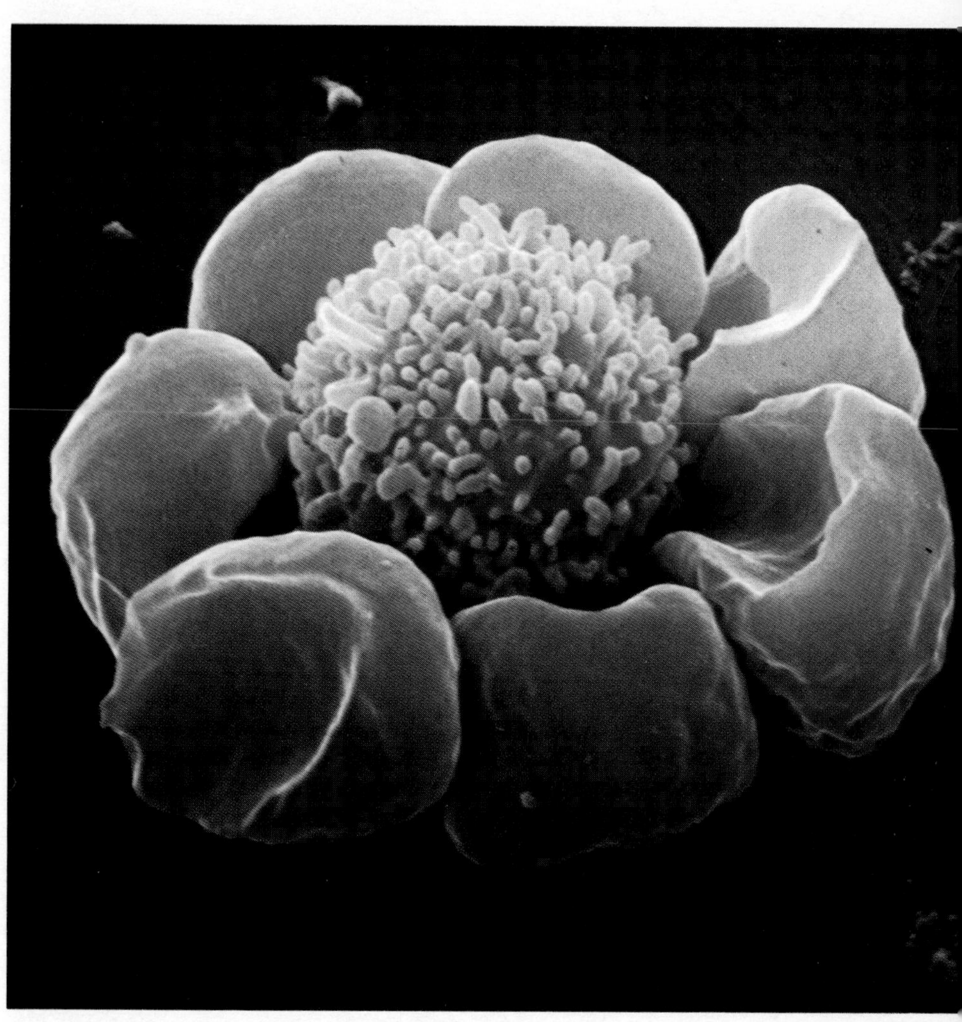

2.1 13,400X

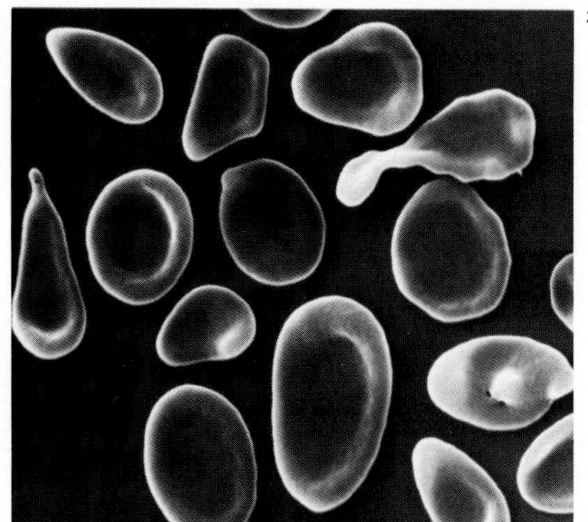

2.2

1,875X

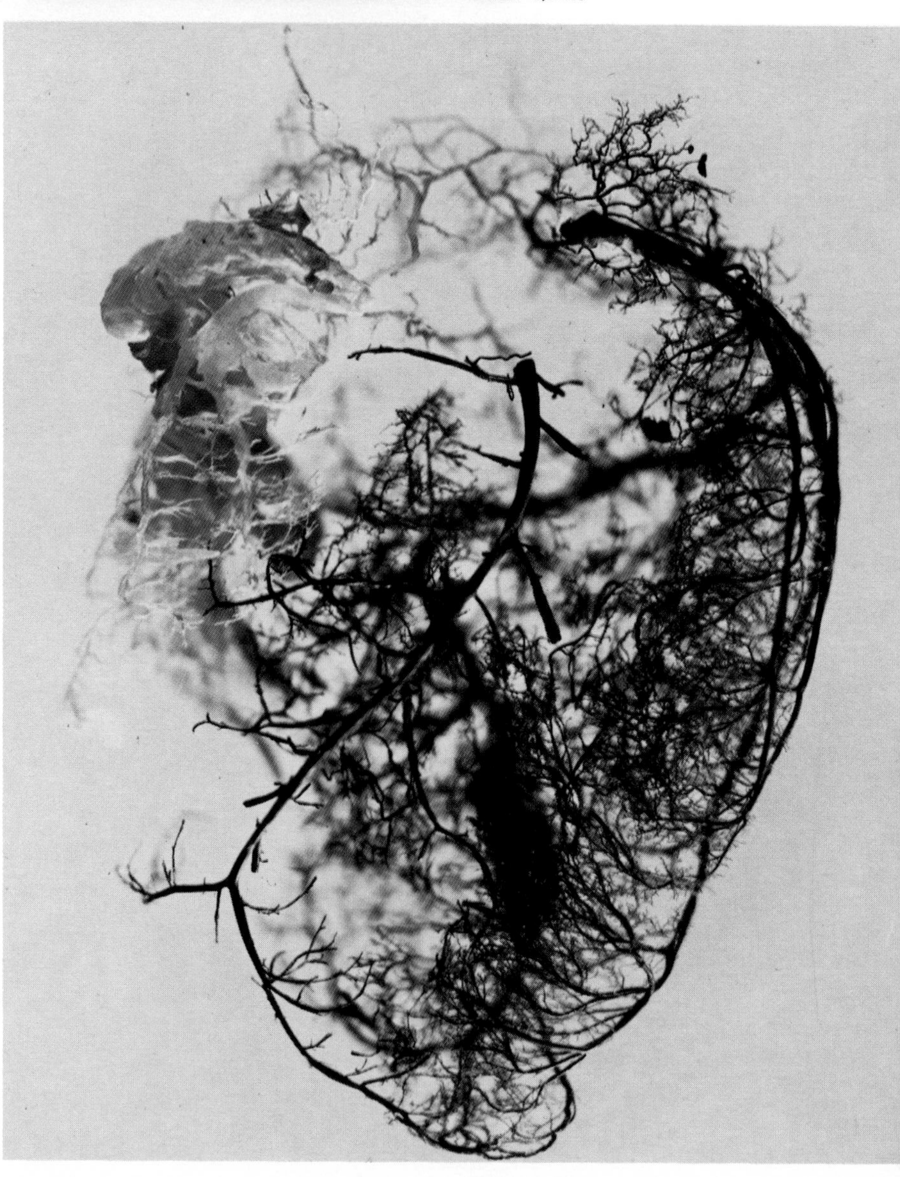

2.2
Both normal and abnormal human blood cells are present in this scanning electron micrograph. The normal cells are flattened, round, biconcave discs. The bizarre shapes of the abnormal cells indicate a genetic disease called Thalassemia, commonly found where malaria is prevalent.

2.3
The human heart is actually a highly modified large blood vessel with an enormous musculature. The continuously active muscles are richly supplied with blood vessels that carry in nutrients and oxygen and remove wastes and carbon dioxide. To prepare the specimen shown, a plastic material was injected into the heart's vessels, and the surrounding tissue was partially eliminated to reveal the network of blood vessels.

2.4
This SEM of a taste bud on the tongue of a frog reveals an array of structures that evolved as specific chemical sensors.

2.5
Fingerlike villi are typically found in intestines, such as that of a mouse, shown in this SEM. Most of the absorption of nutrients from the intestine occurs across the villi membranes. Blood and lymph vessels within the villi carry the nutrients into the general circulation.

2.6
The surface area of a cell is greatly increased by microvilli, which are projections from the cell membrane. This electron micrograph shows microvilli along the apical border of a cell lining the intestine of a cat.

Graziadai/Omikron

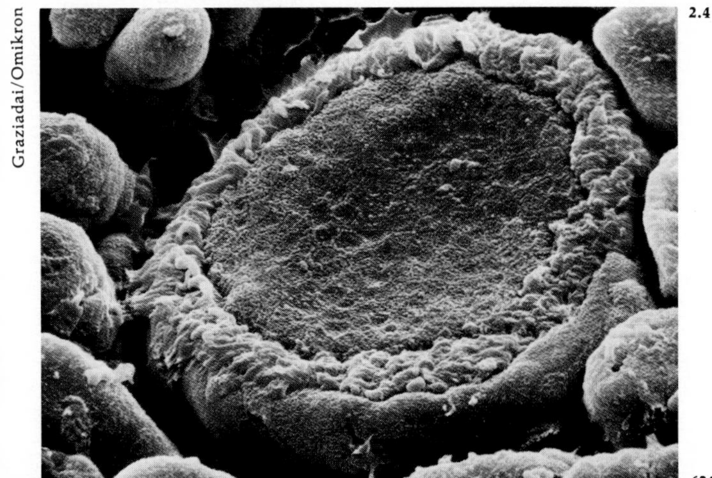

2.4

624X

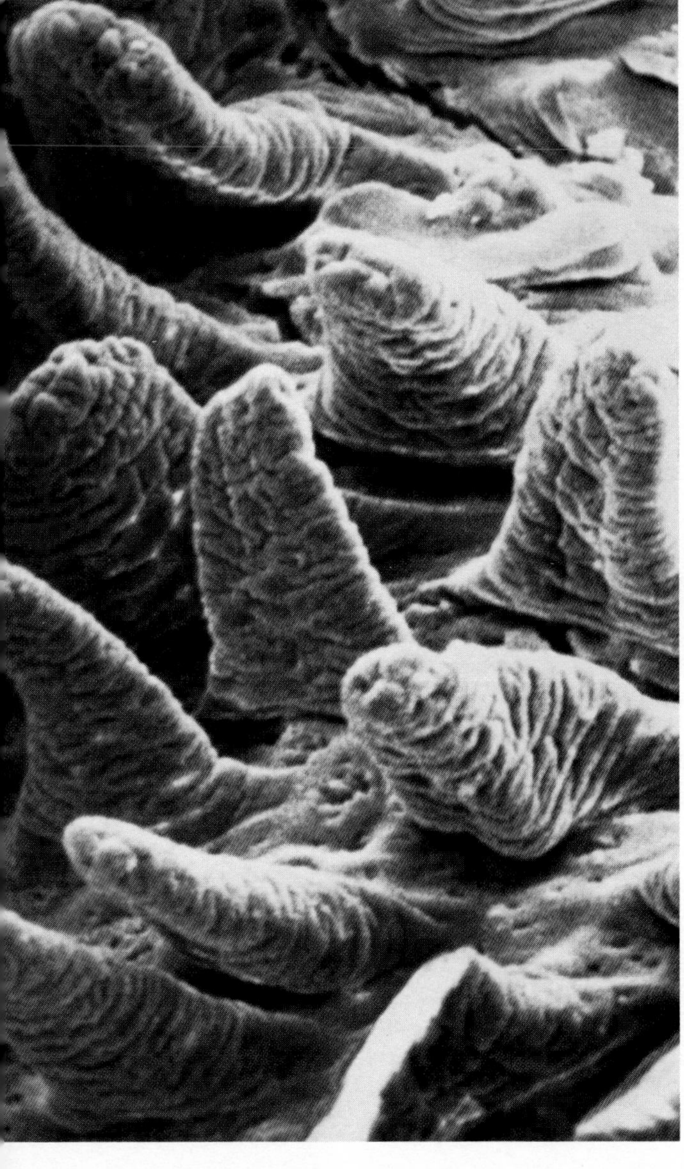

2.5

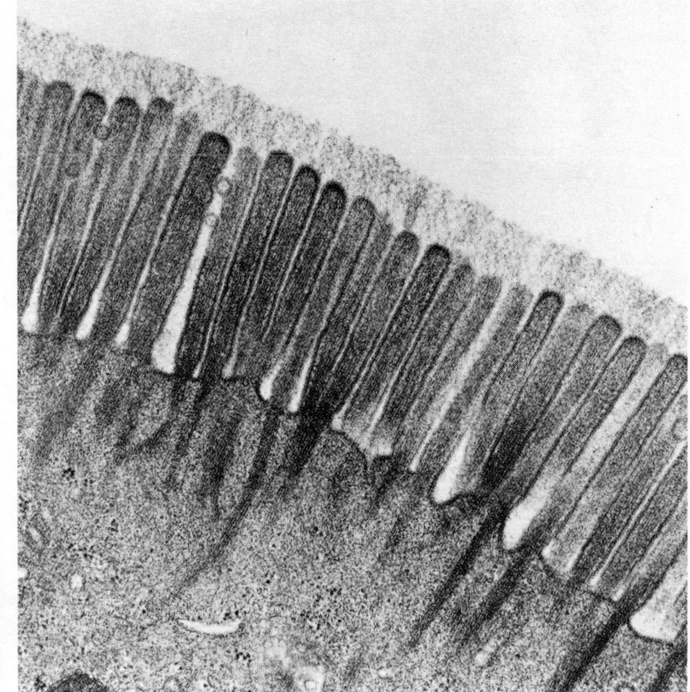

2.6
40,000X

2.7
The tubular structures transporting gases to and from the lungs are lined with ciliated epithelium. The movement of the cilia forces trapped foreign matter out of the lungs.

2.8
An SEM of the eye of a white fly shows the typical structure of an insect eye with its large number of individual light-sensitive receptors arranged in mosaic form. The complexity of the surface architecture suggests the even greater interior complexity.

2.7 8,800X

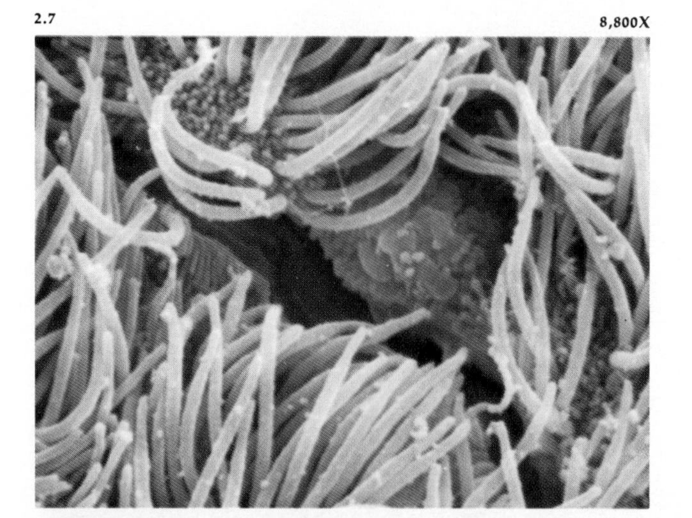

2.8

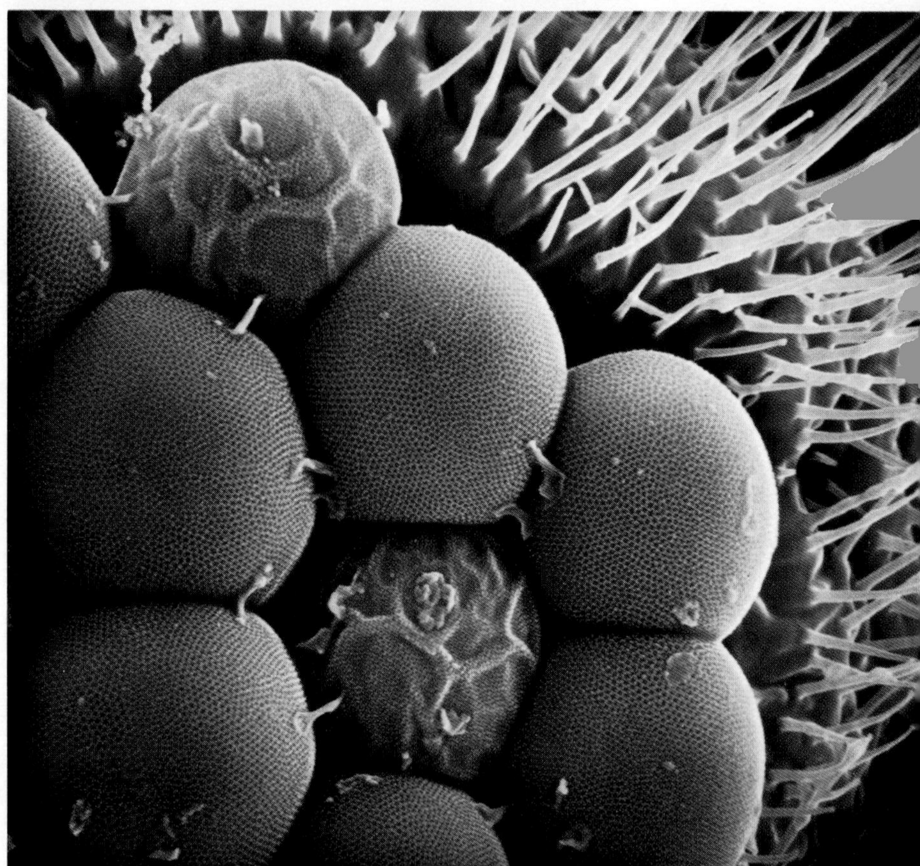

4,060X

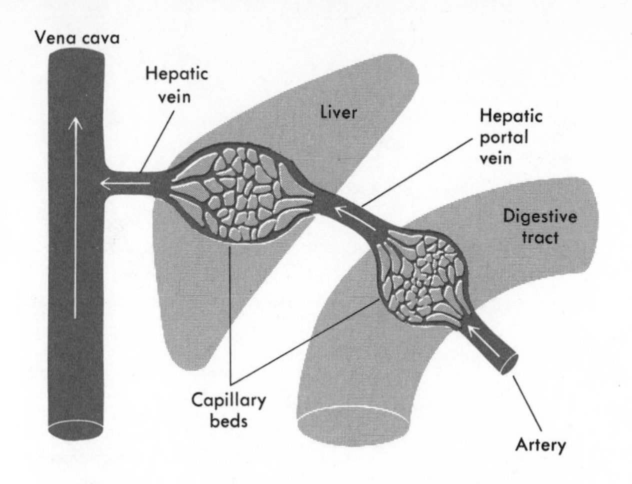

FIGURE 8.9

Mammalian liver. The hepatic portal circulation is unusual because it contains a vein that has capillaries at both ends. Blood flows into the digestive tract, traverses one capillary bed, and is then transported by the hepatic portal vein to the liver. There the blood is distributed through a second capillary bed in the liver, where nutrient regulation occurs. The blood then empties into the hepatic vein, which transports it to the inferior vena cava, leading to the heart.

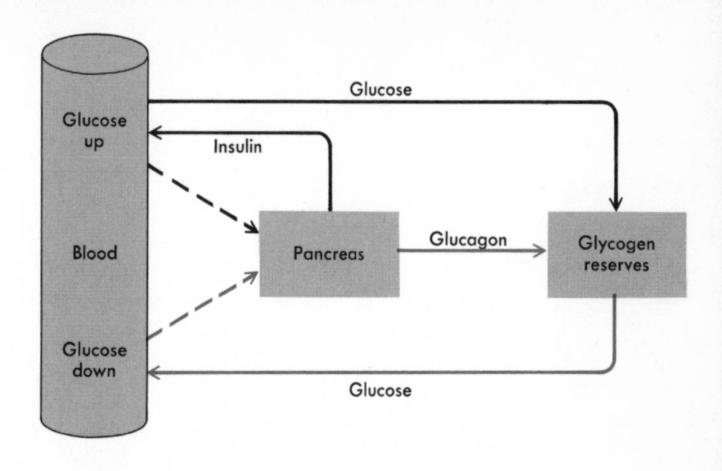

FIGURE 8.10

Glycemic regulation. A decrease in the level of glucose in the blood triggers a signal (colored path) to the pancreas to secrete glucagon, a hormone that acts to convert some of the glycogen reserves stored in the liver to glucose and return it to the blood. An increase in blood glucose sends a different signal (black path) to the pancreas, which secretes insulin and sends it to the blood, where it converts glucose in the blood to glycogen for storage in the liver.

concentration of sugar in the blood is called glycemia, and the process by which it is controlled is **glycemic regulation.** The normal level of glucose is 70 to 120 mg per 100 ml of blood. Levels below 70 mg per 100 ml constitute the condition of hypoglycemia; levels above 120 mg per 100 ml constitute hyperglycemia. Either condition can lead to serious disturbances of the normal functions of cells. For example, hypoglycemia (a result of prolonged fasting) may produce irritability, coma, and finally death.

Blood sugar is derived from three sources: (1) carbohydrate intake, (2) hydrolysis of glycogen stored in the liver (glycolysis), and (3) conversion by the liver of such carbohydrate precursors as amino acids, glycerol derived from fats, and metabolic by-products of glucose metabolism—the last being called **glucogenesis.** The liver also can convert some glucose into fat, which is stored in various deposits throughout the body **(lipogenesis).** The rate and direction of reactions regulating glucose utilization and storage are under a variety of controls. The controls involve specific regulator hormones, such as insulin, glucagon, and norepinephrine (see Chapter 9), and end-product inhibition of key enzymes in the glucose metabolic pathway.

The liver, in concert with other organs, regulates the constancy of blood sugar level. The glucose taken from the circulation by the tissues is replenished by liberation of glucose from the liver into the bloodstream. The liver's capacity to furnish glucose on demand is based on the organ's capacity to convert monosaccharides into glycogen (animal starch), a storage unit for glucose molecules. In this way the liver can retain glucose in the form of glycogen when glucose is in excess. Glycemic regulation is accomplished by a coordinated effort of endocrine glands and regulating centers in the hypothalamic region of the brain. As Fig. 8.10 demonstrates, glycemic regulation also makes use of the feedback principle introduced earlier.

The problem of nitrogen Animals usually ingest far more amino acids than they need. The liver regulates the amount released to the body and breaks down the rest by removing the nitrogenous amino group (a process known as deamination) and converting the remains into glucose, glycogen, or fat. Unlike plants, which can reuse nitrogen, animals are threatened with self-poisoning from nitrogen. When deamination occurs, a toxic substance, ammonia (NH_3), is formed. Although in some organisms the liver releases ammonia into the blood to be transported for excretion, in man and many other animals ammonia is converted by the liver into urea, a less toxic substance. In other groups of animals (birds

Diabetes Mellitus: A Glucose Roller Coaster

Diabetes mellitus is a complex disease characterized in most cases by an insufficient amount or a total lack of insulin, a hormone produced by the pancreas. Some forms of the disease may also result from the body's inability to effectively use the insulin that is produced. Since insulin is involved in the metabolism of carbohydrates, any disorder of insulin production, release, or effect can alter the body's ability to use carbohydrates, and this in turn can alter the metabolism of proteins as well.

There are about three million diabetics in the United States and probably two million other prediabetics. The prediabetics are genetically programmed as diabetics, but no biochemical abnormalities are detectable —yet. The disease is thought to be progressive and proceeds to a chemical diabetes phase. Such patients also show no clinical symptoms, but if given laboratory tests to measure their glucose tolerance, they reveal that they are indeed unable to efficiently handle glucose. At some point the patients pass into a phase in which they show a variety of symptoms associated with insulin lack. This condition is clinical diabetes.

The symptoms of diabetes produce a whole constellation of effects: elevated blood sugar; sugar in the urine, excessive urine production accompanied by thirst to make up for the water loss; weight loss; and weakness. However, not all diabetics show these symptoms; yet they may still show the later complications of the disease. Those complications involve disruptions of blood flow, altered metabolism of the blood vessels, and changes in the retina of the eye, in the kidney, and in the central nervous system. Often these abnormalities can be corrected by administration of insulin or drugs that promote insulin production, but sometimes even insulin treatment fails to prevent the complications associated with the disease. Some of the complications are probably related to some metabolic disturbance whose role is not yet understood.

We know that the genetics of diabetes are complex and that the genes that cause the disease tend to become active late in life. Thus the incidence of diabetes increases dramatically in the over-40 population. Eighty percent of all diabetics are overweight when the disease becomes clinically detectable. This fact suggests that the metabolic disorder had been silently progressing long before the symptoms appeared. Indeed, some investigators think that in many cases the cells that produce insulin, the so-called islets of Langerhans of the pancreas, were congenitally defective.

In order to understand diabetes, one must know how the normal body produces and secretes insulin after a meal. Insulin release is primarily triggered by an increase in blood sugar and to a lesser degree by increases in blood amino acids. Insulin once released has many effects. It accelerates the transport of glucose into muscle and fatty tissues. It promotes the conversion of glucose into glycogen in the liver and muscles. It also promotes the formation of fat in tissue. Finally, insulin promotes the synthesis of new polypeptides, probably working in concert with human growth hormone. Many of these effects of insulin are thought to be due to an effect of the hormone on certain cell membranes, and insulin is known to affect potassium movement across membranes. Given this brief overview of some of the major effects of influence of insulin on normal body functions, is it any wonder that insulin lack in diabetes produces such a wide spectrum of symptoms?

When the diabetic experiences severe insulin lack, absorbed sugars are not transported across cell membranes, and glucose is not converted to glycogen, therefore the blood sugar level rises. Normally blood sugar filtered by the kidney is all reabsorbed into the bloodstream, but since there is a limit to the absorption capacity of the kidney tubules, glucose in excess is flushed out into the urine, and the urine becomes sweet (literally). Since heart muscle, skeletal muscle, and fat tissues need insulin to transport sugar across their membranes, the cells of those tissues literally begin to starve. Deprived of glucose, their fuel of choice, the cells begin to break down free fatty acids and amino acids to provide their energy needs. In the process of breaking down fatty acids, toxic ketone bodies are formed. Ketone bodies are also weak acids, and the diabetic therefore begins to show an increasingly acid urine loaded with ketones. Some of the ketone bodies give a detectable acetone odor to both the breath and the urine. If this process continues without treatment, the patient may actually go into a diabetic coma and even die. Disturbances of amino acid metabolism also occur, and there is often a loss of protein from muscle and bone.

The progression from early stages of diabetes to the clinical, detectable stage may never occur. If development is slow and the disease appears relatively late in life, it can usually be controlled with a special diet or oral drugs that stimulate insulin production. However, in younger persons, the progression to the clinical phase is often very rapid and much more difficult to control. The so-called juvenile form of diabetes usually requires both insulin injections and diet management.

Most diabetics discover their condition in time, and they are usually treated successfully with a combination of drugs and a precisely controlled diet. Their energy intake (calories) and their energy used for daily activity must exactly balance their insulin intake. If they are not careful, they can go from being hypoglycemic (low blood sugar) to hyperglycemic (high blood sugar) in a couple of hours. In a sense, treated diabetics are riding a glucose roller coaster.

Diabetics must test their urine daily, some as often as four times a day, to be sure that little or no sugar is being excreted. If the sugar level in the urine rises, the insulin intake must be adjusted. Insulin level must also be watched. It may become too high if one or more of three conditions develop: too much insulin, too little food, too much exercise. Deprived of glucose, the brain begins to malfunction, first causing decreased irritability and sometimes even convulsions, or insulin shock. Because it can come on rapidly, insulin shock can be very dangerous. Furthermore, it is often preceded by sweating, tremors, dizziness, and staggering, and since these symptoms can easily be mistaken for drunkenness, people are sometimes reluctant to give help. Consequently, most diabetics always carry two items—some high sugar food and a necklace or bracelet that identifies them as diabetics.

The life expectancy of diabetics was once grim. Before 1914, more than 80 percent of diabetic children died by the age of 10. By 1961 that figure had dropped to one percent. Today the life expectancy of a diabetic is about three-fourths of a normal life expectancy.

Diabetics tend to have more health problems than nondiabetics. Blood vessel changes cause a higher incidence of blindness, hardening of the arteries, and coronary artery disease. Poor circulation to the extremities may lead to gangrene. Diabetics are also more prone to infections and kidney disease. A diabetic requiring surgery must be carefully prepared in order to control the blood sugar during and after the operation. Finally, diabetic women who become pregnant have a much higher incidence of spontaneous abortions. However, with today's better modes of treatment, many diabetic women successfully raise families.

With all the developments in recent decades, the diabetic's life style is now much closer to normal than it ever was before. Research continues into basic biochemical problems and genetic causes, as well as into more convenient methods of dispensing insulin.

Someday medical technology may provide us with a method of transplanting or implanting insulin-secreting cells from healthy islets of Langerhans. Or perhaps biomedical engineers can develop an artificial pancreas that will monitor blood sugar and automatically dispense insulin into the body. There is even the possibility that by using recombinant DNA, we can use bacteria to replace the genetic code that diabetics lack—the message that tells the body how to make and secrete insulin. All these possibilities are still far in the future, however. For the present, early detection and current types of treatment remain the best available weapons against diabetes mellitus.

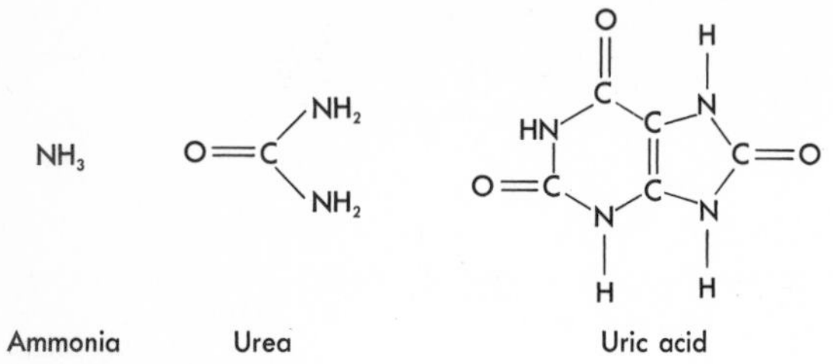

NH_3 — Ammonia

$O{=}C(NH_2)_2$ — Urea

Uric acid

FIGURE 8.11

Structural formulas of three major compounds in which nitrogen—the waste product of amino acid metabolism—is excreted. The principal waste compound functional in mammals is urea. In birds, insects, and reptiles it is uric acid.

and reptiles) nitrogen is converted into uric acid for excretion (Fig. 8.11). In all cases the liver is the treatment plant for nitrogenous wastes. However, without an excretory system these less toxic substances would still accumulate and eventually destroy the organism.

KIDNEYS AND OTHER EXCRETORY STRUCTURES

As organisms moved from marine environments to fresh water and land, special adaptations were required to enable them to survive the differences of osmotic potential posed by their new environments. In regulating water and solute concentrations in the blood, the kidneys can be regarded as the major homeostatic organ in the human body as well as in other higher animals. Homer Smith, a noted physiologist, paying tribute to the importance of the kidneys in regulating homeostasis, commented that "the composition of the blood and the internal environment is determined not by what the mouth ingests but by what the kidneys keep." The kidney is the most complex regulatory organ in the body, as we shall see, but it had its evolutionary beginnings in very simple structures.

Rudimentary excretory structures Even in many simple organisms there is a rudimentary equivalent of the renal systems found in higher organisms. One resembling the tubular structure of the vertebrate kidney is the excretory canal system of the planarian which performs its function through the activity of the **flame cells.** The excretory system of the planarian consists of a set of tubules running the length of each side of the body. The tubules have many branches and dead-end pouches (flame cells), whose tufts of cilia are in constant sweeping motion, not unlike flickering candles—hence the name "flame" cell (Fig. 8.12). Water that must be bailed out is continually entering the flame cells, and the cilia create a current that carries the excess water to the excretory pore, where it is voided.

Most terrestrial organisms rely on the kidneys to control water volume and concentration. The kidneys have an intimate physical and physiological association with the circulatory system. Blood that contains excessive water, salts, and wastes from metabolic reactions can be adjusted so that the concentration of solutes and water is maintained within a normal range. The evolutionary prototype of the vertebrate kidney is easily dis-

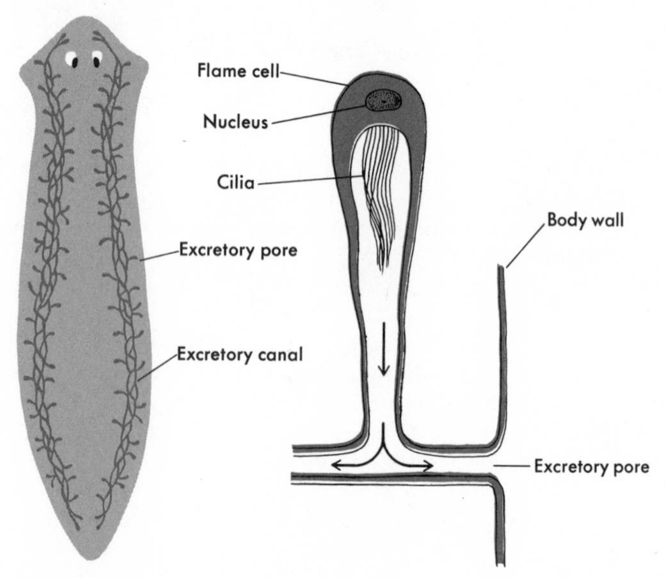

FIGURE 8.12

The flatworm, specifically the planarian, the lowest order of classification in which the rudiments of a tubular excretory system first appear. Two tubules run the length of the body, branching along the way into small bulblike flame cells. Water and waste materials from the tissue fluids enter the flame cells, whose tufts of cilia undulate in a manner suggestive of a flickering flame. With the aid of the ciliary motion, these materials are transported to the excretory pores, where they are excreted.

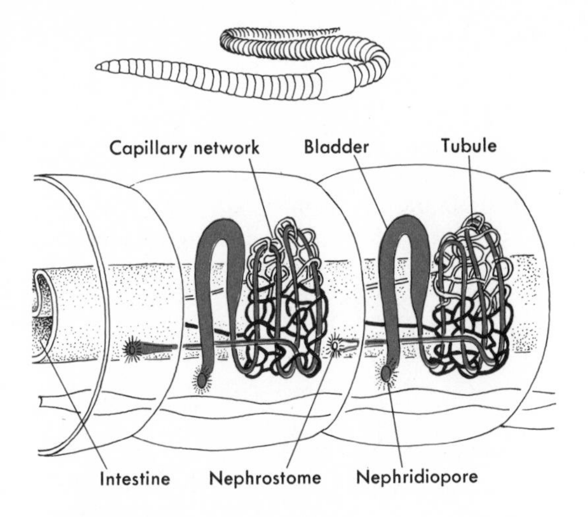

FIGURE 8.13

Excretory system of the earthworm. Each segment in the earthworm (except the first three and the last) contains a pair of coiled tubules, one on each side of the intestine, that empty waste materials into a bladder. The tubule in each segment collects waste from the segment immediately anterior to it by means of a ciliated funnel, the nephrostome, which opens into the tubule. The tubule then passes posteriorly through the membranous septum between the segments and coils around a network of capillaries before emptying into the bladder. The materials are excreted from the bladder through a nephridiopore. The entire structure is called a nephridium. Each segment contains two nephridia, but only the nearer one is shown.

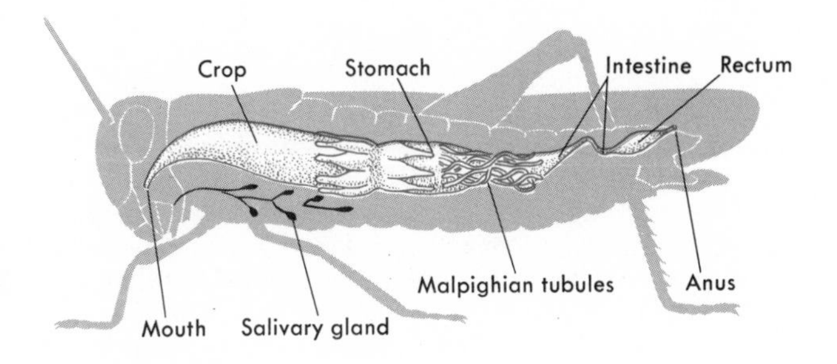

FIGURE 8.14

Excretory structure of the insect. The distal ends of the Malpighian tubules—pouches of the insect digestive tract located between the midgut and hindgut—absorb fluids containing waste materials, which, prior to their entry into the hindgut, undergo a reabsorptive process. The fluid, now concentrated as urine, passes through the hindgut into the rectum. Here additional reabsorption occurs, and the substance is finally excreted with the feces as a dry material.

cernible in the earthworm (Fig. 8.13). This prototype is the **nephridium,** which resembles the vertebrate kidney both structurally and functionally, consisting of a coiled tubule embedded in capillaries. One end of the tubule is exposed to tissue fluid at the **nephrostome** (tissue-fluid opening); the other end of the tubule excretes urine. Cilia sweep tissue fluid into the nephrostome, and materials are exchanged between the fluid passing through the tubule and blood flowing in nearby capillaries. Excess waste products (solutes) are released from the blood into the tubule fluid, and usable substances (salts and other solutes) are passed back into the blood. In this way the fluid flowing through the tubule is altered in composition, as in the blood. At the end the fluid has been transformed into urine.

The insects have evolved a different way of excreting wastes and regulating salt and water. These animals have an open circulatory system, which creates special problems (Fig. 8.14). Their organ for blood filtration is a set of long pouches off the digestive tract at the junction of the midgut and hindgut. These **Malpighian tubules** are bathed in blood. Fluid is absorbed at the blind end of the tube, and as it moves toward the digestive tract, reabsorption of some water and salts occurs. The remaining uric acid and water pass into the hindgut. In the rectum of the insect almost all the water is reabsorbed, and the uric acid is voided with the dry feces.

The renal system in terrestrial vertebrates In terrestrial vertebrates the kidneys are part of a renal system, a group of organs which consists of kidneys, ureters, a urinary bladder, and a urethra (see Fig. 8.15). The function of the system is similar to that of the nephridium of the earthworm, or any other kidneylike organ in other invertebrates: to cleanse the blood and tissue fluids of wastes, while simultaneously adjusting or regulating the amounts of water and other dissolved substances in these fluids. As renal systems evolved and became more complex, they also became more efficient, and the functional units, called **nephrons** in the vertebrate kidney, became grouped structurally into two excretory organs, the kidneys.

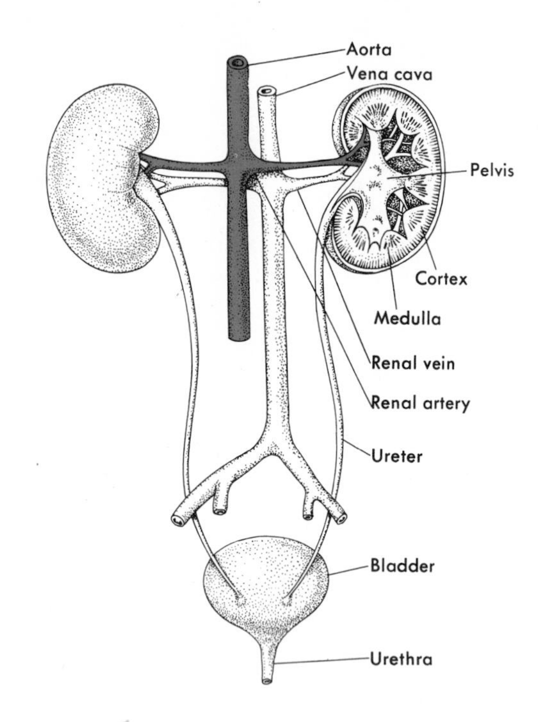

FIGURE 8.15

Renal system in vertebrates. Nephron tubules in the cortex and medulla of each kidney lead into collecting tubules that empty into the pelvis—the central portion of the kidney. Fluids are transported out of the kidneys through the ureters, which empty into the urinary bladder for storage and, later, excretion through the urethra.

Each kidney is very richly supplied with blood and receives a large percentage of the blood forced out of the heart with every beat. Within each human kidney are about one million nephron units. From the blood passing through the kidney, the nephrons remove urea and creatinine (a waste from muscular exertion), SO_4^{--}, acids, some hormones, and a host of other substances, while maintaining a proper balance of water, salts, sugar, and other materials. All of this is done with a high degree of efficiency, within two rather small organs about the size of a man's fist.

The structure of the kidney is well adapted to its function, for within the body of the kidney the large renal artery branches into a sequence of successively smaller vessels that ultimately break up into about one million small tufts of capillaries, thus exposing a tremendous surface area of blood to cleansing (Fig. 8.16).

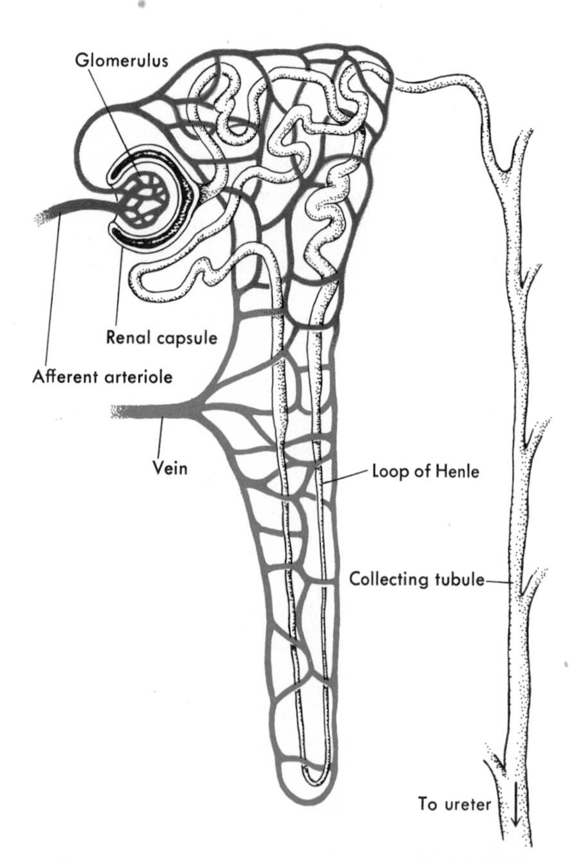

FIGURE 8.16

Human renal system. Each human kidney contains about one million nephrons, the systems of tubules and capillaries that cleanse and adjust the solute and water levels of all the blood in the body. Waste-laden blood enters each kidney through the renal artery, which breaks into millions of thin arterioles to supply each nephron. Water and small waste molecules, as well as some useful molecules, leave the blood by filtering through the pores in the membranes of the capillaries that constitute the glomerulus and enter the nephron tubule by filtering through the membrane of the renal capsule. As the fluid passes through the tubule of the nephron, water, dissolved useful molecules, and some ions are returned to the blood in the quantities that are needed. Waste materials like urea, dissolved in water, pass through the entire length of the tubule and contribute to the formation of urine. Cleansed blood reenters the general circulation through the renal vein.

FIGURE 8.17

The production of concentrated urine in the human kidney depends on the structure of the nephron tubule system, its selective permeability, and the concentration gradient of the surrounding tissues. The initial filtrate moves through the loop of Henle, where sodium ions are pumped out, primarily in the ascending branch. Some of these ions flow into the descending branch and move to be recycled out in the ascending branch, as a means of conserving a gradient in the surrounding tissues. This gradient creates a counter-current system along the collecting tubule. Even though the urine grows more concentrated in the collecting tubule as water passes into the tissues, the increasing osmotic gradient permits the loss of even more water to the tissues. Thus the filtrate becomes much more concentrated at the end of the collecting tubule than it was at the start.

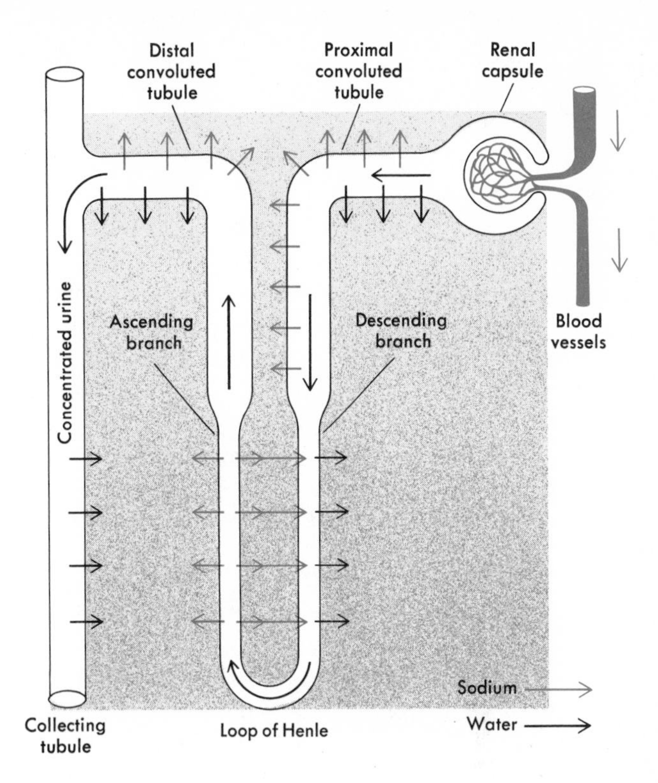

The tufts, called **glomeruli,** are all found in the outer portion of the kidney, the cortex. Each glomerulus is contained within the cup-shaped distal end of a nephron, the **renal capsule,** also called **Bowman's capsule.** Each capsule is a sealed, expanded sac at the end of the nephron, the rest of which is drawn out into a greatly elongated, twisted, and looped tubule in which urine forms. The part of the tubule nearest the Bowman's capsule is twisted, and from its anatomical position (proximal, or near the capsule) it is called the **proximal convoluted tubule.** The proximal convoluted portion of the tubule leads to a loop, the **loop of Henle,** which dips down from the cortex of the kidney into the deeper tissues of the medulla, and then loops back up into the cortex. There it becomes convoluted again and is called the **distal convoluted tubule** (distal means far). It dips back into the medulla and connects with a collecting tubule. Many collecting tubules from many other nephrons terminate in a common collecting duct into which urine drips. The collecting ducts lead to the renal pelvis, a large sac which collects all the urine coming from the kidney and sends it, through the **ureter,** to the **urinary bladder** for temporary storage.

The formation of urine occurs fairly simply (Fig. 8.17). Blood entering the kidney through the renal artery is under quite high pressure—high enough, in fact, to cause water from the blood and dissolved wastes, ions, and other small molecules, which together constitute the renal filtrate, to literally squirt through the pores in the very thin membranes of the capillaries of each glomerulus. The pores are small enough to block the passage of proteins and blood cells, but the blood pressure is high enough to force water out. This process, the separation of materials that leave the blood from those that are retained on the basis of their size in relation to the pores of a membrane, is termed **ultrafiltration.** The ultrafiltration, once formed, enters the capsule portion of the nephron.

The capillaries of the glomerular tuft collect to form small arterioles, which lead again to networks of capillaries, this time surrounding the tubular portion of the nephron. As the blood passes through these capillaries, further and further down the length of the nephron tubule, two other kinds of physiological processes finally lead to the formation of urine and the adjustment of levels of water and other substances in the blood: **selective reabsorption** of needed substances from the filtrate fluid flowing through the tubule, and **secretory excretion** of wastes into the filtrate, which will become urine.

In the proximal convoluted tubule, about 80 percent of the water of the filtrate finally flows back from the tubule into the blood by osmosis, for the pressure of the blood in the capillaries here is not as great as in the glomerulus. Selective reabsorption, accomplished by specific enzymes, takes glucose back from the filtrate, at the rate of about eight trillion molecules per second. By such saving of glucose as required, **glycemic regulation** is accomplished: for the saving is effective only up to a point. When great excesses of glucose are present in the blood and hence in the filtrate, much of it is permitted to pass out of the body in the urine.

As the filtrate passes down the tubule of the nephron, its composition continues to change; other needed substances may be selectively reabsorbed: about three ounces of salt (sodium chloride) are reabsorbed each day as well as some ammonia, a potent base used by the body to regulate pH by counteracting the effects of various acids produced during metabolism. And various wastes are actively secreted, removed from the blood passing near the tubule in capillaries, and added to the filtrate.

What finally emerges in droplets from the ends of each nephron tubule is urine, a solution of wastes and unnecessary substances dissolved in variable amounts of water according to the needs of the body. (For a comparison of the components of urine and other body fluids, see Table 8.1.) Hence the kidney is truly a homeostatic organ, one which regulates within precise limits the vital constituents of the fluids in the body.

As in any homeostatic mechanism, urine composition is controlled through feedback controls. Blood-water concentration is regulated by a brain sensor that monitors the blood's osmotic concentration. If this concentration is slightly diluted, as it is after a person drinks a great deal of water, reabsorption of water is decreased and larger quantities of water are passed into the urine. When water is scarce, reabsorption of water is promoted to conserve as much water as possible.

The homeostatic mechanism that controls the active reabsorption of water involves hypothalamic cells sensitive to changes in the concentration of water in the blood. When the water concentration drops, these cells trigger the release of antidiuretic hormone (ADH) from the posterior pituitary. The hormone stimulates the nephrons of the kidney to reabsorb water. When the water concentration rises to optimal level, the ADH release stops. The entire process takes about one hour.

TABLE 8.1
Comparison of components of plasma, nephric filtrate, and urine, in grams per 100 cc of fluid

MAIN COMPONENTS	PLASMA	NEPHRIC FILTRATE	URINE
Urea	0.03	0.03	2.0
Uric acid	0.004	0.004	0.05
Glucose	0.10	0.10	Trace
Amino acids	0.05	0.05	Trace
Total inorganic salts	0.72	0.72	1.50
Proteins and other colloids	8.00	0.00	0.00

The kidney also regulates blood sodium and oxygen levels. The kidney is sensitive to fluctuations in the oxygen concentration of the blood. When the concentration drops, the kidney, in conjunction with the liver, produces erythropoietin, a hormone which acts to increase the number of oxygen-carrying red blood cells in the bloodstream. Another hormone produced by the kidney is **renin** (released when the kidney is oxygen-deficient) which plays a role in elevating the blood pressure.

SPECIAL PROBLEMS IN WATER REGULATION

Aquatic organisms Aquatic organisms are faced with special problems in regulating water levels in their cells and tissues.

The tissue fluids of the freshwater organisms are hypertonic to their environment. This means that they have more salt and less water per given volume than the environment. Thus salt and other solutes tend to pass out of the organism and water tends to pass in, along a concentration gradient. If this tendency were not counteracted, the fish would lose vital solutes and needlessly gain water. Two regulatory mechanisms prevent this from happening. Excess water is eliminated in dilute urine by the kidneys, and salts are absorbed by specialized cells located in the gills (see Fig. 8.18). The outer covering of the fish is tough and scaly, and relatively impermeable to water.

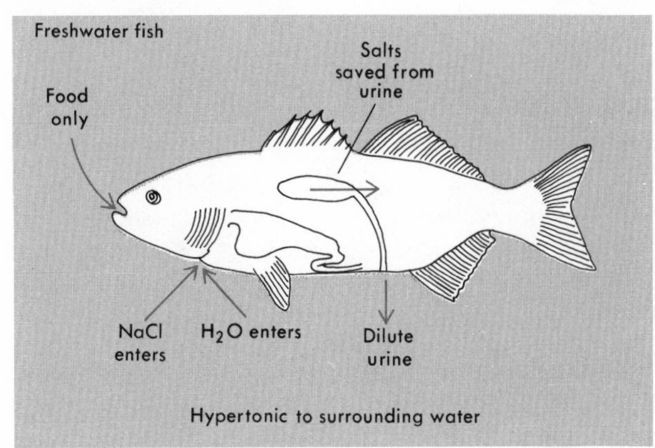

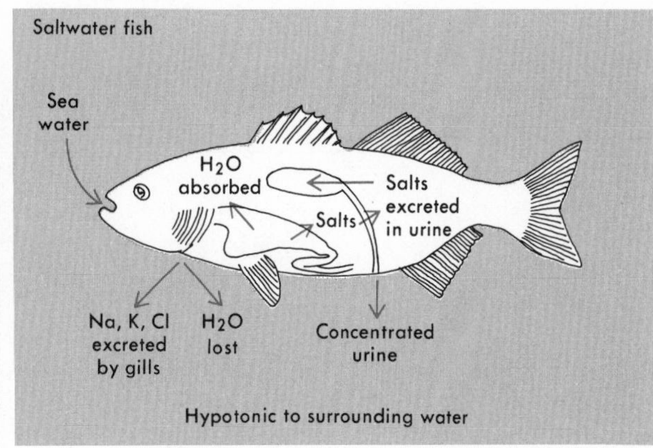

FIGURE 8.18

Osmoregulation. The balance between ion concentration and volume of water is accomplished differently by fish living in fresh water than by saltwater species. Freshwater fish are hypertonic to their external environment, so water has a greater tendency to enter the fish than to leave. Thus the fish seldom drink water, and specialized cells in their gills absorb salts by active transport. Saltwater fish, however, are hypotonic, so they tend to lose water. These species, therefore, drink water to replenish the volume lost through their skin, and through active transport they excrete from their gills the excess salts in their blood.

The saltwater fish have the reverse problem. Their tissue fluids are hypotonic to sea water. This means that they have less salt and more water per given volume than the environment. Thus salts would tend to pass in and water to pass out moving with the gradient. This tendency would of course lead to dehydration. However, these fish save the water from the sea water they swallow and discard the salts. They excrete salts from specialized cells in their gills; they also drink water in large amounts (whereas freshwater fish never do); and they excrete only small quantities of urine, relying instead on the gills to excrete wastes in the form of ammonia. (The kidneys do not perform this function in saltwater fish because, in these species, they are not capable of producing very concentrated salty urine.)

Desert animals Strictly terrestrial species may face severe problems if they happen to live where water is rarely available. The camel, for example, sweats very little during much of the day, as the temperature climbs steeply. Instead, the camel's body temperature fluctuates, rising as the day gets hotter. But when its blood temperature gets dangerously high, threatening damage to the brain or other organs, the camel may begin to perspire, and evaporative cooling will prevent excessive rises in temperature. When the new day begins again, and the temperature begins to rise, the camel's blood is still cool from the preceding night, giving the animal a cool head start on the day. The camel can also tolerate remarkable amounts of water loss from its tissues. A camel that has been in the desert for many days without water may lose 25 percent of its body weight as evaporated water, and it will have an extremely emaciated appearance. But a long drink at a watering place rehydrates the animal. (Despite the old tale, the camel does not store water in its hump.) If a human lost a similar amount of water from his tissues, or suffered such drastic fluctuations in body temperature, it would be fatal.

Some desert animals conserve body water by burrowing deep into the sand during the day; the burrow soon becomes humid as the animal exhales moist air from its lungs, and the saturated air prevents more water from being lost from the lungs by evaporation. Yet another mechanism, employed by the kangaroo rat, is the excretion of a highly concentrated urine; during the production of its urine, most of the water is reabsorbed by the kidneys. This ability to save water permits it to tolerate the drying effects of high temperatures, something that man's renal system cannot do, probably because man's evolutionary history occurred mainly in nondesert environments.

The Body's Defense System

A great variety of potentially harmful foreign substances and infectious organisms invade our bodies daily. Such foreign substances, usually proteins or large carbohydrates, are called *antigens*. Fortunately, over the vast reaches of time, we have evolved a remarkable web of defenses that recognize foreign proteins and destroy them by producing highly specific *antibodies*. The antibodies are produced by our *immune system*. Some of our immune capabilities are inherited. Thus we cannot safely receive blood transfusions from people of differing genetic blood types, whose antigens are foreign and react with our own highly specific antibodies. We can also acquire immunity to foreign agents or organisms by being exposed to them. The exposure triggers the production of antibodies.

Antibodies are large Y-shaped protein molecules called *immunoglobins*. They are produced by specialized white blood cells of the lymphocytic type. Actually there are several types of immunoglobins, e.g., Immunoglobin G, or IgG (the most common), Immunoglobin A, Immunoglobin E, etc. Some immunoglobins are free and circulate in the blood plasma. Others are bound on the surface of cells called T lymphocytes. Obviously, we produce thousands of different types of highly specific antibodies. The specificity to a given antigen is contained on complementary combining sites on the immunoglobin molecule, which interact with part of the antigen molecule. In other words, there is an antigen-antibody interaction that traps and eventually destroys the antigen.

Once the lymphocytes have learned to recognize a foreign substance, they multiply and produce clones, or populations of cells with a memory of that substance. Those cells all have the capability of retaining that memory and of manufacturing antibodies to counteract the antigen. Sometimes such a memory (that is, immunity) is retained for a lifetime, but sometimes there must be a reminder, another exposure to the foreign substance. The second exposure causes a surge of antibody production and a very high level of immunity. For example, a single exposure to an inactivated form of the poliomyelitis virus will trigger the production of some antipolio antibodies, but probably not enough to combat a severe polio virus infection. However, if you receive a booster shot or dose (i.e., a second exposure to the inactivated virus), you will then have full immunity to polio infection. Today we have been able to develop vaccines (harmless inactivated viruses) that can provide immunity against most killer viruses of the past. One formerly deadly and disfiguring virus, smallpox, has just about disappeared.

Now let's go back and review the sequence of events that follows the invasion of the body by a foreign antigen.

When an antigen stimulates activity in the lymphocytes, the manufacturing of antibodies begins. The precise details are not known, but the following sequence is thought to occur (see diagram).

1. The antigen enters the body and is engulfed by phagocytes. The phagocytes, now containing the antigen, interact with the lymphocytes.

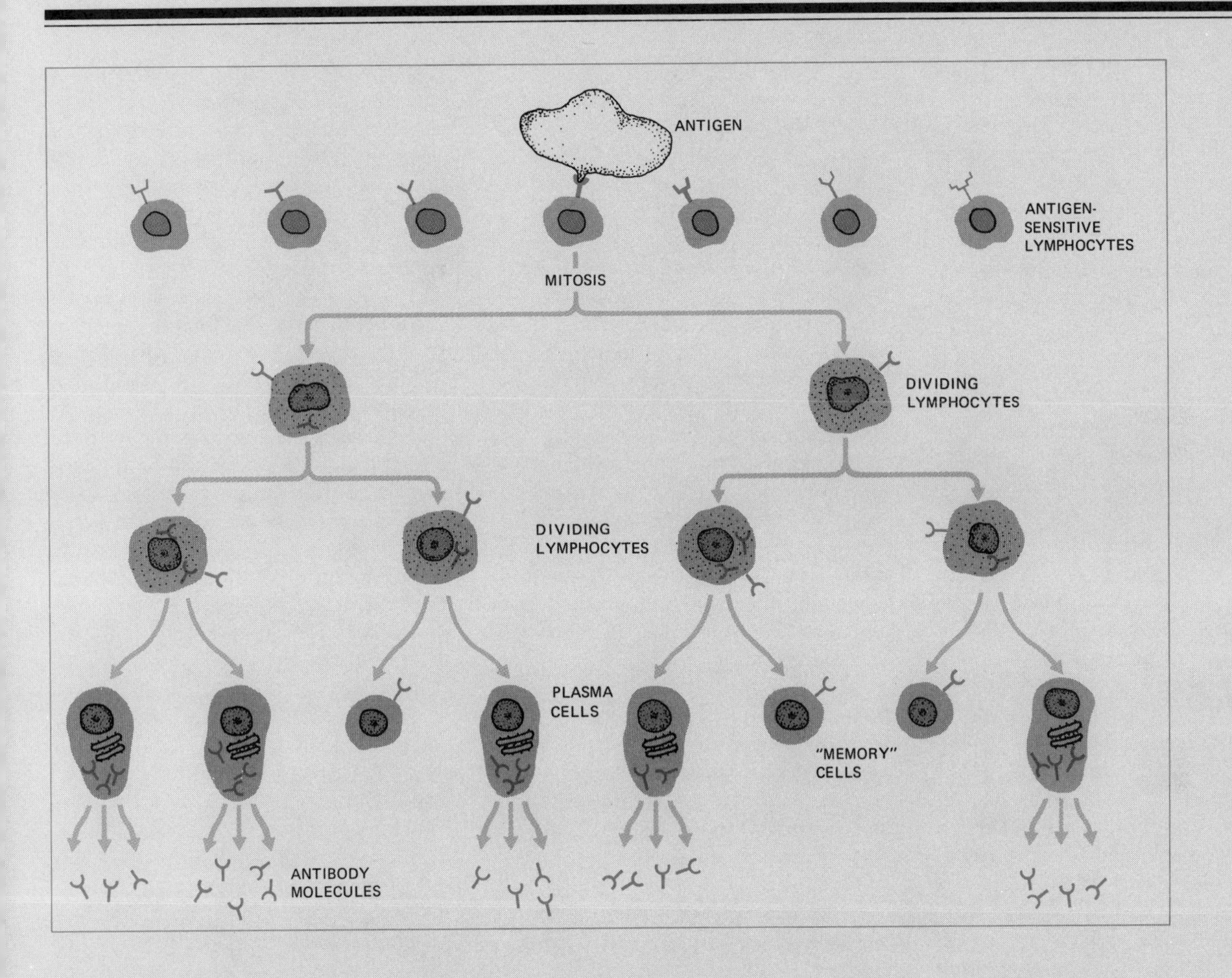

2. The lymphocytes react with the antigen, probably at the same location where the antibody will react with the antigen later on.

3. The lymphocyte is stimulated to divide, producing many identical cells (a clone), or population of cells, all of which can manufacture a particular antibody.

4. After a while, the lymphocytes, or plasma cells, stop dividing and produce antibodies, which are then secreted into the lymph fluid and transported into the bloodstream.

   Up to this point the process may take four to five days.

5. Some of the cells resulting from the division of lymphocytes remain in the blood as "memory cells." Like the original lymphocytes, they are sensitive to the specific antigen. If the antigen returns at a later date, there are many more antigen-sensitive lymphocytes to react to it than there were the first time the antigen appeared. As a result, the second immune response will be greater than the original re-

sponse, and it will occur at a faster rate—one to two days.

After the antibodies have been released into the bloodstream, there are three ways they can affect antigens.

1. They can counteract and neutralize the toxic effects of the antigens.
2. They can coat the antigens to make it easier for the body's phagocytes to destroy them.
3. They can attach themselves to a viral antigen and thereby prevent the virus from infecting other cells.

Responses to transplanted organs When an organ, such as a kidney or heart, is transplanted from one person to another in order to replace a damaged organ, the body of the recipient of the transplant will often "reject" it —that is, the organ will not grow as part of the person's body. It is recognized as "nonself" material. Skin grafts from one part of the body to another part of the same body have been successful. So have organ transplants between identical twins. The body recognizes these tissues as part of its "self." The reaction involved here is called *cellular immunity.* In a cellular immune reaction, no new antibodies are produced. Instead, a certain type of lymphocyte is involved; it is not the same as the lymphocytes that divide into plasma cells and produce antibodies. These special lymphocytes are sensitive to nonself material, and when stimulated by a foreign tissue or organ, they will in turn stimulate other cells in the circulatory system to attach and kill the cells of the transplant.

The cellular immune response can be suppressed by certain chemicals in the transplant recipient, but unfortunately, the patient is then open to other types of infections.

Allergies If an antigen and a special type of antibody unite on certain cells, the union can cause most cells to release a potent vasodilator called histamine. The symptoms produced by the release of histamine vary with the organ affected, but the overall response is dilation of blood vessels, leakiness of blood vessel walls, and swelling and inflammation.

This type of immune response is called an *allergy,* and the antigens are referred to as *allergens.* Such allergens as pollen, dust, and various foods usually enter the body through the digestive or respiratory tract.

Immune Surveillance and Cancer One function of the immune system is apparently to recognize as "nonself" certain abnormal cells within the body and destroy these cells accordingly. This process is called *immune surveillance.*

A cancer is a population of abnormal cells that divides without control, often invading and destroying surrounding tissue. If cancer cells are abnormal and immune surveillance is supposed to recognize and dispose of them, why do we still get cancer? There are several possibilities. First of all, some types of cancer may not be foreign and consequently are not recognized by the immune system. Second, we know that the incidence of cancer increases with age. The immune system's efficiency usually decreases with increasing age and therefore may not be as effective at destroying the cancer cells. Indeed, it has been suggested that many people harbor small clones of potential cancer cells that are kept in control by the immune system. When the immune system becomes less efficient, the cancer clones begin to divide rapidly and soon overwhelm the limited defenses afforded by the immune surveillance system. It is interesting to note that people who have been given drugs that suppress the immune system have a high incidence of some types of cancer. Third, the immune system itself may be responsible. Tumors, like transplants, stimulate a cellular immunity response. Antibodies are not involved in the process, but there is some evidence that they may hinder it by attaching themselves to the cancerous cells.

The antibodies cover the cancer cell's antigenic determinants so that they can't trigger the antibody production that is necessary for the tumor rejection process. Thus the body may develop a tolerance for the tumor and fail to reject it. Another area of cancer research, *tumor immunology,* involves making a weak immune response to cancer antigens vigorous enough to cause significant increases in the production of antitumor antibodies. This type of treatment has been successful with some experimental animal tumors in the laboratory, but medical researchers have not been able to develop a similar treatment for humans.

From John W. Kimball, *Man and Nature* (Reading, Mass.: Addison-Wesley, 1975), p. 396.

FIGURE 8.19

Temperature regulation. The arctic fox withstands very low temperatures by conserving its body heat. The animal's ears and muzzle are relatively short, reducing the area of body surface exposed to the cold. Dense, fluffy fur provides excellent insulation. This fox has curled into a ball and burrowed into the snow to further reduce the area of its body in contact with the freezing air.

PHYSIOLOGICAL REGULATION OF TEMPERATURE

We have seen that homeostatic mechanisms regulate the amount of water, salt, or other substances in an organism's internal environment. In multicellular animals, regulatory mechanisms also permit precise temperature control, despite seasonal or daily temperature fluctuations, or generally high or low environmental temperatures.

Two broad categories of organisms can be distinguished on the basis of temperature-regulating ability: **poikilotherms,** popularly (though inaccurately) called cold-blooded organisms, and **homeotherms,** or warm-blooded organisms.* Poikilotherms, which include most fish, reptiles, and amphibians, have no special physiological temperature-regulating mechanisms. Instead, they rely on the stability of the temperature in their environment or on behavioral mechanisms to ensure that their body temperatures do not become so high that they denature their vital enzyme systems, or so cool that they cause a dangerous slowdown of chemical reactions. Thus a poikilothermic animal may sun itself in early morning to raise its body temperature after a cool night and then seek shade during warmer parts of the day. However, these behaviors cannot control body temperature very precisely.

Birds and mammals belong to the second category of organisms, the homeotherms, which evolved special mechanisms for body-temperature regulation. They regulate temperature by balancing heat gain from the environment and heat production from respiration against heat loss from the surface of the body. Quite precise

---

* The terms poikilotherm (meaning varied heat) and homeotherm (same, or constant, heat) were introduced to counteract the inaccuracy of the long-popular identification of organisms as cold- or warm-blooded. However, these terms have also become less satisfactory in the face of increased knowledge of physiological thermoregulation. There is now a growing tendency in biological literature to use the terms **endotherm** and **ectotherm,** which identify organisms according to the *source* of body heat rather than its constancy. Endotherms obtain heat from their own metabolic activities (endo- means within); ectotherms obtain heat from the environment (ecto- means outside).

temperature control can be thus effected, and mammals may be found living in places where the temperature is as low as −20° F and in temperatures as high as +130° F. Even at these extremes of temperature, mammals can maintain their body temperatures within a few degrees of normal.

When the body temperature drops slightly, as it may when the body is exposed to cool environmental conditions, the hypothalamus mobilizes heat-generating and heat-saving mechanisms. The thyroid is stimulated to release thyroxin, which in turn speeds up cellular respiration, generating heat. The adrenal cortex is stimulated to produce cortical steroids, which promote the availability of glucose for cellular respiration. The appetite may be stimulated to provide more fuel for respiration. Shivering may be promoted, and the rapid movement and respiration of the muscles involved generates heat. A saving of heat is effected by a physical mechanism, vasoconstriction, or narrowing of the blood vessels near the surface of the skin. This action minimizes heat loss by ensuring that the warm blood is diverted from the cool skin surface. In mammals with fur, contraction of small muscles attached to the base of each shaft of hair causes the hair to be raised, thus improving its insulation value and further reducing heat loss. Although the human body lacks fur, the small muscles remain, and they produce the familiar gooseflesh when the body is chilled.

When the body temperature returns to normal, the hypothalamus senses this change and turns these heating mechanisms down. When the body is exposed to a warm environment, the mechanisms of heat production are turned down further, and heat loss is promoted: blood vessels near the surface of the skin dilate, and warm blood moving in them loses its heat to the environment. Such heat loss may be promoted by sweating since evaporation of water from the skin cools the blood. In some mammals that have no sweat glands, other means of cooling by evaporation are employed. For example, a dog pants heavily when warm, and the rich blood supply of its tongue is cooled by the evaporation of saliva. Behavioral mechanisms are also employed, and most animals will seek shade during the hottest part of the day, or take a cooling dip in the nearest pond.

Summary

1. The chemical reactions that typify life occur only under a narrow range of conditions. Living cells must expend energy to continuously adjust and optimize their internal environment to avoid any deviation from the narrow boundaries necessary for life processes. This stabilization of the cell's internal environment is called homeostasis; and without this ability the cell would be a victim of its own metabolic activities. Every living cell performs internal homeostatic regulation, but in multicellular organisms, entire organs interact to regulate the internal environment.

2. All living cells have the same homeostatic problems, such as the maintenance of pH, salt concentration, water volume, nutrient level, and the elimination of toxic wastes. Feedback mechanisms are the key to the cell's maintenance of homeostasis, and feedback control is provided by end-product inhibition.

3. Uncontrolled variation in pH destroys the hydrogen bonds that stabilize the three-dimensional structure of enzymes, consequently altering enzymatic reaction rates and the optimal functioning of the cell. Regulation of pH is a complex process involving mechanisms such as buffers and the elimination of pH-altering wastes such as CO_2 (which forms carbonic acid).

4. Salts are important as components of certain proteins, as well as buffering agents. They help maintain pH and the osmotic concentration of the cell. They are selectively concentrated in or expelled from the cell by active transport at the cell membrane.

5. The cellular regulation of water content is closely linked to the active transport of salts. As the cellular salt concentration rises, water flows passively inward to maintain the osmotic equilibrium. The accumulation of too much water is inhibited in plants by their rigid cell walls. In some unicellular animals, such as the paramecium, contractile vacuoles pump excess water out of the organism.

6. Maintaining a constant level of nutrients is achieved by circulation of digestive vacuoles, by storage of excess nutrients, and by nuclear control of metabolism.

7. With even greater cellular specializations in multicellular animals, certain cells have decreased abilities to regulate their internal environments. The evolution of the liver and kidneys as homeostatic organs enables complex organisms to successfully compensate for changes in their external and/or internal environments.

8. The mammalian liver produces bile, rids the blood of foreign bodies and older erythrocytes, plays a role in the metabolism of nutrients and vitamins, and detoxifies various drugs and poisons that may be in the circulation. The portal circulation of the liver places this organ in the vascular pathway of the nutrient-laden blood from the digestive system. The liver is thus one of the organs involved in carbohydrate balance. The entire system of glucose regulation involves complex feedback interactions between endocrine glands. The liver also regulates the amount of nitrogen formed by deamination of excess amino acids into ammonia.

9. The major homeostatic organ of vertebrates is the kidney. In terrestrial vertebrates they are paired organs, each of which contains about one million functional units called nephrons. Each nephron is surrounded by a unique artery-capillary-venous system. The nephron selectively filters the blood; various ions and nutrients are reabsorbed, wastes are secreted into the nephron tubule, and water is passively and actively reabsorbed.

10. Special problems in water regulation are posed for aquatic organisms and desert animals. Freshwater fish absorb salts through their gills and form dilute urine to eliminate excess water. Saltwater fish avoid dehydration by taking in sea water, and by excreting salts through their gills and small quantities of urine. Desert animals like the camel and kangaroo rat have evolved different physiological and behavioral mechanisms to conserve water.

11. Temperature regulation is achieved in two ways. Poikilotherms, whose temperature varies with that of the environment, rely on behavioral adaptations and on the stability of environmental temperatures. Homeotherms, whose temperature is constant despite environmental fluctuations, have evolved a number of physiological responses that conserve or dissipate heat.

REVIEW AND STUDY QUESTIONS

1. You are thirsty and you drink a glass of water. Explain what happens as an illustration of feedback control.

2. A familiar over-the-counter product is described as "aspirin with buffering." What are the makers of this product suggesting? What does a buffer do in the body?

3. What may happen if an animal takes in too much salt too rapidly? What regulatory mechanisms have evolved that overcome the problem of high salt concentration?

4. What is the main reason that water regulation involves simpler processes in plants than in animals?

5. Many of our body cells need a constant supply of nutrients. Why do we not have to eat constantly to supply such needs?

6. What functions does the mammalian liver perform? The kidneys? The bladder?

7. How is water regulation managed by an ocean fish? A camel? A kangaroo rat?

REFERENCES

Benzinger, T. 1961 (January). "The Human Thermostat." *Scientific American* Offprint no. 129. Freeman, San Francisco.

Bogert, C. B. 1959 (April). "How Reptiles Regulate Their Body Temperature." *Scientific American* Offprint no. 119. Freeman, San Francisco.

Carey, F. G. 1973 (February). "Fishes with Warm Bodies." *Scientific American* Offprint no. 1266. Freeman, San Francisco.

Carlson, A. J., V. Johnson, and H. M. Carert. 1961. *The Machinery of the Body*. University of Chicago Press, Chicago.

Gates, D. M. 1965 (December). "Heat Transfer in Plants." *Scientific American* Offprint no. 1029. Freeman, San Francisco.

Hoar, W. S. 1975. *General and Comparative Physiology*, 2nd edition. Prentice-Hall, Englewood Cliffs, N.J.

Morse, R. A. 1972 (April). "Environmental Control in the Beehive." *Scientific American* Offprint no. 1247. Freeman, San Francisco.

Pitts, R. F. 1974. *Physiology of the Kidney and Body Fluids*, 3rd edition. Year Book Medical Publishers, Chicago.

Schmidt-Nielsen, K. 1959 (December). "The Physiology of the Camel." *Scientific American* Offprint no. 1096. Freeman, San Francisco.

Schmidt-Nielsen, K., and B. Schmidt-Nielsen. 1953 (July). "The Desert Rat." *Scientific American* Offprint no. 1050. Freeman, San Francisco.

Solomon, A. K. 1962 (August). "Pumps in the Living Cell." *Scientific American* Offprint no. 131. Freeman, San Francisco.

SUGGESTIONS FOR FURTHER READING

Bloch, R. M. 1963 (July). "The Social Influence of Salt." *Scientific American* 209(1): 88–98.

The role of the quest for salt in the shaping of history. Includes a brief discussion of the importance of salt to life.

Chapman, C. B., and J. H. Mitchell. 1965 (May). "The Physiology of Exercise." *Scientific American* Offprint no. 1011. Freeman, San Francisco.

Mechanisms for the body's adaption to the demands of exercise.

Fertig, D. S., and V. W. Edmonds. 1969 (October). The Physiology of the House Mouse." *Scientific American* Offprint no. 1159. Freeman, San Francisco.

The mouse's efficient mechanisms for conserving water and energy enable him to live with man.

Heinrich, B., and G. A. Bartholomew. 1972 (June). "Temperature Control in Flying Moths." *Scientific American* Offprint no. 1252. Freeman, San Francisco.

A process that resembles shivering warms the flight muscles of certain insects, enabling them to fly.

Hock, R. J. 1970 (February). "The Physiology of High Altitude." *Scientific American* Offprint no. 1168. Freeman, San Francisco.

Adaptive mechanisms by which animals and man can live above 6000 feet in atmosphere of decreased oxygen.

Irving, L. 1966 (January). "Adaptions to Cold." *Scientific American* Offprint no. 1032. Freeman, San Francisco.

Adaptions include increased metabolism, insulation, and changes in circulation of heat by blood.

Langley, L. L. 1965. *Homeostasis*. Reinhold, New York.

Simple, clear discussion of the nervous and hormonal mechanisms for maintenance of body temperature and weight, blood pressure, respiration, movement, and the composition of blood.

Luria, S. E. 1973. *Life: The Unfinished Experiment*. Scribner's, New York.

Highly readable presentation by a distinguished biologist of the basic concepts of modern human biology. Intended for the layman who wants to gain understanding of the material basis of life.

Schmidt-Nielsen, K. 1959 (January). "Salt Glands." *Scientific American* Offprint no. 1118. Freeman, San Francisco.

Salt-excreting organs in marine birds enable them to drink salt water.

Scholander, P. F. 1963 (December). "The Master Switch of Life." *Scientific American* Offprint no. 172. Freeman, San Francisco.

The vertebrate defense against asphyxia is the diverting of blood from other parts of the body to the heart and brain during extreme stress.

Shepro, D. S., F. A. Belamarich, and C. K. Levy. 1978 (in press). *Human Anatomy and Physiology,* 2nd edition. Holt, Rinehart and Winston, New York.

Smith, H W. 1953 (January). "The Kidney." *Scientific American* Offprint no. 37. Freeman, San Francisco.

Useful explanation of kidney structure, function, and evolution.

Warren, J. V. 1974 (November). "The Physiology of the Giraffe." *Scientific American* Offprint no. 1307. Freeman, San Francisco.

The distance between the giraffe's head and its heart requires a remarkably high blood pressure and very deep breathing.

CHEMICAL CONTROL
chapter nine

Most plants grow upward and will right themselves even when placed horizontally, as shown in this multiple-exposure photograph. This response is induced by auxin, a hormone. Auxin is also responsible for the growth of roots downward.

All organisms must respond to changes in their environment; unicellular organisms have little difficulty in relaying information about external changes to the internal parts of the cell, which can then react. But multicellular organisms have important problems of information transmission, since the coordinated activity of thousands or millions of cells may be involved in the action, and these cells may be quite distant from cells that perceive changes in the environment and control the action. When great speed is essential, as in producing the beautiful coordinated movements of a running antelope, nerves serve as the message carriers; but when the reaction is slower and perhaps sustained, special chemical messengers called *hormones* generally are called into play. Hormones are produced near an information receptor site, then move to a target, the tissue or organ producing a reaction, and alter the target's activity.

Plant hormones are chemically very diverse. They may be involved in producing many different effects, such as bending of the plant toward light, flowering, bending of the roots down and the stems up, the annual thickening of trees, the setting of fruit, and others. Plant hormones are not produced in special organs; they originate in many different cells in different parts of the plant.

Animal hormones are similarly diverse chemically, but they tend to be more complex than plant hormones (Fig. 9.1). They are involved in mobilization of stored nutrient reserves, cyclic reproductive behavior, regulation of metabolism and growth, and many other functions; in short, they serve to maintain homeostasis. However, unlike plant hormones, those in animals are produced in special organs, called **endocrine glands,** which release the endocrine hormones directly into the blood.

FIGURE 9.1

The complexity of a common animal hormone, insulin (A), compared with that of indoleacetic acid (B), a common plant hormone. Insulin is composed of 51 amino acids, each of which is about as complex as the entire auxin molecule. Despite the differences in complexity, both of these molecules serve as regulators of specific activities.

Because the hormones are chemicals that must travel from where they are produced to their target organs, which may be a great distance away, several seconds may elapse before the hormonal information is transmitted. The chemical structures of the hormones are stable and may remain active in the body—like cortisone—for over an hour. This kind of information transmission is therefore more suitable for bringing about reactions in which the time factor is relatively unimportant.

In plants, for example, the reaction time for **phototropism** (a phenomenon in which the plant's stem and leaves bend toward the light source) is rather long (30 minutes), but the hormones are still able to reach the leaves in time to permit them to follow the sun through the course of a day. The rate of hormonal movement in animals is dependent on the flow of the blood and is thus adequate to control such phenomena as the monthly reproductive cycles and the lengthening of bones.

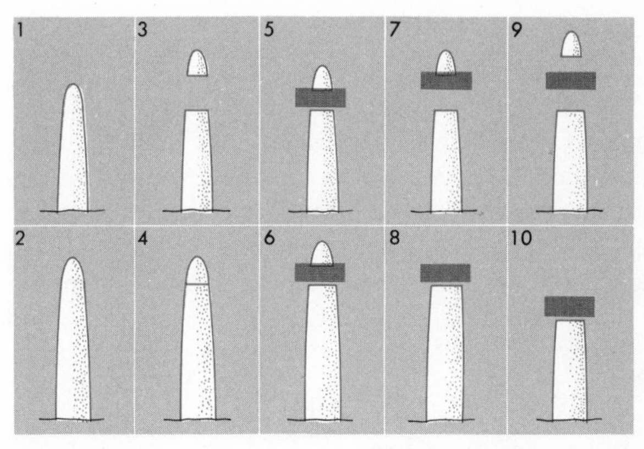

FIGURE 9.2

Demonstration of auxin's effect. Early experiments determined the presence of a hormone, later named auxin, which was produced in the tip of a grass seedling and moved to a lower part of the coleoptile, where it increased the growth of tissues. The intact coleoptile elongates rapidly (1) and may grow perceptibly within a few hours (2). If the tip is removed (3) and then replaced (4), the growth of the coleoptile is relatively undisturbed. If the tip is removed and a piece of gelatin placed between it and the rest of the coleoptile (5), growth is nearly normal (6), because the auxin can move freely through the gelatin. If the tip is removed and placed on the gelatin for a short period of time (7), the hormone diffuses into the gelatin, and the hormone-containing gelatin alone is sufficient to restore growth (8). Plain gelatin, however, that has never been in contact with the hormone source (9), has no effect on growth (10).

Plant hormones

There are several categories of hormones that affect growth in plants. They include the auxins, the gibberellins, the cytokinins, abscisic acid, and ethylene gas.

AUXINS

Why does a stem grow up and a root grow down? Why does a plant bend toward the light? One answer is that a control mechanism is involved. One of the first scientists to become aware of one of these systems of coordination and control was Charles Darwin, who demonstrated by experiments with grass seedlings that the behavior of a young plant stem is directed by the growing tip at its end. Later, other scientists experimented with grass seedlings, as well as other plants, to show that a particular growth substance was involved, whose nature and effects they proceeded to determine.

A grass seedling is a relatively simple structure: At the onset of growth, a **coleoptile**—a sturdy, hollow cylinder—pushes up through the soil first, protecting the young leaves curled up within it. If the tip of the coleoptile is cut off, its growth stops. If the tip is replaced, growth is resumed. Growth also resumes if a thin layer of gelatin is put between the replaced tip and the stump (Fig. 9.2). But if a piece of mica or foil is used instead of the gelatin, growth remains halted. These experiments demonstrate that (1) some factor is produced in the apex, or tip, that promotes growth and (2) the factor is probably a water-soluble chemical that can pass through gelatin, though not through mica or foil.

Later work showed that the growth substance also was directly involved with phototropism, the bending of a plant toward light. The substance, now known as **auxin** (from the Greek *auxein*, to grow), was shown to move from the tip of the coleoptile in greater amounts to the shaded side than to the lighted side. Thus the hypothesis that the plant's curvature was somehow caused by this uneven distribution of the growth substance proved to be true. The shaded side got more auxin and the cells on that side grew longer. When both sides of the stem got equal amounts of light, the bending ceased. This kind of mechanism, in which the response to a stimulus (additional light directed toward the dark side) has the effect of shutting off the response mechanism (the asymmetry of auxin distribution) is another example of negative feedback, described in Chapter 8. Most hormone systems depend on some sort of feedback signals to control the manufacture and release rate of their hormones.

Composition and characteristics of auxin The chemical composition of auxin was not discovered until 1940, when the substance indoleacetic acid (IAA) was isolated as the major auxin in the cells of most plant tissues. By an enzymatic process, IAA is manufactured from the amino acid tryptophan by plant cells in areas of new plant growth. The movement of IAA from the meristematic tissue to regions of other kinds of cellular change —such as cell enlargement and cell differentiation—is *polar;* that is, movement is from the apex to the base. That this polarity is not a result of gravity—that it *is not* merely a downward movement—has been demonstrated experimentally by cutting apart stem segments from a growing plant and reversing their positions so that the part that had been on the bottom was now on the top. Auxin was then applied exogenously—from the outside, by the experimenter, rather than from within the plant itself—and its movement was measured. The auxin still moved from the part that had been on top to the part that had been on bottom—although the positions of these parts were now reversed. This polar movement of auxin thus appears to be the result of active transport by the cells in which the auxin moves.

Since only a very small quantity of auxin is needed to produce the required effects, auxin cannot itself be incorporated as a constituent of stem growth—unlike sugars and starches, which are incorporated. In the sense that auxin works only to cause metabolic changes and is not used directly in the cells, it is a **hormone:** *a substance formed in very small quantities in one part of the plant body* (in this case, the growing tip) *and transported to the other regions* (in this case, the growing parts of the stem) *where it produces its effects.*

Differential responses to auxin Different parts of a plant respond differently to auxin. Root tissue grows best with the lowest concentration of auxin, stems with the highest concentration, buds with a concentration somewhere between the two extremes. Excessive amounts of auxin may inhibit growth. Thus the response of a plant to auxin depends not only on the amount of the substance present but also on the kind of tissue that is reacting to the auxin stimulus. The hormonal system of plants, then, consists of only a few hormones, each of which performs many tasks; this variety of tasks on the part of a single hormone is made possible by the various ways that different tissues respond to the same hormone. Through differential response of plant tissue, auxin can affect a plant in many ways. Two examples are the phenomena of **apical dominance** and **geotropism.**

Apical dominance. As long as the terminal bud of a plant is present, the bud will determine whether the plant will be tall and unbranched, or short and heavily branched (Fig. 9.3). The apical bud produces auxin, which travels down the stem. The amount of auxin produced causes elongation of the main stem but at the same time inhibits lateral buds from growing into branches. If the apical bud is damaged, elongation of the main stem ceases, and the lateral buds elongate, giving the plant a full, branched appearance. The principle of this auxin action is basic to the practice of trimming hedge plants on top to promote fullness (branching), or pinching the tip of a house plant to achieve the same effect.

Geotropism. If a seedling is placed horizontally, auxin will collect on the plant's lower side. In root cells, this extra supply of auxin inhibits elongation of cells on the lower side, with the result that the cells on the upper side of the root enlarge faster and the tissue bends downward. On the other hand, the concentration of auxin on the lower side of the stem promotes elongation

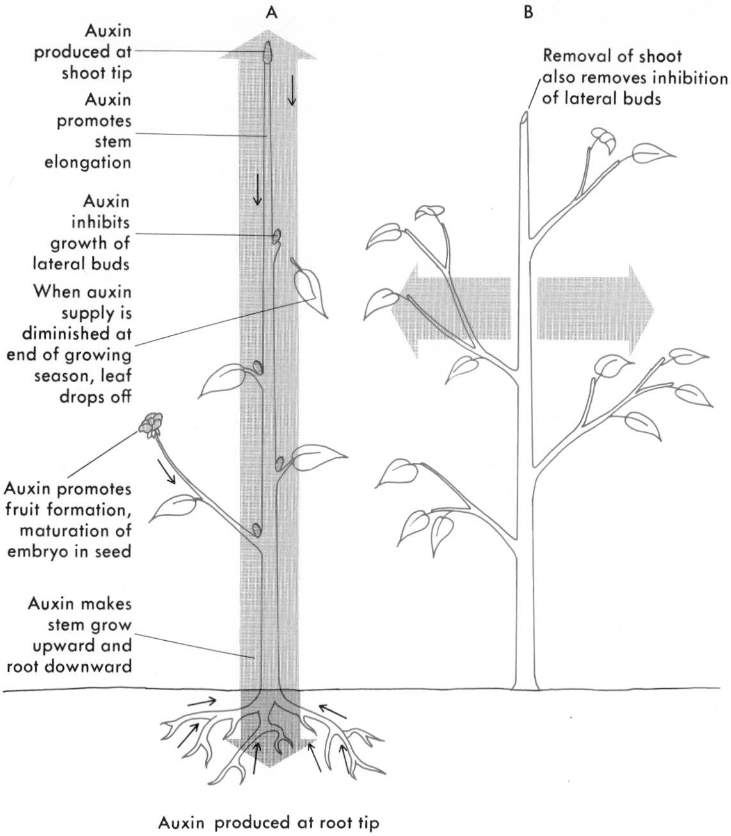

FIGURE 9.3

Multiple effects of auxin. Auxin has many different effects on the growth of a plant, each effect being the result of the sensitivity of the target tissue and the type of response the target tissue can make. (A) Here are shown some of the different responses to auxin, which is produced near the shoot tip (apex) and flows in the direction of the arrows. The overall growth and shape of the plant is controlled by the amount of auxin produced at the apex. If the apex is removed (B), for instance by pruning, the level of auxin in the stem drops. Elongation of the main stem slows or ceases, and the lateral buds, released from their inhibition, elongate markedly, giving the plant a bushy appearance.

on that side, because the cells there enlarge faster, and the stem bends upward. In terms of **geotropism,** which means growth toward the earth, roots are said to be positively geotropic, and stems are said to be negatively geotropic. Geotropism and phototropism are fundamental to understanding how auxins work. Since plant cells have cell walls made of a tough substance, they cannot enlarge (that is, in volume) unless something acts to soften their cell walls. The auxins increase the plasticity of the cell wall, allowing it to "give" and the photoplast to take in more water. As the cell volume increases, the cell wall becomes irreversibly enlarged and the cell becomes permanently elongated.

GIBBERELLINS

Gibberellins, another group of plant hormones, were discovered in the 1920s by Japanese research workers investigating the "foolish seedling" disease in rice. This disease, produced by a fungus, *Gibberella fujikuroi,* caused plants to grow excessively tall and spindly. Gibberellic acid was isolated from the fungus and proved to be the substance causing the disease. Later the gibberellins (the group of more than 40 compounds, which include gibberellic acid) were found to occur naturally as hormones in flowering plants.

Gibberellin, like auxin, promotes cell elongation and seems to stimulate cell reproduction. It also helps to reactivate plant metabolism in dormant plants and seeds and to promote growth.

CYTOKININS

The plant hormone **kinetin** is the most important member of a group of compounds called the **cytokinins.** It was originally isolated from corn, *Zea mays.* Kinetin seems to act antagonistically to auxin. Whereas auxin maintains apical dominance and represses lateral buds, kinetin releases lateral buds from this inhibition. Auxin increases cell elongation; kinetin initiates cell division. Plant cells supplied with auxin will enlarge but not divide, but if kinetin is added, not only do the cells continue to enlarge, but they also divide, leading to controlled growth of cells in a tissue. In addition to its anti-auxin effects, kinetin also has an antisenescence (anti-aging) effect. If a leaf is cut off, color fades because the synthesis of pigments stops. If kinetin is applied to a cut-off leaf, protein synthesis is maintained and the leaf remains green. The mechanism of cytokinin's action is still unknown, but the close relationship of the cytokinin molecule to transfer RNA (see Chapter 15) indicates that it may be directly involved in protein synthesis.

ABSCISIC ACID

Many plants must have a mechanism for slowing down activity to prepare for unfavorable conditions—dry seasons or winter. This slowed activity is called the **dormant period,** a time during which metabolic activity decreases to a very low level. **Abscisic acid** is the hormone that initiates the dormant period in certain plants. It also promotes leaf abscission (leaf drop) and prevents buds from maturing and seeds from germinating during the dormant period. Abscisic acid also plays a role in causing the stomata of plants to close, thus reducing water loss and preventing leaves from drying out during temporary periods of drought.

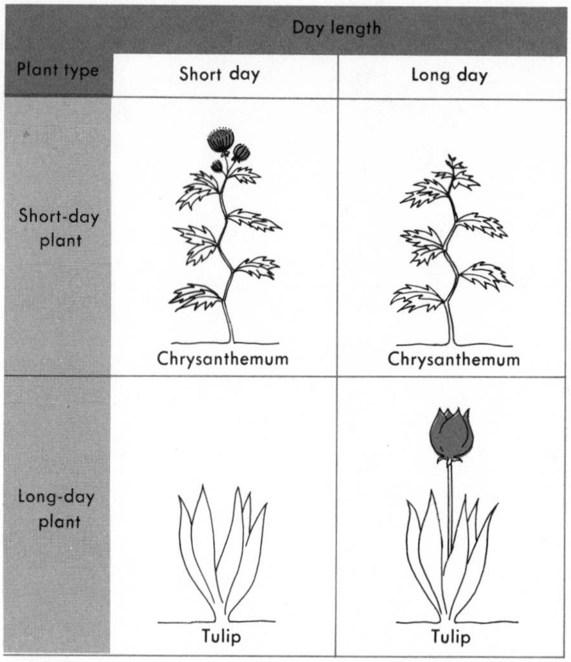

ETHYLENE GAS

The most recently recognized plant hormone is a very simple gas, **ethylene,** which is primarily involved in the ripening of fruit. Since ethylene is often a product of combustion, as well as a common industrial pollutant, it sometimes causes overripening or spoilage. The hormone is very potent, even at low levels of concentration. This fact may be the reason it went long unrecognized as a specific plant hormone. Ethylene prevents young leaves from opening until light inhibits the synthesis of ethylene. It can also promote germination. The interactions of ethylene with other plant hormones, especially auxin, are just beginning to be explored.

PHOTOPERIODISM

The last plant control mechanism that we shall consider is not, strictly speaking, governed by a hormone, but since a hormone called **florigen** is involved in the process, we shall consider it here. **Photoperiodism** is the term used to describe the responses of plants to light periods of varying duration. Flower development is particularly affected by different day lengths, or **photoperiods.** Some plants are **short-day plants,** in which flowering can occur only if plants are illuminated for less than some critical numbers of hours daily. Some are **long-day plants,** in which flowering can occur only if the photoperiod is longer than a certain number of hours. And some plants, known as **day neutral,** flower regardless of photoperiod (see Fig. 9.4). A short-day plant will not flower if it is exposed experimentally to long days (artificial days of different lengths may be created with lights); however, if it is then exposed to only *one* short photoperiod, it will flower even if the long photoperiods are resumed.

Many of the factors in this response were not well understood until recently, and several are still a mystery. First, it has been shown that it is not really the length of the light period that is crucial in determining when a plant will flower, but the period of darkness between two illuminated periods. The receptor of light for the flowering response appears to be the leaf. The cells of the leaf contain a special pigment protein called **phyto-**

	Day length	
Plant type	Short day	Long day
Short-day plant	Chrysanthemum	Chrysanthemum
Long-day plant	Tulip	Tulip

FIGURE 9.4

Effect of day length. Flowering is controlled in many plants by the length of the day (actually by the length of the night, but for historical reasons plants are referred to as short-day or long-day plants). During the lengthening days of late spring and summer, long-day plants begin to bring forth their blossoms, but fall-flowering varieties bide their time making leaves. As the days of autumn shorten, short-day plants like the chrysanthemum turn from the production of leaves to flowering.

chrome, which is partly responsible for the ability of plants to count hours of darkness. When the dark period is proper for flowering for a particular variety of plant, it appears that phytochrome somehow stimulates the production of a hormone, whose existence has been demonstrated although it has never been isolated or collected. This hormone, florigen, moves from the receptor site, the phytochrome-containing cells of the leaf, to the site of activity, the flower-producing cells of the plant, and initiates the production of a flower.

Many other activities of plants are photoperiodically controlled. Small seeds, such as those of lettuce (and of many weeds), will not germinate unless they are exposed to light for the proper duration. This fact ensures that the seeds will germinate only when they are near the surface of the soil and get the light they need.

Animal hormones

Since ancient times, herdsmen have been castrating animals to render them more docile and their meat more tender and tasty. It was not until the nineteenth century, however, that the phenomenon was explained. In 1848, a German physician, A. A. Berthold, performed a classic experiment. First he removed the testes from four roosters. Then he surgically transplanted the testes into the abdomens of two of the caponized (or castrated) roosters. The capons with no testes lost their bright comb and wattle coloring and became peaceful, overstuffed animals. The capons with the transplanted testes, however, behaved as if no caponization had taken place. In appearance and behavior, they seemed like normal, unaltered roosters.

To complete his experiment, Berthold killed the roosterlike capons and examined the transplanted testes. Although he found no new growth of nerves, he did find an extensive growth of blood vessels linking the transplanted testes with the abdominal walls into which they had been placed. Berthold concluded that the testes were producing some substance that passed into the bloodstream and was carried to all parts of the body. It was this substance that was responsible for the retention of masculine characteristics by the two capons, just as it was responsible for the same characteristics in any normal rooster. Berthold found, first, that some body parts released substances into the bloodstream; second, that the bloodstream acted as a transport system for these substances; and third, that these substances affected parts of the body at distances away from where the materials were secreted.

The validity of Berthold's insight had been well established by the early 1900s. Furthemore, his experimental method was generally followed by countless other researchers trying to discover new hormones. First, a body structure suspected of secreting hormones was surgically removed, and then any abnormalties that developed were carefully noted. If a graft of the removed organ in some other body location prevented the appearance of the abnormalities, or if an extract of the organ injected into the animal's bloodstream brought about a "cure," it could be assumed that the structure was a hormone producer.

The hormone-secreting structures of the body constitute what is known as the **endocrine system.** Anatomically it is a most unusual system. Each of the other body systems—skeletal, vascular, nervous, muscular, etc.—consists of parts that are physically connected or chemically similar, or that stem from similar processes of development. The endocrine system, however, consists of a variety of unconnected and unrelated glands, clumps of cells, and individual cells tucked away in various corners of the body. Often the location of a hormone-secreting structure offers no clue to its function. For example, no one could guess from the location of the thyroid gland, in the front of the neck, that it secretes a hormone that controls the rate of oxygen use by all the cells of the body. It was this divorce between location and function that confounded early anatomists, who were accustomed to finding relationships between location and function. The thyroid gland, for example, was at one time assumed to secrete mucus into the throat—or to exist for the purpose of making a woman's neck appear rounded and beautiful.

The disparity between the location of a gland and its function is not the only complication of the study of hormones. In addition, there is often no clearcut anatomical relationship between a hormone and its target

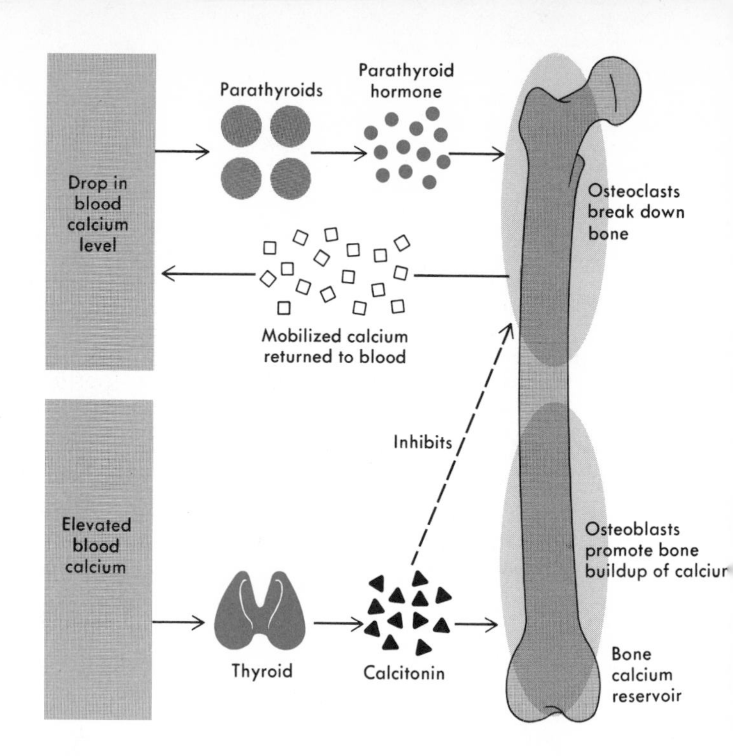

FIGURE 9.5

Hormones and calcium. The maintenance of proper levels of calcium ion in the blood is the result of the activities of two endocrine hormones: parathyroid hormone, secreted by the parathyroid glands, and calcitonin, secreted by the thyroid gland. Parathyroid hormone stimulates calcium release. Calcitonin stimulates calcium deposition and inhibits osteoclasts. Feedback inhibition regulates the calcium levels.

—the particular structures or tissues that a hormone affects. The target of the thyroid hormone may be every cell of the body. The target of another hormone may be a single structure. For example, the hormone secretin, secreted by specialized cells in the upper intestine, stimulates the pancreas to pour out digestive juices. Even though the blood carries secretin to all cells of the body, only the cells of the pancreas and certain liver cells are known to respond. Thus the study of the endocrine system is as much the study of targets as it is of glands.

FEEDBACK CONTROL

Much of the endocrine system is regulated by negative feedback. In effect, the response to the message is to turn off the messenger—as a sleeper turns off the alarm clock that wakes him. Consider, for example, the action of the hormone secretin, mentioned above. When digestion of food begins, acidic digestive juices from the stomach come in contact with specialized cells in the upper intestine. Under the acidic stimulus, cells produce the hormone secretin. Secretin is taken by the bloodstream to the pancreas, which then pours out its alkaline digestive juices into the small intestine. The alkalinity of the pancreatic juices neutralizes the digestive acids that initiated the process and the intestines stop producing secretin. The response to the *message* has turned off the *messenger*. In the meantime, a special enzyme in the blood starts to deactivate the circulating secretin, so that the message rapidly disappears from the bloodstream.

In many situations, the quantity of hormonal secretion and the activity that a hormone induces may function mutually to maintain a delicate balance by negative feedback. For example, the body needs to maintain a constant level of blood calcium. A drop in the blood calcium level stimulates the parathyroid glands (embedded in the posterior surface of the thyroid) to secrete parathyroid hormone (PTH), which causes the release of calcium from bone into the bloodstream. The resulting rise in blood calcium turns off the production (this turning off by the substance itself is the negative feedback) of parathyroid hormone. If the blood calcium level drops again, the parathyroid glands are stimulated again. In this manner, slight elevations or depressions in the blood calcium level are quickly sensed and corrected.

Negative feedback may be supplemented by a second hormone, which acts to reverse the effect of the first. As discussed above, parathyroid hormone acts to maintain blood calcium levels at the expense of bone calcium. Another hormone, calcitonin, has the opposite effect: it inhibits the release of calcium from bone into blood. Thus blood calcium and bone calcium are kept in balance by the action of two hormones, not just one. This interrelationship is presented schematically in Fig. 9.5.

Considering the complexity of relationships between different hormones and the activities they stimulate, it is not surprising to find that very few body activities are regulated by one hormone alone. Consider the metabolism of glucose, for example. All of the following hormones affect it: insulin, glucagon, epinephrine, growth hormone, cortisone—and the list is far from complete.

In addition to regulating body activities, hormones serve to regulate even the endocrine glands themselves. The functioning of certain endocrine glands—the thyroid, the adrenals, and the gonads—is regulated by hormones secreted by the anterior lobe of the pituitary

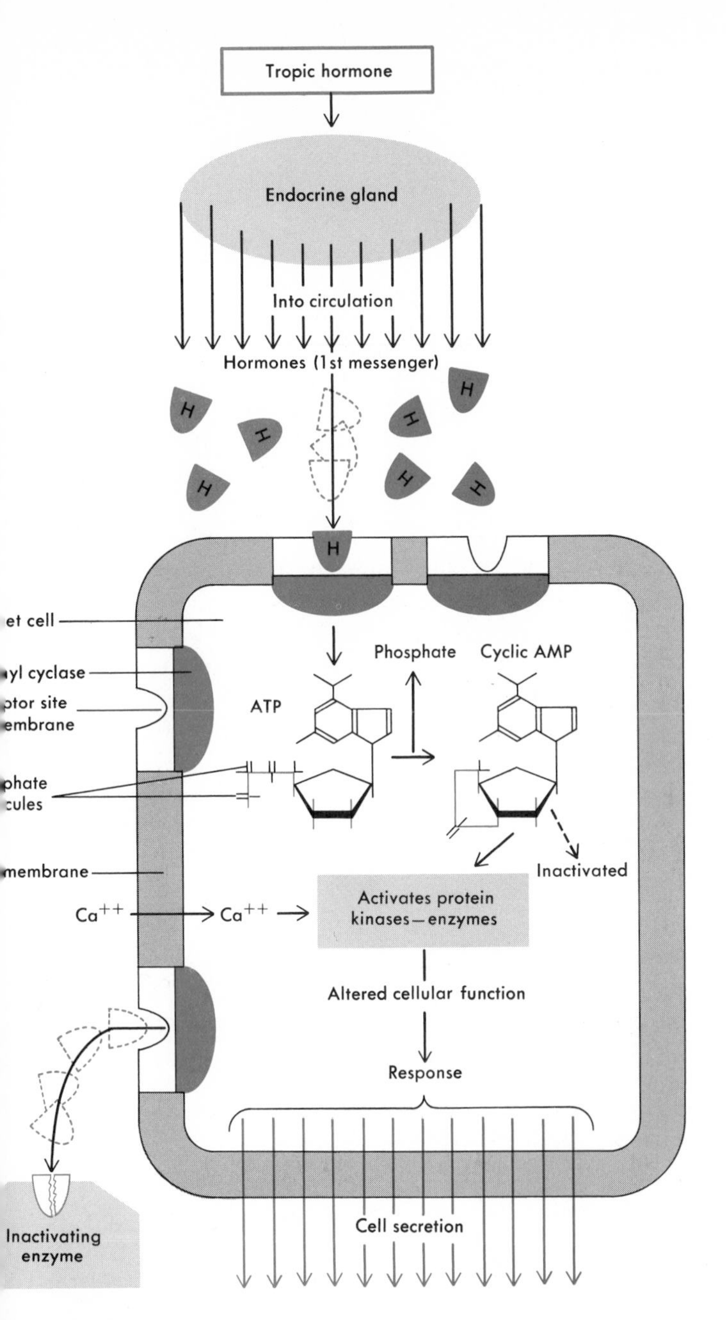

FIGURE 9.6

The two-messenger model of hormone activity. Stimulated by tropic hormones, an endocrine gland secretes its hormones, the first messengers. They travel through the bloodstream to the target cells, where the hormones fit into specific sites on the cell membrane. When in a site, a hormone activates adenyl cyclase, an enzyme that converts ATP into cyclic AMP by removing two phosphate groups and forming a cyclic bonding group at one end of the AMP molecule (see colored portion of molecule). Cyclic AMP is the second messenger. It works with calcium to activate enzymes that alter cell function and bring about target cell response. After completing their function, hormones are inactivated by enzymes in the blood.

gland at the base of the brain. These pituitary hormones, collectively known as **tropic hormones,** typically have a negative feedback relationship with the hormones whose production they stimulate. Thus a pituitary hormone called thyroid-stimulating hormone (TSH) stimulates the thyroid gland to produce thyroid hormone. Thyroid hormone, in turn, shuts off the production of TSH. Consequently, the production of thyroid hormone drops until the resultant drop in the hormone's level in the blood again stimulates the production of TSH. (The actual mechanism involves the **hypothalamus,** the portion of the brain to which the pituitary gland is attached, and which will be discussed later in this chapter.)

HORMONES AND THE SECOND MESSENGER

The question of how hormones produce their effects, once they reach their target cells, has puzzled researchers for many years. The most recent research has led to the concept of a second messenger *inside* the target cell. This second messenger transmits the command of the first messenger, the hormone, which apparently does not pass the barrier of the cell membrane. One such second messenger, *c*yclic *a*denosine *m*ono*p*hosphate (cAMP), has been extensively studied since its discovery in 1958 by Earl W. Sutherland, who received the 1971 Nobel Prize in Medicine for the far-reaching implications of his work. The extensive research on cyclic AMP has demonstrated that many hormones produce their specific effects by altering (increasing or decreasing) the concentrations of this factor within cells. The hormone is believed to combine with a specific receptor site located on the cell membrane of the target cell. This binding then activates a plasma membrane–bound enzyme called adenyl cyclase, whose function is to convert ATP into a cyclic form, 3′,5′-cAMP, which in turn activates cellular enzymes called protein kinases. The protein kinases work in conjunction with increased calcium ion concentrations in the cell to activate yet other enzymes. This in turn causes protein synthesis and thus changes the cell's metabolic activities. Hormone-induced changes in the cell are only indirect effects. A schematic representation of the operation of the two-messenger model of hormone activity is given in Fig. 9.6.

Considering the number of different hormones and the number of different tissue responses they stimulate, it might seem improbable that a single substance—cAMP or a chemical relative—could be the mediator for them all. The difficulty can be resolved, first, in terms of the specific sensitivity of the receptor site on the cell membrane. That site is presumably sensitive to only one hormone; all others pass on without affecting the site. Second, the response is dependent on the nature of the specific cell. A liver cell can produce glucose in response to hormonal prompting, whereas a muscle cell cannot, because muscle cells lack a particular enzyme that is necessary for glucose production.

Not all hormones have been shown to produce their effects by the cAMP route, and recent work has shown that some hormones act to derepress specific structural genes. Thus some hormones that act directly on the cell nucleus do not use a second messenger. Nevertheless, the concept of a second messenger appears firmly established, and other second messengers besides cAMP are being sought. (There appears to be a possibility that a chemically related substance, cyclic guanine monophosphate, may be another such second messenger.)

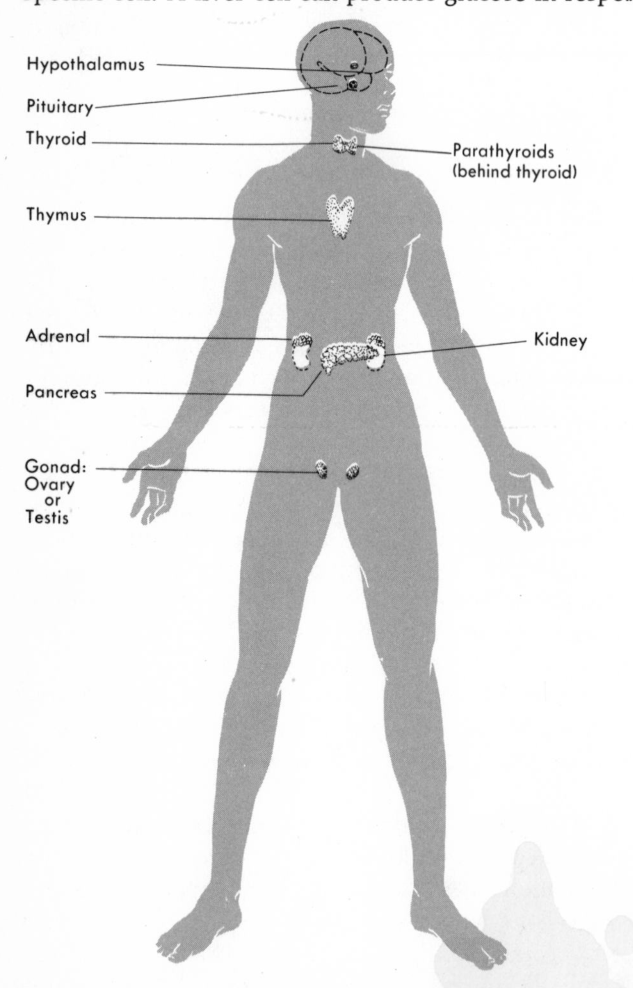

FIGURE 9.7
Location of important endocrine glands.

The human endocrine system

In the remaining sections of this chapter, we will consider the principal glands of the human endocrine system, the hormones they secrete, and the functions performed by those hormones. The location of the glands is identified in Fig. 9.7, and Table 9.1 summarizes this information.

THE HYPOTHALAMUS

For a long time the endocrine system was thought to complement the nervous system, the other bodily control system, but to function essentially independent of it. The endocrine system was thought to control long-term functions, such as growth and maturation, while the nervous system regulated short-term activities, such as muscle movement. The one was assumed to act by chemical regulation, the other by electrical impulses.

Today we know that this hard-and-fast distinction between the endocrine and nervous systems is entirely misleading. With this understanding has come a broadened realization of how the endocrine system does work. As mentioned above, the anterior pituitary gland at the base of the brain secretes hormones that regulate the production of other hormones. Indeed, the anterior pituitary has been called the master gland of the body. In

TABLE 9.1
Principal endocrine glands, their hormones, and hormone function

GLAND	HORMONE	MAJOR FUNCTION	SECRETION REGULATOR
Pituitary, anterior lobe	Growth hormone	Induces growth in bone and muscle	Hypothalamus
	Gonadotropins:		
	Follicle-stimulating hormone (FSH)	Activates follicles in ovary	Blood estrogen; hypothalamus
	Luteinizing hormone (LH)	Stimulates testes in male, corpus luteum in female	Blood levels of testosterone or progesterone; hypothalamus
	Thyroid-stimulating hormone (TSH)	Activates thyroid gland	Blood level of thyroid hormone; hypothalamus
	Adrenocorticotropic hormone (ACTH)	Stimulates adrenal glands	Adrenal cortical hormone in blood; hypothalamus
	Prolactin	Induces secretion of milk, governs parental behavior	Hypothalamus
Pituitary, posterior lobe	Vasopressin (antidiuretic hormone)	Controls water elimination	Osmotic pressure of blood
	Oxytocin	Induces uterine contractions, release of milk	Nervous system
Thyroid	Thyroxin (thyroid hormone)	Regulates metabolism, some phases of development	TSH
	Calcitonin	Stimulates absorption of calcium in bone	Calcium level in blood
Parathyroid	Parathyroid hormone (PTH)	Induces calcium release from bone	Calcium level in blood
Pancreas	Insulin	Lowers blood sugar, induces glycogen storage	Level of blood sugar
	Glucagon	Induces production of glucose from glycogen	Level of blood sugar
Adrenal cortex	Steroids (glucocorticoids, mineralcorticoids, sex hormones)	Control carbohydrate metabolism, salt and water balance	ACTH
Adrenal medulla	Epinephrine, norepinephrine	Raise blood sugar level, cause blood vessel dilation, increase heartbeat	Nervous system
Testes	Testosterone	Induces sperm production, develops and maintains sex characteristics	LH
Ovary, follicle	Estrogens	Induce growth and maintenance of sex characteristics: prepare uterine lining for implantation	FSH and LH
Ovary, corpus luteum	Progesterone	Causes uterine tissue to grow	LH

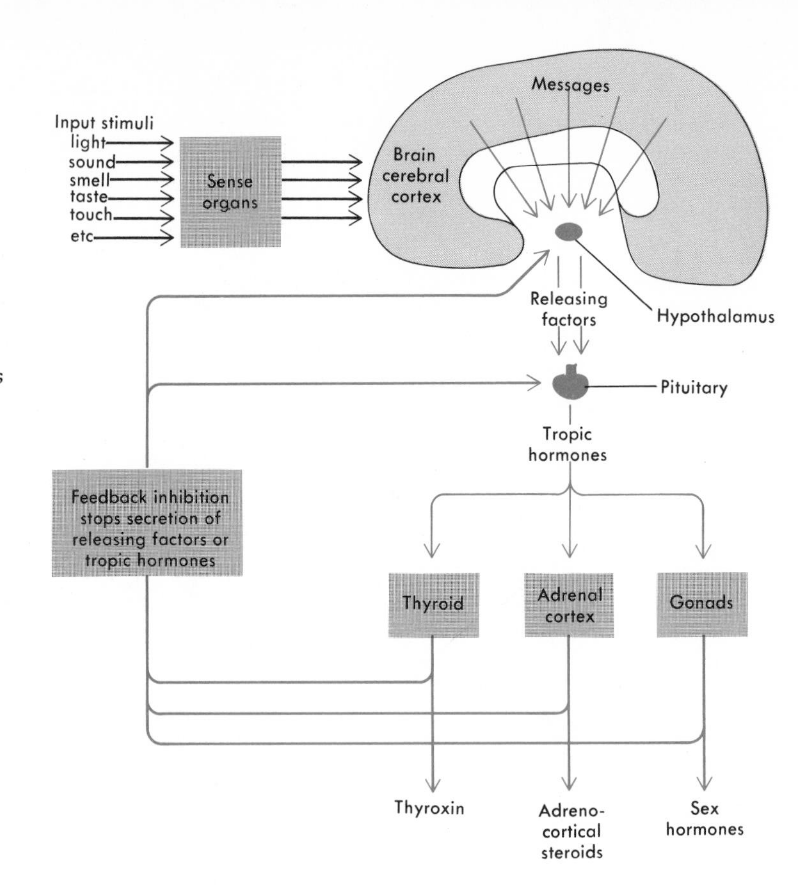

FIGURE 9.8

The relationship between the brain and the endocrine system depends on the hypothalamus. Various stimuli impinging on the sense organs are relayed to the brain, which evaluates and integrates them. The brain sends messages to the hypothalamus through the nerves. The hypothalamus then sends messages to the pituitary in the form of releasing factors, which elicit the secretion of tropic hormones by the pituitary. Tropic hormones in turn cause the secretion of hormones by each affected endocrine gland. Negative feedback operating at the hypothalamic or pituitary level turns off the production of releasing factors or tropic hormones to ensure against excessive secretion.

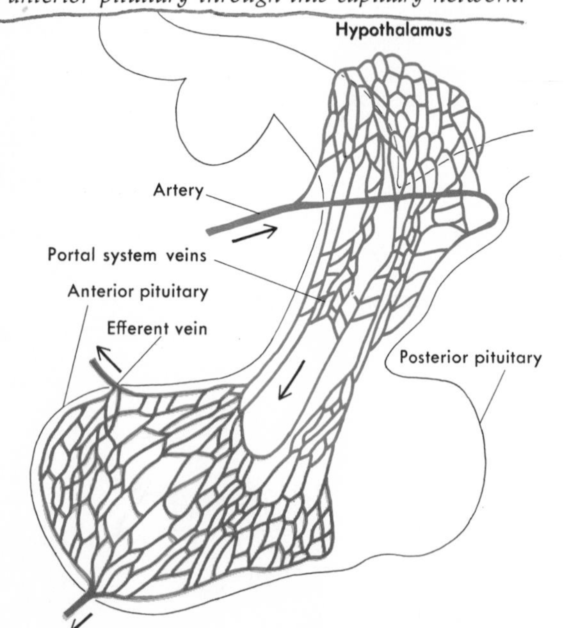

FIGURE 9.9

Capillary blood vessels connect the hypothalamus to the pituitary. The hypothalamus controls the secretions of the anterior pituitary through this capillary network.

the same sense, the brain is the master of the nervous system. The two master structures are physically connected, so it seems logical that their functions should be related. Yet the nature of this relationship had for a long time been a puzzle, for there is no nerve connection between the brain and the anterior pituitary, and hormones and similar substances cannot *normally* pass from the capillaries in tissues surrounding the brain into the brain tissue. Therefore, it appeared that the anterior pituitary could not send a message to the brain either.

Recent research appears to have solved the puzzle, however. It has been discovered that the hypothalamus, the portion of the brain to which the pituitary is attached, is capable of receiving chemical messages via the bloodstream. And within the hypothalamus are specialized nerve cells that actually secrete hormones, usually very small peptide molecules with only three to eight amino acids, in infinitesimally small amounts. These hypothalamic-produced hormones, sometimes referred to as **releasing factors,** are carried directly from capillaries in the hypothalamus to capillaries in the anterior pituitary, where they stimulate the production of the pituitary hormones. Thus the hypothalamus, which is sensitive both to nervous impulses from the rest of the brain and to concentrations of substances in the bloodstream, appears to have emerged as the true master gland for many of the body's hormonal activities (see Figs. 9.8 and 9.9).

POSTERIOR PITUITARY

The posterior lobe of the pituitary gland, an outgrowth of the floor of the hypothalamus, releases two important hormones that are actually secreted by the hypothalamus:

Vasopressin (antidiuretic hormone, **ADH**) is one of the hormones responsible for conserving the body's water supply. In the normal body, the kidneys absorb water and wastes from the bloodstream, then return most of the water back to the bloodstream. The wastes and a small amount of water are excreted. The hormone vasopressin promotes this return of the water from the kidneys to the bloodstream. Inadequate supply of the hormone, such as might occur as the result of a tumor of the posterior pituitary, causes the disease diabetes insipidus (not to be confused with more commonly known but quite different diabetes mellitus). The kidneys of a patient with diabetes insipidus do not adequately return water to the bloodstream. The untreated patient would excrete great quantities of dilute urine and would have to drink gallons of water every day to make up for the loss.

Oxytocin has two known functions involving the contraction of smooth muscle tissue. First, it stimulates the contractions of the uterus during childbirth. Second, it stimulates the contractions of the milk glands in the breasts of a lactating mother. It is these oxytocin-stimulated contractions that expel milk from the breast, and not the baby's sucking, but it is the stimulus of sucking, transmitted to the hypothalamus by nerves, which causes the release of oxytocin from the posterior pituitary. A commercial preparation of oxytocin is sometimes used by obstetricians to bring on labor contractions.

ANTERIOR PITUITARY

The anterior portion of the pituitary is embryologically quite unrelated to the posterior lobe, being an outgrowth of the tissue that forms the roof of the mouth. The anterior pituitary secretes a number of important hormones, many of which stimulate endocrine activities elsewhere in the body.

Growth hormone (GH), or **somatotropic hormone (STH),** stimulates natural child and adolescent growth. Insufficiency of this hormone in the formative years causes the person to be a midget—well formed, but childlike in appearance. This result can occur when the pituitary is injured by disease. In addition, children raised in institutions where they experience difficulty sleeping may suffer stunted growth, because the growth hormone is released mainly during deep sleep. Cessation of growth at the end of adolescence is not due to a lack of growth hormone. Rather, it is attributable to other factors. Adults continue to secrete GH, but its major functions seem to be repair and assistance in the normal utilization of glucose.

Gonadotropins are two (or three—the exact number is still uncertain) hormones that govern the production of sex hormones, sperm, and ova. They are more fully discussed in the chapter on reproduction.

Thyroid-stimulating hormone (TSH) stimulates the production of thyroid hormone by the thyroid gland. Without the stimulus of TSH, the thyroid gland does not function. For this reason, destruction of the anterior pituitary, where TSH is produced, causes many of the same symptoms as destruction of the thyroid. TSH is also involved in the metabolism of fats.

Adrenocorticotropic hormone (ACTH) stimulates the **cortex,** or outer portion, of the adrenal glands to produce certain of their hormones, the **corticoids.**

THYROID

The very important thyroid gland is located at the front and sides of the windpipe, slightly below the larynx. It secretes two types of hormones:

Thyroxin (thyroid hormone) affects the body's use of oxygen—that is, the body's metabolic rate—and also the process of bodily maturation. A child deprived of thyroid hormone develops a condition called **cretinism,** which stunts his mental and physical growth. In lower vertebrates, thyroid hormone is required for proper maturation also; for example, it is the hormone that prompts the tadpole to develop into a frog.

Severe thyroid deprivation in the adult human results in the condition known as **myxedema,** characterized by weight gain, coarsened hair, puffy features, thickened and furrowed tongue, and a general slowing down of bodily and mental processes.

Calcitonin inhibits the movement of calcium from bone into blood by inhibiting the bone-destroying effect of certain cells called osteoclasts, which form an extensive network throughout bone.

PARATHYROID GLANDS

Four small structures, the parathyroid glands, are embedded in the rear surface of the thyroid gland. The glands are known to secrete a single hormone, called parathyroid hormone.

Parathyroid hormone (PTH) reverses the action of calcitonin by stimulating osteoclasts to destroy bone tissue and release calcium and phosphate to the blood. PTH also acts on the kidney, causing it to return excreted calcium to the bloodstream but to dispose of excreted phosphate. The net result is a rise in blood calcium with blood phosphate maintained at a steady level.

Calcium is necessary, among other things, for proper muscle action. An insufficiency of blood calcium may cause extreme irritability of the muscles. Thus destruction of the parathyroid glands, which causes a drop in blood calcium, may throw muscles into spasm—a condition known as **tetany.** The patient, unable to breathe due to the spasm of his respiratory muscles, may die.

PANCREAS

The pancreas is an accessory organ of digestion. It produces certain digestive juices that empty into the upper intestine. Within the pancreas, however, are scattered clumps of cells called **islets of Langerhans,** named after their discoverer, Paul Langerhans. The islets of Langerhans secrete two important hormones:

Insulin, one of the best known of all hormones, influences the metabolism of carbohydrates, fats, and proteins in a variety of tissues and a variety of complex ways. Inadequate insulin secretion gives rise to the disease diabetes mellitus. Diabetes is diagnosed by the presence of sugar in the urine or excessive blood sugar levels. In the absence of insulin, tissues lose their ability to store or use glucose effectively, and the glucose accumulates in the blood. Diabetes is usually controlled by a regular regimen of insulin injections.

Glucagon is a hormone wtih effects opposite to those of insulin. In the liver it works together with the adrenal hormone epinephrine to promote the breakdown of glycogen to glucose. Whereas insulin builds up the stores of liver glycogen, glucagon breaks them down into glucose units again.

ADRENAL CORTEX

Capping the kidneys are two glands, the adrenals, with two portions—an outer layer, the cortex, and an inner, the medulla. The two portions of the adrenals are distinct and different. The cortex produces many substances, collectively known as **steroids,** which are derived from the chemical **cholesterol.** Some are hormones, but some are merely chemical precursors or breakdown products of the hormones. The steroid hormones are classed in three groups: the **glucocorticoids,** which regulate glucose metabolism and affect the body's reaction to infection and allergies; the **mineralocorticoids,** which regulate electrolyte balance in the blood; and the **sex hormones,** which regulate sexual development and function.

The cortical hormones are necessary for life and reproduction. Damage to the adrenal cortex produces **Addison's disease,** characterized by a generalized derangement of carbohydrate metabolism and of blood electrolyte levels. Circulatory and cardiac failure may result. The patient is unable to resist outside stress or infection.

The Use of Diethylstilbestrol in Beef Production

Diethylstilbestrol (DES) is a synthetic female hormone—an estrogen—that mimics the activity of natural estrogens. When injected into or ingested by cattle or sheep, DES has the effect of enhancing weight gain and shorten- the length of time it takes for the animal to reach slaughtering weight.

Many natural and synthetic estrogens (DES included) have been tested and found to cause breast cancer in mice. Thus DES is considered a carcinogen—a substance that produces or leads to the development of cancer. In a September 1976 article in *BioScience* magazine, Thomas H. Jukes estimates that the risk of developing cancer from eating meat containing traces of DES ranges from one case of cancer in 133 years in the U.S. population to one case in 2500 years.

Even though the risk of cancer from the DES in meat appears to be extremely small, the use of DES as a feed supplement is legally regulated by the Delaney clause (1958) of the Food Additives Amendment. The Delaney clause states that "no food additive [to human food] shall be deemed to be safe if it is found to induce cancer when ingested by man or animals." However, the clause includes an additional provision for the use of additives or drugs in food-producing animals. That provision states that a carcinogen may be used in food-producing animals, provided that it does not affect the health of the animal and that no residue can be found in the food products of the animal as determined by a method of analysis approved by the Secretary of Health, Education and Welfare.

In 1954, DES was approved by the Food and Drug Administration for use as a supplement in the diets of cattle or sheep and as an implant that could be inserted in the animal's ear, where it would be distributed by the bloodstream to the rest of the body.

Later in the 1960s, to comply with the Delaney clause, the Food and Drug Administration required that the use of DES be stopped seven days before the animal was slaughtered. The tests in use at that time indicated that DES was not stored in tissues, and that any of the hormone still in the animal's body would be excreted during the seven-day period. The meat would thus be free of the cancer-causing hormone, and there would be no risk to those who consumed the meat.

In November of 1971, two environmental organizations filed a law suit against the Food and Drug Administration and the Department of Agriculture, demanding that the use of DES to fatten livestock be banned entirely. The enforcement of the DES regulations was difficult at best: The Agriculture Department's test for slaughtered meat was not sensitive enough to detect minute amounts of DES in tissue, and the Delaney clause requires that no carcinogen whatsoever remain in the product. It was also difficult to monitor the seven-day limit. Farmers had one chance in 5000 of being caught if they used DES right up to the day of slaughter.

In 1972 and 1973, using a newer, more sensitive method of detection, the Department of Agriculture found DES residues in the edible parts of test animals. The DES showed up even when the dosage was stopped seven days prior to slaughtering. In early 1973, consequently, the use of DES as a supplement and an implant was banned by the Food and Drug Administration. Five manufacturers of DES supplements and implants opposed the ban, and in 1974 the U.S. District Court of Appeals for the District of Columbia overturned the decision on the grounds that the manufacturers of DES were denied a hearing when the ban was proposed.

Before going to a hearing with the manufacturers, the FDA decided it should first resolve the question of the sensitivity and adequacy of the method used to detect drug residues in the food products of animals. Discovering that this was a highly complex and controversial issue that would take a great deal of time to resolve, the FDA decided to move ahead with the proposal of the DES ban and the manufacturers' hearing in 1976.

As of May 1977, the hearing had not been completed, and DES was still being used to promote growth in sheep and cattle.

In a case such as this one, there are many viewpoints that must be considered—in addition to the scientific. Controversies involving biological topics (such as nuclear power, the use of pesticides, etc.) often involve economic, ecological, and political considerations, and a great deal of careful research must be conducted to find the conclusion that is agreeable, safe, and just for all concerned.

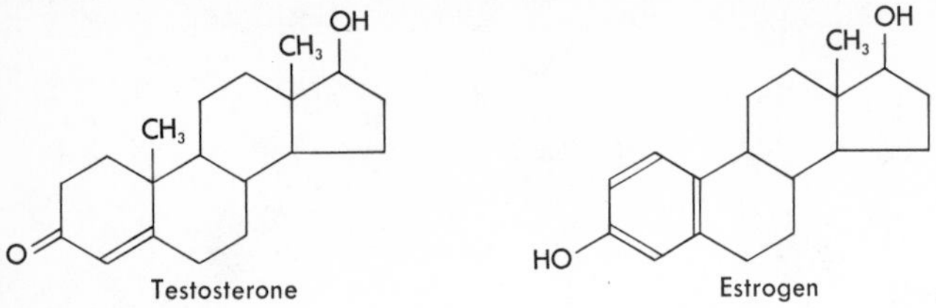

OH
CH_3
CH_3
O
Testosterone

OH
CH_3
HO
Estrogen

FIGURE 9.10

Comparison of testosterone and estrogen. Testosterone, a male sex hormone, is not very different in chemical structure from estrogen, one of the female sex hormones, although their effects are quite different.

ADRENAL MEDULLA

The central portion of the adrenal glands, the adrenal medulla, produces two hormones that are similar to substances produced by certain nerve endings. These two hormones function somewhat similarly and will be considered together.

Epinephrine and **Norepinephrine** are hormones that promote an increase in cardiac activity, elevate the blood pressure, reduce the rate of digestion, increase the flow of blood to the brain and muscles, increase the respiratory rate, and promote the breakdown of liver glycogen to glucose. In short, epinephrine and norepinephrine mobilize the body for action under conditions of sudden stress.

GONADS

The testes produce the male hormone testosterone, and the ovaries produce the female hormones estrogen and progesterone. The role of these hormones in controlling reproduction and producing secondary sex characteristics is considered in Chapter 12, but their secondary functions will be considered here (see Fig. 9.10 for a comparison of their chemical structures).

Testosterone affects the functioning of skeletal muscle, partially accounts for the growth spurt of adolescence, and helps to end that sudden growth by promoting the fusion of the terminal parts of bones to the bone shafts. Testosterone appears to be necessary for sexual desire in men and women (in women testosterone is secreted by the adrenal glands).

Estrogen, one of the two female hormones, is also involved in the growth spurt of adolescence and, like testosterone, helps to end it by similarly slowing bone growth. In addition, it preserves proper muscle tone in the smooth muscles, such as those constituting the walls of blood vessels.

FIGURE 9.11

The feather-like antennae of the male Japanese moon moth contain olfactory receptors sensitive to sex pheromones produced by the glands of the female. The female's scent excites the male, promotes his take-off, and directs him to her vicinity.

OTHER ENDOCRINE ORGANS

Although the stomach and the duodenum are not considered endocrine glands, they contain specialized cells that secrete gastrointestinal hormones. The kidney, blood, and placenta also secrete hormones.

The roles of two glands—the pineal body and the thymus—are not yet fully understood. The **pineal body** is a tiny structure, buried at the center of the brain, which at one time in our evolutionary history may have served as a light-sensitive regulator of certain body cycles. In the mammalian brain, the pineal body appears to be the source of one hormone concerned with inhibiting skin coloration.

The **thymus,** in infants, is a large structure located in the upper chest, but it shrinks to a very small size in adults. Primarily in the early years of development, the thymus "seeds" the blood and lymphatic system with cells known as lymphocytes and produces a substance necessary for the formation of blood proteins called immunoglobulins. Lymphocytes and immunoglobulins form part of the body's defense system against disease and infection. There is some suggestion that the gland may also produce both a growth-stimulating and a growth-inhibiting factor.

Other chemical messengers

The foregoing list has included most of the main hormone classes, at least those occurring in mammals. In addition, there are a number of other substances, some fairly recently discovered, that are similar to hormones in many ways.

Pheromones are one such group of substances. They are released in trace amounts by an organism and, through contact, can induce remarkable behavioral changes in *another* organism, usually of the same species. Pheromones have been studied mostly in insects. They include the attractant that a female moth exudes to lure potential mates from miles around, and the substances that ants of the same swarm or nest secrete to identify and communicate with one another. The effects of the pheromones will be discussed further in Chapter 11.

Prostaglandins—mistakenly named for the prostate gland, which does not secrete them as was originally thought—are substances found in a host of tissues and body fluids. Although not well understood, they appear to influence the contraction and relaxation of the heart and blood vessels; they promote urination and excretion of sodium; they may affect endocrine secretion; they promote uterine contractions in childbirth. Prostaglandins have also been implicated as a cause of headache. In fact, it has been suggested that aspirin works, in part, by blocking the synthesis of the prostaglandins. Although a good deal is known about the particular effects of the prostaglandins, there is as yet no general, unifying concept of how they function or what their purpose is in the body.

In sum, it appears that in both plants and animals, chemical messengers called hormones evolved to help bridge the information gap, bringing messages about various internal and external conditions of the organism to vital centers of action and instructing them to act according to environmental changes, thereby maintaining opimum internal conditions for an organism.

Summary

1. Both plants and animals produce hormones as a means of internal communication and to maintain homeostasis. Formed in various tissues in plants and by the endocrine system in animals, hormones are secreted in minute quantities. Moving from cell to cell in plants and by means of the circulatory system in animals, hormones travel relatively slowly (compared with nerve impulses) and usually affect parts of the organism distant from their point of origin. Hormones regulate such functions as metabolism, growth, and reproduction.

2. Plant hormones regulate plant metabolism, growth, flowering, fruit ripening, and dormancy. Plants produce only a few kinds of hormones, but these effect a multitude of responses, because a single hormone can stimulate different reactions in different tissues.

Auxins, produced in meristematic tissue, cause cellular elongation, photo- and geotropisms, root, stem, and fruit development, and apical dominance.

Gibberellins are produced in stem cambia, seeds, enlarging fruit, and other places where enlargement or awakening from dormancy occur. Alone they promote cell elongation and stimulate metabolism after a dormancy period, and they act synergistically with auxins to promote growth.

Cytokinins, of which kinetin is the most important, act antagonistically to auxins, stimulating cell division rather than elongation and the development of lateral buds rather than apical dominance. They also have an antisenescence effect.

Abscisic acid is the slow-down hormone that acts antagonistically to hormones stimulating growth and development. It initiates leaf abscission and dormancy in certain plants and inhibits bud maturation and seed germination.

Ethylene gas, the substance most recently recognized as a plant hormone, is involved in the ripening of fruit. It also prevents young leaves from opening too soon, and it can promote germination.

Florigen is thought to induce flowering. Florigen production and flowering will occur only if a plant's individual photoperiodic requirements are met and if there are adequate amounts of CO_2 and plant leaves containing the light-sensitive compound phytochrome.

3. Animal hormones are secreted by the endocrine glands. The study of the endocrine system has been complicated by the fact that there is often no apparent functional reason for a gland to be in a certain location and no structural connection between a gland and its hormone's target. Furthermore, glands vary tremendously in size and structure, from small groups of cells like the islets of Langerhans to a large organ such as the thyroid gland.

4. Regulation is achieved by negative feedback mechanisms through which the hormone-induced reaction inhibits hormone secretion or by the action of a pair of hormones which have antagonistic effects.

5. Although the mechanism by which hormones induce target cell activity is not fully understood, scientists have proposed a second-messenger model for certain types of hormones. The first messenger, the hormone, never enters the cell, but makes contact with a receptor site specific to that hormone on the outside surface of the cell membrane. The contact initiates a series of reactions resulting in the synthesis of the second messenger, cAMP, which in turn activates enzymes necessary for the required metabolic activities.

6. The functions of the endocrine and nervous systems are coordinated by the hypothalamus. Sensitive to both nerve impulses and blood chemistry, it monitors body functions and regulates endocrine activity by exerting hormonal control over the anterior lobe of the pituitary gland.

7. The mammalian body contains a number of permanent tissues or organs which secrete hormones: the hypothalamus, the pituitary (anterior and posterior lobes), thyroid, parathyroid, and adrenal glands (cortex and medulla), the islets of Langerhans in the pancreas, and the gonads (ovaries and testes). Other tissues that produce hormones include the stomach, duodenum, kidney, blood, placenta, pineal body, and thymus.

8. Other chemicals with properties similar to hormones are pheromones (whose release affects other organisms) and prostaglandins (whose full effects are still not determined).

REVIEW AND STUDY QUESTIONS

1. What is a hormone? Why is the word "messenger" often included in discussions of how hormones work?

2. What causes the stem and leaves of a plant to turn toward the light? What is this process called?

3. A gardener snips off the growing tip of a plant to promote branching. What makes this practice work?

4. Name two kinds of plant hormones that act antagonistically to each other. What is the effect of each?

5. State two principal differences between the endocrine system and the other systems of the body.

6. What is sometimes called "the master gland of the body"? Why?

7. Name the two types of hormones secreted by the thyroid. Explain what each does in the human body.

8. The disease *diabetes mellitus* results from inadequate secretion of the hormone insulin, and treatment often includes insulin injections. What does insulin do? How is diabetes detected?

REFERENCES

Axelrod, J. 1974 (June). "Neurotransmitters." *Scientific American* Offprint no. 1297. Freeman, San Francisco.

Biale, J. B. 1954 (May). "The Ripening of Fruit." *Scientific American* Offprint no. 118. Freeman, San Francisco.

Jacobs, W. P. 1955 (November). "What Makes Leaves Fall?" *Scientific American* Offprint no. 116. Freeman, San Francisco.

Levey, R. H. 1964 (July). "The Thymus Hormone." *Scientific American* Offprint no. 188. Freeman, San Francisco.

Levine, S. 1971 (January). "Stress and Behavior." *Scientific American* Offprint no. 532. Freeman, San Francisco.

Liddle, G. W., and J. Hardman. 1971. "Cyclic Adenosine Monophosphate as a Mediator of Hormone Action." *New England Journal of Medicine* 285: 560–566.

Pastan, I. 1972 (August). "Cyclic AMP." *Scientific American* Offprint no. 1256. Freeman, San Francisco.

Pike, J. E. 1971 (November). "Prostaglandins." *Scientific American* Offprint no. 1235. Freeman, San Francisco.

Rasmussen, H. 1961 (April). "The Parathyroid Hormone." *Scientific American* Offprint no. 86. Freeman, San Francisco.

Rasmussen, H., and M. M. Pechet. 1970 (October). "Calcitonin." *Scientific American* Offprint no. 1200. Freeman, San Francisco.

Van Overbeek, J. 1968 (July). "The Control of Plant Growth." *Scientific American* Offprint no. 1111. Freeman, San Francisco.

Wurtman, R. J., and J. Axelrod. 1965 (July). "The Pineal Gland." *Scientific American* Offprint no. 1015. Freeman, San Francisco.

Zuckerman, S. 1957 (March). "Hormones." *Scientific American* Offprint no. 1122. Freeman, San Francisco.

SUGGESTIONS FOR FURTHER READING

Arehart-Treichel, J. 1972. "Sperm and Eggs on the Go." *Science News* 102: 108–109.

An interesting report on research on the pituitary sex hormones and their role in male and female fertility. Contradicts the prior belief that the target sex hormones —estrogen, testosterone, etc.—are not actually the source of sex hormone action and deficiency.

Asimov, I. 1963. *The Human Brain: Its Capacities and Functions.* Signet, New York.

A clear, readable description of the functioning of the brain and other systems related to it. Chapters 1 through 5 deal with the endocrine system in man, with separate chapters on pancreatic, thyroid, adrenal, and gonadal hormones.

Butler, W. L., and R. J. Downs. 1960 (December). "Light and Plant Development." *Scientific American* Offprint no. 107. Freeman, San Francisco.

The discovery of the role in flowering of the light-sensitive protein phytochrome.

Galston, A. W. 1964. *The Life of the Green Plant.* Prentice-Hall, Englewood Cliffs, N.J.

Chapter 4 deals with plant hormones, tissue organization, and growth. Chapter 5 includes differentiation, morphogenesis, and photoperiodism.

Robison, G. A., R. W. Butcher, and E. W. Sutherland. 1971. *Cyclic AMP.* Academic Press, New York.

A useful reference volume on cyclic AMP, including summaries of all the research done or being done in this area.

Sutherland, E. W. 1972. "Studies on the Mechanism of Hormone Action." *Science* 177: 401–408.

A reprint of the author's lecture in Sweden on receipt of the Nobel prize for his discovery of cyclic AMP. A clearly written explanation of the function of cyclic AMP, putting forth the view that hormones may be studied at the molecular level.

Wilson, E. O. 1963. "Pheromones." *Scientific American* Offprint no. 157. Freeman, San Francisco.

A review of chemicals used for interorganism communication; their types, structure, and modes of affecting the recipient.

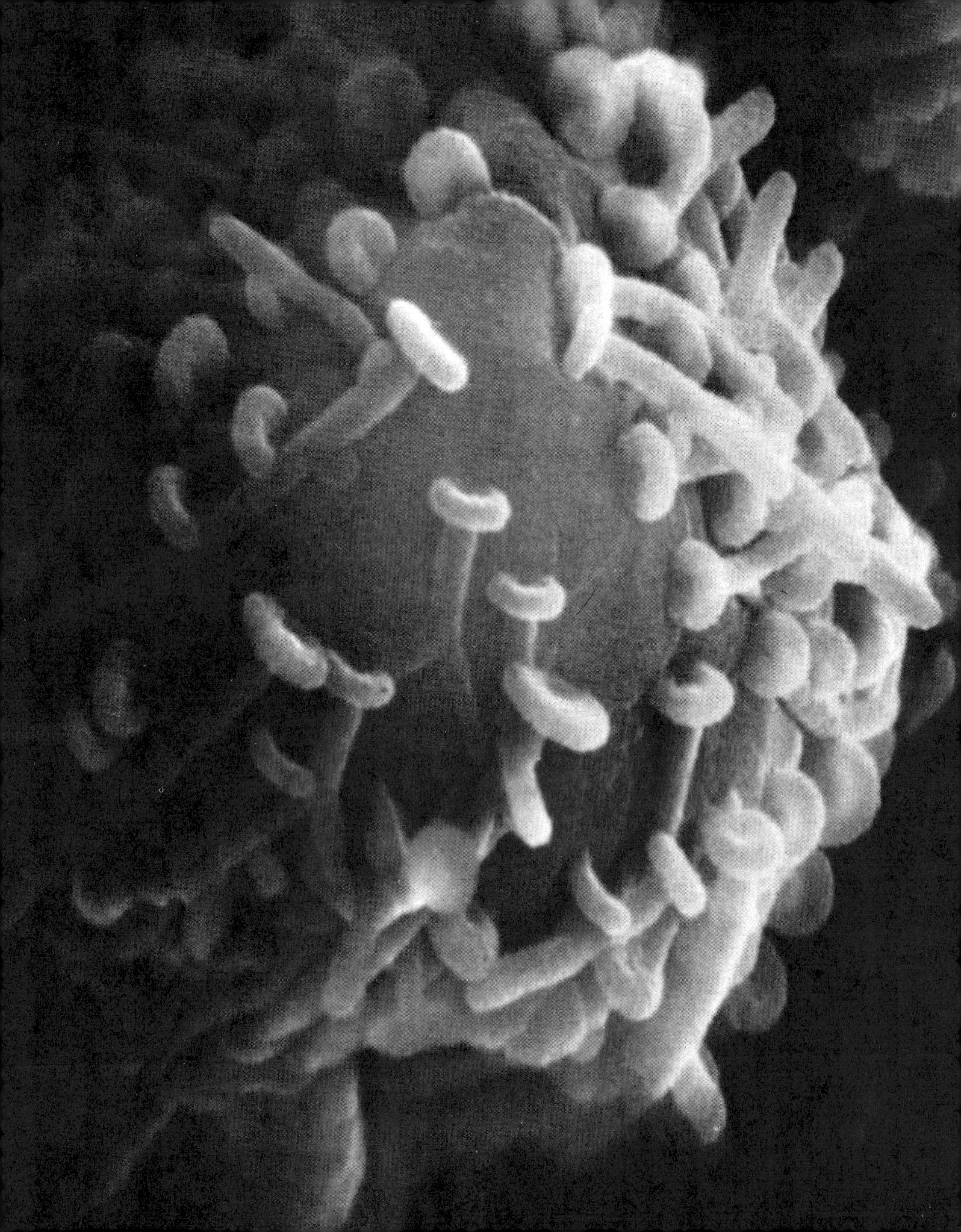

MOVEMENT AND NERVOUS CONTROL

chapter ten

Synaptic knobs and nerve tissue of the sea snail, Aplysia. *As this SEM illustrates, each globular nerve cell, or neuron, receives stimuli from the axons of many other nerve cells. When an impulse arrives, the knob at the end of the axon releases a chemical, and the message is transmitted electrochemically across the synaptic gap to the dendrite membrane.*

6,600X

Most organisms, no matter how simple or complex, utilize movement of some sort to perform the vital life processes of obtaining nutrition, reproducing, and achieving homeostasis. In fact, the capacity for independent movement is one of the characteristics that distinguishes living from nonliving things. Simple organisms like bacteria rely on their environment (wind and water) to carry them from one place to another; plants depend on internal water pressure to turn the broad surfaces of their leaves toward the sun, so they can obtain the energy they need for photosynthesis.

Animal movements are undoubtedly more dramatic. The 60-ton whale breeching the ocean's surface and diving to great depths, the dragonfly capturing its insect prey in midair, the cheetah covering East African grassland at 70 miles per hour—these are familiar examples of the beauty of animals in motion.

Movement is always a response to some input from the environment. In single cells, for example, the stimulus may be chemical or physical. A macrophage (a phagocytic cell) will move toward chemicals released by a dead cell. And protozoa placed in an aquarium will move either toward or away from a light source, depending on the species. Living things move in one integrated process—receiving and processing sensory inputs and reacting to their environments. The more complex the life form, the more complex this process becomes.

Plant movement

Basically, all movement is caused by chemical reactions at the cellular level. But the chemistry of plant cell movement differs from the mechanisms responsible for movement in animal cells. The reason for this difference is expressed in the generally accepted principle that structure tends to be closely related to function. In order to procure nutrients, plants do not require extensive movement, for the soil and water that surround plant life usually abound with nutrients. Thus plant movement can be simple and still be effective.

Some plants, nevertheless, are highly mobile. Algae such as the *Euglena* (a curious unicellular organism with both plant and animal traits) have flagella which are used to propel the organisms through the water. And the seeds produced by plants are often constructed to be moved passively over great distances. Crabgrass seeds have tiny parachutes which enable the seeds to be carried aloft by the wind for considerable distances. Elm seeds are shaped like propellers and can twirl through the air to remote locations. Like animals, many plants must rely on movement in order to reproduce. But whereas animals will generally move actively in search of a mate, plants depend on outside forces, such as wind, water, insects, and other animals, to disperse their seeds.

Most plants remain anchored in the ground at the location where they sprouted, relying on their separate parts to perform the movements that keep them alive. There are two modes in which plant parts may move; one category of movement is called **tropism,** and the other is known as **nastic movement.**

TROPISMS

Tropisms include all the movements of plant parts that are prompted by specific stimuli from the plant's environment. The direction of these movements is always related to the nature of the stimuli that caused them. A tropism may be positive, in which case the plant moves toward the stimulus, or negative, in which case it moves away from the stimulus. Sunlight is an important stimulus toward which most plant leaves exhibit a positive tropism. To absorb as much energy as possible, the leaves of the plant arrange themselves in a way that enables them to expose to the light a maximum of surface area. This response toward light is called a **phototropic** movement.

A

B

FIGURE 10.1

Thigmotropic response in Venus's-flytrap. (A) When a large insect touches specialized trigger hairs on the epidermis, certain cells near the upper and lower surface of the leaf are electrochemically stimulated to change their turgor pressure. (B) The leaf bends and snaps shut, trapping the insect.

Another tropic response involves the direction of growth of the plant parts; for example, the plant's roots grow downward, reaching deep into the soil to areas where moisture may be found, branching into complex systems to increase their surface area for water absorption. Movement either with or against the pull of gravity is called **geotropism.** Some plants are extremely sensitive to touch, and consequently respond to tactile stimuli. One example of this touch sensitivity—called **thigmotropism**—may be seen in the singular feeding apparatus of *Dionaea,* better known as Venus's-flytrap (see Fig. 10.1). Insects are attracted to *Dionaea's* colorful leaf, which contains two sets of trigger hairs. A small insect may land on the leaf without disturbing the trigger mechanism. But a larger insect lighting on the trigger hairs disturbs them enough to activate the closing mechanism. Suddenly the leaf bends into a shape that resembles a clam, trapping the insect inside.

NASTIC MOVEMENTS

Tropisms are movements either toward or away from a stimulus. Nastic movements, on the other hand, refer to plant responses that are not related to the direction of the stimulus. This type of movement is exemplified by *Oxalis*, a common plant resembling a four-leaf clover, whose leaves are extended horizontally during the day. At night the leaves of *Oxalis* begin to fold and droop vertically, finally hugging the stem of the plant. The movement of the leaves could be a reaction to the disappearance of sunlight (a **photonastic** reaction) or a response to the drop in temperature (a **thermonastic** reaction), but the movements themselves are completely independent of any stimulus direction.

The mechanisms of plant movements The phenomena that are responsible for the two forms of plant movement—tropisms and nastic movements—have been identified as **hydrostatic pressure** and **differential cell growth.** Usually rapid movements in plants are caused by sudden changes in hydrostatic pressure, or turgor, within the plant. When the trigger mechanism on the leaf of Venus's-flytrap is activated, cells near the upper surface of the leaf lose water while cells near the lower surface retain water. This causes the leaf to bend quickly, snapping shut over the insect that activated the mechanism. When the seeds of the *Oxalis* plant are ripe, the pod undergoes changes in turgor pressure that cause the structure to act like a catapult, so that it hurls its seeds several feet away from the plant. This action serves to disperse the seeds, thus increasing their chances for survival. Changes in the turgor pressure of flower petals cause the flowers to open and close.

Plant movements like tropisms are due to differential cell growth, the tendency for some cells of a plant to grow more rapidly than others; this process usually results in slower, more permanent movements of cell parts than those produced by changes in turgor pressure. In differential cell growth, one layer of cells grows at a more rapid rate than the layer next to it, causing the more rapidly growing layer to bend. This is why roots are able to bend in response to the pull of gravity; cells which are on the lower side of the plant root, for example, grow relatively slowly, while cells on the upper side grow more rapidly. The effect of this differential growth among root cells is that the root bends and points downward. Root movement occurs slowly, but sometimes differential cell growth is responsible for rapid movements in other areas of the plant. For example, one side of a grape vine tendril grows more rapidly than the opposite side, bending the tendril so that it is always coiled. When the vine comes into contact with a supporting structure, the rate of cell growth in the faster growing layer speeds up considerably, often forming a complete coil which wraps around the support in less than one minute.

Differential cell growth and changes in water pressure are adequate mechanisms for enabling plants to fulfill their nutrient requirements. Plants, generally immobile, are able to obtain nutrients from their immediate environment without having to migrate in search of prey. Animals, however, may have to travel considerable distances to obtain food, and they must also be able to react quickly to alterations in their environment. For this reason, an animal must have a more intricate and specialized system for locomotive activity than a plant —regardless of whether the organism is multicellular or is a single cell.

Animal movement

Animal locomotion is achieved through the action of contractile proteins, usually present in muscle cells. These contractile proteins are molecules that can alter their own length. They are present in animal cells at all levels of organism complexity. Even the simplest organ-

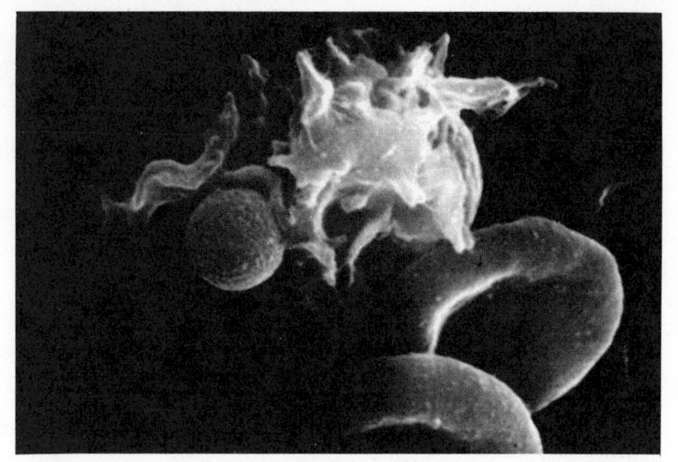

A 5,000X

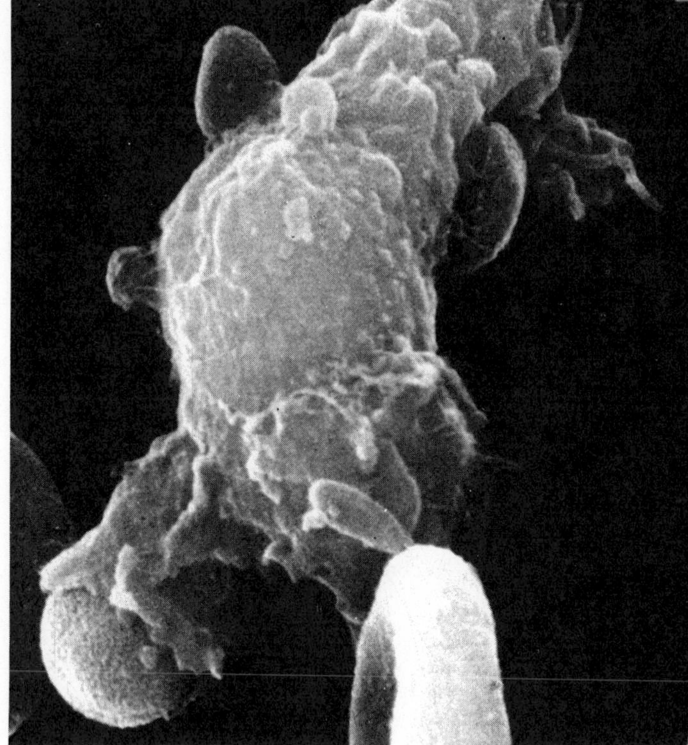

B 7,500X

FIGURE 10.2

Amoeboid movement. Leucocytes (white blood cells) in human blood move to engulf bacteria and other substances. The large irregularly shaped mass in both pictures is a leucocyte. In (A) it is moving toward a yeast particle (the small sphere), and in (B) the leucocyte is beginning to engulf the yeast. Many body cells have exhibited a capacity for this type of movement when isolated and grown in a laboratory culture, although they do not normally engage in phagocytosis.

isms contain a primitive version of the contractile mechanism that gives large organisms movement.

For example, the unicellular amoeba prowls the bottom of muddy ponds in search of prey; its optimum speed is about one foot per day. Movement begins when softer portions of the amoeba's cytoplasm are extended, forming pseudopodia (false feet). These pseudopodia adhere to the muddy terrain and provide a point of contact. After a pseudopod is extended, the bulk of the cytoplasm may be pulled to the point that the pseudopodium has reached, and the process begins again. One theory on the movement of pseudopodia holds that there is a flow of sol-like cytoplasm up through the center of a pseudopodium until the whole organism is moved along. Movement is continued by the conversion of the sol to gel at the front of the pseudopodium and along the sides, and then by the reconversion of the gel to sol. According to another theory, the amoeba can alter its shape and thus accomplish movement because of the operation of two types of structures within its cell, the microtubules and microfilaments. **Microtubules** are tiny rods that seem to function as a skeleton to maintain cellular rigidity. **Microfilaments** appear to contain contractile proteins that provide the means for movement. The proteins in microfilaments seem to be of a variety similar to those found in muscle tissues of higher animals.

Two comparatively more advanced structures for accomplishing unicellular movement also operate through the action of microtubules and microfilaments, whose contractile proteins supply the driving force. The more common of these structures are the cilia—short, hairlike projections which beat rhythmically through the water to propel the organism. The beat of each cilium consists of two types of strokes, more or less comparable to those involved in rowing a boat; the first type is a power stroke that pushes against the water, and the second type is a recovery stroke that moves the cilium back into position. The cilium can reverse its motion almost immediately, halting the organism and then thrusting it into the opposite direction.

The second type of motile structure is the flagellum —a single strand longer than a cilium that protrudes from part of the organism. The flagellum can move backward and forward like the cilium. It also can curl and, with a propellerlike motion, drive the animal forward.

Some cells of complex higher animals have retained cilia, flagella, and pseudopodia as means of locomotion. In humans, the phagocytic white blood cell engulfs bacteria and other foreign substances with the aid of pseudopodia (see Fig. 10.2). The sperm cell moves by means of flagella, and ciliated cells line parts of the respiratory system. As multicellular animals became more complex,

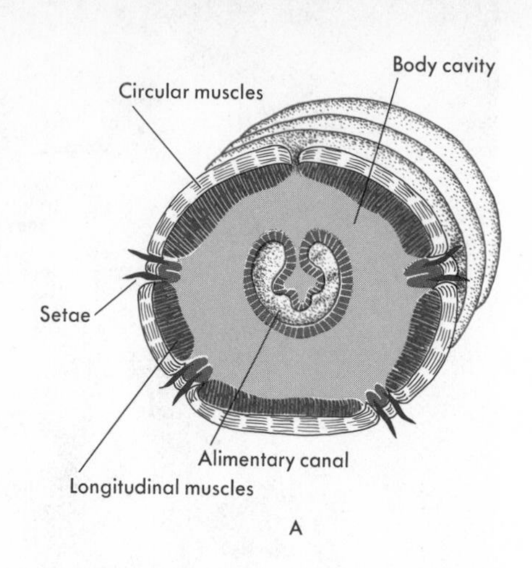

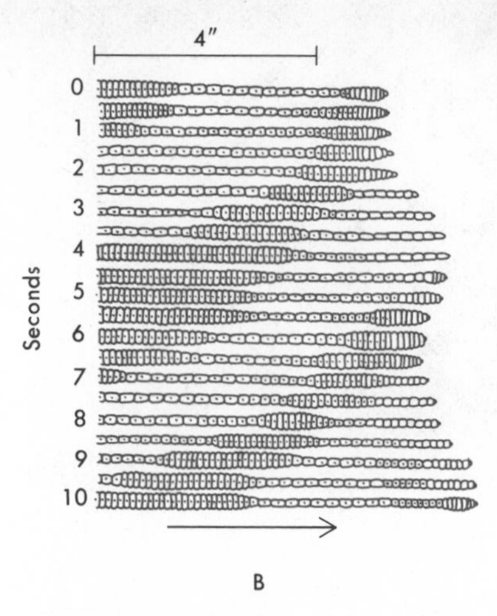

FIGURE 10.3

Muscles in earthworm locomotion. (A) A cross section through the body of an earthworm shows an outer layer of circular muscles, whose contraction stretches the animal, and an inner layer of longitudinal muscles, whose contraction shortens it. The four pairs of setae serve to grip the animal against the stratum over which it is moving. The body cavity, in conjunction with the two pairs of muscles, serves as the hydrostatic skeleton of the earthworm, and the alimentary canal is the passage in which food travels through the organism. (B) The shape and progress of an earthworm over a 10-second period demonstrates an anterior-to-posterior wave of localized longitudinal muscle contractions, followed by an anterior-to-posterior wave of localized circular muscle contractions. This sequence moves the animal forward and enables the setae to take a new hold for the next series of contractions.

they developed specialized muscle cells with contractile properties.

To translate the force of muscular contractions into movement, the muscles must push against some medium of the environment. To walk, man pushes his feet against the ground; to fly, a sparrow flaps its wings against the air. The methods whereby an organism translates force into movement are directly related to the organism's complexity. Multicellular organisms may be classified from the relatively simple to the more complex on the basis of skeletal structure. It is the contraction of the organism's muscles in conjunction with the animal's particular skeletal features that make possible the functions of locomotion. Organisms with the simplest skeletons possess a simpler system of muscular contraction, and organisms with more complex structures utilize a more intricate system of contraction. The three kinds of skeletons found among animal species are the hydrostatic skeleton, the exoskeleton, and the endoskeleton.

SKELETAL SYSTEMS

Hydrostatic skeleton Relatively simple animals like the earthworm have a most unusual skeleton—a mass of body fluid known as the **hydrostatic skeleton.** The earthworm possesses two types of muscles—circular muscles and longitudinal muscles—which operate against its fluid skeleton. Together, the muscles manipulate the fluid back and forth. First the circular muscles contract, forcing the fluid through the length of the body, much as toothpaste is forced out of a tube when the tube is squeezed. This movement of the fluid stretches the longitudinal muscles so that the body is extended. Under the body are bristles called **setae** that hook into the ground to maintain the new position. Then the longitudinal muscles contract. This action pulls the fluid back and also pulls the rear of the body forward (Fig. 10.3).

The earthworm's system of muscular contraction, operating in association with a hydrostatic skeleton, is one of the simplest forms of locomotion system found among animals. In the earthworm, the circular muscles and longitudinal muscles act as an **antagonistic pair,** in which each set works in opposition to the other. The body is thus capable of extension and contraction through the transmission of the muscular force by the fluid skeleton.

Exoskeleton The next higher order of skeletal sophistication, found among the more advanced animals, is the **exoskeleton** of molluscs and arthropods. In this system of locomotion, the force of muscular contractions is transmitted by the exoskeleton—a structure with a rigid outer shell. Exoskeletons may be hinged with antagonistic muscles positioned on each side of the hinge. While one muscle contracts and thus bends the hinged part down, the other muscle contracts to pull the part up. This is how clam and oyster shells are held together so that their contents are often difficult to get at.

The exoskeletons of flying insects, such as the housefly, possess a specialized adaptation for flight. The fly's wing operates on a principle similar to a lever, with the fulcrum located at the point where the wing is attached to the body. Muscles are arranged as antagonistic pairs on either side of the fulcrum. One muscle of each pair contracts and thus pulls the wing up. Then the muscle relaxes while the other member of the pair contracts to pull the wing down. One unusual feature of these muscles is that each serves to stimulate the other. The relaxation of one set of muscles serves as the stimulus for the contraction of another.

There are, however, disadvantages for organisms that possess an exoskeleton. One problem encountered is that the organism's growth cycle must include a molting period to discard the old shell so that the animal may grow larger. During the time required to grow a new exoskeleton, the animal is especially vulnerable to predators. The soft-shell crab which appears so often on seafood menus is actually a crab that was caught during one of its molting periods and so does not possess a mature, hardened shell.

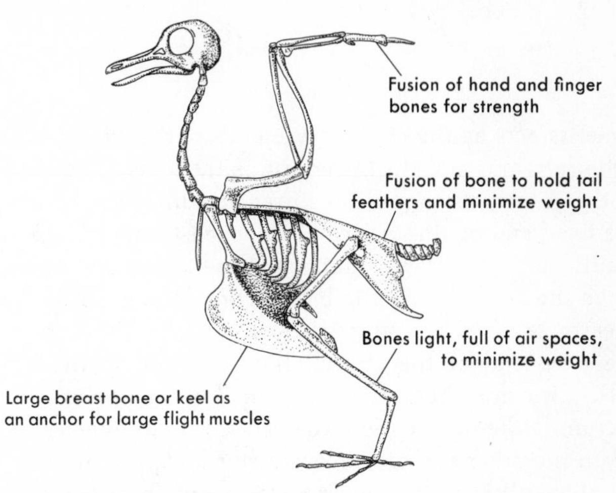

FIGURE 10.4

The endoskeleton of a bird contains a number of bones that are structurally and functionally adapted for flying.

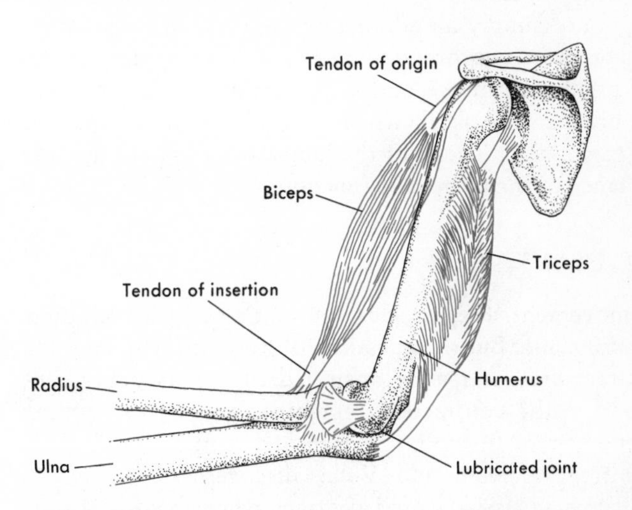

FIGURE 10.5

Antagonistically paired muscles. The movement of the forearm is controlled by a pair of antagonistic muscles, the biceps and the triceps. The main body of the biceps muscle is located on the front of the upper arm. (You can feel it tighten if you crook your arm and tense your muscles.) When the biceps contracts, it pulls on the radius bone of the forearm and causes the arm to bend at the elbow. The main body of the triceps muscle is located on the back of the upper arm. (You can feel it if you straighten your arm slowly but firmly.) When the triceps contracts, it pulls on the ulna of the forearm in a direction opposite to the biceps and causes the arm to straighten.

Endoskeleton The lever principle of the exoskeleton is also operative in the function of the **endoskeleton**—the most advanced of skeletal structures. The endoskeleton, found in vertebrates, consists of bones and cartilage. The bones are arranged to perform as levers, which are manipulated by antagonistic pairs of muscles attached to them. As with the muscles of the exoskeleton, one muscle of the pair contracts to bend the hinged part; the other muscle contracts to straighten it.

The human elbow provides a good example of how the endoskeletal system operates (Fig. 10.5). The hinge is found between the upper arm bone, the humerus, and the two bones of the lower arm, the radius and ulna. The humerus acts as the stationary member; the elbow joint is the fulcrum; and the lower arm is the lever. The muscle that contracts to pull the lower arm up is the biceps. The fixed end of this muscle is attached to the humerus, and the end that moves is attached to the forearm bones above the fulcrum. As the biceps contracts, it raises the forearm lever. The antagonistic muscle is the triceps. The fixed end of this muscle is also found in the humerus; its movable end is on the forearm below the fulcrum. When the triceps contracts, it is in effect pulling on the other end of the lever, thus lowering the forearm. Note that these two muscles work in opposition. When one is contracted, the other is automatically stretched. The skeleton serves to transmit the enormous force of the muscular contraction.

It is important to note, however, that the endoskeleton is not simply an adjunct of the muscular system. It functions in two other respects—first as a support for the organism, and second as protection for softer internal organs. The skull, which encases the brain, and the rib cage, which encircles the lungs, demonstrate the importance of the protective function.

SKELETAL MUSCLES

In movement, the muscles contract slowly, developing great tension. But the questions of how and why muscles contract are not readily answered. It is reasonably certain that the contraction represents an electrochemical phenomenon. In the late eighteenth century two Italian scientists, Galvani and Volta, discovered by accident that muscle tissue would contract when exposed to an electric shock. One day Galvani left a pair of freshly dissected frog legs on a table next to his rather primitive electric generator (electricity was a scientific novelty then). When they were accidentally struck by an electric spark, the muscles contracted abruptly. This led Galvani to believe that muscular activity was somehow governed by electrical impulses.

Today the theory of muscular contraction as an effect of electrical stimulation is widely accepted (see Fig. 10.6). Electronic stimulation of muscles is a common laboratory procedure. Two of the major findings, how-

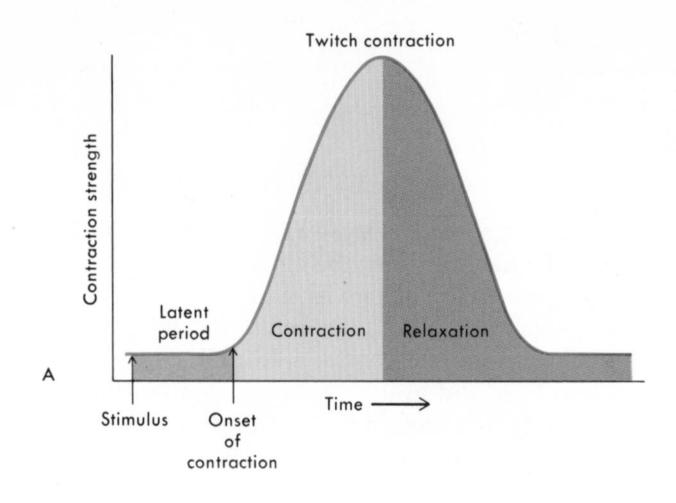

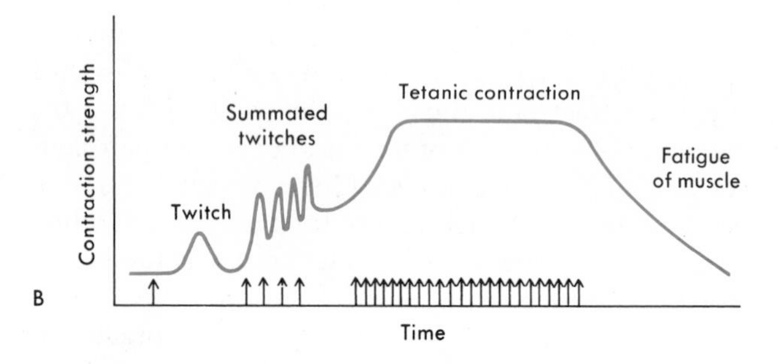

FIGURE 10.6

Muscular activity. (A) A muscular contraction initiated by a single brief stimulus is called a twitch. The twitch contraction occurs in three stages: latent period, contraction, and relaxation. When a muscle is stimulated by a nerve impulse or electric shock, the muscle does not immediately respond. The time before the onset of contraction is called the latent period. When contraction begins, the strength of contraction builds up gradually. At peak strength of contraction, the relaxation period begins, and contraction strength gradually diminishes. Within a certain range of stimulus intensity, the strength of a twitch response is proportional to the strength of the stimulus. (B) If a muscle is restimulated before the end of a twitch, the muscle will contract again, but this time the strength of contraction will be roughly proportional to the intensity of the initial and second stimuli together. Thus there is a summation of the effects produced by the stimuli. If stimulation occurs with increased frequency, summation continues with a gradually decreasing amount of relaxation between twitches. If stimulation is very frequent, relaxation does not occur, and the muscle remains in a state of contraction called tetanus. In experiments with an isolated muscle, tetanus can be maintained for a while, but then the muscle becomes too fatigued to sustain contraction. In normal muscles, tetanus is responsible for muscle tone.

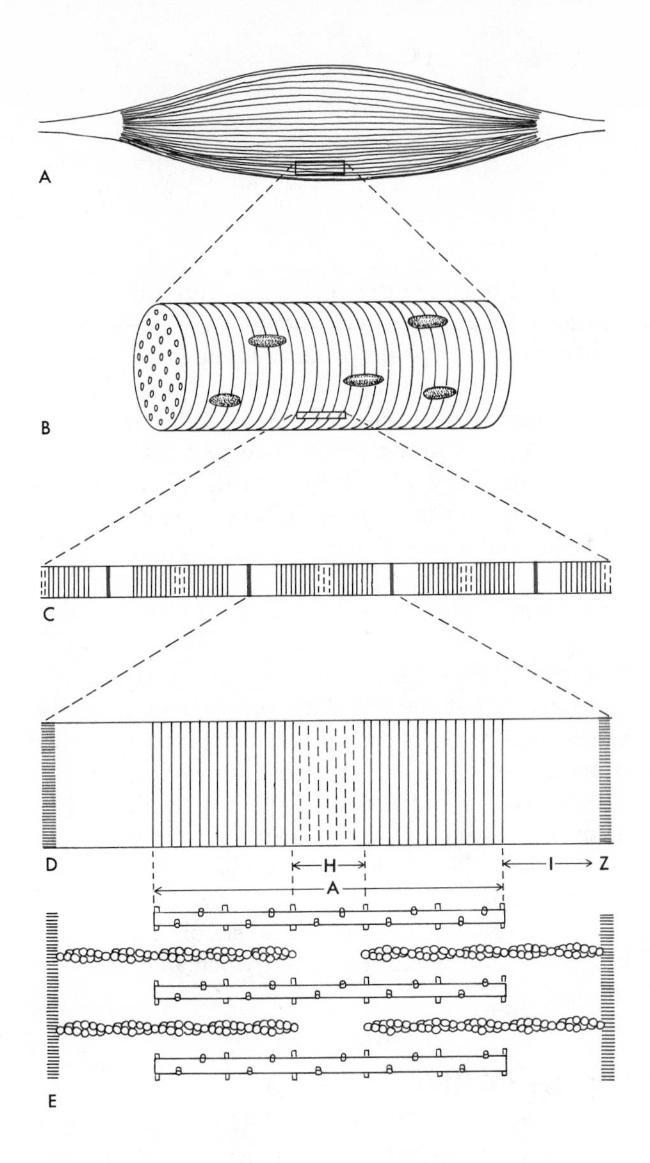

FIGURE 10.7

The structure of striated muscle in diagrams showing increasing detail. A muscle (A) is composed of muscle cells or fibers (B) that appear striated under a light microscope. Each fiber contains a bundle of myofibrils (C) associated with cell nuclei and mitochondria (the oval structures in B). A single myofibril has a characteristic repeating banding pattern consisting of a Z-line, an I-band, an A-band containing a central H-zone, another I-band, and another Z-line (D). Electron microscope photographs have revealed that banding patterns are caused by the arrangement of thin and thick protein filaments within the myofibril (E). The thick filaments, composed of myosin, have protuberances called cross-bridges, which fit into special sites on the thin filaments, composed of actin. During contraction the paired actin filaments move toward each other, narrowing the H-zone and the distance between Z-lines.

ever, seem to be contradictory. The response of a bundle of muscle fibers increases as the strength of the electrical stimulus increases. Yet a single muscle fiber responds only when the stimulus has attained a particular strength —the **threshold**—and continues to respond at the same intensity even when the strength of the stimulus is increased (until the stimulus becomes so intense that it damages the fiber). The explanation for this apparent inconsistency is that each fiber alone has an *all-or-none* response. The greater the strength of the stimulus reaching an entire muscle, however, the larger the number of fibers that are stimulated.

Electrical impulses are carried to muscle fibers by nerve cells (motor neurons) that are connected to a number of the fibers. Together, a neuron and the fibers connected to it make up a **motor unit.** The action of the motor unit—the firing of the stimulus and the contraction of the fibers—occurs as a single, integrated function. Each muscle fiber is surrounded by a membrane—the **sarcolemma**—which is strongly polarized; that is, the inner surface of the membrane is negatively charged relative to the outer surface. This difference in electrical potential is maintained by a transport mechanism that keeps positively charged sodium ions moving out of the cell. When an electrical impulse arrives, however, it changes the permeability of the membrane. Now the positively charged sodium ions rush back through the fiber's cell membrane, and the membrane reverses its electrical potential. It is this reversal that in some unknown way initiates contraction.

Structurally, every skeletal muscle fiber is made of several smaller fibers called **myofibrils** (see Fig. 10.7). Myofibrils are in turn composed of protein filaments. These filaments consist of **myosin** and **actin;** myosin filaments are longer and thicker than actin, and the two have a regular pattern of light and dark bands that makes skeletal muscle appear striped, or striated. During muscular contraction, myosin and actin filaments slide past each other. Sites known as cross-bridges appear to form and break between actin and myosin filaments. This process requires energy, which is probably supplied by the conversion of ATP to ADP at each cross-bridge site.

Muscular contraction In order for the ATP molecule to release its energy, the bond holding the terminal phosphate group must be broken. This is achieved by an enzyme called an ATP-ase. Research by H. E. Huxley shows that the protruding heads of the myosin molecules are somehow changed so that they serve as an ATP-ase. Like other protein molecules, myosin can function both as a structural element and, with some modifications, as an enzyme. Recent research has demonstrated that the activation of myosin is provided by an influx of calcium ions, made possible when an electrical stimulus depolarizes the membranes within the muscle fiber.

It takes considerable time to form enough cross-bridges to sustain a high degree of tension. Therefore a muscle that is contracting rapidly cannot develop as much tension as one that is contracting slowly; so much energy is being expended in the process of contraction that there is not enough available to form additional cross-bridges. The skeletal muscles of vertebrates are usually relatively shorter than those in other animals. Shorter muscles are more economical: the longer the muscle, the more energy it must employ to detach and build cross-bridges.

OTHER TYPES OF MUSCLE

Much of the early research in the field of muscle physiology was concentrated primarily on the skeletal, or striated, muscle of vertebrates. However, smooth muscle and cardiac muscle have in more recent years also been extensively studied. Each type has special properties and functions. (The reader may wish to review at this time the discussion of muscle cells in Chapter 4 and, in particular, Fig. 4.17.)

Smooth muscle In vertebrates one finds a type of muscle which is not striated. This is the smooth muscle located in the walls of the intestine and blood vessels, the bladder, the female uterus, and the iris of the eye. Smooth muscle differs from striated muscle in both structure and function.

Smooth muscle cells are thin, with tapering ends. No regular pattern appears in the fibers. This suggests that even though smooth muscle contains actin and myosin, the sliding mechanism of contraction may be different. Smooth muscle is capable of contracting, though more slowly than striated muscle. In fact, it seems that the smooth muscles lining internal organs and vessels have no relaxation phase. Contraction of these muscles is continuously maintained over long periods of time. Since smooth muscles are innervated by that part of the nervous system that is not directly under voluntary control, they cannot be easily moved, as can a skeletal muscle of the arm, for example.

Cardiac muscle Heart, or cardiac, muscle is striated like skeletal muscle, but the fibers are organized differently. Instead of occurring in parallel bundles, cardiac muscle fibers are branched so as to form a network of tightly compressed cells. Contraction is effected by the sliding together of actin and myosin filaments, and it is initiated by specialized cells of the heart's pacemaker, as discussed in Chapter 7. The fact that cardiac muscles contain a large number of mitochondria should not be surprising in view of the constant energy demands made on that tissue.

Nervous control

Muscles, as we have seen, are activated by neural impulses, the electrical impulses through which, with unbelievable speed, we communicate information to virtually all parts of our bodies. The direction of evolution was markedly influenced by organisms' ability to obtain and process information—quickly and accurately—through nervous impulses. The basic conducting unit for a nervous impulse is the nerve cell, or neuron.

THE NEURON

Among the phyla, there is a very large variety of types of neurons. Indeed, even with the human organism, neuronal types show a considerable amount of variation. However, all neurons perform essentially the same function—they send bioelectric signals. Despite differences

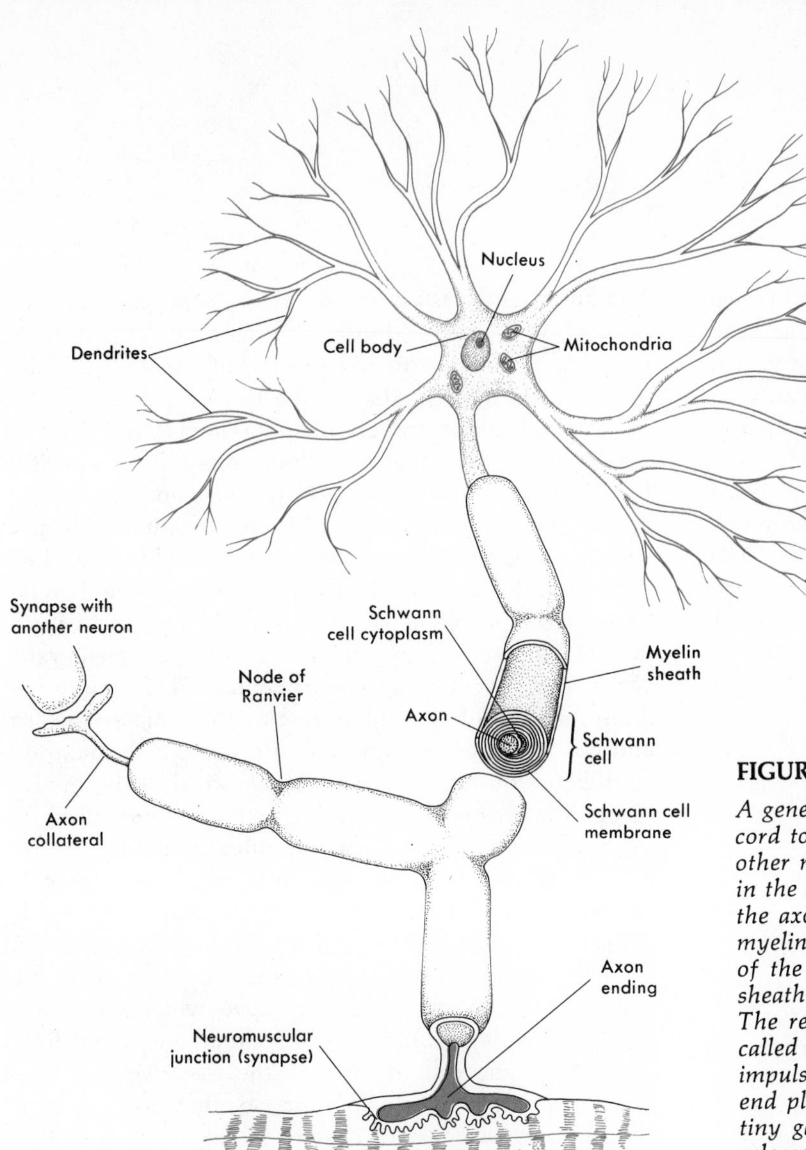

FIGURE 10.8

A generalized motor neuron, extending from the spinal cord to a muscle. The dendrites receive impulses from other neurons, in this case sensory neurons or neurons in the spinal cord. The impulse is then transmitted along the axon. Motor axons are frequently surrounded by a myelin sheath, which consists of a tightly wound spiral of the cell membranes of Schwann cells. The myelin sheath is not continuous for the full length of the axon. The regularly spaced interruptions in the sheath are called nodes of Ranvier. The sheath serves to speed up impulse transmission. The axon terminates in a motor end plate, which is separated from the muscle by a tiny gap called the neuromuscular junction. Chemicals released by the end plate cross the gap and stimulate the muscle to contract.

in their architecture and size, neurons share certain common characteristics (see Fig. 10.8). The neuron has a nucleated cell body, which is a bulbous portion that contains the nucleus and cytoplasmic organelles necessary for metabolic activity. Thin, multibranched projections, called **dendrites,** extend from one end of the cell body, and a relatively long and thicker projection, called the **axon,** extends from the other. In vertebrates, the axon may be wrapped along its length by a sheath of fatty material called myelin. The myelin sheath—interrupted at intervals of 1 to 2 mm at points called the **nodes of Ranvier**—serves to insulate the axon and hastens the conduction of nerve impulses.

Although the neuron is the basic structural and functional unit of the nervous system, other cells—the **glial cells**—support and protect the neurons.

THE NERVE IMPULSE

The membrane of the neuron is polarized by electrical charges. A voltage potential of from 60 to 70 millivolts, called the **resting** or **membrane potential**—the voltage differential between the membrane's negatively charged interior and its positively charged exterior—exists across the membrane while the cell is in its resting state. Just as the muscle fiber reacts only when the strength of the stimulus is at or greater than a minimum threshold level, so too does the neuron. The intensity, duration, and rate of the stimuli determine whether or not the nerve will respond. Although subthreshold stimuli do cause an electrical change, this change is not of sufficient magnitude to evoke an impulse. And like the muscle fiber, when the neuron *does* fire, the impulse is transmitted at maximal strength, no matter how strong the stimulus, thus following the all-or-none principle (see Fig. 10.9). After having fired an impulse, the neuron has a **refractory period** of approximately one millisecond (one thousandth of a second) during which it is incapable of responding to another stimulus—regardless of strength.

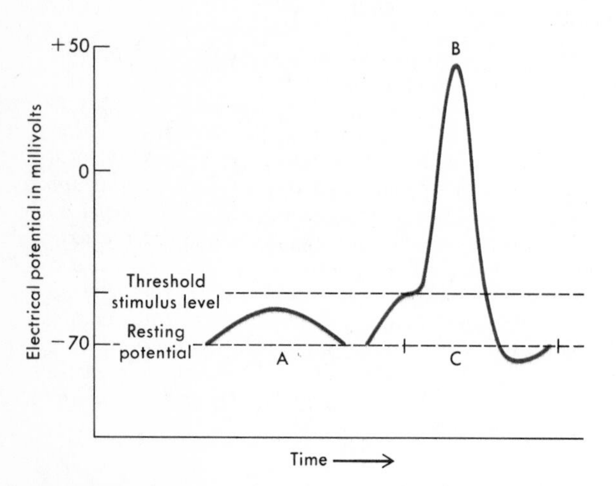

FIGURE 10.9

The nerve impulse. Neurons exhibit an all-or-none response to stimuli. They will not fire (carry impulses) unless stimuli are stronger than a minimum threshold level. If stimuli are above the threshold level, neurons fire at one intensity and speed, irrespective of the strength of the stimuli. The diagram shows the typical action potential of a firing neuron (B), the refractory period of a single firing (C), including the compensatory overshoot, and the effect of a subthreshold stimulus (A), in which the potential of the neuron rises slightly above its resting potential. Although the neuron does not fire, it is much closer to the threshold level and may be readily triggered by a slight additional stimulus.

The Sodium-potassium pump How is the impulse conducted? We have said that the inside of a neuron is negative in relation to the outside of the cell. Equally important is the fact that in the neuron's resting state, its internal concentration of potassium ions (K^+) is much greater than that of sodium ions (Na^+). Physical forces of diffusion would tend to cause sodium ions to diffuse through its pores (which are sometimes referred to as **gates,** or valves, in the neuron membrane), while potassium diffuses out. However, the membrane is relatively impermeable to sodium ions, and it tends to accumulate potassium ions. In fact, the neuron membrane can actively, at the expense of cellular ATP, pump sodium ions out of the cell as fast as they diffuse inward and can pump potassium ions into the cell to make up for the loss through diffusion. However, the cell can also respond to chemical, physical, or electrical stimuli in its external environment. The stimulus must be strong enough to affect the cell—it must be above the cell's threshold for response. When an adequate stimulus is delivered to the neuron, the permeability characteristics of the membrane are changed, and it appears that the sodium gates in the membrane open wider for a moment, permitting rapid influx of sodium ions; their presence inside the cell neutralizes the negative charge—depolarizes the cell membrane so that its electrical charge is reversed. The change in electrical potential, called the **action potential,** is the impulse or message sent by a neuron. This impulse spreads down the nerve fiber (the axon) in a wavelike form. Rapidly—in 5 to 10 milliseconds—the sodium is pumped out and the neuron is repolarized and ready to carry a second signal or impulse.

Transmission of the neural impulse is not at all like sending telegraph signals over a wire: the neuron's message is self-propagating, and it continues to spread down the axon, and from neuron to neuron, after the stimulus has been removed. A stimulus cause depolarization at one end of the neuron—the dendrite end; because the depolarization is an electrical change, it can "shock" the region of the axon next to it. In myelinated neurons the action potential skips rapidly from one node of Ran-

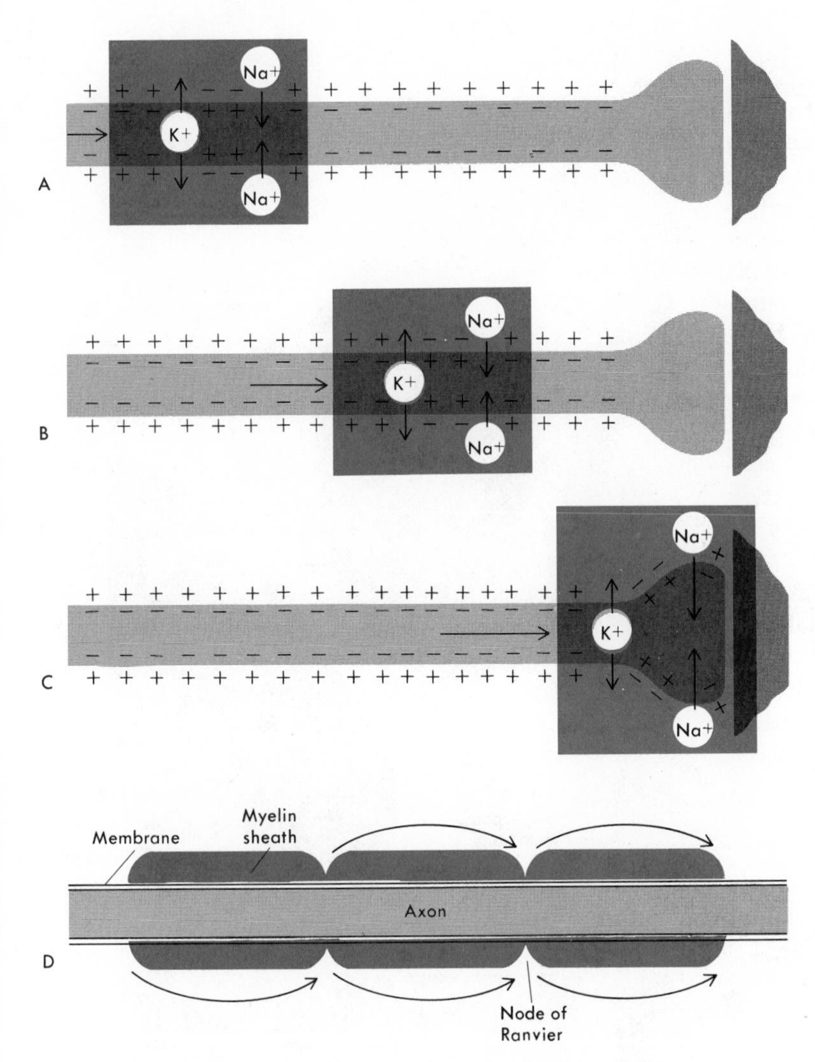

FIGURE 10.10

Gradient changes in a nerve membrane during the propagation of a potential. (A) All nerve membranes are polarized; that is, the inside is negatively charged in relation to the outside. The membrane maintains this potential by actively pumping Na^+ ions out of the cell and K^+ ions in. In response to a stimulus (energy change) a portion of the membrane becomes depolarized and loses its impermeability to Na^+. Sodium ions flow inward and make the inside positive. To restore the negative inner charge, K^+ ions are permitted to flow outward. (B, C). The excited depolarized section of the membrane affects the adjacent segment, which in turn becomes depolarized, and so on down the membrane. Alteration of the Na^+-K^+ gradients causes the potential to be propagated quickly along the entire length of the membrane. The nerve cell is specialized for this type of electrical transmission. (D) This schematic shows how an action potential moves along a myelinated nerve. The potential does not travel by sequentially depolarizing the entire length of the axonal membrane, but rather by jumping from one node of Ranvier to the next. It can actually jump two nodes at a time. The advantages of this type of conduction are increased velocity of propagation and a conservation of energy needed to return the potential to its normal state, since only the nodes are depolarized, not the entire axon.

vier to the next. A schematic representation of the transmission of the neural impulse is given in Fig. 10.10. The process in unmyelinated neurons is similar, except that the action potentials are generated along the entire length of the axon, not only at the nodes. This takes somewhat longer—hence, myelinated neurons transmit more rapidly than unmyelinated.

Synapse Each axon terminates in a bulbous structure called the **synaptic bulb.** The electron microscope reveals an infinitesimal separation—the **synaptic cleft**—of from 200 to 500 Å between the synaptic bulb of an axon and the dendrite to which it must transmit its message. This junction between the axon of one neuron and the dendrite of another is called the **synapse.**

Naturally, the physiologists who first observed this separation wondered how impulses could be transmitted across it. In the 1950s it was discovered that this was achieved by a chemical transmitter substance, such as acetylcholine or norepinephrine. Vesicles containing the chemical transmitter empty into the synaptic cleft whenever the synaptic bulb is stimulated by an action potential (Fig. 10.11). One chemical transmitter, acetylcholine, diffuses in from 1 to 2 milliseconds across the cleft to stimulate the dendrite membrane. The permeability of the dendrite membrane to potassium and sodium ions is thereby altered, and an action potential is generated. Once the membrane is activated, acetylcholinesterase—an enzyme present in the synapse—hydrolyzes the acetylcholine so that it will not keep the associated neuron depolarized and refractory. It is because of the synapse—and specifically because chemical transmitters are contained only in the synaptic bulb of axons—that action potentials pass from axon to dendrite and not in the reverse direction.

Not all synapses, however, serve to stimulate the adjacent dendrite to fire. Certain synapses contain inhibitory substances that halt impulse transmission. In the brain, for example, an overall adjustment of the impulse is possible as it travels from the site of stimulation. In short, neural responses may be modulated through discharges across inhibitory synapses.

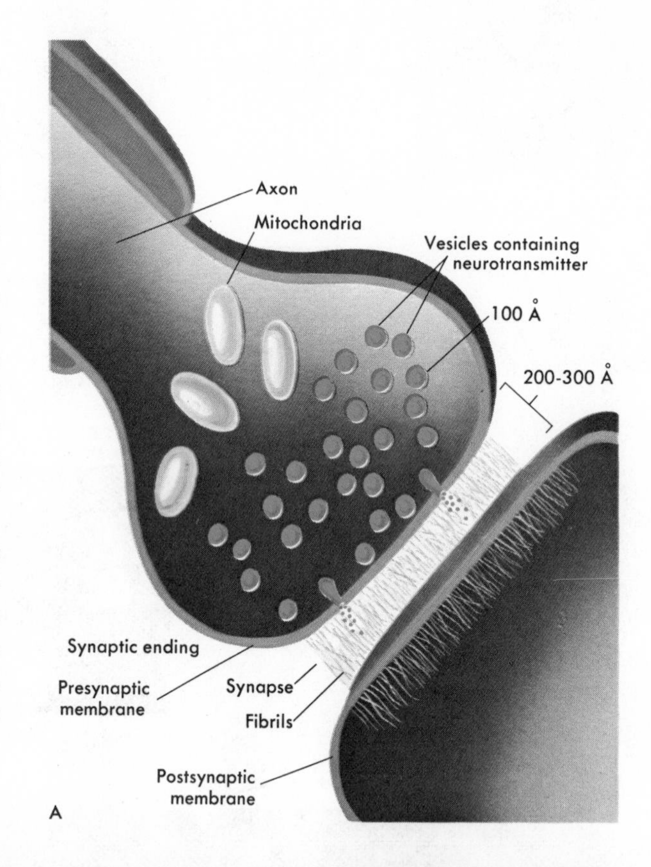

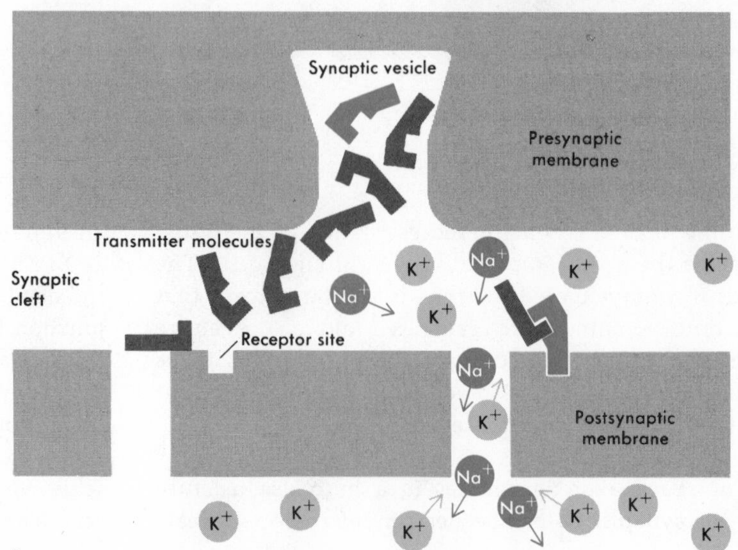

FIGURE 10.11

Impulses transmitted from one neuron to another at a synapse, a gap between the neurons. (A) The impulse travels down the axon of the first neuron to a synaptic ending, where the arrival of the electrical impulse stimulates the vesicles there to release a chemical transmitter into the synapse. The chemical diffuses across the synapse to the postsynaptic membrane on the dendrite of the second neuron and stimulates the neuron to transmit an impulse. (B) The mechanism of synapse transmission involves chemical transmitters, which may fit into specific sites on the postsynaptic membrane, thereby opening channels in the membrane that permit the free entrance of sodium ions (Na^+) into the neuron. The entrance of sodium ions initiates the impulse transmission process.

Conditioning for Athletes at the Cellular Level

When Frank Shorter crossed the finish line and earned his gold medal in the 1972 Olympic Games marathon, he looked as though he'd been out for only a casual jaunt. But after an all-out 400-meter race, a sprinter appears to have completed a rather painful ordeal. The intense exertion of the competitive weight lifter can leave him almost literally breathless. If all these individuals are in proper condition for their event, what accounts for the differences in appearance afterwards? Another way of stating the same question: Why are long-distance athletic events performed at slower rates than the sprint events?

The answer lies in the amount of time the particular event takes, the condition of the athlete, and the intensity of the exercise. Is the athlete performing with maximum effort? At the cellular/biochemical level, one can answer the question by determining which of the three possible metabolic pathways is being used to produce the ATP that in turn provides the energy to run the body.

The modern coach and competitive athlete—or even the recreational athlete—can use knowledge of the operation of these energy-producing pathways to improve performance. The adjustments the body makes through training are nothing short of astonishing. What is equally amazing is that the metabolic pathways can be selectively improved.

During slight or moderate activity, the body oxidizes glucose (from glycogen) and produces ATP almost exclusively through the Krebs cycle and electron transport systems. This is an efficient way to oxidize glucose and obtain maximum energy from it. However, this route requires that oxygen be available as the "final electron acceptor." That is, oxygen (O_2) must be available within the mitochondrion, where the electron transport system is operating. To get to the mitochondrion, oxygen must be transported by the blood from the lungs to the cells and their mitochondria. During slight or moderate activity, the rate at which oxygen becomes available to mitochondria is adequate to support the production of ATP. However, with increased activity and increased requirements for ATP, the rate of oxygen transport is inadequate to support the ATP production by the Krebs cycle and electron transport systems. Under conditions of heavy exercise, two anaerobic (not requiring oxygen) ATP production routes are called upon. Muscle phosphocreatine (PC) and ADP react to form creatine and ATP in a one-step, single-enzyme reaction. This reaction can promptly produce ATP within the muscle and requires no oxygen. However, the phosphocreatine stored in the muscle is limited and is rapidly depleted.

The other important anaerobic pathway is the glycolytic pathway, which results in the production of pyruvic acid and ATP. During slight or moderate activity, the pyruvic acid produced by this pathway enters the Krebs cycle and is completely oxidized to carbon dioxide and water. During strenuous activity, the rate at which glucose can be metabolized through the glycolytic scheme is not limited by the availability of O_2 as an electron acceptor. It speeds up considerably with the increased production of pyruvic acid. The pyruvic acid serves as the electron acceptor and is reduced to lactic acid. Lactic acid accumulates in muscles and is responsible for the pain and tiring of muscles during sustained, rapid muscular activity.

The buildup of lactic acid and increase of pain may severely limit the duration of rapid muscle activity. However, when strenuous activity is reduced, the lactic acid is converted back to pyruvate, which enters the Krebs cycle and is completely oxidized to CO_2 and H_2O via the electron transport system. As lactic acid is drained off, muscle pain diminishes, and a burst of rapid muscle activity is again possible.

Runners will recognize in this description of metabolic events an explanation of the phenomenon of "second wind."

Anyone who jogs, swims, skis cross-country, or bicycles has probably noticed that hard breathing does not begin as soon as exercise starts. A lag time of a few minutes passes before the body senses the need for more oxygen. But the 100-meter sprinter, who uses little oxygen for the intense work done in just a few seconds, must get the energy from other sources—the anaerobic pathways. The phosphocreatine and lactic acid routes provide most of the energy for sprints and other sports where quick bursts of energy are required.

World-class marathon runners maintain a speed of about five minutes per mile for more than two hours, and yet at the finish of a race their lactic acid levels are only moderately elevated. Marathoners maintain a pace well below maximum capacity so that the body's need for oxygen is about equal to the amount inhaled—a "steady state" condition. Instead of using an anaerobic pathway, they use the aerobic Krebs cycle to produce ATP.

It is no secret that training improves performance, and now physiologists have provided some of the reasons. Today the training is tailored for the athlete and the event. The sprinter must increase the capacity of the anaerobic energy pathways, but the distance runner emphasizes development of aerobic capacity. The middle-distance runner has perhaps the hardest job of all—to develop both anerobic and aerobic routes. The 1500-meter runner, for example, uses both methods of producing energy about equally.

Interval training has provided an enormous boost to improving all three pathways. To increase the capacity of the PC/ATP and lactic acid pathways, for example, an athlete no longer runs all-out sprints for as long as possible. To decrease fatigue, the runner may do perhaps six "sets," each of which consists of 10 seconds of maximum running followed by 30 seconds of rest. The total is 60 seconds of exercise, but each 10-second interval can be run more intensely with more benefit to conditioning. The body recovers somewhat during each rest period. All three pathways can be improved by interval training, depending on the length of the run and the time of the rest.

At the cellular level, several changes occur to increase the body's ability to generate energy. Anaerobically, larger amounts of PC and ATP can be stored by muscle cells. The enzyme creatine kinase, which forms ATP from PC, is more active. The body's ability grows toward maximum performance.

Two types of skeletal muscle fibers—red, or "slow twitch," and white, or "fast twitch"—are involved primarily in aerobic or anaerobic metabolism, respectively. In training designed to increase anaerobic capacity, the size of the white fibers increases, as does the amount of stored glycogen. The activity of the glycolytic enzymes goes up.

On the other hand, training for distance events emphasizes long-distance workouts at less than maximum intensity to raise the athlete's *aerobic* capacity. As a result, the amount of myoglobin (an oxygen-binding protein similar to hemoglobin) and the number of mitochondria of the red muscle fibers increase. More glycogen is stored, and the activity of enzymes used in the Krebs cycle and the electron transport system rises.

Recovery from exercise, like exercise itself, is a two-step process. The ATP/PC system, which is so rapidly depleted, is restored most quickly—in two or three minutes. Metabolism of lactic acid takes about an hour. Formerly researchers thought the accumulated lactic acid was synthesized directly into glycogen, but now they know that it is oxidized to carbon dioxide and water. An athlete can hasten lactic acid recovery, or recovery from oxygen debt, not by rest, but by light exercise, which increases oxygen intake and blood circulation.

However, the lost glycogen stores are not immediately regenerated. An athlete needs two to five days to restore muscle and liver glycogen. Restoration takes five days on a low-carbohydrate diet, two if the diet is high in carbohydrates. The benefits of the high-protein "training table" meals are largely myth.

And what is the reward to the athlete from all this metabolic juggling? To the nonathlete, the activity, discipline, and pain probably appear masochistic, a waste of time, or just plain crazy. No one, after all, has proved that athletes really do live longer. An athlete might tersely mutter something about the "quality of life." The professional athlete has very obvious monetary rewards, and marathoners like Frank Shorter may well be addicted to the sport.

George Sheehan, a marathoner with decades of running behind him, takes a philosophic tone in suggesting what many different types of athletes feel who have been in training for several years. In his book *On Running,* he writes, "The runner does not run because he is too slight for football or hasn't the ability to put a ball through a hoop or can't hit a curve ball. He doesn't run primarily to lose weight or prevent heart attacks. He runs because he has to. . . . He is fulfilling himself and becoming the person he is."

RECEPTOR CELLS

Origin of the nerve impulse Thus far we have been considering the structure of the neuron and its function, as well as the nature of the action potential. Yet every action potential must begin somewhere, as an effect in response to some cause. This cause, or instigating factor, is called the **stimulus.** We shall now consider what a stimulus is and precisely how a neuron is affected by it.

All stimuli have in common the fact that they are forms of energy. Yet sources of energy are numerous, and there are various forms of energy, such as light, chemical, and heat energy. As multicellular organisms evolved, there also evolved a means for converting various types of energy into one basic form for utilization. Only with such a mechanism would the organism be able to integrate the various signals it receives from its environment and to process them rapidly. It is the receptor—usually a specialized epithelial cell (or a modified neuron), often with a modified hair-like cilium—that has evolved as an energy transducer. The receptor and the sensory neuron act together to convert into an electrochemical impulse all forms of energy to which the organism responds.

After aeons of time, specialized organs and structures evolved to contain the receptor cells and sensory neurons and to provide greater accuracy of stimulus reception. Eventually, organisms were able not only to sense various stimuli but also, as a result of the combined action of their sensory organs, to actually perceive the environment. Thus organisms could see and hear, smell and taste. Yet despite this specialization and development, probably no organism perceives its environment accurately or completely; this would have entailed extraordinary development of sensory apparatus. Instead, organisms have evolved means of perception that suffice to ensure *their* survival—what the earthworm sees, what the bumblebee sees, and what a human sees are all highly different modes of perception. The earthworm senses only the presence of light; the bee probably sees a blur. And although the man sees objects, he has no way of knowing how much he does *not* see. Assuredly, since the human eye is receptive to wavelengths of light that encompass only a very limited portion of the radiation spectrum, man is missing quite a lot.

Vision Vision in many of the lower animals is restricted to the ability to detect light and in some cases the intensity or direction of the light source. Protozoa, as we have seen, react to the presence of light, usually moving away if it is intense. The flatworm (planaria) has specialized light-sensitive clumps of cells on either side of its head—these are two dots that give the front end of this organism its oddly facelike appearance. Such organisms appear to be able to survive without more extensive visual information about their surroundings.

Higher animals must have more precise information if they are to contend with a varied environment. The organ that specializes in conveying visual information —including visual *images*—is the eye, an organ that Charles Darwin celebrated as the masterpiece of evolution.

Two very different types of eyes have evolved independently among animals. These are the arthropod or common eye and the vertebrate eye, which is often called the camera eye.

As typified by the human eye, the **vertebrate eye** works on the principle of a camera. Essentially a spherical, multilayered structure, the human eye consists of a front layer of transparent connective tissue called the **cornea;** a highly vascularized ring of muscular pigmented tissue called the **iris,** in whose center is an aperture, the **pupil,** through which light may pass; and situated just behind the pupil, the **lens**—the structure that focuses light rays on a light-sensitive area called the **retina.** The pigmented iris serves to block most light rays so that the small beam of light that does pass through the pupil can be focused distinctly on the retina. The human eye—through changes in the position and shape or curvature of the lens—is capable of a relatively high degree of resolution, so that it can register sharp images of objects both near and far.

The evolution of receptor cells with a strong sensitivity to light reaches an apex of complexity in the **retina** —a structure containing several layers of neurons and, in its innermost layer, the light receptor cells. In hu-

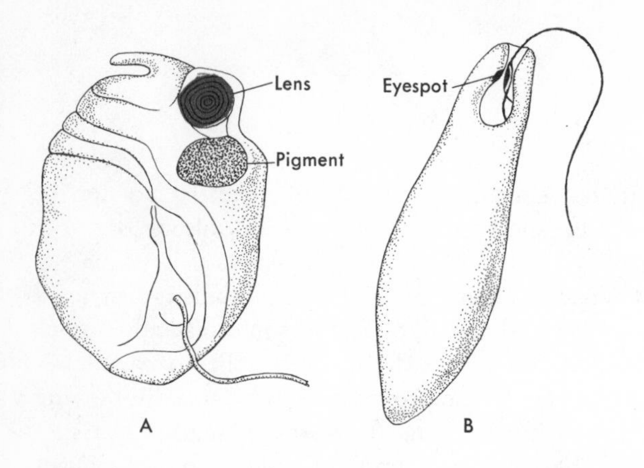

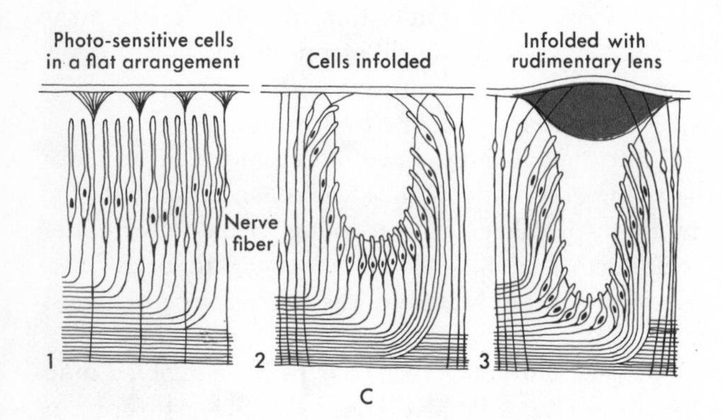

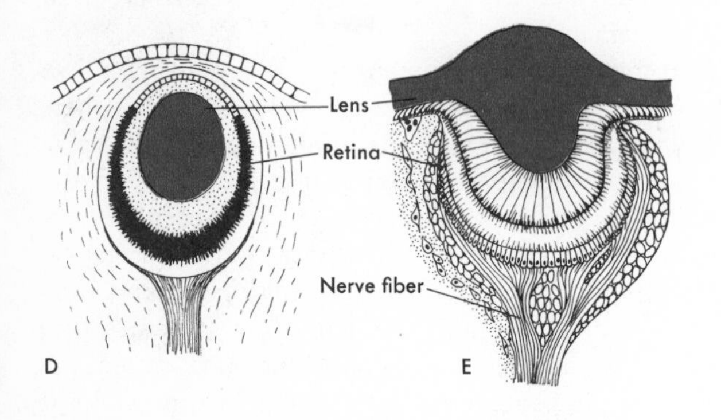

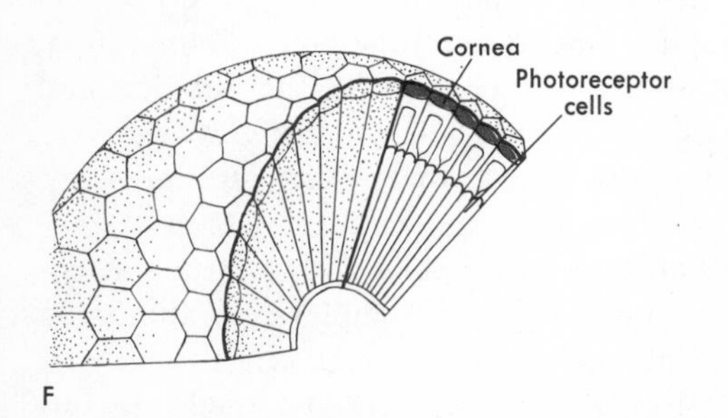

FIGURE 10.12

Independently evolved light-sensitive organs in unrelated types of organisms. The frequency of this occurrence attests to the importance of the ability to respond to light. Simple photosensitive systems consist only of light-sensitive areas. More complex ones have lenses that concentrate, or focus, light. The diagrams show the simplest type of photoreceptive system found in unicellular organisms. Pouchetia *has a rudimentary lens (A), but* Euglena *has only a pigmented light-sensitive area (B). Systems are found in starfish consisting of specialized cells arranged in various configurations (C). Snails (D) and spiders (E) have true eyes, each with lens and retina. The compound eye is characteristic of insects (F). It is composed of thousands of cone-shaped units called ommatidia. Each ommatidium contains a cornea, a lens, photoreceptor cells, and a rhabdom, which conducts light to the photoreceptor cells. Pigment cells surrounding the photoreceptor cells prevent light from leaking into other ommatidia. Each ommatidium receives a pinpoint image, and the whole eye perceives a mosaiclike image of the world around it. The human eye (G) perceives only a single image. Light passes through the cornea and is focused by the lens onto the retina. Photoreceptor cells in the retina called rods and cones are sensitive to different types of light and send impulses along the optic nerve to the brain, which interprets them as vision. There are about 7 million cone cells in a human eye, and about 100 million rods.*

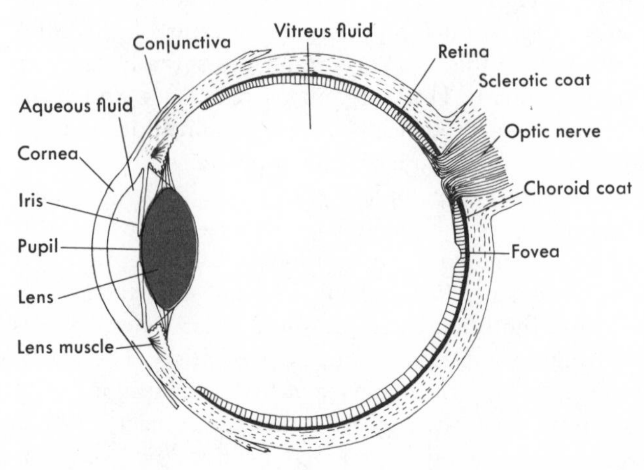

mans, there are two types of light receptor cells. The **rod cells,** arranged along the periphery of the retina, are functionally suited to operate in dim light but incapable of registering color. Some nocturnal animals rely entirely on rods for their vision, as we do in very dim light. Cells of the second type, **cone cells,** are situated in the human eye toward the center of the retina in a spot called the **fovea centralis.** These cells require brighter light, but they are capable of registering color and of delivering sharper images. Of the 100 million or so receptor cells in the human retina of each eye, only 7 million are cones.

Both rod and cone cells are long and slender and, in terrestrial vertebrates, contain visual pigments. In man **rhodopsin** is the pigment present in the rods. Three other pigments are present in the cones—each with maximum sensitivity to green, red, or blue—but they work much like rhodopsin. The rhodopsin molecule is composed of **opsin,** a protein, and a slightly altered molecule of vitamin A called **retinene** to which the opsin is joined. Rhodopsin becomes unstable and splits when light is absorbed, and the opsin undergoes structural modification. It is these alterations in the rhodopsin structure that generate a modification in the electrical potential of the cell membrane. In this way an action potential is initiated in a sensory neuron. Since impulses may be transmitted along a path of sensory neurons at a rate of 500 per second, continuous perception is possible. Just a few of the variants of light-sensitive organs that have evolved in many types of organisms are shown in Fig. 10.12.

With respect to vision, man is especially well adapted: of the vertebrates, he and only a few other primates are capable of color vision, whereas most mammals see only in black and white and varying shades of gray. In addition, the frontal position of the two eyes gives man and certain other animals a distinctive stereoscopic or 3-D vision.

The **arthropod eye,** found in insects and crustaceans, is composed of numerous independent units called **ommatidia**—long tubelike structures whose sides are darkly pigmented, perhaps to prevent entry of light from unsuitable angles or internal reflections of light. The photosensitive cells are located at the bottom, and because of the compound nature of the eye (consisting of many facets), the organism may perceive a mosaic image of its environment, each ommatidium registering a piece of it. However, it is also possible that the insect brain is capable of deciphering the mosaic of signals from the compound eye and making a clear image from it. Moreover, the photosensitive pigments in arthropods perceive color, adapting the eyes of these animals to meet the de-

FIGURE 10.13

The head of a housefly. Most adult insects have two bulging compound eyes on either side of the head, which give a wide field of vision in all directions. It is not certain whether the insect's brain can decipher a single clear picture from the mosaic image produced by its compound eye.

FIGURE 10.14

Differences in vision. Insect vision detects most of the spectrum of electromagnetic radiation that we do, but with a shift toward the shorter wave lengths. Red is invisible to an insect, but ultraviolet can be seen as a color. The same flower has been photographed in regular light (top) and under ultraviolet light (bottom) to show this difference in vision. The evolutionary trend in flower development has favored patterns and coloration that can attract insects to the gamete-containing center of the flower, since the main structural and functional purpose of a flower is to effect pollination, and insects are the main agents in this process. Thus nectar guides, invisible to our eyes but appearing as a dark center to eyes that can see ultraviolet, direct the insects to the reproductive center of the flower.

mands of their many varied activities—obtaining nutrients from the nectaries of flowers, or preying on other insects. An interesting example of such adaptation is described in Fig. 10.14.

Hearing All organs of hearing are composed of specialized ciliated epithelial cells, called **hair cells,** whose cilia are modified to form long thin projections. Sound waves reaching the hair cell, whether through the air or through liquid, cause the cilia to move and thereby generate an action potential in adjacent sensory neurons.

In insects, a membrane called the **tympanum** is situated on both sides of either the abdomen or the legs. It is stretched across an opening that is a modification of the tracheal passages of gas exchange. Sound waves cause the tympanum to vibrate, and the energy in the vibrations is transmitted as a nerve impulse by hair cells attached to the base of the membrane.

The vibrating membrane is also found in the vertebrate ear, although in vertebrates the outer tympanic membrane (the eardrum) is not directly associated with any sensory cells. Instead, the actual organ of hearing—the **cochlea**—is located deep inside the skull.

The cochlea converts sound waves into vibrations in the viscous liquid that fills that organ. These liquid vibrations are received by the basilar membrane, which consists of many tiny hair cells. The consequent movement of the hair cells generates an electrical impulse, which is received and transmitted by a neuron.

The human ear is a relatively complex structure consisting of three parts (see Fig. 10.15). The outer ear includes the **pinna,** or ear flap, and the **auditory canal,** which leads from the pinna to the tympanic membrane. The middle ear includes three bones, the **hammer, anvil,** and **stapes,** and the **Eustachian tube**—a passage leading from the middle ear cavity to the pharynx for equalizing air pressure. The inner ear is a chamber divided into an upper and a lower section. The upper section of the inner ear contains three **semicircular canals,** within which is the equilibrium-sensing apparatus. The lower section houses the cochlea—the actual auditory organ composed of three fluid-filled tubes in which the energy in sound waves is transduced into vibrations in the

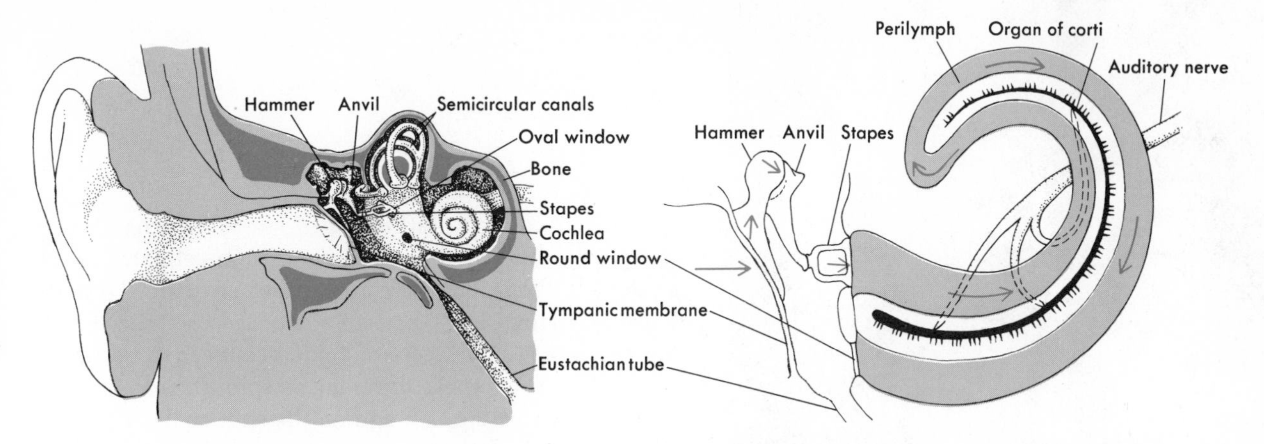

FIGURE 10.15

Human ear. The human ear has two functions: the perception of sound, accomplished in the cochlea, and the perception of spatial orientation, accomplished in the semicircular canals. These two sensing organs, the cochlea and the semicircular canals, make up the inner ear. The outer and middle ears function in the transmission of sound to the cochlea. In the process of hearing, air vibrations pass through the pinna into the auditory canal and cause the tympanic membrane (eardrum) to vibrate. Eardrum movement is transmitted through the middle ear to the oval window by the hammer, anvil, and stapes. Oval window vibrations generate resonance waves in the liquid that fills the canals of the cochlea. These waves cause the basilar membrane to move and thus cause the cilia of the hair cells to vibrate. Cilia movements generate nervous impulses, which are transmitted by the auditory nerve to the brain. The brain perceives spatial orientation and maintains balance with the three semicircular canals. Each canal is oriented in a different plane, and impulses from the three canals are interpreted by the brain as the orientation of the body in three dimensions.

liquid. Two of these tubes—the **vestibular canal** and the **tympanic canal**—are connected at one end; the **cochlear canal** lies between them. The vestibular canal is connected with the middle ear through a membrane called the **oval window,** against which, on the other side, lies the stapes. The tympanic canal ends in a membrane called the **round window.** However, it is from the cochlear canal that the **organ of Corti**—the sensory portion of the cochlea—leads into the auditory nerve. This nerve conducts action potentials (transduced from fluid vibrations of the hair cells in the organ of Corti's basilar membrane) to the brain. Sounds of different pitch stimulate hair cells in different areas of the cochlea, and thus the organ of Corti can discriminate tone as well as loudness.

The auditory ranges of organisms vary according to the number of hair cells in the basilar membrane. Man, for example, can hear sounds of up to 20 kiloherz (20,000 cycles per second); cats hear up to 40 kiloherz; bats—creatures with probably the greatest sensitivity to high pitch—hear sounds of over 100 kiloherz. Bats, in fact, employ their extraordinarily developed sense of hearing to locate objects. They emit brief high-frequency squeaks whose echoes from objects in the bat's path are heard by the animal, whereupon the presence of these objects is sensed.

Smell and taste Whereas the auditory and visual senses depend on stimulation of sensory receptors by energy *waves,* smell and taste involve specialized epithelial cells called **chemoreceptors** that are stimulated by ions and molecules. Since both senses share a common stimulus source, smell and taste are interdependent. When we eat spaghetti, for example, the chemoreceptors on our tongues convey information concerning how salty, acid, and sweet the dish is. Yet much of our sense of the spaghetti's flavor actually derives from molecules released by spices in the sauce that stimulate chemoreceptors within our noses. Thus chemoreceptors in man and all other mammals are located on the tongue and at the back of the nose. The receptors, or taste buds, on the tongue convey only four different sensations of taste—saltiness and sweetness on the front, sourness (acidness) on the peripheries, and bitterness at the rear. Other animals, however, have chemoreceptors located on various other parts of their bodies. Some insects, for example, have them on their feet and on the hairs along their legs. The taste buds in fish are located on the skin of their heads.

FIGURE 10.16

Chemoreceptors. Snakes have a good sense of smell through their nostrils, but that sense is enhanced by special chemoreceptors located in cavities on the roof of the mouth. The receptor cells are stimulated by minute particles picked up on the snake's flickering tongue; thus this green grass snake is literally both smelling and tasting its environment.

It is thought that odors are perceived when molecules of the odorous substance fit into particular sites on the receptor cell's membrane. Since seven classes of odor are known to be perceptible by man, seven different types of receptor sites are hypothesized, each structurally unique and designed to accept, like a lock and key, only molecules with a particular configuration. It is indeed probable that a chemical reaction occurs at the receptor site, and this generates the action potential that is transmitted to the brain. Certain parts of the brain and some sensory regions in the blood vessels also contain chemoreceptors sensitive to changes in pH, oxygen, and carbon dioxide levels.

Touch, temperature, and pain There are several types of receptors that sense contact with environmental surfaces and that enable the organism to perceive the temperature of its surroundings. There is the hair cell—the specialized ciliated epithelial cell similar to that found in the ear—and there are numerous unmyelinated nerve endings. In addition, there are various structures—called **Pacinian corpuscles**—that are composed of a nerve ending surrounded by layers of folded connective tissue. In most vertebrates these different receptors are located on various parts of the face and on the skin of the arms and legs. Insects have touch receptors on their feet, and fish appear to have them in their fins. Fish also have a structure called the **lateral line system**—an arrangement of specialized hair cells situated inside a canal-like depression running longitudinally down the side of the body. In addition to sensing water temperature and pH, these cells are also responsive to changes in the velocity of water passing over them and hence convey information concerning the speed at which the fish is moving.

The sensation of pain is not very well understood. There do seem to be pain receptors in the epithelium, but apparently the message they carry is not very specific. Moreover, any sensory modality, if stimulated strongly enough, evokes pain. Pain is a highly personal sensation, in which psychological factors may be at least as important as physiological ones. Recent studies suggest that males tolerate pain better than females, and young men better than older men. Cultural differences have also been suggested.

In addition to receptors subject to stimuli from the external environment, there are specialized sensory cells called **proprioceptors** that receive stimuli from organs and systems of the internal environment and, by monitoring their functions, serve as the information input for all of the homeostatic mechanisms. Two of the proprioceptor systems include the stretch receptors, such as those in the cardiac blood vessels that help maintain constancy of blood pressure, and the chemoreceptors, such as those in the carotid body that promote the constancy of blood pH.

A third proprioceptor system is one through which animals obtain their sense of movement and posture, or their **kinesthetic** sense. In the kinesthetic system, sensory receptors are located within the connective tissue of joints, in the tendons that attach muscles to bones, in ligaments, and also within specialized muscle fibers that contain thin receptor nerve endings in their middle. The joint receptors are sensitive to movement of the joint through an arc greater than 10° and thus register information concerning the degree to which the joint is bent. Receptors in tendons and ligaments measure the extent of muscular movement. Nerve endings with intrafusal fibers transmit impulses during relaxation of the fiber's contactile elements and so register the state of muscular contraction in that region. This permits the brain to continuously monitor and regulate posture and the position and movements of the limbs.

Finally, there are the sensors of equilibrium—in vertebrates, the **otoliths,** crystals of calcium carbonate situated atop hair cells located in the upper portion of the inner ear. Movements of the head cause movements of the otoliths, thereby causing the hair cells to transmit action potentials that relate the position of the head—the organism's sense of **static equilibrium.** The animal's sense of bodily movement—its direction and speed—is conveyed by the semicircular canals, the organs of **dynamic equilibrium** that, like the chambers housing the otoliths, are also located in the upper portion of the inner ear.

In summary, sensory receptors—whether they occur as nerve endings in the skin or function in a sensory organ like the eye—convey information about the organism's external and internal environment. To be effective, this information, conveyed as electrochemical impulses, must be delivered to specific areas in the organism, where the impulses may be interpreted and coordinated in order to achieve a response. The organ that has evolved to assume the major role of integrating and coordinating the overwhelming number of impulses received per unit of time is the brain. It is in the brain, for example, that sensory impulses "make sense." For it is in the brain that sensory areas evolved to accept impulse transmissions from particular sensory organs. Thus nerve impulses from the retina travel directly to the brain's visual center—an area whose function is specifically to receive visual stimuli and interpret them in a way that registers an awareness of the image formed on the retina. In short, it is the particular *circuit* of neurons leading from the eye that determines the *visual* perception of their impulse—because that circuit of neurons leads only to the brain's visual center and nowhere else.

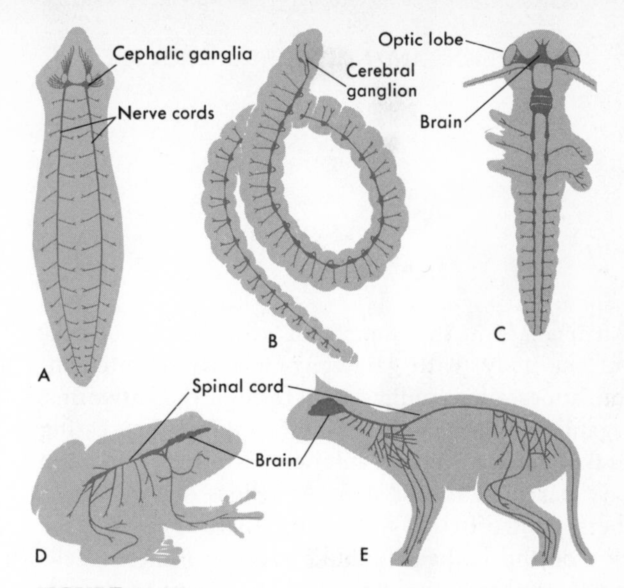

FIGURE 10.17

Several representative animals with successively more complex nervous systems. (A) The flatworm has a bilateral ladderlike nervous system. The cephalic ganglia are a primitive form of cephalization, or brain development. (B) The earthworm has a single central nerve cord running down the ventral side of its body. (C) Arthropods (including segmented animals with exoskeletons, such as insects, crayfish, centipedes, etc.) have paired ventral nerve cords and a dorsal brain. (D) The simple vertebrate, in this case an amphibian, has a single dorsal nerve cord surrounded by bone and a well-developed dorsal brain encased in a bony skull. (E) Higher vertebrates are basically similar to lower vertebrates, but they have more highly developed brains.

Central nervous system in vertebrates

In vertebrates, a brain and the spinal cord constitute the central nervous system, or **CNS.** Although the spinal cord is capable of responding independently of the brain, it is largely subordinate to the brain in the overall function of the central nervous system.

The course of the brain's evolutionary history can be posited from observations of its development in the lower invertebrates (see Fig. 10.17). Coelenterates, such as the hydra, have *no* brain. Instead, a network of nerve cells radiates throughout the hydra's body to form an interconnecting circuit called a **nerve net.** Although the nerve net serves to conduct impulses rapidly enough from any receptor site, the system lacks integration and coordination. If the hydra receives a tactile stimulus at one point on its body, rather than confining its response to that one area, the hydra's entire body flinches and withdraws.

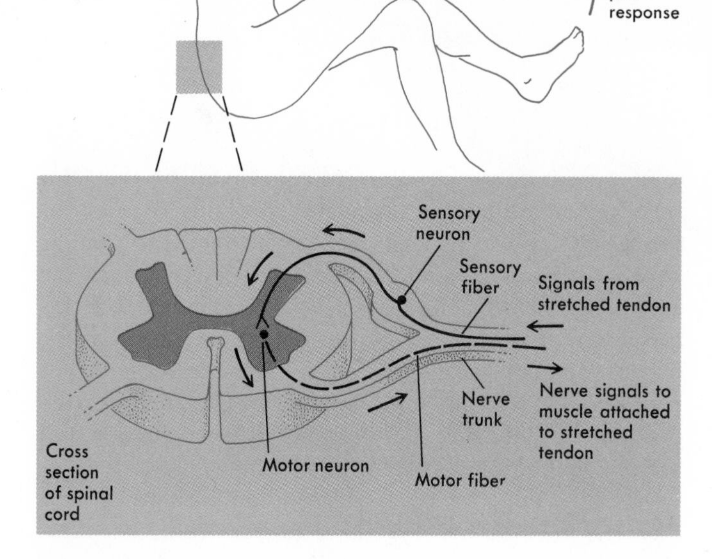

FIGURE 10.18

The knee-jerk reflex, an example of the simplest type of reflex arc. The stimulus to the tendon at the knee sends a signal along a sensory neuron to the spinal cord. There the axon of the sensory neuron synapses with the dendrites of a motor neuron, and the signal returns to the muscle attached to the tendon and causes the knee-jerk response.

An increase in the number of neurons in a specific area of the body, with greater complexity of interconnection, appears in a rudimentary form in the flatworms, the organisms in the phylum Platyhelminthes possessing bilateral symmetry and an anterior-posterior orientation of body parts. In the phylum Annelida, one of whose members is the oft-cited earthworm, there appears in the anterior of the body a bulblike cluster of nerve cells called the cerebral ganglion, which functions to coordinate the ganglia found in each of the body segments. By contrast, the set of ganglia in arthropods and molluscs is highly developed, and it actually functions somewhat like a brain in that it receives much of the sensory information and extensively coordinates motor activity. Yet not even in the arthropods and molluscs is the brain completely indispensable, since a headless insect can manage to survive for a period of time.

Only in the higher vertebrates does the brain assume almost complete dominance over the vital functions of the organism. Twelve **cranial nerves** connect with the brain, and there are **spinal nerves** that connect with the spinal cord. A **nerve** is a bundle of many neurons tightly wrapped with tough connective tissue. Without such protection the delicate neuron fibers (axons) could be broken easily. Although sensory input along the spinal nerves can travel up the spinal cord to the brain, the spinal cord itself is also capable of assuming a coordinative function, as in the **reflex arc.**

REFLEX ARCS

The knee-jerk reflex elicited by a tap on the knee is an excellent example of a response effected through the simplest type of reflex arc (Fig. 10.18). Only *two* neuron are involved in this reflex. The sensory neuron extends from a stretch receptor in the tendon to the spinal cord, where the cell body of the neuron is located just above the dorsal surface of the cord in a structure called the **dorsal-root ganglion.** The dorsal-root ganglion houses the cell bodies of sensory neurons that enter the spinal cord. Here the axon synapses with the dendrites of a **motor neuron** whose cell body is located *in* the ventral portion of the grey matter of the spinal cord. The axon of the motor neuron exits from the ventral part of the cord and leads all the way to an effector cell—the same fiber that was attached to the stretched tendon—which responds to the stimulus with a sudden contraction, or jerk. Since only two neurons are involved in the knee jerk, this reflex is termed a **monosynaptic** reflex arc. Such simple connections permit unconscious control of many bodily activities, such as breathing and blinking. There are more complex reflex arcs involving, for example, intercalary neurons that carry information up and down the spinal cord.

AUTONOMIC NERVOUS SYSTEM

Whereas the reflex arc does not involve the brain directly, the autonomic nervous system does. The autonomic system is simply that part of the central nervous system that automatically regulates muscles of internal organs (viscera) not usually subject to voluntary control, in response to information coming from receptors in these organs, as well as from the brain. Thus the autonomic system also involves the pathways provided by the cranial and spinal nerves.

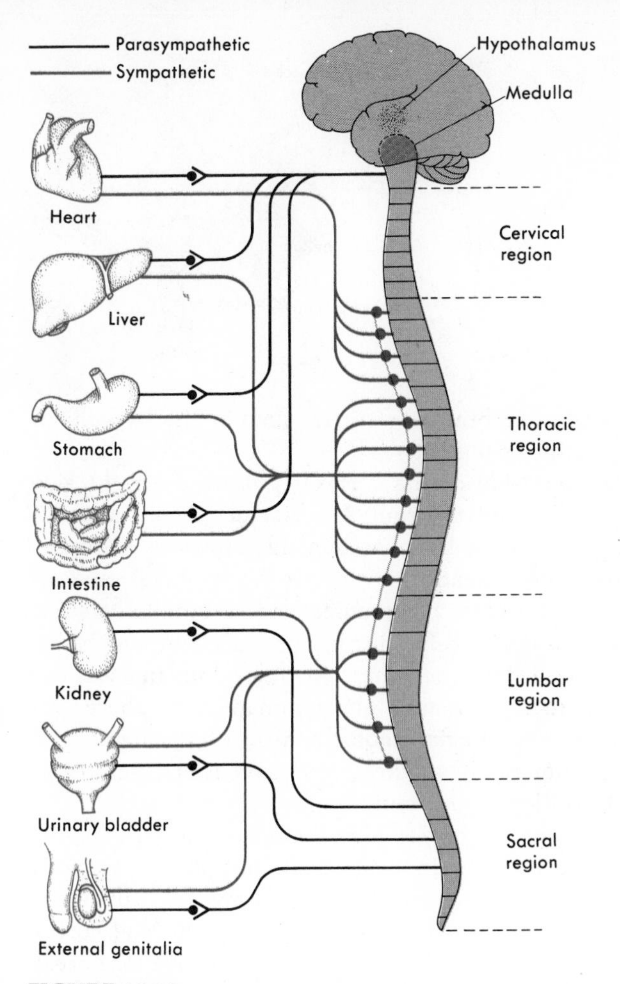

FIGURE 10.19

Human nervous system. The autonomic nervous system is made up of the sympathetic and the parasympathetic systems. Most vital organs of the body are regulated by both systems, and for any given organ the effects of the two systems are opposite. The sympathetic system releases norepinephrine from nerve endings; the parasympathetic system releases acetylcholine. Sympathetic nerves emanate from the spinal cord in the thoracic and lumbar regions, and the parasympathetic nerves originate from the medulla in the brain and the sacral portion of the spinal cord.

The autonomic nervous system actually comprises two individual systems that often operate as antagonists in the same organ (Fig. 10.19). The **parasympathetic** system regulates the animal's homeostatic mechanism when the organism is at rest; the **sympathetic** system takes over in times of emergency or stress to prepare the animal for "flight or fight." Although many organs and glands are innervated by both systems, each produces its own effect by causing the release of a different chemical transmitter. In a stressful situation, for example, the sympathetic system prompts the adrenal glands to release epinephrine and norepinephrine—substances that increase the blood pressure, the flow of blood to the skeletal muscles and brain, the heart rate, the rate of breathing, and the capacity of the body to take in a greater volume of air. When stress disappears, however, it is the parasympathetic system that dominates. It causes the release of acetylcholine, the substance that signals all the rapidly functioning organs and glands, sparked by the sympathetic system, to slow down. Central control (i.e., the brain) superimposes its influences over autonomic reflexes in the parasympathetic and sympathetic systems.

THE VERTEBRATE BRAIN

Among the lower vertebrates, the brain consists of three main divisions: hindbrain, midbrain, and forebrain (see Fig. 10.20). The hindbrain consists of two major structures, the medulla oblongata and the cerebellum. Through the **medulla oblongata**—really the enlarged end of the spinal cord—pass most of the sensory nerves leading *to* the brain, as well as parts of nearly every motor neuron leading from the brain to effector cells. This makes the medulla by virtue of its location, the initial relay center of the brain. The medulla also contains several vital nerve centers, concentrations of specialized neurons that regulate respiration, blood flow, and heart rate. Although of a much smaller size in frogs and fish, the **cerebellum** appears as a large outgrowth of the medulla in birds and mammals. Its large size in these animals provides for them the possibility of a greater number of synapses and therefore more capacity for coordinating information. As a result, birds and mammals are able to run and fly—functions requiring rapid muscle movement and more intricate control of the musculature. More primitive vertebrates do not have as well-developed motor-coordinating mechanisms.

Whereas the hindbrain has evolved to a much greater size in the human, the midbrain has not. However, the midbrain has lost most of the integrative function it assumes in the lower vertebrates, in whom it operates to integrate visual impulses with other impulses from the motor nerves and the autonomic nervous system that pass into it from the hindbrain. In the higher vertebrates, the midbrain serves mainly as a relay sta-

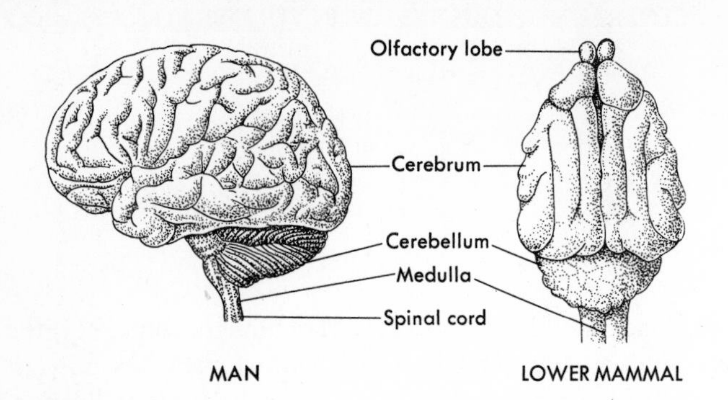

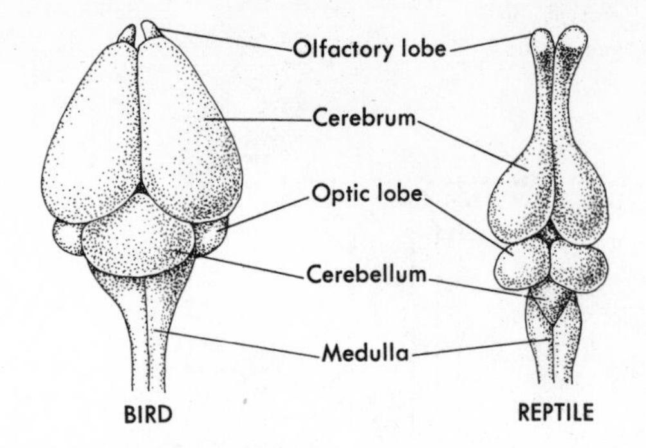

FIGURE 10.20

The comparative structure of vertebrate brains and the human brain. Note the relative sizes of the cerebrum in the different vertebrate classes. An increase in the size of the cerebrum in proportion to the rest of the brain corresponds to an increased ability to learn. Other areas of the brain include the cerebellum, which coordinates muscular activity; the medulla, which relays messages to and from the spinal cord and controls visceral functions; optic lobes, which integrate visual inputs; olfactory lobes, which determine the sense of smell; the thalamus, which integrates sensory inputs; and the hypothalamus, which regulates emotions, sensations, and many visceral functions.

tion through which impulses move from the hindbrain to the forebrain, although it does retain the function of regulating the reflex action of the irises and eyelids.

It is the forebrain, consisting of several distinct areas, that has changed extraordinarily through the course of evolution. The thalamus in higher vertebrates serves an integrative function for impulses arriving from all the sensory systems. It coordinates the data received and transmits information to other parts of the forebrain and to the spinal cord. It also serves as a major relay center for pain. The hypothalamus incorporates centers for the regulation of hormone release as well as the regulation of water balance and body temperature. It has also been determined, through electrode implantation, that the hypothalamus contains areas that control sensations and emotions. In it are located the centers that regulate appetite, thirst, pleasure, pain, rage, passivity, and sexual desire.

The third major area of the forebrain is the **cerebrum,** in man the largest portion of the brain, whose convoluted surface is so disproportionately expanded that it appears to spill over the other parts of the forebrain. The cerebrum consists of an outer rind, the cerebral cortex, formed of five layers of neurons (the grey matter) and an inner core formed of millions of nerve fibers (the white matter). The cerebral cortex regulates voluntary motor activities, interprets sensory inputs to achieve meaningful perception, and performs mnemonic and cognitive functions. It also exerts some sort of control over lower brain structures, such as the hypothalamus and the medulla.

The left side of the cortex regulates the right side of the body, and the right side performs its duties for the left side of the body. In humans, however, there has also been specialization in the two hemispheres, and although both halves communicate via a bundle of some 200 million fibers, called the corpus collosum, one hemisphere dominates the other. (In most humans the left hemisphere is dominant.) The dominant hemisphere specializes in speech, fine motor control of the hand, and abstract thought. The other hemisphere is artistic, rhythmical, emotional, and mute.

Although the purely motor and sensory areas of the cortex are reduced in man, each part of the body is nevertheless represented there by its own control center —everything from the toes to the thumb. The cortex achieves its apex of distinction, however, in those areas that provide for such functions as memory and thought; for it is precisely these areas that serve to distinguish man from all other forms of animal life.

MODERN VIEWS OF THE BRAIN

It is no longer sufficient to study the brain only anatomically. As Sir John Eccles, a distinguished Australian physiologist and Nobel prizewinner, has said, "It is not enough to think of the brain as a large collection of units or components. We must pay more attention to the patterns of organization in which the units are joined together. We must study the incredibly detailed connectivity that is present at birth and keeps changing throughout as in the learning process."

Indeed, the study of brain function has accelerated dramatically within the last 50 years. In the 1930s W. R. Hess, a Swiss physiologist, perfected a method for inserting fine electrodes into the brain tissue of freely moving animals. Thus the effect of electrical stimulation on various parts of the brain could be observed, and the functional areas of the brain could be more accurately mapped. During the last few decades, brain studies have

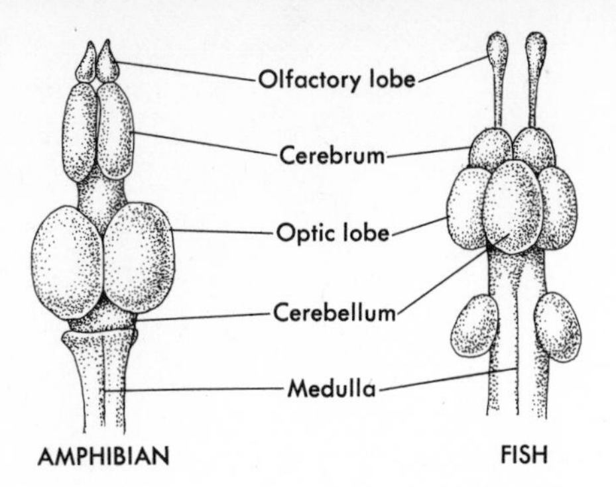

been facilitated by the development by Ralph Gerard of the microelectrode, whose tip is only a few thousandths of a millimeter wide, and which can be inserted into a single nerve cell to record events. Recently, researchers have been able to refine their studies of brain chemistry with the use of microinjection needles whereby they can inject chemicals into specific centers deep in the interior of the brain. In addition, further insight into brain function has been provided by progressively more refined surgical techniques.

It is not surprising that physiologists and others have begun to observe the functioning of the nervous system and the brain in new ways and have attempted to specify some of the behavioral aspects of neurophysiology. Robert Doty, for example, tends to look at the brain in terms of six categories of functions: sensorial, effective, attentive, motivational, mnemonic, and, in the higher primates, cognitive. Discussing the nervous system in terms of these functions is a useful way of placing in perspective some of the more recent brain studies. A study of the attentive functions of the brain might focus on its ability to select particular inputs and ignore others. Research has shown, for example, that a cat bombarded with a series of clicks produces a characteristic electrical response to each one—until a mouse appears in the field of vision, and then the electrical response in the cat's brain disappears. A study of motivational functions might focus on the capability of electrical stimulation or surgery to radically change behavior. Loss of a certain portion of the hypothalamus, for example, turns a tame animal into a vicious one. Study of the mnemonic (memory) functions might investigate whether or not the learning of visual discriminations can be transferred from one hemisphere of the brain to the other, or might seek to explain the loss of memory following electroconvulsive shock. The basis for memory, the permanent storage of information, is also increasingly studied. There are some indications that RNA and protein synthesis may be involved, though in what capacity has not been determined.

Given the pace and significance of modern brain research and the fact that experimental animals are likely to yield results that are relevant to human physiology (Doty states that "the mouse and man have the same types of neurons, and that in general the major principles of neural function vary but slightly in the mammalian series"), it seems likely that the next decade will bring a new understanding of the functioning of the neurological bases of human behavior.

Summary

1. Movement is a principal characteristic of living things, and virtually all organisms require some form of movement to accomplish the functions of nutrition, reproduction, communication, and homeostasis.

2. The cellular chemistry determining movement in plants differs from that in animals. Plant movements are generally of two types: *tropisms,* which include directed movements in response to specific stimuli from the environment; and *nastic movements,* which are responses not related to the direction of the stimulus—such as the folding of leaves at night.

3. Animal movement is more complex, and basically it tends to result from reversible changes in configuration and length of contractile proteins within the cell. In unicellular animals microfilaments and microtubules serve to initiate and guide the streaming cytoplasm or to form cilia and flagella.

4. In terrestrial multicellular animals, muscle cells contain the contractile proteins. The muscles mobilize the animal's appendages of locomotion by their specific connections to the skeletal system of the animal. Contraction of these muscles moves the animal.

5. Animals may have a hydrostatic skeleton, an exoskeleton, or an endoskeleton. In the last, antagonistic muscles are attached around a joint and under nervous stimulation certain muscles contract, causing the bones to move on the lever principle.

6. A muscle fiber contracts when a specific threshold value of stimulation is exceeded. Each fiber responds with a complete contraction. This is the all-or-none principle of response for individual fibers. As the strength of the stimulus increases, more muscle fibers respond, giving graded response to stimuli of different intensities.

7. The contraction of a muscle fiber is stimulated by a depolarization of the membrane surrounding each muscle fiber (the sarcolemma) by an electrochemical impulse from a nerve that supplies the fiber. This depolarization temporarily alters the gradients of Na^+ and K^+ ions on either side of the membrane of the muscle cell, stimulating the contracile proteins to slide together.

8. The neuron is the basic structural and functional unit of the nervous system. It is a cell specialized for the conduction or propagation of impulses (in the form of action potentials) from specialized receptor cells to specialized effector cells.

9. All living cells maintain a resting potential, setting up an ionic gradient by actively transporting Na^+ ions out of the cell and pumping K^+ ions inward. An action potential is initiated in a nerve cell when some stimulus (change in energy) that is greater in magnitude than a specific threshold value causes a localized depolarization of the neuron. This energy change disrupts the resting potential of that area, so that Na^+ ions are allowed to flow into the cell. This depolarization causes the neuron to fire in an all-or-none fashion, and a transient change of potential is propagated down the membrane of the neuron to the synaptic ending of the axon. Here the action potential triggers the release of a specific chemical, such as acetylcholine, which diffuses across the synapse, where it depolarizes another neuron. In myelinated neurons, transmission of the impulse is faster, since the action potential skips from one node of Ranvier to another.

10. Receptor cells are usually specialized epithelial cells with a modified hairlike cilium, and they *transduce* energy from the environment into an electrochemical impulse. Examples of receptor cells are the rod and cone cells of the vertebrate retina, and the hair cells of the organ of Corti in the ear. The latter transduce mechanical vibrations of the basilar membrane into action potentials that the brain translates as sound.

11. Proprioceptors receive stimuli from organs and systems of the internal environment.

12. The brain is the organ that integrates and processes the multitude of sensory information that is constantly being transmitted by our external and internal receptors. The central nervous system comprises the brain and spinal cord. The autonomic nervous system, that part of the central nervous system that regulates all the processes not under voluntary control, comprises parasympathetic and sympathetic components. These individual systems act as an antagonistic pair. In general, the sympathetic system excites or stimulates the organ it innervates, and the parasympathetic inhibits it.

13. The vertebrate brain reflects an evolutionary trend toward development of the hindbrain and the forebrain, whereas the midbrain is mainly a relay station between the other two sections. The cerebral cortex of the forebrain has evolved to the greatest extent in the higher vertebrates such as dolphins and man. It is this area that functions in memory and thought and makes us distinctive from all other forms of life.

REVIEW AND STUDY QUESTIONS

1. What is the essential difference between tropism and nastic movement? Give an example of each.

2. Describe the movement of an amoeba. Of an earthworm.

3. What are the three types of skeleton? Name one or more animals in each category.

4. What is meant by the term "all-or-none response" in muscular contraction?

5. Identify the kind(s) of stimulus and the kind(s) of receptors involved in each of the five senses.

6. Trace the route of the impulse in the knee-jerk reflex from stimulus to response.

7. Name and explain the difference between the two parts of the autonomic nervous system.

REFERENCES

Amoore, J. E., J. W. Johnston, and M. Rubin. 1964 (February). "The Stereochemical Theory of Odor." *Scientific American Offprint* no. 297. Freeman, San Francisco.

Cohen, C. 1975 (November). "The Protein Switch of Muscle Contraction." *Scientific American* Offprint no. 1329. Freeman, San Francisco.

Eccles, J. 1965 (January). "The Synapse." *Scientific American* Offprint no. 1001. Freeman, San Francisco.

Evarts, E. V. 1973 (July). "Brain Mechanisms in Movement." *Scientific American* Offprint no. 1277. Freeman, San Francisco.

Guillemin, R., and R. Burgus. 1972 (November). "The Hormones of the Hypothalamus." *Scientific American* Offprint no. 1260. Freeman, San Francisco.

Huxley, H. E. 1965 (December). "The Mechanism of Muscular Contraction." *Scientific American* Offprint no. 1026. Freeman, San Francisco.

Jacobson, M., and R. K. Hunt. 1973 (February). "The Origins of Nerve-Cell Specificity." *Scientific American* Offprint no. 1265. Freeman, San Francisco.

Llinas, R. R. 1975 (January). "The Cortex of the Cerebellum." *Scientific American* Offprint no. 1312. Freeman, San Francisco.

MacNichol, E. G., Jr. 1964 (December). "Three-Pigment Color Vision." *Scientific American* Offprint no. 197. Freeman, San Francisco.

Melzak, R. 1961 (February). "The Perception of Pain." *Scientific American* Offprint no. 457. Freeman, San Francisco.

Murray, J. M., and A. Weber. "The Cooperative Action of Muscle Proteins." *Scientific American* Offprint no. 1290. Freeman, San Francisco.

Sperry, R. W. 1964 (January). "The Great Cerebral Commisure." *Scientific American* Offprint no. 174. Freeman, San Francisco.

SUGGESTIONS FOR FURTHER READING

Axelrod, J. 1974 (June). "Neurotransmitters." *Scientific American* Offprint no. 1297. Freeman, San Francisco.

Communication of nerve cells by means of chemical messengers released from nerve-fiber endings.

Dowling, J. E. 1966 (October). "Night Blindness." *Scientific American* Offprint no. 1053. Freeman, San Francisco.

Report on experiments with rats that clarify the relationship of vitamin A and vision in dim light.

Eccles, J. C. 1958 (September). "The Physiology of Imagination." *Scientific American* Offprint no. 65. Freeman, San Francisco.

Demonstrates that patterns of electrical waves in the brain correspond to creativity.

Girdano, D. D., and D. A. Girdano. 1975. *Drugs: A Factual Account*, 2nd edition. Addison-Wesley, Reading, Mass.

A thorough examination of drug use and abuse, including the historical, social, and legal impact of drugs in our society, as well as the physiology and pharmacology of drugs.

Hess, E. H. 1965 (April). "Attitude and Pupil Size." *Scientific American* Offprint no. 493. Freeman, San Francisco.

Interest and emotion affect pupil dilation as well as the brightness of light.

Katz, B. 1961 (September). "How Cells Communicate." *Scientific American* Offprint no. 98. Freeman, San Francisco.

Coordination of the activities of cells in multicellular animals by chemical messengers.

Olds, J. 1956 (October). "Pleasure Centers in the Brain." *Scientific American* Offprint no. 30. Freeman, San Francisco.

Experimentation on rats showing that electric currents in certain sections of the brain can satisfy basic drives.

Schneider, D. 1974 (July). "The Sex-Attractant Receptor of Moths." *Scientific American* Offprint no. 1299. Freeman, San Francisco.

A report on the extremely delicate receptor system of the male silk moth.

BEHAVIORAL PATTERNS

chapter eleven

A male baboon abandons himself to sensual pleasure as a dutiful female removes insects and dirt from his coat. Each individual baboon's social status is reflected by the amount of grooming he receives, but all members of a troop give and receive some grooming attention as a reinforcement of social bonds.

In 1973, the Nobel Prize Committee, acknowledging for the first time the major research advances in our understanding of behavior and sociobiology, awarded the Nobel prize to three leaders in the field. These three men—Konrad Lorenz, Nikolaas Tinbergen, and Karl von Frisch—had explained how elaborate patterns of behavioral responses to the environment reflect both learned and innate (genetically programmed) responses. They showed that organisms possess remarkably complex, subtle, versatile patterns of communication that work via the nervous system to control behavior. Furthermore, they showed that such natural phenomena have great adaptive biological significance and reflect a pattern of phylogenetic history, or evolution.

All these men, but Lorenz in particular, made their initial observations outside the laboratory, and their original insights came from detailed observation of organisms' responses in their natural habitat. Instead of viewing behavior narrowly as an input-output mechanism, they developed a new science, in which the relationship between an animal's makeup, its experience, and its early development all play a major role in the making of the animal's behavioral repertoire. This science is called **ethology.**

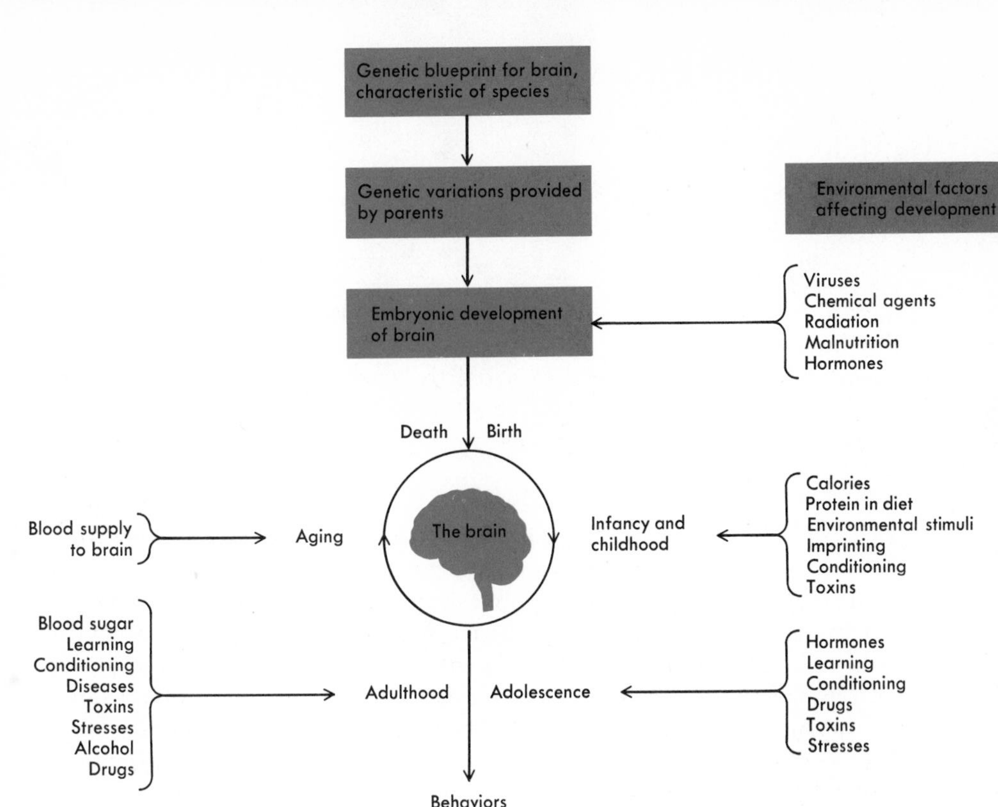

FIGURE 11.1

A schematic overview of some of the major factors involved in behavior. The general character and function of the brain are determined by genetic makeup and embryonic development. Some behavior is established genetically with the development of neural circuits. Other behavior is in part genetically determined but requires some form of learning during a critical period in the organism's development. Finally, some forms of behavior are continually modified by learning and reflect the adaptability of the nervous system.

Innate behavior

All behavior may be classified generally as either innate or acquired (learned), although many behavior patterns may result from a combination of both innate and acquired characteristics. Innate behavior is inherited and is necessary for survival. It is a stereotyped behavior shared among all members of a species (or for reproductive behaviors, among all members of a sex in a species). Innate behavior is demonstrated even by those animals that were isolated from birth from their own kind and therefore could not have learned the behavior from observation. Moreover, innate behavior is instinctive and occurs automatically in response to certain stimuli. Innate animal behaviors involve approach-avoidance behaviors (taxes), reflex behavior, fixed action patterns, and releasers.

APPROACH-AVOIDANCE BEHAVIORS

Like the tropic response in plants, **taxes** consist of movements directed either toward or away from the direction from which the stimulus is received. The types of tactic responses are **phototaxes, chemotaxes,** and **thermotaxes**—responses to light, chemicals, and temperature.

The paramecium that encounters a noxious chemical agent first swims backward; it reverses ciliary movement to drive itself away from the stimulus and then turns to one side and swims forward in a new direction (see Fig. 11.2). This type of response, generally termed an **avoiding reaction,** is specifically an example of negative chemotaxis.

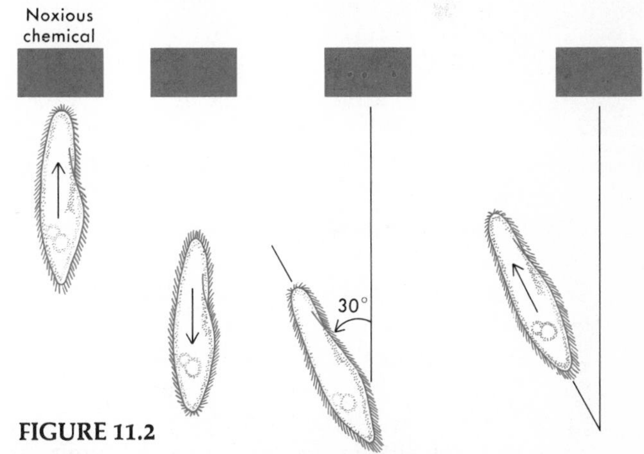

FIGURE 11.2

Avoidance behavior. When confronted with a noxious chemical, a paramecium exhibits an avoidance reaction. Reversing the direction of ciliary action, the paramecium backs up and then moves forward again at an angle. It repeats this pattern until it has successfully avoided the noxious substance.

A single stimulus may elicit both positive and negative taxes, depending on its intensity. The *Euglena*, a unicellular organism containing the green pigment chlorophyll, exhibits positive phototaxis by swimming toward light of moderate intensity, although very strong light elicits a negative phototactic response.

REFLEX BEHAVIOR

More complex than the taxes are those innate responses that are called reflexes. The reflex arc in vertebrates consists of a fixed circuit involving sensors, sensory neurons, motor neurons and effectors. These relatively simple neural connections are literally obligate command circuits, which always produce a standard response if activated. However, if the organism has a brain, signals from higher centers superimpose their influence over these simple reflex arcs, and responses may be modified. Reflex behaviors are purposeful and often have to do with avoidance or withdrawal from some noxious stimulus.

FIXED ACTION PATTERNS

Whereas simple reflexes usually involve only a few neurons, there are other behavior patterns that involve the entire organism (brain, muscles, glands) and therefore produce behaviors that are difficult to analyze. These so-called **fixed action patterns,** formerly called instincts, are often said to be internally triggered by some hormone (e.g., the sex drive) or by receptors signaling an internal change (e.g., thirst or hunger).

Fixed action patterns are characterized by several distinct features. They are usually inherited or genetically determined responses that occur in the same way in all members of a species, including those that have been raised in isolation. They generally consist of a constellation of responses involving greater complexity and coordination than simple reflexes. Often these fixed action patterns are elicited by one or more environmental stimuli. Despite these criteria, the study of "instinctive" behavior has always posed a delicate problem to biologists, psychologists, and other contenders in the so-called nature-nurture controversy. Because the role of environment and experience is sometimes unclear, scientists now tend to avoid the term instinct altogether, preferring to talk about species-specific behaviors, or fixed action patterns.

Releasers Stereotyped behaviors are usually initiated by external environmental stimuli—signals that evoke complex behavior patterns. Such environmental signals are called **innate releasing factors,** or **releasers.** These releasers may consist of physical features of the environment, such as the intensity of light or water temperature, or they may involve some feature of another animal, such as color, shape, odor, movement, or sound.

In animals, innate releasing factors and fixed action patterns (responses) are most often seen in intraspecific communication (between members of the same species), and they can best be demonstrated in the highly stereotyped behavior of the three-spined stickleback, a small fish studied by Nikolaas Tinbergen in research that culminated in a Nobel prize award. As an ethologist, Tinbergen first observed three-spined sticklebacks in their natural habitat, recording in detail all aspects of their mating behavior, courtship, territorial defense, and aggressiveness. Analyzing these observations, the scientist shrewdly guessed and then demonstrated that a male stickleback, with his characteristic red belly, would attack any red-bellied male in his territory and would even attack a self-image in a mirror. This led to the idea that the red belly, when viewed by a male, released a fixed action pattern—attack! Tinbergen subsequently tested his hypothesis by means of plastic models with and without red bellies, and the resident male stickleback attacked only the red-bellied model. On the other hand, a round, distended belly—such as that of an egg-laden female stickleback—evoked a totally different response: an elaborate courtship behavior that triggered in the female behavioral circuits that caused her to follow the male to the nest he had built and to enter it. Her tail protruding from the nest acted as another innate releasing factor, inducing the male to direct rhythmic thrusts at the female's tail, massage her abdomen, and thus prod her into laying her eggs (Fig. 11.3).

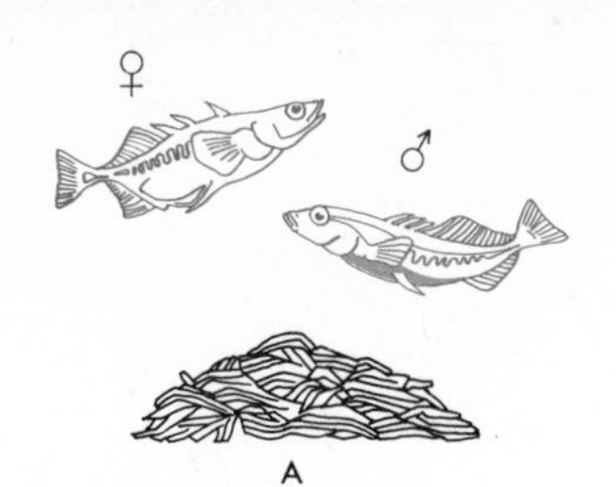

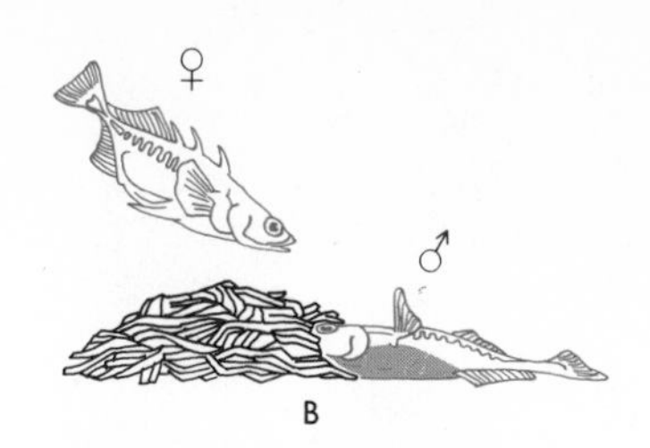

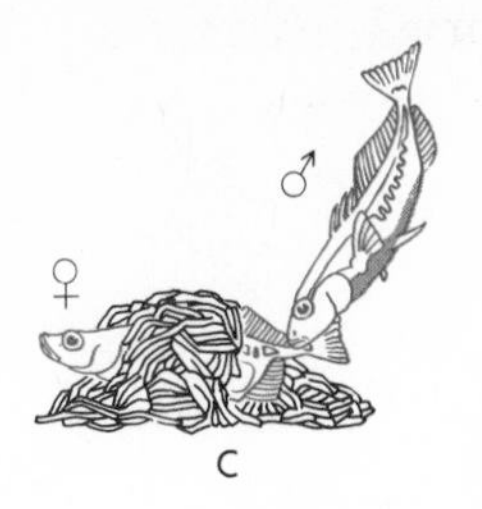

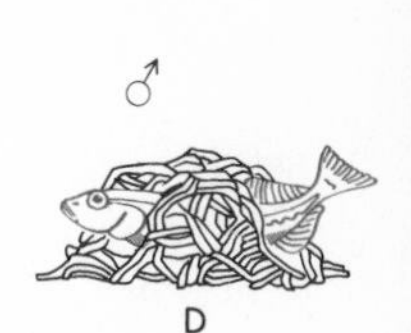

FIGURE 11.3

Successful completion of mating by the stickleback. (A) The male incites the female in ritualized courtship display of his red belly, which acts as a releaser of behavior from the female ready to lay eggs. (B) The male indicates his nest and lies on his side at its entrance, as inciting ritual continues. (C) After the female enters the nest, her projecting tail stimulates the male to prod her with his snout. That behavior in turn triggers the female to lay her eggs. (D) The male enters the nest after the female leaves. He fertilizes the eggs and remains to fan the nest with his fins and guard it from predators.

Other innate releasing factors have been described for other species. For example, a baby gull pecks at the red spot on its parent's beak, and this releases parental feeding behavior—the regurgitating of food into the young one's mouth. Sometimes releasers serve to inhibit normal behavior. The cheeping of the baby turkey inhibits aggressive pecking behavior of the female turkey. If her ears are stopped up, the mother will peck her progeny to death.

Many behavioral patterns are obligate, or command, stereotypical responses to the innate releasing factor. Such behaviors are seen primarily in organisms low on the phylogenetic ladder and not seen clearly in man. Human behavior instead is far more complex, and it is largely dependent on learning. Nonetheless, we do seem to retain some obligate—though usually well-masked—innate behaviors.

Learned behaviors

Learned behaviors may include the simplest modifications of innate behavior, as well as the complexities of human reasoning. There are, of course, various types of learning that occur at different levels of complexity. They involve phenomena called imprinting, habituation, conditioning, and trial and error, as well as higher forms of learning, such as concept formation and reasoning.

IMPRINTING AND EARLY ENVIRONMENTAL LEARNING

In many of the lower species, as well as in man, the experiences of the newborn during the earliest period of its life are often critical determinants of subsequent adult behavior. The timing of such experiences is critical as it determines what the effects on the organism will be. Some of these experiences are very subtle, and certainly in man they are not perceived consciously at all.

For an animal to behave in a certain way in response to a particular stimulus—for the behavior to be "wired in" so that it occurs automatically whenever the stimulus is confronted—the response to the stimulus must be learned within a relatively brief, critical period of time early in the animal's life. Learning of this type is called **imprinting,** and it results in the rapid development of a strong, stable preference for some type of stimulus early in life. It is a form of learning that establishes relatively permanent neural circuits in the brain, and these circuits control some subsequent behaviors (see Fig. 11.4). The behavior of a gosling or a duckling in relation to its mother illustrates the phenomenon

FIGURE 11.4

The effect of imprinting. During a critical period shortly after hatching, ducklings form a strong attachment to the first moving object they see and follow it everywhere. Under normal circumstances, as you see here, that object is the mother. The survival value of imprinting is tremendous because it ensures that the young will stay within the protective care of the mother.

rather well. The newly hatched bird follows its mother. However, a gosling hatched from an incubated egg and removed from the incubator into the care of a human will follow the person as if he or she were the actual mother. It is known that the critical period for imprinting this following pattern in chicks and ducklings is approximately 36 hours from the time of hatching. After this period, this particular behavior can never be imprinted. The implication is that if a newly hatched chick is given a surrogate mother during the critical period, it will not imprint successfully as a member of its species. Usually, of course, it does. The types of stimuli involved in the process of imprinting are mostly visual, chemical, and auditory. The baby chick gets its mother "wired in" by seeing her, by sensing her characteristic odor, and by hearing her walk around.

Early deprivation The vital importance of appropriate learning experiences early in the animal's life is quite apparent from studies performed on how the animal is affected when deprived of these experiences. In these studies the early environment of the animal is limited, and so are its experiences. The behavior of the deprived animal is then compared with the behavior of a control group raised under normal and satisfactory conditions. The studies reveal, for example, that the quantity and nature of the food provided the youngster during this early period has a decided effect on its future behavior. When deprived of food during infancy, the adult animal tends to hoard more food than its counterparts in the control group. Thus deprived rats tend to hoard more food pellets as adults than do control animals fed consistently. Moreover, it has been suggested that protein deprivation coupled with cultural deprivation can affect humans during the first four years of life to prevent full functioning of the brain.

In addition to food as an environmental component, exposure to other animals seems to be critical. Early isolation quite markedly affects later social behavior. Harlow has found that monkeys who do not receive normal mothering by a member of their species do not demonstrate normal sexual or parental behavior as adults. Similarly, chicks raised in isolation tend to isolate themselves when introduced into the flock, and they do not respond to the food calls of the mother hen.

Extra experience Another method for dramatizing the importance of the early environment is to enhance it—enrich it so that it is better than normal. Studies indicate that a rat that is placed in a stimulus-rich environment is more curious and more adept at maze learning than a rat raised in an unstimulating environment. Moreover, the favored rat develops more neural connections and a larger brain.

HABITUATION

The tendency of an animal to learn *not* to respond to certain stimuli that originally evoked a response is defined as **habituation.** Indeed, most neurons in the brain and spinal cord habituate to—or 'turn off"—when exposed to any monotonous, repetitive signal. For example, ducklings freeze motionless at the sudden appearance of any object overhead, but they gradually become accustomed to objects seen frequently, such as falling leaves and friendly birds. However, ducklings never become habituated to the relatively infrequent appearance of a hawk.

CONDITIONED LEARNING

The conditioned reflex is in large part a form of approach-avoidance learning. In the conditioning process, an innate reflex, one normally elicited by an associated stimulus, is evoked by a previously neutral stimulus. The neutral stimulus calls forth the reflex simply because the neutral stimulus has become associated with the original stimulus that *normally* evoked the reflex.

I. P. Pavlov, the famous Russian physiologist, first described the conditioned reflex in a classic experiment. He repeatedly blew meat powder into the mouth of a dog that had been restrained in a harness, and recorded the amount of salivation—a reflex that occurs naturally whenever food is taken into the mouth. After a while, he began to ring a bell just prior to placing the food in the dog's mouth, repeating this operation at successive

Sociobiology: Extrapolating from Animal to Human Behavior

One of the most recent developments in the scientific community is sociobiology—the study of the biological basis of social behavior in all organisms from insects to fish to baboons to man. It involves information gathered not only from biological areas, such as genetics and ecology, but from psychology and anthropology as well.

Among the many hotly debated questions involving the value and possible political implications of sociobiology is one we will examine here: Can information on animal and insect behavior be validly applied and projected to the behavior of human beings?

The "peck order" shown in certain animal communities is based on the "rights" of certain individuals to be aggressive toward others. The dominance hierarchy of a baboon troop, for example, usually has a dominant male and three or so subordinates who comprise a sort of central governing body within the troop. Outside this central governing body, the dominance hierarchy continues further, with young males approaching prime usually standing higher in the order than females and aging males. The females, in turn, stand higher than the young juveniles and infants.

It would seem that a social structure built upon the absolute right of aggressive dominance over subordinates could hardly be a peaceful one. Yet for animal societies based on peck orders, peace generally reigns. Furthermore, it would seem obvious that the overthrow of a superior by a subordinate could occur only by violent means. Yet this, too, proves not to be the case. Rarely, for example, is a dominant male deposed by a major battle; instead, he falls from power gradually as he ages and fails to meet routine challenges and tests of his ability to carry out his responsibilities. Studies of baboon troops over a period of several years reveal a gradually changing structure, with dominant males being replaced slowly by younger animals who are later replaced by those younger still. The result is a strong and continuing social order allowing high efficiency in ensuring adequate food for all and maximum protection for the weaker.

However, one might observe a baboon troop for a long time without detecting the peck-order framework on which it is constructed. The reason for this is that in nature the animals forage widely for food and there is generally enough for all. Thus the competitive situations necessary to demonstrate dominance and subordination do not occur often. Rarely, perhaps, only one or two females may be in estrus (i.e., receptive for mating), in which case the males may compete for her. Or an ethologist, wishing to study the peck order, may feed the troop by scattering the grain in a small area or, perhaps, by throwing it to only one animal. Under these artificially imposed conditions the troop must condense into a smaller area, and chances for contact between individual animals are greatly enhanced. Though mere threats are usually enough to prevent large-scale violence, fights under such conditions can and do occur.

From Jeffrey J. W. Baker and Garland E. Allen, *The Study of Biology*, 3rd edition (Reading, Mass.: Addison-Wesley, 1977), pp. 1115–1118.

Thus population density, the number of individual organisms per unit of space, definitely affects behavior. Precisely *how* it does so has recently been the object of much research, research involving many different species of organisms. With overpopulation clearly the most serious problem facing mankind, the findings of this research have obvious significance.

One of the earliest studies of this type was carried out by John B. Calhoun of Rockefeller University in New York. Calhoun first worked outdoors with the wild Norway rat. He confined a population in a quarter-acre enclosure, where the animals had no escape from the behavioral consequences of population density increase. The animals were spared, however, the normally accompanying overpopulation conditions of starvation and disease, for they were given an abundance of food and places to nest, in addition to being protected from predation. Under these "ideal" conditions, the observed reproductive rate predicted an adult population of 5000 individuals. Actually, it numbered 150! Investigation revealed an extremely high infant mortality rate; even with only 150 adults in the enclosure, the stress of enforced social interaction led to disruption of maternal behavior. The result was death for most of the young.

Calhoun later moved his study inside and performed more sophisticated investigations dealing with the effects of population density on behavior. Again infant mortality climbed, with levels of from 80 to 90% not uncommon. Females stopped building nests for their young; eating habits were changed; normally peaceful reproductive behavior patterns were disrupted and fights often resulted. Sexual deviations became common, as did cannibalism, total social disorientation, and high adult mortality rates. A startling number of these deaths occurred in females as a result of either pregnancy or giving birth.

Results such as these are troublesome, for they seem uncomfortably close to the human situation today. In heavily overpopulated Latin America, for example, despite the fact that most such deaths undoubtedly go unreported, pregnancy and childbirth deaths are high. Chile, for example, has 271.9 deaths per 100,000 births, as opposed to Sweden's 11.3. . . .

But how valid *are* extrapolations of such results to the human situation? It is difficult to judge. The results of Calhoun's studies concerning the effects of high population density on behavior are admirably supported by studies of other species both higher and lower on the evolutionary scale than the Norway rat. Most important, they seem to hold true for primate societies. When primate species normally peaceful in the wild are kept under high-population-density conditions in a zoo, violence leading to maiming and death becomes quite common. Under such conditions, animals of all ranks in the peck order are thrown into close proximity, and there is no escape. Subordinates direct their aggression toward animals of the next lower peck-order status, and a chain reaction of violence results. But it is erroneous to ascribe the *cause* of such behavior to density. Confinement in artificial surroundings would be an equally, or more important, variable. The problem is that there has been relatively little direct and systematic observation of animals under conditions of great density in the wild. Thus problems exist at the simplest level of observation in attempting to determine the true effects of density on animal behavior.

Again, the sticky question: Can we extrapolate to humans? . . . Is there a relationship between the abuse of infant monkeys by their mothers when kept in crowded conditions and the increase in the "battered baby syndrome" (child beating and torture) reported in areas of high human population density?

It can be truthfully said that we simply do not know. After all, Calhoun's rats, though overcrowded, were relatively "wealthy," being kept well supplied with food and water. In humans overcrowding tends to be associated with poverty and poor education. If nothing else, the example points out neatly how fallible simple-appearing cause-effect relationships may often be. Thus some scientists, such as molecular geneticist and Nobel laureate Joshua Lederberg, urge more caution in extrapolation from crowding experiments in lower animals to humans. Others feel that limiting density by limiting population growth is absolutely essential to prevent war, poverty, and the social pathology that seems to be associated with overcrowding.

intervals. Eventually, he rang the bell without following with the powder. The dog salivated anyway. In Pavlovian terms, the food is an **unconditioned stimulus,** one which naturally elicits the response. The bell is a **conditioned stimulus,** a neutral stimulus that elicits the response only after a process of conditioning.

More recent studies in conditioned learning, such as those by B. F. Skinner, have shown that by using automated cages that pair a reward (e.g., a food pellet) with the performance of some act, one can evoke, hasten, and prolong learning. This type of learning is called **operant conditioning.** In all operant conditioning, the reward **reinforces** a specific behavior and causes the animal to learn to perform it again. However, with the passage of time, learned or conditioned reflexes, if not reinforced, will fade away and be forgotten. This effect is known as **extinction.** Reinforcement of conditioned learning can also occur by punishment for failure to perform. In general, however, painful or unpleasant stimuli bring about conditioned avoidance behavior.

Much of human learning, intellectual and practical, is conditioned learning. Some recent researchers have shown that we can condition even many visceral responses, such as changes in blood pressure or heart rate, through a variety of conditioning techniques.

TRIAL-AND-ERROR LEARNING

Perhaps the oldest form of learning is that acquired from trial and error. Laboratory experiments have made possible the controlled study of this form of learning, but it also occurs in the natural world, of course, as well as in the laboratory. A blue jay may learn that eating a monarch butterfly (full of milkweed juices) will induce vomiting in a short period of time. The jay soon makes the association of the particular pattern and colors of the monarch butterfly with the sick reaction, thus learning to avoid that source of unpleasantness. For man, trial-and-error learning is a basic tool in survival. Often it takes no more than one trial for us to learn to avoid a particularly painful situation. And conversely, pleasurable rewards can also result in learning by trial and error.

HIGHER LEARNING

Maze learning More complex decision making is seen in maze learning, a kind of trial-and-error learning that is often used to study learning in the laboratory.

In maze learning, an animal is taught to proceed directly to a food-laden goal area that is not at first visible. The animal must learn to make a series of discriminations at various choice points within the maze, thereby wending its way through to the goal. The simplest mazes are *T*- or *Y*-shaped, involving only one choice point. Increased complexity can be obtained by increasing the number of choice points.

Watching an earthworm learn to solve a simple T-maze, where reward consists of being returned to the home container and punishment involves a mild electric shock, almost never ceases to amaze. Among the invertebrates, however, the *Formica* ant is the champion maze runner. It is capable of learning within only 25 trials how to traverse without error mazes consisting of as many as six blind alleys. Among the vertebrates, only mammals behave in a manner that seems to suggest some form of primitive reasoning. They appear to make mental trials—pausing at each choice point to ponder the alternative responses.

Insight learning and concept formation Much of higher learning involves a more calculated approach to the solution of a problem than random trial and error. Insight is based on the extrapolation of past learning to a new and immediate situation. One of the simplest techniques for determining whether or not an animal displays insight learning is posing a barrier or some other device between an animal and some food that it can smell or see. The animal cannot approach the food directly but must somehow get around the barrier. Animals who can figure out how to circumvent the barrier are said to employ insight. Chimpanzees, for example, will stack boxes directly under an out-of-reach banana and then climb up to get it. This use of tools by an animal (see also Fig. 11.5) illustrates insight learning. Laboratory-reared rats and dogs also appear to have a capacity for some insight learning. Man, of course, is

FIGURE 11.5

Insight learning and concept formation produce complex behavior like the use of crude tools. Fifi, a five-year-old chimpanzee, strips down a blade of grass and inserts it into a termite mound to fish for insects. A few species of birds are known to use tools in a similar manner.

best at problem solving requiring insight and abstraction and the formation of concepts. Concept learning occurs, for example, when a child—or a chimpanzee—learns to group similar items together. If the child can identify a poodle and a Saint Bernard as members of a common species, we say he has mastered the concept "dog." Insight and concept learning may be regarded as the establishment of new neural circuits that produce more complex behavior.

CURIOSITY AND LEARNING

Reward, punishment, pleasure, and satiation motivate animals to a given behavior following a given experience, but as one goes up the evolutionary ladder, one finds that curiosity and manipulative behavior are programmed into many animals. The curiosity drive has been measured experimentally in monkeys, who will do work and even tolerate shocks in order to open a peephole that will permit them to see a new object. Curiosity enhances the chance of trial-and-error learning. In fact, emotional or neurotic animals are often much less curious than normal ones, showing far less exploratory behavior. Curious behavior ensures that animals expand into new environments. Monkeys will also show a manipulative behavior with their hands for no other reward than the pleasure of latching and unlatching hooks.

DISPLACEMENT BEHAVIORS

Animals frustrated by their environment will often present some bizarre behavior pattern totally alien to the behavior usually evoked by a sensory input. For example, a rat that will normally not attack a mouse may do so when bullied by a larger and more aggressive rat. The attack on the mouse constitutes **displacement behavior.** Although many displacement behaviors may seem to have no real constructive function, they help to alleviate tension and frustration.

Animal communication and social behavior

Intraspecific communication among animals involves the generation by one individual of a given species of chemical or physical signals that are detectable by the sense organs of other animals of the same species. These stimuli in turn convey a very specific type of information to the recipient of the signal. Most animals engage in some form of communication, often in conjunction with reproductive behavior, care of offspring, navigation, or migration. Reception of communication signals is accomplished through the tactile, chemical, optical, and acoustical systems.

TACTILE SYSTEMS

Animals engaged in tactile communication must be in physical contact or within very close range of each other. Tactile signals are usually transmitted through specific

FIGURE 11.6

Communication behavior. Animals communicate through a complex system of body language, in which subtle postures and movements convey meaning. Here a dominant wolf, standing over a low-ranking pack member, indicates aggression in upright ears and an intense, direct stare. The low-ranking wolf inhibits that aggression by assuming a submissive posture—rolling on its back, flattening its ears, and pulling its lips back. The wolf standing to the side is ambivalent. It holds its ground, but anxiety draws ears and mouth down and back, and it avoids direct eye contact.

bodily interaction, such as one animal's grasping or simply jostling another. This type of communication sometimes finds expression in ritualized fighting—two animals engaging in what seems to be mortal combat although the animals seldom injure each other. Work with primates has shown that cuddling, stroking, and caressing all constitute important tactile signals that affect both behavioral development and immediate responses to other individuals in the species.

CHEMICAL SYSTEMS

Chemical signaling is the most widespread form of communication found among animal species, a fact probably related to a chemical system's inherent capacity for acting across great distances. Chemical substances employed specifically for social communication, called **pheromones,** are secreted in minute quantities by one animal to influence in a specific manner the behavior of another animal of the same species—often at a distance.

Pheromones released by certain species of ants indicate sources of food. On its trek back from a feeding site, a worker fire ant secretes a pheromone that marks the trail. As other workers locate the source, they too secrete the pheromone along the same trail, thus reinforcing it. When the food source is exhausted, however, returning workers no longer mark the trail, and since the volatile pheromone's scent disappears quickly, the marked trail also disappears. Higher organisms, including the primates, are known to rely on unconscious responses to chemicals in the air. Female monkeys (and humans) emit a pheromone sex attractant called copulin, which evokes sexual arousal in the male monkey.

Chemical signals may also convey alarm. For example, when a certain species of minnow is wounded, it releases a substance that other members of its school sense as a warning to remove themselves from possible danger. Many animals also secrete repellents that protect themselves or their nests from predators.

Despite a number of suggestive studies on the role of pheromones in human behavior, that role is probably minor. However, one provocative finding is that the adult woman's ability to detect the smell of certain synthetic musks depends on the presence and concentration of the female hormone estrogen in her bloodstream. Women who have had their ovaries removed cannot smell musk until it is present in a concentration that is about a thousand times the level that their hormone-laden sisters can detect.

OPTICAL SYSTEMS

Animals having well-developed optical organs usually use optical signals that still serve quite well in some circumstances although they are often not effective over extensive distances. For example, optical signals are fre-

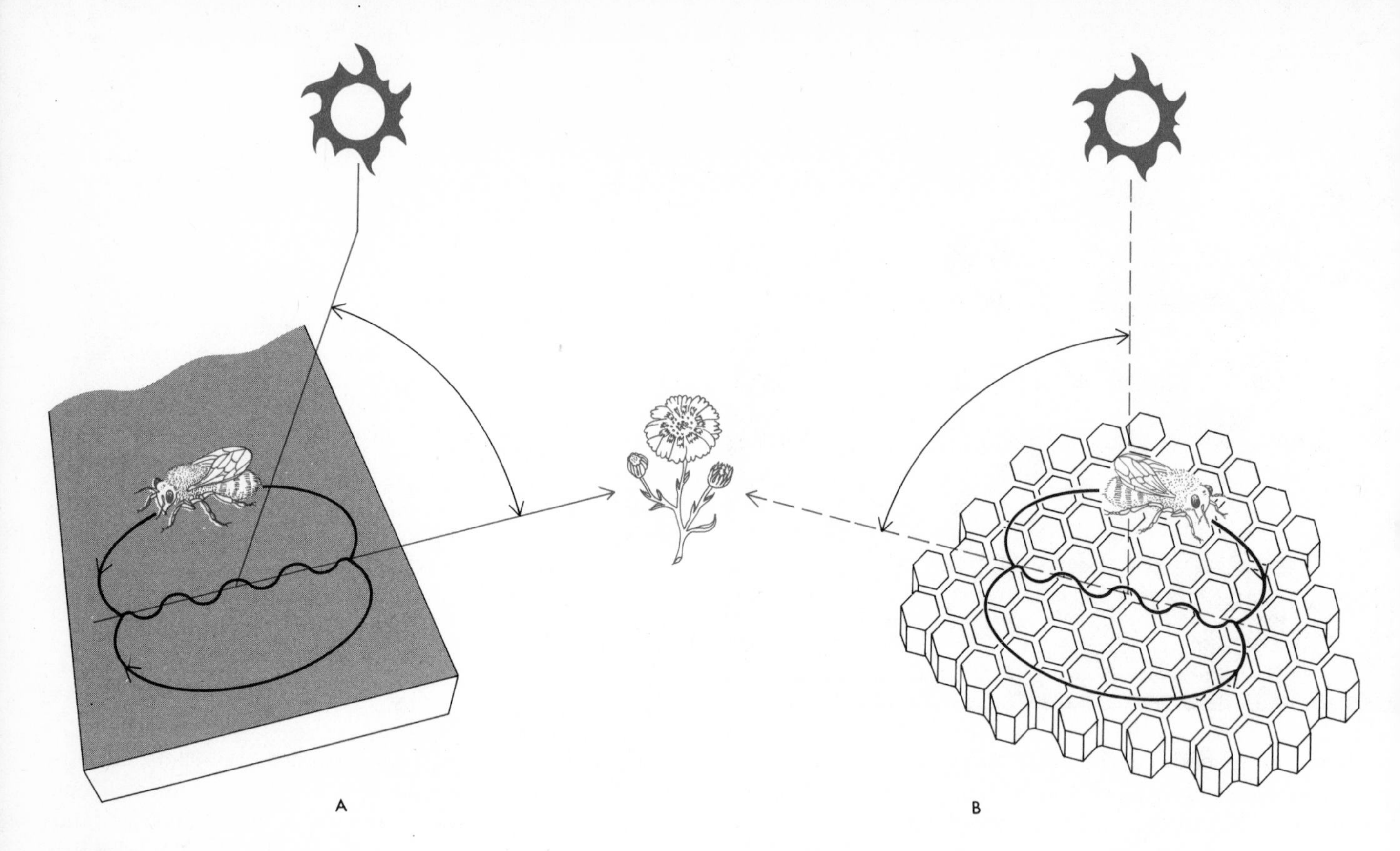

FIGURE 11.7

The tail-wagging dance of the honeybees. To indicate to fellow workers the whereabouts of a food source, the honeybee performs a tail-wagging dance that locates the food in terms of the direction of the sun. The pattern of the dance is a flattened figure eight with a straight section between the two loops. If the bee performs the dance outside the hive, the straight section points directly at the food source. If the bee performs inside the hive, it orients its dance by gravity. The vertical represents the direction of the sun, and the angle between the vertical and the dance's straight section indicates the angle between the sun and the food source. The relative speed of the dance indicates the distance of the food.

quently used as a means for species recognition. Fish swim in schools and remain together by visually identifying the specific physical patterns of other members. Many kinds of herds and flocks also remain intact through the mutual visual recognition of their members.

Visual signals are used to communicate the direction and sometimes the distance of food sources. The intricate dance of the honey bee is an example of this kind of communication (Fig. 11.7). Finally, visual cues are utilized by many species to signal alarm. These signals usually involve the flashing of some brightly colored part of the body. Birds, for example, may signal with tail or wing feathers, mammals with fur or hair. Many species of deer have a distinctive white patch of fur beneath the tail, which they flash as they flee from danger.

ACOUSTICAL SYSTEMS

A form of communication that is effective over relatively great distances is sound, which animals may produce by voice, chewing, walking, and flying. Organisms employ sounds as signals of alarm and also as a means of communicating their positions.

Some birds employ different sounds to indicate the type of predator they perceive. Others issue alarms that even indicate the proximity of the predator. For example, as danger draws near, the catbird's alarm changes from a comparatively brief note to what sounds like a long meow. Some animals—chiefly those that are social in their habits—have an all-clear signal that alerts companions to the departure of the enemy.

The uncanny ability of bats to fly safely in absolute darkness, even when blinded, can be attributed to a special type of acoustical system known as **echolocation.** The bat's larynx produces pulses of very high-frequency sound. When these pulses (sound waves) strike an object, they are reflected back to the bat as echoes, which are picked up by the bat's very large ears. The amplified

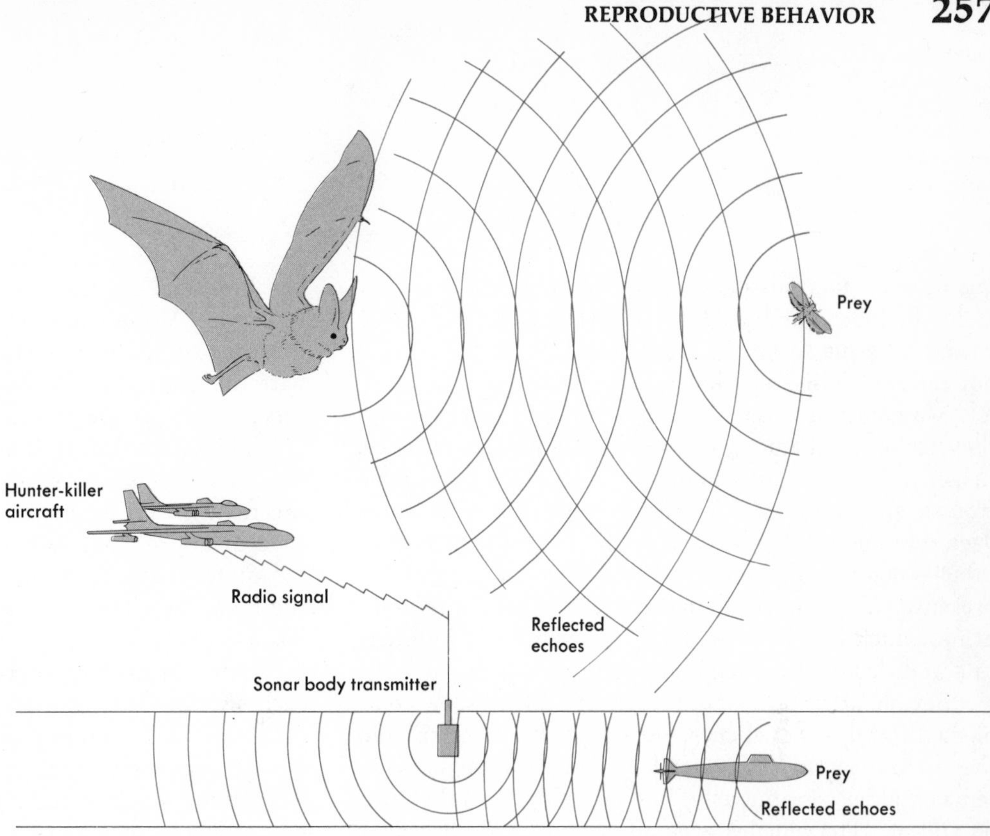

sensory information relayed to its brain enables the bat to alter its course to avoid an obstacle—or to zero in on its next meal (Fig. 11.8). Modern submarine-hunting sonar systems are of course echolocation systems that operate on essentially the same principles.

Acoustical systems are also used to keep herds and flocks together. Canadian geese produce a loud honking cry to remain together in the air. Porpoises use a variety of clicks and whistles at frequencies above the limits of human hearing to convey much information on the position or presence of predators.

One problem with auditory cues is the similarities in sound. Nonetheless, as many as twenty species of crickets may be singing a mating song in a given locale, yet the female always finds a mate of the right species. Electronic analysis of male cricket calls shows a distinctive pulse rate for each species.

Reproductive behavior

In reproductive behavior, the processes of communication reach an apex of sophistication. Indeed, it is this behavior on which survival of the species depends. The complexities of communication in relation to reproduction are understandable, because the process in each species involves a male and a female locating each other at a specific time, recognizing each other as potential mates, and effecting the union of their gametes.

FIGURE 11.8

Echolocation. The bat has uncanny ability to fly in absolute darkness and to obtain remarkably efficient results in hunting nocturnal insects. It catches 40 to 60 percent of those it attacks. Man has attempted to copy the echolocation system of the bat with a sophisticated electronic system for hunting submarines. When the sub is located, the hunting plane or ship can launch a sound-homing torpedo.

SEXUAL ATTRACTION

The varieties of communication employed for the purpose of sexual attraction include all the types of signals discussed so far—chemical, tactile, visual, and acoustical.

Chemical cues are used by many animals and are found in nearly all insect species. In most cases, it is the male that is attracted by a chemical releaser produced by the female. The female gypsy moth, a well-known example, releases a sex attractant from scent glands near her reproductive organs. An infinitesimal quantity of this substance (10 micrograms) can draw males in huge numbers from a distance of over 4,000 yards. It seems to be the signal (that is, the odor) and not the female herself that lures.

Mammals use scents to attract the opposite sex, and they are unusual among the vertebrates for this practice. In general, sexual attractants are secreted by the male from special scent glands. For example, the male musk deer rubs his hind legs—where the scent glands are located—against trees to alert females in the area of his presence. Cats and dogs emit sex attractants with their urine, which is then sniffed with great interest by other cats and dogs in the neighborhood.

Sexual attraction and recognition are also accomplished through visual cues. During the breeding season, the marine worm *Odontosyllis* surfaces at night. The females glow continuously, but the males literally turn on and off. The females generally arrive at the breeding site first, and the glow of their large gathering attracts the males. Visual cues may also take the form of bright plumage or, as in many species of the butterfly, bright patterns on the male insect's wings. Such colorations serve to attract the female of the species.

Even acoustical cues are employed as sexual attractants. The male *Aedes aegypti* mosquito is attracted to the female by the sound of her wings during flight—a tone of approximately 500 cycles per second. That the mosquito is attracted only by the tone is illustrated by the fact that a tuning fork emitting the same frequency has a similar effect. Auditory cues are also used by male birds, which emit a song-like signal that is unique to each species. The male alligator or crocodile attracts his mate with a resounding roar. The frog calls forth with a croaking noise. Among such mammals as cats and dogs, however, it is the female that emits auditory signals to the male.

FIGURE 11.9

Mating behavior. During estrus, the urine of the lioness has a characteristic odor that attracts every male in the vicinity. She may act stand-offish or blatantly seductive. Her lovestruck mate accompanies her everywhere and will mate with her every 20 to 30 minutes for hours at a time. In general, their courtship is affectionate, but when the lioness's estrus ends—so does the romance.

COURTSHIP AND MATING

Courtship rituals—elaborate forms of communication—are performed by the males of many species. Their purpose is to arouse the female's mating instinct and thereby induce the female to copulate. Courtship rituals may involve any or all of the cues—auditory, tactile, visual, and chemical.

Tactile cues The function of tactile stimulation in sexual arousal is well illustrated by the mating behavior of the male octopus. First he caresses the female with one of his sucker-clad arms and, when she becomes receptive, reaches into his body cavity with a modified arm and removes the spermatophore, a package that contains his sperm. He then deposits the spermatophore in the female's body cavity at a location near her oviducts.

Tactile cues are also an important component of the sexual behavior of many mammals, involving direct stimulation of the genital areas for arousal. For example, rabbits ovulate in response to tactile stimulation associated with copulation (and judging from the number of rabbits, it's a pretty efficient system).

Auditory cues A great number of animals use auditory signals to announce both territory and sexual receptivity. The eerie bellow of the Olympic elk, the high-pitched whistle of the spring peeper, the summer morning's songs of birds and evening chirping of insects are all examples of the sound of sex. And so is that cacophony of bleating frogs in the night. The male frog inflates his large vocal sacs and then deflates them to produce the characteristic song in chorus with his companions. Investigators have recorded an extensive variety of frog songs, which identify the species and allow for localization.

Visual cues Visual communication is involved in the most elaborate courtship displays. Among the reptiles, brightly colored bodies are displayed, and the colors serve to attract the female. Male salamanders and lizards, for example, are often brightly colored. The male canyon lizard raises himself off the ground into a vertical position to obtain the maximum color effect. While the female observes this display, the male approaches and seizes her with his jaws, maneuvering her into the mating position.

FIGURE 11.10
Visual and auditory cues. The sage grouse, a large game bird, is a native of the western plains. In early spring, males gather on chosen mating grounds to begin their unique courtship display. Strutting pompously, wings raised and tail outspread, the cock inflates two large sacs on either side of his neck. Groups of spectator females are attracted by the guttural booming and croaking call produced when the sacs are drawn in and air is expelled from the throat.

Similar behavior is involved among the courtship patterns of the males in certain species of birds, the male birds of paradise being a particularly well-known example. Courtship and mating behaviors among mammals often involve gestures instead of brilliant coloring.

Parental behavior

Parental behavior may be defined as any function performed either by one or by both parents that facilitates the survival of their offspring. This generally consists of attentive or inattentive behavior toward the egg and, later, treatment of the young.

A

B

C

D

FIGURE 11.11

Nest-building behavior. In order to protect their eggs, various species of birds construct unusual nests, sometimes in out-of-the-way places. (A) The cactus wren prefers the thorny cactus as protection for its gourd-shaped nest of sticks, thorns, and grasses. (B) The barn swallow's nest, a bowl-shaped mass of mud pellets and straw lined with feathers, is attached to a sheltered projection in caves or barns and other buildings. (C) The male weaver bird builds a nest of intricately woven vegetable matter, from which he hangs upside-down as he advertises for a mate with a buzzing, chirping call. (D) The cattle egret builds a loose mass of twigs and branches high in the tree tops.

Parental behaviors involve innate defense of brood and territory in many species, behaviors that are actually triggered by a variety of sensory signals (visual, tactile, auditory) that release appropriate behavior on the part of adults. Because many signals given by the young can inhibit aggressive behavior, parents will even tolerate from their offspring certain abusive behaviors that would normally evoke attack. Nevertheless, some species, among them certain spiders and fish, will eat their own young.

BEHAVIOR TOWARD THE EGGS

Certain animal species are oviparous: they lay eggs from which the young hatch at some later time outside the parent body. The majority of marine fish deposit their eggs on the ocean surface, leaving the newly born fish to fend for themselves. Most marine and freshwater invertebrates also abandon their eggs immediately after fertilization, as do many of the terrestrial invertebrates. However, the parents often do select an appropriate and safe place in which to deposit the eggs. Many reptiles, for example, lay their eggs inside burrows of sand or soil, frequently concealing the eggs with a protective covering. Other species, such as the earthworm, encase their eggs in a cocoon.

Many species, however, go to greater lengths to protect their eggs and their offspring. They build nests, either prior to mating or after, and lay their eggs there (Fig. 11.11). Among birds, the nest is built by the male and/or the female, and often parents alternate in incubating the eggs and in foraging for food to feed the young.

FIGURE 11.12

Maternal behavior. Elephants are born into close-knit family units, each headed by a matriarch. All members of the family take part in the care of small calves, generally treating them with tolerance and kindness. The young elephant is nurtured for the first ten years of its life, sometimes longer, and learns from the herd and its mother necessary survival skills and acceptable social behavior.

Community cooperation in the tending of eggs is practiced by penguins in their harsh Antarctic environment. During the mating season the penguin loses feathers on a small patch located on the underside of its abdomen. Because this is a highly vascularized area, the brood patch, as it is called, is warm and can be used for incubation of the egg. It also appears that the brood patch is an erogenous zone that stimulates a pleasure center in the penguin's brain. Upon seeing an untended egg, a penguin will place it over her feet and under the brood patch, and she will tend it until another penguin comes to relieve her. A community effort prevents the eggs from perishing quickly, as would otherwise happen at such low temperatures.

BEHAVIOR TOWARD THE YOUNG

Once the young are born, they are either left to fend for themselves or, in many species, nurtured by their parents. One of the most important parental functions is the feeding of the young. Birds, for example, feed their nestlings, a function often shared by both male and female parents. Mammals possess unique feeding behavior, nourishing their young with milk secreted from the mothers' **mammary** glands.

A second behavioral pattern directed toward the young is protection. Predators abound and the young must be shielded. One form of protection involves audible signaling. For example, some birds sound calls of alarm in times of peril, and the chicks respond by scurrying for a hiding place. Protective behavior also takes the form of retrieval—the rounding up of toddlers that have strayed away from the nest area. Protection usually takes the form of physical action. For example, when rat pups wander away, the mother picks them up with her mouth and returns them to the nest.

Another type of parental behavior, although not so critical as feeding or protection, is **grooming**—generally a mammalian function where the parent cleans the young and rids it of parasites and dirt. Grooming cements the social bonds between parent and offspring. The langur monkey, for example, strokes her infant and licks it, even while the baby is asleep.

In some animals parental behavior is highly developed and is exercised over extended periods of time. With predatory cats, such as the lion and cheetah, the mother usually undertakes a long training period during which the novice is taught how to hunt. In cheetahs, this takes about 14 months. Of all animals, man has the

FIGURE 11.13

Auditory cues. Social animals stay together in groups by using and responding to various cues. Members of this school of dolphins always keep within hearing distance of others in the group. Schooling is a common phenomenon among aquatic animals and has a strong selective value. It gives individuals the protection of numbers and maintains intraspecific contacts that facilitate reproduction.

longest childhood; as an heir to human culture, he has the most to learn.

Migration and navigation

Migrations are movements of animals to a specific area and their subsequent return to their place of origin. The time of the day or the year in which these movements occur is predictable, for these are behaviors that occur automatically in response to certain environmental stimuli, such as changes in day length, changes in the tides, lunar cycles, etc. Nevertheless, migratory movements involve an inherent ability to navigate, a function exemplified by the ability of certain animals to return (called the **homing instinct**) to an area from which they have been removed.

Migration has a vital role in species survival. Migrating animals are able to remain in consistently favorable environments; they migrate to northern areas in the summer and to southern areas in the winter, so they are almost always in favorable surroundings. Migration also allows species to obtain food sources from several environments, rather than being limited to just one.

Migratory activity is often related to reproductive behavior. Large crustaceans, such as the lobster, migrate in clusters to shallow waters during their mating season and then swim out to deep water to spawn—a type of movement called **benthic migration** (from the Greek, meaning "bottom of the sea")—and spend the winter there, where food is plentiful. The most impressive migrations are performed by birds. These animals generally migrate from a northern breeding area in the summer to a southern area in the winter. The migratory routes usually provide good areas for feeding—a reason that northbound routes during the spring differ from southbound routes in the fall. The golden plover, for example, migrates along a looplike route. In the fall it moves from Arctic Canada to southern South America by way of New England and Labrador, where late summer berries provide the plover with adequate food. In spring, those berries are gone, and thus the plover's return trip north follows a different route.

Other animals that migrate for purposes of reproduction are certain species of fish, amphibians, aquatic reptiles, and aquatic mammals. Among aquatic reptiles, certain sea turtles feed off the coast of Brazil and, every two to three years, swim approximately 1400 miles to Ascension Island, where they spawn. Among aquatic mammals, the most extensive migrations are seen in the whales. The grey whale, for example, migrates in the winter from the waters of the Arctic Ocean south to the sheltered lagoons of Baja California, in Mexico, where it breeds.

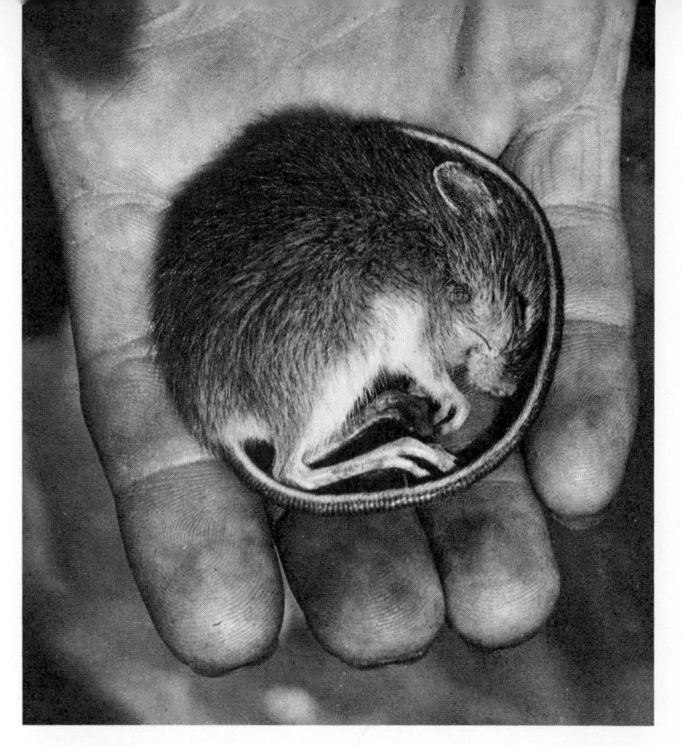

FIGURE 11.14

Seasonal rhythms. Certain animals like this jumping mouse respond to the onset of winter by entering a state of hibernation. Heart, breathing, and metabolic rate slow down, body temperature drops, and stored layers of fat are the only source of energy. This rhythmic alternation of hibernation and activity is apparently regulated by an innate biological clock.

NAVIGATIONAL CUES

Successful migration and navigation must of necessity involve sensory cues. Those used are chemical, visual, and auditory. A famous example of navigation by means of chemical cues is that of the Chinook salmon of the northwestern United States. The salmon follows its nose from the Pacific Ocean to the small stream in which it was born and where it will spawn a new generation. The vital role of chemical cues in the behavior of salmon has been demonstrated experimentally by plugging their nasal sacs. Their success in returning to their natal stream in significant numbers was inhibited.

The sensory device used most extensively in navigation is the visual cue. Many species make use of one or more celestial bodies for directional purposes. The starling, for example, determines direction by the angle of the sun. Green turtles, after nesting or hatching in the sand, reach the sea by heading toward the brightest portion of the horizon (which is the seaward portion).

Biological clocks

Since the origin of life there have been two highly predictable variables that have not drastically changed. First, each 24 hours the sun rises and sets; hence there is a 24-hour cycle. It is not surprising to find that a 24-hour rhythmicity has been incorporated into our metabolic machinery. Second, because the earth is tilted on its axis, all portions of the earth show cyclical changes in day length—short days in winter, long in summer. This rhythmicity has also been incorporated behaviorally, as in seasonal sexual changes.

CIRCADIAN RHYTHM

For many years it was thought that the daily rhythms exhibited by living things were the result of direct influences by the physical environment. Hence statements arose of the sort alluding to the chirping of crickets at nightfall and the singing of birds at sunrise, implying that the organism is totally dependent on its environment to activate its behavior. We now know, however, that organisms do not simply respond to environmental cues but contain within their systems very accurate mechanisms for measuring time. In other words biological rhythms are innate. The environment does indeed act on the organism, but only to help maintain its internal clock at the "correct" time. It was Franz Halberg of the University of Minnesota who, in 1959, designated these daily rhythms as **circadian**—from the Latin "circa" meaning "about" and "dies" meaning "day."

Experiments have been performed on various species that manifest rhythmic behavior in an attempt to confirm the innate nature of such behavior. The fruit fly *Drosophilia*, for example, emerges from its pupal case—and into adulthood—only at dawn. When eggs are laid under static laboratory conditions and allowed to develop there, the egg population does not exhibit rhythmic behavior. Adults emerge at all times of the day. Yet a single brief flash of light is sufficient stimulus to initiate rhythmicity so that future generations of flies emerge at intervals approximating a multiple of the original 24-hour period. This rhythmicity occurs even after 25 generations of flies are reared under static conditions. The evidence therefore strongly supports the conclusion that the intervals are determined by an innate property of the *Drosophila*.

Experiments have also been performed on the fiddler crab—an animal whose body darkens as the sun rises but turns pale as the sun sets. When the crabs are experimentally deprived of the more obvious clues signaling the passage of day, they continue to darken and lighten over each 24-hour period.

It has recently been suggested that there are also circannual (approximate yearly) rhythms. Cycles consisting of periods of hibernation alternating with periods of activity are thought to be just such a phenomenon. The annual hibernation cycle of the squirrel, for example, consists of a period that extends *regularly* from 324 days to 329, measured from the start of one hibernation to the onset of the next. This suggests an innate instinctual rhythmicity. In addition, experiments performed on the warbler indicate that migratory behavior is also a circannual response. Even when the birds were positioned in their normal wintering region, they still displayed a migratory urge at the appropriate time of the year.

Summary

1. All organisms—plant and animal—exhibit behavior of some kind. Behavior affects an organism's relationship with its environment and can be classified as either innate or acquired (i.e., learned). Innate behavior is genetically determined and often has survival value for either the individual or its species. Innate responses are automatic and specific to a given stimulus. Unicellular and simple multicellular organisms exhibit approach and avoidance behaviors, or negative and positive taxes, in response to such environmental factors as temperature, light, and chemical and physical factors.

In higher animals with developed nervous systems, the least complicated form of innate behavior is the simple reflex, in which a stimulus evokes a reaction through a single reflex arc. Fixed action patterns, or instincts, are also innate and automatic, but they involve the entire organism and may include a very elaborate series of activities in response to a number of stimuli. Instinctive behavior is triggered by releasers.

2. Acquired behavior is learned during an organism's lifetime. The learning of certain critical determinants of adult behavior during the earliest period of life is called imprinting. An organism may become habituated to certain stimuli so that it does not respond to them, and it may learn to respond in various ways to certain other stimuli. When an organism learns to exhibit an automatic reflex in response to a once neutral stimulus, it is said to have learned a conditioned reflex.

The ability of an organism to learn and to use simple forms of reasoning can be determined by different types of experiments. Learning by trial and error involves the least complicated form of reasoning—learning that reward is associated with one type of behavior. The ability to solve mazes requires learning a sequence of responses in order to receive reward. The ability to find a reward out of sight or smelling range requires insight, and more complicated problems require the ability to understand abstract concepts. Curiosity is an important motivating force in learning.

3. Most organisms communicate with other organisms, usually in conjunction with reproduction, caring for offspring, or navigation. Communication may be accomplished by tactile, chemical, visual, or auditory signals.

4. Many elaborate instinctive behavior patterns have evolved in conjunction with reproduction. Chemical, tactile, visual, and auditory cues affect such activities as sexual attraction, recognition, courtship, mating and gamete release.

5. Instinctive parental behavior patterns help the young survive and develop into healthy adults. Although many lower animals do not protect their eggs, they often select or make a safe place for them. Nests, burrows, and cocoons serve a protective function. Higher organisms protect first their eggs and then their young until the latter can survive by themselves. Mammals protect and feed their young.

6. Many animals exhibit seasonal or time-oriented behavior stimulated by innate biological clocks. The best known is migration to favorable environments for feeding and/or reproduction. Chemical, visual, and auditory mechanisms aid in migrational navigation. Daily rhythmic behavior patterns are called circadian rhythms.

REVIEW AND STUDY QUESTIONS

1. Why was the award of a Nobel prize to three ethologists in 1973 considered especially noteworthy?

2. What are the characterizing features of fixed action patterns?

3. What are releasers? Give an example of a releaser and explain how it works.

4. A now-classic picture shows Konrad Lorenz followed by a brood of ducklings. What behavior principle does the picture demonstrate?

5. Explain what is meant by extinction in conditioned learning.

6. Why is trial-and-error learning described as "perhaps the oldest form of learning"?

7. Give two examples of pheromones and describe the circumstances in which each is used.

8. A common problem in our world of rapid transportation has been called "jet lag." What biological basis is suggested as explanation for this difficulty in adjusting to time changes when traveling?

REFERENCES

Bennet-Clark, H. C., and Ewing, A. W. 1970 (July). "The Love Song of the Fruit Fly." *Scientific American* Offprint no. 1183. Freeman, San Francisco.

Brown, R. A., Jr., J. W. Hastings, and J. D. Palmer. 1970. *The Biological Clock: Two Views.* Academic Press, New York.

Comfort, A. 1971. "Likelihood of Human Pheromones." *Nature* 230: 432–433.

———. 1971. "Communication May Be Odorous." *New Scientist* 49: 412–414.

Hölldobler, B. 1971 (March). "Communication Between Ants and Their Guests." *Scientific American* Offprint no. 1218. Freeman, San Francisco.

Maier, R. A., and B. M. Maier. 1970. *Corporative Animal Behavior.* Brooks/Cole, Belmont, Calif.

Marler, P., and W. J. Hamilton III. 1966. *Mechanisms of Animal Behavior.* Wiley, New York

McClintock, M. K. 1971. "Menstrual Synchrony and Suppression." *Nature* 229: 244–245.

Menaker, M. 1969. "Biological Clocks." *Bioscience* 19: 681–692.

Michael, R. P., and E. B. Keverne. 1968. "Pheromones in the Communication of Sexual Status in Primates." *Nature* 218: 746–749

Pengelley, E. T., and S. J. Asmundson. 1971 (April). "Annual Biological Clocks." *Scientific American* Offprint no. 1219. Freeman, San Francisco.

Sauer, E. G. F. 1958 (August). "Celestial Navigation by Birds." *Scientific American* Offprint no. 133. Freeman, San Francisco.

Tinbergen, N. 1952. "The Curious Behavior of the Stickleback." *Scientific American* Offprint no. 414. Freeman, San Francisco.

Wenner, A. M. 1964 (April). "Sound Communication in Honeybees." *Scientific American* Offprint no. 181. Freeman, San Francisco.

Wilson, E. O. 1975. *Sociobiology: The New Synthesis.* Belknap/Harvard Press, Cambridge, Mass.

SUGGESTIONS FOR FURTHER READING

Dethier, V. G., and E. Stellar. 1970. *Animal Behavior,* 3rd edition. Prentice-Hall, Englewood Cliffs, N.J.

Covers animal behavior in terms of its evolutionary heritage and neurological basis. Heavily illustrated with clear and useful diagrams.

Hailman, J. P. 1969 (December). "How an Instinct Is Learned." *Scientific American* Offprint no. 1165. Freeman, San Francisco.

A study of the feeding behavior of sea gull chicks indicates that an instinct is not fully developed at birth.

Johnson, C. E., ed. 1970. *Contemporary Readings in Behavior.* McGraw-Hill, New York.

Interestingly written articles on a wide range of topics in the field of behavior.

Lorenz, K. 1952. *King Solomon's Ring.* Thomas Y. Crowell, New York.

This extremely interesting analysis of animal behavior—from instinct to intelligent and social behavior—includes examples such as the jackdaw, turtledove, fighting fish, and stickleback. The writing is lively as well as informative.

Manning, A. 1972. *An Introduction to Animal Behavior,* 2nd edition. Addison-Wesley, Reading, Mass.

Selective coverage is given to types of behavior and the motivation behind each. The author analyzes behavior both physiologically and psychologically and also includes a brief discussion of behavioral evolution.

Thorpe, W. H. 1973 (August). "Duet-Singing Birds." *Scientific American* Offprint no. 1279. Freeman, San Francisco.

Examination of the remarkably precise song of the male and female of certain tropical species, which permits them to maintain close communication in dense foliage.

Todd, J. H. 1971 (May). "The Chemical Languages of Fishes." *Scientific American* Offprint no. 1222. Freeman, San Francisco.

Examination of the exquisitely sensitive organs of smell of certain fishes, as well as the behavioral activities involving those organs.

Van der Kloot, W. G. 1972. *Behavior,* 2nd edition. Holt, Rinehart and Winston, New York.

A brief introduction to the principles of behavior, biological as well as psychological. Well illustrated with clear diagrams and informative photographs.

THE CONTINUITY OF LIFE

part three

REPRODUCTIVE PROCESSES

chapter twelve

After the female wolf spider lays her eggs, she binds them in a sac of silken thread and attaches it to her spinnerets for safekeeping as she roams about and runs down her prey. When the young spiders hatch, they climb to their mother's back. This female, glaring at the camera with three rows of eyes, will continue to carry her newly hatched young until their first molt.

The ability to reproduce its own kind is one of the unique characteristics of a living organism. Since every organism eventually faces individual death, the survival of a species and the general continuity of life depend on reproduction. In organisms that reproduce sexually—combining the genetic possibilities of two individuals—reproduction also provides for the rapid introduction of new characteristics and hence the variability that is necessary for the processes of evolution to operate.

Reproductive processes vary in complexity, from the simple fission of the unicellular bacterium to the complex series of events that constitutes a mammalian pregnancy. Specialized reproductive structures also vary greatly, being virtually nonexistent in some simple organisms, and highly specialized in such groups as the flowering plants and the higher vertebrates. Even the participants in the reproductive process vary. Some organisms reproduce sexually, and some asexually. We shall begin our discussion of the reproductive processes with an examination of what is perhaps the simplest reproductive event—the asexual reproduction of unicellular forms of life.

Reproduction among unicellular organisms

ASEXUAL METHODS

Fission The simplest form of asexual reproduction is fission, a process that involves the splitting of a parent cell into two independent and equal daughter cells. Prior to division, all genetic material is replicated. If the cell has a visible nucleus (as does the amoeba), it divides into two portions. If the cell does not have a nucleus—if it is a procaryotic cell, like a bacterium—the nuclear material is equally, if not conspicuously, divided, and a cell wall then begins to form near the center of the parent cell, eventually dividing it into two new cells. Each of these daughter cells becomes a separate individual and grows to normal size.

Bacterial cells generally divide in this manner—some at the remarkable rate of once every 20 minutes. But even though fission is an especially rapid means of reproduction, it has certain disadvantages. Because the daughter cells receive genetic material from only one parent, variation from generation to generation does not occur as part of the normal reproductive process. This means, of course, that new adaptive characteristics are not normally introduced. They do occur randomly, however, by genetic accidents called mutations. Given the rapid reproduction of bacteria and the large number of individuals produced, such mutations tend to occur often enough to produce new characteristics in a bacterial population—which may by accident produce a mutant bacterium and then a bacteria strain that is resistant to penicillin or some other antibiotic, for example.

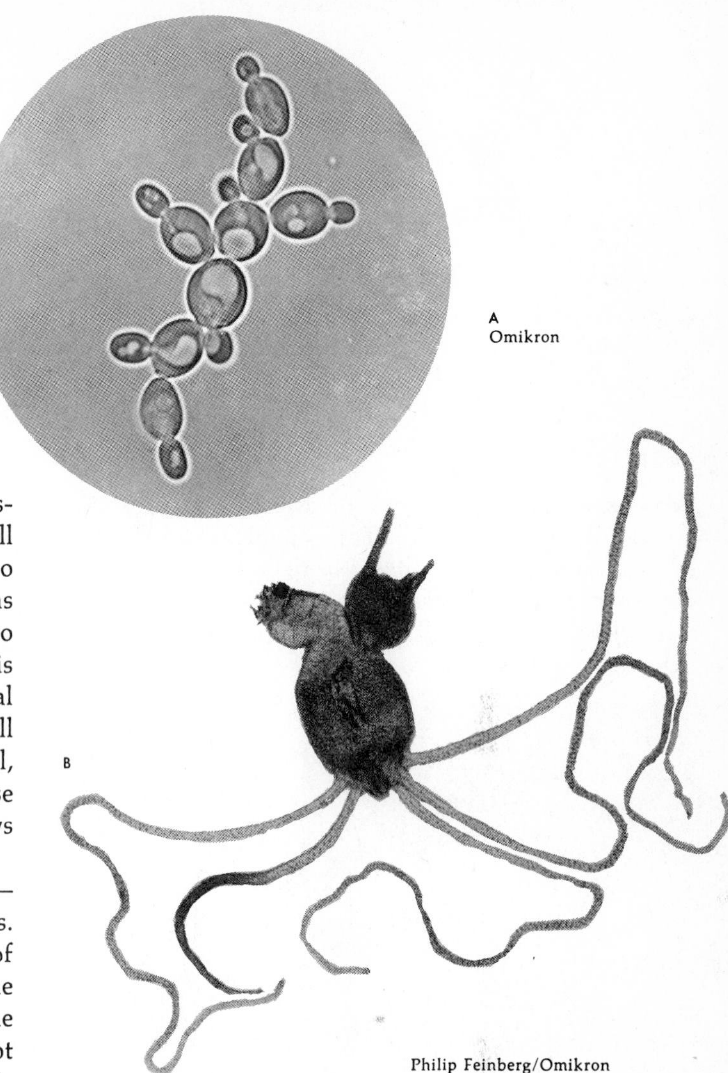

FIGURE 12.1

Budding. Not only unicellular but simple multicellular organisms can reproduce by budding. (A) When a yeast cell attains a certain size, the nucleus moves to the outermost portion of the cell, divides, and is pinched off from the cell along with a small share of cytoplasm. (B) The hydra, a multicellular organism, develops buds on its outer surface. The buds eventually break off and form independent organisms.

Budding Fission is not the only method of reproduction among unicellular organisms. Yeasts and some other microorganisms typically reproduce by budding. The cell produces a microversion of itself—a kind of bud that develops on one of its outer surfaces. The bud eventually breaks off as a new cell (Fig. 12.1). As in other forms of asexual reproduction, the daughter cell is a genetic duplicate of its parent.

Sporulation A group of unicellular parasitic animals, the Sporozoa, engage in a third kind of asexual reproduction. The sporozoan, established within the tissues of its host, becomes multinucleate, and proceeds to undergo a form of multiple fission, dividing into many smaller cells or spores. These burst forth and may eventually undergo another round of sporulation (or they may enter a cycle of sexual reproduction). The best

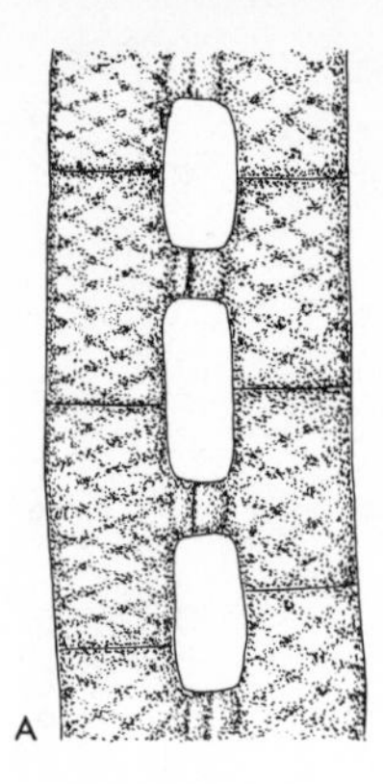

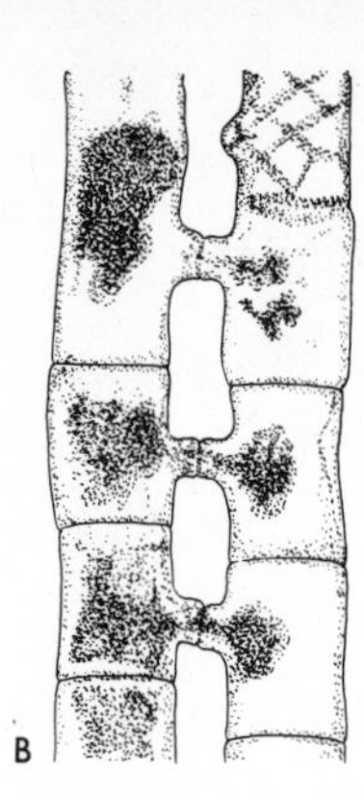

FIGURE 12.3

Conjugation variant. Some colonial algae, such as Spirogyra, *engage in a form of conjugation. (A) Two filaments align themselves side by side, forming bridges between opposite cells. (B) The contents of the cells of one filament then migrate across the bridge to the other and form an oval zygote (or zygospore). The zygote secretes a cystlike covering that enables it to survive in cold weather to produce a new filament in the spring.*

known sporozoan is the *Plasmodium,* whose species causes malaria in man. Sporulation occurs most often as part of the reproductive cycle of multicellular plants and fungi.

SEXUAL METHODS

Some unicellular organisms occasionally engage in a form of sexual **conjugation.** For example, two paramecia of the same species may partially fuse along one surface and then exchange genetic material before separating. Subsequently each will engage in fission, the usual asexual means of reproduction of the paramecium. The occasional sexual spree introduces new genetic combinations that would not appear if the animal reproduced only asexually. It is important to note that conjugation is not properly a form of reproduction, for no additional individuals are created. Conjugation has an adaptive value insofar as it leads to the production of individuals with new characteristics. In fact, paramecia are most likely to engage in sexual conjugation in times of environmental stress, when new adaptive characteristics may be most beneficial. In Fig. 12.2 two paramecia are shown during the process of conjugation.

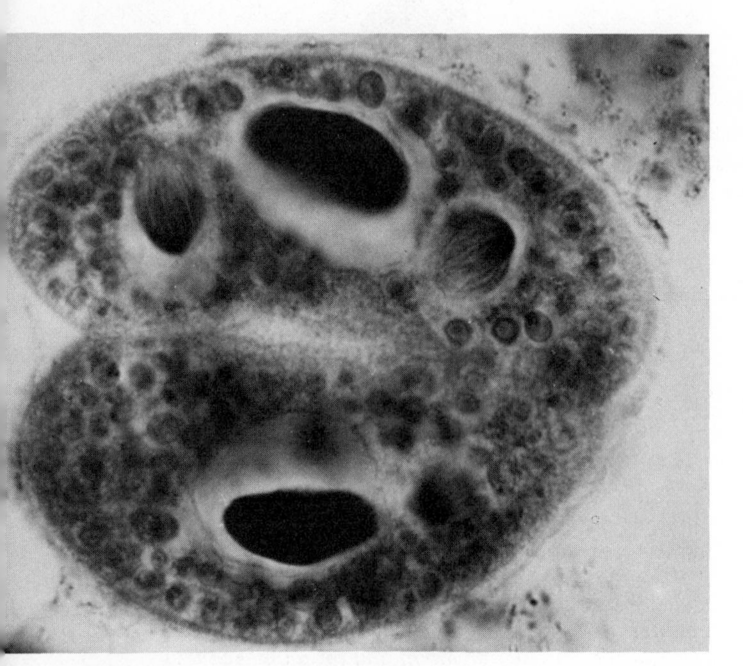

Courtesy of Carolina Biological Supply Company

FIGURE 12.2

Conjugation. The cytoplasm of two paramecia in conjugation merges along the ventral surface. The nucleus of each cell then undergoes a process of division and reorganization, shown in the cell on the top. One of two gametes thus formed is traded to the partner. When the paramecia separate, the nuclear material in each fuses. Although no additional individuals have been created, the two original organisms have acquired new genetic combinations.

Several unicellular plants can also reproduce sexually. *Chlamydomonas,* for example, is a simple green alga which sometimes produces unspecialized sex cells, or gametes, that combine to form a new individual. Other colonial algae, such as *Spirogyra* and *Volvox,* sometimes reproduce sexually by means of specialized gametes (see Fig. 12.3).

Reproduction in multicellular plants

VEGETATIVE REPRODUCTION

One of the ways in which gardeners typically cultivate new plants is by using cuttings of stems or underground

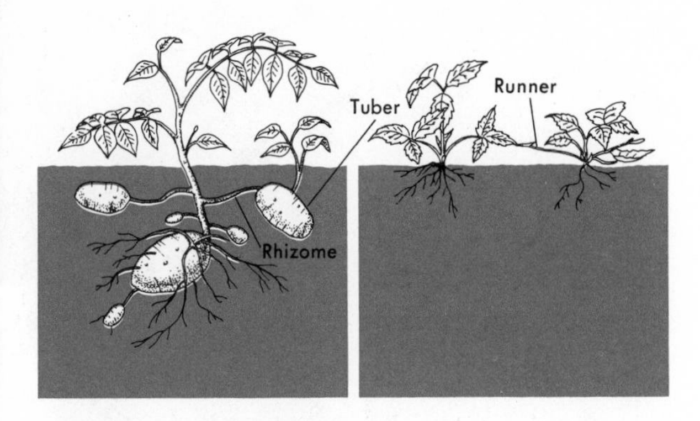

FIGURE 12.4

Vegetative reproduction. New shoots and root systems develop from the rhizome of the potato plant (left) and the runner of the strawberry (right).

stems called bulbs or rhizomes. Some plants, like the African violet, can even be grown from a piece of a leaf. Although such plant parts are not specialized for reproductive purposes, they can be made to propagate new individuals, either in the garden or spontaneously in nature. This form of reproduction is termed **vegetative reproduction.** Because it involves only one parent, it is a form of asexual reproduction and results in new plants with the same genetic makeup as the parent plant.

The two structures most often involved in vegetative reproduction are **rhizomes** and **runners,** or stolons (Fig. 12.4). Rhizomes are modified underground stems, common to ferns and some flowering plants, such as the iris. In some plants, the terminal portion of the rhizome is enlarged to form a **tuber.** In the familiar potato plant, the tuber is the edible portion, that is, the potato. Underground rhizomes send up stems that develop into new plants very near to the parent. Another form of vegetative reproduction involves runners, above-ground horizontal stems. Where the runner touches the ground, it sends out roots from which new plants develop. The strawberry plant, certain grasses, and cacti take advantage of this structure for reproductive purposes. Other modified stems involved in reproduction include bulbs (onions, daffodils) and corms (gladiolus). In bulbs, food is stored in leafy scales, whereas in corms it is stored in the stems, which are therefore engorged.

LIFE CYCLES

We have seen that some unicellular organisms that reproduce asexually are also capable of sexual conjugation. In the plant kingdom alternation between asexual and sexual reproduction is a characteristic pattern; in fact, it can be observed in every one of the plant phyla. Most plants in the course of their life cycles alternate between an asexual generation that reproduces by spores (that is, a **sporophyte** generation), and a sexual generation that reproduces by gametes (a **gametophyte** generation). Despite the fact that this is the common way of considering the alternation of generations in plants, note that it is actually an extended process of sexual reproduction, as indicated in the next section.

Sporophytes and gametophytes The sporophyte and gametophyte generations can be distinguished from one another not only on the basis of their method of reproduction, but also by their chromosome numbers. The gametes of every species contain a certain number of chromosomes. This number (represented as n) is designated the *haploid* number, and it is characteristic of the gametophyte generation (which is therefore also called the haploid generation). The gametophyte generation produces haploid gametes. When two gametes fuse to form a new individual, the resulting zygote, or fertilized cell, has twice the haploid number of chromosomes ($2n$) and so is called *diploid.* The organisms that develop from zygotes constitute the diploid generation. The diploid generation produces spores and thus is also called the sporophyte generation. Spores are produced by meiosis, a process of cell division (which will be described in the next chapter) in which the chromosome number is reduced from diploid ($2n$) to haploid (n). Thus spores (also called meiospores) are haploid, they grow by simple cell division, and they develop into another haploid, or gametophyte, generation. The two generations generally differ in appearance—just as the pollen grains differ from the flower—but they are nevertheless forms of the same organism. In some phyla the sporophyte generation is more often observed; in others

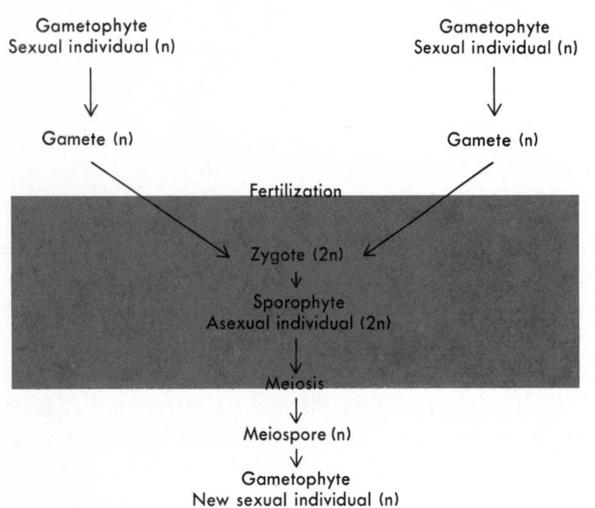

FIGURE 12.5

Alternation of generations in plants. Alternation between sporophyte and gametophyte generations is the characteristic reproductive pattern of multicellular plants. In some groups, the sporophyte generation is more conspicuous, in others the gametophyte. The shaded portion of the diagram represents the diploid part of the life cycle.

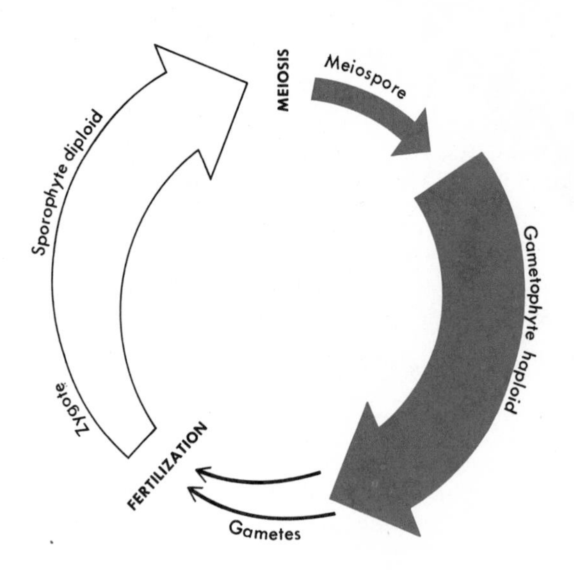

FIGURE 12.6

The reproductive cycle of a typical bryophyte. The gametophyte generation, shown in color, is dominant.

it is the gametophyte generation that is more prominent (see Fig. 12.5).

Life cycle of mosses (bryophyta) In the mosses, the gametophyte, or haploid, generation is more prominent; this is the velvety green cover that we are accustomed to finding on damp rocks (see Figs. 12.6 and 12.7). In some species one gametophyte plant will produce both eggs and sperm, whereas in others the different plants are either male or female. The sperm-producing organ is the **antheridium,** and the egg-producing organ is the **archegonium.** These reproductive structures appear only in the gametophyte generation.

When the egg cell of a female moss plant is fertilized by a male sperm, in some species it grows into a small diploid sporophyte plant—a thin stalk with a sporangium, or spore-producing organ, at the tip. In the sporangium, the process of cell division called meiosis produces haploid spores. Under favorable conditions, these will develop into a new gametophyte generation.

FIGURE 12.7

Reproductive structures of bryophytes. Gametes from a male bryophyte and a female bryophyte fuse to produce the sporophyte generation, which can be seen growing out of the gametophyte plant.

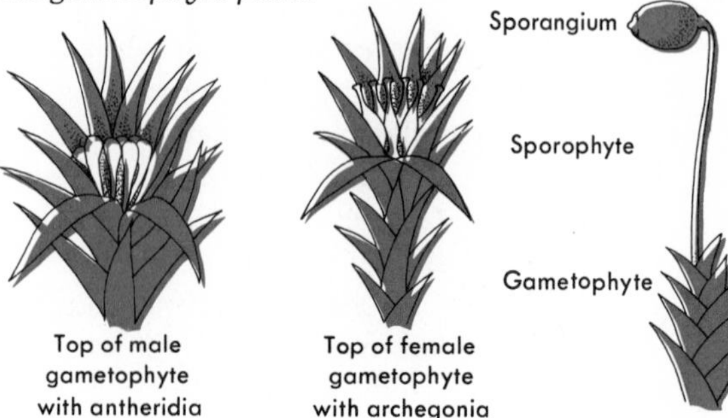

Life cycle of ferns (pteropsida) In ferns the generational roles are reversed. The familiar feathery fern plant is the sporophyte generation (see Figs. 12.8 and 12.9). On many ferns one can see sporangia clustered in small brown dots, called **sori,** on the underside of the fern leaf. The spores grow into small, flat, heart-shaped plants called **prothallia**—these are the gametophyte generation. Fertilization takes place within the prothallium, as sperm cells are released from the antheridium and swim to the egg-containing archegonium. The resulting zygote develops into another sporophyte generation—the recognizable fern.

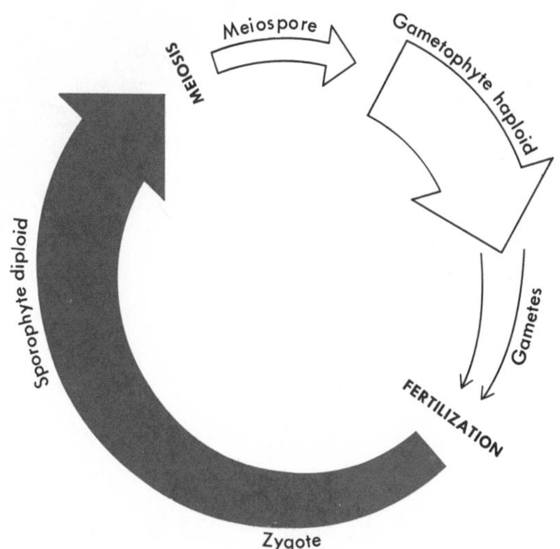

FIGURE 12.8

Reproductive cycle of a typical fern. The sporophyte generation, shown in color, is dominant.

FIGURE 12.9

Reproductive structures of ferns. Sori are found on the underside of the sporophyte leaf of the fern. The arrangement differs according to the species; two variations are shown.

REPRODUCTION IN SEED PLANTS

With the evolution of the seed plants—the flowering and cone-bearing plants—several new adaptations to terrestrial life have evolved. Sperm cells no longer depend on water as a medium through which to reach the egg cell. Instead, sperm-producing pollen grains (the male gametophytes) are carried by the wind, birds, insects, and other means; they reach the female gametophyte contained within the ovule by growing a special tube, the pollen tube, that allows for the passage of the sperm to the egg. The fertilized egg (zygote) enters a seed stage—protected by an outer coat, and nourished by a food supply—until such time as conditions favor its growth. Of course, many millions of pollen grains must be produced to ensure fertilization, and a large number of seeds must be dispersed for even a few new plants to take root.

The conspicuous forms of the flowering plants and conifers are the sporophyte generation. In these plants the gametophytes are reduced to structures completely dependent on the sporophyte plant.

REPRODUCTIVE STRUCTURES IN FLOWERING PLANTS

The reproductive structures of the angiosperms (Fig. 12.10) are familiar to us as flowers. In the center of a circle of petals, or **corolla,** are one or more female sex organs called **pistils,** or **carpels.** At the base of the pistil is the ovary, which contains the ovules. Extending dorsally from the ovary is a **style,** at the top of which is a structure called the **stigma,** which functions to catch pollen grains. Around the pistil is a circle of male sex organs called **stamens.** Each stamen consists of an **anther,** the spore-forming organ, and a **filament** for attachment. The spores produced in the anther are haploid, and these differentiate into pollen grains, the male gametophyte. The female gametophyte, also haploid, develops from a large spore that undergoes meiosis within a specialized ovule in the ovary. Around the base of the flower is a set of **sepals,** known collectively as a **calyx.**

Pollination is the transfer of pollen grain from the anther to the stigma (either that of the same flower or of

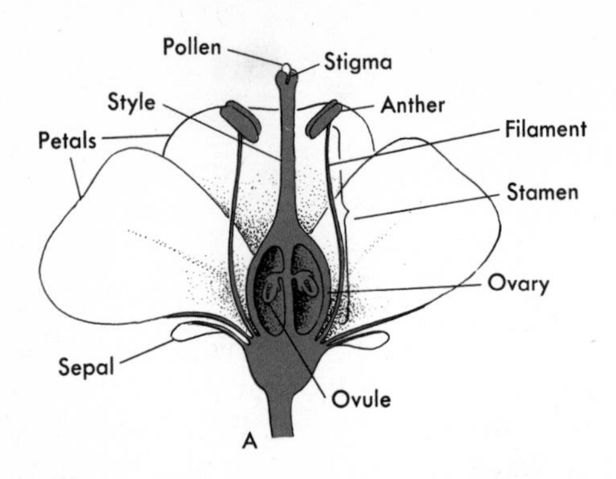

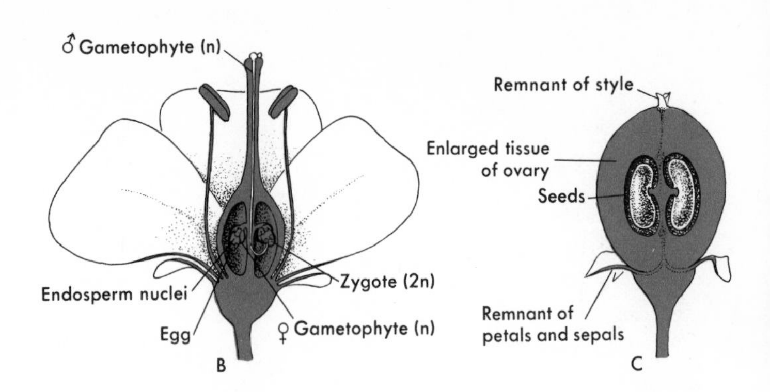

FIGURE 12.10

Reproductive structures of angiosperms. The flower (A) is part of the sporophyte generation of an angiosperm. The female gametophyte is enclosed in the ovules, and the male gametophyte grows to meet it (B). Subsequently, a seed (C) is formed.

another flower). This can be accomplished by means of wind or by insects or birds who are attracted by the color of the flower or the sweet solution or nectar within. The pollen grain, positioned on the stigma, will then grow a pollen tube that reaches down the style to the ovary. Here it encounters a female gametophyte enclosed in the ovule.

Usually two sperm cells, which result from the meiotic division of a diploid spore cell within a chamber of the anther, pass down and meet haploid cells developed within the embryo sac. One of the sperm unites with an egg cell to form a zygote, which eventually grows into the plant embryo. The other sperm cell unites with a pair of cell nuclei that have also been produced by the female gametophyte. The result of this union grows into tissue called **endosperm** that provides food for the growing embryo. (Ordinary flour is mostly ground wheat endosperm—the wheat embryo or wheat germ is usually removed so that the flour will keep longer.) Endosperm and embryo are usually covered with a protective coat, and the result is the **seed.** Sometimes the seed is encased in a **fruit** that facilitates dispersal. A seed is defined as a mature ovule; a fruit is defined as a mature ovary. The dandelion seed is surrounded by a dry fruit that can be carried by the wind. Other seeds may be encased in fleshy fruits that are eaten by animals; the seeds are then expelled by the animal and may germinate at some distance from the original plant.

Reproduction in conifers is similar to that in flowering plants. The female gametophytic generation develops on the inner surfaces of the scales of the female cones, the large, typical cones. Small cones produce the male gametophytes, which differentiate and are borne by the wind as pollen. As in flowering plants, the pollen grain produces a pollen tube through which the sperm may reach and fertilize the egg. The zygote develops into an embryo, the ovule into a hard seed cover. The seed is eventually released from the plant to germinate and form a new sporophyte plant. No fruit can form in conifers because pistils are lacking.

Reproduction in multicellular animals

As might be expected, animals, and in particular lower invertebrates, continue to show the same reproductive modes encountered among unicellular organisms. Thus budding is the normal reproductive method in both porifera (sponges) and certain coelenterates (relatives of the jellyfish). Among sponges, the buds typically remain attached to the parent, so that what develops is an interconnected colony of sponge animals. In the hydra, the buds break from the parent and grow into mature individuals.

Another form of asexual reproduction found in certain animals is **parthenogenesis,** a process whereby egg cells develop into young without having been fertilized by male sperm. Parthenogenesis is found primarily in arthropods and generally as an alternative to fertilization. Certain plant-eating insects called aphids produce several successive generations of fatherless young before a generation that reproduces sexually appears. However, a few species seem to reproduce by parthenogenesis alone: the white-fringed beetle is not known to produce in any other way, for no males have ever been found. Laboratory experiments have shown that certain stimuli, such as pinpricks or radical environmental changes, can induce parthenogenesis in species where it does not normally occur. This suggests that the egg is like a loaded bomb that in some species can be triggered by a variety of mechanical or chemical stimuli.

SEXUAL REPRODUCTION IN ANIMALS

Most higher animals rely on sexual reproduction to propagate the species. Usually the species is divided into male and female individuals, but individuals of some species—chiefly among annelids (e.g., earthworms) and molluscs (e.g., certain snails)—have developed the reproductive structures of both sexes. Such animals are said to be **hermaphrodites** (after Hermes and Aphrodite, the Greek god and goddess of love). In some cases the hermaphrodite engages in self-fertilization. This results in a virtual genetic duplicate of the parent. However, self-fertilization is a distinct advantage to a parasitic animal like a tapeworm, which may be the sole member of its species to find its way into a particular host. Under other circumstances, the tapeworm may seek a mate. Often two hermaphrodites will couple and cross-fertilize. Earthworms, for example, lie together, the heads pointing in opposite directions, and engage in a sticky trade of sperm cells.

Fertilization In sexual reproduction that involves more than one individual, there is a general overriding problem—ensuring that the sperm reaches the egg and fertilizes it. A **zygote** is the diploid cell that results from the fusion of egg and sperm nuclei. In many species, particularly aquatic species, fertilization is *external:* both the egg and sperm leave the parents before fertilization occurs. Generally, the female lays the eggs and the male swims over them, depositing his sperm. Other

FIGURE 12.11

The reptiles were among the first land-dwelling higher organisms. One adaptation that enabled them to survive successfully in a terrestrial environment was the tough-shelled egg with its fluid-filled interior. The modern box turtle laying her eggs is a descendant of that first hard-shelled egg producer.

aquatic animals—and nearly all land animals—rely on *internal* fertilization, involving the introduction of sperm cells directly into the female reproductive tract.

Internal fertilization is much more efficient for terrestrial animals, for in the absence of water the chances for successful external fertilization are very greatly reduced. In many species, the male has developed a special copulatory organ, a penis or penile structure, to deposit sperm within the reproductive tract of the female. In animals that lack this structure—amphibians and most birds, for example—the transfer of sperm is made by the juncture of the external openings of the male and female reproductive tracts. In insects and in all internal fertilizing vertebrates except the mammals, internal fertilization precedes the laying of eggs.

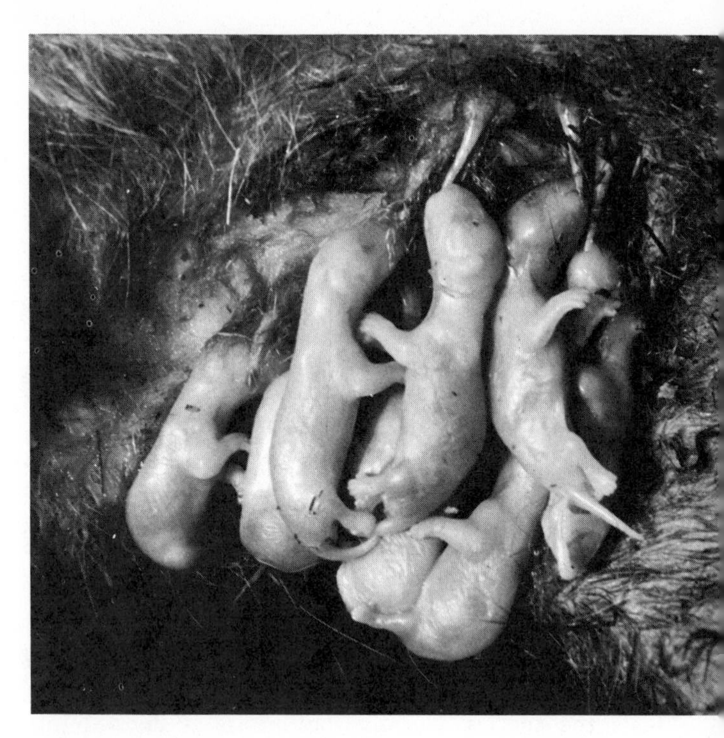

FIGURE 12.12
Young marsupials are born at a very immature stage of development and must crawl unaided up their mother's abdomen to the safety of her pouch. These naked baby opossums have fastened onto the teats in their mother's pouch, where they will remain until they are strong enough to emerge and begin living independently.

Development of the zygote Once the egg is fertilized, the next problem for the zygote is surviving a series of complex developmental processes until hatching or birth. A number of successful strategies have evolved which provide the developing embryo with the protection and moisture it needs. In aquatic animals, the water itself provides moisture, so their eggs do not need a hard shell; instead, the eggs are covered with a gelatinous substance, such as is found on caviar (fish eggs). Land animals, however, either return to the water to lay their eggs or produce eggs containing structures that prevent the egg from drying out. Most of the amphibians (the earliest terrestrial vertebrates) return to the water; reptiles and birds, however, are able to lay their eggs on land, because their eggs have a hard external shell and membranes, including the tough **amnion,** which prevent drying.

In some species, the eggs develop in or on the body of one of the parents—not always the female. Certain male amphibians bear eggs in the folds of their moist skins, and many species of fish carry their eggs in the mouth. In many animals, the complete development of the fertilized ovum takes place inside the female's body. In rattlesnakes and most sharks, for example, the eggs develop internally, and the young are born alive. In these animals, the embryo, though inside the female's body, draws on food supplies stored inside the egg.

Mammals have developed a unique variant of this internal development of the embryo. In the majority of mammal species, the embryo is attached to its mother by a special organ, the placenta, that develops from the wall of the womb, or uterus. The placenta permits the selective interchange of nutrients and wastes between the mother and the developing embryo. Not only is the embryo protected, but the mother is freed from all nest-sitting responsibilities, and she can remain mobile. In larger mammals, the embryo develops inside the mother for a considerable period of time—up to 21 months for an elephant. Upon birth the mammalian young is nourished by milk from his mother's mammary glands.

The young of the marsupials (e.g., opossum, kangaroo, koala bear) are born at a very immature stage and are subsequently suckled in a pouch on the mother's abdomen (Fig. 12.12). Here they are protected and nourished until they are ready to live independently.

Human reproduction

In animals that practice external reproduction, reproductive structures are rather simple, consisting only of a **gonad** (a gamete-producing organ) and a duct to carry the gamete, either egg or sperm, to the environment where fertilization takes place. In many animals that reproduce on land, a copulatory organ, or penis, is used to deposit sperm into the female reproductive tract. In animals that bear living young, the female reproductive system must also provide an internal environment within which the zygote can develop. The human reproductive system includes structures that fulfill all these specialized functions (see Fig. 12.13).

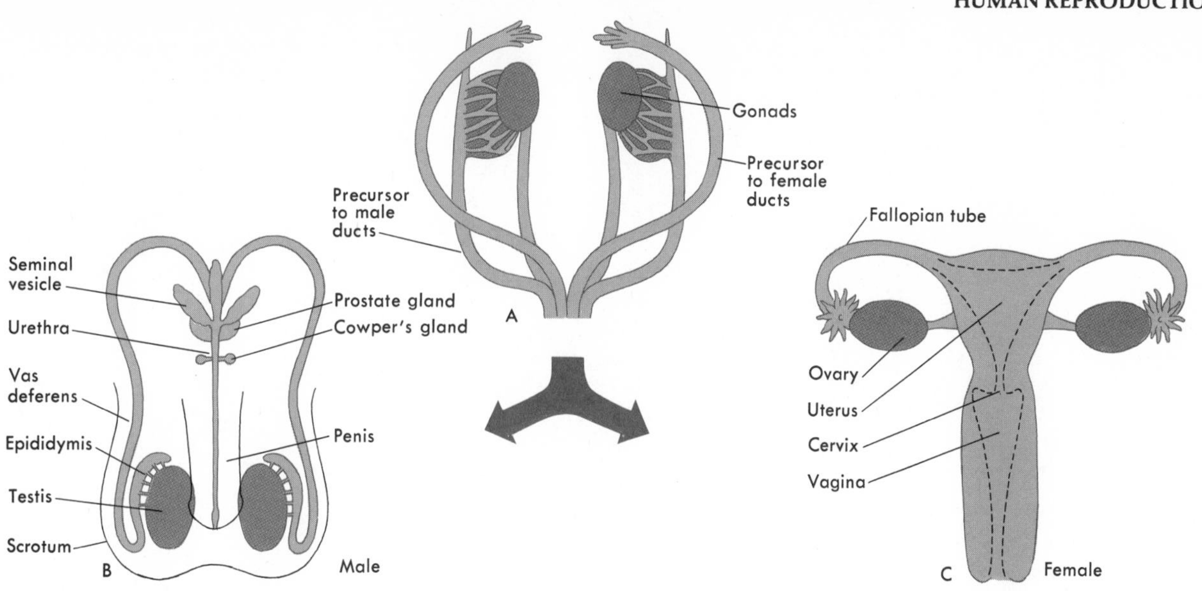

FIGURE 12.13

Human reproductive structures. Diagrammatic representations of the reproductive structures in the embryo (A) and in the mature male (B) and female (C) demonstrate developmental relationships. Although the two mature systems are quite different, they are derived from the same structures in the embryo.

Human sexual behavior has its basis in biological principles, but its many manifestations are in large part products of our various cultures, values, and individual tastes. Only in the past twenty years have serious scientists begun to systematically explore human sexual behavior. Studies have been and are being undertaken by biologists, psychologists, and medical researchers. The pioneer work of Alfred Kinsey in the late 1940s and the more recent work of the Masters and Johnson clinic and others have provided insight into the physiology of some human sexual responses. Because sexual behavior is complex and variable, an adequate discussion would be beyond the scope of this text. However, several recent paperback books have dealt with the topic in some detail, and some of them are included in the suggested readings at the end of this chapter. In this text, the discussion will be limited primarily to reproductive structures and their function.

FEMALE REPRODUCTIVE STRUCTURES

The internal and external organs of reproduction in both sexes are called the **genitals,** or genitalia (from the Greek, "to give birth"). The gonads are the organs that produce the gametes—the ovum, or unfertilized egg, in the female, and the sperm in the male. In females the gonads are internal; they are called **ovaries** because of the cells they produce.

The ovaries, approximately 1½" long and 1" wide and almond-shaped, lie within and toward the base of the abdominal cavity. Although there is no opening from the ovaries to the rest of the internal sex organs, they are attached to the uterus by broad ligaments.

Estimates suggest that there are at birth anywhere from 30,000 to 300,000 potential egg cells within the ovaries, each contained in a sac-like **follicle;** however, only 350 to 450 of these ever mature. At the onset of sexual maturity, hormones begin to induce the eggs to mature within the follicle and to be released at the rate of approximately one every four weeks from alternate ovaries. The eggs are released into a **fallopian tube,** which serves as an oviduct—a passageway through which the ovum travels to the uterus. Each tube is approximately 5½" long and is filled with cilia that help move the ovum toward the uterus. Fertilization, if it occurs, takes place during this passage through the fallopian tube, usually at its distal end.

The embryo is retained in the uterus for the course of its development. The **uterus** is a pear-shaped muscular organ with a lining of vascular tissue. It is about

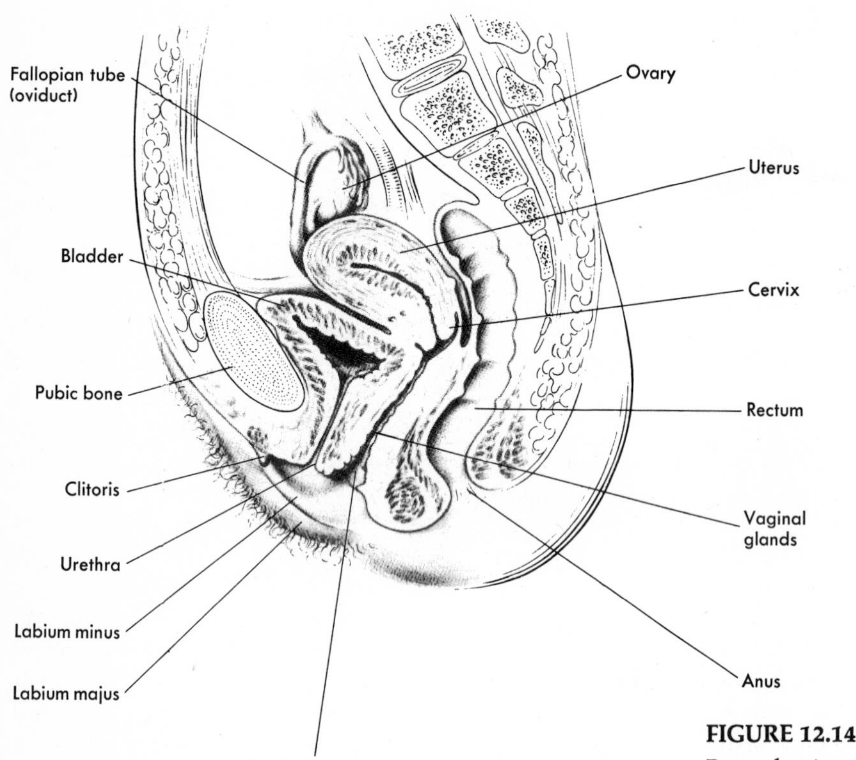

FIGURE 12.14
Reproductive system in the human female.

3½″ × 2½″ × 1½″ in its normal state, but its capacity is impressive: during the final stages of pregnancy it may measure 15″ × 10″ × 8″. This increase in size is due not only to the elasticity of the muscle fibers, but also to growth in the size of each fiber through the incorporation of additional proteins. After birth, the uterus returns to a size close to that before pregnancy.

The muscular part of the uterus is called the **myometrium.** The lining of the uterus, a spongy lining filled with vascular tissue, is the **endometrium.** This tissue undergoes changes each month in preparation for possible reception of a fertilized egg. If the ovum is not fertilized, the lining is shed and menstruation occurs. The cervix, or neck of the uterus, connects with the vagina and normally has a very small opening. The cervix also undergoes many tissue changes during the menstrual cycle and pregnancy. During birth it dilates, or expands, opening wide enough to allow the head of the infant to pass through and into the vagina. The vagina, a well-lubricated tube about 3″ long, stretches many times its size during birth. It is sometimes called the birth canal; it is also the specialized receptacle for the male penis (Fig. 12.14).

Development and release of the ovum The cyclical nature of human female sexuality is demonstrated by menstruation, a monthly discharge of blood and tissue. However, not until this century was menstruation put into its proper perspective as one phase in a complex series of events that surround the development and release of the ovum in preparation for its fertilization (Fig. 12.15).

Because the only outward sign of the menstrual cycle is the shedding of the endometrial lining in the show of blood, we usually measure the cycle from the first day of menstruation (Day 1). The cycle technically

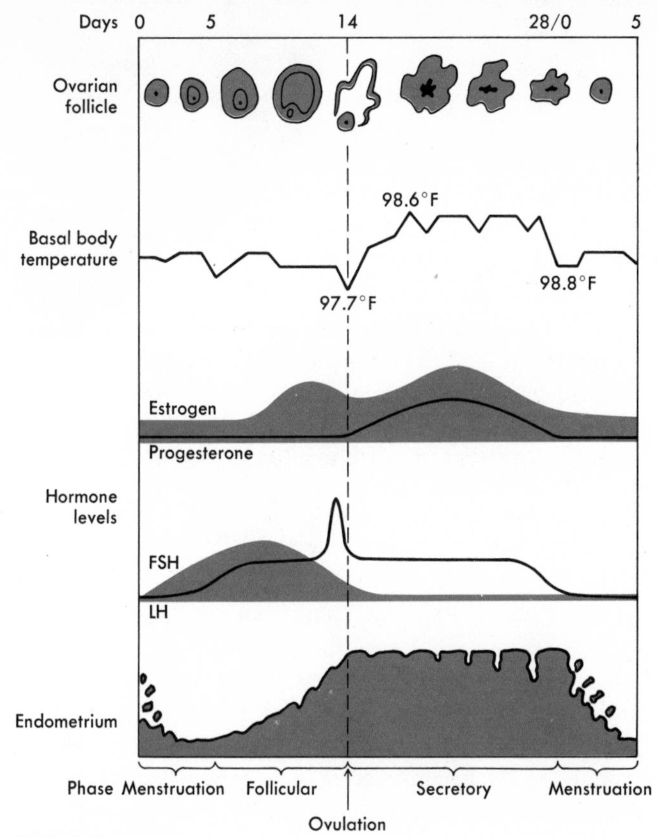

FIGURE 12.15

Combination diagram of the menstrual cycle. The bottom segment indicates the changing condition of the endometrium in response to changes in hormone levels, shown in the next segments. The remainder of the diagram relates body temperature and development of the ovarian follicle to those changes. Note particularly the multiple changes before and after ovulation, which occurs on Day 14.

begins about Day 5, when the pituitary gland, a small gland located at the base of the brain, secretes a hormone called the **follicle-stimulating hormone** (FSH), which causes a few follicles in the ovaries to begin developing their immature ova. During this part of the cycle, called the follicular phase, the developing follicles secrete a female hormone, **estrogen.** The secretion of estrogen increases as the follicles grow, rising rapidly to a high point on about Day 14. Because of the attendant proliferation of tissue in the endometrium, this part of the menstrual cycle—from the beginning to Day 14—is also known as the proliferative phase. Toward the end of this phase; one of the developing follicles matures more rapidly than the others and is designated the **Graafian follicle.** The other follicles eventually shrink and disappear.

On or about Day 14, the rise in the level of estrogen causes the pituitary to stop secreting FSH and to secrete **luteinizing hormone** (LH) in a sudden surge, which stimulates the Graafian follicle to release the mature ovum—an event called **ovulation.** As the ovum begins its journey through the fallopian tube, the now-emptied follicle becomes filled with a fatty yellow substance, which gives the organ its new name, **corpus luteum** (Latin for "yellow body"). The corpus luteum continues the secretion of estrogen and begins the secretion of **progesterone,** another female hormone, which causes the endometrium to become spongelike and secrete a mucuslike fluid. This part of the cycle, from Day 14 to Day 28, is called the secretory phase (sometimes the luteal phase).

If the ovum is not fertilized, the corpus luteum soon degenerates and discontinues its secretion. Without hormone support, the uterus sheds its endometrial lining of blood, tissue, and mucus. Progesterone's presence in the blood indirectly inhibits the sloughing of the endometrium and the development of other follicles. Without it, menstruation occurs. If fertilization does take place, cellular division begins in the zygote as it moves down the fallopian tube toward the uterine cavity, where it will implant itself in the richly prepared endometrium.

All normally ovulating women are familiar with the physical effects of the menstrual cycle. By the end of the secretory phase, the uterus is engorged with blood and the breasts are often enlarged. As a consequence, both organs may be tender. Just prior to and during menstruation, the uterine contractions necessary to expel the endometrial lining are often felt as cramps of varying intensity. Some women also experience changes in emotional sensitivity, commonly referred to as premenstrual tension.

For fertilization to occur, the vagina must be prepared to receive the sperm-bearing penis. The vagina responds to sexual stimulation in several ways. It secretes a small amount of lubricating substances. Its upper portions enlarge and expand while the lower third contracts, and sometimes it undergoes rhythmical waves of contraction. These changes provide more room and lubrication for the penis, and they enhance the movement of the sperm toward the dilated cervix, the opening to the uterine canal.

CONTRACEPTION

For reasons ranging from population control through family planning to maternal health, couples may often wish to avoid having that sperm and egg meet. Methods of contraception are becoming increasingly sophisticated and effective, offering several alternatives to those wishing to prevent conception. Oral contraceptives—"the pill"—contain small amounts of synthetic hormones similar to estrogen and progesterone. When taken daily, these pills provide sufficient hormonal content in the blood to inhibit pituitary secretion of FSH and LH, thus preventing ovulation. The pill also induces changes in the endometrium so that it is less receptive to implantation of the zygote, even if ovulation does take place. The pill offers almost 100 percent contraceptive effectiveness when taken properly. Of all methods of birth control currently available, only sterilization has a higher rate of effectiveness. The long-range safety of the pill, however, remains a subject of considerable discussion.

Another method of contraception involves the intrauterine device (IUD)—an oddly shaped plastic device, sometimes wound with copper or impregnated with progesterone—which is placed in the uterine cavity. The precise contraceptive mechanism is still unknown, but the IUD appears to alter the endometrial lining's receptivity to the zgote and to interfere with the migration of sperm. Although its does not prevent ovulation or fertilization, the IUD is nevertheless nearly as effective as the pill. A principal drawback, however, is that some women cannot adjust to their use with both comfort and safety.

The diaphragm, used with or without jellies or creams that kill sperm (spermicides), is a dome-shaped latex rubber cup that fits over the cervical opening and prevents sperm from entering the uterus. The condom is a rubber sheath that fits over the penis to achieve the same objective. These two mechanical methods are reasonably effective if the materials are not defective and are used properly.

The "rhythm" method is based on an attempt to predict the time of ovulation by charting the woman's basal body temperature. Theoretically, if intercourse is avoided around the expected time of ovulation, fertilization will not occur. However, the frequent and unpredictable irregularities in the menstrual cycle and the inability to precisely pinpoint ovulation are limitations to the effectiveness of this method. (Indeed, people using the rhythm system are often called "parents"!)

MALE REPRODUCTIVE STRUCTURES

The male reproductive system consists of four types of structures: the gonads (or **testes**), which produce sperm and secrete the male hormone testosterone; a system of ducts through which the sperm passes; some accessory glands whose secretions constitute the seminal fluid in which the sperm is suspended; and a specialized organ, the penis, which becomes erect (engorged with blood) and enlarges during sexual excitement so that when placed within the vagina, it is close to the cervical opening (Fig. 12.16). The testes, or testicles, are located outside the body cavity in a thin-walled pouch called the scrotum. The temperature of the testicles within the scrotum is a few degrees lower than the temperature within the body; sperm are particularly sensitive to heat. However, for some mammals heat is not a problem, and the testicles remain within the body. In others they descend into the scrotum prior to birth, at sexual maturity, or during the mating season.

The sperm passes from the testes through a system of ducts. The **epididymis,** located on the outside of the testes and partially surrounding them, collects the sperm. The **vas deferens** conveys it from the epididymis to the ejaculatory duct, which contracts and expels the sperm into the urethra. The urethra carries the sperm through the penis and into the vagina of the female.

Associated glands, such as the prostate, the seminal vesicles, and Cowper's glands, add certain fluids to the ejaculatory duct as a vehicle for the sperm cells. The resulting alkaline **seminal fluid** is a slightly viscous, opaque liquid that aids the sperm in traveling through the female reproductive tract and reaching the egg alive. Among the components of the fluid are sugars that nourish the sperm and activate them; a chemical buffer that protects them against the excessively acid environments frequently found in the reproductive tract and

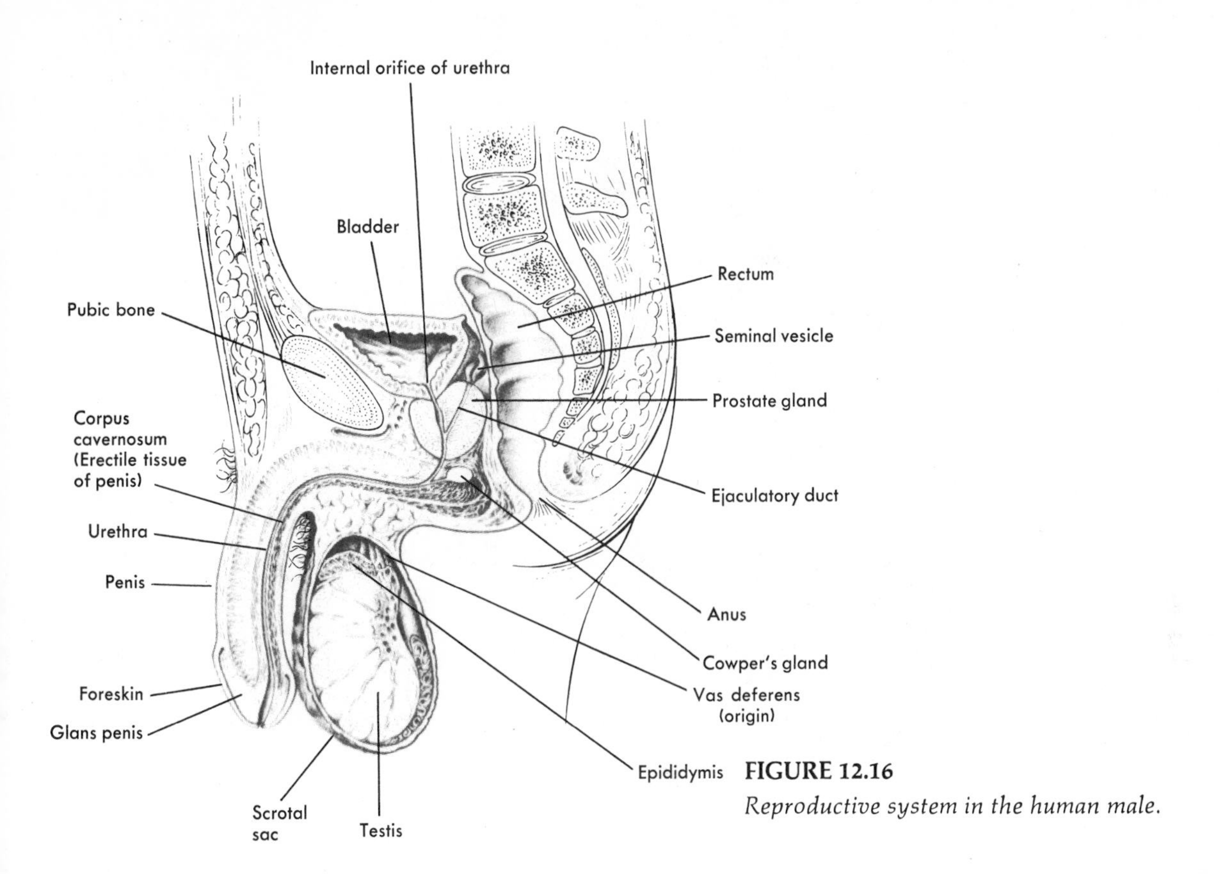

FIGURE 12.16
Reproductive system in the human male.

protects against basic environments as well; and a lubricant for the female reproductive tract.

The seminal fluid is discharged by the penis. Erection occurs because the blood vessels leading from the penis become constricted, and the blood, unable to flow out, fills the spongy tissue.

This arousal response in the male is usually much more rapid than the female arousal response, although the intensity of the response, as well as the potential sex drive, is equal for males and females. Most males after orgasm (usually simultaneous with ejaculation of semen) also show rapid detumescence (return to the normal unaroused state) and have a protracted postorgasmic refractory phase during which repeated arousal is difficult. Females, on the other hand, are capable of repeated orgasms, and they experience relatively slow detumescence.

Ejaculation in males and orgasm in both males and females can occur without relationship to reproduction. During the stage of sleep known as REM (for "rapid-eye-movement"), young males sometimes experience spontaneous ejaculation of semen ("wet dreams"). Both males and females can induce orgasm by self-arousal (masturbation) or mutual stimulation.

In a healthy male, one ejaculation will forcefully emit between 70 and 100 million sperm per cc (and an ejaculation is usually around 5 cc). Yet very few sperm reach the upper end of the fallopian tubes, where fertilization takes place—so few that sterility is defined clinically as a condition in which there are fewer than 35 million sperm per cc. Some sperm move too slowly; some move in the wrong direction; some simply die, usually within 10 seconds of emission. Although in most cases only one sperm will unite with the ovum, a number of sperm must be present for fertilization to occur. Surrounding the ovum is a special layer of cells that can be dissolved only by the combined enzymes of many sperm. Upon entry of the head of a sperm (which contains the genetic materials), the egg immediately responds by forming a fertilization membrane that cannot be penetrated by other sperm. Fertilization occurs as

Birth Control Methods

The accompanying chart compares the most common methods of birth control in a number of ways, including two categories of failure rate. The *theoretical failure rate* is based on hypothetical perfect use of the method. The more realistic (and usually higher) *actual failure rate* is based on records of actual use of the method over a period of time. The difference between the two rates points out that for any method to be effective it must be applied correctly and consistently.

METHOD	APPROXIMATE FAILURE RATE *		HOW IT WORKS	POSSIBLE SIDE EFFECTS	PHYSICIAN INVOLVEMENT
	Theoretical Failure Rate	Actual-Use Failure Rate			
STERILIZATION	Tubal ligation: 0.04 Vasectomy: less than 0.15	Tubal ligation: 0.04 Vasectomy: 0.15	Female (tubal ligation): Fallopian tubes are cut, tied, blocked, or removed to prevent eggs from reaching the uterus. Male (vasectomy): Small piece of vas deferens is removed to prevent sperm from getting from testes to prostate.	None known	Doctor required for both operations
ORAL CONTRACEPTIVES ("The Pill")	Less than 1.0	2–5	The pill, taken daily, contains a synthetic progesterone and estrogen, which prevent ovulation from occurring. Should it occur, the progestin and estrogen make enough changes in the uterine environment to make it hostile to implantation.	*Bad:* Possible thromboembolism (blockage of major blood vessel by blood clot)—uncommon in healthy young women; may increase risk of gallstones, raise blood pressure. *Good:* Lessens risk of benign breast tumors, ovarian cysts, and perhaps cancer of uterine lining; menstrual cycles more regular; lessens premenstrual tension and depression; may help acne.	Doctor's prescription required
CONDOM (Prophylactic, "rubber")	1.0 (if used with spermicidal jelly)	5.0 (if used with spermicidal jelly)	Sheath of rubber, latex, or animal membrane is placed over the erect penis before ejaculation; semen is caught in sheath and doesn't reach vagina unless some is spilled as man withdraws or condom ruptures.	None	None

*measured in pregnancies per 100 women per year

METHOD	APPROXIMATE FAILURE RATE *		HOW IT WORKS	POSSIBLE SIDE EFFECTS	PHYSICIAN INVOLVEMENT
	Theoretical Failure Rate	Actual-Use Failure Rate			
INTRAUTERINE DEVICE (IUD)	1–5	6	Small plastic device inserted into uterus by doctor; precise reason it works is not known—thought to prevent pregnancy by disturbing uterine environment.	Can cause excessive bleeding, cramps; if woman does get pregnant with IUD, chance of miscarriage is 30–50% more than in normal pregnancy.	Doctor required for insertion of IUD into uterus
DIAPHRAGM AND CERVICAL CAP	3	20–25	A domed rubber cap, fitted to the vagina, forms a physical barrier that covers the cervix and prevents sperm penetration; spermicidal cream or jelly must be used; can be inserted up to 2 hr prior to intercourse and must be left in at least 6 hours after.	Spermicidal cream or jelly may cause vaginal irritation.	Doctor required for fitting diaphraghm to the individual
CHEMICAL CONTRACEPTIVES	3 (best average is with aerosol foams)	30 (best average is with aerosol foams)	Spermicidal creams, jellies, foams, or suppositories that immobilize and kill sperm must be inserted into vagina 5–15 min before ejaculation so they can disperse throughout the vagina.	Chemicals in the contraceptive may irritate vagina.	None
WITHDRAWAL (Coitus interruptus)	15	20–25	Male withdraws from vagina before ejaculation.	None	None
RHYTHM	15	35	Planned abstinence from intercourse during the female's fertile period after ovulation; to determine fertile period, she must keep track of her temperature daily, the length of her menstrual cycles to determine when ovulation may take place.	None	Supervision by a doctor advised

METHOD	APPROXIMATE FAILURE RATE*		HOW IT WORKS	POSSIBLE SIDE EFFECTS	PHYSICIAN INVOLVEMENT
	Theoretical Failure Rate	Actual-Use Failure Rate			
DOUCHING	40 pregnancies per 100 women per year	40 pregnancies per 100 women per year	Female must flush all semen out of vagina *immediately* after intercourse with a syringe full of water and/or other spermicidal liquid.	None	None
MORNING AFTER PILL (Diethylstilbestrol)	No data available	No data available	High dosage of estrogen taken within 72 hours of intercourse will prevent pregnancy; usually administered only in emergency cases, such as rape or incest.	Nausea, vomiting, bleeding, thromboembolism; if a woman is pregnant by another intercourse and takes the morning after pill, the diethylstilbestrol may cause cervical or vaginal cancer if the baby is a girl.	Doctor required for administration of pill

The methods cited are by no means the final word in birth control. Because emphasis on controlling population growth is so strong today, researchers are constantly searching for safer, more effective, and more carefree methods of birth control. Some methods, still in the developmental stages, that have received publicity are the following.

1. An antipregnancy vaccine that would cause a pregnant woman to menstruate, "washing" the fertilized egg out of the uterus.

2. A birth control pill for men that would decrease the production of the male sex hormone and, as a result, sperm output. This pill would be supplemented by a monthly shot of testosterone, which would maintain the man's sex drive.

3. A plastic ring inserted into the vagina that would prevent pregnancy by releasing a steady dose of estrogen and progesterone into the bloodstream through the mucous membranes of the vagina. The dosage of estrogen and progesterone, though enough to prevent pregnancy, would be lower than the dosage in the Pill, therefore avoiding the Pill's potentially dangerous side effects.

Additional up-to-date information on birth control methods can be obtained from Planned Parenthood–World Population chapters located throughout the United States.

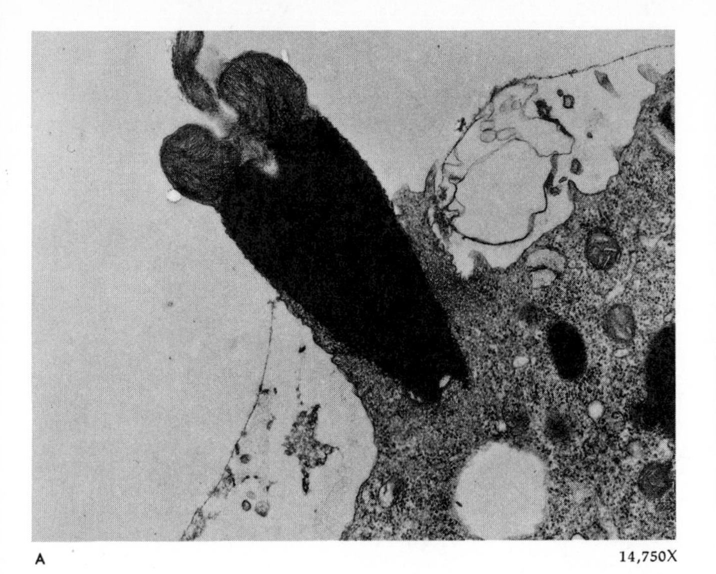
A 14,750X

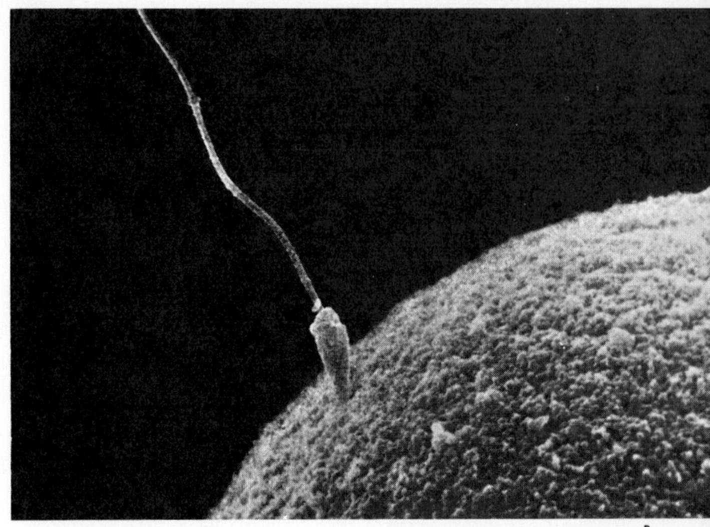
B

FIGURE 12.17

The moment of fertilization shown in a transmission electron micrograph (A) and a scanning electron micrograph (B). At the point where the head of the sperm contacts the egg, the membranes of the two cells fuse, and both sperm and egg are enclosed in one continuous membrane. As a result of rapid changes in the membrane, the egg surface is no longer receptive to other sperm. Other changes that occur within the egg include the triggering of the remaining meiotic divisions, increases in metabolic rates, and protein synthesis.

the haploid nucleus of the sperm fuses with the haploid nucleus of the egg to create the zygote—and potentially a new individual (see Fig. 12.17).

Formation of sperm and male hormones The testes in the human male have two essential functions—the formation of sperm cells and the production of male hormones. Two types of tissue within each testis perform these functions. **Seminiferous tubules,** highly convoluted tubes, are the site of sperm production. Under the influence of the pituitary hormone FSH (the same hormone that acts on the female follicles), specialized cells lining the wall of each tubule undergo sperm formation, or **spermatogenesis,** until mature sperm are developed. Between the seminiferous tubules are narrow spaces containing **interstitial cells.** Another pituitary hormone, called luteinizing hormone (LH) in females and interstitial cell-stimulating hormone (ICSH) in males, stimulates the interstitial cells to produce and secrete testosterone. Several masculinizing hormones, or **androgens,** are produced by the testes, but testosterone is the most potent of these; it influences the appearance of the male's secondary sex characteristics, including the maturation of the seminal vesicles, the growth of facial hair, and other changes associated with puberty. A feedback control system appears to govern the level of these gonadotropic hormones in the blood. A summary of the principal effects of these hormones is given in Table 12.1.

TABLE 12.1
Gonadotropic and sex hormone effects

HORMONE	ORIGIN	PRINCIPAL EFFECTS	CONTROL
FSH	Pituitary	Stimulates growth of ovarian follicle; estrogen production; maturation of seminiferous tubules and sperm production	Hypothalamus
LH	Pituitary	Stimulates progesterone production; ovulation; maturation of interstitial cells	Hypothalamus
Estrogen	Ovary (follicle)	Produces and maintains female sex characteristics; thickens lining of uterus; inhibits FSH production	FSH
Progesterone	Ovary (corpus luteum)	Thickens lining of uterus; inhibits LH production	LH
Testosterone	Testes	Produces and maintains male sex characteristics; stimulates sperm production; inhibits LH production	LH

Summary

1. The ability to reproduce is one characteristic of living things. This process in unicellular organisms may be accomplished by fission, budding, sporulation, conjugation, or sexual reproduction.

2. In multicellular plants, propagation may occur vegetatively, usually by means of runners or rhizomes. Most multicellular plants, however, have life cycles involving a sporophyte and a gametophyte generation. In the mosses the gametophyte generation dominates; in the ferns and seed plants the sporophyte generation is most conspicuous. In these higher plants the gametophytes are part of the flower or cones.

3. Some multicellular animals propagate by budding or parthenogenesis, but most rely on sexual reproduction. In some species the individuals are hermaphrodites, having the reproductive structures of both sexes. In different species fertilization may be either external or internal. External fertilization occurs most readily in an aquatic environment. Internal fertilization is characteristic of most land animals, since it increases the probability that the male and female gametes will be able to meet.

4. Ensuring the survival of the embryo also presents different problems on land and in water. In an aquatic environment eggs need not have elaborate protective coverings. Fish and amphibians lay their eggs in the water. Birds and reptiles lay eggs surrounded by membranes and a hard shell that protect the embryo and provide moisture and nutrients. In mammals internal development offers the embryo the same protection, while freeing the mother from nest tending.

5. The genitalia are the organs of reproduction; the gonads are those organs that produce gametes. In the human female the gonads are internal and are called ovaries; in the human male the gonads are outside the body cavity and are called testes.

6. The sex hormones play an important role in maintaining sex characteristics and fertility. In the female the interaction of FSH, LH, estrogen, and progesterone causes the ripening of the ovum, its release from the follicle into the fallopian tube, and the preparation of the uterus for the reception of the fertilized egg. If the ovum is not fertilized, the lining sloughs away and a new menstrual cycle is begun. In the male the testes produce both sperm, in the seminiferous tubules, and androgens, in the interstitial cells. LH stimulates the production of the androgens.

7. The sperm, produced in the testes, pass through a system of ducts to the urethra. Seminal fluid, resulting from the secretions of associated glands and containing millions of sperm, is ejaculated from the penis, an organ of spongy erectile tissue through which the urethra passes.

8. The human egg is fertilized internally as sperm are deposited in the vagina close to the cervix and make their way into the fallopian tubes. Only one sperm can fertilize an egg because the egg forms a fertilization membrane impenetrable to other sperm.

REVIEW AND STUDY QUESTIONS

1. What is conjugation? Why is it not a true form of reproduction? Why is it nonetheless usually discussed in connection with reproduction?

2. What are the names of the two structures that are most often involved in vegetative reproduction? How does each one function?

3. Explain what is meant by the term "alternation of generations" in plants.

4. What reason or reasons can you give for the fact that a very large proportion of the plants we are familiar with are flowering plants?

5. In terms of evolutionary importance, compare sexual and asexual reproduction.

6. What besides the process of reproduction itself do you think would be included in the term "perpetuation of kind"?

7. What is the essential difference between parthenogenesis and hermaphroditism?

REFERENCES

Austin, C. R. 1965. *Fertilization.* Prentice-Hall, Englewood Cliffs, N.J.

Golanty, E. 1975. *Human Reproduction.* Holt, Rinehart and Winston, New York.

Hardin, G., ed. *Population, Evolution, and Birth Control,* 2nd edition. Freeman, San Francisco.

Hutt, S. J., and C. Hutt, eds. 1973. *Early Human Development.* Oxford, London.

Pengelley, E. T. 1974. *Sex and Human Life.* Addison-Wesley, Reading, Mass.

Rowlands, I. W., ed. 1966. *Comparative Biology of Reproduction in Mammals.* Academic Press, New York.

Van Gelder, R. G. 1969. *Biology of Mammals.* Scribner, New York.

Van Tienhoven, A. 1968. *Reproduction of Vertebrates.* Saunders, Philadelphia.

SUGGESTIONS FOR FURTHER READING

Harrison, R. G., and W. Montagna. 1973. *Man,* 2nd edition. Prentice-Hall, Englewood Cliffs, N.J.

Includes a highly factual yet readable review of human reproduction and behavior.

Odell, W. D. and D. L. Moyer. 1971. *Physiology of Reproduction.* Mosby, St. Louis

A discussion by two physicians of the human reproductive organs and their associated hormones. Excellent illustrations, including electron micrographs, clear line art.

Reuben, D. 1969. *Everything You Always Wanted to Know About Sex.* David McKay, New York.

Popularly written summary of all facets of human sexuality from a physiological and cultural point of view.

Wilmoth, J. H. 1967. *Biology of Invertebrata.* Prentice-Hall, Englewood Cliffs, N.J.

Clearly organized treatment of each invertebrate phylum, including a brief coverage of reproduction and embryology.

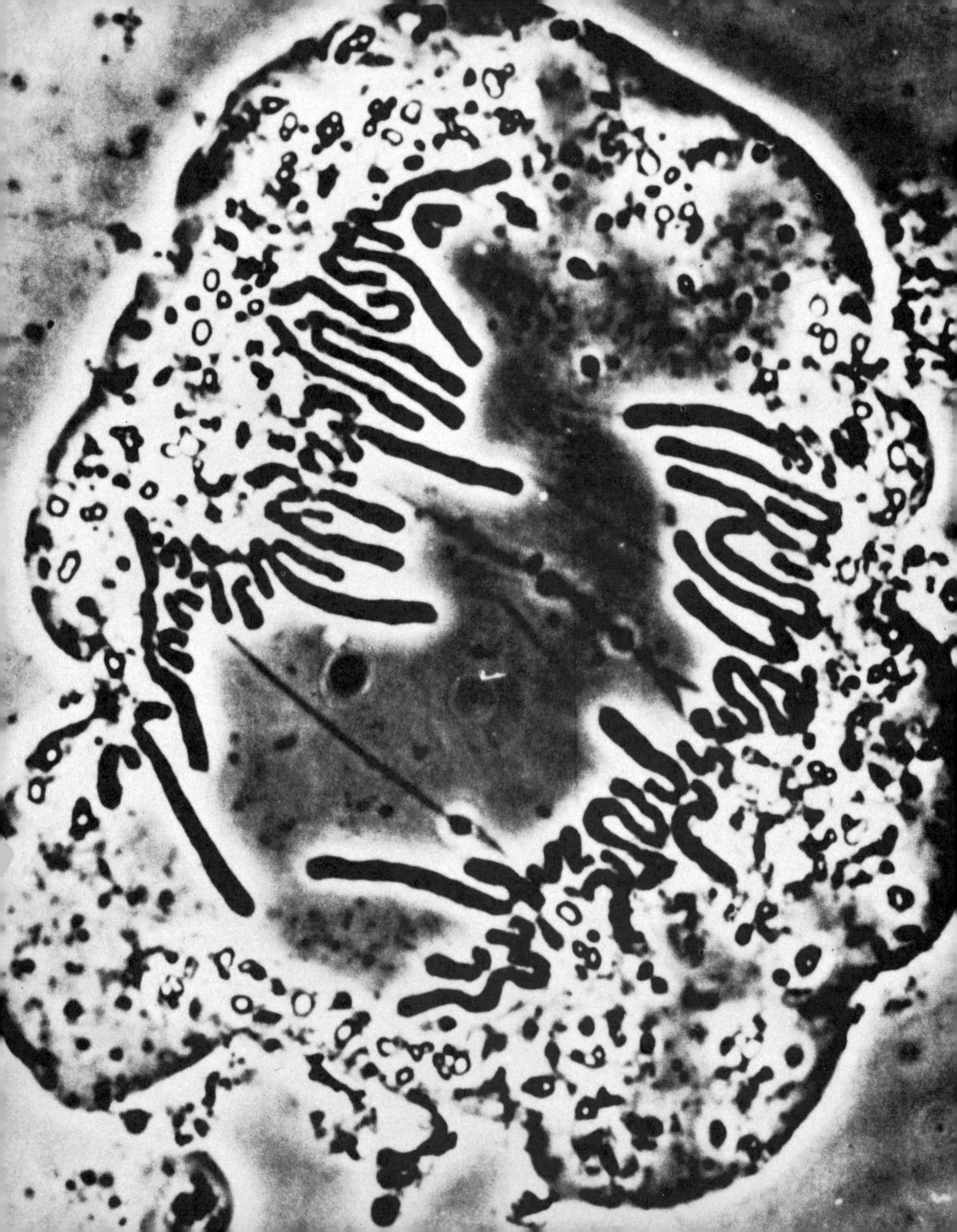

CELLULAR DIVISION
chapter thirteen

Growth, repair, and reproduction depend on the ability of cells to divide. The ceaseless dance of the chromosomes as they duplicate and separate into new cells has two variations—mitosis for cell replacement, body growth, or asexual reproduction, and meiosis for the formation of gametes. Shown here is the late anaphase stage of mitosis in a living endosperm cell of the African blood lily (Haemanthus katherinae).

Living organisms have a unique combination of properties. Not only do they maintain and repair themselves; they also grow and reproduce themselves. The maintenance process alone is remarkable. Life is in fact a race between replacement and death. Each of us, in a lifetime, will produce a total of 40 pounds of skin. Every 36 hours, we must replace the entire population of cells lining the intestines. An average red blood cell lasts 120 days; a white cell usually lasts only seven days. We are constantly renewing various kinds of cells in our bodies. At the heart of cellular replacement is the process of cell division, which makes possible the necessary maintenance and repair of tissues.

Cell division actually consists of two related processes; nuclear division (**mitosis** and **meiosis**) and division of the cytoplasm (**cytokinesis**). Usually these two processes occur in the same cycle, although in some instances—e.g., cancer—nuclear division may occur in the absence of cytokinesis. Mitosis is the form of cell division that occurs in normal growth and replacement of somatic tissues. In mitosis the number of chromosomes remains constant. Meiosis, a specialized form of mitosis, is found only in tissues involved in sexual reproduction and gamete or spore formation. Cellular divisions during meiosis reduce the number of chromosomes to half the number that is characteristic of the species. Thus all gametes of higher plants and animals are haploid cells.

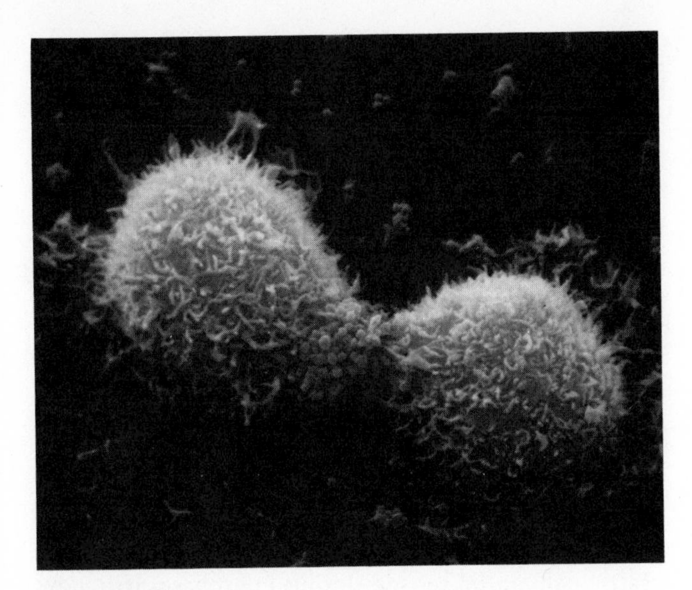

FIGURE 13.1
Cells dividing in tissue culture. The stimulus to stop dividing, which normally prevents uncontrolled cell division in a tissue, does not seem to operate on cancer cells.

Both cytokinesis and mitosis have recently been of considerable interest to cancer researchers because improved understanding of normal growth may contribute to the understanding of seemingly uncontrolled growth, or cancer (see Fig. 13.1). Ordinarily, cells undergo division and multiply, but not indefinitely. Some people are bigger, some smaller, but all stop growing at some point. Organs—the heart, kidneys, eyes, and the others—reach a certain size in the adult and then stop growing. Cell division usually continues at a proper rate to ensure that dead or injured cells are replaced by new cell division. Apparently there are control mechanisms—consisting of chemical signals—that start and stop normal cellular division. But in cancer the signal system fails, leading to an uncontrolled reproduction of cells. Many now hope that the study of mitosis will lead to a cure for or prevention of cancerous growth.

The second process that we will study in this chapter, meiosis, is responsible for the production of sperm and egg cells, or **gametes.** But, perhaps most important, the process is critical to evolution because it permits tremendous variation in the progeny of sexually reproducing organisms. Even when such organisms have relatively long life spans, their populations can evolve rapidly enough to survive through even quite rapid environmental shifts. In the next chapter we shall see how such variations arise; our discussion of molecular genetics will explain the chemical mechanisms of inheritance.

Cell headquarters

Inside every cell is a membrane-bound structure that contains almost all of that cell's DNA. This organelle is the nucleus, and it serves as cell headquarters. It contains information molecules enabling it to exert control over routine cellular activities and behavior. The information or instructions contained within the nucleus may be transmitted from generation to generation, and offspring cells thereby retain the basic properties of parent cells.

Early cell biologists sought to determine how essential the nucleus was for normal cell life. Nuclei can be removed from the cells of plants and animals by microsurgery without major harm to the cytoplasm. Once a nucleus is removed, a cell may continue to function for a time as if nothing had happened. But when it uses up its "stock" of molecules made on instructions from the nucleus, it dies. For example, when the nucleus is removed from an amoeba, the enucleated amoeba seems to act in normal fashion, but gradually it stops feeding, its movements diminish, its shape changes, and it dies.

Clearly the information inside the nucleus is vital. But how is information stored in the nucleus, and how is it transmitted from generation to generation? We now know that the information in a cell, necessary for the synthesis of all cellular components, is stored in the macromolecule DNA, which is found in dark-staining nuclear material called **chromatin.** Normally chromatin is a very diffuse collection of long, thin, twisting fibers, but during cell division, it forms rodlike structures called **chromosomes.** Each chromosome appears to have only a portion of the total information in a cell. In order for a cell to function correctly, it must have an entire set of chromosomes. The chromosomes in a set appear to be arranged in pairs; each member of a pair is called a **homolog** of the other member. There is thus a degree of duplication of information by each homologous chromosome in a pair. But as we shall see, the information is not precisely duplicated. One might liken the situation to that of the house-builder with blueprints that include detailed instructions for building a kitchen but allow two choices of color for the cabinets.

The number of chromosomes in the set varies tremendously among different types of organisms. The guinea pig, for example, has 62 chromosomes in every cell; the crayfish has 200; the potato, 48; and the human being, 46. When new cells replace dead or injured ones in the body of an organism, they function just like the ones they replace, because in mitotic cell division the entire chromosome set is duplicated, and a complete set of chromosomes is transmitted to each new cell.

But a different type of cell division operates in the production of gametes, sperm and egg, necessary for sexual reproduction. For here the nuclei of two cells become united at fertilization. A second kind of division, meiosis, makes sure that each gamete gets only one-half the usual number of chromosomes in its nucleus, so that the zygote formed at fertilization has only one whole set of information.

Mitotic division

There is no certain knowledge of what *initiates* mitosis or how it is stopped, but it appears to be triggered in part by the ratio of nuclear material to cytoplasmic volume. It takes place when nuclear volume is high relative to cytoplasmic volume. Even if this is not the cause, we understand the events of mitosis well enough to describe them clearly. As a matter of convenience, we consider mitosis as occurring in a series of phases, each of which can be examined separately. But the cell's life cycle, of which mitosis is only a part, is actually a continuum (see Fig. 13.2); we examine it by stages, or phases, only for teaching purposes.

THE STAGES OF MITOSIS

Interphase Interphase extends from the last actual cell division to the beginning of the next. It is the period in which the cell is not dividing. For a long time, the cell was thought to be resting during interphase. It is now known that this is not true.

In interphase, a cell metabolizes—produces protein, grows, uses nutrients, and, most vital, synthesizes DNA for a new set of chromosomes. Before it divides again, it must **replicate** the chromosomes so it will have *two* full sets available for the next division, one for itself and one for its daughter cell.

During interphase the nucleus can be seen as a distinct structure bounded by a membrane, but no chromosomes are visible as such. Instead, there is what appears to be a tangle of threads. For a long time, scientists believed that chromosomes did not exist at all during interphase. The present evidence, however, indicates that they are there, but they are thin, elongated, and wrapped around each other so that they cannot be distinguished individually with a light microscope.

Also seen near the nucleus in animals cells during interphase (but not in most plant cells) are two tiny bodies, called centrioles. In many cells the centrioles replicate during interphase.

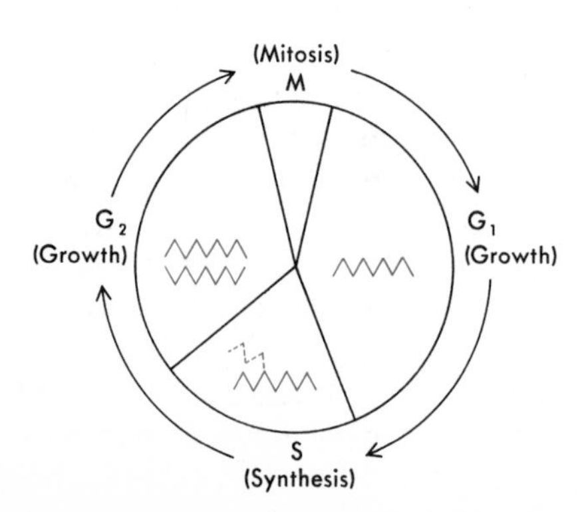

FIGURE 13.2

The life cycle of a cell represented as a clock circle to emphasize the small proportion of time spent in mitosis. The remainder of the time, interphase, consists of three periods of activity in most cells. During the first growth period (G_1) the cell builds materials especially for its own maintenance. In the synthesis phase (S) the major steps in DNA synthesis and replication take place. During the second growth period (G_2) the cell completes DNA replication and synthesizes proteins in preparation for division in the period of mitosis (M) that follows. The drawing within each segment suggests the status of DNA in that period.

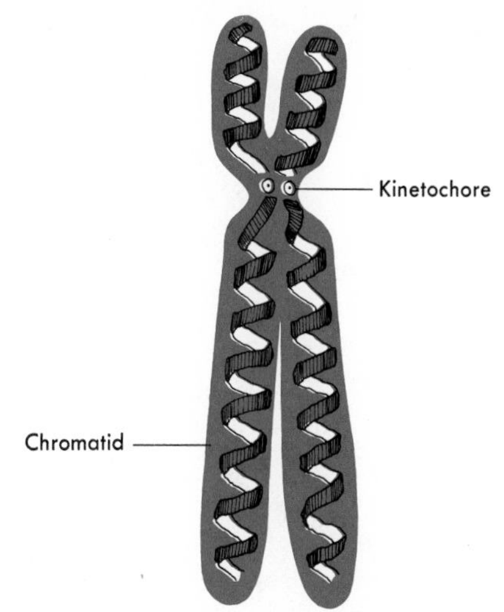

FIGURE 13.3

Late prophase chromosome in a cell that is about to divide mitotically. The original chromosome has replicated itself, and the two chromatids are held together at a granular region called the kinetochore. During the next stage in mitosis, metaphase, the chromatids will separate from each other at the kinetochore, and each will then become a single chromosome.

Prophase In prophase, the earliest stage of actual mitosis, the chromosomes now become visible—first as long, thin threads. As prophase progresses, the threads shorten, thicken, and become rodlike.

In the late stage of prophase, each chromosome can be seen as two components (each resembling a chromosome), called **chromatids,** which are joined together by a small body, a **kinetochore** (or **centromere**), sometimes near the end and sometimes near the center of the chromosome (Fig. 13.3). The chromatids are exactly alike—as indeed they should be, for during chromosome replication in interphase, one serves as the model for the production of the other.

During prophase in animals, the replicated centrioles begin to separate, and each one takes up a position on either side of the nucleus. The centrioles seem to be the site of synthesis of fibrils, actually microtubules, which stretch across the nucleus forming a structure called the **spindle apparatus.** The spindle-fiber microtubules seem to be of two types: some stretch from centriole to centriole uninterrupted; others are attached at their center to a kinetochore that holds two chromatids together. The centrioles at either end of the spindle apparatus are often referred to as the poles of the spindle; from them also arise, in addition to the microtubules of the spindle, **astral rays,** short microtubules that make the poles look like a pincushion (Fig. 13.4). In plants the spindle forms despite the absence of centrioles, so the role of the centriole in spindle formation is not quite clear.

As the spindle forms, the nuclear membrane begins to disappear. The formation of the spindle also seems to affect the chromosomes. The chromosomes, which until now have been randomly distributed inside the nucleus, begin to orient themselves and to move toward the middle of the spindle. At the very end of prophase, each chromosome—which is clearly composed of two chromatids—has joined its own individual fibril in the spindle.

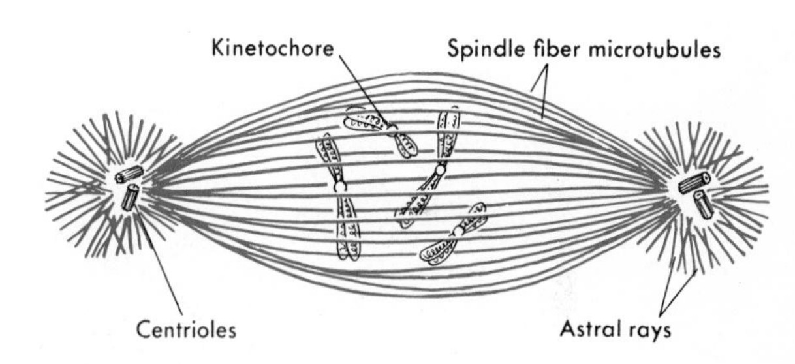

FIGURE 13.4

An artist's conception of the spindle apparatus of an animal cell during mitosis. Note the raylike microtubules around the centrioles, forming the astral rays. Each chromosome is attached to a spindle fiber by its kinetochore.

Metaphase Metaphase is a brief stage, but it is particularly useful for cytologists interested in studying chromosomes. During metaphase, the chromosomes are lined up, easily visible in the center, or equator, of the spindle.

During metaphase, each kinetochore, which has joined together the two chromatids of a chromosome, undergoes division. As this happens, each chromatid stops being one of a pair on a chromosome and becomes a single chromosome in its own right. Now, for a moment, the cell contains double the normal complement of chromosomes so that half can be passed on to the daughter cell.

Anaphase As anaphase begins, the once-paired chromatids that formed each chromosome begin to move. The

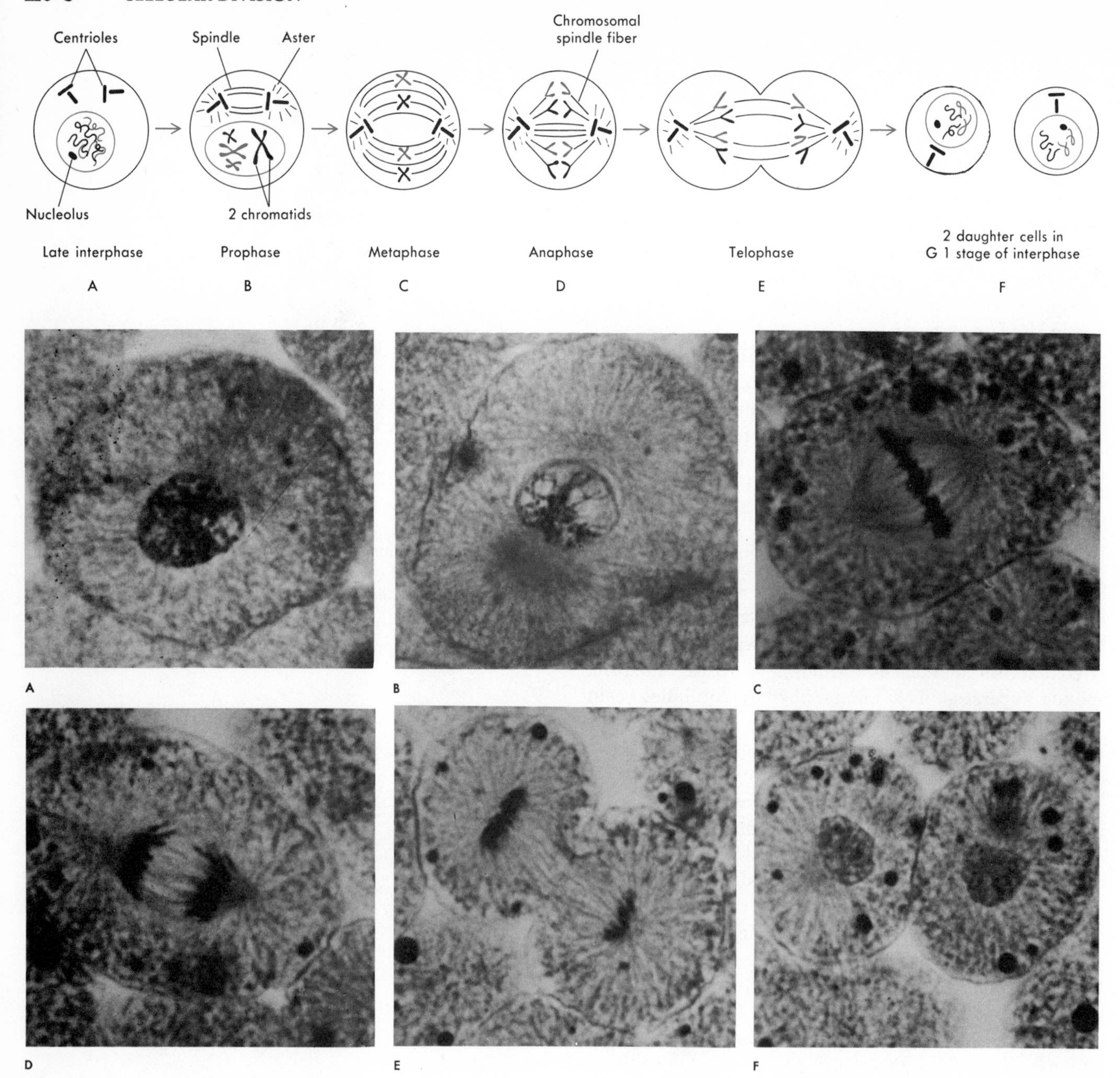

FIGURE 13.5

Mitosis in a whitefish cell. (A) Interphase, during which DNA replication occurs; the chromosomes, in the form of chromatin, remain attenuated twisted threads. Just before prophase begins, the centrioles of an animal cell separate. (B) Prophase is signaled by the appearance of chromosomes. By the end of prophase, each chromosome is clearly composed of two chromatids, joined at the kinetochore, and the nuclear envelope has disappeared. (C) In metaphase the chromosomes align along the equator of the spindle. (D) In anaphase the kinetochore region seems to divide, separating the chromatids. Each chromatid (now a chromosome) appears to be pulled or pushed away from the equator toward the poles of the spindle. (E) In telophase the movement of chromosomes has been completed, and cytoplasmic division begins. (F) At the end of telophase the daughter cells have formed with nuclear envelopes intact, cytoplasmic division is completed, and each cell enters interphase.

chromosomes act as if they are being pulled apart lengthwise—and one chromatid moves toward one pole of the spindle, while the other moves toward the other pole.

Early in anaphase, the chromosomes can be seen as two groups, equal in number and appearance, still only a short distance apart. Late in anaphase, there is greater separation as the two groups near their respective poles on the spindle.

What accomplishes the separation? What can explain the chromosome movement? There is still do definitive answer to the mystery, although there are several theories.

According to one theory, the fibrils pull the chromosomes. Proponents point out that although the kinetochore, which connects chromosome to fibril, seems to move steadily during anaphase, the arms of the chromosome (the portions not connected directly to the kinetochore) seem to drag along behind. A second theory suggests that the chromosomes may be pushed toward the poles instead of being pulled by the part of the fibrils between the centriole and the kinetochore. The idea that the chromosomes are pushed seems consistent with the way the microtubules in the spindle apparatus change length. The microtubules, which are quite rigid rodlike assemblies of protein subunits, elongate by adding subunits at either end, and they shorten by dropping subunits off. On the other hand, recent support for the idea that the chromosomes are pulled has come from the discovery of actin in nonmuscle cells. But whether they are pushed or pulled, they do move apart—until telophase is reached.

Telophase In the last stage of mitosis, telophase, the two sets of chromosomes have arrived at their respective poles. Now they begin to spread out to become, soon, as indistinguishable as they normally are during interphase. Gradually, the spindle vanishes. Nuclear membranes appear, forming about each of the two new nuclei. Late in telophase the centrioles duplicate themselves in animal cells; telophase ends when the new nuclei show all the characteristics of interphase.

This sequence of events, illustrated in Fig. 13.5, does not represent the whole story of cell division, which, along with nuclear mitosis, may include cytokinesis, or division of the cytoplasm.

CYTOKINESIS

Cytokinesis is common but not an invariable occurrence. In some plants, mitosis may take place without cytoplasmic division. The result is a body with several nuclei but few or no cellular partitions, such as in the multicellular algae *Nitella* and *Caulerpa*. Moreover, mitosis without cytokinesis is responsible for the giant multinucleate cells characteristic of some kinds of cancer.

When cytokinesis occurs, it often begins late in anaphase and is completed during telophase. In animals, cytokinesis usually begins when a cleavage furrow forms around a cell. It develops in the equatorial region of the spindle and grows progressively deeper, until finally it severs both spindle and cell, producing two cells.

Even now, almost nothing is known about the mechanism of cleavage furrow formation. One theory is that a special ring of cytoplasm develops around the exterior of the cell and contracts and pinches. A second theory is that the furrow grows inward as the result of new membrane formation within the cell. Still a third holds that some fibrils may glue themselves to the cell surface and pull inward. A factor working against the last theory, however, is that in some instances, cytokinesis actually takes place after rather than during mitosis, at a time when the spindle no longer can be seen.

In plant cells, which are surrounded by rigid walls, cleavage furrows never appear. Instead, the cytoplasm is divided by a **cell plate** and a new cell wall. The cell plate seems to be formed from vesicles, or membranous sacs, which are made by the Golgi apparatus; these membranous sacs fuse near the center of the spindle apparatus, which is still present at telophase, to form a flattened structure called the **phragmoplast.** As the phragmoplast grows by the addition of new membranes from the Golgi vesicles, it spreads toward the edges of the cell, finally forming a cell plate, a flattened, double membrane which cuts the cell in half. In the space between the two membranes of the cell plate, a new cell wall forms, probably from the contents of the Golgi

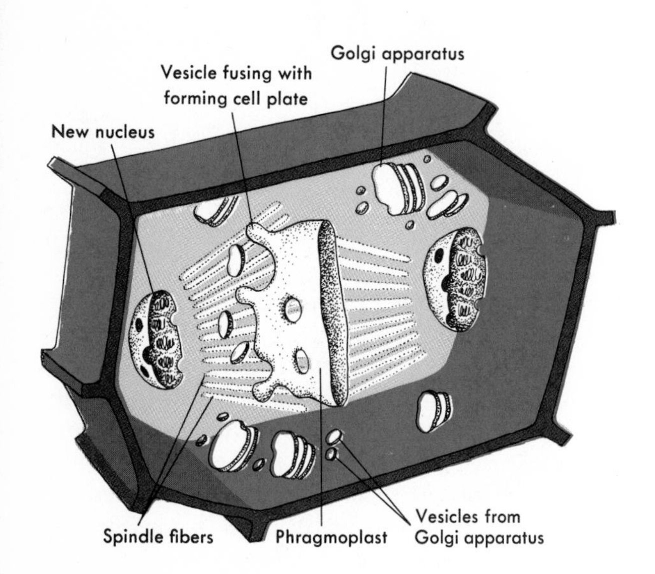

FIGURE 13.6

Two major differences between plant and animal mitosis. First, centrioles are rarely found in plant cells, and the spindle poles form without them. Second, cytokinesis (cytoplasmic division) and the production of a new cell wall for each daughter cell occur by the formation of a phragmoplast from fused vesicles made by the Golgi apparatus. The phragmoplast grows toward the inner edges of the cell in telophase, finally forming a cell plate, a double membrane with a space between, which separates the two cells. The wall that forms in the space appears to be made of materials carried in the vesicles that formed the phragmoplast.

vesicles whose membranes contributed to the formation of the cell plate (Fig. 13.6).

THE IMPORTANCE OF MITOSIS

Mitosis is a basic life process—for growth, for dead cell replacement, for found healing, sometimes for body part regeneration, and in many lower plants and animals for the simple asexual production of new individuals.

Commonly among unicellular organisms—for example, protozoans, bacteria, yeasts, and many algae—reproduction is accomplished by **fission:** the nucleus goes through mitosis and the entire organism then divides to form two separate organisms (Fig. 13.7). **Budding** is another form of asexual reproduction accomplished by mitotic division. For example, an adult hydra, a small freshwater animal, grows part of its body by repeated mitotic division. The new part that forms, called a **bud,** takes the shape of a new, small hydra. Eventually the small hydra breaks off from the parent body and becomes an offspring. Many other organisms similarly reproduce by budding. The wine- and beer-making industries are dependent on this process to fill their beer "wort" or wine "must" with billions of active young yeast cells that can ferment and make alcohol.

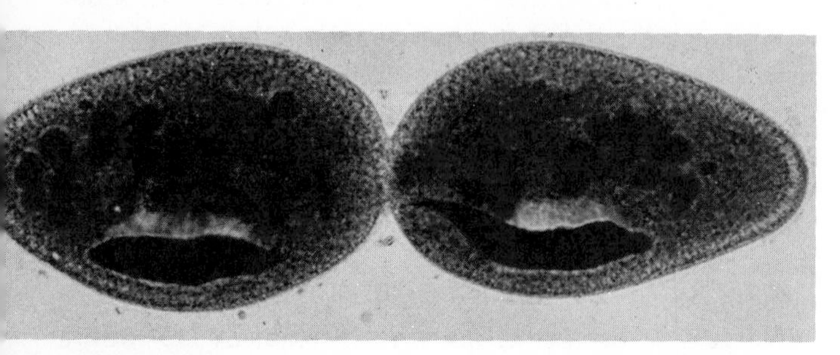

Courtesy of Carolina Biological Supply Company

FIGURE 13.7

Single-celled organisms generally reproduce asexually by fission. When an individual reaches a certain size, it divides, creating two daughter individuals, each exactly like the parent. Here fission in a paramecium is almost completed. The two daughter cells are connected by only a tiny amount of protoplasm.

Meiotic division

Sexual reproduction involves the merging of two sex cells called gametes. If the resulting fertilized cell is to have the normal number of chromosomes in its nucleus, each sex cell must come to the meeting with only half the normal chromosome complement. A special type of *reduction* cell division, meiosis, evolved that ensures continuity of chromosome number from generation to generation.

Recall that each human body cell contains 46 chromosomes. Actually, there are 23 pairs—23 different types of chromosomes, two of each type. The normal number (46) is called a **diploid** number, indicating the pairing arrangement of the 23 types.

In a gamete, there are 23 types of chromosomes, but just one of each type. This reduced number is called the **haploid** number.

Meiosis, which is responsible for giving each gamete a haploid number of chromosomes, starts with a cell containing the diploid number. Although the cell undergoes two successive divisions, the chromosomes are replicated only once, resulting in four daughter cells, each containing the haploid number.

Research in recent years has increased our understanding of how faulty meiosis may affect development in humans. In the next chapter we will examine some of the abnormalities that occur because of the presence in human cells of single extra chromosomes.

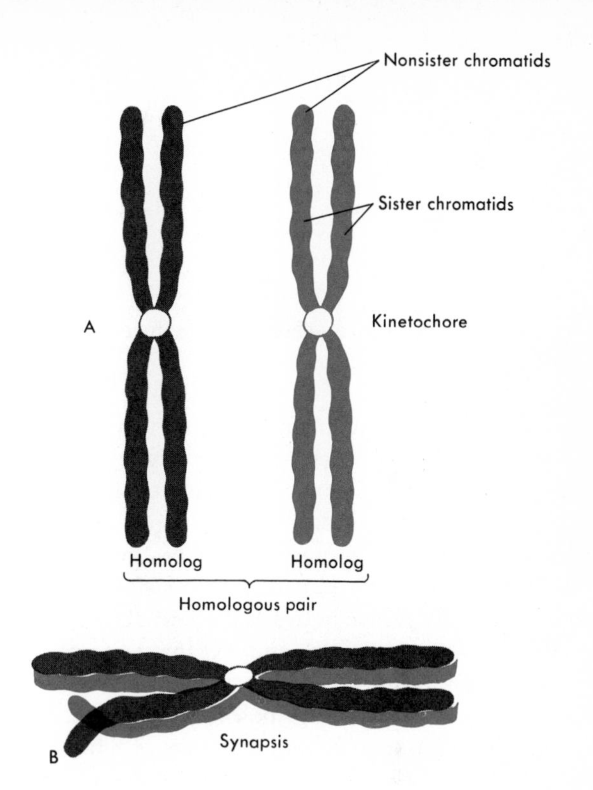

FIGURE 13.8

(A) In a diploid organism there are two (homologous) chromosomes of each type. The types may be distinguished by their shape and size and by the position of the kinetochore, which may be found at different distances from the end of the chromosome. (B) During meiosis the homologous chromosomes come together in synapsis, during which they may cross over or exchange parts of chromatids.

THE FIRST DIVISION OF MEIOSIS

Prophase 1 The first stage of meiosis is, up to a point, very much like the beginning of mitosis.

As in mitosis, replication of chromosomes has occurred during interphase. As prophase 1 begins, the chromosomal mass in the nucleus changes, and chromosomes become visible when viewed with the light microscope. Gradually, too, the spindle forms and the nuclear membrane disappears.

But a distinctive feature of meiotic prophase 1 is that as the chromosomes continue to take shape, the two members of each pair gravitate toward each other until they lie side by side, not fused but intertwined, in a process called **synapsis.** At this point, each chromosome consists of two chromatids, so that a synapsed chromosome pair has *four* chromatids. The two chromatids of a single homolog are called **sister chromatids.** Chromatids belonging to the two different homologs are called **nonsister chromatids** (see Fig. 13.8). Because a synapsed chromosome pair has two chromosomes, it is often referred to as a **bivalent.** Then the two chromosomes of a pair move apart slightly and, as this happens, some of the chromatids cling to each other at one or more sites, where they may exchange parts in a process called **crossing-over.** The site of crossing-over is called a **chiasma** (plural: **chiasmata**). A crossing-over between sister chromatids will not cause genetic change, since neither the amount nor the function of genetic material is altered. However, a chiasma between nonsister chromatids, which come from separate parents, will result in a new combination of genes on the homologous chromosomes and therefore more variety in offspring. (There will be further discussion of crossing-over in Chapter 14.) Toward the end of prophase 1, each synaptic pair of chromosomes moves as a unit to become oriented on the spindle.

Metaphase 1 In this second stage of meiosis, the chromosome pairs (containing four chromatids) are arranged around the spindle and are attached to spindle fibrils by their kinetochores. However, in meiosis, the *chromatids* constituting a chromosome do not separate during the first phases of division; instead, each of the homologous *chromosomes* in a bivalent (each chromosome still composed of two chromatids) will be separated. During the first meiotic metaphase, each kinetochore becomes attached to a spindle fiber.

Anaphase 1 The first anaphase in meiosis also differs markedly from its counterpart in mitosis. In mitotic anaphase, there is a division of the single kinetochore of each chromosome, whereupon the chromatids become free to move toward opposite poles of the spindle. But in meiosis, there is *no kinetochore division.* This is the fundamental difference between the first stage of meiosis (as well as crossing-over) and the process of mitosis. The chromosomes of each synaptic pair move away from each other toward opposite poles without the chromatids' having separated. If both members of a homologous pair travel to the same pole, the outcome will be two gametes with an extra chromosome each and two gametes that are short one chromosome each. In the next chapter we will consider some of the results of such abnormal variations.

There is no constraint on the homologous chromosomes to move toward a *particular* pole. And since each chromosome in a homologous pair is slightly different, one having come from the mother and the other from the father at conception, this essentially random movement assorts the chromosomes. As we shall see, the assortment, or mixing up of maternal and paternal chromosomes during meiosis, greatly contributes to the varied appearance of offspring.

In contrast now to mitotic anaphase, during which 46 chromosomes (in humans) move to each pole, only 23 do so during meiotic anaphase. The chromatids of each chromosome do not yet separate.

Telophase 1 As this stage begins, the chromosomes have arrived at the poles of the spindle. At each pole, there is a haploid set of chromosomes, but each chromosome is composed of two chromatids attached to one kinetochore.

The spindle disappears, new nuclear membranes appear, and the chromosomes uncoil and fade from view. Except for the fact that each of the two new nuclei formed in meiosis has half the number of chromosomes originally present in the parent nucleus, telophase in mitosis and meiosis is much the same.

Interkinesis Following the telophase stage in mitosis, interphase occurs. In meiosis, however, telophase 1 may be followed directly by the second sequence of divisions, or it may be followed by a short period called interkinesis. Whereas in interphase, genetic material is replicated and new chromatids are formed, replication does not occur during interkinesis; each chromosome is *already* composed of two chromatids.

THE SECOND DIVISION OF MEIOSIS

After interkinesis (or directly after telophase 1), a second sequence of divisions, much like that of mitosis, begins.

Prophase 2 In each daughter cell produced by the first division of meiosis, a spindle forms, and the nuclear membrane disappears. The chromosomes coil and begin to move toward the middle of the spindle.

Metaphase 2 As each chromosome arrives at the spindle, it becomes attached, by its kinetochore, to a fibril. As in mitotic metaphase, the kinetochores divide, and the chromatids are now free to move away from each other. The division of the kinetochore is the fundamental difference between metaphase 2 and metaphase 1.

Anaphase 2 Now the chromatids of each chromosome move apart, one proceeding toward one pole of the spindle while the other moves toward the other pole. When the chromatids are visibly separated from one another, they are once again called chromosomes.

Telophase 2 As in mitotic telophase, the chromosomes arrive at their poles, the spindle vanishes, and new nuclear membranes appear.

The new nuclei are unlike those formed during mitotic telophase. *Two* haploid cells with replicated chromosomes have resulted from the first meiotic division. And in the second division, each of these cells divides again—to produce four haploid daughter cells, with the chromosomes of the parent cell distributed among them.

The sequential steps of meiosis are summarized in Fig. 13.9.

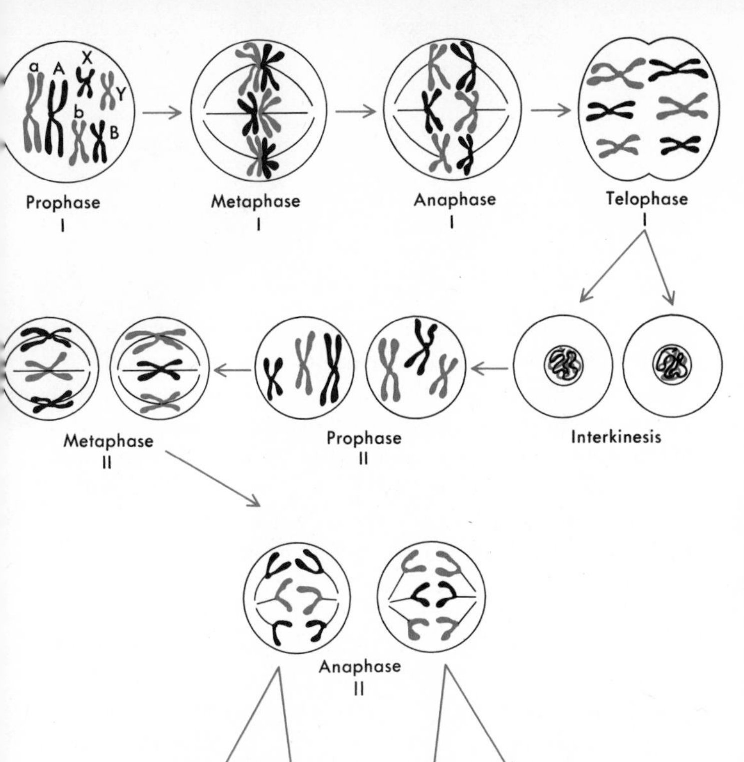

FIGURE 13.9

Meiosis. Prophase 1: In this prolonged stage, in which several substages are recognizable, homologous chromosomes (shown in color and black, each homolog in a pair labeled with a letter) seek each other out and come together in synapsis. Each chromosome is clearly composed of two chromatids. Metaphase 1: The bivalents, each composed of two homologous chromosomes (four chromatids) line up on the equator of the spindle. Anaphase 1: The homologous chromosomes are pulled or pushed to opposite poles. Note that the chromatids of each chromosome have not separated. Note also that the distribution of the homologs, shown in color, is random. Telophase 1: Two new cells result. The nucleus of each contains one chromosome from each homologous pair, but distribution of the homologs, represented in color and black, is random. Interkinesis: No replication occurs between the two divisions of meiosis. The second round of meiotic divisions is similar to mitosis. Anaphase 2: The identical chromatids composing each chromosome are separated from each other, as in mitosis. Telophase 2: Four haploid cells result. Note the random distribution of colored and black homologs.

Comparison of mitosis and meiosis

Mitosis and meiosis are forms of nuclear division which are similar in many respects; however, mitosis occurs only where the new cells resulting from division must be identical structurally and functionally—in cell replacement in a multicellular organism, production of additional cells for growth, or production of new organisms by fission. Meiosis occurs *only* when organisms produce haploid cells—gametes and meiospores—special reproductive cells with *one-half* the normal number of chromosomes.

A graphic comparison of the various stages of mitosis and meiosis appears in Fig. 13.10.

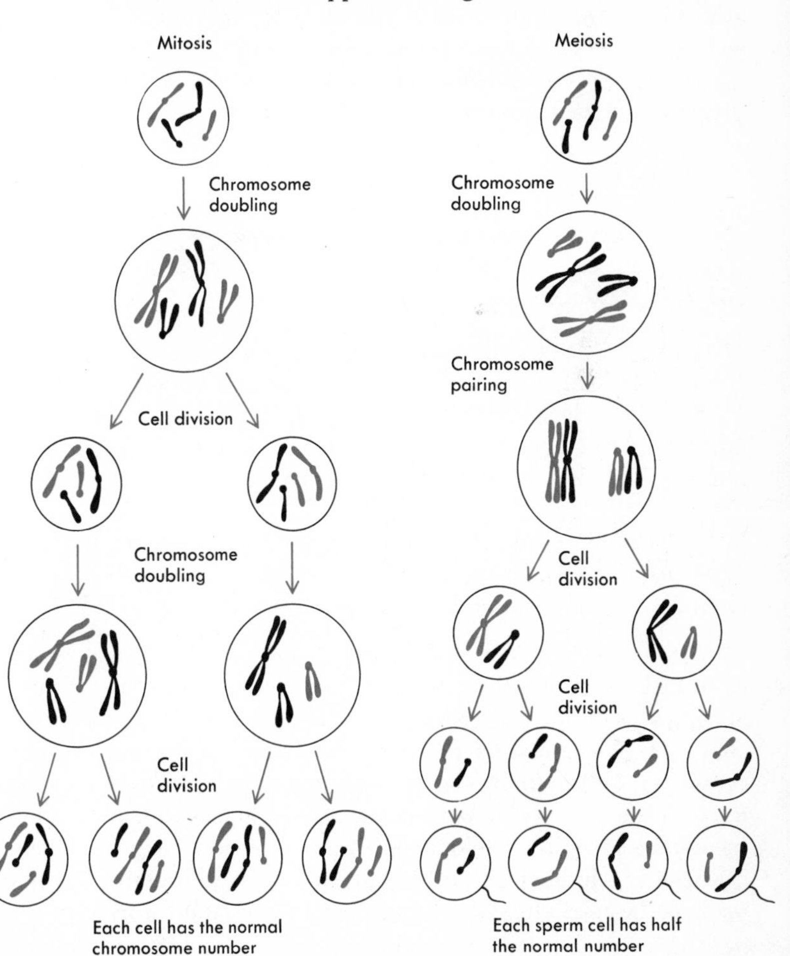

FIGURE 13.10

Comparison of mitosis and meiosis. Black chromosomes came from one parent, colored from the other.

Both types of nuclear division begin the same way: DNA replication occurs, and a new chromosome forms for every one that was present. In both types of division, each new chromosome formed remains attached to its parent by a region called the kinetochore; while attached in this way, each chromosome (parent and daughter) is called a chromatid. In meiosis, homologous chromosomes pair, forming bivalents composed of four chromatids; in mitosis no pairing of homologous chromosomes occurs.

In mitosis, two daughter cells, identical in chromosome type and number to the parent cell from which they came, are formed after the chromosomes duplicate once, and subsequent movements separate the chromatids. In meiosis four cells with half the chromosome number of the parent cell form.

In meiosis, chromosome duplication occurs once, but chromosome movements occur twice. During the first meiotic anaphase, homologous chromosomes in each bivalent are separated from each other, resulting in two daughter cells, each with half a set of chromosomes. At this stage, however, each chromosome is still composed of two chromatids stuck together at their kinetochore. The second meiotic anaphase separates the chromatids of each chromosome, resulting in four daughter cells, each containing one homologous chromosome from the homologous pairs of the diploid parent cell, and therefore a haploid, or half, set of chromosomes.

The importance of the information contained in the chromosomes, in units we now call genes, was appreciated only late in the nineteenth century. It was the work, forgotten for many years, of an Austrian monk which led to the concept of genes as "atoms of heredity." We shall consider Mendel's work and the principles of Mendelian genetics in the next chapter.

Summary

1. Growth, dead cell replacement, regeneration, and certain kinds of asexual reproduction depend on cells' ability to divide mitotically. During the four stages of mitosis, a cell's genetic material, replicated during interphase, is divided equally: Each new daughter cell gets the same number and kind of chromosomes as the original cell. When chromosome division is nearly finished, cytoplasmic division occurs, and two complete cells with identical sets of chromosomes are produced.

2. Plant and animal mitosis differ in two ways. Most plant cells have no centrioles, and the spindle apparatus is formed without their presence at the poles. In animal cells, spindle formation appears to be directed by the centrioles, and cytoplasmic division occurs by constriction of the cytoplasm between the new nuclei which have been formed by telophase. In plant cells, the rigid cell wall precludes this mechanism; instead, a phragmoplast (a large, flattened, double membrane sac) forms near the equator of the spindle apparatus. It reaches the walls of the cell, forming a cell plate which separates the two new cells. A cell wall forms between the membranes of the cell plate, probably from materials contained in the Golgi vesicles.

3. Sexual reproduction and certain asexual plant reproduction stages are dependent on the formation of haploid cells from diploid cells by meiosis. Meiosis requires two cell divisions. During the first division, homologous chromosomes synapse. During synapsis genetic material may be exchanged between homologous chromosomes by crossing-over. At the first meiotic anaphase, homologous chromosomes are separated from each other. The second division is similar to mitosis: the chromatids of the double-stranded chromosomes are separated as the cells divide. The net result of meiosis is four haploid cells, each with a set of unique chromosomes.

REVIEW AND STUDY QUESTIONS

1. Why are the processes of cell division of primary significance to cancer researchers?

2. Name the stages of mitosis and state briefly what happens in each.

3. Name the stages of meiosis and state briefly what happens in each.

4. How many chromosomes are there in a typical cell in the human body? How many are there in a human sperm?

5. Describe the process of crossing-over and explain its significance.

6. Define each of the following and distinguish one from another: chromatin, chromatid, chromosome.

7. In what ways are mitosis and meiosis alike, and in what ways are they different?

REFERENCES

Huettner, A. F. 1949. *Fundamentals of Comparative Embryology of the Vertebrates,* revised edition. Macmillan, New York.

Mazia, D. 1974 (January). "The Cell Cycle." *Scientific American Offprint* no. 1288. Freeman, San Francisco.

———. 1961 (September). "How Cells Divide." *Scientific American* Offprint no. 93. Freeman, San Francisco.

Swanson, C. 1969. *The Cell,* 3rd edition. Prentice-Hall, Englewood Cliffs, N.J.

Swanson, C., et al. 1967. *Cytogenetics.* Prentice-Hall, Englewood Cliffs, N.J.

Wilson, G. B. 1968. *Cytogenetics.* Van Nostrand-Reinhold, New York.

———. 1966. *Cell Division and the Mitotic Cycle.* Van Nostrand-Reinhold, New York.

SUGGESTIONS FOR FURTHER READING

Cohn, N. S. 1964. *Elements of Cytology,* 2nd edition. Harcourt Brace Jovanovich, New York.

This clear and interesting description of the structure and function of cells includes useful information on mitosis and meiosis.

Garber, E. D. 1972. *Cytogenetics: An Introduction.* McGraw-Hill, New York.

A short but in-depth review of the various normal chromosomal characteristics of mitosis and meiosis, plus the abnormal or unusual characteristics of chromosomal behavior.

Giese, A. C. 1973. *Cell Physiology,* 4th edition. Saunders, Philadelphia.

An extremely valuable book for the interested biology student. Of particular importance is the section devoted to growth and cell division, including information on kinetics and mechanisms of mitosis.

Levine, R. P. 1968. *Genetics,* 2nd edition. Holt, Rinehart and Winston, New York.

Includes in Chapter 2 a concise, accurate treatment of mitosis and meiosis, accompanied by excellent photographs and drawings.

Loewy, A. G., and P. Siekevitz. 1970. *Cell Structure and Function,* 2nd edition. Holt, Rinehart and Winston, New York.

A good comprehensive description of cell structure and function. Chapter 4 contains a clear explanation of mitosis and meiosis, with excellent visual representation of the processes.

Novikoff, A. B., and E. Holtzmann. 1970. *Cells and Organelles.* Holt, Rinehart and Winston, New York.

Excellent discussion of the mitotic structures (spindle, chromosomes, kinetochore, and others) and some mechanisms of assembly and movement.

Patt, D. I., and G. R. Patt. 1975. *An Introduction to Modern Genetics.* Addison-Wesley, Reading, Mass.

Emphasis on the molecular aspects of genetics and on developmental genetics. Clear, understandable, and readable.

PATTERNS OF INHERITANCE

chapter fourteen

Scientists at one time believed that patterns of inheritance involved an equal blending of parental characteristics in offspring that were always intermediate between the two parents. The science of genetics has disproved that belief, showing that inherited traits are determined by information-carrying units called genes. This herd of horses, zebroids, and mules demonstrates the variation in characteristics that can be inherited from horse, donkey, and zebra parents.

That like begets like may have been one of primitive man's earliest observations. Possibly this recognition that characteristics could be transmitted from one generation to another led to the discovery that by selective breeding certain traits could be perpetuated and others caused to disappear. Thus about 10,000 years ago, people may have begun the purposeful selection and breeding of docile sheep that would be easy to keep in herds near an encampment, dogs that preferred the companionship of men, and domesticated cereal grains that were more productive than wild varieties and easier to harvest.

The careful selection of crop plants and stock animals and the methodical breeding of the selected parents have been responsible for tailored crops and tailored animals. We have developed fruits that are sized, shaped, and timed to mature uniformly so that they are easy to harvest and pack by machine. We have bred cows that produce milk very efficiently and dogs whose bodies are especially well adapted for specific tasks. For example, by selective breeding we produced the basset hound, a dog built low to the ground so that it can sniff a trail.

But man, the most curious animal, has never been content merely to be able to manipulate. The history of science teaches us that he has always tried to understand, perhaps in the hope of more facile and effective manipulation. Thus geneticists have sought and learned the answers to many questions of heredity. **Genetics,** the disciplined study of heredity, can now explain why offspring, instead of looking and functioning exactly like either parent, may be a mosaic of characteristics inherited from both, or may even resemble grandparents more than parents. These heritable characteristics appear to occur in some discrete form as units. The units, called genes, are *information* carriers, segments of molecules of DNA in the nuclear bodies called chromosomes, in which are encoded instructions for the performance of every chemical reaction in a cell. Thus are produced the visible characteristics and structures of every cell and, naturally, of the organism of which they may be part.

What we now know of genes—how they work and interact with one another and the environment, what can go wrong with them, and how some genetic errors can be corrected—has grown from the work of an Austrian monk, Gregor Mendel.

The origins of Mendel's work

In 1677, the Dutch lens-maker, Anton van Leeuwenhoek, peering through his "magic looking-glass"—a home-made microscope—discovered living sperm in the seminal fluid of man and various animals. But his microscope was of poor resolution and the mistaken idea arose that each human sperm must contain a tiny creature—a kind of "little man." Once the sperm became implanted in the female womb, the tiny creature was nourished by the mother, who presumably made no contribution to the child other than that of providing a warm, moist, nutritious environment. Another 150 years would pass before an actual human ovum was seen, and then it was theorized that the egg contained the "little man," and the function of the sperm was simply to trigger the egg into developing. By the mid-nineteenth century, however, the cell theory had become accepted. Sperm and ova now were recognized to be special cells and were called **gametes** (from the Greek word meaning marriage).

There was at this time a widely held theory that each parent contributed a single cell to each offspring, and that somehow hereditary material became mixed or blended. According to this theory, for example, a blue pigeon mated with a black pigeon would produce a steel-gray pigeon of a faintly bluish tinge. It was believed that hereditary material, once blended, could not be separated into its original components; thus all progeny of the steel-gray pigeon must likewise be gray.

But several significant difficulties prohibited complete acceptance of this theory. For one thing, it could not account for the fact that characteristics might disappear for a generation, even several generations, and then suddenly reappear. More important, if the blending theory were valid, there would be difficulty in accounting for the variety of organisms. Especially to Charles Darwin, who was to conceive of evolution on a grand scale, the thought of gray pigeons for all eternity was untenable. For how then could evolution be possible? For evolution there had to be variety, and blending would produce monotonous uniformity instead.

FIGURE 14.1

Gregor Johann Mendel (1822–1884). An Augustinian monk living in Austria, Mendel was the first to propose an accurate explanation of hybridization. Mendel crossbred different varieties of garden peas through many generations and derived fundamental laws of heredity by observing the distribution of traits in the offspring. A report of his work and his conclusions was published in 1866, but it was ignored and forgotten until its rediscovery by three different researchers in 1900.

It was Abbot Gregor Mendel of Brunn (Fig. 14.1) who showed that inheritance involved something far different from blending. Mendel is often pictured as a gentle, elderly monk who whiled away his quiet hours in his garden. Actually, Mendel was a former schoolmaster who had been educated in physics and mathematics. Undoubtedly he enjoyed gardening as a hobby, but it was also an opportunity for disciplined empirical study. Mendel was particularly interested in crossbreeding plants and recording the results. And he became intrigued by the idea that it might be possible not just to crossbreed and wait to see what the results were but rather to predict the results, to anticipate what the **hybrids**, or crossbred offspring, might be like.

In the garden of the monastery at Brunn, beginning in the 1840s, Mendel raised and cross-fertilized selected varieties of garden peas for many generations. By 1865, when he read a paper to the Brunn Society for the Study of Natural Science, he was able to report the exact frequency with which certain hereditary traits could be expected to occur—with some allowance for variations due to chance—in successive generations. He had reduced the mysteries of cultivating hybrids to a few statistical laws and had provided a foundation for a science of genetics.

MENDEL'S EXPERIMENTS

The pea plant Mendel's choice of the garden pea was no accident. The pea was readily available, grew reliably (nearly all the seeds he planted germinated), and was relatively simple to cultivate. Furthermore, it had a short generation time: the plants flowered and produced seed (embryos of the next generation) in less than one summer. Another advantage was that several different varieties of plants existed, each with its own distinctive characteristics. The plants that Mendel had available differed in seed form and color, in stem length, in flower color, and in the form and color of the pod.

Moreover, the flower of the pea plant is so constructed that it is naturally self-fertilizing (see Fig. 14.2). The pollen from another flower cannot reach the stigma; the ovules of a flower are fertilized only by its own pollen. Thus the pea plant can be purebred, so that its characteristics remain unchanged from generation to generation. But if crossbreeding is wanted, an immature flower can be opened readily, so that its anther can be removed and the stigma later touched with pollen from another plant. In other words, deliberate cross-fertilization is possible, but accidental crossbreeding is not. Consequently, the pea afforded Mendel unusual control over crossbreeding, enabling him to record his experiments with great accuracy.

In a typical experiment, Mendel used pollen from flowers of a strain (purebred line) that produced wrinkled seeds to fertilize the flowers of a strain that produced round seeds. Conversely, he used pollen from plants producing round seeds to fertilize flowers of plants producing wrinkled seeds. This cross-fertilization resulted in complete uniformity of the offspring. Every one of the several hundred offspring plants produced round seeds, whether the pollen had come from a round-seed or wrinkled-seed plant.

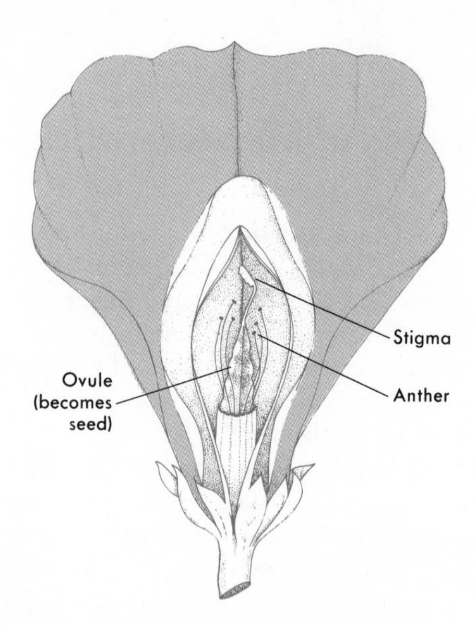

FIGURE 14.2

A pea flower is particularly suited for genetic experiments, because fertilization normally occurs before the flower opens. In order to crossbreed, Mendel opened each bud before pollination and removed the anthers. He then dusted the stigma of each flower with pollen from another plant. Thus he could fertilize the stigma of a red flower with pollen from a white flower.

In the following year, Mendel planted 253 of the round seeds and allowed the resulting pea plants to fertilize themselves. He got 7,324 seeds from the plants of the second generation and found that 5,474 were round and 1,850 were wrinkled—a ratio of 2.96, or almost 3 round to 1 wrinkled seed. And when he made several other crosses between plants having other single-characteristic differences, he got the same overall result.

For convenience, crosses such as these are now usually represented symbolically. The plants used for the first cross, wrinkled seeds and round, are termed the parental, or P, generation. Symbolically, the first cross is written:

P Round × wrinkled

The round offspring of this cross are termed the first filial generation, or F_1. Symbolically:

P Round × wrinkled
F_1 All round

These F_1 plants are called **monohybrids.** The self-crosses in the F_1 generation produce a new generation of plants, termed the F_2 generation:

P Round × wrinkled
F_1 Round × round
F_2 3 Round : 1 wrinkled

For easy recall, the F_1 are the children of the P (parental) generation, and the F_2 are the grandchildren of the P generation.

It seemed to Mendel that if there are two alternative parental characteristics, such as round and wrinkled seeds, and if only one—in this experiment, round—appears in the first filial generation, then that characteristic should be termed the **dominant** characteristic. But the characteristic that does not appear in that generation, wrinkled, will turn up again in one-quarter of the members of the second filial generation. This characteristic Mendel termed **recessive.**

The dominant and recessive relationships were found to exist in at least six other traits that Mendel tested. When, for example, Mendel crossed red-flowered peas with white-flowered peas, the first generation of offspring bore red flowers only (dominant). When this generation was left to fertilize itself, the second generation consisted of 705 red-flowered plants and 224 white-flowered, approximately a 3 to 1 ratio (see Fig. 14.3).

Experiments such as these led Mendel to his brilliant insights. The hereditary characteristics of peas, he deduced, must be carried and transmitted as discrete units, which he called factors (and which we now call **genes**).

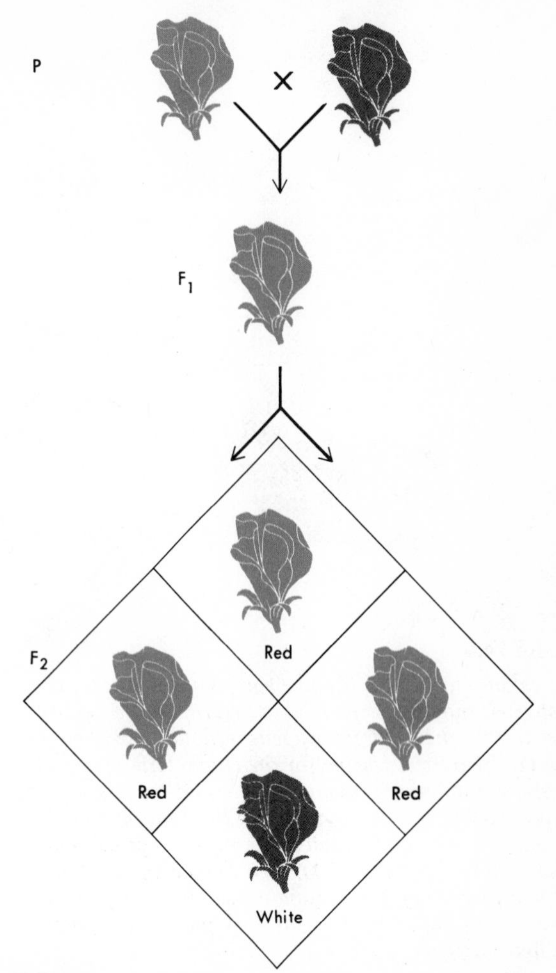

FIGURE 14.3

A cross similar to the ones that Mendel performed with single traits, such as flower color, and from which he derived his principle of segregation. The parental generation (P) consists of a pure red flower crossed with a pure white flower. The offspring generation (F_1) are all red. Self-crosses among the red F_1 flowers yield a ratio of three red flowers to one white flower in the F_2 generation.

He also postulated that each pea plant must have two hereditary units for each characteristic, one received from the pollen and one from the egg that gave rise to the seed from which the plant germinated.

A new plant, Mendel reasoned, receives one gene for each characteristic from its male parent and one from the female. In the first generation of hybrid offspring, only the characteristic of the dominant gene (round seed or red flower) appears, even though the recessive gene (wrinkled or white) is also present in every plant.

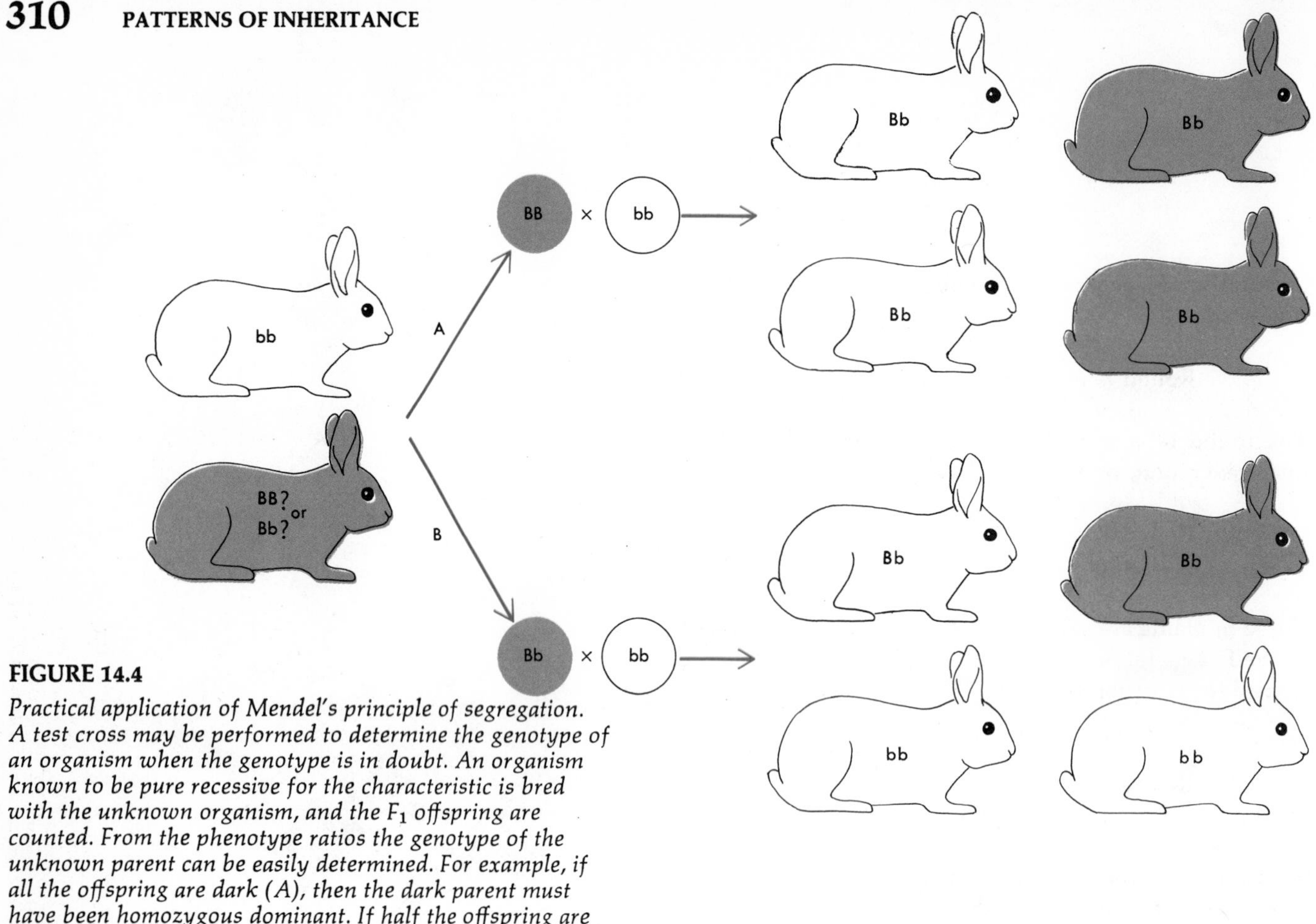

FIGURE 14.4

Practical application of Mendel's principle of segregation. A test cross may be performed to determine the genotype of an organism when the genotype is in doubt. An organism known to be pure recessive for the characteristic is bred with the unknown organism, and the F_1 offspring are counted. From the phenotype ratios the genotype of the unknown parent can be easily determined. For example, if all the offspring are dark (A), then the dark parent must have been homozygous dominant. If half the offspring are light and half dark (B), then the dark parent must have been heterozygous.

Then, with self-fertilization of the first generation, several kinds of seeds are produced, of which only one of every four has a pair of recessive genes. Thus the recessive characteristic becomes apparent in only one-quarter of the members of the second filial generation (F_2).

Since Mendel's time, a number of terms have been introduced in order to conveniently refer to and symbolize genes. For example, for the characteristic flower color we have mentioned two genes, one for red and one for white; these two are called **alleles,** or alternative gene types, for a characteristic. In some cases there may be three or more alternatives; each alternative is called an allele. The two genes that *are* present in a given individual are termed **homologous;** that is, they are genes related to a particular characteristic, such as flower color. The two homologous genes for flower color might be two genes for red flowers, or one for red and one for white, or two for white. If there were three alleles—for example, a gene for red, one for white, and a third gene for blue flowers they could be present in homologous pairs, such as red-white, red-red, red-blue, white-white, etc., in different plants. Whatever pair of alleles were present in a plant, the two would be homologous for the characteristic of flower color.

Now let us examine Mendel's crosses carefully. A red-flowered plant from a pure strain of pea producing only red flowers has a pair of dominant genes that can be denoted by capital letters, R/R. On the other hand, a white-flowered plant from a pure strain of white-flower producers has a pair of homologous genes denoted by the letters r/r. After cross-fertilization, all plants of the first generation, receiving one gene for flower color from each parent, will have the genes R/r and will show the dominant red-flower characteristic. This combination of genes, or any other that might be present, is termed the **genotype** of the plant; either alternative visible expression of the genotype, in this case red flowers, is called the **phenotype.** That is, we say this plant has the genotype R/r and the phenotype red flowers.

FIGURE 14.5

Dihybrid cross, such as those from which Mendel derived his principle of independent assortment. In the (P) generation, purebred plants producing round yellow seeds were crossed with purebred plants that produced wrinkled green seeds. All the F_1 plants produced round yellow seeds. These seeds were germinated and their flowers self-crossed. The results of the self-cross were F_2 plants producing round yellow, round green, wrinkled yellow, and wrinkled green seeds.

When flowers on plants from hybrid seeds of the F_1 generation self-fertilize, each flower (which produces two types of egg and two types of pollen: R eggs and r eggs, R pollen and r pollen) produces seeds with three different possible combinations of these genes or genotypes: R/R, R/r, and r/r. The dominant gene is present in two of these genotypes; when the seeds have grown into mature plants which produce flowers, it will be found that both genotypes with the dominant gene will produce the phenotype red flowers. Only the genotype r/r, having no dominant allele, will produce white flowers.

MENDEL'S LAWS

By such work—crosses involving single characteristics—Mendel arrived at what has come to be called his *first law*, or the *Principle of Segregation*. Briefly, it states that every phenotype is the result of the inheritance of genes in pairs. The genes *do not* fuse or blend in successive generations, but they may segregate, or separate, from each other and recombine in different ways.

Mendel arrived at a second principle by experimenting with **dihybrid** crosses, using pea strains differing in two alterative characteristics, such as seed color and shape. For example, he mated the plants of one strain whose seeds were round and yellow with those of another strain whose seeds were wrinkled and green. All seeds of the first generation of offspring were round and yellow. This result was similar to that with monohybrid crosses, which had indicated that yellow and round are the dominant, wrinkled and green the recessive characteristics.

But then, the second generation of 556 hybrid offspring produced 315 yellow-round seeds, 101 yellow-wrinkled, 108 green-round, and 32 green-wrinkled—a ratio of 9:3:3:1 among the four types (Fig. 14.5).

Yet there was nothing in such dihybrid experiments to contradict the results obtained in the monohybrid experiments. New combinations of characteristics appeared, but the ratio of round to wrinkled was still 3:1, as was the ratio of yellow to green. It was simply that the round and yellow characteristics and the wrinkled and green characteristics which were paired in the first filial generation were inherited in different combinations by the second filial generation. Noting this, Mendel arrived at what is now known as his *second law*, although in his original paper, Mendel had considered this the only "law" determined by his work; the principles of dominance and segregation were steps along the way. The *Principle of Independent Assortment* states that genes for a given characteristic recombine independently of the genes for any other characteristic.

Mendel's choice of such a simple system to study was most fortunate. As it happens, many—perhaps even most—human characteristics (and many in lower plants and animals) are the result of more complicated inheritance. The language of human genes has, one might say, various adjectives and adverbs for each phenotypic character. Thus skin color may occur in a very wide range of shades, from yellow to black. In human genetics, with its degrees of intensity, many phenotypic characteristics are the result of not just two paired homologous genes but of many genes—that is, the result of **multifactorial inheritance.**

Even so, Mendel was far ahead of his time. Although his experiments and deductions were published in the *Journal of the Brunn Society of Natural Science,* they were ignored by the scientific community for 35 years. Even Charles Darwin, passionately concerned with genetic problems of how traits were passed on and changed over the course of time, did not realize the significance of Mendel's work. However, there is evidence that Darwin was aware of it, because a book in Darwin's personal library, in which Mendel's work is discussed, was thoroughly underlined.

Eventually the abbot's research was to become the foundation for the science of genetics. By the turn of the century, 16 years after Mendel's death, biologists were ready to rediscover, accept, and carry on his findings.

Modern genetics

It is now possible to describe Mendel's work using the terminology that has arisen since his time—and this will be helpful in understanding the progress that has been made.

There is uncertainty as to whether Mendel knew anything about the processes of cell division, but his insights and conclusions correlate with what is now known about chromosomes and how they behave during meiosis.

In a diploid nucleus, as we have seen, there are two of each type of chromosome. The two chromosomes, similar in structure and carrying genes for the same characteristics, are said to be homologous. Each chromosome in a pair contains a gene for each characteristic. A diploid cell, therefore, has two of each type of gene, which may or may not be identical.

During meiosis, the two members of a chromosome pair separate and the gametes contain only a single chromosome of each type and therefore only one of each homologous gene. What Mendel had inferred from his breeding experiments fits remarkably well with what can be seen by microscopic examination of nuclei.

MODERN REPRESENTATIONS

Let us recapitulate the terminology that we have learned and add some that will enable us to discuss the inheritance of human characteristics.

Genes for a given characteristic are inherited in pairs and are said to be homologous with each other. Usually there are two alleles, or alternative forms of a gene, one of which may be dominant and the other recessive. The genes are symbolized by one or a few letters, usually taken from the name of the characteristic, such as R for the dominant gene for red flowers and r for the recessive gene for white flowers.

An organism may inherit a homologous pair of genes consisting of the *same* genes, either two dominant or two recessive. It is then said to be **homozygous dominant** (RR) or **homozygous recessive** (rr). Or each gene in a homologous pair may be a different allele, such as one dominant and one recessive, and the organism is said to be **heterozygous** (Rr).

Whatever combination of genes is inherited is referred to as the genotype, and the visible characteristics that they produce are referred to as the phenotype. But it is impossible from the phenotype alone to tell the genotype of all individuals, since there is often more than one genotype for a given phenotype.

Thus a systematic way of determining the possible genotypes of parents and offspring should prove useful. Today we use the system called the Punnett square, developed by the English geneticist, L. C. Punnett, who first used it for the analysis of genotype distributions (Fig. 14.6).

Using the Punnett square A Punnett square is constructed by showing the possible kinds of gametes a male parent can produce by meiosis, representing them along a horizontal line. The possible kinds of gametes

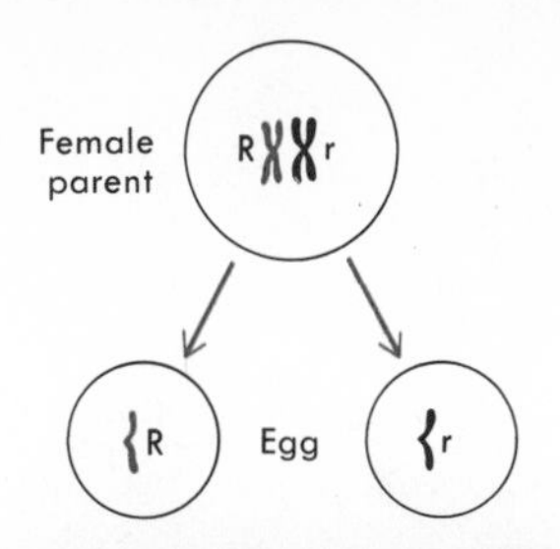

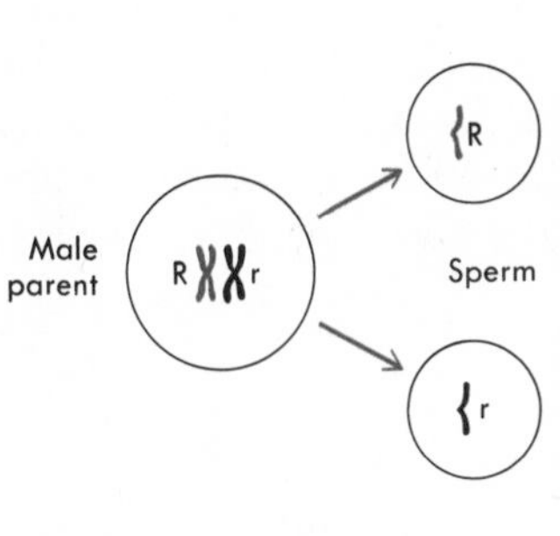

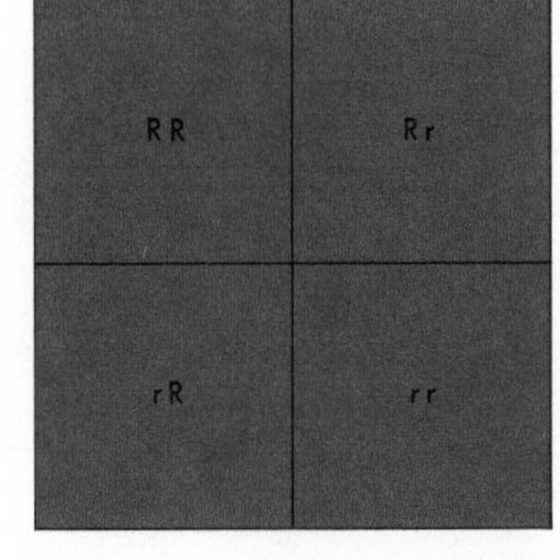

FIGURE 14.6

The Punnett square, a useful device for organizing the different gamete types that may be produced by meiosis. The crosses may be indicated in the boxes, and the resulting genotypic ratios are simply counted.

the female parent can produce by meiosis are represented in a vertical column. Then squares are formed thus:

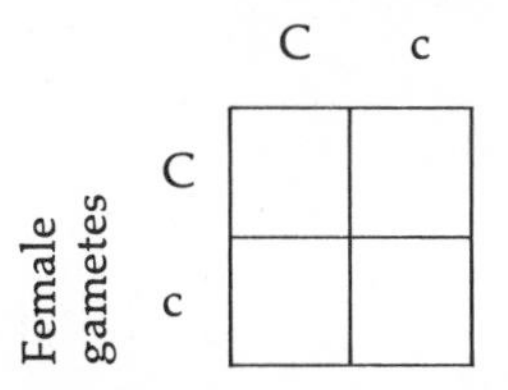

Each box then can be filled in with the allele from the female gamete followed by the allele from the male, so that each indicates the genotype for one possible zygote combination:

Male gametes

Female gametes	C	c
C	CC	Cc
c	Cc	cc

It is then apparent that the ratio of genotypes is 1:2:1 (1CC:2Cc:1cc), and the phenotypic ratio is 3:1 (3 dominant: 1 recessive).

In human genetics, the same principle can be applied when only two alleles, one dominant and one recessive, are involved. For example, a young couple has just had an albino child as their firstborn. (An albino lacks the pigment melanin and therefore has no color in skin, eyes, or hair; see Fig. 14.7.) What are the probabil-

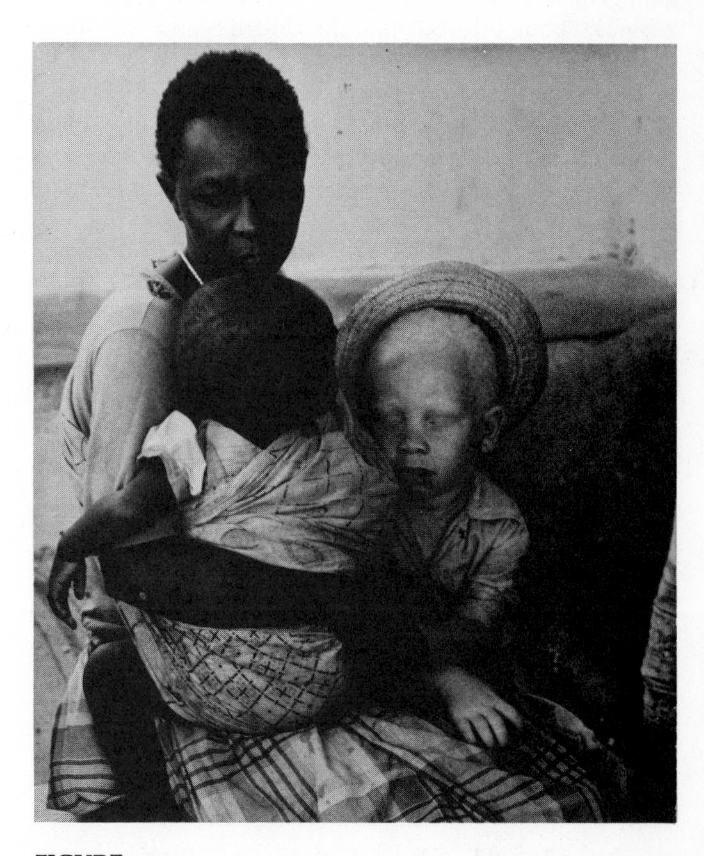

FIGURE 14.7

A black mother, who is a carrier of the albino trait, poses with her two children. One is an albino, an individual lacking an enzyme necessary for the production of pigment in skin, hair, or eyes. He has inherited a recessive gene for the albino trait from each parent. The black child in her arms may be either heterozygous or homozygous for pigmentation genes.

ilities that succeeding children may be albinos? For each gene that influences the formation of the brown pigment melanin in skin, hair, and iris of the eye, there is an allelic gene. Whereas the dominant allele A instructs cells how to produce an enzyme required for melanin formation, the recessive allele a does not. When one of each allele is present, skin pigmentation will be normal; the dominant gene permits cells to make all the required enzyme. The recessive phenotype, albino, will be expressed only when no dominant allele is present—when the person is homozygous recessive. Few people are albinos, about 1 in 20,000 in the United States, yet about 1 in 71 carries the recessive gene and its dominant allele—and is thus heterozygous dominant.

The normally pigmented couple whose first child has been an albino are both heterozygous and have one gene of each kind. Let us represent this with a Punnett square.

Male gametes

Female gametes	A	a
A	AA	Aa
a	Aa	aa

Note that the probability of any child being albino is 1 in 4 ($\frac{1}{4}$). Suppose that one parent, the father, is heterozygous and thus a carrier for albinism, while the mother is homozygous, herself an albino. What could we predict about their potential offspring? Again a square shows the probability:

Male gametes

Female gametes	A	a
a	Aa	aa
a	Aa	aa

The probability is 1 in 2 or $\frac{1}{2}$ that any child will be albino.

A Punnett square can also demonstrate the probabilities involved in dihybrid crosses. If a Punnett square is constructed, using R for the dominant allele for round and r for the allele for wrinkled seed, and G for the allele for yellow and g for the allele for green seed, it would look like this:

RrGg
(round yellow male parent)

Sperm

RgGr (round yellow female parent) Eggs	RG	Rg	rG	rg
RG	RRGG	RRGg	RrGG	RrGg
Rg	RRGg	RRgg	RrGg	Rrgg
rG	RrGG	RrGg	rrGG	rrGg
rg	RrGg	Rrgg	rrGg	rrgg

If we examine the genotypes that we have listed in the Punnett square, we will find that there are only four phenotypes represented: round green, round yellow, wrinkled green, and wrinkled yellow. And now, if we count the individuals of each phenotype, we will find that they occur in the following numbers: 9 round yellow: 3 wrinkled yellow: 3 round green: 1 wrinkled green. This ratio, 9:3:3:1, is typical for the combinations of dominant and recessive characteristics in a dihybrid cross with *any* dominant and recessive characteristics.

A simpler procedure Punnett squares can be used for crosses involving still more genes and characteristics, but they eventually become cumbersome. A much simpler method is based on the principle that the likelihood of a number of events occurring independently but simultaneously is equal to the product of the likelihood that each will occur separately.

Consider, for example, the Punnett square we have just constructed for a Mendelian dihybrid cross. Sixteen possible combinations of gametes can result when two F_1 offspring heterozygous for both characteristics are crossed. The combinations included nine genotypes (RRGG, RRGg, RRgg, RrGG, RrGg, Rrgg, rrGG, rrGg,

rrgg), and these determined four different phenotypes in the ratio of 9:3:3:1.

To figure out how many of the 16 combinations could produce the wrinkled yellow phenotype, we could proceed this way:

Since wrinkled is recessive, in a monohybrid cross it would be expected to appear in $\frac{1}{4}$ of the F_2 offspring. Since yellow is dominant, it would be expected in $\frac{3}{4}$ of the F_2 offspring. Multiply the two values together, $\frac{1}{4} \times \frac{3}{4}$, and the result is $\frac{3}{16}$, indicating that three of the 16 combinations can be expected to produce a wrinkled yellow phenotype.

Similarly, to figure out how many combinations could provide round yellow phenotypes, the probabilities of $\frac{3}{4}$ for each dominant character can be multiplied together—$\frac{3}{4} \times \frac{3}{4}$—to produce $\frac{9}{16}$, indicating that nine of the 16 combinations would lead to round yellow phenotypes.

The same formula can be applied to trihybrid, tetrahybrid, and still more complex crosses. Consider, for example, a cross involving seed shape, seed color, and flower color. To determine what proportion of the F_2 offspring would have red flowers and wrinkled yellow seeds, we could use the probability for each characteristic in a monohybrid cross. That would be $\frac{3}{4}$ for red flowers, $\frac{1}{4}$ for wrinkled seeds, and $\frac{3}{4}$ for yellow seeds. When the three probabilities are multiplied together—$\frac{3}{4} \times \frac{1}{4} \times \frac{3}{4}$ —the result is $\frac{9}{64}$, indicating that nine of the 64 possible combinations in a trihybrid cross would produce red flowers and wrinkled yellow seeds.

Non-Mendelian genetics

Not all genetic traits are described by Mendel's principles. Some exceptions include codominance of multiple alleles, incomplete dominance, polygenic inheritance, environmental influences on gene expression, expressivity and penetrance, and sex-linked, sex-influenced, and sex-limited genes. However, the following discussion will explain how such deviations from Mendel's principles can be understood as modifications of his principles.

SEX INHERITANCE

Late in the nineteenth century a German biologist, Walther Flemming, treated the cell nucleus of a salamander with a synthetic red dye and found that certain rodlike bodies absorb the dye and become relatively easy to see. And although the bodies are actually colorless unless dyed, they were misnamed chromosomes from the Greek word for color.

Over the years, 46 chromosomes were identified in human cells (and of course, specific numbers of chromosomes were found in other species). It became apparent that chromosomes exist in pairs and that each pair has a different size and shape from others in the nucleus. Twenty-two pairs in the human cell were numbered by length for reference purposes.

The 23rd pair, however, presented a problem. Sometimes the two members of the pair resembled each other; sometimes they did not. The larger of the two was called X, and the smaller was labeled Y. With the discovery that males have one X and one Y chromosome while females always have two Xs, it was not long before it was suspected that these particular chromosomes have something to do with sex determination. The chromosomes of the 23rd pair came to be called **sex chromosomes.** The rest of the chromosomes are known as **autosomes.**

And, indeed, sex determination was soon revealed decisively around 1910 by Thomas H. Morgan of Columbia University. Experimenting with *Drosophila* (the fruit fly), Morgan revealed the principle of sex determination, which was later applied to human genetics.

Although somatic cells have 23 pairs of 46 chromosomes, the reproductive cells, ova (eggs) and sperm, receive half as many chromosomes as the result of meiosis. Each ovum carries 22 autosomes and an X chromosome. Sperm cells, however, are of two kinds. All have 22 autosomes, but about half carry an X chromosome, and the other half carry a Y chromosome.

The sex of an offspring depends on which type of sperm penetrates and fertilizes the egg. If it is an X-carrying sperm, the resulting zygote will be endowed with two Xs, and the embryo will be female. If a

Y-carrying sperm does the fertilizing, resulting in an XY combination, a male will develop.

As a result of present-day insights into gene functioning, it is known that everyone embodies in his genetic material the potentialities of both sexes. Males have the genes for female as well as male characteristics; females have them for male as well as female characteristics. However, it appears that the expression or activity of the genes for male or female characteristics is controlled by the genes on the sex chromosomes. In other words, when an XX combination is inherited, the "female" genes are permitted expression, and the physical and physiological characteristics of a female result. Similarly, the inheritance of an XY combination results in the structures and physiological characteristics of a male.

Evidence for this has come from laboratory experiments with chicks, for example. When male hormones are injected into young female chicks, they develop the comb and spurs of the male. Their latent maleness has been stimulated. Actually, studies of embryonic development have shown that *early* in development, the *reproductive* structures are the same for both sexes. They contain cells that can produce testes and others that can produce ovaries. A little later, the organs of one sex, testes or ovaries, continue to develop in the individual, but those of the other sex do not. The sex chromosomes apparently determine which sex organs will be expressed and therefore which sex hormones will be produced in quantity, and it is these hormones which in turn greatly influence the expression of genes controlling sex characteristics.

Nothing in nature is invariably perfect. Sex chromosome abnormalities, although rare, do occur. For example, the two sex chromosomes, instead of separating normally during meiosis, may adhere to each other—a malfunction called nondisjunction of the chromosomes—so that both go to one gamete while the other gamete receives none. If a human ovum without its normal X chromosome is fertilized by an X-carrying sperm, the resulting 45-chromosome zygote will be a female, who will develop Turner's syndrome. Her sex organs will remain infantile, and she will fail to mature sexually at adolescence. She will be physically smaller than the average female and may also be mentally retarded. The single X chromosome she has received from her father has been sufficient to give her a female body but not enough to provide adequate female hormones to allow her to mature properly.

If an ovum without its normal X chromosome is fertilized by a Y-carrying sperm, there is not even a chance for development. It appears that some genes on the X chromosome are essential for life. Without such genes, the embryo may begin to develop but soon dies.

If an ovum carrying two X chromosomes is fertilized by a Y sperm, the resulting XXY combination will produce an abnormal male with Klinefelter's syndrome. He will have male sex organs, but they will be abnormally small and without the cells needed for sperm production, and he will be sterile. His muscular development may be somewhat feminine, and there may be some degree of breast development. In effect, with an extra X, some genes for femininity are allowed a degree of expression. Klinefelter's syndrome received some publicity during a recent international athletic event; one participant, seemingly a woman and registered as such, was found to be a man with Klinefelter's syndrome. Since then, many international athletic groups have begun to require chromosome analyses for some women entrants.

If a double-X ovum is fertilized by an X sperm, the 3X combination often produces a relatively normal female. But the 3X combination does appear to carry with it an increased likelihood of mental illness.

Just as an unusual event during meiosis may lead to abnormally endowed ova, so it may produce abnormal sperm, some with both X and Y chromosomes, some with neither. When an XY sperm fertilizes a normal X egg, the resulting XXY combination leads to Klinefelter's syndrome. When a sperm carrying neither X nor Y fertilizes a normal X-carrying ovum, the result is Turner's syndrome.

Another abnormal sex chromosome combination only recently studied has led to legal controvery. The combination—XYY—has been found among 5 to 12 percent of the male prison population tested, but is found among only 0.25 percent of the civilian male population. These statistics led to the argument that the

XYY combination was responsible for mental disturbances that led to crime, and that men with this abnormal inheritance should not be held legally accountable for their crimes. But the XYY combination has also been detected in chromosome studies of men leading normal lives, so the evidence is not clear-cut. Although the genotype may increase the probability for abnormal personality, abnormal behavior is not inevitable.

How can the XYY combination arise? During the second division of meiosis, the Y chromatids may fail to separate (the malfunction of nondisjunction), to that both Ys go to one sperm while the second sperm receives none. The fertilization of a normal X ovum by a YY sperm produces the XYY zygote.

Barr bodies Unlike the human male, the female appears to be a mosaic of sex-linked traits, with some cells in her body following the instructions of recessive alleles on one of her X chromosomes, while other cells follow the instruction of dominant alleles on the homologous X chromosome.

It appears that early in the life of all of the cells (except the gametes) in the female body, one of the two X chromosomes coils itself into a tight spiral and becomes genetically inactive; the resulting chromosome is called a **Barr body.** In some of the cells the inactive chromosome is the X from the mother; in others it is the X from the father. Thus it is possible for some cells to express the dominant trait while others are simultaneously expressing a recessive trait.

Sex linkage As we have seen, the two sex chromosomes are an odd pair. The X chromosome, one of the largest in the cell, is about three times the size of Y. It appears to be much more active in body functioning, containing many more protein-ordering genes.

The Y chromosome does have a few genes that are alleles of genes at one end of the X chromosome, but most of the genes on the X chromosome have no alleles on the Y chromosome. Thus these genes will always be expressed in the male, even if they are normally recessive alleles, because there are no corresponding dominant alleles on a second (X) chromosome as there are in the female. Because their expression is dependent on the sex of the individual, such genes are called **sex-linked.** The Y chromosome also carries a few genes which have no corresponding alleles on the X chromosome. These Y-linked **holandric** genes cause the development of certain characteristics (such as hair in the ears) only in the male. Those genes which are homologous on both the X and the Y chromosomes are called incompletely sex-linked.

There are hundreds of thousands of individual characteristics which distinguish human beings from each other, but only 46 chromosomes. It stands to reason that the hundreds of thousands of genes present in a human must be distributed in the 46 chromosomes present. Thus many genes must always be inherited together; in fact, Mendel's laws and his ability to predict the outcome of dihybrid crosses were fortuitously based on characteristics, such as flower color and seed shape, whose genes are on different chromosomes. But because of the large number of genes and the relatively small number of chromosomes, most human characteristics are inherited in groups—certain characteristics always appearing with certain others.

Of major importance in the study of linkage groups and the principles of Mendelian genetics was the research of Thomas H. Morgan. In his extensive experiments with the fruit fly, *Drosophila melanogaster*—an organism evidently chosen for its prolific breeding, short generation time, and easy care—Morgan concentrated on the sex chromosomes. He found, for example, that when he bred a particular strain of fruit fly, only the male ever had white eyes; he concluded that the genes for eye color in the fruit fly must be carried on the same chromosomes which determine sex. By 1915 Morgan had discovered about 20 traits in the fruit fly that he called sex-linked because of their transmission on the sex-determining chromosomes.

Since then, geneticists have investigated various linkage groups in humans, but because of the particular defects caused by genes on the sex-chromosomes, most interest has centered on sex-linked traits.

Color blindness is an example of a sex-linked trait. A gene on the X chromosome governs the production of

Is There a "Criminal Chromosome"?

In 1961 a human male was found showing an *XYY* chromosome complement, a result of nondisjunction of two *Y* chromosomes (presumably during spermatogenesis in the subject's father). In December 1965, Dr. Patricia Jacobs and her colleagues at Western General Hospital in Edinburgh (Scotland) published a cytogenetic study of male inmates in the hospital's security ward. These inmates all had records of what was considered violent criminal behavior. Of the 197 men surveyed, 7 showed the *XYY* chromosome condition.

There was at the time very little data available on the frequency of *YY* nondisjunctions; but the related *XX* nondisjunctions were known to occur in about 1.3 out of every 1000 births, or about 0.1% of the time. Thus the appearance of *XYY* in 3.5% of the inmates of a penal institution appeared to Dr. Jacobs and others to be a significant departure from the norm. They hypothesized that there might be a cause-effect relationship between presence of the extra *Y* chromosome and a tendency toward aggressive and violent behavior. As a result of this early work and the publicity it received, many surveys were carried out to screen for *XYY* males. Most of these surveys were done on inmates in various kinds of prisons or mental institutions. More examples of *XYY* males were uncovered. The frequency of *XYY* males in at least some penal institutions appeared to be significantly higher than the frequency estimated for the general population.

Given the apparently higher frequency of the extra *Y* chromosome in prison populations, geneticists and psychologists put forward the hypothesis that the extra *Y* chromosome causes males to have a greater tendency toward violent and antisocial behavior. Such behavior, they argued, could lead to various kinds of criminality. The precise means by which an extra *Y* chromosome could produce a "tendency to violence" was a matter of some speculation. Some workers suggested that presence of the extra *Y* chromosome might increase the amount of male hormone secreted, thus causing an increased level of general aggressive behavior. Others suggested that the extra *Y* chromosome might increase the rate of development, especially around puberty, making *XYY* boys grow faster, appear larger than normal for their age, and become hyperactive social misfits. A third possibility suggested was that the extra *Y* chromosome specifically affected brain development, acting on "violence centers" supposed to exist in areas such as the hypothalamus or amygdala. While no biologist has any idea about the mechanisms that might be involved, by the mid-1960s a number of researchers maintained that there might be a direct link between possession of an *XYY* chromosome complement and a person's chances of becoming a criminal.

The *XYY* story became another example of a theory of biological determinism: a theory attempting to explain a social phenomenon (in this case aggressive behavior and criminality) in strictly biological terms. The opposing hypothesis is that in the vast majority

From Jeffrey J. W. Baker and Garland E. Allen, *The Study of Biology*, 3rd edition (Reading, Mass.: Addison-Wesley, 1977), pp. 545–548.

of cases, aggressive and criminal behavior is caused by unfavorable environments (family problems, slum neighborhoods, racial or ethnic prejudice, economic deprivation, etc.).

By the late 1960s the *XYY* story had been widely circulated through leading newspapers and magazines in much of western Europe and the United States. In France and Australia *XYY* defendants in two murder trials were given light sentences (in one case acquitted) on the grounds that their violent behavior was beyond their control, rooted as it supposedly was in the genes. A report appeared in 1968 that Richard Speck, who in 1966 killed 8 nurses in Chicago, was *XYY*. (That the report was later shown to be false got very little publicity). The result of much of this publicity was to gradually convince the public that biologists, particularly geneticists, accepted as valid the hypothesis that an extra *Y* chromosome caused an increased tendency toward violent and criminal behavior. In some circles the extra *Y* was called the "criminal chromosome."

One of the chief critics of the theory on scientific grounds is Dr. Digamber S. Borgaonkar, of the Johns Hopkins University School of Medicine. Dr. Borgaonkar made an exhaustive study of most of the *XYY* cases reported, examining both the data and the methods used to obtain the data. His conclusion is that most of the studies were carelessly executed. For instance, he found that the data were often so unreliable as to be virtually meaningless, and that the suggestion of any relationship between an extra *Y* chromosome and criminality was consequently unsupportable.

The criticisms Dr. Borgaonkar has directed against the *XYY* work provide several guidelines to data collection and analysis:

1. One of the important assertions of the *XYY* work is that *XYY* males are more disposed toward violence than *XY* males. Dr. Borgaonkar found from analyzing papers reporting on behavior traits of *XYY* males within penal institutions that at least in these circumstances, *XYY* males were on the whole more cooperative than their *XY* counterparts. Thus the claim that *XYY* males are more aggressive cannot be considered valid *without specifying the environment involved.*

2. Few physiological or psychological traits have been found that actually distinguish *XYY* males from other males. Only height appears to be distinctive (*XYY* males are on the average slightly taller than *XY* males). Other traits—skeletal structure, electroencephalograms (EEG's), electrocardiograms (EKG's), skin traits, etc.—all seem to be average. Hormonal levels are no different for populations of *XYY*'s and *XY*'s. The IQ of *XYY* males appears to be about the mean for inmates of penal institutions. No significant differences in personality traits distinguish *XYY*'s from *XY*'s. In short, by all significant physical or physiological criteria that might affect behavior, *XYY* males rate about the same as *XY* males.

3. Methodologically, the techniques of collecting and analyzing behavioral data about *XYY* males appeared to Borgaonkar, on the average, to be very unreliable and nonrigorous.

a) All the studies lacked either a "blind" or "double-blind" procedure. In a "blind" experiment, an investigator interviewing a subject to determine behavioral and personality traits does not know what is being tested for (i.e., that the patient is suspected of displaying violent behavior); the patient may know, however, the purpose of the study. In a "double-blind" experiment, neither investigator nor subject would know what relationships were being sought. "Blind" and "double-blind" procedures help ensure that neither investigators nor subjects will find more of what they are looking for than is actually there.

b) Virtually none of the studies of *XYY* males had been conducted with matched control groups against which data on the behavior of the *XYY* subjects could be compared. That is, in testing the hypothesis that an extra *Y* chromosome is a significant cause for criminal behavior, it is necessary to eliminate other variables (such as poor socioeconomic status, bad family life, etc.) that may also have profound effects in molding personality. Most of the studies compared the behavior of *XYY* males to control groups of randomly-chosen *XY* males not matched for social class, family background, economic status, and the like. Thus, several variables are introduced simultaneously: the presence of different chromosome complements, different social class and various family backgrounds. When two or more variables are present it is impossible to say which may be the more important cause for a particular phenomenon.

c) Researchers placed much reliance for descriptions of violent or criminal behavior on sources such as police records, legal documents, or records from correctional institutions. Not only are these descriptions likely to be highly variable; they are also subjectively biased in very specific ways. For example, there is no standard definition of what constitutes "violent" behavior. Is swearing at a prison authority or police officer evidence of a tendency toward violence? Or, does physical violence have to be involved? Moreover, police and prison administrators are likely to classify as violent any behavior openly disrespectful of or hostile to their own authority. Yet mere resistance to authority is not by itself adequate indication of a propensity to violence and criminality.

d) There was an element of selection involved in the subjects who were investigated in most of the studies. Only a small fraction of violent behavior actually comes to the attention of authorities and is recorded. Much more needs to be known, Dr. Borgaonkar argues, about the kinds of violent behavior displayed by people who do not come to official notice. Is it the same, less, or more than that displayed by those who are actually caught, convicted, and placed in penal institutions to become readily available objects for study? To claim that *XYY* males are more violent than *XY* males requires knowing what kind of violence *XY* males in the general population perpetrate. In fact, Dr. Borgaonkar reports that penal institution records indicate that *XYY* males have committed no more violent crimes than the *XY* males in prison with them. In the absence of data about the differences between the kinds of crimes that arouse official attention and those that do not, it could be hypothesized that *XYY* males are simply more open and honest than *XY* males and thus get caught more readily. In other words, if there is any genetic basis to the argument at all, it is that perhaps the extra *Y* chromosome is an "honesty-determining," rather than a "criminal-determining" chromosome!

In concluding his study, Dr. Borgaonkar points out that some behavioral and developmental conditions (such as Down's syndrome, or "Mongolism") are definitely hereditary in nature. He does not deny the role of heredity in determining, in a general way, some broad patterns of personality development. On the other hand, he emphasizes that behavioral problems once thought to be largely hereditary (certain kinds of epilepsy or abnormalities in EEG) have been strongly linked to social class and family stability. He cautions against assuming a genetic cause for something as specific as behavior, when the obvious environmental influences that can affect such behavior have been largely ignored. He writes:

Inadequate understanding of the phenomena and the premature conclusions about the XYY phenotype, which have been reported with distressing frequency, have produced remarkably simplistic views of the interactions between XYY genotypes and the almost infinitely varied environments with which they interact. We should always keep in mind that even the demonstration of a genetic contribution to poor impulse control warrants only the conclusion that in certain environments some persons with particular genotypes will respond by developing certain behavioral problems more frequently than others. However, this does not preclude the possibility that *in some other environments persons with the very same genotypes (i.e., XYY) may well manifest socially adaptive behaviors.**

The *XYY* case has raised other questions about research on human subjects, especially where negative aspects of an individual's makeup are the main focus. The issue came to light dramatically in 1974 and 1975 at the Harvard University Medical School in Boston. A large project screening for *XYY* babies had been in progress at Boston Hospital for Women (formerly Boston Lying-In Hospital) from 1968 to 1975. The researchers heading the project, psychiatrist Stanley Walzer and geneticist Park Gerald, both from the Harvard Medical School, wanted to identify *XYY* males born in the hospital and follow their personality and behavioral development through adulthood. Walzer and Gerald explained that the purpose of their study was to identify *XYY* genotypes early in a child's life, so that psychiatric counseling could be provided to help overcome personality problems if and when they arose. Their research was funded by the Crime and Delinquency Division of the National Institutes of Mental Health.

Dr. Jonathan Beckwith, also of the Harvard Medical School, and Dr. Jonathan King of Massachusetts Institute of Technology mounted an extensive campaign, beginning in 1974, to have the *XYY* project closed down. They argued that parents participating in the project (that is, who had agreed to have their children studied) were not adequately informed of what the project was about. They claimed that parents did not understand the stigma that might be attached to their child, even if he appeared perfectly normal in his behavior, if he were known as an *XYY* type. (A well-known American geneticist appalled his colleagues at a professional meeting recently by claiming that he "wouldn't invite an *XYY* home to dinner.")

Given the misinformation the public has received in considerable doses about "criminal chromosomes," Beckwith claims that Walzer and Gerald did not take adequate steps to inform the parents of how participation in the project might affect their child's future. Moreover, Beckwith and King argue that given what most people now think about the *XYY* genotype, the hypothesized relationship between an extra *Y* chromosome and aggressive behavior could become a self-fulfilling prophecy. In other words, parents who know their children are *XYY* will treat them differently, perhaps pushing them almost unconsciously toward violent behavior. The medical school's Human Studies Committee found that Walzer and Gerald's work did comply with their requirements that (1) informed consent be properly obtained, (2) the patients' rights be protected, and (3) the benefits of participating in the study outweigh the risks.

Beckwith and King do not agree. They argued that the committee was composed of established doctors who had a stake in protecting themselves and their colleagues from challenges to the fundamental nature of the research. Beckwith and King have aimed their criticism of the Boston project largely at the moral and ethical implications of the *XYY* research. However, they would not necessarily have done so if the scientific basis of the work were not also faulty. This is an important point. Of course the rights of patients should be protected in all such research projects, no matter how valid the scientific basis of a piece of research. But when the research methods and the data are themselves questionable, the harm can be considerably greater. Beckwith and King argued that the patient's rights are considerably more in jeopardy when the conclusions from supposedly scientific work are erroneous. To be stigmatized for life is bad under any circumstances; to be stigmatized erroneously is worse.

As a result of the pressure brought to bear on the project by Beckwith, King, and others, the *XYY* project in Boston was discontinued in the spring of 1975. Some people think Walzer stopped screening *XYY*'s because he finally realized that the risks outweighed the benefits. Walzer denies this. "I hope no one thinks I don't still believe in my research," he declares. "I do. But this whole thing has been a terrible strain. My family has been threatened. I've been made to feel like a dirty person. I was just too emotionally tired to go on." Walzer agrees that talk of a "criminal chromosome" is nonsense, but he still thinks there is enough evidence of certain learning difficulties in *XYY* children to justify an early identification leading to corrective therapy. King and Beckwith, on the other hand, claim that the potential harm to individual people is far greater than the potential good, and that under such conditions stopping the screening project is justifiable.

---

* D. S. Borgaonkar and Saleem A. Shah, "The XYY chromosome male—or syndrome?" *Progress in Medical Genetics* X, 10 (1974), pp. 135–222.

cones, small bodies in the retina of the eye, which are sensitive to green light. A recessive allele of the gene produces cones that fail to detect green clearly enough. The result is that green objects are frequently confused with red objects.

Since a boy has only a single X chromosome, he has a far greater chance of being color-blind than a girl. About one boy in 12 is in fact color-blind, but only one girl in 250 is affected.

A boy will be color-blind if he receives a recessive allele on his single X chromosome, since there is no homologous position on his Y chromosome for a dominant normal allele, which would permit the production of normal cones. A girl inheriting one recessive allele on one X chromosome and a dominant normal allele on the other X chromosome will have normal vision. She is called a **carrier,** however, since she could pass her recessive allele on to a son or daughter, despite her own normal vision. But in order for her daughter to be color-blind, the carrier woman would have to have children by a color-blind man. And the probability of each of her sons being color-blind is $\frac{1}{2}$, regardless of the genotype of their father.

Hemophilia, the bleeders' disease, is another illustration of sex-linked inheritance. Hemophilia is the result of a deficiency of a blood factor required for normal clotting. In a normal person, when any kind of wound causes blood to leak out of a blood vessel, the blood quickly responds to the danger by clotting, and the clot seals the wound. In a hemophiliac, however, clotting may be delayed for hours, allowing excessive and even fatal loss of blood.

Hemophilia occurs in about one male in 10,000, but it occurs in only one female in 100 million. As with color blindness, a boy needs to receive only the recessive gene on the one X chromosome he inherits from his mother to have hemophilia. If a girl receives one such gene, she will be a carrier, but she needs to inherit two of the genes to have hemophilia, and there is little chance of that. Because of the Barr body phenomenon, however, she may show barely detectable signs of her affliction. In a woman heterozygous for the gene for hemophilia, the X with the normal gene can be functioning in some cells, producing clotting factor, while the other X chromosome may be functioning in other cells, which therefore cannot produce the clotting factor. Although she appears to be normal, extremely sensitive tests can detect a slight delay in her blood-clotting time.

Some sex-linked traits can be found only in males because the responsible recessive gene on the X chromosome has such a lethal effect. For example, pseudohypertrophic muscular dystrophy affects only boys at about the age of 10, producing wasting of muscle and death in the early teens. There is little chance that a woman carrier of the gene would ever bear a female child from a man with the disease, since an afflicted male, from whom the second X chromosome must come, will never reach reproductive age.

On the other hand, there are some sex-linked characteristics that occur more often in females than in males. When an X chromosome gene is dominant, a female can receive it from either mother or father and thus has double the chances of inheriting the gene as does a male, who can get it only from his mother. One such trait is a defect in dentine, a protein in teeth, that leads to premature erosion of the teeth. A man with defective dentine may pass the trait to his daughters through his X chromosome, but he cannot pass it to his sons, to whom he gives only the Y chromosome.

In addition to sex-linked genes, there are genes that are sex-limited or sex-influenced. Although these genes are not on the sex chromosomes, their expression is dependent on the sex of the individual.

Sex-limited genes, unlike sex-linked ones, may be carried on *any autosome* yet are normally expressed in only *one* sex. Thus a woman is normally beardless although she carries the genes required for a beard. In fact, the character of the beards of her sons will depend on her genes as well as her husband's. For example, a man may have a relatively soft beard, and yet his son, inheriting a dominant gene from his mother, may have a tough, wiry beard.

An example of a **sex-influenced** gene is that responsible for the ordinary male type of baldness. Although such baldness is not unknown in women, it is far more common in men. It is believed to result from a gene

TABLE 14.1
The A-B-O blood antigens and antibodies

BLOOD TYPES	ERYTHROCYTE ANTIGENS	PLASMA ANTIBODIES	TRANSFUSION RELATIONSHIPS	
			DONOR TO	RECIPIENT FROM
O	none	anti-A anti-B	O,A,B,AB	O
A	A	anti-B	A,AB	O,A
B	B	anti-A	B,AB	O,B
AB	A,B	none	AB	O,A,B,AB

which is *recessive in women* but dominant in men. A man can be bald when he has only a single gene for baldness since that gene will be dominant. In a woman, however, since the gene is recessive, two of them must be present before she can achieve the dubious distinction of becoming bald.

CODOMINANCE

Allelic genes do not invariably have a dominant-recessive relationship. Sometimes both genes may be expressed simultaneously. They are then considered to be **codominant.**

Such codominance is illustrated by human A-B-O blood antigen-antibody types, which are heritable. A blood antigen is a protein molecule that can provoke production by the body of antibodies, other proteins which chemically bind to the antigens and inactivate them. Usually this mechanism is used as protection against foreign substances that may enter the body.

Human red blood cells contain antigens, and it is because they do that blood transfusions must be done carefully, using properly matched blood. There are many known antigens in human blood, but those that are most familiar to us are the A and B antigens. Some people have only A antigen and are known as type A. Others have only B antigen and are known as type B. A small number have both A and B and are classified as type AB. Some people have *neither* antigen and are classified as type O.

Blood plasma may also contain antibodies. If red cells containing an antigen are mixed with plasma containing the corresponding antibody, the red cells are affected: they **agglutinate,** or clump together. The reaction can be serious, sometimes even fatal.

Every individual has all the antibodies in his plasma that can be present without clumping his red cells. Type O individuals have antibodies *a* and *b*, which clump antigens A and B; type A individuals have *b* antibodies; type B have *a* antibodies; and type AB people have neither kind of antibody (see Table 13-1). To ensure that no clumping is induced, it is necessary to check the blood type of both recipient and donor before a blood transfusion is administered.

The blood transfused will usually be of the same type as the recipient's blood. However, when the same type is not available, another type may be used, so long as the plasma of the recipient and the red cells of the donor are compatible. For example, type O can be given to anyone since its red cells contain no antigens and so will be compatible with any plasma. However, a person with type O blood can receive only type O blood since his plasma contains antibodies for both A and B. A person with type AB blood, whose plasma has no antibodies, can receive blood of any type, but his blood cannot be donated except to other type AB individuals.

Three allelic genes are responsible for blood types. One of these, I^A, produces A antigen. An allele, I^B, produces B antigen. A third allele, a recessive designated i, produces neither of the antigens. I^A and I^B are dominant over i, but neither I^A nor I^B is dominant over the other; they are codominant, so that a person who receives both of them will have type AB blood.

When i and i are both inherited, the blood type is O. When I^A is inherited with I^A or with i, the blood type

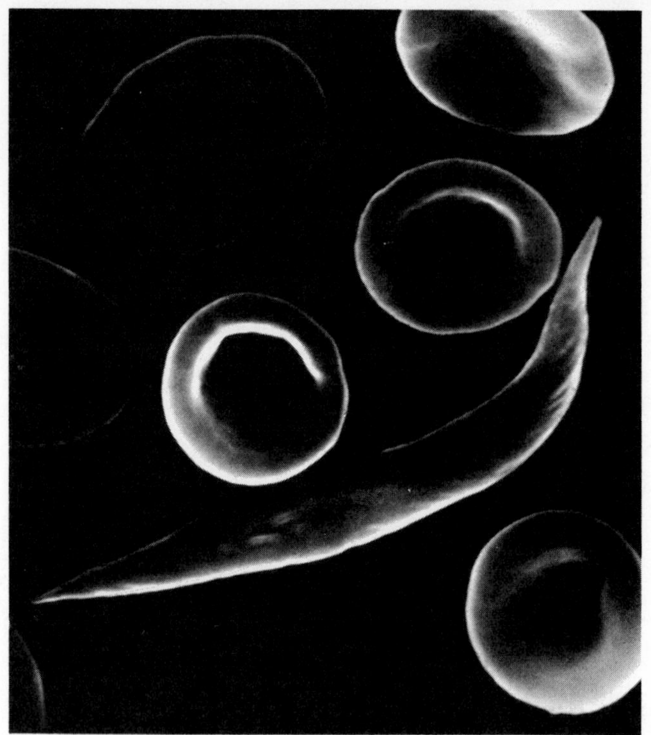

3,900X

FIGURE 14.8

A red blood cell characteristic of sickle-cell anemia surrounded by normal red cells. The biconcave disc shape of the normal cell is distorted by defective hemoglobin into an elongated crescentlike shape. The abnormal cell cannot transport oxygen efficiently.

is A. When I^B is inherited with I^B or with i, the blood type is B. When I^A is inherited with I^B, the type is AB.

INCOMPLETE DOMINANCE

In some instances, allelic genes may be only partially expressed, and there may be a blending of characteristics. Only two alleles may be present, but instead of only two phenotypes, a third intermediate type appears. For example, in some cattle, when a red bull is mated with a white cow, the offspring are roan, a shade between red and white. Both the gene for red and the gene for white are being partially expressed. Similarly, a cross of a white snapdragon with a red snapdragon produces pink instead of the expected red flowers.

Sickle-cell anemia is a human disease in which such incomplete dominance is involved. In sickle-cell anemia, there is a defect in hemoglobin, the vital pigment in red blood cells which serves to transport oxygen to the tissues. For normal production of hemoglobin, a normal hemoglobin gene is required. There is an allele that can alter the hemoglobin structure. The alteration is very slight, but it is enough to cause the hemoglobin molecules to aggregate in long chains, especially when the blood oxygen level fails. The hemoglobin chains distort some of the red cells so that they become sickle-shaped and as a result are unable to perform efficiently (Fig. 14.8).

When two sickle-producing genes are inherited, the sickling is so great that very severe sickle-cell anemia develops and may lead to early death. But a heterozygous individual will fall in between the normal person and the homozygous defective individual. Some of his hemoglobin will be normal, some will be abnormal. He has the sickle-cell trait, but he will suffer only at times from mild anemia and most of the time can lead a normal life.

MULTIFACTORIAL INHERITANCE

Most human traits are probably the result of **polygenic** or **multifactorial inheritance.** In this type of inheritance there are many nonallelic gene pairs called **polygenes,** each of which contributes in some way to the production of the characteristic. The reason is that most human characteristics are the result of complicated interactions between different chemicals, which may be produced in different parts of the body.

Thus most traits—height, weight, intelligence, etc. —are determined by many genes, as well as by the environment. The classical questions of whether nature or nurture is primarily responsible for the expression of a particular trait continue to be asked, but the geneticist tries only to determine how much of the difference observed in the expression of a trait is due to inheritance.

Because of multifactorial inheritance, people are not either light-skinned or dark-skinned, as they would be if it were a matter of only dominant and recessive genes —nor can they be classified simply as light, medium, or dark, as they could be if it were a matter of intermediate or codominant inheritance.

Skin coloration is largely dependent on the presence of a brown pigment called melanin, the distribution of the pigment, and also the presence of blood. The

FIGURE 14.9

Human inheritance for height is polygenic (determined by more than one gene), but stature is also influenced by environmental factors. Although a typical suit of armor from the Middle Ages would be quite a bit too small for the average American male today, genetic selection alone could not account for such a difference in only a few hundred years. In all probability, improved nutrition and hygiene are more directly responsible.

pinkishness in a light-skinned individual comes from blood circulating in vessels near the skin surface, as evidenced by blushing when the vessels dilate and carry more blood, and by blanching when the vessels contract and carry less. The brownish component of the color comes from the pigment melanin, and the darkness of the brown hue depends largely on the distribution of melanin in the skin. In light shades of skin, melanin is present in only one of the five cell layers of the epidermis. In very dark brown skin, all five layers contain melanin. This complex interaction of factors in skin color is thought to be the result of the inheritance of at least four pairs of genes; each pair is nonallelic to the others. Thus each occupies a different position along the chromosome:

$$\frac{S_1 \quad S_2 \quad S_3 \quad S_4}{S_1 \quad S_2 \quad S_3 \quad S_4}$$

The intensity of the skin color is dependent on how many of the polygenes are recessive or dominant.

EPISTATIC GENES

Even when many genes are involved, it is possible for one of the pairs to block the effects of others. For example, an individual may have the genes for melanin deposition, but if he has two recessive genes for the production of tyrosinase, an enzyme essential for the production of all the melanin in his body, he will have no tyrosinase and therefore will be unable to form melanin and so will be an albino. In that case, the genes for the expression of any brown pigment in the skin are said to be **epistatic** (controlling the expression of other genes) to the polygenes for melanin distribution.

Many other characteristics in humans are multifactorially determined. Such characteristics vary across a fairly wide range, with most people having midrange values and only a few having extreme values. Any time there is such a continuous distribution of values, there is strong evidence of multifactorial inheritance. And other genes, such as the epistatic ones, or *environmental* factors, or *both* are involved in producing the melange of characteristics of varying degrees found in humans. Indeed, environment has very considerable effects on the expression of genes.

ENVIRONMENT AND GENE EXPRESSION

Genes may set limits. They may, for example, determine the upper limit for an individual's possible height. But the environment may determine how tall the individual actually will be (see Fig. 14.9). In most advanced countries of the world—as the result of better food, shelter,

FIGURE 14.10
The effects of environment on the phenotypic expression of genes. The Himalayan rabbit is normally homozygous for the gene responsible for the production of black pigmentation. However, when reared at normal laboratory temperatures, the rabbit grows white fur. If a patch of fur is plucked and the new hairs are permitted to grow back at lowered temperature, they grow in black. Thus a quality of the environment, temperature, may have a strong effect on the expression of genes.

and health care—children now are usually taller than their parents. A favorable environment may allow fuller expression of genes. This can be seen in plant growth: in good soil, with suitable amounts of rain and sunshine, and without blights or other sickness, plants grow taller than others of similar inheritance growing under less propitious conditions.

One classic experiment to demonstrate the interplay between genes and environment involves the Himalayan rabbit (Fig. 14.10). The rabbit normally is white with black ears, nose, feet, and tail. But if a patch of fur is removed from the back of the rabbit and an ice pack is applied at the bald site, the new fur that grows in will be black rather than the original white. Actually, the rabbit's extremities—ears, nose, feet, and tail—are ordinarily black because the gene for black color in the rabbit expresses itself only when the temperature is relatively low, as it usually is at body extremities. The new fur on the back is black because the ice pack gave the gene a chance to express itself there.

A human disease, phenylketonuria (PKU), offers another illustration of the gene-environment relationship. In 1934, a Norwegian physician and biochemist, Asbjorn Folling, was approached by the mother of two severely retarded children. Although she had made repeated attempts to try to find out the cause of her children's mental condition—and also of a peculiar musty odor they gave off—no physician had been able to help. Folling became interested. Testing the children's urine with ferric chloride (which changes color in the presence of certain foreign substances), he was astonished to see the urine specimens turn blue-green. He had never before encountered this in such testing. Three months later, Folling identified the responsible material—a by-product of phenylalanine, an amino acid present in proteins. Another phenylalanine by-product discovered in the children's sweat explained their strange odor.

When Folling tested children in nearby institutions for the mentally retarded, he found that many had the same condition. It seemed to Folling that because of some inherited error of metabolism, normally harmless phenylalanine—common in milk, meat, and other proteins—was being turned into a toxic substance. Subsequently, other researchers have been able to show that the metabolic error involves the gene-caused lack of a liver enzyme needed to metabolize phenylalanine. Lacking that enzyme, a child with PKU is unable to dispose properly of phenylalanine, and the ordinarily harmless substance builds up to abnormal levels in the blood and becomes toxic. If the child is left untreated, the brain is irreversibly damaged; by age 2 or 3, IQ may fall below 50 and often close to 20. Although, as yet, the missing enzyme cannot be supplied, physicians have learned to manipulate the environment through a diet which minimizes the intake of phenylalanine, reducing the chance for it to accumulate in toxic quantities.

Thus, the presence of a substance, phenylalanine, in the environment influences the expression of a gene. When the phenylalanine is removed from the environment, the effect of the gene expression—mental retardation—does not appear.

There have been instances noted in which genetic peculiarities, though once of no consequence, have become dangerous through environmental changes brought about by advances of civilization. For example, thousands of Afrikaners descended from one Dutch immigrant who arrived in South Africa in the seventeenth century have a genetic trait which makes it deadly for them to use barbiturates (sleeping pills).

Genes, therefore, do not act in a vacuum. Their effects may be modified by the external and internal environment. Their effects may be expressed in different ways or may not be expressed at all.

EXPRESSIVITY AND PENETRANCE

As we have seen, a range of intensity of inherited effects is possible because of polygenic inheritance; but it has also been shown that a large range of effects—from no perceptible effect to intense expression of a characteristic—can result when only *one* pair of genes is inherited. The intensity of the characteristic is referred to as its **expressivity;** the percentage of individuals who have inherited the genes for the characteristic and who actually show it is called its **penetrance.** The distinction between these two measurements is clearly shown in Fig. 14.11. The variation appears to be the result of several factors. A dominant gene may not produce its effect at all because of the presence of an epistatic gene. Other genes may be inherited which can contribute in some way to the expression of a gene, or the environment may in some way alter the expression of a gene.

Variability can be seen in an inherited defect called syndactyly, or webbing of the fingers. The characteristic appears to be inherited as a simple dominant; that is, only one gene pair is involved. Yet the expressivity of the characteristic is quite variable. In some people with at least one dominant gene for syndactyly—which can be shown with a pedigree or with a family tree—the fingers are entirely normal. In other individuals there may be incomplete webbing of two and sometimes even three fingers.

Sometimes, a variable external environmental condition, such as diet, helps to explain differences in gene expression; in other cases, there seem to be internal causes that affect penetrance and expressivity.

One example is the eye condition, blue sclera, in which the white of the eye (the sclera) appears bluish. The gene for blue sclera behaves as a simple dominant, and theoretically, anyone who possesses the gene, in either heterozygous or homozygous form, *should* show the phenotype. But in only about 90 percent of people with the gene is there a detectable sign of it. The penetrance is thus 90 percent. And among those showing the phenotype, the expressivity is extremely variable. In some cases, the sclera is so darkly blue as to appear almost

FIGURE 14.11

Penetrance and expressivity. Penetrance refers to the percentage of individuals who have inherited a trait and exhibit it. Expressivity refers to the degree to which the trait appears. Each member of this Spanish family has partially formed hands, so penetrance is complete. The mother has four fingers, but her children have as few as two, showing a variation in the expressivity of the trait.

black. In other cases, there are varying shades, ranging all the way to very light blue.

Since it is not yet possible to pin down causes, we can say that the effects of some genes penetrate to different degrees; i.e., they show through (penetrate) the other visible characteristics of the body to different degrees, depending on the density or opacity of other characteristics encoded by other genes. An analogy illustrates this phenomenon. The gene for blue sclera, for example, may be considered somewhat like a light source of constant intensity. The intensity of the transmitted light does not vary. But if filters of different densities are placed between any light source and an observer, the observer will perceive variations in intensity. With a dense enough filter, the light will not penetrate at all; it seems to the observer to be extinguished.

Similarly, the sclera gene is not responsible for penetrance. The gene remains the same, but the penetrance may be determined by differences in the individual's body (reflections of the entire **genome**—all the genes in an individual) in which the sclera genes reside.

ERRORS IN TRANSMISSION

There are two levels at which genetic changes may occur. There may be a change in a single gene that alters in some way the function of the gene. Such changes, called mutations, arise through various chemical errors or damage to DNA. The second level is chromosomal.

Mutations Living systems, as we have noted, are *mutable,* or subject to change in their heritable characteristics. Because they are, they can adapt or change from generation to generation to function optimally in a slowly changing environment. Almost all **mutations,** or changes in genetic material, are nonviable, however, because they upset the finely balanced inherited characteristics of an organism rather than help it. Such nonviable mutations are called **lethals;** when inherited they can produce death of the fetus or of an infant shortly after birth. But the effects of such lethal genes may not appear until several generations after the mutation occurs if the mutant allele is recessive. For sexual reproduction, which provides an organism with two copies of every gene, can mask the recessive allele with a dominant normal one. Two well-understood mutants of this type, lethal when inherited as a homozygous recessive and not too harmful when inherited heterozygously along with a dominant gene, are those which produce the blood disease sickle-cell anemia and Thalassemia.

Although it may appear at first to be strange, the persistence of mutations which produce sickle-cell anemia and a similar disease called Thalassemia are believed to be the result of a successful adaptation to the environment. They flourish in regions where malaria is common. Individuals who inherit a single sickle-cell or Thalassemia gene have a greater than usual resistance to malaria. In some areas of Africa, 40 percent of the population are sickle-cell–gene carriers. On the Mediterranean island of Sardinia, Thalassemia genes have been found to be prevalent among inhabitants of lowland areas which at one time were malarial, but few people in the mountains, where malaria has never flourished, have Thalassemia genes. Unfortunately, although a single sickle-cell or Thalassemia gene provides added protection against malaria, those who inherit two such genes may die of severe anemia.

Many of the genes that cause disease, some investigators believe, originally became common in some populations as a means of meeting environmental needs. In regions where such genes are no longer of adaptive value, they tend to disappear. There is some evidence that in regions where there is little malaria—in the United States, for example—the incidence of sickle-cell genes among blacks has begun to decline.

Chromosomal aberrations The second level at which genetic material may be influenced involves a change in chromosome structure. Aberrations may occur in the choreographic movements of chromosomes during meiosis and therefore are passed on to the next generation via gametes.

There are several types of chromosomal aberrations —inversion, deletion, translocation, and duplication— that may alter gene expression (Fig. 14.12). The processes involved are changes in gene position on the

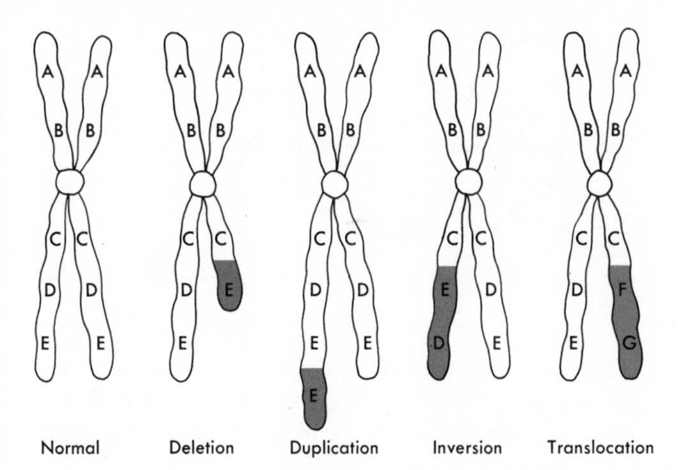

FIGURE 14.12

Examples of chromosomal aberrations. Deletion occurs when a portion of a chromatid (D) breaks off and disappears. The broken end (E) may fuse again with the chromatid, leaving a middle part missing. Duplication results when a portion (E) that should have separated during cell division remains attached as an extra unit. In inversion, part of a chromatid (DE) breaks off, turns over, and fuses again. Translocation occurs when portions of two nonhomologous chromosomes break in crossing over and then fuse again out of place. In the drawing above, segment FG comes from another (nonhomologous) chromosome.

chromosome, removal of portions of a chromosome, or changes in linkage groups.

Chromosome displacements or translocations sometimes have tragic results. For example, in humans, the chromosome pairs numbered 13–14–15 are catalogued by the letter D. Pairs 21 and 22 are catalogued by the letter G. Sometimes one chromosome of the G pairs is displaced in what is called a D/G translocation. One of the No. 21 chromosomes may cross over to, and join up with, one of the No. 15 chromosomes—in effect, riding piggyback and giving that one No. 15 chromosome a lopsided look.*

A person who has such a translocation is a **balanced carrier** for mongolism—balanced in the sense that the total amount of genetic material on the chromosomes in each cell is normal although the number of chromosomes is only 45. But the balance is disturbed when in the next generation an offspring inherits from a carrier a lopsided 15 chromosome and a 21, plus a 21 from the other parent. The total inheritance, then, is three 21 chromosomes, two in normal position and one on the lopsided 15. The three 21 chromosomes determine that the resulting child will be mongoloid.

CROSSING-OVER

One of the simplest types of chromosomal changes is crossing-over. In this phenomenon, old linkages between genes on homologous chromosomes are broken, and new linkages are established. But the changes are not truly mutations, since neither the amount nor the function of genetic material is altered.

Crossing-over takes place during the first meiotic division, when homologous chromosomes are synapsed. It appears that the chromatids of a chromosome may break and exchange parts. For example, in two homologous chromosomes lying side by side in synapsis, one may carry gene P at one end and gene Q at the other end. The other may have alleles p and q at corresponding sites. If there should be breaks in the chromatids at matching points and if there should be an exchange of fragments between the two chromatids before the breaks are repaired, the result of that crossing-over could be a chromatid carrying genes p and Q and another carrying P and q. As a result, when four gametes are produced by meiosis, no longer do two of them carry genes P and Q

* The reference here to the No. 21 chromosome is an example of a conscious continuation, for practical reasons, of a known error. When after long search the cause of Down's syndrome (mongolism) was identified as a chromosomal aberration, it was thought that the chromosome involved was No. 21. Later research established that No. 22 was in fact the aberrant chromosome, but the descriptive identification of the condition, "trisomy 21," has become so entrenched in medical writing on the subject that there has been general agreement to continue to use it.

while the other two carry p and q. Instead, each gamete now carries different genes. In one, there is PQ; in another, Pq; in the third, pQ; and in the fourth pq.

Crossing-over can add to variability, increasing the genetic combinations possible with a cross and introducing new combinations of visible characteristics.

For investigators, crossing-over serves another useful purpose. It produces new phenotypes that may be used to map the locations of genes on chromosomes.

The principle of mapping is simple. A break can appear at any point along the length of a chromosome. It is as likely to occur in one place as another. The possibility that a break may occur at some point between two linked genes on a chromosome increases as the distance between the genes increases. The farther apart their locations and the greater the length of chromosome between them, the more room there is for breakage.

In fact, the frequency of crossing-over between any two linked genes is proportional to the distance separating them. Noting this frequency, the biologist can determine distances between genes; that is, he can map gene locations. Crossing-over frequencies do not indicate absolute distances but rather suggest relative intervals between genes. The basic unit of relative distance between genes on a chromosome has been defined as the distance within which crossing-over takes place 1 percent of the time.

Investigators have been able to draw chromosome maps (particularly for certain insects and some types of bacteria) showing many genes, their order of occurrence on chromosomes, and the relative distances between them. Consider a hypothetical case in which two genes, A and F, are known to be linked. They have been mapped, and the distance between them has been found to be 15 units. Now another gene, X, is found to be linked to F, crossing over with F 7 percent of the time.

There are two possible sequences of A, F, and X on the chromosome. One possible sequence would be:

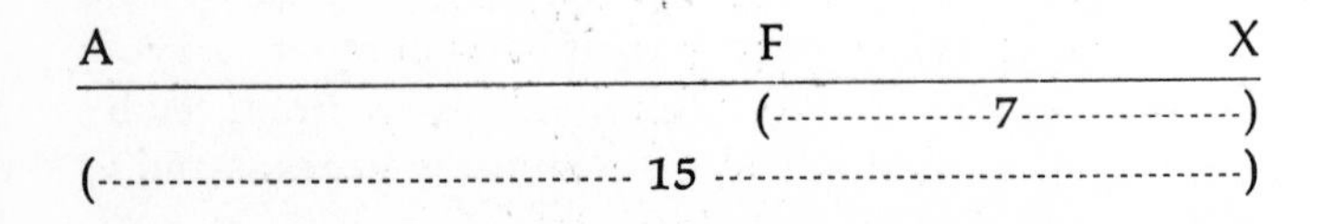

But there could also be an alternative ordering:

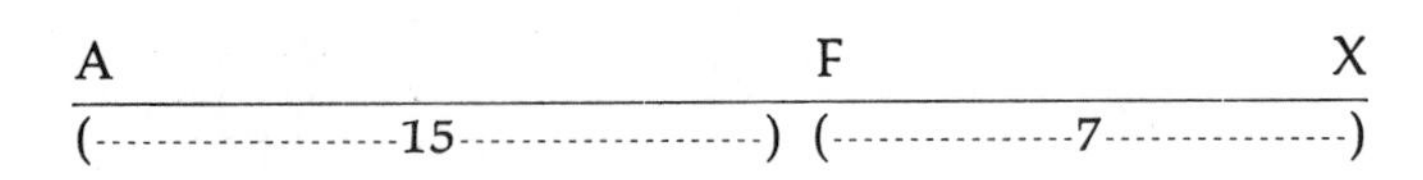

The proper choice can be made by determining the percentage of the time that X crosses over with A. If the crossing-over occurs 22 percent (15 + 7) of the time, then the second alternative must be the correct one. But if X crosses over with A 8 percent (15 − 7) of the time, then the first sequence must be the right one.

It thus becomes possible to define the locations and relative distance between many genes on a chromosome by establishing the crossing-over frequency between each gene and at least two other "landmark" genes.

There is another scientific advantage to be gained from establishing crossing-over frequencies for mapping. Crossing-over helps to indicate which genes control a trait. A gene can then be defined as a point, or **locus,** on a chromosome that controls one trait. If two traits (such as lack of pigmentation in the eyes and in the skin) prove to be invariably linked, the assumption can be reasonably made that they must be functions of a single gene. If, however, they do cross over, they must be functions of different genes.

What we have seen thus far represents a body of hard-won knowledge that can be referred to as classical genetics. Classical genetics did not depend on an understanding—it had none—of the chemical nature of the gene. It was to be the province of molecular genetics—which came afterward—to analyze the gene, to view the hundreds on hundreds of nucleotides within it, and to fathom the mysteries of how it functions.

But if, for classical genetics, the gene was largely an abstract concept, the success of classical genetics in developing theories to explain heredity—the transmission of characteristics from generation to generation, how mutations arise, and how mutations may lead to abnormalities in some instances but may make possible evolutionary development in other instances—has been of tremendous importance.

Summary

1. Although it was once believed that inherited characteristics were a blend of parental characteristics, Gregor Mendel disproved this theory in his experiments with peas in the mid-nineteenth century. Mendel found that each phenotypic characteristic he studied (e.g., seed color, shape) was determined by a pair of hereditary units, today called genes, and that there were two types of genes, dominant and recessive. He observed that successive generations of plants exhibited either the dominant or the recessive trait, and he reasoned that the recessive trait was expressed only if two recessive genes were present.

2. Besides the dominant-recessive gene relationship, Mendel's work was the basis for several other fundamental concepts in genetics. An organism's genetic makeup is called its genotype. Each characteristic is determined by a pair of genes; the pairs may be homozygous or heterozygous. An organism's genotype is expressed in its outward appearance, or phenotype.

3. Mendel proposed two laws of genetics. The first, the law of segregation, states that in gamete formation homologous gene pairs separate, and each gene is distributed to different gametes. The genes may then recombine in different ways after fertilization. The second, the law of independent assortment, states that two traits are inherited independently of each other. Although the first law has not been disproved, twentieth-century experiments have shown that the second law is true only for characteristics whose genes are on different chromosomes.

4. Genotype and phenotype frequencies resulting from a given cross can be determined by a Punnett square or by probability mathematics.

5. Through the discovery and study of sex chromosomes in the fruit fly and in man, modern geneticists have learned about gene linkage and sex-linked characteristics. They found, in exception to Mendel's second law, that certain characteristics are inherited together. The genes for these traits are located on the same chromosome and are said to be linked. Certain genes located on sex chromosomes are sex-linked. In humans, recessive sex-linked traits, such as hemophilia and color blindness, appear mostly in males. Sex-related characteristics whose genes are located on the autosomes may be either sex-limited or sex-influenced.

6. Although Mendel studied only genes with a dominant-recessive relationship, genes may express themselves or affect other genes in a variety of ways. Allelic pairs may exhibit codominance, in which both alleles express themselves phenotypically. They may also show incomplete dominance, in which heterozygous genotypes produce phenotypes which appear to be mixtures of the two alleles present. Many phenotypes are multifactorial, with several allelic pairs contributing to the outward appearance. Some genes are epistatic to a whole series of other genes; their presence can effectively inactivate other genes for a given phenotype.

Environmental factors, such as temperature and diet, will also affect genotype expression. In sum, the expressivity and penetrance of a gene or genes is dependent on both genetic and environmental influences.

7. Although crossing-over, the exchange of genes between homologous chromosomes, commonly occurs during meiosis, sometimes there are random changes in genetic material. These changes are called mutations if they occur on the molecular level and chromosomal aberrations if they occur on the chromosomal level. Chromosomal aberrations include inversions, deletions, duplications, and translocations of whole chromosomes or chromosome segments. Although a genetic change may be beneficial and may generate qualities which favorably affect an organism's adaptation to its environment, more commonly, random genetic changes are detrimental to the viability of an organism.

REVIEW AND STUDY QUESTIONS

1. Imagine yourself magically transported to a farm in eighteenth-century England. The farmer tells you that he and his father before him have selectively bred sheep that produce wool of exceptionally high quality. The farmer, of course, knows nothing about genetics. How would you go about explaining to him why selective breeding works?

2. Why was Mendel's choice of the pea plant a particularly good choice?

3. What is Mendel's Principle of Segregation?

4. What is Mendel's Principle of Independent Assortment?

5. Because they are both heterozygous for albinism, Mr. and Mrs. Smith know the probability that they will produce an albino child is 1 in 4. Since they already have one albino child, they assume that their next child will be normally pigmented. Are they right? Explain your answer.

6. Why are there many more color-blind boys than color-blind girls?

7. What is the meaning of the statement that someone is a carrier of a genetic trait?

8. Explain the difference between expressivity and penetrance.

REFERENCES

Dishotsky, N. I., W. D. Loughman, R. E. Mogar, and W. R. Lipscomb 1971. "LSD and Genetic Damage" *Science* 172 (3982): 431–440

Grundbacher, F. J. 1972. "Human X Chromosome Carries Quantitative Genes for Immunoglobin M." *Science* 176 (4032): 311–312.

Herrnstein, R. 1971. "I.Q." *The Atlantic Monthly* 228(3): 43–64.

Heston, L. L. 1970. "The Genetics of Schizophrenic and Schizoid Disease." *Science* 167(3917): 249–256.

LaDu, B. N., Jr. 1971 (June). "The Genetics of Drug Reactions." *Hospital Practice*, pp. 97–107.

Moore, J. A. 1970. *Heredity and Development*, 2nd edition. Oxford, New York.

Rimoin, D. L. 1971 (February). "Genetic Defects of Growth Hormone." *Hospital Practice*, pp. 113–124.

Winchester, A. M. 1972. *Genetics: A Survey of the Principles of Heredity*, 4th edition. Houghton Mifflin, Boston.

SUGGESTIONS FOR FURTHER READING

Bodmer, W. F., and L. L. Cavalli-Cforza. 1970 (April). "Intelligence and Race." *Scientific American* Offprint no. 1199. Freeman, San Francisco.

A discussion of studies attempting to find a basis in genetics for behavioral differences between races and social classes. The authors conclude that no good case can be made for such studies on either scientific or practical grounds.

Clarke, B. 1975 (August). "The Causes of Biological Diversity." *Scientific American* Offprint no. 1326. Freeman, San Francisco.

Examination of differences in genetic traits within the same species, with emphasis on causes.

Goldsby, R. A. 1971. *Race and Races.* Macmillan, New York.

A biologist's successful attempt at objectively discussing race as a purely genetic differentiation. Accompanied by excellent photographs and art work, the text presents the biological, historical, and psychological evolution of races.

Levine, R. P. 1968. *Genetics*, 2nd edition. Holt, Rinehart and Winston, New York.

A clear, comprehensive, but minimally technical introduction to the nature, transmission, and action of genetic material.

Stahl, F. W. 1969. *The Mechanics of Inheritance*, 2nd edition. Prentice-Hall, Englewood Cliffs, N.J.

A clear presentation of the basic concepts of genetics. Includes useful illustrations, references, and test questions.

Stent, G. S. 1971. *Molecular Genetics: An Introductory Narrative.* Freeman, San Francisco.

Comprehensive introduction to the concepts and development of molecular biology. Well-written and fascinating reading for beginning or more advanced students.

Sturtevant, A. H., and G. W. Beadle. 1962. *An Introduction to Genetics.* Dover, New York.

Originally published in 1939, this is the classic work on the results of Drosophila *genetics.*

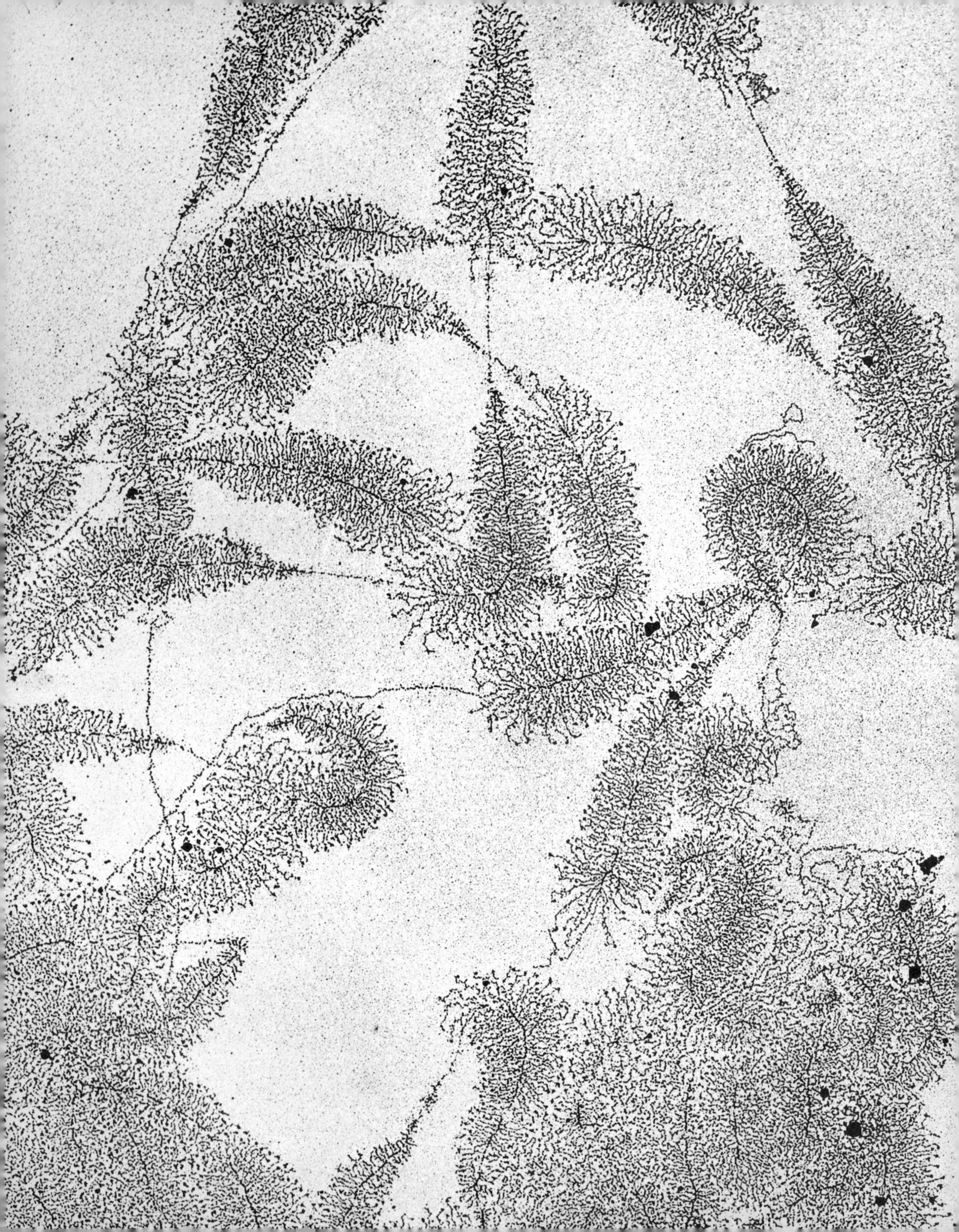

MOLECULAR GENETICS AND PROTEIN SYNTHESIS

chapter fifteen

This historic electron micrograph shows the beginning of the process of transferring genetic information. The backbone of each carrot-shaped structure is the nucleic acid DNA, consisting of genes strung like beads along a necklace. The hairlike filaments are strands of the nucleic acid RNA; short fibers are in early stages of synthesis, and long fibers are nearly completed strands. When the double-helical DNA molecule opens, each chain can act as a template for the synthesis of RNA. A special form of RNA, called messenger RNA, carries the genetic code from the DNA to the ribosomes, where protein synthesis takes place.

The field of genetics has known two bursts of rapid growth in the twentieth century. The first came shortly after the turn of the century with the rediscovery of the ideas and experiments of Mendel. The second, beginning after World War II, continues to this day.

By 1940, a great deal of information had been accumulated about the abstract hereditary unit, the gene. Practical use could be made of information derived from genetic studies. Breeding procedures could be formulated, and records could be made showing the transmission of heritable traits from generation to generation. However, little was known about the chemistry of the unit called the gene.

Classical geneticists thought the gene was a theoretical entity existing on the chromosomes that was somehow responsible for the transmission and expression of phenotypic traits. In essence, there were three questions to be answered. First, what are genes made of chemically? Second, how are they duplicated and transmitted to offspring? And finally, how is the coded information of the genes interpreted by a cell or organism into its phenotype? Despite the fact that the first question seems the easiest to answer, information leading to the answer to the last question was obtained first.

The one-gene-one-enzyme hypothesis

The reaction by which a substance is manufactured in a cell is catalyzed by an enzyme. How, then, do we explain the synthesis of enzymes themselves? If there must be enzymes to form enzymes, it is not clear where the first enzyme could have come from. There is a certain similarity in this dilemma to the chicken-and-egg impasse. Something else in the cell must direct the assembly of amino acids—the components of enzymes—in the proper sequence and make possible their replication in daughter cells.

The first conception that genes might be the "something else in the cell" responsible for enzyme formation was offered as far back as 1908 by an English physician, Sir Archibald Garrod. Garrod had become interested in certain congenital diseases which he called "inborn errors of metabolism." Among them, for example, was a rare disorder, alkaptonuria, characterized by the excretion of a chemical, alkapton, in the urine. The substance caused the urine to turn dark on exposure to air, a sign that made the disorder easily recognizable soon after birth.

Garrod theorized that the diseases stemmed from failure of a specific chemical reaction to take place and that the failure resulted from a lack of a "ferment" (now called an enzyme), which in turn resulted from a defect in a single gene.

Garrod, like Mendel, was far ahead of his time. Finally, with work beginning in 1941 for which they were to receive the Nobel prize, George W. Beadle and Edward L. Tatum (Fig. 15.1) were able to confirm Garrod's thesis. They showed indirectly but conclusively that *enzymatically* catalyzed reactions leading to the production of vitamins, amino acids, and the building blocks of nucleic acids were *directed by genes*—specific, identifiable genes. Just as Garrod had thought, *genes control the synthesis of enzymes.* Later this principle was extended—genes control the synthesis of *all* proteins.

FIGURE 15.1

Beadle and Tatum. George W. Beadle (1903–) an American geneticist and former chairman of the White House Conference on Health, studied the nature of inheritance in bacteria with Edward L. Tatum (1909–) at the Rockefeller Institute for Medical Research. They received the 1958 Nobel prize in Medicine and Physiology for their discovery that genes transmit hereditary characteristics by controlling enzyme synthesis, thereby regulating cellular chemical reactions.

As the result of such work, it was becoming clear what genes *do*—but not what they *are* or *how* they operate. The answers to these questions came later.

Perhaps the implications of work such as Garrod's —that there must be a definite relationship between genes and enzymes—should have been clear to geneticists earlier. Yet clarification did not come until Beadle and Tatum presented their one-gene-one-enzyme hypothesis, which holds that each gene has one primary function—to direct the formation of a specific enzyme. It is by controlling the enzymes that catalyze the chemical reactions of cells—reactions which in turn determine phenotypic characters—that genes exert their influence.

Beadle and Tatum worked with *Neurospora crassa,* a red bread mold. The organism offers many advantages for study. It is easy to culture in the laboratory, and its life cycle is short—only 10 days. Furthermore, although it sometimes reproduces sexually, it also readily reproduces asexually, so that new strains can be produced without the genetic change that would result from sexual recombination.

The two scientists irradiated asexual spores to produce mutations. Each mutation affected the normal nutritional capability of a mold. It could not make for itself all the nutrients a normal mold can produce. They then germinated the spores on a complete nutritional medium —agar mixed with every nutrient the molds might need to grow.

With a large colony thus established from each spore, Beadle and Tatum went about finding the mutants among all the colonies and establishing their identities. Samples of each colony were transferred to a minimal nutrient medium, which lacked several nutrients. A mutant which had lost the ability to make one of the missing nutrients would not grow on it. The mutants could thus be recognized. When a sample from a colony would not grow on the minimal medium, the researchers could be sure that the colony was a mutant. With the mutant strains thus determined, the next step was to find out which supplements in the complete medium not present in the minimal medium were needed. Patiently, the two men placed samples of each mutant colony in a series of various media until they determined in each case which compound—a specific vitamin or amino acid, for example—the mutant molds had lost the ability to synthesize.

When they had finished, Beadle and Tatum had established that in every case of lost ability to synthesize one particular nutrient, the trait was *inherited* in a pattern indicating that only *one* gene was involved. The work of the two men had profound impact on genetics. If each gene had but one function, it provided hope that ultimately the function of each gene could be determined.

We can begin to understand how genes work, by relating the one-gene-one-enzyme principle to Mendelian principles. Consider flower color, for example. A red flower (or any other color flower) gets its color as the result of the production of a pigment molecule in reactions catalyzed by an enzyme. When simple Mendelian dominance is responsible, the dominant gene carries information for the production of an enzyme (or, as we shall see, a family of closely related enzymes) that causes red pigment to be formed from constituents in the cells of the flower. In either the homozygous dominant or heterozygous individual, there is at least one working copy of the gene. In the homozygous recessive case, there is no working copy (both genes are defective). Consequently, no pigment-producing enzyme is made, and the white (absence of pigment) coloration appears. The same principle applies to albinism in humans.

Incomplete dominance, or blending, becomes understandable in terms of gene-directed enzyme production. For example, in snapdragons with red flowers, the homozygous dominant makes a large amount of pigment-producing enzyme. The hue is therefore saturated—that is, the density of pigment molecules is high. In the heterozygous case, the single working copy of the gene produces less enzyme, and there is not enough pigment to saturate the hue. The flower is pink. In the homozygous recessive case, there is no working copy of the color gene, no pigment at all is made, and a white, pigment-deficient flower results.

Codominance, too, is easily explained. In human A-B-O blood groupings, type A blood has A antigens, and type B has B antigens (all antigens are proteins). Heterozygous A or B individuals (i.e., with only one A or one B gene) produce either A or B antigen, respectively. Heterozygous AB individuals make both proteins and thus show both A and B antigenic reactions. Type O blood results when an individual inherits homozygous recessive genes. There are then *no* enzymes for A *or* B antigen production, and the blood contains no A or B antigens.

The fundamental problem

By applying essentially classical genetic techniques, scientists had made progress toward understanding how genes determine the phenotype—namely, by somehow directing the synthesis of specific enzymes. The primary question remained unanswered, however: What is the genetic material?

Even as late as 1950, H. J. Muller, a Nobel prize winner and one of the world's most distinguished genetic researchers, wrote: "the real core of genetic theory still appears to lie in the deep unknown. That is, we have as yet no actual knowledge of the mechanism underlying that unique property which makes a gene a gene—the ability to cause synthesis of another structure like itself, in which even the mutations of the original gene are copied. . . . We do not know of such things yet in chemistry."

But they were to become known. Even as Muller was writing, a base for getting at the knowledge was being laid.

THE PHYSICAL SCIENTISTS ENTER THE PICTURE

In 1945, a little book, *What is Life?*, appeared. It was written by Erwin Schrödinger, an Austrian physicist, and it challenged physical scientists to consider the unsolved physical-chemical problems of genetics. Although their knowledge of biology was generally limited, Schrödinger encouraged them with the observation that "the obvious inability of present-day physics and chemistry to account [for events in living organisms] is no reason at all for doubting that they can be accounted for by those sciences." He added that organisms are large in comparison with atoms; surely they must obey physical laws.

Schrödinger also pointed out that, unlike ordinary chemicals in reaction mixtures—and certainly, gene chemicals must participate in some sort of reaction if they are to transmit their information—genes are remarkably stable. Although they influence much of the chemistry of a cell, they do so without themselves being changed from generation to generation to generation. Perhaps, Schrödinger suggested, genes are stable because chromosomes are crystalline structures. Perhaps the relatively large chromosome crystal, if such it is, is made up of a number of structurally similar parts, and the arrangement of these parts determines the hereditary code.

A fundamental tenet of Schrödinger's argument was that there must be laws of physics that govern genetics. And enticed by the possibility of discovering new laws by which living things operate, many physical scientists moved into the field of biology.

Their activities were to revolutionize biological science. Classical genetics had conceived the gene, and **molecular biology** was analyzing it, in the hope of dispelling some of the mystery of the nature of life itself. Molecular biologists were to find within the gene—a gross unit for a trait—a group of about 1000 pairs of four chemicals, called **nucleotides,** arranged in distinctive order, with this order ultimately determining the order of component amino acids in structural proteins and enzymes.

The physicists made use of the electron microscope to reveal details of structure that could not be seen with ordinary optical instruments. They brought to bear X-ray crystallography, a technique for deducing a molecule's structure by taking X-ray photographs of it from different angles.

They also changed the organisms used for study. Mendel had investigated selected characteristics in a dozen or so generations of peas. Morgan had studied the common fruit fly, which has a generation time of 14 days and can be raised in large quantities in a small

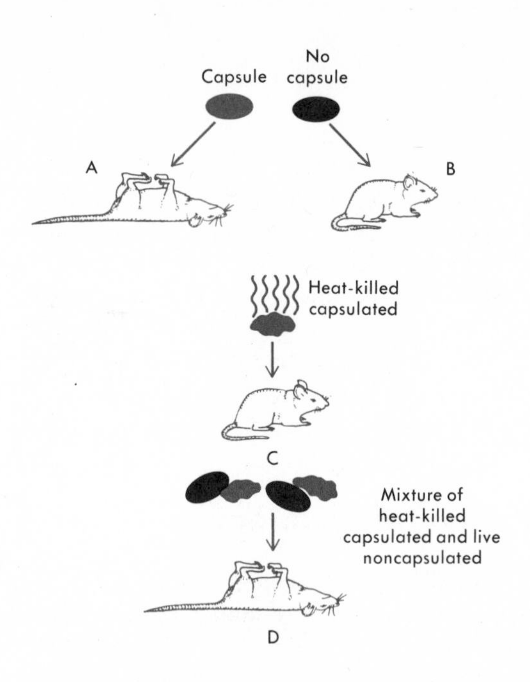

FIGURE 15.2

The work of Avery, MacLeod, and McCarty with pneumococci, pneumonia-causing bacteria, demonstrated that DNA is the agent that carries genetic information. (A) Pneumococci that cause disease are surrounded by capsules resistant to the body's disease defenses. (B) A mutant, noncapsulated pneumococcus strain and (C) heat-killed capsulated bacteria do not cause disease. The experimenters found, as they expected, that capsulated pneumococci injected in mice caused death but noncapsulated and heat-killed bacteria did not. They further discovered to their surprise that (D) a mixture of dead capsulated and live noncapsulated bacteria caused death. Autopsy showed capsulated bacteria in the dead mice. Somehow the dead bacteria's ability to make capsules had been transferred to the live bacteria. The scientists isolated and identified the transferring agent as DNA.

laboratory. The physicists-turned-geneticists turned to bacteria and viruses, which are, indeed, less complex than higher organisms (having fewer components) and which breed so fast (a generation in an hour or even less) that billions can be raised overnight for study the next morning.

Bacterial reproduction is rapid and precise. A bacterium, in each generation period, duplicates itself so that daughter cells are identical to each other and to the parent cell. If the organism is to duplicate its cell structures, it must do so using the simple materials present in a growth medium. The heart of the problem of cell self-reproduction lies with the proteins, which are of paramount importance. Not only do proteins make up much of cell structure; they also form the enzymes, the catalysts for all of a cell's chemical reactions. Without them, a cell could neither reproduce nor maintain itself.

The molecular biologists soon discovered how enzymes, as well as other proteins, are produced. Proteins, both structural and enzymatic, are composed of about 20 different amino acids. The sequence of the amino acids distinguishes one protein from another. The mechanism determining the sequence, the way in which a particular amino acid is introduced at a particular point in a polypeptide chain and another at a second point, was soon made clear.

Because there are so many diverse yet specific proteins, it was assumed for years that genes must be proteins. Indeed, proteins were the only molecules known to have such diversity. But then, in 1944 at the Rockefeller Institute, Oswald T. Avery, Colin MacLeod, and Maclyn McCarty made a significant discovery while working with a bacterium, pneumococcus, that is a cause of human pneumonia (see Fig. 15.2). The bacterium is a pathogen—able to cause disease—because it forms a polysaccharide capsule outside its cell wall, and the capsule serves to protect the organism from destruction by human body defense mechanisms. But some mutant pneumococci lose their ability to form the capsule and are no longer pathogens. The three scientists were able to obtain pure DNA from pathogenic pneumococci. When they added just a tiny amount of that DNA to a culture of nonpathogenic mutant pneumococci, the latter regained their ability to form the capsule. Moreover, their progeny had the capsule-forming ability. Apparently, a defective gene had been replaced. No protein was involved, and yet genetic information was transferred.

This led to the question whether DNA might really be the carrier of heredity. Hardly anyone could believe it, largely because of a mistaken notion about the nature of DNA. In fact, as far back as 1924, Robert Feulgen had found DNA to be a major component of chromosomes, and it had been suspected of playing some role in hereditary processes. But DNA was thought to be simply a *uniform* polymer composed of nucleotides containing bases called adenine, guanine, cytosine, and thymine. Such a molecule could no more carry information than could a sentence composed of just a single letter.

But then came a key finding that disposed of the uniform polymer idea and was to lead to the Nobel prize–winning model of DNA built by Watson and Crick.

Late in the 1940s Erwin Chargaff had begun to adapt paper chromatography—a newly invented technique for the separation and quantitative analysis of chemicals—to the identification of nucleic acids. When Chargaff extracted DNA from calf thymus nuclei, he found that its four nucleotide bases were not present in the equal proportions that would be expected if DNA really was a uniform polymer. Instead, the DNA contained 28 percent adenine, 28 percent thymine, 24 percent guanine, and 20 percent cytosine. Chargaff went on to similarly analyze samples of DNA extracted from many different kinds of nuclei. He found consistently that the amount of adenine equalled the amount of thymine and the amount of guanine equalled the amount of cytosine. These relationships did not solve the puzzle of DNA's structure, however, because he found that the proportions of the four bases varied within wide limits. Thus it became clear that DNA was not a uniform, monotonous polymer.

By 1950 it had begun to look, indeed, as if DNA was the genetic molecule, and that genetic information might be coded in the sequence of the four bases, as sentences have information encoded in the order of 26 symbolic elements, or letters. It also began to appear that the precise *sequence* of the four bases in the DNA molecule controlled the *sequence* of amino acids in the enzyme over which a gene had control.

The DNA molecule was known to be composed of a few relatively simple building-block compounds, the nucleotides, bound together in sequence. Each nucleotide is made up of three constituents: a phosphate group, a five-carbon sugar called deoxyribose, and one of the four organic nitrogen-containing bases. Two of the bases, adenine and guanine, are known as purines and have a double-ring structure. The other two, thymine and cytosine, are pyrimidines and have a single-ring structure. The phosphate group and base are bonded to the sugar.

FIGURE 15.3

Watson and Crick with DNA model. James D. Watson (1928–), the whiz kid of American biologists, received his Ph.D. at 22 and a Nobel prize for work that was completed at 24. Watson and the English biophysicist Francis H. C. Crick (1916–) began studying the structure of DNA at Cambridge University in 1951, and in only two years they were able to publish their conclusions. For this work they shared with Maurice H. F. Wilkins the Nobel prize in Physiology and Medicine.

Knowing the chemical nature was not enough to explain the genetic properties of DNA, however. The crucial next step was determining the spatial arrangement of the atoms within the molecule.

THE WATSON-CRICK MODEL

On a late winter day in 1953, two excited young men, James D. Watson, 24, and Francis Crick, 36, dashed out of Cambridge University's Cavendish Laboratory and into the Eagle, a pub traditionally favored by Cambridge scientists. As the two talked intensely over drinks, friends stopped to learn the cause of the excitement. Finally, Crick burst out exultantly: "We have discovered the secret of life!"

On that day the two scientists, then unknown outside the Laboratory, had finally worked out DNA's now famous double-helical, spiral-staircase structure (Fig. 15.3).

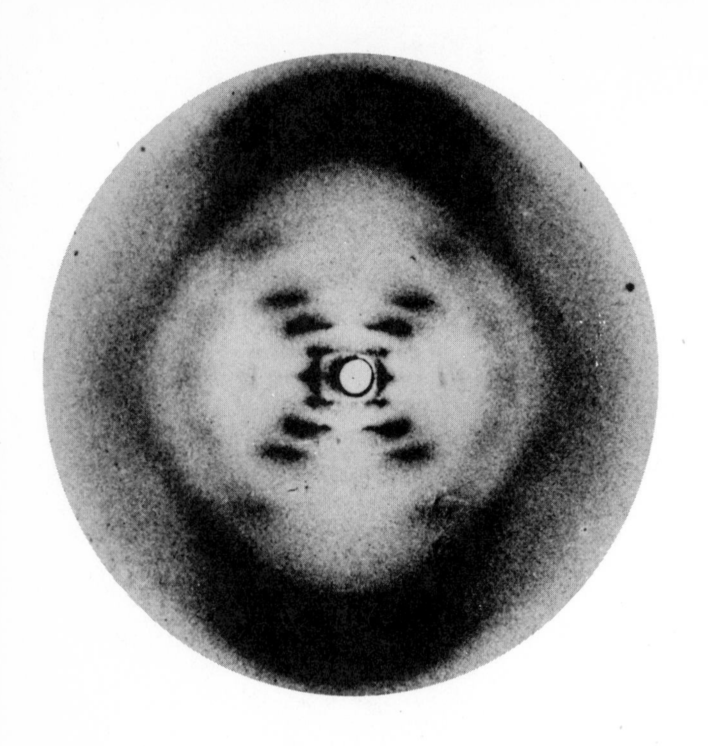

FIGURE 15.4

This now classic photograph of DNA by the X-ray diffraction process, taken by Rosalind Franklin, was an important part of the evidence on which Watson and Crick based their model of DNA structure. In The Double Helix, *Watson's narrative of the events leading to the breakthrough, he acknowledged that his "initial impressions of her, both scientific and personal . . . were often wrong. . . . The X-ray work she did at King's is increasingly regarded as superb. . . . Later . . . she took up work on tobacco mosaic virus and quickly extended our qualitative ideas about helical construction into a precise quantitative picture, definitely establishing the essential helical parameters* and *locating the ribonucleic chain halfway out from the central axis." Rosalind Franklin died in 1958 at the age of 37.*

A common pattern accounts for a large proportion of major new scientific developments. Many researchers work to gain information, and there is a pooling of the information through the various media of scientific communication. Then one or two scientists with deep insight come along and put the fragments together. That was the pattern that lay behind the creation of the DNA model. Watson and Crick had done no laboratory research themselves, but they made brilliant use of the work of many others.

For example, while Watson and Crick were working at Cavendish, important work was being carried on at King's College, London. There Maurice H. F. Wilkins and his colleagues—most notably among them, Rosalind Franklin—were applying the techniques of X-ray crystallography to DNA crystals to produce photographic plates with distinctive patterns of dark dots. From the positions of the dots, the skilled observers could infer much about positions and arrangements of atoms (see Fig. 15.4).

Watson and Crick used data supplied by Wilkins and Franklin. They also applied the technique of molecular model-building, which had been pioneered by Linus Pauling in his work that disclosed the physical structure of proteins (Fig. 15.5). Molecular model-building is a technique in which plastic or metal representations of atoms are stuck together like a child's Tinker-Toy to

FIGURE 15.5

Pauling with protein models. Linus C. Pauling (1901–   ) received the 1954 Nobel prize in Chemistry for his studies of molecular structure. His technique of building models and his X-ray crystallography studies led to the discovery that protein molecules are wound in a helix configuration. Watson and Crick later adopted the model-building technique in working out the double helix structure of DNA.

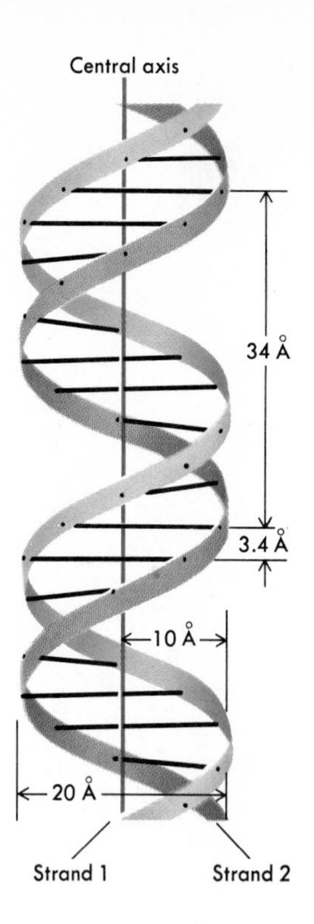

FIGURE 15.6

Diagrammatic representation of a DNA molecule identifying the periodicities of 3.4 Å, 20 Å, and 34 Å observed in X-ray diffraction photographs. The first measurement, 3.4 Å, is the distance between nucleotide base pairs; 20 Å is the width of the DNA molecule; and 34 Å is the length of one complete turn of the double helix. Hydrogen bonds connect the complementary base pairs that constitute the "rungs."

make models of molecules. Watson and Crick made many false starts; the pieces would not fit.

The King's College X-ray diffraction analyses showed three major periodicities, or regularly recurring measurements in crystalline DNA: one of 3.4 Å (angstroms), a second periodicity of 20 Å, and a third of 34 Å.

Watson and Crick were certain of two things. First, the 3.4 Å periodicity represented the distance between two successive nucleotides in the DNA chain. And second, the width of the chain was 20 Å. However, the 34 Å periodicity was at first difficult to understand. Finally they theorized that the DNA molecule might be twisted or coiled. Then the 34 Å period might correspond to the distance between adjacent turns of the coil. In that case there would be exactly 10 nucleotides per turn of the coil if the nucleotides were 3.4 Å apart. Now one more serious problem had to be faced in reconciling their model with the data. The model of a twisted molecule composed of a single chain of nucleotides would have only half the density DNA was known to have. They deduced that DNA was a *two-stranded* molecule, made of a pair of chains of nucleotides wound around an imaginary cylinder in a helix.

When Watson and Crick began to tinker with their model, trying many arrangements of bases, sugars, and phosphates, they found that the arrangement that best agreed with all the known properties of DNA was one in which the two nucleotide chains wound around an imaginary cylinder like a twisted ladder, in a helix, with adenine-thymine and cytosine-guanine base pairs representing the steps and sugar-phosphate side chains representing the upright supports, and with the base pairs held together by hydrogen bonds (Fig. 15.6).

Moreover, Watson and Crick established, each cross rung could not be composed of two purine or two pyrimidine bases. There was not enough room for two purines with their double-ring structures. There was too much room for two pyrimidines with their single rings, but not enough room for three. Invariably, a rung had to be made up of one purine and one pyrimidine.

With this information, Watson and Crick could go further. Although adenine and cytosine were of the right size to make a rung, it became clear that they could not be lined up in a way to allow hydrogen bonding between them. The same was true for guanine and thymine. There were only two possible base pairings: adenine with thymine, and guanine with cytosine. Thus Watson and Crick had solved one of the last critical and difficult aspects of the structure of DNA by reference to the earlier finding of Chargaff that the ratio between adenine and thymine in any DNA molecule is always almost one, and so is the ratio of guanine to cytosine. These ratios occur because of the pairings.

Significance of the model The Watson-Crick model tells us, of course, that DNA consists of two strands or polymers composed of four nucleotides. One strand is a complement of the other, much in the same way that photographic positives and negatives are complementary. There are no restrictions on the sequence of nucleotide bases in the double helix, and the DNA double helix may be hundreds of thousands of base pairs long. For example, there are 200,000 such pairs in the DNA core of one virus particle.

Since any sequence of nucleotide bases is possible in a DNA strand, the virtually limitless variations can make for a molecular language. It is possible that the sequence of bases along a strand somehow can spell out the order of amino acids in proteins. The hereditary material thus carries a set of instructions for making a cell's structural and enzymatic proteins.

Moreover, if the hereditary information is carried in a strand as a sequence of four kinds of bases, then every DNA molecule with its two strands carries two complete sets of information. In their first report on the double helix, which appeared in the British journal *Nature,* Watson and Crick were somewhat playful. "It has not," they wrote, "escaped our notice that the specific pairing we have postulated immediately suggests a possible copying mechanism for the genetic material."

DNA REPLICATION

After mitosis, each cell manufactures its full quota of ribosome "factories," which can produce protein when mitosis is over and vegetative activities resume. The blueprints in the chromosomes are duplicated exactly, and one set goes to each daughter cell. It is also now understood how the chromosomal blueprints are reproduced before mitosis distributes them.

Shortly after the appearance of their first report on the double helix, Watson and Crick proceeded to describe the possible mechanism by which DNA could be copied and distributed to daughter cells.

Since a DNA molecule is double-stranded, they envisioned that each strand could serve as the template for the formation of a complementary strand. If the two strands were separated—in effect, unzipped—by rupturing the hydrogen bonds between the paired purines and pyrimidines, there would be complete information in each strand for producing a new partner. Each purine and pyrimidine base could attract a complementary free nucleotide and hold it in place on the parental template strand with hydrogen bonds. Then the nucleotides could be bonded to each other to form a new strand. Thus each of the two separated original strands could generate a new partner, and there would be two complete double-stranded molecules exactly like the original molecule. But each new DNA molecule would contain one old strand, from the parental DNA molecule, and one completely newly synthesized strand. Thus part of the parental DNA was said to be conserved, and this hypothetical mechanism came to be called semiconservative DNA replication.

It is the semiconservative replication model that is commonly accepted now. Evidence to support this concept came from many researchers. Some of the vital groundwork that supported further research was performed by a team of researchers headed by Arthur Kornberg in 1957 (Fig. 15.7). Kornberg and his associates learned how to synthesize DNA in a test tube. They extracted from bacteria the synthesizing machinery (enzymes) and some DNA (to act as a primer) and added to them ATP (as an energy source) and the four nucleotides. The nucleotides were labeled with ^{14}C in order that their progress in the chemical reactions could be traced. After permitting the reaction to proceed, they found that new DNA was indeed formed, and that it was radioactive. The radioactive nucleotides had been incorporated in the new DNA. Kornberg's work, which won a Nobel prize, showed that DNA *could* be formed in vitro (in a nonliving system), and that the new DNA was a *copy* of the primer. This did not, however, provide incontrovertible evidence that copying of DNA proceeded exactly as Watson and Crick envisioned.

Such evidence came from work in 1958 by M. S. Meselson and F. W. Stahl at the California Institute of Technology (Fig. 15.8). Meselson and Stahl first grew many generations of *Escherichia coli* bacteria in a nutrient medium in which nitrogen was supplied only as the heavy isotope ^{15}N instead of the normal isotope ^{14}N. After all the ^{14}N in the bacterial DNA was replaced by ^{15}N, the nitrogen in the nutrient medium was switched to the normal, lighter ^{14}N isotope. At regular intervals Meselson and Stahl removed cell samples, extracted their DNA, and by high-speed centrifugation separated DNA of different densities. They found that when cells

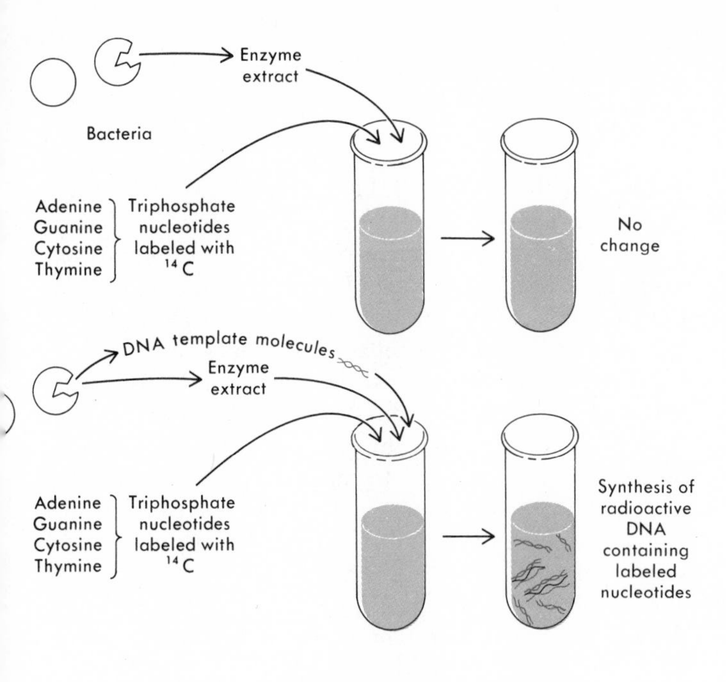

FIGURE 15.7

Arthur Kornberg and his associates set out to prove that DNA-synthesis enzymes could make copies of existing DNA. Obtaining protein extracts including all cellular enzymes from bacterial cells, they mixed them in various combinations with ^{14}C-labeled radioactive nucleotides. If enzymes or primer DNA were not present, or if enzymes and DNA but not all of the four DNA nucleotides were present, no synthesis occurred. However, if enzymes, DNA, and the four nucleotides were all present, radioactive DNA was synthesized. When Kornberg chemically analyzed the labeled DNA, he found it had the same nucleotide composition as its DNA template, indicating that the structure of the template and the new DNA were the same. Kornberg also analyzed the bacterial protein extract and isolated the synthesis-catalyzing enzyme, which he called DNA polymerase. For this work Kornberg received the Nobel prize.

containing only heavy DNA (^{15}N in the purine and pyrimidine bases of both strands) were permitted to divide mitotically once in the ^{14}N medium, the DNA of daughter cells contained half ^{15}N and half ^{14}N. This, of course, was to be expected if the two parental chains containing heavy DNA separated and served as templates for new complementary chains containing only ^{14}N. Each new DNA molecule would have one heavy chain from the parent and one newly produced light chain.

Meselson and Stahl went further, treating the new DNA molecules to break the hydrogen bonds and separate the two strands. And when the separated chains

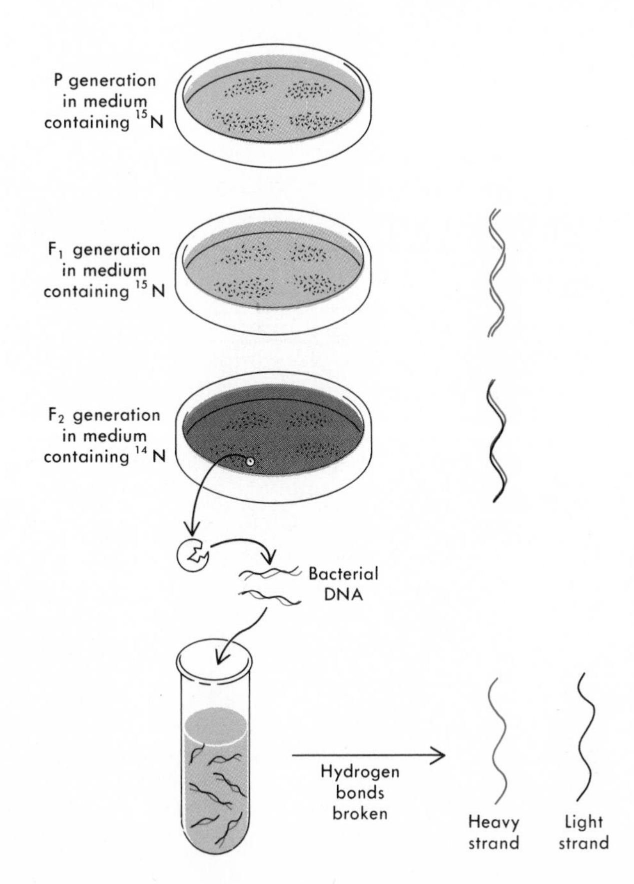

FIGURE 15.8

The work of M. S. Meselson and F. W. Stahl proved that DNA replication follows the semiconservative model proposed by Watson and Crick. After growing several generations of Escherichia coli *bacteria on a medium containing ^{15}N, a heavy isotope of nitrogen, the scientists started growing the bacteria on a medium containing ^{14}N, the normal isotope of nitrogen. From the first bacterial generation (F_2) grown in the ^{14}N culture, they extracted DNA and found that it was lighter than DNA from bacteria grown for generations on ^{15}N medium, yet heavier than DNA from bacteria grown for generations on a ^{14}N medium. Separating the strands of F_2 DNA by breaking the hydrogen bonds between nucleotide base pairs, they found that one strand was heavy and contained only ^{15}N and that the other strand was light and contained only ^{14}N.*

were analyzed, they were found to consist of one heavy and one light strand, showing that the N isotopes had not been distributed randomly, but each was localized in just one of the strands, as the Watson-Crick theory indicated it would be.

Function of genetic material

Earlier we learned that genes are responsible for coding for enzymes in accordance with the one-gene-one-enzyme hypothesis. Watson and Crick thought that a full explanation of this problem would be obtained with relative ease, but actually it took molecular biologists 13 more years to figure out precisely how DNA accounted for the structure of enzymes.

THE CODE

With only four letters, the four bases, to serve as an "alphabet," how could DNA select from among 20 amino acids to produce complex proteins? At first blush, the problem may seem insuperable. Yet a conventional language like English can be communicated by means of even a four-symbol alphabet, as it is by telegraphy.

With the four-letter DNA alphabet, there can be only four one-letter words (A, T, G, C). But there can be 16 two-letter words (AA, TT, GG, CC, AT, AG, AC, TG, TC, TA, GC, GA, GT, CA, CT, and CG). And there can be 64 three-letter words, more than enough to symbolize the 20 amino acids used in protein synthesis.

Crick and two colleagues devised a code that called for each triplet (called a **codon**) of nucleotide bases in the DNA double helix to specify one amino acid. But punctuation is also essential in a language to separate the symbols. Suppose there is a sequence of nucleotides, such as GCA, TGG, TCG, ATT, etc. The sequence seems readable enough with commas in between. But there are no commas in the DNA language. Many elaborate schemes were suggested to try to get around the need for punctuation, but as it turned out, they were unnecessary. The code is extremely simple. There are nucleotide triplets—codons—each coding for a specific amino acid, as shown in Table 15.1. And the code letters for the triplets—in groups of three, from message beginning to end without commas or any kind of punctuation—are read and understood. The only sorts of punctuation are **initiator** and **terminator** codons, which represent the start and finish of a set of instructions, or gene.

To illustrate the dramatic degree of miniaturization achieved in the DNA code, Beadle has pointed out that the nucleus of a single human cell contains 5 billion nucleotide pairs, enough to make 1.7 billion letters, or 340 million words averaging five letters each, which would be enough to fill 1000 printed volumes, 500 words per page and 600 pages per volume. Moreover, if one were to stack all the DNA strands of the nuclei of the egg cells that have given rise to all humans currently on earth (some 3 billion), they would fit into a cubical box only $\frac{1}{8}''$ on a side. The tremendous amount of information contained in genes is of vital importance in determining every structural and functional quality of every organism.

SITE OF PROTEIN SYNTHESIS

Most of the DNA in a cell is located in the nucleus (or in the nuclear region of a bacterial cell which has no nucleus per se). But protein synthesis occurs in the cytoplasm of a eucaryote and away from the nuclear region in a procaryote. Thus DNA's information must be transmitted out of the nucleus to the sites of cellular protein synthesis.

Beginning in the late 1930s, investigators had become aware that large amounts of a nucleic acid called ribonucleic acid, or RNA, could be found in the cells of such organs as the pancreas in vertebrates and the silk gland in silkworms. These are cells which are extremely active in producing proteins. On the other hand, in cells such as those in muscle tissue, which secrete no protein, little RNA was to be found.

In addition, microscopic studies had established that almost all cellular RNA is in the cytoplasm, concentrated there in small particles. The particles were found to contain a considerable amount of protein along with the RNA. By the 1940s, Jean Brachet, who along with T.

TABLE 15.1
The genetic code

FIRST NUCLEOTIDE	SECOND NUCLEOTIDE: U	C	A	G	THIRD NUCLEOTIDE
U	UUU, UUC } Phenylalanine UUA, UUG } Leucine	UCU, UCC, UCA, UCG } Serine	UAU, UAC } Tyrosine UAA stop UAG stop	UGU, UGC } Cysteine UGA stop UGG Tryptophan	U C A G
C	CUU, CUC, CUA, CUG } Leucine	CCU, CCC, CCA, CCG } Proline	CAU, CAC } Histidine CAA, CAG } Glutamine	CGU, CGC, CGA, CGG } Arginine	U C A G
A	AUU, AUC, AUA } Isoleucine AUG (start) Methionine	ACU, ACC, ACA, ACG } Threonine	AAU, AAC } Asparagine AAA, AAG } Lysine	AGU, AGC } Serine AGA, AGG } Arginine	U C A G
G	GUU, GUC, GUA, GUG } Valine	GCU, GCC, GCA, GCG } Alanine	GAU, GAC } Aspartic acid GAA, GAG } Glutamic acid	GGU, GGC, GGA, GGG } Glycine	U C A G

Caspersson had been studying the RNA-protein particles, suggested that they were the sites of protein synthesis. The RNA protein particles have since been termed ribosomes.

By the 1950s, around the time that Watson and Crick were developing their model, it was well established that the ribosomes were indeed the sites of protein synthesis. By then, radioactively labeled amino acids had become available. A rat could be given an injection of a labeled amino acid, and when, shortly thereafter, RNA-protein particles were extracted from tissue samples, radioactive atoms could be found in the ribosomes, indicating that the labeled amino acid was being incorporated in the polypeptide chains being constructed there.

But if the information about the synthesis of proteins resides in DNA in the nucleus, there must be some way in which the information is copied and the copies or messages are sent to the ribosomes. And it now seems clear that RNA, a substance much like DNA, is *both* the message and the messenger that brings the message to the cytoplasm.

MECHANISM OF PROTEIN SYNTHESIS—TRANSCRIPTION

Although DNA and RNA are both nucleic acids, there are differences between them. In DNA the sugar of the backbone is deoxyribose; in RNA it is ribose, which is very similar to deoxyribose but has one more oxygen atom. Another difference is that RNA contains the base uracil rather than the thymine found in DNA, and so the RNA pyrimidine bases consist of cytosine and uracil, whereas those in DNA are cytosine and thymine. Finally, DNA is usually double-stranded, but RNA usually has only a single strand.

Despite these differences, DNA can serve as a template, or reusable pattern, for the synthesis of RNA, and the synthesis can proceed in much the same way as that for new DNA. Synthesis occurs during the time that a

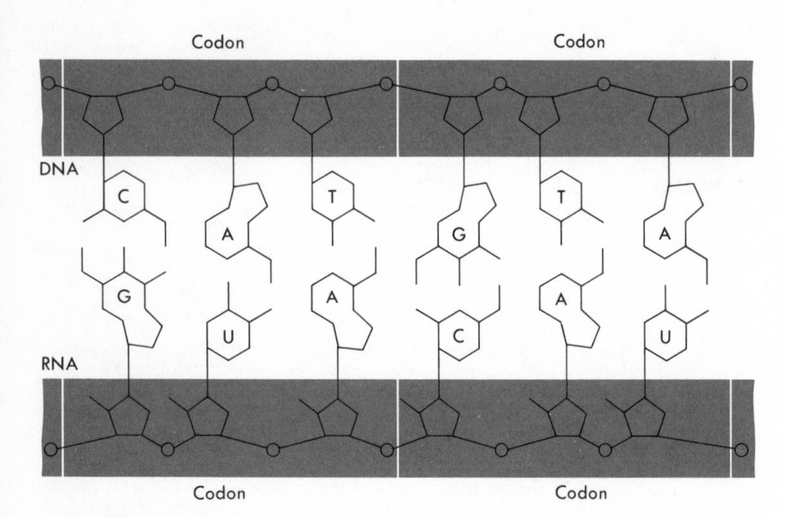

FIGURE 15.9

The first step in cellular protein synthesis is the formation of a strand of messenger RNA from a DNA template. One strand of DNA unwinds slightly, and RNA nucleotides pair up with bases on the DNA strand. In the presence of the enzyme RNA polymerase, the RNA nucleotides bond together, forming a molecule of mRNA complementary to the DNA template strand.

cell is making any sort of protein, and several, even all, of the genes can serve as templates simultaneously.

Although we do not know how RNA synthesis actually takes place on the DNA template, most researchers working on the problem feel that DNA is somehow unwound so that the normally twisted double-helical DNA molecule reveals its two strands. Only one strand serves as a template—that is, has useful information encoded. The other has no useful information for protein synthesis, but it can be useful during repair, when it can be a copy of the information-containing strand.

After the DNA molecule opens up to expose the working strand, the process of **transcription,** or copying, begins. Ribonucleotides—which are very similar to deoxyribonucleotides—are assembled into a strand of RNA called **messenger RNA,** or **mRNA** (Fig. 15.9). The ribonucleotides contain the bases adenine, guanine, and cytosine, but in place of thymine the fourth kind of ribonucleotide contains uracil. As in DNA replication, mRNA synthesis involves pairing of complementary bases, with the exception that RNA's uracil pairs with DNA's adenine. Thus the order of nucleotides on a strand of mRNA is a copy, in complementary form, of the codons in a molecule of DNA. Once formed, the newly made mRNA moves from the DNA in the chromosome to the protein assembly sites in the cytoplasm.

A visible indication of this process—transcription—appears in the giant chromosomes of certain flies, such as the fruit fly. A "puff" or protuberance forms around each gene during its transcription into RNA. The puff, as electron microscope studies show, is composed largely of untwisted and extended DNA in the genes. It can also be seen in most cells during interphase, when a cell is vegetative and nondividing—actively secreting or building up more proteinaceous parts of itself as it grows—and the chromosomes become indistinct individually and appear as granular or fibrous **chromatin** in the nucleus. The granular or fibrous appearance results from the untwisting of many gene regions of each chromosome to permit formation of mRNA. During mitosis, when protein synthesis shuts down, the DNA condenses with large amounts of protein, and the rodlike mitotic chromosomes appear.

MESSENGER RNA

After mRNA is made, it must leave the nucleus in a eucaryote (in a procaryote, ribosomes are situated close to the DNA so they can grab mRNAs as they are formed). Electron micrographs show clearly that the nuclear membranes of eucaryotes are porous and that the pores offer a path for communication between nucleoplasm and cytoplasm. In fact, one recent electron micrograph purports to show an mRNA fragment actually squeezing through a pore in a membrane.

Once in the cytoplasm, the mRNAs associate with ribosomes, the actual organelles of synthesis. The ribosomes can be thought of as assembly plants, relatively stable organelles capable of producing proteins on order. The order comes in the form of mRNA transcripts of the

DNA message from each gene. Each transcript may be used by many ribosomes; in fact, many ribosomes may stick to the mRNA at one time, all using its instructions, and giving it the appearance of a string with beads on it. This complex of mRNA with many ribosomes is called a **polyribosome** (Fig. 15.10). In time, the instructions become illegible and they are, in effect, discarded; i.e., an enzyme hydrolyzes them.

The cell has a regulating mechanism (which we will discuss shortly) by which it can control the amount of a particular protein that it produces. If synthesis is to continue, new mRNA copies must be sent to the ribosomes. If the relatively unstable mRNA is not replaced, the production of a particular protein stops.

FIGURE 15.10

Electron micrograph of polyribosomes. Note the beads-on-a-string appearance of the spherical ribosome particles stuck to a strand of mRNA.

THE ADAPTOR: TRANSFER RNA

Once mRNA molecules bring orders for protein synthesis to the ribosomes, the amino acid building blocks must be assembled to make the proteins specified by those orders. In 1958, when Crick considered the problem of assembly, he observed that a first, but naive, idea might be that the RNA would take up some shape that might provide 20 different cavities, one for the particular shape of each of the 20 amino acids. But that seemed implausible on physical and chemical grounds. "One would expect, therefore," Crick wrote, "that whatever went on to the template in a specific way did so by forming hydrogen bonds. It is therefore a natural hypothesis that the amino acid is carried to the template by an adaptor molecule, and that the adaptor is the part which actually fits on to the RNA."

This hypothesis would require 20 adaptors, one for each amino acid. It was anybody's guess, Crick admitted,

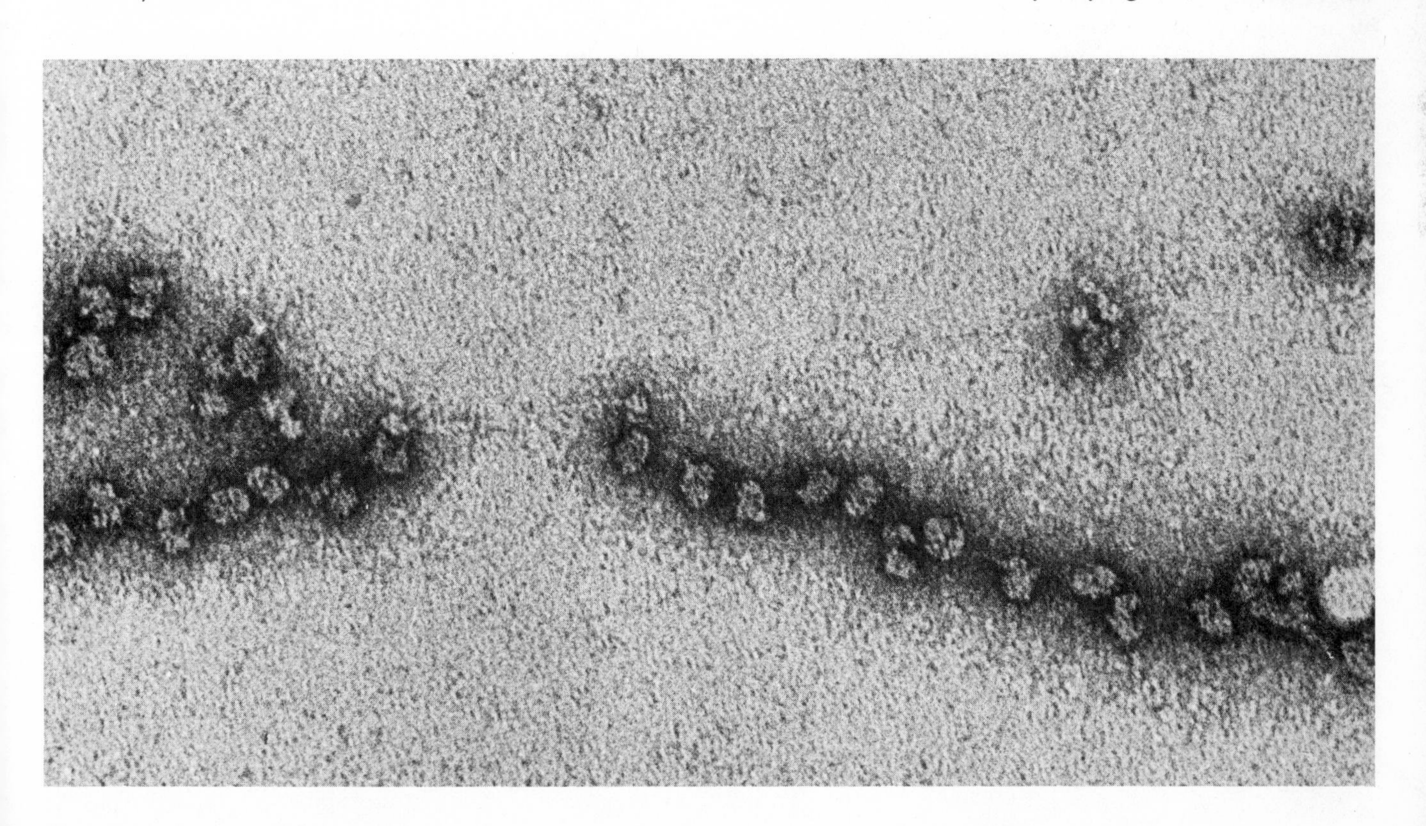

what kind of molecules such adaptors might be, but it seemed to him likely that they would contain nucleotides. That would permit them to join to the RNA template by the same pairing of bases as is found in DNA, or between DNA and mRNA when the latter is synthesized. Crick also suggested that a separate enzyme would be needed to join each adaptor to its own amino acid, and "enzymes, being made of protein, can probably make such distinctions more easily than can nucleic acid."

In the light of an adaptor theory, the amino acid assembly might proceed in the following way. *Each* amino acid molecule is provided with a different adaptor. The adaptor contains an **anticodon,** which is a nucleotide triplet *complementary* in its nucleotide sequence to the codon, the nucleotide triplet in the mRNA that codes for the amino acid. Once equipped with adaptors, the amino acids move to the ribosomes and to the messenger template where they are held in the proper places by hydrogen bonds between complementary purines and pyrimidines of adaptor and mRNA molecules. Then, once lined up properly along the mRNA, the amino acids are joined to each other by peptide bonds and are thus freed from their adaptor bonds and joined to the polypeptide chain.

Although this was a speculative scheme, evidence soon was found to support its validity. Almost at the same time the scheme was being formulated, a previously unknown type of RNA was found, first in eucaryotic cells, then in bacteria. It proved to be a much shorter RNA molecule, only 80 nucleotides long. Soon investigators were able to determine that the first step in the utilization of an amino acid for protein manufacture is its activation by an amino acid–activating enzyme so it becomes more amenable to chemical reaction. The enzyme catalyzes the attachment of the amino acid to the shorter type of RNA—called **transfer RNA,** or **tRNA**—and it is by means of this amino acid (or aminoacyl) tRNA complex that an amino acid reaches its destination in the polypeptide chain. It was soon determined that there is a species of tRNA for each of the 20 amino acids.

MECHANISM OF PROTEIN SYNTHESIS—TRANSLATION

Now we can understand the complete process of how protein sequences are determined by the messenger RNAs, the transfer RNAs, and the ribosomes (see Fig. 15.11). Aminoacyl tRNAs collide at random with the exposed codons on mRNA molecules attached to ribosomes. When an anticodon of an aminoacyl tRNA happens to match, it is fitted into place in the growing polypeptide chain, and the ribosome moves down one codon for a repetition of the performance. In this way, the *primary* structure, that is, the order of bases on the double-stranded DNA molecule, is copied into the primary structure of an mRNA molecule, which is in turn directly responsible for the primary structure (the order of amino acids in the polypeptide) of the growing protein.

SYNTHESIS CONTROL

In these ways, insights into the synthesis machinery of a cell on a molecular level were obtained. But questions about the mechanisms that regulate synthesis remained. Does a cell express, at all times, all of its protein-producing capabilities? That would not be economical any more than it would be to have every factory in a national economic system produce everything it can every day.

Consider, for example, a situation in which many bacteria find themselves. They are living and multiplying contentedly on some given nutrient medium. Then a useful sugar is introduced into the medium. Somewhere in their DNA blueprints there may be instructions for producing an enzyme capable of digesting or using the newly introduced sugar. It would certainly not be economical if they were to be always expending the energy and consuming the materials needed to produce the sugar-digesting enzyme, and indeed they do not—but as soon as the sugar is introduced they turn on the specific enzyme-producing processes. Thus there remained the question of how bacteria start the production process, and how the process is turned off when the sugar is all gone and no more enzyme is needed.

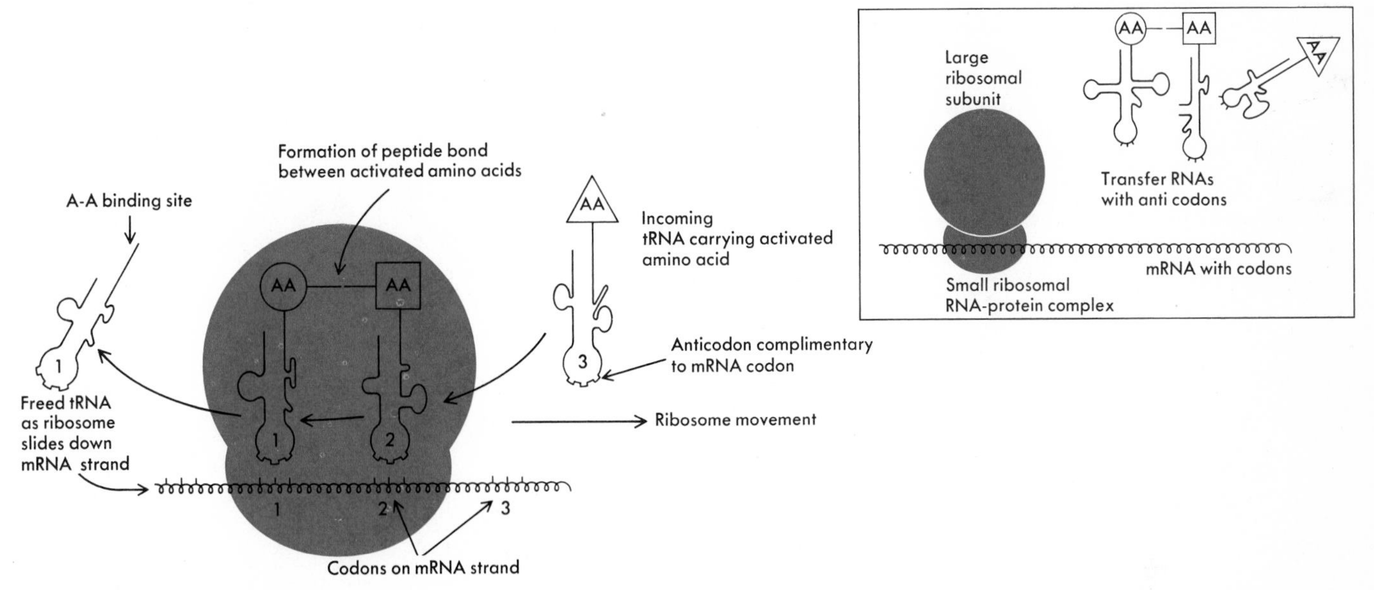

FIGURE 15.11

In the upper right corner are shown the major RNAs involved in protein synthesis. The ribosome attaches to the mRNA strand and slides along, exposing new codon triplets to anticodons of tRNA held in place on the large ribosomal subunit. This temporary viselike positioning allows time for the activated amino acids of tRNA 1 and tRNA 2 to form a peptide bond. The first unit, tRNA 1, is freed as the ribosome moves, and tRNA 2 occupies its site on the large ribosomal subunit. In so doing, it frees a binding site for occupation by incoming tRNA 3 with its activated amino acid, which will then be added to the growing peptide chain. This process continues until a terminator codon signals it to stop.

A model for these controls—or gene regulation—has been devised by François Jacob and Jacques Monod. They noted that in normal *Escherichia coli* bacteria, an enzyme is produced for the utilization of lactose only when the enzyme is actually needed. Jacob and Monod found a mutant strain of the bacteria which produced the enzyme *continuously*. They were able to establish that the recessive mutant gene—which they called a **regulator gene**—is not a gene that blueprints enzyme structure but rather regulates the *activity* of the gene that determines the structure of the enzyme.

Since the mutant allele of this regulator is recessive, Jacob and Monod postulated that the *normal* allele must produce a substance that can turn off, or *repress*, the activity of the gene that determines the enzyme structure—which they called a **structural gene**—and that the repressor substance is one the mutant allele cannot produce.

Jacob and Monod found another mutant strain of *E. coli* which produces the enzyme for lactose continuously. In this strain the mutant allele is dominant rather than recessive. They determined that this gene is different from both regulator and structural genes. When the mutant allele of this third gene is present, the structural gene keeps functioning whether or not repressor substance is produced by the regulator gene. Evidently the mutant is insensitive to the repressor substance. Jacob and Monod gave the name **operator gene** to the gene that turns the structural genes on or off.

They went on to find that such a regulator-operator-structural gene system exists for other bacterial activities, and they proposed a model to account for the functions of the three types of genes (Fig. 15.12). An operator gene, the model indicates, acts to switch on or off the activity of a structural gene (or genes) next to it. When the structural genes are switched on, they produce mRNA, which moves to the ribosomes in the cytoplasm to serve as a template for protein (in this case, enzyme) synthesis. Often one operator gene controls a series of structural genes next to it on the chromosome. Such a collection of operator gene and structural genes is called an **operon.**

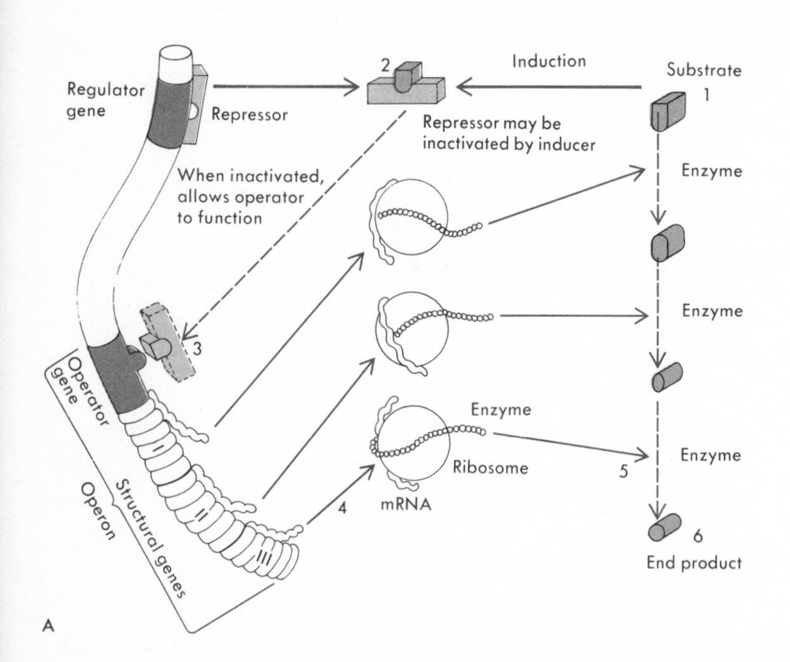

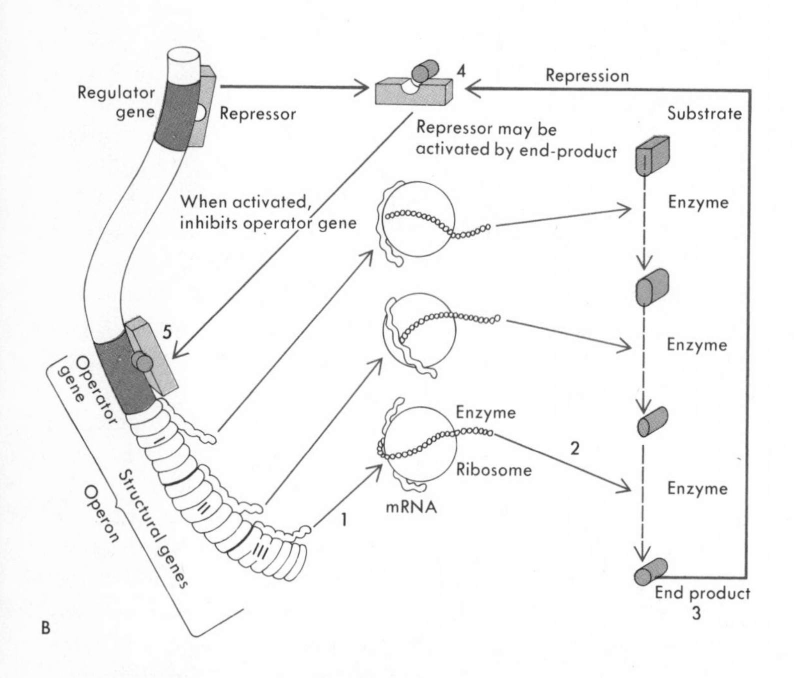

FIGURE 15.12

The Jacob-Monod model of regulation of genes suggested two possible modes of control. (A) In induction, the regulator gene makes a repressor substance that can normally attach to the operator gene and turn it off. However, when a substrate (1) is present, it binds with the repressor (2), preventing it from attaching to the operator gene (3). Thus the operator is free to turn on the structural genes (4), and the enzymes of a biochemical pathway may be fabricated (5,6). When the substrate is used up, there are no substrate molecules left to inactivate the repressor. The repressor is again free to bind with the operator and turn it off. (B) In repression, the regulator gene makes a repressor molecule that cannot itself inactivate the operator gene. The enzymes of the biochemical pathway are thus manufactured (1), and substrate molecules are converted (2) to end-product molecules (3). End-product molecules combine (4) with repressor molecules, and the repressor–end-product combination inactivates the operator (5), stopping further synthesis of enzymes of the biochemical pathway.

According to the model, therefore, control works in the following way. A particular substrate may nearly always be available in the cell's cytoplasm. The product that can be made from it is needed in some definite amount. As the organism uses up the product and stores fall below a critical level, the repressor, which previously has been activated by the presence of adequate amounts of the end product, becomes inactivated. Now the operator gene, no longer repressed, is free to turn on structural genes and start up the enzyme-producing machinery. When enough **end product** has been produced, the repressor is activated and the machinery turned off.

GENE REGULATION IN EUCARYOTIC CELLS

Gene transcription, in which DNA manufactures complementary RNA molecules that in turn synthesize specific proteins, clearly appears to be a highly regulated process. The brilliant studies of Jacob and Monod on *Escherichia coli* have done much to explain genetic control in the procaryotes, but no similar regulatory mechanism has been conclusively found in complex nucleated eucaryotic cells.

Among the many differences between eucaryotic cells and bacteria is the fact that, in addition to DNA and bits of RNA, eucaryotic chromosomes contain large amounts of basic (i.e., alkaline) histone proteins and acidic nonhistone proteins. A considerable amount of recent research has been directed toward determining how these proteins are involved in gene regulation. Much of the evidence suggests that nonhistone proteins play a key role, but precisely what that role is has not yet been worked out. Several models have been proposed, each supported by some data, but no model for gene regulation in eucaryotic cells is as yet so well supported as the operon system of *E. coli.*

Molecular genetics of genetic change

ERRORS: MUTATIONS

Despite the precision with which DNA replicates, the mechanism is not infallible. Errors or *mutations* may occur as the result of some external influence. Environmental influences known to produce mutations include high-energy radiation, such as from X rays or gamma rays, and many chemicals, such as mustard gas (used during World War I) and related compounds, peroxides, epoxides, nitrites, some purines and some pyrimidines.

Mutations on the gene level, as *contrasted* with those on the chromosomal level (discussed in Chapter 14), are called point mutations. They are somewhat analogous to typographical errors in a printed message, because they involve additions, deletions, or substitutions of nucleotides.

Consider, for example, the genetic message

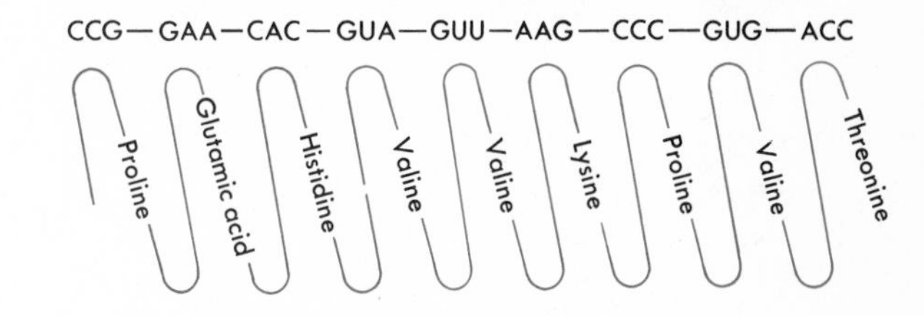

The message is read from left to right in triplets and the nine triplet words may be instructions for the synthesis of a segment of a protein chain.

If a single letter is added or subtracted, it will change the way the message is read, altering triplets to the right of the error and producing a corresponding change in the protein chain segment.

For example, if the twelfth letter, the A in GUA, should be *deleted*, the message after the omission, as well as before it, would have to be read in triplets. It would become

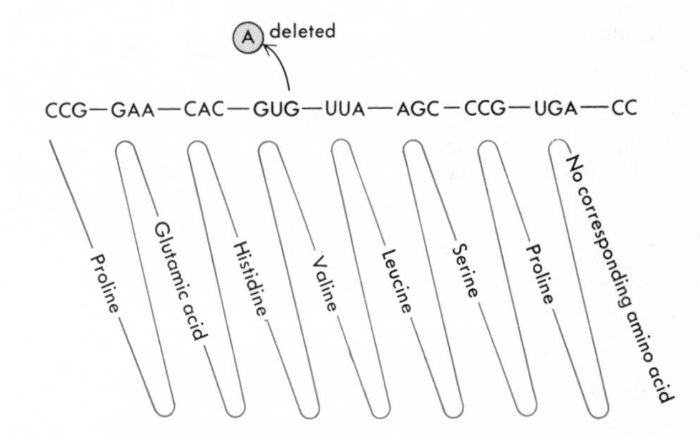

Similarly, the *addition* of a single letter could produce marked changes in the meaning of the instructions and the assembly of amino acids into a protein chain.

In **base substitution,** one nucleotide replaces another by mistake so that, for example, in place of a required UUA triplet, a UGA triplet may be used.

Although some errors have little significance, other point mutations may cause serious disability. The only difference between either Thalassemia or sickle-cell hemoglobin and normal hemoglobin, for example, lies in

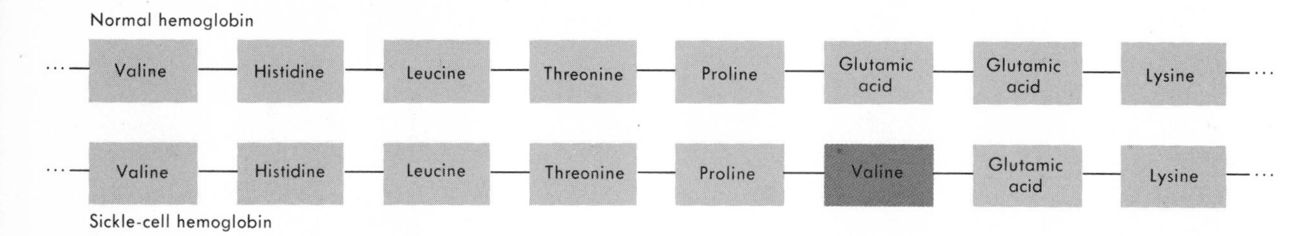

FIGURE 15.13

Sickle-cell hemoglobin results when only a single amino acid is substituted in the polypeptide. This alteration is the result of a mutation that changes one nucleotide in a single codon in the section of DNA that codes for the production of hemoglobin.

one amino acid in the molecule—the result of a mistake in a single nucleotide in the gene responsible for the production of hemoglobin (Fig. 15.13).

ERROR REPAIR

In a sense, DNA is characterized by redundancy, which gives it the potential for rectifying errors. The two strands of the molecule are complementary images of each other, and one can serve, as it does in replication during mitosis, as a template for the other, permitting reconstruction of the mate strand if damage has occurred.

Mistakes often can be repaired successfully. For example, ultraviolet light ordinarily is deadly for many bacteria. Yet investigators have found that some organisms survive the damage to their DNA inflicted by exposure to ultraviolet radiation or X-rays, because they have the ability to repair the damage readily.

A model proposed for the repair mechanism suggests that the process is one of cut and patch, in which one or more enzymes constantly travel along a bacterial DNA molecule, much as a test car travels over railroad tracks testing for structural faults. If an error in DNA is encountered, the enzyme cuts the polynucleotide chain in two places, one ahead of and one behind the faulty nucleotide. That nucleotide and perhaps a few adjacent ones are removed from the chain. Additional enzyme activity follows, and with the intact, error-free, complementary DNA strand serving as a template, a new section of nucleotides to fill the gap in the cut chain is assembled and bonded into place.

The future of molecular genetics

Molecular genetics is a relatively new field, and its future will undoubtedly provide us with many benefits. However, there are some risks, and appropriate precautions must be taken. With growing knowledge of gene structure and new techniques of gene manipulation, it may become possible to replace damaged genes in human patients with genes manufactured in laboratories. Although that goal remains distant, a number of important developments have already pointed toward it.

A step in the direction of correcting a genetic defect was taken when mouse cells lacking a specific enzyme were cultured and chick red blood cells containing the enzyme were added to the culture. The hope was that the two kinds of cells would fuse and that the DNA coding for the enzyme would be transferred from the chick red cell nuclei to the mouse cell nuclei and become established there. Examination of the fused cells showed that they had all the characteristics of the mouse cells and had gained the ability to synthesize the previously missing enzyme. Moreover, analysis of the enzyme itself showed it to be identical with the chick enzyme, suggesting that the mouse cells had, indeed, acquired the chick gene for the enzyme. If it should become possible to apply this technique to the multitude of human genetic diseases, mankind would benefit greatly.

The discovery of recombinant DNA and an understanding of some of the processes involved arose from the efforts of many scientists. **Recombinant DNA** is a special form of DNA with the capacity to separate into fragments, which then reassemble in new arrangements. Consequently, researchers recognized its potential as a tool that could be used to transfer genes and their genetic functions from one organism to another. The human congenital disease, galactosemia, is the result of lack of an enzyme essential for digesting galactose, a sugar. *E. coli* bacteria possess a gene for the enzyme.

The Recombinant DNA Debate

NEW STRAINS OF LIFE—OR DEATH

RECOMBINANT DNA RESEARCH:

A FAUSTIAN BARGAIN?

SECRET GENETIC RESEARCH DANGEROUS

TINKERING WITH LIFE

These headlines, which appeared in *The New York Times, Science, The Denver Post,* and *Time,* reflect the controversy surrounding the relatively new recombinant DNA research.

Scientists literally *are* tinkering with genetic material to create new "hybrid" DNA molecules from the DNA of two different species. DNA from the bacterium *Escherichia coli,* or *E. coli,* has been most commonly used in this research. In addition to its single large circular strand of DNA, *E. coli* has several smaller circular DNA units called plasmids. Into the plasmids—we can think of them as vehicles—geneticists have inserted small DNA fragments—or passengers—from other species.

The technique is actually quite simple. *E. coli* cells are split open, and the cellular material is centrifuged to separate the lighter-weight plasmids from the chromosomes. The plasmid DNA is then opened at a very specific location by a specific enzyme, an endonuclease. Placed in this mixture are additional pieces of foreign (passenger) DNA, which were detached from the parent molecule by the endonuclease enzyme.

Both passenger and vehicle DNA have "sticky ends" and are attracted to each other. The enzyme DNA ligase chemically links the two fragments. The "hybrid" DNA plasmid that has been created probably never existed in nature. It represents in one unit a sequence of genes not previously found in living organisms in nature. If the hybrid DNA plasmids are added to a vessel containing *E. coli* cells, under the right conditions they will be taken into the *E. coli* cells. Thereafter, when *E. coli* doubles its DNA prior to cell division, the new plasmids are also reproduced. Whole populations of *E. coli* cells can be grown containing the foreign DNA, which may give *E. coli* new properties. If the foreign DNA provides for the synthesis of enzymes that in turn make a vitamin not normally synthesized by *E. coli,* a new type of *E. coli* with important new properties has been manufactured.

As petroleum supplies decrease, the cost of nitrogen-containing fertilizers made from them will continue to soar. In nature, the roots of legumes—plants such as peas and beans—support bacteria that are actually fertilizer factories. They take free nitrogen from the air and convert it into soluble compounds that nourish plants. What if other bacteria could be transformed through recombinant DNA research so that they acquired such nitrogen-fixing ability? Might they then be induced to live in the roots of corn, carrots, squash? The world would thereby have less need for artificially prepared nitrogen fertilizers.

The potential benefits of recombinant DNA research are limited only by the imagination of the scientist, but what are the potential risks involved in such research? Critics of recombinant DNA research are opposed to the use of *E. coli* as a vehicle because *E. coli* normally lives in the intestines of many animals, including humans. If a transformed *E. coli* escaped from the laboratory and entered human intestinal tracts, great harm could result. An *E. coli* strain with the capacity to produce insulin, for example, might manufacture the hormone in large quantities in the human gut. An infected individual, as a result, could go into insulin shock and die.

What if antibiotic resistance were incorporated into an infectious bacterium like *Streptococcus* or *Staphylococcus?* What might happen if a normally harmless bacterium were given the ability to produce deadly poisons? The potential for a new type of chemical or germ warfare is frightening.

In 1973, geneticists presented recent advances in recombinant DNA research at the Gordon Research Conference on Nucleic Acids in New Hampshire. Scientists at the conference, on hearing the details of some of the experiments, worried about the possible dangers. They wrote to the National Academy of Sciences and published that letter in *Science* and *Nature* to express "deep concern" and to ask that a study committee be established to evaluate the dangers inherent in the recombinant DNA research.

In 1974, the leading recombinant DNA researchers, members of the study committee that the National Academy of Sciences had set up, asked the world's scientists to join with them in ceasing to perform two types of experiments. They would no longer incorporate into bacterial plasmids antibiotic resistance or the ability to form toxins. Nor would they incorporate DNA from cancer-causing viruses into bacterial plasmids or other viruses.

Those reassurances did not quell the doubts of many critics of the research. The debate, which until then had been carried on mostly in scientific journals or at conferences, grew hotter and began to be covered by major newspapers. City governments in Ann Arbor, Michigan, Palo Alto, California, New Haven, Connecticut,

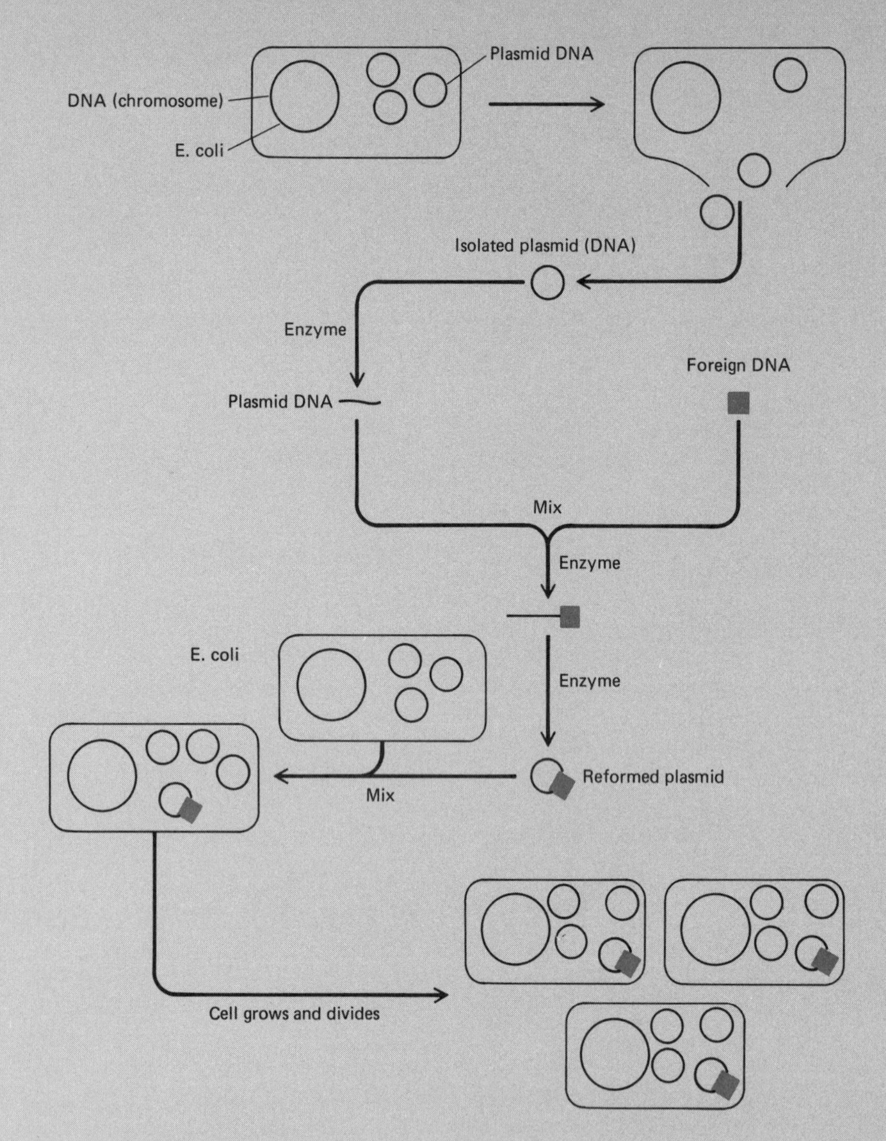

and Cambridge, Massachusetts, considered restricting or banning recombinant DNA research. Academic freedom and social responsibility seemed to be on a collision course.

Robert Sinsheimer of the California Institute of Technology expressed the philosophic concerns of many when he stated that scientists "must recognize that to reshape man is not a beguiling laboratory experiment but an enterprise that involves the ultimate in value judgment."

In 1976, the National Institutes of Health (NIH) issued guidelines spelling out precautions to be taken in recombinant DNA research. Those controls are voluntary, however, and they apply only to laboratories receiving federal funding. Carroll Williams of Harvard University has expressed his concern: "Scientists are racing for advantage and priority in a hotly competitive field and they are likely to do what they can to win the race. . . . One scientist told me, 'If we don't get the containment facility we want, we'll just reclassify our experiments from a higher to a lower security requirement.' " Sinsheimer further comments, "What we are doing is almost certainly irreversible. Knowing human frailty, these structures will escape, and there is no way of recapturing them."

Recombinant DNA research has continued, despite the controversy. Recently published results indicate that scientists are proceeding cautiously. Roy Curtiss at the University of Alabama has developed a weakened strain of *E. coli,* which is very sensitive to ultraviolet light and is incapable of producing thymine. It can be killed by sunlight and detergents, and it cannot grow unless thymine, one of the building blocks of DNA, is supplied in the culture medium. The probability of its survival outside the laboratory is extremely small. Therefore other experiments using this weakened bacterium can be performed with less danger to the researcher and the public.

In an article in *Science* in 1977, John Abelson reported that a "worst case" experiment is to be performed with utmost care by NIH. They will try to transform *E. coli* into a pathogen. If they are *unsuccessful,* other related experiments may be performed with less concern. Abelson also noted that researchers have been the guinea pigs in recombinant DNA research. (So far there have been no illnesses reported as a result of this work. The fear remains, however, that if cancer is developing in the researchers, it will not show up for years.)

The middle course—protecting the public while allowing research to continue—is apparently being followed. The World Health Organization has encouraged scientists to cooperate in designing experiments and in assessing risks, to perform meaningful experiments quickly and safely, and to share the results.

The fears of NIH biochemist Maxine Singer that DNA research might be banned altogether have not materialized. Hers has been a moderate and concerned voice throughout the controversy. Perhaps, then, her comments should be given special note. She has written that recombinant DNA research "must continue; nothing less than science is at stake."

Investigators succeeded in growing a colony of bacterial viruses which had acquired and integrated the bacterial gene for the enzyme into their own DNA. When enzyme-deficient human cells then were treated with the bacterial viruses, a few of the cells regained the ability to produce the enzyme, apparently as the result of incorporating the bacterial gene into their own DNA.

Thus, there is some hope that a sickle-cell human could be infected with a virus that might carry genetic information to his blood-forming cells which would enable those cells to make proper hemoglobin, thereby "curing" sickle-cell anemia. Similar techniques might be equally suitable for infecting albinos with genes for the production of tyrosinase, thus making possible normal pigmentation. Many other heritable defects, the result of point mutations, might be similarly treatable.

In an important breakthrough, Nobel prizewinner H. Gobind Khorana announced in 1976 that he and his team, working over a number of years at both the University of Wisconsin and the Massachusetts Institute of Technology, had succeeded in implanting an entirely synthetic gene in a virus, a bacteriophage, where it performed its intended function of synthesizing RNA polymerase. This brillant work brings us closer to making genetic engineering a reality.

Genetic engineering may have many beneficial uses. In some laboratories scientists are now studying the molecular basis of thought and memory. They are working with a relatively simple animal, the flatworm, which has certain definite behavioral characteristics. Behavioral mutants are being isolated and genetically mapped to determine what the molecular basis for behavioral changes may be. In other words, a start is being made in trying to identify molecules that may account for thinking and remembering, but it is, of course, a long way from flatworms to man. Furthermore, some of the work on the chemical basis for memory is highly suspect and requires further experimentation, so we are far from the stage at which we might be optimistic about implanting genes for genius.

Genetic manipulation of crop plants and stock animals may replace the tedious process of crossbreeding. Perhaps it will someday become possible for technicians to collect or to construct the best genes possible and to infect plant seeds or the gametes of stock animals with them to make more and superior vegetables and meat. A very hungry world may then be more effectively fed while the problem of curbing the human birthrate is being resolved.

However, there is another side to this story, one that warns of danger in the future of molecular genetics. Responsible voices are increasingly being raised against such molecular experimentation. There are fears that the experimenters may not be able to control the agents of gene transfer, which of necessity are organisms with the ability to reproduce themselves. Some leading biologists, at a conference in 1975, even took the unprecedented step of proposing rules to limit genetic research. One outcome of that move was the issuance by the Director of the National Institutes of Health of a set of safety guidelines.

Beyond safety alone, furthermore, there are social, ethical, and philosophical ramifications to molecular genetics. As one nonscientist participant in the 1975 conference warned, "As crucial as your research seems to you to the achievement of progress, you should be prepared for the eventuality that the public may not agree."

Summary

1. After World War II the field of genetics turned from the study of phenotype characteristics and frequencies in macroscopic organisms to the chemical and physical analysis of microscopic organisms and submicroscopic molecules. New techniques, such as electron microscopy and X-ray crystallography, greatly increased the molecular geneticist's ability to probe the mysterious world of genes and heredity.

2. In 1953 James Watson and Francis Crick deciphered the structure of DNA. They postulated that the DNA molecule consists of two strands of polymerized nucleotides. The strands are held together by hydrogen bonding between paired purine and pyrimidine bases and are wound around each other in a double helix, with one complete turn for every 10 base pairs. Watson and Crick further proposed the now universally accepted semiconservative DNA duplication mechanism, in which the DNA molecule "unzips" to serve as a template. The result is two daughter molecules exactly like the original and each containing one strand from the parent molecule and one newly assembled strand.

3. The genetic code contained in the sequence of nucleotides in DNA transmits blueprints and control mechanisms for protein synthesis. Each word or codon consists of three bases and represents a specific amino acid. As George Beadle and Edward Tatum demonstrated, each gene directs the formation of one specific protein, an enzyme. It is now understood that all proteins—not just enzymes—in a cell are manufactured according to instructions in genes.

4. Protein synthesis begins in the nucleus when one strand of a DNA molecule serves as a template for the synthesis of a complementary strand of mRNA. Messenger RNA then migrates to the cytoplasm, where it becomes associated with several ribosomes, forming polyribosomes.

Also in the cytoplasm are molecules of tRNA. Each has its own anticodon, which is complementary to an mRNA codon, and each becomes bonded to only one specific kind of amino acid. As a ribosome moves down an mRNA strand, molecules of tRNA whose anticodons complement mRNA codons momentarily line up opposite the mRNA strand. Their amino acids are transferred—in the presence of enzymes—to a growing polypeptide chain attached to the ribosome. Finally, the completed protein, its component amino acids in the sequence stipulated by the genetic code, detaches from the mRNA and the ribosome.

5. François Jacob and Jacques Monod have provided a model for gene regulation in procaryotes: the operon control system. The enzyme-producing activities of a structural gene can be switched on and off by the action of an operator gene. Structural genes and their operator gene, located near each other on a chromosome, are called an operon. Regulator genes control the activity of operons by producing substances which can inhibit the activity of operator genes. This model can account for the observed phenomena of substrate induction and end-product repression of gene activity.

6. In eucaryotic cells, no similar regulatory mechanism has been conclusively found, but current evidence suggests that the nonhistone proteins found in the chromosomes of eucaryotes play a key role in gene regulation.

7. Occasionally mutations occur, causing heritable alterations of genetic material. There is some evidence that some kinds of mutations are repaired by repair enzymes.

8. The field of molecular genetics has opened the door to an understanding of the mechanisms of formation of cellular structures and characteristics. Such an understanding may permit genetic engineering that may result in cures for certain kinds of inherited diseases. However, there are risks involved, even when proper precautions are taken, and responsible voices are being raised against some kinds of molecular experimentation.

REVIEW AND STUDY QUESTIONS

1. What is the meaning of the term "molecular biology"?

2. Explain briefly the particular breakthrough achieved by the following scientists: Beadle and Tatum; Chargaff; Kornberg; Meselson and Stahl; Khorana.

3. In a frequently related anecdote (included in this chapter) Crick said of the work he and Watson had just done, "We have discovered the secret of life!" What did Crick mean?

4. Describe the double helix structure of DNA.

5. How are incomplete dominance and codominance explained in terms of the one-gene-one-enzyme theory?

6. What particular fact pointed toward a genetic code made up of three-part units?

7. Distinguish between messenger RNA and transfer RNA.

8. What is recombinant DNA?

REFERENCES

Barry, J. M. 1964. *Molecular Biology: Genes and the Chemical Control of Living Cells.* Prentice-Hall, Englewood Cliffs, N.J.

Benzer, S. 1962 (January). "The Fine Structure of the Gene." *Scientific American* Offprint no. 120. Freeman, San Francisco.

Brown, D. D. 1973 (August). "The Isolation of Genes." *Scientific American* Offprint no. 1278. Freeman, San Francisco.

Crick, F. H. C. 1966 (October). "The Genetic Code: III." *Scientific American* Offprint no. 1052. Freeman, San Francisco.

———. 1962 (October). "The Genetic Code." *Scientific American* Offprint no. 123. Freeman, San Francisco.

———. 1954 (October). "The Structure of the Hereditary Material." *Scientific American* Offprint no. 5. Freeman, San Francisco.

Hartman, P. E., and S. R. Suskind. 1969. *Gene Action,* 2nd edition. Prentice-Hall, Englewood Cliffs, N.J.

Jacob, F., and J. Monod. 1961. "Genetics' Regulatory Mechanisms in the Synthesis of Proteins." *Journal of Molecular Biology* 3:318–356.

Levine, L. 1973. *Biology of the Gene,* 2nd edition. Mosby, St. Louis.

Nirenberg, M. W. 1963 (March). "The Genetic Code: II." *Scientific American* Offprint no. 153. Freeman, San Francisco.

Sobell, H. M. 1974 (August). "How Actinomycin Binds to DNA." *Scientific American* Offprint no. 1303. Freeman, San Francisco.

Stein, G. S., J. S. Stein, and L. J. Kleinsmith. 1975 (February). "Chromosomal Proteins and Gene Regulation" *Scientific American* Offprint no. 1315. Freeman, San Francisco.

Stent, G. S. 1971. *Molecular Genetics.* Freeman, San Francisco.

Watson, J. D. 1976. *Molecular Biology of the Gene,* 3rd edition. Benjamin, New York.

SUGGESTIONS FOR FURTHER READING

Cohen, S. N. 1975 (July). "The Manipulation of Genes." *Scientific American* Offprint no. 1324. Freeman, San Francisco.

Report on methods deevloped by which genetic information is transferred from one organism to another

Friedman, T., and R. Roblin. 1972. "Gene Therapy for Human Genetic Disease." *Science* 175(4025): 949–955.

Analysis of the prospects for using isolated DNA segments or mammalian viruses as vectors in gene therapy. The authors also cover the techniques and ethico-scientific criteria the therapy should satisfy.

Ingram, V. M. 1972. *The Biosynthesis of Macromolecules,* 2nd edition. Benjamin, New York.

A review of the biosynthesis of RNA and proteins, including the model of genetic control of protein structure.

Miller, O. L., Jr. 1973 (March). "The Visualization of Genes in Action." *Scientific American* Offprint no. 1267. Freeman, San Francisco.

Excellent examples of electron microscopy on transcription of genes. The pictures resemble diagrams of genetic and biochemical data.

Rosenberg, E. 1971. *Cell and Molecular Biology: An Appreciation.* Holt, Rinehart and Winston, New York.

An excellent general introduction to the subject, lucidly written, with elegant illustrations to accompany the text. Good for the beginning student.

Watson, J. D. 1968. *The Double Helix.* Atheneum, New York (paperback, Signet, 1969).

A narrative of the discovery of the structure of DNA, with great insight into the ways scientists work.

Wolfe, S. 1972. *Biology of the Cell.* Wadsworth, Belmont, Calif.

A solid presentation with good illustrations. This text is excellent for students who plan to do advanced work in cell biology.

DEVELOPMENT
chapter sixteen

Development consists of the orderly sequential changes that an organism undergoes during its life cycle. This turtle emerging from the egg has gone through its most dramatic transformations during the embryonic stage, but development will continue. The turtle will grow and mature according to the interaction between its unique genetic inheritance and its external environment.

All living things are capable of reproduction—that is, they can make more individuals like themselves. Except in the more primitive forms of life, the process consists of more than merely growing larger and splitting into two new individuals, each like the parent. Indeed, in multicellular plants and animals, there evolved two types of special reproductive cells, the sperm and the egg, which unite to create the start of a new individual. The characteristic form that emerges is genetically programmed through controlled cellular division and differentiation of cells, which form aggregates of distinct shapes. The processes of cellular division, growth, and the structural and functional shaping of new individuals of the species—which may occur simultaneously or separated in time—constitute development. Development is an orderly progression of changes that occur in the life cycle of an organism. It includes growth, the many aspects of maturation, the regeneration or repair of tissues, and the creation of a new organism from part of the parent.

Development proceeds only after a reproductive unit has been formed by some kind of cellular division in the parent. We have seen that the type of reproductive unit formed depends on the organisms. Most plants produce both gametes and spores. The latter are haploid reproductive units or cells that need not unite sexually with other cells in order to begin developing into new individuals. Other organisms, those that reproduce by sexual means, produce only gametes, either egg or sperm cells (in some cases both).

Sexual reproduction, which allows for the recombination of genes from two parents, produces organisms that differ slightly from one generation to the next, thereby accelerating the course of evolution. Because of their importance in producing most of the plants and animals with which we are familiar and because asexual development has fewer complexities, we shall restrict ourselves to a study of the processes of development which follow the union of the sperm and egg in plants and animals. Most of the dramatic changes occur very soon after fertilization takes place, during embryonic development, but it should be remembered that embryology is only one part of the study of development. The schematic diagram in Fig. 16.1 indicates the cycle of development in higher animals.

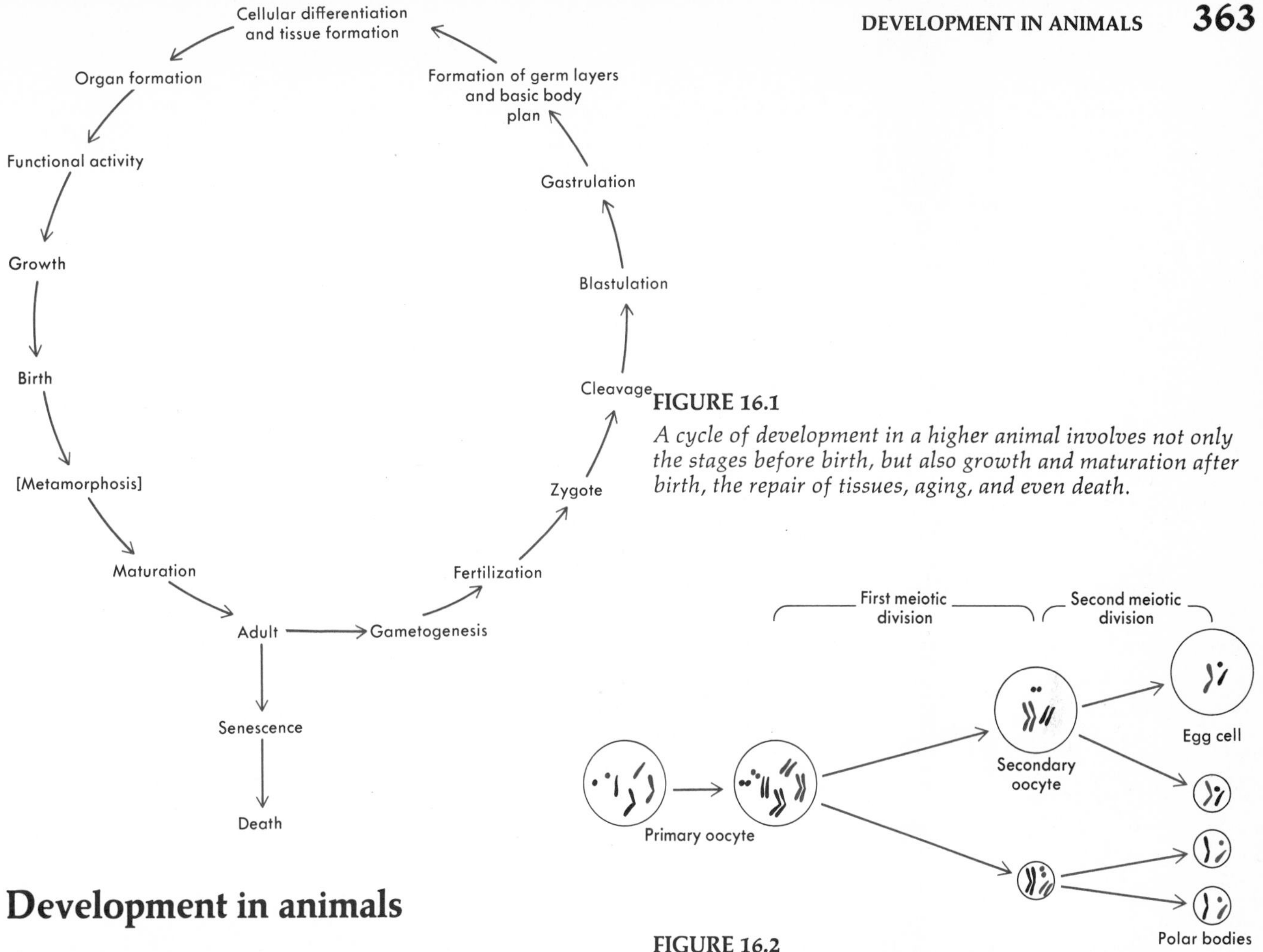

FIGURE 16.1

A cycle of development in a higher animal involves not only the stages before birth, but also growth and maturation after birth, the repair of tissues, aging, and even death.

FIGURE 16.2

Oogenesis. The primary oocyte undergoes two meiotic divisions. The process results in the formation of only one functional gamete and three associated polar bodies, thus conserving the cytoplasmic material to nourish the developing zygote.

Development in animals

In animals, two types of gametes form. Eggs, haploid cells formed in the ovary of the female, are generally large and incapable of self-movement; the sperm, haploid cells produced by the male gonad, are small and highly motile. Although the sperm brings to the new individual a complement of genetic information necessary to trigger the precisely programmed events of development that follow, many of the early events are dependent on the structure of the egg alone. The structure of the egg, in turn, is dependent on a sequence of events that occur during its formation.

EGG AND SPERM FORMATION

Oogenesis, the formation of the egg, begins in the ovary. During the development of females, a few diploid primordial germ cells undergo mitotic divisions, increasing manyfold the number of primary oocytes, cells destined one day to become mature eggs. These diploid egg cells grow and accumulate nutrients and materials such as RNA and enzymes prior to undergoing the final steps of maturation into haploid egg cells. At last, in sexually mature individuals, some primary oocytes undergo meiotic divisions. During meiosis, as we have seen in Chapter 13, the two sets of homologous chromosomes are separated from each other, resulting in four cells, each haploid, or with half the number of chromosomes contained in a normal body cell. However, during oogenesis, the cytoplasmic divisions that accompany the chromosome movements are unequal. As a result, there are three tiny nucleated cells, or polar bodies, containing very little cytoplasm, and one very large haploid egg, containing large amounts of cytoplasm and nutritive yolk, which will later be fashioned into an embryo. The polar bodies are then cast off, and they subsequently disintegrate (Fig. 16.2).

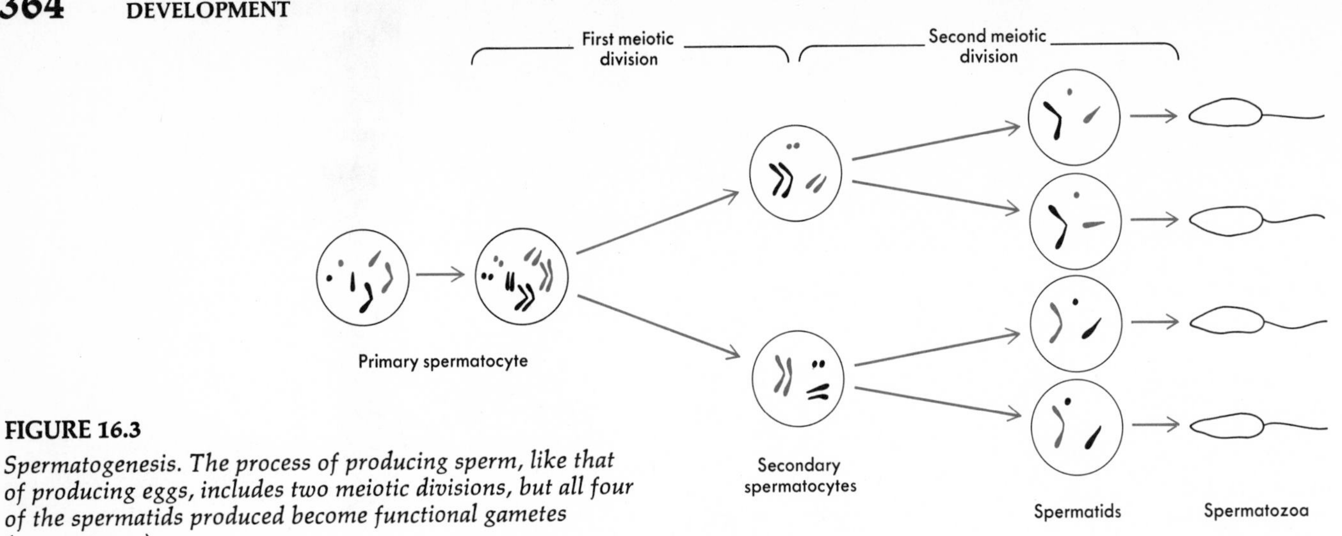

FIGURE 16.3

Spermatogenesis. The process of producing sperm, like that of producing eggs, includes two meiotic divisions, but all four of the spermatids produced become functional gametes (spermatozoa).

Spermatogenesis, which occurs in the testes, is a process similar to oogenesis. As in oogenesis, meiosis occurs, resulting in four haploid cells called spermatids. The spermatids, of equal size, are eventually transformed into fully mature sperm. Each sperm, when mature, appears to be virtually a mobile nucleus, which consists of a haploid set of chromosomes bearing the father's genetic contribution to the new organism, as well as machinery for delivering the nucleus to the egg, penetrating it, and activating the egg to begin development (Fig. 16.3). Very little cytoplasm is present in the sperm, and as a result, the period of motility may be brief, because nutrients are quickly exhausted. In mammals, for example, nutrients to prolong motility must be provided in the seminal fluid.

FERTILIZATION

Following the release of the sperm cells in the vicinity of a mature egg, the sperm move toward the egg in the female reproductive tract by means of their own flagella.

As a sperm reaches an unfertilized egg, a very rapid series of events occurs. The acrosomal filament, contained in the acrosome body at the tip of the sperm, establishes contact with the membrane of the egg. At first there is no visible reaction, but suddenly the egg responds, often producing an elevated cone of cytoplasm that seems to envelop the sperm and enfold it within the cytoplasm of the egg. The acrosomal membrane makes contact with the broken egg membrane and fuses with it, becoming a part of the membranous structure of the resulting fertilized egg. Also the egg may quickly undergo changes at its surface, producing from its cortex a fertilization membrane that separates from the egg surface, leaving the egg surrounded by a fluid, in which it floats, and a tough outer membrane that prevents the entry of other sperm. A similar series of events may be caused artificially by pricking the egg with a needle or irritating it with certain chemicals. In some organisms, such as bees or aphids, activation of the egg normally begins without the presence of sperm. The activation, called parthenogenesis, results in a female organism.

Following this entry of the sperm, or **syngamy,** the sperm and egg nuclei move toward each other, and their chromosomes become attached to the fibers of a spindle apparatus formed from part of the sperm. As the chromosomes come together on a common spindle, we may say that **fertilization** has been completed when two haploid sets of chromosomes mix. The completed fusion of the two cells forms the **zygote,** or first cell of the new individual. In some developmental patterns, such as that of a frog, the zygote rotates under the influence of gravity so that its densest part (the yolk-filled vegetal pole) lies beneath its lighter part (the cytoplasmic animal pole). In any other orientation, normal development is blocked. In all organisms, the zygote now begins to develop into an adult by sequential gene action.

PROCESSES OF ANIMAL DEVELOPMENT

The subsequent shaping of the organism from the plastic zygote entails three types of processes, which proceed through the interaction of the genetic program, contained within the zygote nucleus, with its surroundings. These processes may occur separately in time or together in any combination. Indeed, it is in large part sequential responses of the cell and its organelles to the intracellular and extracellular events following fertilization that determine the particular shape of the organism that develops. These events may be summarized as follows:

1. **Differentiation** is the production of new characteristics in the cell, or new substances that determine the functional fate of the cell. For example, separate parts of the embryo will differentiate to form muscle, bone, and blood cells. Some changes are irreversible, restricting the cell or part of the developing organism to a particular developmental pathway. This is known as **determination.**

2. **Morphogenesis,** or literally the generation of shape, is caused by growth of some cells and not of others, by cellular movement, and in some cases by cellular death. It is morphogenesis that, for example, transforms the spherical zygote into a fully formed organism.

3. **Growth,** or the increase in mass of the developing organism, is the result of cellular division by mitosis, followed by synthesis of additional materials in the cells.

CLEAVAGE STAGES AND LATER CELL DIVISION

Blastula The fertilized zygote, a very small sphere, seems to rest for a moment. In some organisms, such as the frog, the zygote, now floating within the fertilization membrane, contains enough stored nutrients to carry it through all the steps of development. In humans and

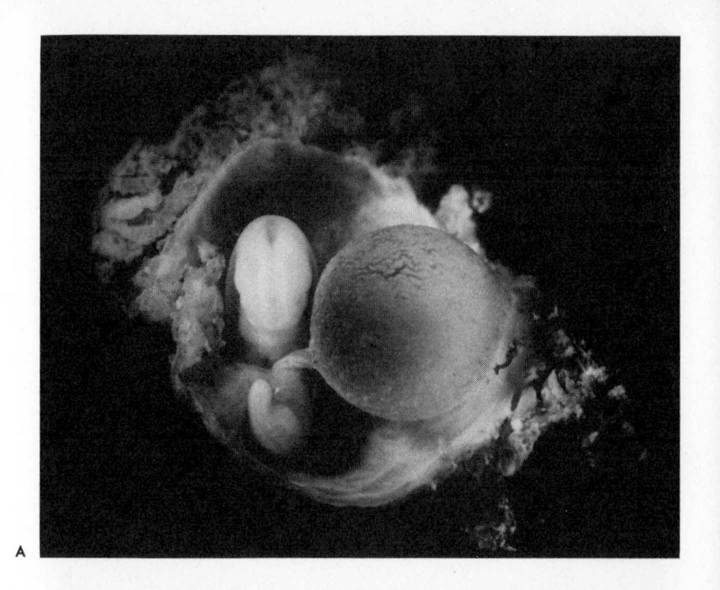

A

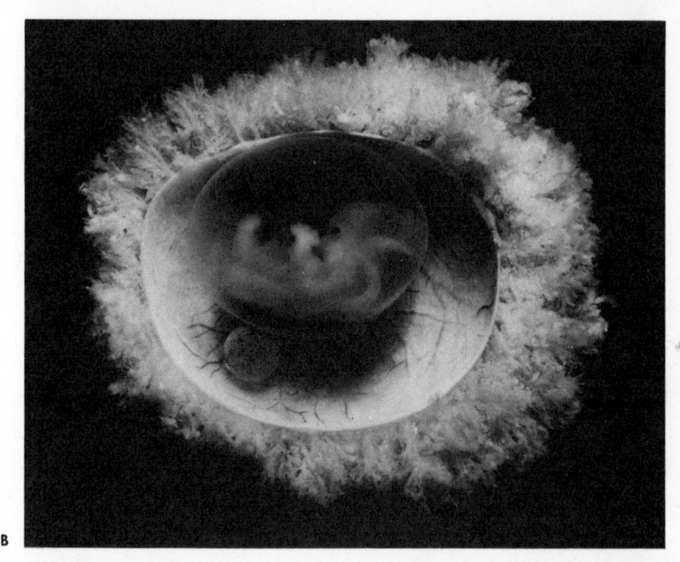

B

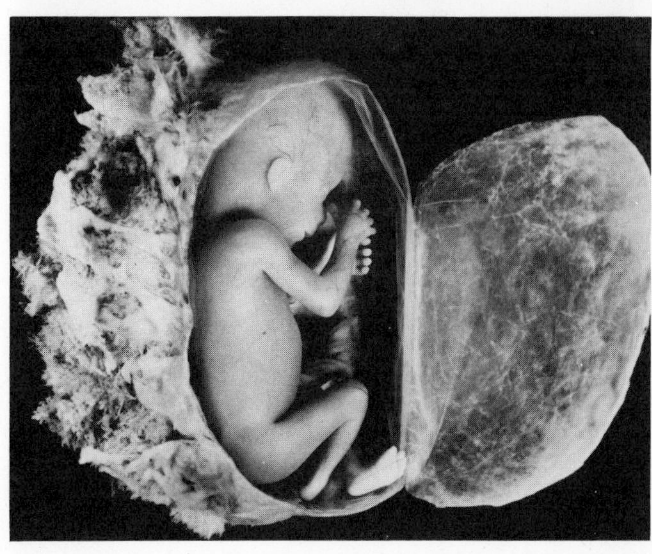

C

FIGURE 16.4

Approximately seven days after fertilization, the hollow ball of cells containing the human embryo implants itself in the uterine lining. As the inner cell mass develops into the new human being, other cells form the amniotic membrane and the placenta, which will nurture the growing organism. The placenta acts as an intermediary between the bloodstreams of mother and child, allowing nutrients to diffuse in and excreted wastes to be carried out. (A) At 28 days, the embryo is approximately 5 mm long and is still attached to a large floating yolk sac. (B) At seven weeks, the embryo measures 18 mm; during this period facial features, fingers, and toes become clearly recognizable. (C) By twelve and a half weeks the fetus, now approximately 93 mm long, can coordinate nerves and muscles in response to stimuli.

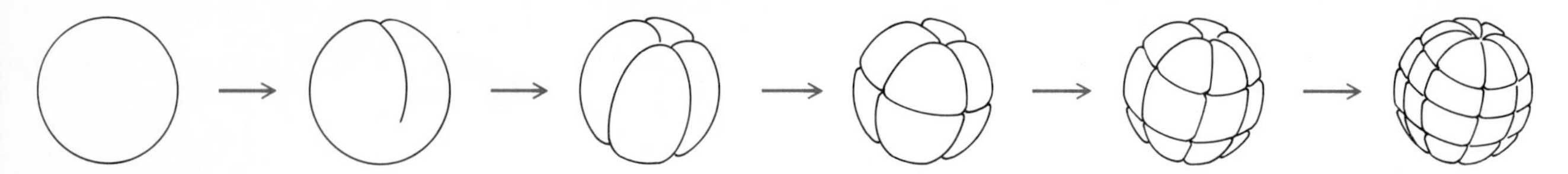

FIGURE 16.5

In the early cleavage stages the overall size of the embryo changes very little. Instead, a single large cell develops into a large number of smaller ones. In this way the ratio of surface to overall volume is increased, permitting the increased respiration necessary to the actively growing embryonic cells.

most other mammals, the zygote is wholly dependent on its mother for nurture and support. Still within the fallopian tube, the zygote rolls or is swept downward into the uterus, where it implants or attaches itself to the uterine tissues of the mother. There it will remain until parturition (birth), deriving from its mother its nutrient and oxygen supply.

Once the zygote is in position, cleavage begins. Cleavage is a specialized type of mitotic division in which the daughter cells do not grow between successive divisions; the result of cleavage is at first a hollow ball, or **blastula.** With each successive cleavage these cells get smaller (Fig. 16.5). The initial cleavages seem to be highly organized. In a frog's egg the first cleavage, which is always vertically oriented, cuts the zygote into two smaller cells. The next is also vertically oriented and at right angles to the first, producing a four-cell stage. With the third division the visible beginnings of differentiation occur. This third division is always horizontal and at right angles to the first two. However, since the cleavage does not occur at the equator of the embryo, the result is eight cells of unequal size. This distinction in size is a forerunner of the myriad distinctions that arise as the cells of the embryo become more and more specialized and their fates are determined. It is in large part the distribution of yolk materials within the egg that, after the second cleavage, makes symmetrical cleavages impossible.

Although at first the cleavage mitoses occur regularly, with each cleavage involving the entire embryo, soon the cells of the blastula are quite small, and cleavages become asynchronous. That is, the cells at the animal, or cytoplasmic, pole begin to cleave more often than those at the yolk-filled vegetal pole, and a skin of very small cells covers the top of the embryo. Then another remarkable change begins. Some of the cells on the upper surface begin to move.

These movements, in part, signal the beginning of the next developmental changes in the embryo, which constitute **gastrulation,** or the formation of **gastrula** with a new internal cavity, or **archenteron,** which is also called the primitive gut or alimentary tract.

Gastrula As the turmoil of cellular movement begins at the surface, a pore or opening, the blastopore, forms on the part of the embryo destined to become the tail. The moving cells of the upper surface of the embryo turn under at the blastopore, and as they move into the hollow of the sphere, they form a tubelike second layer, like a long stocking. As this tube elongates and fills the interior of the embryo, the outer portion elongates somewhat, too. Thus is laid down the basic tube-within-a-tube body plan that is characteristic of most animals. In invertebrates this morphogenetic event, the sculpting of a tube-within-a-tube from a hollow sphere, is the result of slightly different events. In the starfish, for example, a small dent, or invagination, appears on one side of the hollow blastula, instead of a pore. The dent becomes deeper and deeper, and the gastrula takes on the appearance of a soft, hollow rubber ball into which someone has pushed a finger. But despite the fact that the developmental events in invertebrates are different from those in vertebrates, the basic result is the same—a tube-within-a-tube body plan.

As a result of gastrulation the embryo has two distinct layers of cells. The inner layer, which constitutes the wall of the archenteron tube, is called the endoderm (inner skin); the cells which remain on the outer surface constitute the ectoderm, or outer skin. At this point the embryo may be called diploblastic, or two-layered. In higher organisms the diploblastic stage is transient. In many lower organisms, however, development seems to stop here, although differentiation of cells in the blastula-like or gastrula-like shape will occur.

At some time during gastrulation in higher animals, a third layer of cells—the mesoderm, or middle skin—begins to form, and the embryo may now be called triploblastic. Each of these layers—the ectoderm, mesoderm, and endoderm—is called a germ layer, for each becomes partitioned and carved in various ways to form all the body structures—ears, eyes, muscles, glands—of the adult.

The mesoderm arises differently in various kinds of animals. In invertebrates, such as the earthworm, a pair of cells, the pole cells, are pushed into the blastocoel (the cavity within the blastula) during gastrulation, and they divide many times to form the many cells of the mesoderm. In vertebrates, such as the frog, cells on the inner surface of the rim of the blastopore begin to divide and proliferate to form part of the mesoderm. They are joined by other cells that arise by cell division at the sides of the archenteron.

Cellular divisions within the new mesoderm produce a mass of tissue, which fills the space between the ectoderm and the endodermal archenteron tube. As it grows, the mesoderm forms a long hollow space within the mass of cells. This **coelom,** or body cavity, separates the mesoderm into two layers. The outer layer, which becomes attached to the inner surface of the ectoderm, is called the **somatic mesoderm.** The ectoderm and somatic mesoderm together will later become the body wall of the adult, consisting of skin and muscles. The inner layer of mesoderm, called the **visceral mesoderm,** attaches itself to the endoderm (that is, to the archenteron tube) and, will among other things, become the musculature of the intestines. Thus the tube-within-a-tube structure is still maintained, permitting the intestines in the adult the freedom of movement necessary for their functioning. (For some variations in coelomic structure, see Fig. 16.6.)

Bridging the coelomic cavity there remains a thin double sheet of mesoderm called the **mesentery,** which holds the digestive structures freely suspended within the adult body cavity. Blood vessels and nerves also form within the tissue of the mesentery to connect the gut with the circulatory and nervous system.

Increasingly, the organism becomes more complex, and the organs are formed. A vertebrate organism becomes elongated and fishlike in appearance as its nervous, muscular, and skeletal systems develop.

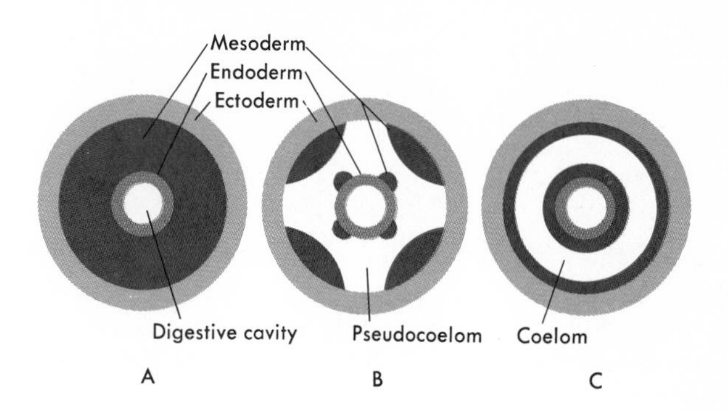

FIGURE 16.6

Differences between acoelomate (A), pseudocoelomate (B), and coelomate (C) body types depend on whether there is free space between the ectoderm and endoderm. In (A) the space is filled with mesoderm, so there is no coelom, or body cavity. In (B) there is a functional body cavity, but it is called a pseudocoelom because it is bounded partly by ectoderm and partly by endoderm. A true coelom (C) is entirely bounded by mesoderm. These body types constitute one basis for taxonomic classification.

Neurula The beginning of **neurulation,** or the formation of the nervous system in vertebrates, is marked by the appearance of a flattened plate of ectoderm that runs along the back of the embryo. Soon it reaches from the head to the tail. The outer surface of the cells contracts and the sides of the plate rise up, leaving a groove between the newly risen skin. Eventually the skin that forms the neural ridges grows together, leaving a hollow tube, the neural tube, beneath the fused ectoderm. Within the tube, the remaining fluid (cerebrospinal fluid) circulates between the brain and spinal cord. As the organism becomes more and more mature, the front end of the tube swells and changes shape to form a brain, and the rest of the tube becomes the spinal nerve cord. Other parts of the nervous system form from other ectodermal tissues.

In the frog, the embryo becomes a self-supporting, free-swimming larva that initially is more fishlike than froglike. Later this larva, the tadpole, undergoes a metamorphosis in which its body is reshaped into a form that is somewhat more suitable for terrestrial life. The fishlike tail is reabsorbed, small fore- and hindlimbs form, and the organism begins to look more like an adult frog.

These events, which we have but briefly described here, have been studied by generations of biologists. But it is only within the past few decades that the underlying causes of these events of development have been discovered. More complete understanding awaited the additional knowledge of the gene and of DNA's role in protein synthesis, as well as the powerful new technology that has grown up around molecular biology. Many things about development are not fully understood yet, but we can discern the mechanisms that control the many facets of development.

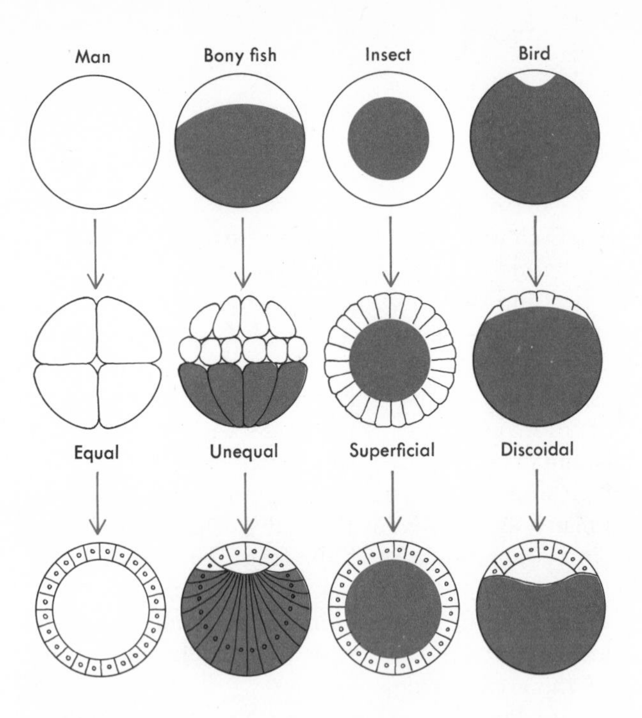

FIGURE 16.7
Influence of yolk content and distribution on cleavage and blastulation in four egg types. In the human egg, which contains no yolk, cleavage is equal and produces a coeloblastula. The yolk (color) in the other three eggs affects the cleavage patterns and thus modifies the structure of the blastula and the later embryology of the species. For example, yolk interferes with the rate of cleavage, so yolky eggs divide more slowly.

Interacting systems in animal development

THE QUALITIES OF THE EGG

Much of the developmental pattern followed by an organism is the result of the properties of the egg, which are in large measure the result of the synthesis of components in the mother's body and their fabrication into the structure of the egg.

Yolk, which serves as a nutrient source to the embryo, varies in amount. Almost no yolk is found in the human egg (since the embryo derives nutrients directly from the mother rather than from the egg). In birds' eggs, on the other hand, the egg consists mostly of the yellow yolk, with a tiny disc of cytoplasm perched on one side of the yolk. Between these extremes is the frog's egg, which consists of a yolky vegetal pole, approximately the lower half of the egg, and a cytoplasmic upper animal portion.

The nutrient substances of the yolk are evidently supplied by the mother. The yolk does more than just furnish energy and substance to the embryo, however. The pattern of yolk deposition also influences the cleavage of the embryo (see Fig. 16.7). Yolk is an impediment to cleavage. In a human egg, for example, the cleavages pass completely through the egg, producing equal-sized cells, since there is little yolk to impede cleavage. In a frog's egg, in which about half the zygote is filled with yolk, only the first few cleavages pass entirely through the cell. As the divisions continue, the interference of the yolk slows cellular divisions on the lower half of the egg while divisions continue more rapidly at the top. This differential rate of division leads to differential growth and is an important factor in the movements of cells during gastrulation. The large number of cells overgrow the available surface at the animal pole, and having only one possible space to fill, the blastocoel, they enter the blastopore and move into the center of the embryo.

Egg cortex The amount of yolk present is a visible physical property of the egg, which markedly influences the course of development. But there are other, more subtle, properties of the egg that can have great effects. The egg cortex, a gelatinous colloidal layer about 3μ thick, sometimes contains concentrations of pigment, as in the frog's egg. Upon fertilization, part of the cortex moves, carrying with it some of the pigment granules.

Portfolio 3: Reproduction and Development

3.1
This scanning electron micrograph shows sperm and eggs of the sea urchin. From the large number of sperm swarming around the egg, only one will successfully penetrate the membrane and fertilize the egg.

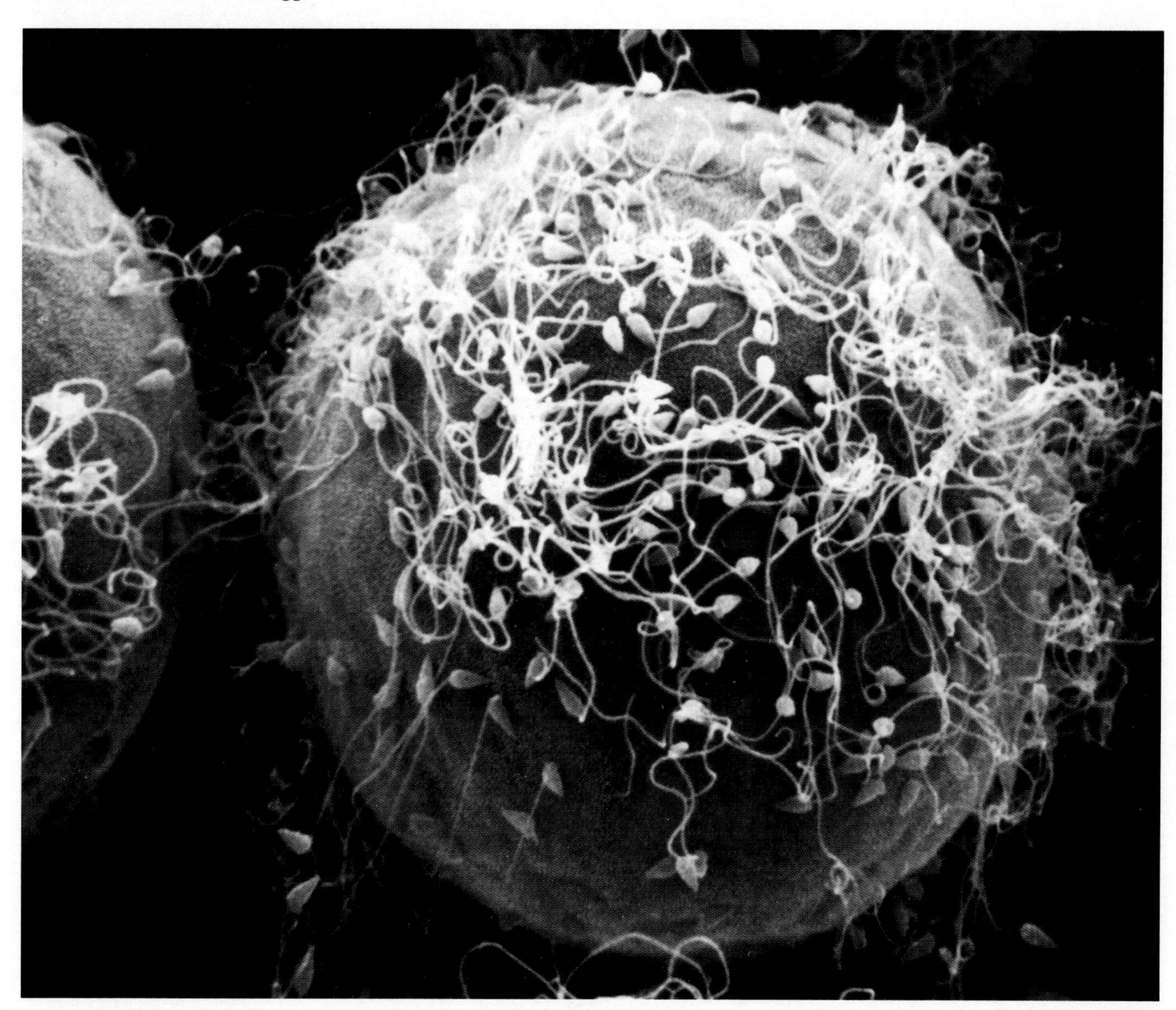

3.1 **2,415X**

3.2

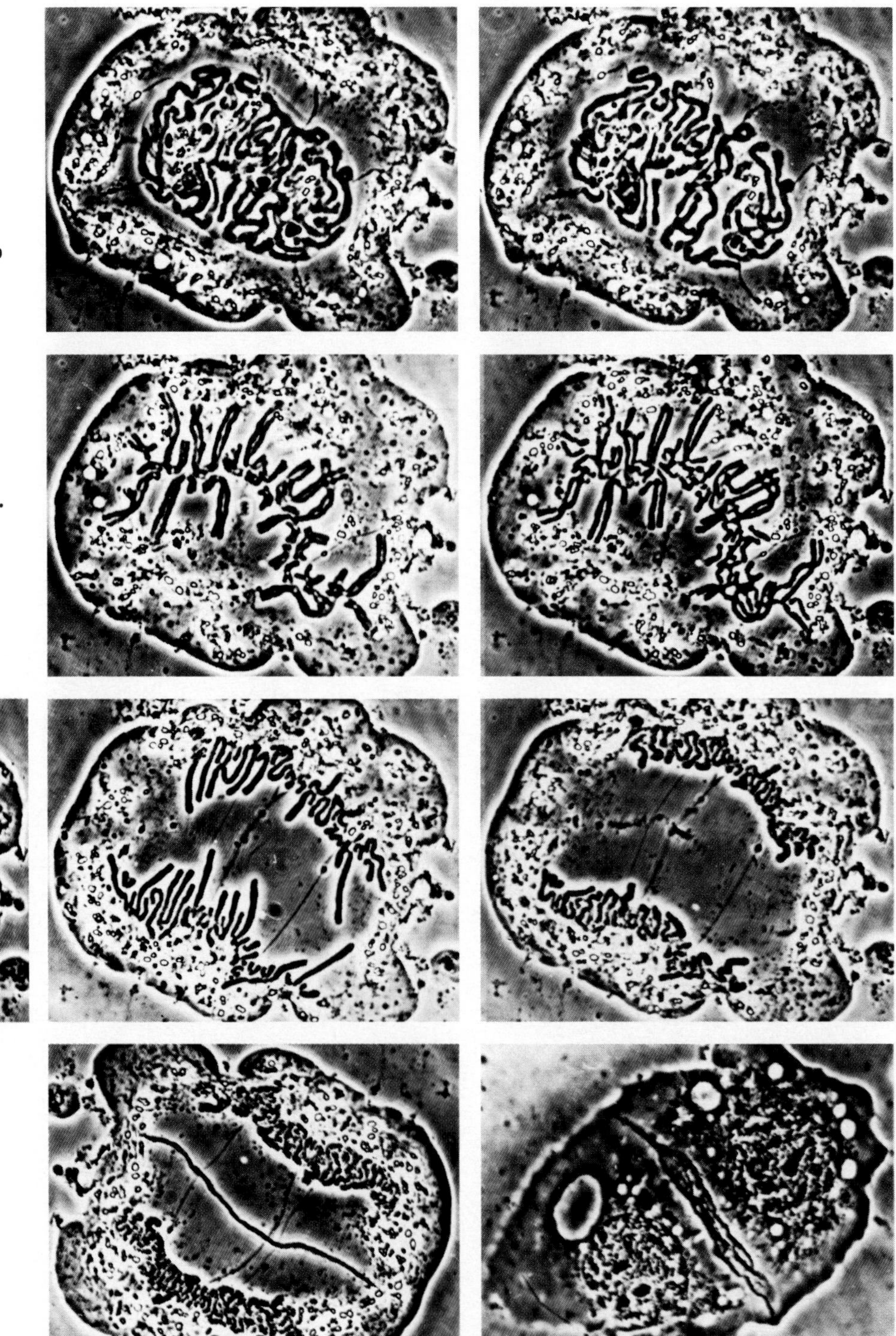

3.2
Light microscopy and time-lapse photography reveal the stages of mitosis in a living cell of the African blood lily. The photographs represent frozen moments of a continuous, dynamic process that may require up to an hour to complete.

3.3
Reproduction by binary fission of a ciliate, *Didinium nasutum,* **is shown in this sequence of scanning electron micrographs. In B and C, additional girdles of cilia are developing. In D the daughter cells are about to separate.**

3.3

A 925X

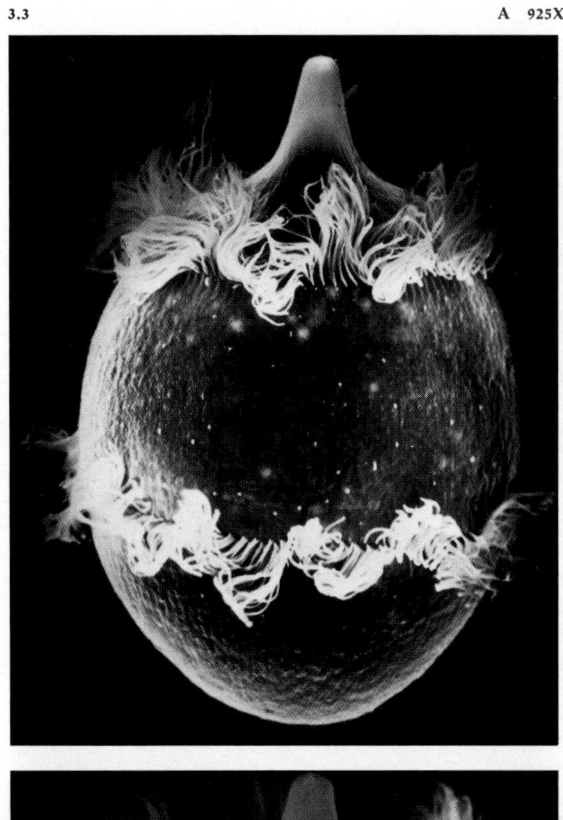

B 935X

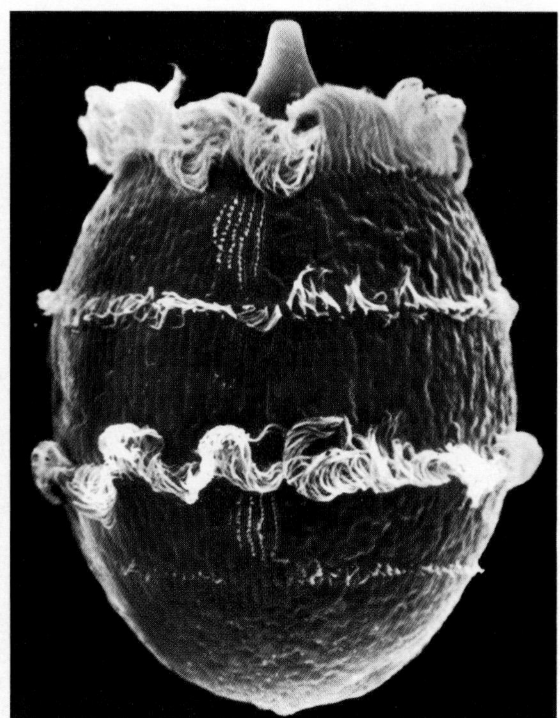

C 830X

D 700X

3.4
These micrographs show longitudinal (A) and transverse (B) sections through the tip of a growing shoot. Arrows point to dark areas of cell division that produce small, densely packed cells. Growth is accomplished as the cells increase in number and enlarge. Specialization of cell function (differentiation) gives rise to leaves and vascular tissue.

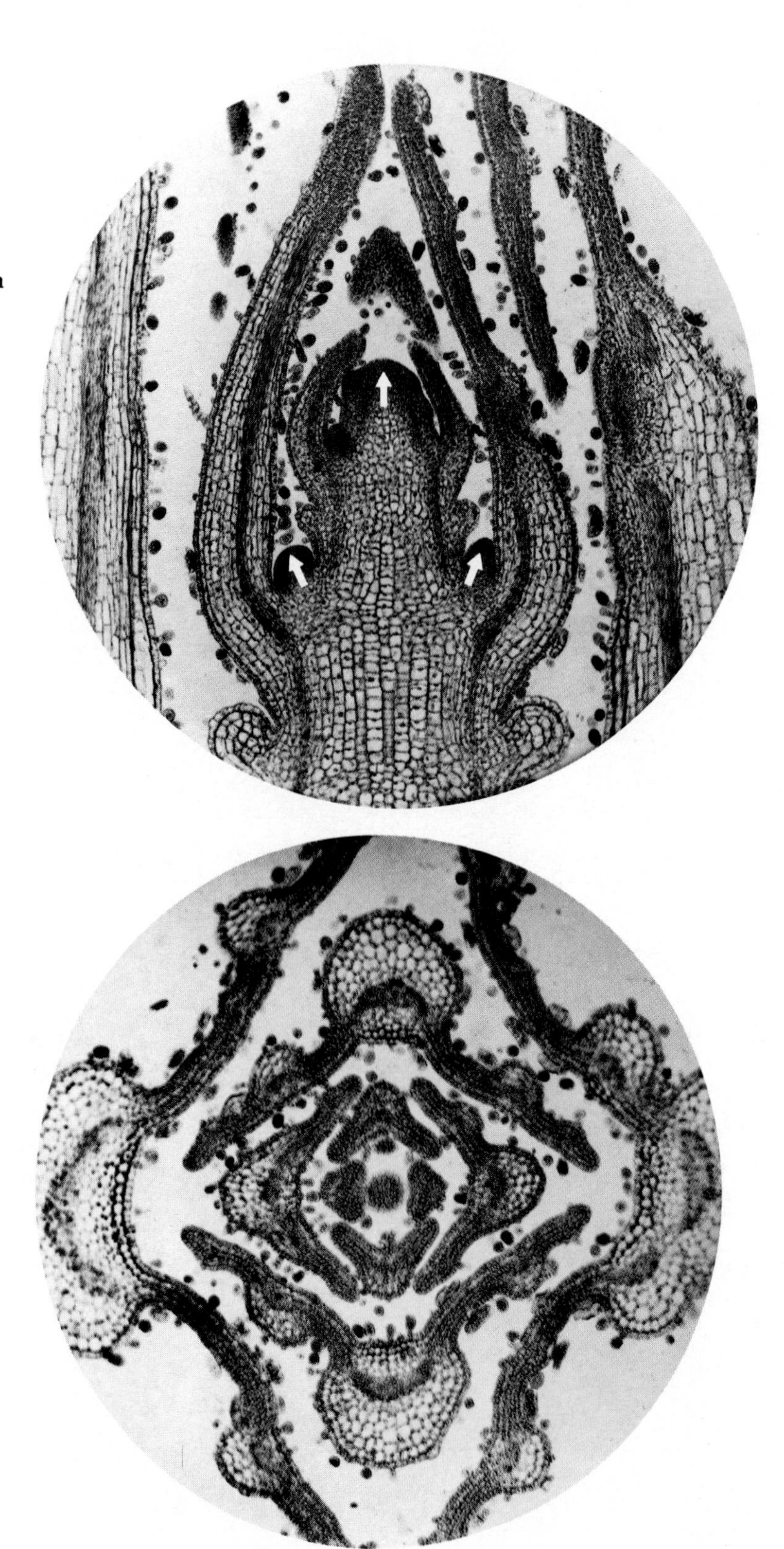

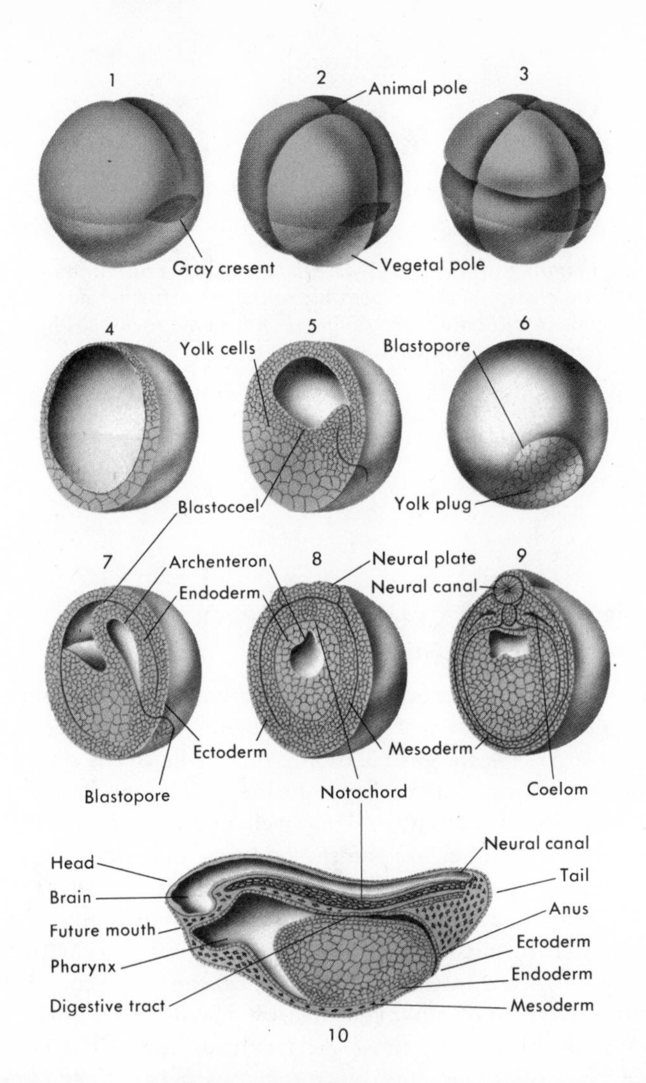

The result is a gray crescent-shaped region near the junction between the animal and vegetal poles. Although this change occurs before cleavage begins, it seems to direct or mark the egg physically, because the first cleavage always passes through the gray crescent, cutting it in half (Fig. 16.8). Later, the primary organizer of the embryo, the dorsal lip of the blastopore, always forms near the middle of one edge of the crescent. More striking than the fates of natural markings near the surface of the egg are those of artificial dye marks placed experimentally on the egg. Experiments have shown that

FIGURE 16.8

Early development of a frog. After the fusion of sperm and egg pronuclei, the zygote begins a series of cleavages, in which cells divide but do not separate. In (1) the first cleavage furrow has appeared, starting the 2-cell stage. The second and third cleavages (2, 3) produce 4-cell and 8-cell stages. Cleavage continues until a hollow ball of cells, the blastula (4), is formed and develops into a gastrula with three germ layers and a primitive gut (5–7). The neural folds then undergo rapid cell division, moving together to form the neural tube (8, 9). A later frog embryo (10) already shows the basic features of the tadpole it will become. The photographs show the 2-cell, 8-cell, and 32-cell stages, the late yolk plug, and finally the formation of the neural groove, from which the neural tube will form.

Photos courtesy of Carolina Biological Supply Company

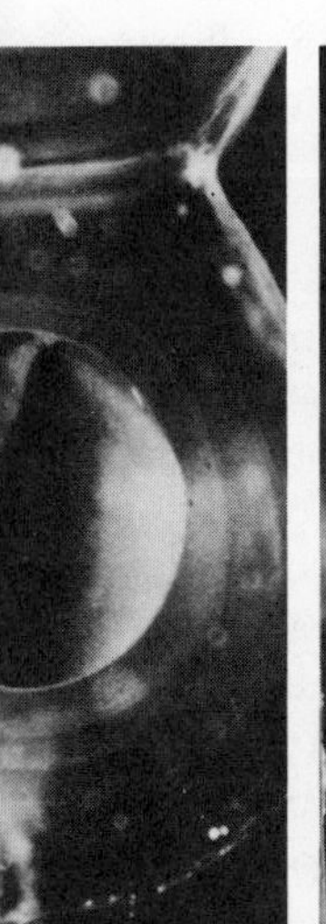
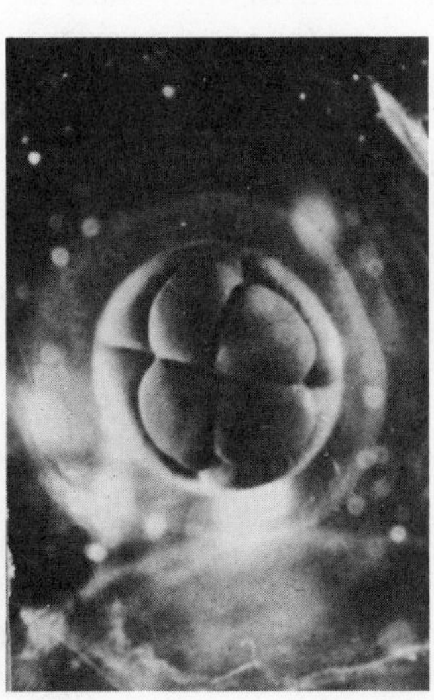

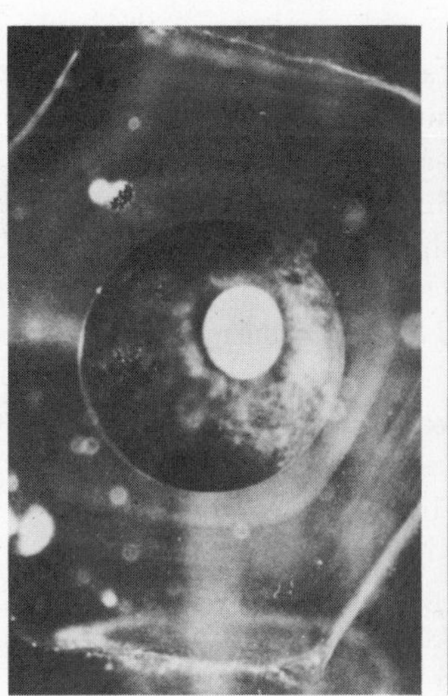
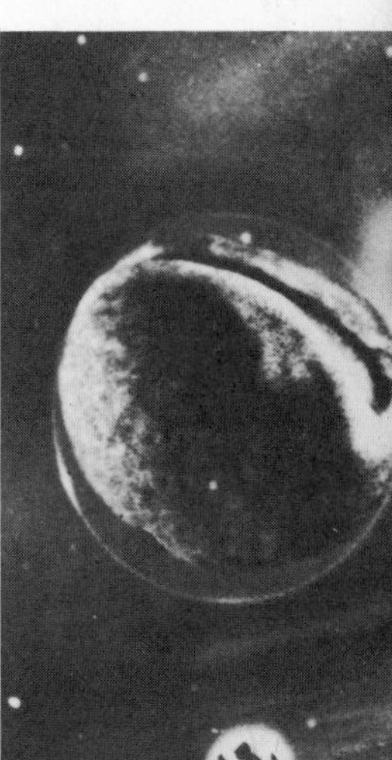

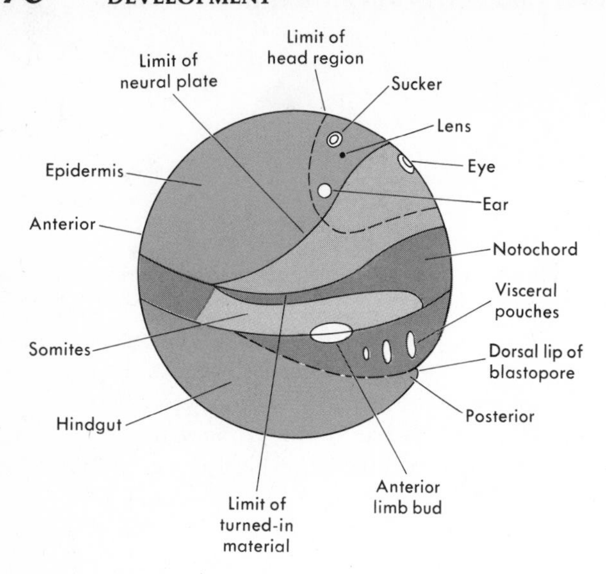

FIGURE 16.9

Fate mapping. In the early gastrula stage, differentiation is well under way, and it is possible to tell what further development will take place in many embryonic regions. This is a fate map of Amphioxus, *a lancelet.*

marked portions of the uncleaved surface always become incorporated into particular structures of the adult. Thus the egg cortex is very highly organized and seems to direct the further development of the egg, but the method by which the direction is accomplished is not fully understood. The marking technique, called fate mapping (Fig. 16.9), has been extremely useful in describing the events of development, but it has not revealed much concerning the mechanisms of development. Instead, the field of molecular biology has supplied some answers.

Egg RNA Long before the era of molecular biology brought an understanding of the mechanisms of protein synthesis, embryologists studying the structure of the egg noted that prior to fertilization the egg nucleus expanded, forming a germinal vesicle. It is during this brief period that much of the egg's program for the production of necessary proteins and enzymes is established. The egg chromosomes unwind and disentangle, forming lampbrushlike structures that tremendously increase the surface area of the information-carrying DNA. Much mRNA, containing copies of genetic instructions for the subsequent production of proteins, is synthesized, as well as tRNA and ribosomal RNA. These substances stand ready in the egg, assembling at the proper times the necessary enzymes and structural proteins. Because RNA is not again synthesized during subsequent cleavages, it may be said that early developmental events are in large measure the result of specific properties of the egg.

CELL MOVEMENT AND THE PROCESS OF MORPHOGENESIS

As development proceeds, factors other than those built into the egg must be responsible for the complex events. As we have seen, gastrulation begins when cells on the surface of the embryo begin to move. The mechanisms of cellular movement are not well understood. Equally important, neither are the reasons that cells adhere and interact in particular ways to form the new tissues. In the starfish, for example, studies have shown that the cells that form the blastopore depression of the gastrula may be forced into the hollow space, the blastocoel, by the movement of other cells nearby. Having been pushed into the blastocoel, those cells extrude long filaments, which grow across the blastocoel, reach the inner surface of the ectoderm on the opposite side, and, adhering to that surface, pull the cells of the depression further inside to form a tube.

The mechanism of cellular movement in vertebrate embryos is just as poorly understood. Perhaps the rampant cellular division that occurs at the animal pole during gastrulation expands the ectoderm so much that it buckles or folds in some way. This is problematical, for why should the folding arise only at the blastopore, and no wrinkles appear all over the surface of the embryo? We can only guess at the answers. It is thought that cells may adhere only to certain cells, but not to others. Experimenters have dissociated embryonic kidney cells and watched as these cells reassembled themselves into kidneylike structures. Cells also seem to stop moving or stop dividing when they touch the surfaces of other types of cells. These two factors—stimulation and inhibition by contact—doubtless influence the direction of cellular movement. They may also partially explain why the paths of movement seem to be so greatly affected by the changes at the lip of the blastopore.

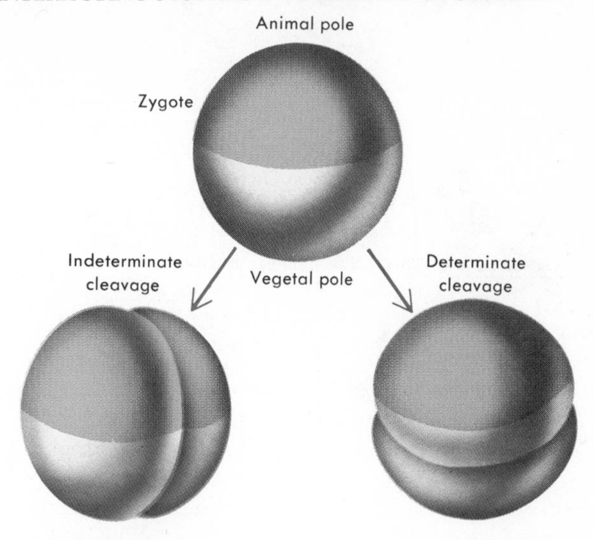

FIGURE 16.10

Determinate and indeterminate cleavage. Suppose that an essential material that influences development is concentrated at the animal pole of the zygote. If the initial cleavage is longitudinal, dividing the material equally between the two daughter cells, the cleavage is indeterminate. If it is latitudinal, giving all or most of the material to one daughter cell, the nuclear material in the daughter cells will be subjected at this early stage to different cytoplasmic influences, and the cleavage is therefore determinate. Indeterminate cleavage is characteristic of higher forms of animal life, primarily the vertebrates.

DIFFERENTIATION

The movements characteristic of gastrulation are morphogenetic; that is, they are responsible for shaping the developing organism. However, other changes occur besides those which replace the hollow ball of cells with a tube-within-a-tube. For the cells themselves begin to alter their shape, structure, and function, some of them differentiating into cells typical of lung tissue, others into liver, muscles, blood, and so on. The processes of differentiation do not begin at the same time in all organisms, nor is differentiation suddenly completed. Many cells do not become fully differentiated until quite late. In some vertebrates and in invertebrates, cellular determination, or restriction of further modification, is less fixed. In most vertebrates, however, the functional and structural nature of cells is usually unalterably fixed some time during development.

In some types of embryos, cellular determination begins very early, after only one or two cleavages have produced two or four cells. (The reason for such early determination is demonstrated graphically in Fig. 16.10.) Removal of one of these cells, for example, results in a contingent of cells in part of the embryo that fails to develop, and the ensuing adult lacks some part of its body.

In the vertebrates, however, determination often occurs somewhat later. In the late nineteenth century, August Weismann showed experimentally that the two blastomeres (cells) resulting from the first cleavage of a frog may be separated, and each will regulate its development and form into a normal adult frog. Blastomeres separated after the next few divisions develop normally, but those separated thereafter lack the ability to form whole organisms. Weismann proposed that as development proceeded, cells became differentiated as the result of the differential distribution of genetic material. That is, the nucleus became different from cell to cell. The nuclear differences resulted in changes in the structure and function of the cells, thereby differentiating them.

This proposal seems to be a reasonable explanation. We do know that the genes are responsible for the specific proteins that compose the structural and functional parts of cells, and it would be reasonable to assume that any alteration of genetic material during cellular division resulted in the differentiation of the cell. This hypothesis, however, was not definitively tested until the middle of the twentieth century.

In the 1950s Robert W. Briggs and Thomas J. King of the Institute for Cancer Research tested the genetic capability of nuclei from the cells of very late frog blastulas. These nuclei were carefully removed and placed inside frogs' eggs that had been activated artificially (without involvement of sperm) and then enucleated (had their own nuclei removed). When the enucleated eggs were implanted with nuclei from late

Stages of Human Development from Fertilization to Birth

FERTILIZATION AND THE FIRST MONTH

The life of every one of us began very simply and in a tiny way—a sperm cell 1/800th of an inch long fertilized an ovum (egg) 1/175th of an inch in diameter and weighing 1/20th of a millionth of an ounce. Included in the genetic material of both germ cells—sperm and egg—were the complete instructions to guide the development and growth into a recognizable human baby.

After 36 hours,* the fertilized egg (zygote) divides for the first time. The zygote then travels from the fallopian tube where it was fertilized into the uterus, where, between days 6 and 12, it proceeds to embed itself in the uterine lining. The embryo, now 1/100th of an inch long and consisting of hundreds of cells, draws nourishment from the uterine tissues and continues to grow.

As the embryo implants itself in the uterine lining, a mass of tissues and blood vessels—some from the mother, some from the embryo—begins to form. For the next 9 months this mass, called the *placenta,* nourishes the embryo, provides it with oxygen, and removes the waste products (CO_2, urine) that accumulate. The embryo is connected to the placenta by the umbilical cord, which contains one vein that carries nutrients and oxygen from the mother to the embryo, as well as two arteries that carry waste products out of the embryo and into the placenta to be disposed of through the mother's lungs, kidneys, etc. The vein and arteries in the umbilical cord are part of the embryo's circulatory system, not the mother's.

With the nutrients provided through the placenta, the embryo continues to develop. At days 11 to 13, a bulge develops where the spinal cord will eventually be. At day 17, blood cells begin to form, and on day 18, a heart forms and begins to pulsate by the 24th day.

Days 18 to 28 see a great many "blue print" developments in the embryo, now 2/25ths of an inch long:

1. The foundations for the nervous system (brain, spinal cord, and nerves) and eyes are laid.
2. The first of three kidneylike structures begins to form. The first two structures are the basis of the ducts in the male and female reproductive systems (fallopian tubes, vas deferens, etc.). The two initial structures are eventually displaced, but the third and final structure becomes the basis for the kidneys that must last for a lifetime.
3. The foundations for 40 pairs of muscles form (28 days).
4. The liver is recognizable, and lung "buds" are formed.

By day 30, the end of the first month of its existence, the embryo has come a long way. It has increased 40 times in size to become 1/4 of an inch long. Its weight has increased 3000 times, and it now has a trunk and a head. The plans are laid for its organs, and it has set up a way to get nutrients and oxygen from its mother. It is in the first month that the embryo is most susceptible to outside influences, such as diseases, drugs, and radiation. If something affects embryonic development in the early stages, the result may be deformities that can never be altered. Through all this remarkable activity, the mother, who gains about a pound, usually doesn't even realize that she is pregnant.

MONTH 2

In the fifth week, arms and legs begin to sprout, and the stomach and esophagus form. The embryo, now called a fetus, is 1/3rd of an inch long and weighs 1/1000th of an ounce.

The sixth week brings the following events: (1) face muscles, teeth, and hands begin to form; (2) the eyes are pigmented; and (3) fingers, toes, and gonads appear, although the sex of the half-inch fetus can't be determined.

During the seventh week, around day 46, the gonads develop into ovaries or testes, and a microscopic examination of the fetus would reveal its sex. By now, the fetus is 4/5ths of an inch long, from the head to the base of the spine, and a distinguishable neck now connects the large, forward-drooping head to the trunk. The heart is now complete.

In the eighth week, mammary glands are identifiable, and tastebuds form, among other developments. The fetus is 1 1/4 inches long and weighs 1/30th of an ounce, fifty times what it weighed at the end of the first month. All the major organ systems have begun, and it looks more identifiably human.

---

* All times (days, hours) are approximate.

MONTH 3

At nine weeks, the fetus is 1½ inches from head to base of spine and weighs 1/7th of an ounce. The head is no longer bent forward on the chest, and the fetus has fingernails, toenails, and hair follicles on its skin. The skeleton and muscles develop and lend a more human look to the fetus.

In the eleventh and twelfth weeks, the fetus is 3 inches long and weighs ½ ounce. Toothbuds form, and sockets develop in the jaws for the teeth. The mother should be careful to take in enough calcium and minerals throughout her pregnancy to ensure that the fetus will have enough calcium for its teeth. Many internal organs (stomach, liver, trachea, lungs, etc.) continue to develop toward their final form.

The first three months (called the first trimester) are by far the most critical in the pregnancy. At the end of them, the fetus has all the organs it will ever have or need. It is complete, yet unable to exist outside the uterus on its own. In the remaining six months, the organs develop to the point where they can sustain a live baby.

THE SECOND TRIMESTER: MONTHS 4, 5, and 6

Month 4—the 13th through the 17th weeks—represents a period of tremendous growth for the fetus. It lengthens from 3 inches to 6 inches (head to toe) and gains 3½ ounces (from ½ oz. to 4 oz.) It is during this month that the fetus develops lips and finger pads—with fingerprints! In the female, the oöcytes form, representing all the potential ova the female will ever have. The heart beats regularly and circulates 25 quarts of blood a day.

Although by the fifth month the fetus has reached a height of 8 inches (head to toe) and a weight of ½ a pound, it still can't survive outside the uterus. Its lungs, skin, and most of its organs are not yet developed enough to support it independently.

The fetus develops a covering of hair all over its body during the fifth month. Called lanugo, this covering remains on the baby for the rest of the pregnancy, but it usually falls off just before birth. Hair can also be seen on the head, eyebrows, and eyelashes.

From about the nineteenth week on, the mother can feel the fetus moving. Although it has always moved, the mother cannot feel its movement before this time. The fetus kicks, stretches, hiccups, and sleeps; it can be awakened by loud noises or sudden movements.

During the sixth month, the mother may gain as much as a pound a week, and the fetus begins to look more and more human. It is taller and more erect (10–12 inches from head to toe), and weighs 1½ pounds. The skin, lacking fatty deposits to fill it out, is red and wrinkled, the eyes are structurally complete, and the lids, closed since Month 3, reopen. Ossification (the changing of cartilage to bone) begins and continues until after birth.

THE THIRD TRIMESTER: MONTHS 7, 8, and 9

For the first time, if the fetus is born early, it will have a chance to survive on its own. It gains about five pounds and grows to about 20 inches. In preparation for its survival outside the uterus, the fetus stockpiles calcium for its bones and teeth, iron for its blood, and protein for its growth.

If born in the seventh month, a fetus has a 10 percent chance of survival. It is possible to survive now (and not earlier) because the brain has developed sufficiently to direct breathing and swallowing and to regulate the body temperature. The brain has also developed the centers for sight, smell, and hearing. The nervous system has advanced, and the fetus shows some basic reflexes, such as sucking and grasping (fingers and toes). Although the digestive and immune systems are not completed, survival chances are now up to 70 percent.

As the fetus makes ready for its head-first grand entrance at the end of the ninth month, it is 19–20 inches tall and weighs 6–8 pounds. It has acquired antibodies from the mother's blood to protect it from various illnesses, its lungs are ready to breathe, and its digestive system is ready to take in its first meal.

Two to three weeks before the delivery the placenta regresses and fetal growth stops, in preparation for birth.

*****

Bernard Shaw once said, "Except during the nine months before he draws his first breath, no man manages his affairs as well as a tree does."

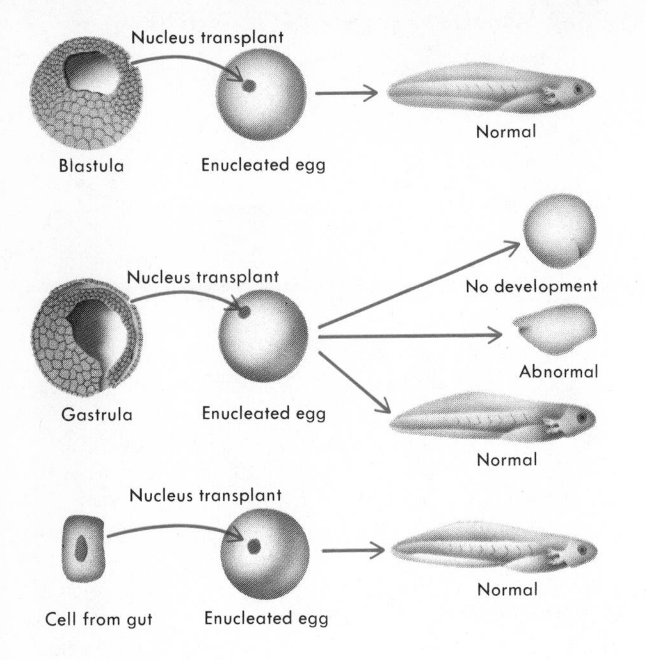

FIGURE 16.11

Constancy of genetic information. Successive experiments with enucleated eggs have revealed that normal development may be obtained even using the genetic information within the nucleus of certain cells from a relatively late stage of development. This result indicates that the genetic composition of cells is constant, but that their activity may change as development progresses. Certain gastrula nuclei fail to guide normal development because their activity has become restricted through differentiation.

blastula cells, a significant number of them developed entirely normally (Fig. 16.11). Clearly the implanted nucleus had all the genetic information necessary to direct the development of the egg into a frog. Even older nuclei extracted from the cells of later stages permitted normal development, as demonstrated by experiments in which the transplanted nuclei came from the intestinal cells of tadpoles. Thus the hypothesis of Weismann was discarded. Truly, all the nuclei in an organism have all the genetic information necessary to form the whole organism again.

Further insight was provided by the work of Jacob and Monod, which we discussed in Chapter 15. These men have shown that, in some organisms, materials present in the cellular cytoplasm can either initiate or inhibit the genetic production of proteins. Developmental biologists believe that as cleavage proceeds, each nucleus finds itself in a slightly different and distinctive cytoplasmic environment. Microenvironmental influences due to polarities in the gradients of certain chemicals, pH, temperature, light, and gravity are established. As we have seen, the substances in the egg cytoplasm—such as RNA, yolk, or materials in the relatively rigid cortex—are not uniformly distributed, and cleavage of the spherical egg results in a little more yolk (or other substances) in one cell, and a little less in another. Later, when protein synthesis is resumed, each cell, with a composition slightly different from the next cell, might produce proteins different from its neighbors. This further alters the cytoplasm of the cell, and in turn further alters the types or amounts of proteins produced. In the process, differentiation occurs to make them different from their neighbors.

Thus we can, hypothetically at least, account for many of the events of development. Differentiation, or the change in the structure and functions of the cells of an embryo, results in part from initial heterogeneous composition of the egg, and in part from alterations caused by the interactions between the nuclei of new cells and their cytoplasm. Morphogenesis, which is partially responsible for alterations in the shape of the embryo, is the result of movements of cells that occur either because they become amoeboid or because differential growth pushes many cells away from the site of rapid growth. Differential adhesive capability, or inhibitions to further division when two cells adhere, further shape the embryo.

But these factors alone are not totally responsible for the molding of highly specialized organs, such as eyes, ears, and glands, from the embryonic tissues. As the embryo grows and becomes more complex, groups of neighboring cells become more alike and form tissues. These tissues seem to spur the development of other tissues, a process often termed **induction.**

INDUCTION

Induction in developing embryos—the formation of specific types of tissues in response to the presence of other tissues—begins early. During gastrulation the cells in the region of the blastopore seems to have a distinct inductive influence. This tissue, referred to as the dorsal lip of the blastopore, appears to be essential for shaping the egg into a gastrula. Its removal halts further development, whereas removal of other tissues has no such effect. More striking yet have been transplantation experiments in which the dorsal lip tissue is removed from a donor embryo and transplanted into a host gastrula. The host, with two dorsal blastoporal lips—its own and that of the donor—forms two embryos, joined together like Siamese twins. Somehow the lip tissue organizes the surrounding cells and marshals their movements and differentiation to form the new organism. Curiously, dead dorsal lip cells retain the ability to induce the formation of a neural tube in an embryo, as can the chemical dye, methylene blue. This suggests that cells are so responsive at this stage that almost any environmental change can trigger this phase of development. Possibly the mechanism is direct interaction with the DNA in cell nuclei, which regulates the kinds of proteins that are produced and, in turn, the kinds of enzymes and structures.

Many other inductive events occur in shaping the embryo. After the neural tube—the forerunner of the brain and spinal cord—is laid down, the remainder of the nervous system is produced. From the spinal cord arise nerves—bundles of neurons which connect with the appropriate muscles. From the brain arise sensory nerves and components of sensory structures, such as the eyes. These events also apparently depend on induction, the development of one kind of tissue caused by the influence of another that is already present. The individual neurons of a nerve grow, for example, from small spindle-shaped cells situated near the edges of the neural tube. These cells extrude long thin processes of cytoplasm, which become the axon of each neuron. The tip of each forming axon is amoeboid, and it sends out thin strands of pioneering fibers that penetrate the limb-bud tissue that will form the bones, muscles, and other tissues of the limbs. Thus the nerve cells seem to respond only to the cells of the tissues that they will supply, and their growth is directed toward them. If the limb-bud is removed from a developing vertebrate embryo before the process of innervation is complete, a tangled mat of neurons forms around the scar marking the removal of the limb-bud. Somehow the information that there is no tissue to innervate is fed back to the developing neuron cell bodies, and all the neurons, with no destination, disintegrate. Conversely, implantation of an extra limb-bud during these stages induces the formation of additional neurons. Chemical feedback is evidently the process whereby the body of the nerve cell is informed of conditions prevalent at a great distance from it.

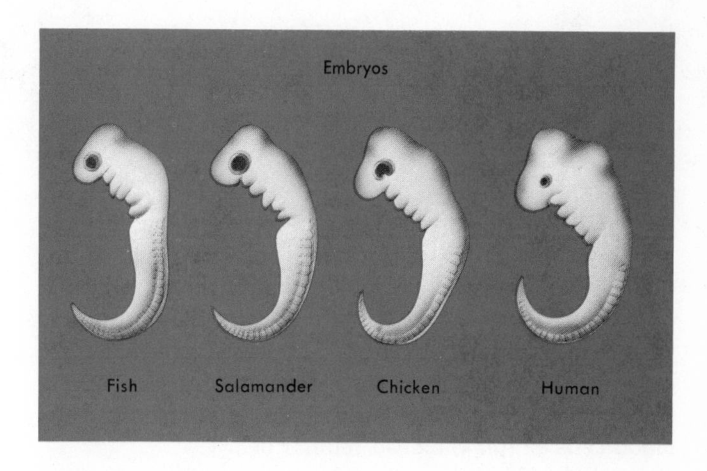

FIGURE 16.12

The process of development is not totally efficient. The resemblance of these early vertebrate embryos indicates that structures retained from the evolutionary past—for example, a tail—may appear at some stage in the embryo and then be eliminated or converted to some other structure. In the bird, the tail bones will fuse together during development, whereas in the human organism, they will be almost completely lost, only the vestigial coccyx remaining. Cellular death plays an important role in the processes of conversion and elimination of tissues.

CELLULAR DEATH IN MORPHOGENESIS

So far we have considered developmental mechanisms that involve the dynamics—growth and other changes—of live cells. In some parts of a vertebrate, however, selective cellular death also plays a role in development (see Fig. 16.12). In the developing wing of a chick, for example, cellular death in part of the limb-bud, the precursor of the wing, is responsible for the formation of the digits (rudimentary fingerlike protrusions that point up the common body plan of all vertebrates, whether

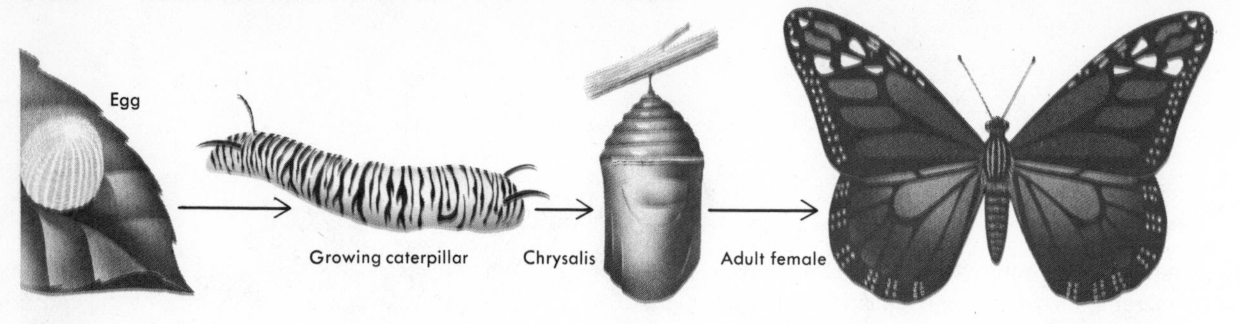

they swim, walk, or fly). The cells between the developing fingers begin to die, leaving only the cells of the digit alive. The mechanism by which these particular cells destroy themselves is not understood. It is thought that their lysosomal enzymes are released and that these cause the cells to die. Once cellular destruction within the limb-bud has become widespread, phagocytic white blood cells engulf the dead cellular debris.

Cellular death also plays a role in the shaping of the vertebrate gonads, or reproductive structures. The sex of an individual is of course determined genetically, but early in development the gonads are undifferentiated in that they contain primordia for both testes and ovaries. One primordium persists and develops further, and the other dies or degenerates. Sometimes, however, developmental "mistakes" occur, and an organism genetically of one sex has gonads intermediate in structure between male and female, or even more like the sex opposite its genetically inherited one. Here, too, lysosomal enzymes are probably responsible for the breakdown of unnecessary cells. But what triggers the release of these enzymes? Although the process is not completely understood, biologists think that hormones may signal the self-destruction of some cells and not others.

METAMORPHOSIS

The processes of development do not end when the organism first becomes self-supporting. In some insects, such as moths and butterflies, the zygote develops into a free-living caterpillar, which can forage, eat, and support itself completely. But the caterpillar is only a premature moth or butterfly. It soon ceases to chew on leaves, encases itself in special hardened cuticle tissue, and rests, while inside its pupal case (the cuticle) its entire anatomy is altered (see Fig. 16.13). After this pupal stage a winged adult emerges, leaving the hollow cuticle behind. Many kinds of cellular transformation, including cellular death, occur to effect such a startling metamorphosis of a rather homely "worm" into a lovely winged moth. Although other insects undergo a transformation process, it may not be quite so radical as that of the caterpillar. The organisms merely undergo a series of molts, or shedding of the outer skin of the body, emerging each time a little larger and more nearly like the adult of their species.

FIGURE 16.13

Metamorphosis. One type of metamorphosis is commonly observed in the butterfly. The egg gives rise to a caterpillar, which, after feeding and growing, pupates within a chrysalis it spins about its body. Within the chrysalis the tissues undergo fundamental reorganization. The metamorphosed caterpillar emerges as the mature, breeding butterfly.

In insects, the stimulus for metamorphosis and molting seems to be distinctly hormonal. Two hormones, known as the **juvenile hormone** and the **growth and differentiation hormone,** have been found. These two chemicals seem to be directly responsible for maintaining the juvenile characteristics long enough for the immature organism to become able to feed itself, and then for initiating the growth and development that transform the organism into an adult at the proper time.

Such metamorphosis, however, is not at all restricted to the invertebrates; some vertebrates also undergo remarkable changes in anatomy. The frog, for example, begins its free-swimming life well adapted to this environment, for it has at first a fishlike tail, lacks arms and legs, and has functional gills, which enable it to respire in the water. As it gets older, however, the form of its body changes, and it is able to walk and breathe on land. These metamorphic changes involve the destruction of many cells and tissues in every part of the juvenile frog's body. The gills of the juvenile are lost, and the adult's lungs become functional; the size and shape of the jaws change; new limbs, the arms and legs, appear; and so on. Although these changes in the frog are well described, the physiological basis of amphibian metamorphosis is yet unexplained. It is known, however, that thyroid hormones largely effect the changes.

Development in plants

In plants, development of an adult organism from one or two cells has many similarities to development in animals. In both, a zygote undergoes cellular divisions. In animals, the earliest divisions are called cleavage, a kind of carving of the cytoplasm into smaller and smaller parcels—cells which then go on to form more cells typical of adult tissues and organs. In plants, the cell divisions involve ordinary mitosis, which makes new cells of more or less normal size.

In both plants and animals, cellular divisions eventually lead to growth, although plant embryos may arrest their growth during periods of dormancy. Growth of the organism may entail both an increase in the number of cells as a result of mitosis, and enlargement of certain cells, a frequent occurrence in plants. Growth may continue in plants indefinitely; some towering sequoias have been growing for more than a thousand years.

Differentiation occurs in both plants and animals. In animals, cells become more and more determined with time, as they become highly and irreversibly specialized. A few animals, salamanders and starfish, for example, can regenerate parts. The salamander can replace a lost limb in six weeks. In plants, regeneration is much more common. In fact, live differentiated cells in the adult can often "dedifferentiate." That is, if cultured properly, they can produce a whole new organism by processes similar to the formation of an adult from a zygote.

Morphogenesis occurs in both plants and animals. In animals, morphogenetic movements—actual migrations of cells around the developing embryo—help to shape the new animal. In plants, the cells are rigidly fixed in place by intercellular cement, which binds the cells together. Plants compensate by having zones of active cell division, the meristems, in position for further development.

The embryonic plant or animal needs a supply of nutrients to provide energy until the organism can become self-sustaining. In animals, yolk from the egg may serve this purpose, or the embryonic animal may live inside its mother and derive its nourishment from her body. In some plants, a sperm fertilizes the polar bodies to produce endosperm, a living tissue that sustains the embryo. In the flowering plants, the mature embryo also has seed leaves, or cotyledons, to nourish it.

The processes of development in plants and animals—the changes in cellular size, shape, function, and destiny—result from complex interplays between the nucleus and cytoplasm of individual cells, interplays between and among different cells, some cellular movements, and even the death of some cells. In animals, it is relatively easy to imagine that as cleavage carves up the cytoplasm of the egg, nuclei resulting from successive divisions find themselves in different cytoplasmic surroundings. Since the cytoplasm of these cells is different, biologists have theorized that different kinds of interactions between nuclear DNA and components of the cytoplasm could yield different paths of development in different cells. The causative factors in plant development are less well understood. It appears that later developmental changes—which occur well after the forces of morphogenesis have established the shape of the plant—are caused by hormones produced by cells in different parts of the plant. It is not clear, however, how the initial cellular changes, which produce a multicellular embryo from a unicellular zygote, occur.

GERMINATION

In the flowering plants, the egg is fertilized by pollen entering the ovary. After all the mitotic divisions have produced an embryo, the seed containing that embryo is released from the parent plant. This relatively dry seed now can remain dormant for months or even years. Winter represents the typical dormant period, after which the warmth and moisture of spring awaken the suspended embryo. This period of renewed development is called **germination** and is particularly characterized by a massive absorption of water. Filling cell vacuoles, the water soon occupies much of the space within the cells, increasing their size as well as the size of the entire seed by as much as 100 percent. The seed's enzymatic activity also increases, as does its metabolic rate, allowing for synthesis of new protoplasm, renewed cellular division, and absorption of still more water.

Eventually, the expanding embryo breaks through the seed coat and emerges as a seedling, a tiny rudiment of the adult plant to be. This mature embryo is made up of

a **cotyledonary node,** to which one or two cotyledons are attached;
the **epicotyl,** just above the cotyledon, consisting of the beginnings of a bud and tiny leaves, which will constitute most of the shoot, the above-ground portion of the plant;
the **hypocotyl,** at the opposite end of the axis, which will become part of the primary root, or part of the stem; and
the **radicle,** which will become the primary root.

The radicle, the first part of the embryo to emerge from the fractured seed coat, immediately grows downward, as a result of geotropism, regardless of the embryo's position at the time of germination. Root hairs as well as secondary roots soon grow from this structure. These roots hold the plant securely in position within the soil, or substrate, and also absorb water.

In some plants, such as the common garden bean, the upper portion of the hypocotyl arches upward and penetrates the surface of the soil, taking the epicotyl and cotyledons out with it. In these plants, the lowermost portion of the shoot is actually the uppermost portion of the hypocotyl (Fig. 16.14). In other plants, such as the garden pea, the hypocotyl does not arch; the cotyledons remain beneath the ground, and the epicotyl grows up, pushing through the surface. This entire above-ground shoot, then, is of epicotyl origin.

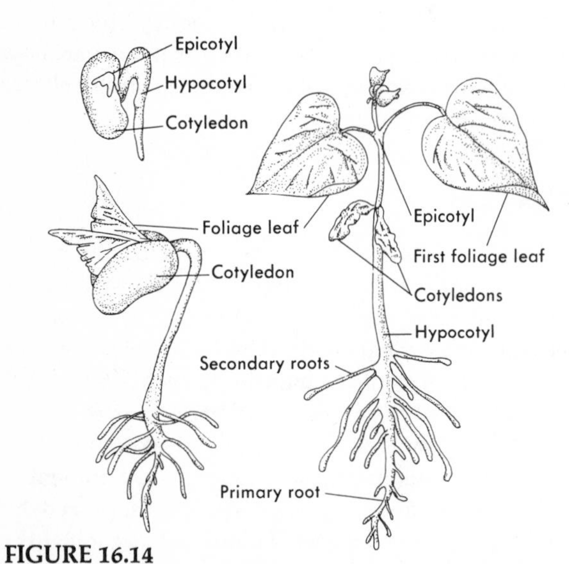

FIGURE 16.14

Plant development. In the germination of the common garden bean, the epicotyl develops into the upper stem and first foliage leaves, and the hypocotyl grows into the root and lower stem. The cotyledons nourish the young plant until it is able to obtain nourishment for itself, and then they wither and drop off.

GROWTH AND THE MERISTEM

The slow initial growth is followed by a longer period of more rapid growth. Perennial plants continue such growth throughout their lives. The upward and downward growth, resulting in the extension of the roots and the shoot, is accomplished through cell multiplication, cell elongation, and vacuolation. These functions are continually and simultaneously performed by the meristems, a busy section of cell formation located at the tips of the root and the shoot.

Roots The root tip is covered by parenchyma cells, collectively called the root cap. The root cap works through the soil, sometimes by secreting a lubricant and by shedding cells to further reduce friction. Root cap cells do not divide. They are replaced by those provided by the root apical meristem. Most of the new cells form farther up the root, however, at what might be termed the rear of the meristem section. As the root grows, the meristem remains at the tip, leaving the new cells, which increase the length of the root, behind it. The mitotic activity of these new cells decreases with the increasing distance between them and the meristem, which is forever moving farther away from them. After they cease dividing altogether, these cells undergo an elongation process, stretching out and absorbing water into their vacuoles.

FIGURE 16.15

The apical meristem of the horsetail (Equisetum). *At the tip of stem and roots is a dome-shaped area of embryonic cells called the meristem. Meristematic cells, which are continually dividing and increasing, are responsible for the growth of the shoot.*

Eventually the water-filled vacuoles fill practically the entire cell, forcing the protoplasm against the cell walls. This vacuolation, as it is termed, is a cause of cell growth quite different from the cytoplasmic increase common in animal cells. Only a few millimeters long, the zone of cell elongation is directly behind the zone of cell multiplication.

Stems The same process is responsible for the growth of the stems, for there is a meristem area located at the stem tip. In the stem, periodic changes in the apical meristem are responsible for production of **nodes,** which give rise to leaf primordia; nodes are not present along roots. The intervening sections of stem between the nodes are called **internodes.** Nodes and internodes alternate along the length of a stem.

Stem growth results mainly from the elongation of cells in the internodes. As leaf primordia develop on the stem tip (at the apical meristem region), they first cluster, forming a bud that contains, in addition to the rudimentary leaves, sections of not yet elongated stem. In perennials, plants that winter over from year to year, this bud opens and the internodes stretch out during the springtime. Meristematic growth ceases in the leaves once they are fully formed. By this time, however, some meristematic tissue has arisen at the attachment of the leaf and the node, and an axillary bud develops, producing branch stems (Fig. 16.16).

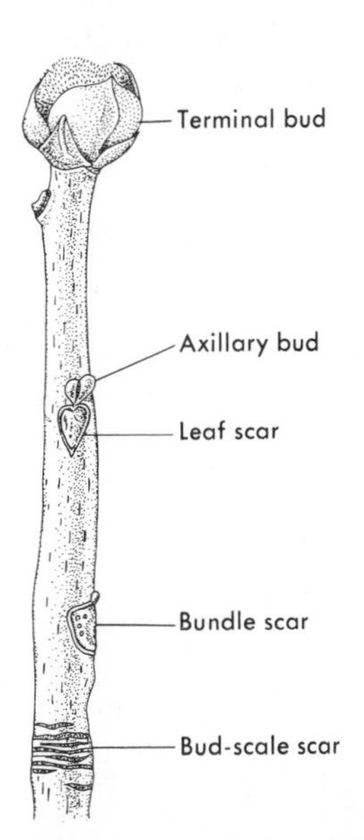

FIGURE 16.16

The growing end of a dormant winter stem reveals its development history. The length of stem between the bud-scale scar and the terminal bud represents one year's growth. Leaf scars and their associated bundle scars show where the petioles were attached to the stem and where vascular bundles were connected to the leaves. The axillary bud will develop into a branch stem in the next growing season, and the terminal bud will produce further stem growth, leaving behind a bud-scale scar similar to the one below it.

DIFFERENTIATION

The apical meristems produce fundamentally identical cells. As in animals, there must occur a process of differentiation, providing the different types of cells needed for formation of a mature organism—vessel cells, sieve cells, etc. What a cell will become is determined by its position in the root or stem when it is elongating.

Roots The process of maturation leading to a fully differentiated cell begins, in fact, during the period of elongation. Cells in the meristematic zone of elongation become arranged in three concentric layers along the cylindrical structure of the root, according to the following plan of distribution.

1. The outer layer, or **protoderm,** quickly becomes the epidermis.

2. The **provascular cylinder,** or inner core, is destined to become the stele, or central vascular tissue, with an outer layer of cells that will be the pericycle layer. The cells of the central vascular tissue are particularly elongated.

3. An area of **ground tissue** separates the other two and becomes the cortex and the endodermis.

Stems There is a slight variation in the differentiation of a stem. Here a second ground tissue section is formed inside the provascular cylinder, becoming the pith—that soft, spongy tissue characteristic of many young branch and stem centers.

Subsequent increase in the diameter of stems and branches is a function of the **cambium,** a thin layer of cells that continually produces xylem cells from its inner face and secondary phloem in the outward direction.

The epidermis and cortex break and fall off as a result of the increase in diameter. As this happens, the cork cambium, a new, lateral meristem, is formed along the outer perimeter from some of the old cortex cells. This cork cambium generates a secondary protective material called, not surprisingly, the cork, which takes the place of the lost epidermis and cortex. In some plants, such as the cork oak (*Quercus suber*), the amount of cork produced is quite large and is harvested commercially, eventually finding its way into wine bottles.

REGENERATION

A remarkable capability of plant life not shared by most animals is the ability to regenerate, that is, to grow back a part that has been removed or to grow a full plant from a section removed from another plant.

The cuttings used by horticulturists for plant propagation are an example of this process. A callus develops at the point of the cut on the donor plant. The piece that was removed often produces roots and branches, and a typical development pattern ensues. Tissues of carrot plants cultured in a laboratory experiment demonstrate the process of regeneration. The newest individual cells are further isolated, and they are observed to reproduce and reach an embryo stage, which eventually gives rise to mature carrot plants. This capacity of plant cells, known as **totipotency,** is a tremendous boon to horticulturists, botanists, and agriculturists, since it allows them to duplicate a plant that has been found hardy or productive merely by removing, planting, and cultivating some of its tissue. A plant of comparable if not identical qualities will develop from the cutting or cell.

Summary

1. Development is the orderly progression of changes that occur in the life cycle of an organism. These changes include growth, maturation, regeneration or repair, and the creation of a new organism.

2. The events of animal development begin with the formation of gametes. Oogenesis results in three polar bodies and a single large haploid egg. Spermatogenesis results in four haploid sperm. Fertilization—or the joining of egg and sperm—produces a zygote. The three processes through which the zygote forms an embryo are differentiation, morphogenesis, and growth. The cleavage stages are the blastula, a hollow ball of cells, and the gastrula, which forms the tube-within-a-tube body plan. Gastrulation is followed in higher animals by the formation of mesoderm surrounding the coelom, or body cavity. In the vertebrates, neurulation forms the neural tube.

3. The properties of the egg are critical in determining the course of development. The egg yolk serves as a nutrient source to the embryo and influences development by impeding cleavage. The egg cortex contains materials that play a key role in directing further development. Much of the egg's program for the production of necessary proteins and enzymes is established in a brief period prior to fertilization, when the germinal vesicle forms.

4. Morphogenesis—the shaping of the embryo—is accomplished by cellular movements that are the result of amoeboid activity of differential growth. Cellular death also contributes to morphogenesis.

5. Cellular differentiation results from the initial heterogeneous composition of the egg and from alterations caused by interactions between the nuclei of new cells and their cytoplasm. The genetic composition of the nuclei is not altered, but the activity of the genes is affected. One important mechanism of differentiation is induction —the formation of specific tissues in response to other tissues, either through contact stimulation or inhibition.

6. The process of development does not cease at birth. In certain organisms, metamorphosis involves the complete transformation of anatomical features from the larval to the adult stage.

7. Plant development also includes morphogenesis, differentiation, and growth. Even though plant development may be arrested during dormancy, the period of growth may continue throughout the lifetime of the plant. Regeneration is a common property of plants. The mechanisms of plant development are less well understood than those of animal development.

8. The period of germination in the plant seed is one of renewed development that is characterized by the massive absorption of water. The embryo that emerges is composed of the cotyledonary node, the epicotyl, the hypocotyl, and the radicle.

9. Extension of the root and the shoot is accomplished through cell multiplication, cell elongation, and vacuolation, functions that are performed by the meristems at the tips of the root and the shoot. Stem growth in diameter results from meristem activity and from elongation of the internodes.

10. Cellular differentiation is determined by the position of the cell in the root or stem when the cell is elongating. Cells in the meristematic zone of elongation of roots become arranged in the outer protoderm (which becomes the epidermis), the provascular cylinder (which becomes the stele), and the ground tissue (which becomes the cortex and the endodermis). Stem differentiation involves the formation of the pith from a second ground tissue inside the provascular cylinder, as well as the formation of the other layers. Subsequent increase in the diameter of stems and branches is a function of the cambium.

REVIEW AND STUDY QUESTIONS

1. What is included in the term "development"?

2. Name and describe briefly the three processes of development. Do all three processes occur in both animals and plants?

3. Why is yolk present in an egg? What effect does the amount of yolk have on an egg's development?

4. Distinguish each of the following: blastula, gastrula, neurula.

5. What is a fate map? What purpose does it serve?

6. From his experiments with frogs, Weismann proposed—incorrectly—that as an organism develops, its genetic material is distributed unequally among the cells, which become differentiated as a result. How was Weismann's hypothesis refuted?

7. What performs for plants the function that yolk performs for animals?

8. What is regeneration? Citing examples, make a generalization about the power of regeneration in terms of a comparison between animals and plants.

REFERENCES

Balinsky, I. 1975. *An Introduction to Embryology*, 4th edition. Saunders, Philadelphia.

Ballard, W. W. 1964. *Comparative Anatomy and Embryology*. Ronald Press, New York.

Barth, L. J. 1964. *Development: Selected Topics*. Addison-Wesley, Reading, Mass.

Ebert, J. D. 1970. *Interacting Systems in Development*. Holt, Rinehart and Winston, New York.

Edwards, R. G. 1966 (August). "Mammalian Eggs in the Laboratory." *Scientific American* Offprint no. 1047. Freeman, San Francisco.

Gurdon, J. B. 1968 (December). "Transplanted Nuclei and Cell Differentiation." *Scientific American* Offprint no. 1128. Freeman, San Francisco.

Jacob, F., and J. Monod. 1961. "Genetic Regulatory Mechanisms in the Synthesis of Proteins." *Journal of Molecular Biology* 3:318–356.

Pastan, I. 1972 (August). "Cyclic AMP." *Scientific American* Offprint no. 1256. Freeman, San Francisco.

Trinkus, J. P. 1969. *Cells into Organs: The Forces That Shape the Embryo*. Prentice-Hall, Englewood Cliffs, N.J.

SUGGESTIONS FOR FURTHER READING

Bonner, J. T. 1963. *Morphogenesis*. Atheneum, New York.

A comprehensive introduction to the development of living organisms that uses examples from a variety of living forms, touching on the physics and chemistry of development, patterns of growth and morphogenic movements, polarity and symmetry, and patterns of differentiation.

Bulmer, M. G. 1970. *The Biology of Twinning in Man*. Oxford University Press, New York.

A compact, beautifully clear though rather technical account of the phenomenon of twinning that examines the various types of twins and reviews the vexed question of why twinning has persisted in man.

Huettner, A. F. 1949. *Fundamentals of Comparative Embryology of the Vertebrates*, revised edition. Macmillan, New York.

A reasonably simple presentation of vertebrate embryology, with an especially clear account of the development of amphioxus. Very good illustrations.

Rugh, R., and L. B. Shettles. 1971. *From Conception to Birth*. Harper & Row, New York.

A detailed description of embryonic development, day by day and week by week. Excellent photographs.

Willier, B. H., and J. M. Oppenheimer, eds. 1974. *Foundations of Experimental Embryology*, 2nd edition. Prentice-Hall, Englewood Cliffs, N.J.

A collection of important papers focusing on current research in embryology.

ORGANISMS AND THEIR ENVIRONMENT

part four

EVOLUTION
chapter seventeen

Basic to understanding the general concept of evolution is acceptance of the premise that all beings alive today and all those who lived at any time in the past have descended from some common ancestor—the first life form.

In 1860, the year after the publication of *On the Origin of Species,* Charles Darwin wrote to his friend, the geologist Charles Lyell:

> I have noted in a Manchester newspaper a rather good squib, showing that I have proved "might is right," and therefore that Napoleon is right and every cheating tradesman is also right.

This ironic "squib" no doubt rather cleverly poked fun at the moral outrage generated in the Christian world by Darwin's revolutionary theory of evolution. In *Origin of Species,* Darwin set forth and carefully documented the idea that life had not been created and placed on the earth in unchangeable perfect forms according to a divine plan. Life had existed for billions of years and had been constantly changing and developing. Theologians of the time claimed that God's creatures were perfect and their offspring would have perpetual life, but Darwin claimed fossil evidence showed that numerous species had been sufficiently imperfect or maladapted to have become extinct. These species had succumbed to the pressures of a natural law that to the Christian mind seemed far from divine. It determined that neither the meek nor the pure in heart inherited the earth; instead, the survivors were the most fit, those best able to defend themselves, breed, and raise their young in a hostile, changing world.

To certain members of the scientific community the elements of Darwin's arguments were not unfamiliar, but to the layman and the clergyman Darwin's theory was probably more shocking than was Copernicus's theory of a sun-centered universe when it was proposed about 300 years before. Copernicus denied man the central spot in the universe, but it seemed to nineteenth-century theologians that Darwin wanted to depose God. Religious objections to the theory of evolution continued on into the twentieth century, and the furor did not die down in America until after the famous Scopes "monkey trial" in Tennessee in 1925. But by that time geneticists had discovered chromosomes, genes, and their functions and thus were able to demonstrate the existence of the hereditary mechanisms implied in Darwin's theory.

FIGURE 17.1
The young Charles Darwin examining fossils in South America during the voyage of the Beagle. *His extensive observations led to the conviction that evolution over enormous periods of time was a valid explanation of biological diversity. An inveterate collector, Darwin kept a voluminous mass of carefully compiled data that played an important part in the rapid acceptance of his theory of natural selection as the mechanism of evolution.*

Evolution is the modification and development of species through the hereditary transmission of slight variations from generation to generation. The study of evolution has revealed much about the history of life and the earth, as well as about the variety and quality of life in the past and in the present. We can only speculate about the direction of evolution in the future. As new technology has developed, man has gained the power to quickly and radically alter the forces that shape the direction of evolution. These forces generally act very slowly, but for certain rapidly reproducing organisms, man has already modified the forces sufficiently to effect significant evolutionary changes. For example, strains of DDT-resistant insects and penicillin-resistant bacteria have evolved since the beginning of continuous widespread use of DDT and penicillin, posing greater problems of virulence.

Technology has now advanced to the point that man is on the verge of being able to shape his own course of development. The potential for change ranges from the catastrophic to the magnificent, and the effect and extent of man's influence will depend on the skill and foresight with which he uses his technology.

This chapter, then, will discuss the development of evolutionary theory and the environmental and genetic factors that shape evolution itself.

Historical perspectives

The development of a scientifically based evolutionary theory took place in roughly 250 years, a rather short period when one considers the tremendous changes in scientific thinking that occurred during that time. The development can be divided into two periods. The first, which began in the late seventeenth century, consisted of the events leading up to and including Darwin's proposal of the theory of evolution in 1858. The second, which extended into the 1930s, consisted of the discovery and elucidation of the genetic mechanisms that underlie Darwin's theory. The main characters in this scientific drama were Jean Baptiste de Lamarck, Charles Darwin, Gregor Mendel, Hugo De Vries, and Thomas Hunt Morgan, but many other figures, including Thomas Jefferson, played bit parts.

Darwin is credited with having proposed the accepted theory of evolution, but he did not work in a vacuum. Many elements of this theory had already been published and accepted by other scientists. In fact, Darwin's own grandfather had published a paper suggesting the existence of an evolutionary process about 70 years before. Darwin's achievement was to pick out the valid aspects of other theories, add insights and observations of his own, and synthesize an accurate and well-documented theory. It was in the 150 years before Darwin that scientists made the discoveries and conclusions that paved the way for Darwin's work.

The period was characterized by monumental upheavals in scientific thought, as man's sense of time and space expanded from a human to a cosmic scale, and scientists increasingly explained natural phenomena with scientific rather than theological arguments. In the late seventeenth century, Western scientists generally accepted some form of the biblical account of the creation of the heavens, the earth, and life, and the small scale of time and the universe that this account implied. Scientists thought that the universe was big but oriented to man, that life on earth had been created in six days about 6000 years before, that the major events in early earth history were recorded in the Bible and the writings of the ancients, and above all, that the universe was created and ruled by God. His ways were unknowable, and therefore his works could be catalogued and wondered at but not analyzed or understood. These ideas, which seriously limited the scope of scientific inquiry, were constantly reinforced by the very influential church. It was still a time when scientific societies could require members to retract theories on the basis that they were too heretical.

Discoveries and speculation in the fields of geology and paleontology (the study of fossils) led to the questioning of the religious views of the creation of earth and life. At the turn of the seventeenth century, these were young fields, but the success of Newton's laws of physics in explaining the motion of the stars and planets had inspired scientists to look for natural causes in other phenomena. Agents of geologic change—earthquakes, volcanoes, landslides, etc.—were well known, and the presence of fossils of ocean-dwelling organisms in the mountains was evidence that tremendous changes had taken place. Nevertheless, the understanding of geologic mechanisms progressed slowly, perhaps because the creation of the earth was so explicitly spelled out in the Bible. Early geologists spent considerable energy trying to show that their theories did not contradict the Scriptures.

During the eighteenth century there was a running argument in geology between proponents of two opposed philosophies. The catastrophists supported the theory that the earth was formed by a series of catacylsmic events, and the uniformitarians believed the currently held view that geologic change occurs mainly in gradual day-to-day processes according to constant natural laws. The major obstacle to understanding geologic forces was the scientists' notion of time. The concept of the earth as a few thousand years old favored the theory of fast-acting cataclysmic development. Time sense gradually increased, however, with such discoveries as the one that many French and German mountains were in fact old well-worn volcanoes, much different in appearance from active ones.

The first man to fully realize the age of the earth and the forces that had shaped it was the uniformitarian James Hutton, who published his theories in 1788. Applying the Newtonian concept of constantly operating natural laws, Hutton concluded that geological transformation was a constant self-perpetuating process of stratification, upheaval, and erosion. The process created a "succession of worlds," of which there was "no vestige of a beginning—no prospect of an end." Hutton had grasped the age and power of the natural forces and had introduced a new concept as well, namely, that change involved the generation of new worlds, not just the degeneration of old ones.

Hutton's theory was rejected and nearly forgotten by the catastrophists, whose views were more commonly accepted at the time. However, uniformitarianism reemerged in 1830 with the publication of the first volume of Charles Lyell's *Principles of Geology*, a book that profoundly influenced Darwin.

If geology was primitive in 1700, the study of fossils was even more so. (Fossils are the hardened remains or imprints of early life-forms preserved in rock.) Fossils had been found, but there was much dispute as to what they were and where they had come from (Fig. 17.2). The main issues of eighteenth-century paleontology were, first, the possibility that species could have died out or become extinct, and second, the possibility that, like the earth, organisms could have changed with time. Again, the major obstacles to the resolution of these issues were an inaccurate view of the age of the earth and religious views about the creation of life. It was commonly believed at the time that God had created all

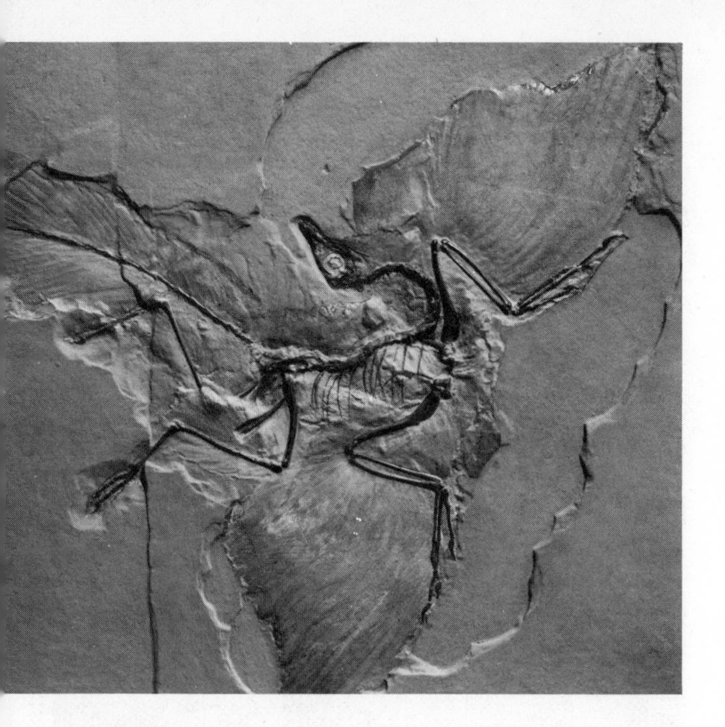

FIGURE 17.2

The fossil Archaeopteryx, *whose sensational discovery in the 1860s drew eminent collectors from everywhere. The Darwinists triumphantly heralded the feathered reptile as the evolutionary link between reptiles and true birds. By careful examination of such fossils, we can obtain valuable information on the environmental conditions and ecological relationships of a particular geological era, as indicated by the position of the fossils within a geological formation. More direct methods of dating have also been devised by the use of carbon 14 or potassium-argon analysis.*

creatures and that his wisdom had ensured that each type was perfectly suited to its role in life and would exist forever. The discovery of fossils proving the existence of organisms unlike any known organisms was naturally disturbing, but extinction was not accepted as a valid explanation until the controversy about mammoths reached its end in 1799.

In the early 1700s the fossil remains of giant elephantlike animals called mammoths were discovered in New York, Virginia, and Siberia. Larger than any elephant alive, they were found in areas much colder than the habitats of normal elephants. Speculation on their species and origin began.

Some said they were the same species as modern elephants; others, that they were a different extinct species. Thomas Jefferson got into the act in the 1790s and proved himself a better statesman than paleontologist. He thought they were a different species with different climatic requirements, but assuming that nature would not have "permitted any one race of her animals to become extinct," he suggested that some of these beasts still existed in unexplored regions of the earth. In 1799 Georges Cuvier, now considered the father of vertebrate paleontology, finally settled the matter, proving by comparing differences in anatomy that the mammoths formed a separate extinct species.

Although the concept of extinction was accepted by 1800, the concept of evolutionary change was not. The brilliant and respected Cuvier believed, as did many others, in progressivism, the theory that the divine plan for life on earth was a series of successively more advanced worlds, which God filled with successively more advanced forms of life. Each world was destroyed by a geologic cataclysm, and few if any survivors lived to flourish in the next age. The final world, the culmination of this progressive series, was the present age of man. Cuvier believed that this final world had begun about 6000 years earlier.

In sum, many paleontologists believed that life was an immutable expression of divine wisdom, and few had grasped the uniformitarian geologists' concept of the extreme age of the earth. However, a few men—including Charles Darwin's grandfather, Erasmus Darwin, and Jean Baptiste de Lamarck—believed that new forms of life had evolved from ancient ones.

THEORIES OF EVOLUTION

An early theory of how evolution operates was proposed by Jean Baptiste de Lamarck, an invertebrate paleontologist. He had studied fossils of primitive sea animals, and perhaps their simple structure had suggested the possibility of development potential, whereas the mammoths had seemed to Cuvier so highly developed that they could only have been put on earth in their monumental form by God. At any rate, Lamarck recognized that organisms might change over time, and Cuvier did not. Lamarck also saw great significance in the fact that plants and animals seemed suited physically and in behavior to their type of life, and he attributed this suitability to natural forces, not to the wisdom and foresight of God.

FIGURE 17.3
According to the Lamarckian theory of evolution, living organisms have evolved through the inheritance of acquired characteristics. Thus the giraffe lengthened its neck by frequently stretching to reach foliage in the trees. The resulting modifications in the bones and muscles of the neck were transmitted to offspring, and succeeding generations had longer and longer necks. Existing data discredit this theory in favor of the process of natural selection.

Lamarck published his theory of evolution in 1809, the year Charles Darwin was born. Lamarck saw that animals could be fitted onto a ladder of progression from a simple marine form to man. He held that living forms were slowly changing and gradually becoming more complex. According to his theory, the mechanism of change was the *inheritance of acquired characteristics.* An individual of a species might develop certain parts of his body through extensive use or allow other parts to dwindle or shrink through disuse. Physical characteristics acquired during a lifetime could then be inherited by offspring. Characteristics that adapted an individual to its environment were used and retained (see Fig. 17.3). Characteristics that had no particular adaptive value were not used and gradually disappeared. For example, we can use Lamarck's theory to make a hypothetical scheme for the evolution of New World monkeys.

Suppose an ancient monkey ancestor with a short tail took a fancy to bananas and found that he could use his tail to steady himself in the trees. Suppose he used his tail constantly, and as a result it became more muscular and even slightly longer. These tail characteristics would then be inherited by his children, whose tails in turn would be lengthened and strengthened even further through use. The final outcome of this sequence might have been the New World monkeys with their long prehensive tails. On the other hand, the brother of that early ancestral monkey might have been less adept in the trees and found adequate food supplies on the ground. Suppose he never used his tail and it became weaker and shorter. Again these tail characteristics would be passed on to his offspring, who, like their father, fed on the ground and did not use their tails. The race sired by these monkeys might have been the tailless apes like the gorilla and the chimpanzee.

This scheme is invalid, of course, because Lamarck erred in his assumptions about inheritance. It is certainly true that physical characteristics change with use and disuse, but these characteristics cannot be inherited, because they are not encoded in the genetic material passed to the next generation. Lamarck's theory was effectively challenged by a German scientist named August Weismann, who raised more than 20 successive generations of mice and cut off their tails when they were born. The last generation was born with tails just like all the others.

DARWIN AND NATURAL SELECTION

It was about 50 years after Lamarck had published his theory that Charles Darwin publicly proposed the theoretical key to the evolutionary process. Whereas Lamarck's ideas were mostly speculation, Darwin had spent a good part of his life making observations and compiling a tremendous body of scientific information that was later used as evidence supporting his theory. He began his observations at the age of 22 aboard the British ship H.M.S. *Beagle.*

In 1831 the British navy sent the *Beagle* on a five-year voyage around the world to chart the shores of South America and the Pacific islands. Darwin collected and classified the flora and fauna at sea and ashore. The trip was an opportunity to get a panoramic view of the life and geography of many parts of the world and to acquire a depth of knowledge that few scientists could equal. Significantly, Darwin took with him a copy of *Principles of Geology* by Charles Lyell.

Everywhere Darwin looked he saw evidence of change and variation. The ostrichlike birds called rheas in northern Argentina looked quite different from those at the tip of the continent. The vegetation on the Argentine pampas appeared to have changed radically since the arrival of European civilization, for European plant species were eliminating native species. In the fossil beds near Cape Horn, Darwin discovered the remains of giant armadillos very different from live armadillos. Using Lyell's book, Darwin was able to recognize geological changes as well.

After passing through the Straits of Magellan, the *Beagle* headed up the western coast of South America and then west to the Galapagos Islands, 600 miles off the coast of Ecuador. There Darwin saw even more striking evidence of change. On the islands (which did not seem particularly old geologically) lived animals and plants distinctly different from those on the continent. There were unique species of reptiles, including giant land tortoises weighing up to 500 lbs, and land and marine iguanas, lizardlike animals 3 to 4 ft long. Furthermore, Darwin noticed slight differences between animals on different islands. He observed 14 varieties of finches that were generally alike except for the shape of their bills. Each bill type corresponded to specific feeding habits. Each of the varieties was found only on certain islands.

By the end of the trip Darwin was convinced from his observations of the validity of evolution as a means of explaining biological diversity. Furthermore, he had been greatly impressed by Lyell's book and had accepted the uniformitarian view of geological change and Lyell's theory that changes in the environment had caused extinction, migration, and modification of early forms of life. He did not yet understand, however, the mechanisms that shaped the direction of evolution.

Darwin's readings in animal breeding and human populations provided the key. He knew that new varieties of domestic animals could be bred by mating animals with desired characteristics. There appeared to be a lot of variation within species of domestic animals, and many new breeds were possible. When breeding animals, man selected the qualities that would appear in the next generation. What selective force acted in nature?

In 1838 Darwin read Thomas Malthus's essay on human population. Malthus held that man multiplied geometrically, while his food supplies grew only arithmetically. The result was a tremendous struggle for food and for life itself. Darwin realized that struggle was the selective force in nature that determined the traits of the next generation. Only the most fit could survive to breed. He called this process, the elimination of the less fit, **natural selection.**

Darwin wrote an essay on evolution in 1842 and enlarged it in 1844, but he did not publish his ideas, because he was collecting more and more evidence for his theories. But then in 1858 a naturalist named Alfred Russel Wallace asked Darwin to look over a manuscript he had written. Darwin discovered that Wallace had independently arrived at the same theory of evolution. On July 1, 1858, Darwin and Wallace jointly presented papers on evolution at a meeting of the Linnaean Society of London, and in 1859 Darwin published his classic *Origin of Species.* The book is an encyclopedic treatment of evolution, but significantly it omits any discussion of man. It was not until 1871 that Darwin was ready to publish his ideas on human evolution in *The Descent of Man.*

The elements of the Darwin-Wallace concept of evolution by natural selection are as follows:

1. Living organisms tend to multiply geometrically, producing many more offspring than can possibly survive.
2. The number of individuals of any species remains relatively constant.

TABLE 17.1
Evolutionary developments

ERA	PERIOD	EPOCH	CHARACTERISTIC SPECIES ANIMALS	PLANTS	CLIMATIC AND GEOLOGICAL EVENTS	YEARS BEFORE PRESENT (MILLION)
	Quaternary	Holocene	Modern man		Glaciations	.01
		Pleistocene	Early man		Glaciations	2.5–3
		Pliocene	Large carnivores			7
Cenozoic	Tertiary	Miocene	Abundant grazing mammals			26
		Oligocene	Large running mammals	Angiosperms dominant	Rocky Mountains raised and eroded	37
		Eocene	Modern types of mammals			54
		Paleocene	First placental mammals; modern birds			65
	Cretaceous		Climax of reptiles and ammonities, followed by extinction	First flowering plants; conifers dominant		135
Mesozoic	Jurassic		First birds; first true mammals; many reptiles and ammonites	Cycadeoids (flowering cycads)		180
	Triassic		Labyrinthodont amphibians; first dinosaurs	Abundant cycads and conifers		225
	Permian		Chondrostean fish, cotylosaurs dominant; widespread extinction of marine animals; including trilobites; many insect groups		Allegheny mountains formed	280
	Pennsylvanian (Carboniferous)		Great coal swamps; first reptiles		Continental glaciation in southern hemisphere	310
Paleozoic	Mississippian		Sharks and amphibians	Seed ferns and lycopods	Appalachian orogeny	345
	Devonian		First amphibians; fishes very abundant			400
	Silurian		Euryptids, ostracoderms	First terrestrial plants		435
	Ordovician		Brachiopods and first fishes; marine invertebrates dominant	Marine algae abundant		500
	Cambrian		First abundant fossils of marine life; trilobites and brachiopods dominant			600
Precambrian	(Period divisions not well established)		Relatively few and primitive fossils		Continental glaciations	

3. Therefore, there is a struggle for survival in which many organisms die.
4. Individuals of the same species vary slightly in their physical and behavioral characteristics.
5. Some variations are particularly beneficial in a certain environment, making an organism more fit to survive. These variations are called adaptations. Examples are the white coat of an arctic animal or the ability of a desert-dwelling organism to live with little water. The more fit organisms with favorable or adaptive variations survive in greater numbers and produce more young than those with less favorable variations.
6. Offspring tend to inherit the characteristics of their parents. Since individuals with favorable variations produce more offspring than the less fit, in the next generation there is a greater proportion of individuals with favorable variations. In successive generations advantageous traits continue to be preserved and accumulated by the natural selection of those most fit to live and reproduce.

In the light of present-day knowledge, the Darwin-Wallace theory of evolution is not without weaknesses, but it successfully identifies and explains natural selection as the process that shapes evolutionary change. One problem with the theory is that it does not explain why or how offspring can inherit the favorable variations of their parents. In 1859 genes and chromosomes were unknown. It remained for the biologists in the following 70 years to reveal the genetic basis of evolutionary development.

GENETIC MECHANISMS OF EVOLUTION

In the second phase of the development of evolutionary thought, the major discoveries were the mechanics of inheritance and the sources of variations within species. Gregor Mendel first understood the basic laws of inheritance from his experiments with garden peas (discussed in Chapter 14). Mendel presented a paper summarizing the results of his experiments in 1865, but it went unnoticed and was quickly forgotten. In 1900 Mendel's work was rediscovered by three different geneticists working independently: Hugo De Vries, Carl Correns, and Erich Tschermak von Seysenegg. Mendel's work provided a scientifically based explanation for Darwin's assumption that offspring can inherit traits of their parents, and it also demonstrated how variations between individuals could arise.

When he rediscovered Mendel's work, De Vries was studying variations in the evening primrose. He found that daughter plants occasionally exhibited characteristics radically different from those of the parents. He called these new traits mutations, a term that has come to mean spontaneous changes in genes or chromosomes. Although De Vries's mutations turned out to be only an unusual form of genetic segregation, the term he coined became accepted.

In 1910 Thomas Hunt Morgan discovered the first true example of De Vries's mutations. He published a paper in *Science* magazine that began: "In a pedigree culture of *Drosophila* [the fruit fly] which had been running for nearly a year through a considerable number of generations, a male appeared with white eyes. The normal males have brilliant red eyes." The find was one of Morgan's many contributions to genetics. He also studied gene linkage, crossing-over, and chromosomal aberrations, which together with mutations are recognized as key sources of genetic and evolutionary variation.

Since Darwin, the theory of evolution has undergone many changes, although the modern concept still relies on Darwinian natural selection. We now recognize that evolution is a balance between variation and the molding of variation into adaptive channels. Four processes—recombination, fertilization, mutation, and chromosomal aberration—produce genetic variations that are then modified over generations by natural selection and reproductive isolation. The balance may also be affected by three accessory processes. Migration of individuals takes place from one population to another. Hybridization between closely related species can change the amount of variability within populations. And chance, acting in small populations, may alter the way in which natural selection shapes the direction of evolution. These principles are illustrated in Fig. 17.4 and will be examined in the following sections of the chapter.

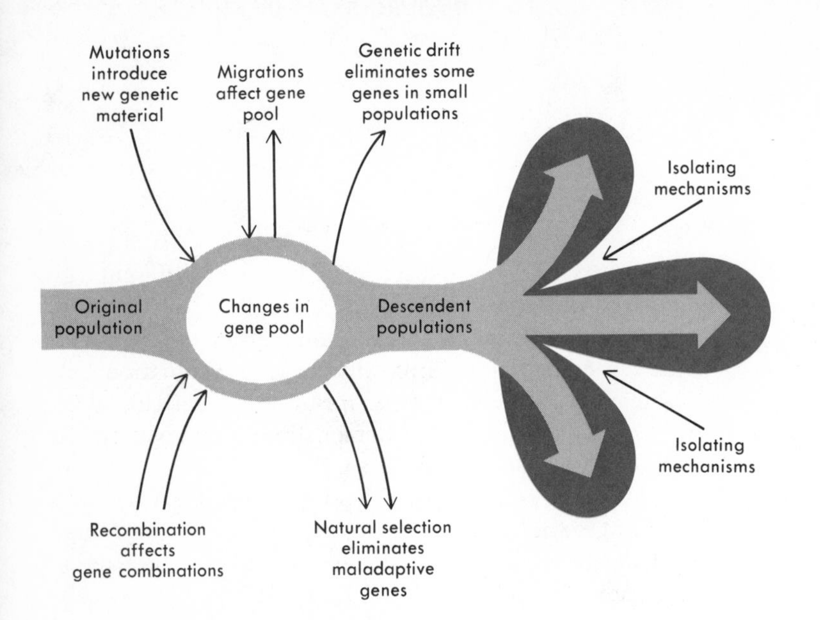

FIGURE 17.4

Evolution as a process not of changes in an individual during his lifetime, but rather of changes in the genetic makeup of populations over many generations. Changes introduced into the gene pool through mutation, recombination, and migration will be eliminated or maintained by the process of natural selection. If groups within an initially unified population are separated by some means and prevented from exchanging genes, these groups will diverge gradually and evolve into new species with distinct gene pools.

Sources of variability

Members of any population of sexually breeding organisms differ from one another to a greater or lesser degree. Some differences, like height or hair color, are immediately obvious. Others, like the inability to produce the hormone insulin, would require a medical examination. Nevertheless, all variations are produced by two kinds of factors: environmental influences, such as nutrition, disease, age, or use; and genetic makeup. Each organism has a genetic constitution (genotype) and a physical makeup (phenotype). The genotype remains constant throughout life, but the phenotype is always changing from birth to death.

Genetically determined phenotypic traits like eye color do not change, but many variations, such as weight or musculature, are the product of an individual's activities or environment. Some purely phenotypic traits, such as injury-produced paralysis, can be very damaging; others can be adaptive and beneficial. For example, a species of buttercup that lives in the water at the edge of ponds can be induced by its environment to produce different kinds of leaves. On the surface it puts out broad whole leaves, but underwater it produces dissected or many-segmented leaves. The significance of the two leaf variations is that each in different circumstances provides an optimum balance of the two essentials for photosynthesis—light and water. On the surface there is plenty of light, but water loss is a problem. Since big leaves have a smaller surface volume ratio than dissected leaves, big leaves provide less space where evaporation can occur. Underwater, light is the limiting factor. Since dissected leaves have more surface area than the big ones, they provide more space with which to absorb the diffuse underwater sunlight.

Traits that are exclusively phenotypic cannot be ininherited and therefore cannot be part of evolutionary development. Lamarck believed they could, of course, but there is no way externally induced variations can become part of a genotype. In fact, the only inheritable characteristics are those encoded in the chromosomes carried by eggs and sperm. **Somatic mutations,** gene changes in cells of the body, cannot be inherited because they are local and do not affect gamete genotypes.

The raw material of variation and of evolutionary change is the variety of genes and gene combinations that can occur within a species. Genetically determined variations arise in a number of ways.

Recall from Chapter 13 that eggs and sperm are formed through meiosis. During the first meiotic division, synapsis occurs between homologous chromosomes, and genetic material is exchanged. Genes located close together on the same chromosome are said to be linked because they usually are not split up during this exchange. However, many genes are separated as some cross over to another chromosome. An infinite number of gene arrangements can come from this sort of genetic scrambling. When a new organism is created by the fertilization of an egg, the union of genetic material from the two parents is called recombination. Whereas linkage and crossing-over scramble the genes of one parent, recombination mixes the genes of two parents, producing new genotypes and phenotypes.

Recombination, linkage, and crossing-over are mechanisms extremely important in the generation of new variants to be put through the sieve of natural selection. They occur each time spermatogenesis, oogenesis, and fertilization occur, and in large populations they can generate almost limitless gene combinations. One might think that these mechanisms merely hash around the same old genes without generating anything new. However, the rearrangement of genes can produce startling effects, because genes influence one another. Certain combinations can be highly adaptive; others can be neutral or maladaptive within the context of a specific environment.

Another major mechanism that produces new variants is gene mutation. Mutations are caused by such environmental factors as radiation, heat, mechanical trauma, and chemicals. These forces are naturally part of the environment, and they generate mutations at low but constant rates that can be calculated experimentally.

Limitless mutations are possible, but most are detrimental to the fitness of an organism. Furthermore, the greater the phenotypic effect produced by a mutation, the more detrimental that mutation tends to be. It is not surprising that mutations are usually not adaptive. In a line of organisms bred primarily for fitness, as wild species are, any alterations would be likely to reduce fitness. Because mutations are generally maladaptive, the majority are quickly selected out and are thus insignificant from the point of view of evolution.

Under certain circumstances, however, mutations are adaptive and can become raw materials for evolutionary change. Sometimes a mutation modifies the expression of other genes. For example, if the combination of the mutation and the other genes produces traits more adaptive than those generated without the mutation, and if the mutation and other genes become linked, then the new adaptive qualities can be inherited by offspring. In other instances, mutations may be adaptive in a particular environment. For example, when a certain mutant of the fruit fly *Drosophila funebris* is raised at 16 or 29°C, it is less fit than nonmutants, but in a 25°C environment, it is more successful than the normal flies.

Another variation-generating process is chromosome aberration, the alteration of gene sequence on a chromosome and/or the alteration of the number of the chromosome complement. As with gene mutations, these must occur in the gametes to be inherited. The possible chromosome structural aberrations are deletion, duplication, inversion, and translocation. There are two types of chromosome number changes: **aneuploidy** and **euploidy.** Sexually reproducing organisms have two sets of chromosomes, one from each parent. Each set contains n chromosomes, and thus the total chromosome complement is $2n$. Aneuploidy occurs when chromosomes are lost or gained, so that the total complement is less or more than $2n$—for example, $2n + 1$ or $2n - 2$. Euploidy occurs when complete sets of chromosomes are gained or lost, and the chromosome number is a multiple of n other than 2—for example, $3n$ or n.

Since chromosome aberrations involve changes in a whole sequence of genes, they tend to produce dramatic effects that are usually detrimental to survival. If aberrations are not lethal, they may cause an individual to be infertile, because unmatched pairs of chromosomes disrupt synapsing and division during meiosis and thus cause abnormal gamete production. Nevertheless, some chromosome aberrations have undoubtedly produced favorable variations and become part of evolutionary development.

Not all mutations are acted on immediately by natural selection. Deleterious mutations may be carried for generations because they are recessive and masked by the dominant allele or because they are closely linked to a gene of particularly high selective value. The sum of these unfavorable mutations not yet eliminated by natural selection is called the **mutational load.**

Recombination, linkage, crossing-over, mutation, and chromosomal aberration can generate limitless variations, but much of the potential genetic variability of a population is unexpressed. Adaptive and maladaptive recessive traits and gene combinations often do not appear in phenotypes unless inbreeding occurs. Moreover, many gene varieties, including damaging ones, are retained in a population because heterozygous individuals have a distinct selective advantage. This phenomenon, known as **heterozygote superiority** or **hybrid vigor,** is fairly common, and an example can be seen in certain Negro populations in Africa. As we have seen in earlier chapters, sickle-cell anemia is a genetically induced blood disease, the result of a genotype homozygous for two mutant alleles. Homozygotes have severe anemia and usually die before reproductive age. One might expect that if all homozygous individuals died, the mutant allele would soon be eliminated. However, because heterozygotes exhibit a resistance to malaria, they have a distinct advantage in the malarial regions of Africa. Therefore the mutant allele is retained, despite its potentially lethal effects in the homozygous condition.

Diversity in populations

We have seen how new genotypes and new variations arise among individuals of a population. But evolution is not a process that occurs in individuals—it occurs in populations over many generations. The relationship of individual genotypes to the genes held collectively by a population, as well as the mechanisms by which a population's collective characteristics change during evolution, are shown in the study of population genetics.

An individual's traits are determined by his genotype, and in somewhat the same way the character of a population is determined by its **gene pool,** the sum of all the genes held by the members of the population. The frequency of a given phenotypic characteristic in a population depends directly on the dominance of the genes that determine it and the frequency of those genes within the gene pool.

For example, let us make the unlikely assumption that only one pair of alleles determine coat color for a population of mice. Suppose that a dark brown color is determined by the dominant allele B and a tan color by the recessive gene b. Suppose further that the frequency of B in the gene pool is 80 percent. Since B and b are the only genes determining coat color, the frequency of b must be 20 percent.

How do these frequencies affect the appearance of the population as a whole? We can determine genotype and phenotype frequencies of the mouse population by using a Punnett square. Assuming that coat color is in no way sex-linked, and that 80 percent of the population's eggs and sperm will contain B and 20 percent will contain b, we can set up the axes of the square and determine the genotype frequencies in the first filial generation:

		Eggs	
		0.8 B	0.2 b
Sperm	0.8 B	0.64 BB brown	0.16 Bb brown
	0.2 b	0.16 Bb brown	0.04 bb tan

In the first fiilial generation there will be 64 percent homozygous dominant, 32 percent heterozygous, and 4 percent homozygous recessive mice in the population. Therefore 96 percent will be brown and 4 percent tan.

Tan mice would be rather rare, and one might expect that the tan trait would disappear after a few generations. We can see if this starts to happen in the second filial generation.

First, we must calculate the number and kind of eggs and sperm the first filial generation can produce. BB individuals (male and female) are 64 percent of the population and will produce 64 percent of the gametes. 100 percent of their gametes will contain the gene B. Therefore, they will produce 0.64×1 or 0.64 B gametes. By the same reasoning, Bb individuals make up 32 percent of the population and will produce 32 percent of the gametes, half of which (0.16) will carry B and half (0.16) b. The bb individuals will produce 4 percent of the gametes, all of which will carry b. Adding up the kinds of gametes:

0.64 B	0.16 b
0.16 B	0.04 b
0.80 B	0.20 b

These are the same frequencies as in the parental generation. Thus the genotypes and phenotypes in the second filial generation will be exactly the same as in the first. Tan mice will not be eliminated. If nothing happens to alter the gene frequencies, tan mice will continue to appear at the same low frequency for generations.

In 1908 G. H. Hardy and W. Weinberg independently established a mathematical relationship for gene and genotype frequencies within a population, which has been a valuable tool in the study of population genetics and evolution. Called the Hardy-Weinberg law, the equation is expressed

$$p^2 + 2pq + q^2 = 1.$$

It is an algebraic expression of the relationships we have been using in the Punnett square calculations. Literally, it means that the total incidence of a pair of alleles in a population equals the sum of all the homo- and heterozygous genotype frequencies for those alleles. In less technical terms, it states that the original genetic variability of a gene pool will be maintained at the same frequencies from generation to generation. However, it is obvious that the gene frequencies in a gene pool would not remain the same if any genes were added or lost. The Hardy-Weinberg law holds true only in ideal, genetically stable conditions.

For the Hardy-Weinberg genetic equilibrium actually to occur, five requirements would have to be satisfied to ensure that a population's gene pool frequencies would remain constant.

1. The population must be very large so that chance cannot affect gene pool frequencies.
2. There must be random breeding within the population.
3. There must be no emigration out of or immigration into the population; that is, the population must be effectively isolated.
4. There must be no mutations, or the mutation rate must equal the reverse mutation rate.
5. All haploid and diploid genotypes must have equal selective value.

Population size is significant because random events can radically change gene frequencies in small populations. In a population of 10, for example, if one individual has gene A and that individual is suddenly killed, the frequency of A drops immediately from 10 percent to zero. Random gene frequency fluctuations in small populations are called **genetic drift.** Large populations are protected against genetic drift because many organisms carry each type of gene, and the fate of a gene is never dependent on one organism. Large populations can and do exist, and thus the first requirement can be satisfied.

Random breeding implies that an individual is equally likely to breed with each member of the opposite sex in the population. If certain organisms are excluded from breeding or others breed only among themselves, genes will be lost or gene frequencies altered. Because such factors as personal preference or proximity do have these effects on breeding patterns, a random breeding situation could never occur.

Migration and immigration would have obvious effects on the gene pool, resulting in the loss or gain of genes. The absence of all emigration and immigration could occur in a population in nature.

Mutations occur naturally at low but steady rates, and thus there could be no population in which there was no mutation. The balancing of mutation with reverse mutation would be highly unlikely, because reverse mutations are relatively rare. The effect of mutation in changing gene frequencies and disrupting genetic equilibrium is called **mutation pressure.**

The fifth requirement, equal genotype fitness, would also be impossible. Although genotypes may be adaptive or maladaptive in subtle ways, variation implies varied fitness. The forces in the environment that eliminate unfavorable variations, thereby altering gene frequencies, are called **selection pressures.**

Although some of these five conditions for a Hardy-Weinberg equilibrium could occur, most are improbable or impossible. Gene frequencies in a population do not remain the same. In fact, evolution can be defined as frequency change in the gene pool. Over generations, the processes of genetic drift, restricted breeding, migration, mutation pressures, and selective pressures act on the gene pool. As the genes that are adaptive are retained and others are eliminated, the characteristics of a population are continually modified. Thus the significance of the Hardy-Weinberg law is that it cannot hold for all genes in all populations in all generations; in essence, it proves the fact of evolution.

SELECTION PRESSURE

Selection pressure is a major cause of gene pool modification, and it was this selection pressure that was critical in Darwin's concept of evolution. Mutation, genetic drift, and migration are significant to evolution because they give rise to new gene pool frequencies and thus to new variations. Selection pressure alters gene pool frequencies by eliminating nonadaptive genetic traits.

The effect of selection pressure on gene and genotype frequencies can be demonstrated mathematically. As an example, let us use the mouse population discussed earlier, in which the frequency of the brown coat color gene B was 80 percent and the frequency of the tan coat color gene b was 20 percent. Suppose this time that the brown color is less adaptive in some way, and that selection pressure has reduced the frequency of B to 70 percent. Then the frequency of b will be 30 percent, and we can use a Punnett square to determine genotype frequencies in the first filial generation:

		Eggs	
		0.7B	0.3b
Sperm	0.7B	0.49BB brown	0.21bB brown
	0.3b	0.21Bb brown	0.09bb tan

Comparing these results with those of our earlier Punnett square calculation, we see that selection has reduced the proportion of BB individuals from 64 percent to 49 percent in only one generation. The proportion of bb individuals has risen from 4 percent to 9 percent. If selection pressure continues to act on the gene pool in this manner, the proportion of BB individuals will become very small, and tan mice will eventually predominate.

Although natural selection is a constant force, its pressure on the gene pool may be very slow in a constant environment. However, if an organism's environment changes, selection pressures will increase in proportion to the environmental change. Selection will strongly favor those who are able to adapt to the new conditions, and their genes are much more likely to become the predominant ones in the gene pool.

There are three types of selection in populations: **stabilizing selection, directional selection,** and **disruptive selection.** Each is dependent on environmental conditions.

Stabilizing selection occurs when the environment is relatively constant for a long period of time. It favors average individuals and tends to keep the gene pool genetically constant by eliminating extreme variations. For example, a severe storm killed a large number of a flock of sparrows. When examined, the dead birds proved to have abnormally long or short wings in comparison with those of the average sparrow. Selection had acted against extreme variation in wing length.

When the environment is changing in a particular direction—for example, becoming colder or including more predators—directional selection occurs. Organisms with traits that help them adapt to the new conditions are favored. Others, whose traits are at the opposite end of the variation spectrum, become the victims of selection pressures against them. The character of the population as a whole shifts in one direction. Directional selection could consist of the replacement of one allele by another in the gene pool, but it generally involves more than a pair of alleles.

A classic example of this type of selection is the evolution of the legs of the horse (Fig. 17.5). The earliest horse ancestor was a small animal about two feet high. Because there was constant selection pressure for individuals who could run swiftly to avoid predators, long

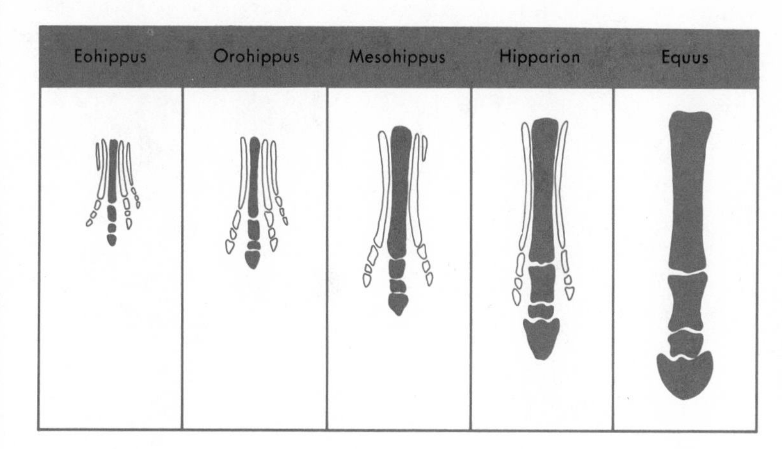

FIGURE 17.5

Directional evolution as illustrated by the changes that occurred in the forefoot of the horse over a period of 60 million years. Adapting to changing environmental conditions, the horse evolved from a terrier-sized, five-toed creature who browsed on leaves and soft plants to the relatively large modern horse, Equus caballus. *Increasing adaptation for life on open plains was achieved through lengthening of the forelimbs and reduction of five toes to a single hoof.*

legs adapted for running were distinctly advantageous, and horse evolution moved steadily in that direction.

In a more modern example, biologists have discovered bacterial strains with many times the resistance to antibiotics of usual bacterial strains. In the laboratory it has been possible to evolve highly resistant strains by introducing into their environment a strong selection pressure in the form of increasing amounts of an antibiotic. The first step in the experiment was to subject normal colon bacteria to a high dose of an antibiotic. Most bacteria died, but mutants with resistance to the particular antibiotic survived and were transferred to a new medium with more concentrated antibiotic. The experiment was repeated, and eventually through directional selection a strain evolved that could withstand 250 times the dose tolerated by the original normal bacteria.

Disruptive selection occurs when the environment changes in different ways in different parts of a population's distribution. Selection pressure favors those organisms in each area which are adapted to the new changes and selects against average individuals. Thus two or more diverging adaptive groups arise out of one population. For example, in the Sacramento Valley in California a population of sunflowers was studied over a 12-year period. The genetically unified population was hybrid between two species, and it occupied an area 300 feet long and 20 feet wide in a ditch. In five years the population had split into two subgroups with an area dividing them in which very few sunflowers grew. At the bottom of the ditch, on a site that was relatively damp, was a group that looked generally like the original hybrids, although it had some of the characteristics of one of the hybrid parents. Five hundred feet away on a higher, drier site was the other group, which generally had the characteristics of the other hybrid parent. Divergent selective pressures from drier and wetter environmental conditions induced a division of the hybrid population and a divergence of characteristics.

Thus selection pressures shape populations according to the demands of a specific environment. A sunflower's ability to live in wet soil becomes a selective advantage in a ditch, whereas it is less adaptive in a drier site only 500 feet away. As the genes for certain traits in specific areas and climates are accumulated in a population's gene pool, the population becomes better adapted to its world.

ADAPTATIONS

Genetically controlled traits that improve an organism's survival and breeding potential are called **adaptations.** Some adaptations are slight; others are highly specialized and have evolved over long periods of constant selective pressure. However, all are beneficial only in the context of an organism's specific life style and environment. The variety and quality of adaptations in nature are both astounding and intriguing.

Adaptations affect many different aspects of life. The angler fish's wormlike appendage over his mouth with which he "fishes" for prey, the arctic hare's white warm coat, the salmon's ability to find a spawning ground in the river where she was hatched, the kangaroo's pouch in which she protects her young until they are able to live on their own, the various forms into

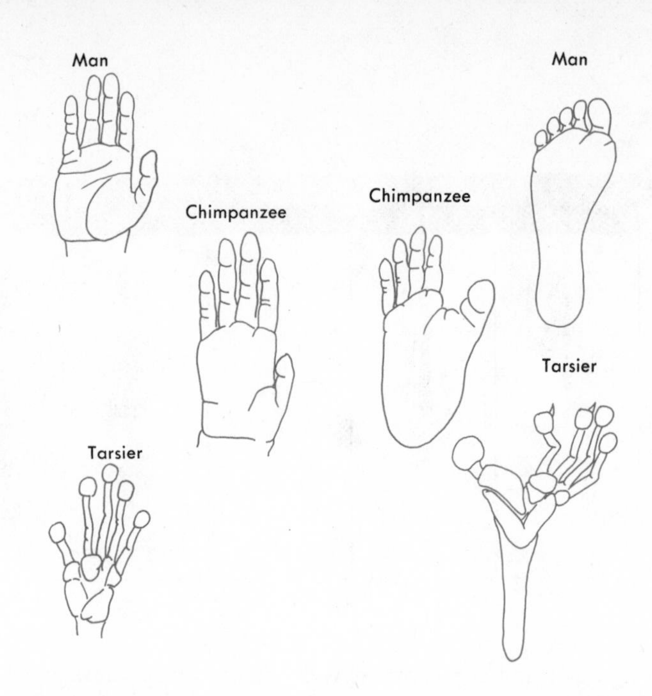

FIGURE 17.6
The evolution of the primate hand and foot as a reflection of adaptation to a variety of functions. The tarsier, a jumper, has disks on the tips of its digits that help the animal to hold on after leaping. Although all primates are able to grasp with their fingers, the chimpanzee is much more dexterous than the tarsier in manipulating individual digits and the thumb. Man's larger and more complex brain has enabled him to perfect this dexterity and to use his opposable thumb in a precision grip that can serve countless purposes. His big toe, however, is no longer opposable. It has been strengthened and aligned with the other digits to provide a specialized solid base for walking and standing.

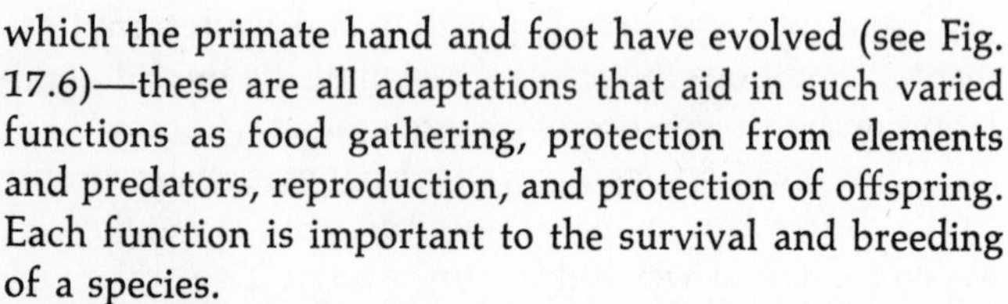

which the primate hand and foot have evolved (see Fig. 17.6)—these are all adaptations that aid in such varied functions as food gathering, protection from elements and predators, reproduction, and protection of offspring. Each function is important to the survival and breeding of a species.

The need to find adequate food supplies has resulted in a multitude of adaptations throughout the plant and animal kingdoms. One of the most unusual adaptations is that utilized by the woodpecker finch *(Camarhynchus pallidus)*, one of the 14 species of finch Darwin discovered on the Galapagos Islands. The finches that first arrived on the Galapagos Islands found adequate food supplies, but as the population grew and competition for existing food supplies increased, some of the finches turned to new forms of food to supplement the normal source, seeds. One of these, the woodpecker finch, eats grubs that live in the bark of trees. However, it does not have the strong beak and long tongue of the woodpecker, so it extracts its food by using a probe—a stick or a cactus spine.

Many organisms living in harsh climates have special traits that help them to survive. In the desert, for example, there are cacti whose unusually thick tissues provide water storage area. The kangaroo rat solves the water shortage problem in a different way. Its body chemistry is regulated so that it can derive the only water it requires from its dry food, and consequently it never drinks at all.

Another common type of adaptation makes males and females of the same species more easily recognizable to each other, thereby ensuring that mating will occur and perpetuate the species. In many species of birds the female chooses a mating partner. The male's brightly colored and distinctive markings are a factor in helping the female to recognize and respond to one of her own kind, although they make the male more visible to predators as well. Since one male may impregnate several females, the male of the species is more biologically expendable.

The most visible and dramatic adaptation is protective coloration, external markings that help an organism to evade possible predators. There are several forms of this adaptation. Camouflage, or **cryptic coloration,** is an appearance that blends in with the organism's usual surroundings, making it difficult to see (Fig. 17.7). Female birds are usually a rather dull brown or grey, in sharp contrast to the sometimes gaudy finery of their mates. Females spend a lot of time on the nest during the breeding season, and their dull coloration protects them and their young from being seen by predators.

An interesting example of the importance of camouflage and its effects on populations can be seen in the English land snail *Cepaea nemoralis.* The snail's range includes meadows and different kinds of woodlands. Its shell may be yellow, pink, or brown, and in addition it may or may not be marked with black bands. In the meadows and woods without undergrowth, where the

FIGURE 17.7

Cryptic coloration. The sundial flounder blends almost perfectly into its sandy background. Such cryptic coloration is an adaptive characteristic that helps animals escape predators. Some animals can contract or expand their pigment cells to become lighter or darker, depending on their surroundings.

FIGURE 17.8

Batesian mimicry. The monarch butterfly is a highly toxic species because of its diet. It collects the nectar from the poisonous milkweed flower and concentrates this poison in its tissues. A predator that eats a monarch gets sick and subsequently learns to avoid its color and patterning. The viceroy, which is smaller but similarly marked, is not protected by any toxic juices of its own, but its mimicry of the monarch's coloration scares off would-be predators.

background is fairly uniform and green, there is a very high proportion of yellow unbanded snails. Here birds easily find and eat any banded dark snails. However, in woods with underbrush, there is a much higher proportion of snails with banded shells, which blend with the mottled forest floor. Here it is the lighter snails that are discovered and eaten. The banded and nonbanded forms have existed since ancient times, so this situation is not a passing phase in snail evolution. Camouflage indeed provides a great selective advantage, but if an organism inhabits more than one environment, selection pressure will act to maintain color variation within the population, despite the disadvantage arising from the right color in the wrong environment.

Another type of protective coloration is warning or **aposematic coloration.** Organisms that have unusually effective defenses or are particularly unpalatable to the taste advertise their presence with brilliant or distinctive markings. Predators who have had previous experience with these organisms quickly learn to recognize the markings and steer clear. For example, bees with their painful stings, skunks with their potent scent, and monarch butterflies with their unpleasant taste all have conspicuous markings that warn prospective predators. Organisms similar in appearance to an aposematic animal acquire some protection by virtue of being confused with and mistaken for the unpleasant animal. Mimics are fairly common in nature, and two types of mimicry, Batesian and Mullerian, have been observed. In **Batesian mimicry,** a palatable, otherwise unprotected organism closely resembles an aposematic one. The classic example of Batesian mimicry is the resemblance of the viceroy butterfly *(Limenitis archippus)* to the distasteful and distinctively marked monarch butterfly *(Danaus plexippus)*, as illustrated in Fig. 17.8.

The effectiveness of Batesian mimicry was demonstrated by experiments with the viceroy, the monarch, and Florida scrub jays *(Cyanocitta coerulescens)*. Young jays were allowed to feed on viceroys, and the jays did so with no hesitancy. Then milkweed-feeding monarchs were introduced into the jays' environment. Initially the jays tried to eat the monarchs, but one painful vomit-inducing experience was enough to bring on consistent avoidance thereafter. When 50 monarchs had been introduced, viceroys were again fed to the jays, but the jays avoided them as consistently as they had the monarchs.

In **Mullerian mimicry** two aposematic species resemble each other. Each species is actually distasteful or has effective defenses in its own right, but the confusion of the two by prospective predators reinforces the negative association generated by their appearance. Furthermore, if the two species looked different, young predators would ultimately learn to avoid each separately. With Mullerian mimicry, predators learn to avoid only the type of markings, and fewer individuals of each species are lost during the learning period.

Because plants and animals living in the wild must be well adapted to their environment, their characteristics usually have some obvious adaptive value. The cactus's spines keep away thirsty animals who might otherwise use the succulent cactus tissues as an easy source of water. The lion's claws and teeth enable it to catch and tear apart its prey. However, some characteristics seem to be exceptions to this rule. Although the spots on the back of the leopard frog *(Rana pipiens)* do serve as camouflage, the markings are not consistent throughout its range. Frogs from New York are quite different from those in Florida. Could the differences possibly be a product of different selective pressures in the two areas? It seems unlikely, and differing characteristics like these, which seem to have no adaptive significance, are called neutral or nonadaptive traits.

Because adaptation is at such a premium in wild populations, the existence of nonadaptive traits is surprising and their origin a matter of controversy. It is possible that there is no such thing as a nonadaptive trait. Characteristics that seem to be of no particular selective value may, in fact, be found of significance if the organism's life activities are studied in detail.

For example, a house mouse has a specific number of whiskerlike bristles over each eye. Although there seems to be no particular adaptive significance in this fact, bristle number is remarkably constant, and mice with fewer bristles are less viable than normal mice throughout the species. In laboratory studies to determine the selective value of bristle number, normal mice and mice from whom bristles had been removed were run through mazes. The latter were far less successful than the normal mice, indicating that the bristles were of value, probably because they help mice to navigate the dark narrow passageways in which they live.

In other cases it has been found that nonadaptive traits are the secondary effects of a genotype that has great adaptive value. For example, the possible colors for onion bulbs—white and red—seem to have no adaptive significance. However, only red bulbs contain protocatechnic acid and catechol, two chemicals that are poisonous to the spores of the smudge fungus *Colletotrichum.* If the smudge fungus is introduced into the environment, red-bulbed onions with fungus resistance will have a great selective advantage over the white-bulbed variety.

In both instances, mice and onions, what appeared to be nonadaptive traits proved to be either adaptive or associated with an adaptive trait as a secondary characteristic. Therefore neither was really neutral. It has been suggested that, if nonadaptive traits actually do exist, they become established because they are closely linked genetically with a trait of great adaptive value, or that they arise through genetic drift. If the latter, then chance is the agent of their increased frequency in the population, rather than selection pressure.

Speciation

In previous sections we discussed the mechanisms by which a single species becomes modified during evolutionary development. The generation of new species from pre-existing ones is a process called **speciation.** We shall now consider the nature of a species and the means by which new species can arise.

Organisms are generally classified in one species because of their similar appearance. However, because organisms do vary, it is sometimes difficult to establish an organism's species by physical comparison alone. The most precise definition of species is a group of similar organisms that (1) occupy the same or contiguous regions, (2) live at the same geologic time, and (3) reproduce sexually in nature with members of the same group but not with members of other groups, and produce fertile offspring. The third point means that under normal conditions a species must be isolated reproductively from all other species.

As there is variation between species, there is also variation within species. Some species subgroups may have very distinctive characteristics and are classified as subspecies or races. No matter how distinctive they are in appearance, groups are classified as subspecies if they breed with other members of the species as a whole. The concept of subspecies can be explained in terms of **gene flow,** the transfer of genes from one population's gene pool to that of another. If two populations interbreed, gene flow occurs between them, and because of this fact, the populations are considered to be subspecies rather than separate species.

During the course of evolution, speciation has occurred in two ways. The first and by far the slower (but more common) way is through a process of division, isolation, and modification of subpopulations derived from a single population. The second way is by hybridization, the generation of a new species by interbreeding between two closely related species. Hybridization occurs in nature only under unusual circumstances. It is important to keep in mind, however, that a new species is created only when reproductive isolation has been established under normal conditions.

DIVISION-ISOLATION-MODIFICATION

Speciation is essentially the establishment of a new gene pool, isolated by lack of reproductive exchange and different from all other gene pools. The first step is the creation of new gene pools from the gene pool of a single population. If this step is to be accomplished, the single population must be divided in such a way that the new subgroups will not interbreed. Several processes may be involved.

As noted in the discussion of selection pressure, environmental changes modify selection pressures. If the environment changes in different ways in different sections of a population's range, migration and disruptive selection will act on the population, dividing it into groups adapted to each new set of conditions. If these groups move very far apart or are separated by some barrier, they will become geographically isolated. Geographical isolation can be achieved without environmental changes if a few individuals or seeds migrate to a relatively inaccessible spot like a new island or valley. Once isolation is complete, the gene pools of the subpopulations are completely separated, and the populations are said to be **allopatric** (having different distributions).

In rare cases groups can become isolated in the same range. In certain species of insects the mother lays her eggs on or near a specific food source; offspring tend to mate with others raised on the same food source and to lay their eggs there. Examples of these specialized preferences can be seen in subgroups of the American two-winged fly, *Rhagoletis pomenella.* The two types of larvae feed on apples and blueberries, respectively. Interbreeding between the two groups probably never occurs in nature, even though the only apparent difference between the two groups is size.

Whether subpopulations become separated by geographical or other mechanisms, they must be divided so that they do not breed together. The next and last step in speciation is the evolutionary modification of the groups so that they become genetically distinct.

The establishment of new groups in itself often gives rise to gene pool frequencies different from those in the original population. If a new population is sired by only a few individuals, genetic drift plays a significant role in gene pool modification. If a new population arises by means of disruptive selection and isolation, the majority of individuals will have traits adaptive to their new environment and therefore will differ somewhat from the parent or other offshoot populations in other environments. Once isolated, the population's gene pool will become modified as selection acts on new traits arising from recombination, mutation, and chromosome aberration.

If isolation is not complete and interbreeding between populations occurs at the border of their ranges, subspecies develop. For example, the common frog *Rana pipiens* can be found from Canada south to Panama on the eastern half of North America. Frogs from Quebec are distinctly different in their markings from Pana-

FIGURE 17.9

Subspeciation in the North American leopard frog Rana pipiens. *Ranging from Canada to Panama, these frogs are considered one species, but populations from different regions have sufficiently different markings to be classed as distinct subspecies.* Rana pipiens sephenocephala *(top) from the southern United States differs from* Rana pipiens pipiens, *a more northern subspecies, in that it has a white spot on its eardrum and variant markings on its belly and legs.*

manian frogs, and intermediate types from intermediate areas can be distinguished as well. Although different in appearance (Fig. 17.9), frogs from neighboring areas do occasionally mate and produce viable offspring. For that reason, all the different varieties are considered subspecies of *Rana pipiens.* Frogs from distant ranges do not meet in the wild, but it is interesting to note that if they are mated in a laboratory, the offspring do not survive beyond the early stages of embryological development. This fact indicates that the genetic makeup of distant subspecies is different enough to result in genetic dysfunction when their gametes are combined. If there were no intermediate subspecies, the distant subspecies would be considered separate species of frog, because when mated they are unable to produce viable offspring.

If isolation between divergent populations is complete, or if a subspecies moves in a course of particularly radical adaptive modification, genetic barriers arise that inhibit successful intergroup breeding. If this continues, it first inhibits and then prevents gene flow. When gene flow is blocked, speciation is complete.

This idea can be expressed in a more organic way. The barriers to gene exchange are called reproductive isolating mechanisms. The test of speciation occurs in nature if and when two newly evolved populations become **sympatric** (occupying the same range). If reproductive isolating mechanisms block successful reproduction between sympatric populations, the populations are distinct species.

Adaptive radiation The diversification of one species into a variety of species with relation to ways of maintaining life and reproducing successfully is the process called **adaptive radiation.** Because of limitations on food and nesting grounds, etc., only a certain number of organisms can occupy a habitat. If two or more species with the same requirements for survival try to occupy a single area, the less-adapted species will be eliminated. As noted earlier in this chapter, Darwin observed this kind of species elimination in Argentina, where plants introduced from Europe were driving out the native species. If there is competition for a habitat, it is a great selective advantage for subpopulations to develop new

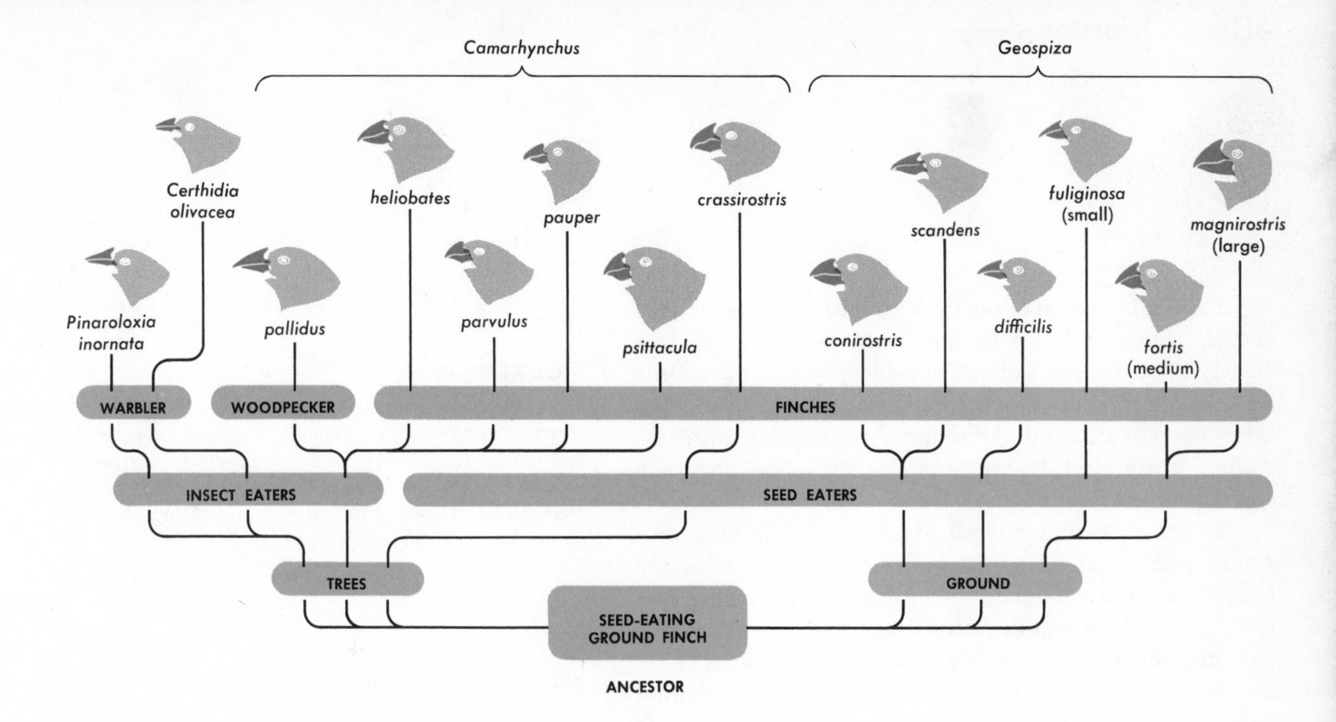

specialized habits and characteristics, which later may become the basis of new species differences.

Note that evolution can be studied at two distinct levels. At the microevolutionary level, mutation, migration, genetic drift, and natural selection are significant forces in effecting changes of gene frequencies. At the macroevolutionary level, adaptive radiation occurs via reproductive isolation and hybridization.

One of the best-known examples of adaptive radiation is that of the 14 species of finch that Darwin discovered on the Galapagos Islands. Though very similar in body markings, they differed considerably in the shape and size of their beaks (Fig. 17.10). Probably originally descended from a single pair of finches, the various species each developed specialized feeding habits. As a result, some of the separate species were able to occupy the same islands without competing for food.

Speciation probably occurred as small groups of individuals flew to a new island where, isolated for many generations from the original population, they developed new feeding habits and new adaptive beak types.

REPRODUCTIVE ISOLATING MECHANISMS

Reproductive isolating mechanisms prevent successful interspecies breeding. They have been classed in two groups: **prezygotic mechanisms,** which inhibit mating and egg fertilization, and **postzygotic mechanisms,** which negatively affect the viability and fertility of hybrids. The summary of these mechanisms given in Table 17.2 is expanded in the explanatory paragraphs that follow.

FIGURE 17.10
Darwin's finches. The 14 species of Darwin's finches evolved by a process that initially involved divergence of populations isolated on different islands. Later, when sympatry occurred (that is, when the various species came to occupy the same range), these differences were intensified, because natural selection favored the evolution of divergent feeding and nesting habits that minimized competition. Beak differences reflect diverse eating habits and function in species recognition.

TABLE 17.2
Reproductive isolating mechanisms

Prezygotic mechanisms
1. Isolation by habitat
2. Seasonal or temporal differences in mating patterns
3. Behavioral differences in courtship
4. Structural differences in reproductive organs that disrupt mating
5. Inhibition of fertilization
Postzygotic mechanisms
1. Hybrid inviability or weakness
2. Hybrid infertility
3. Hybrid breakdown

1. *Isolation by habitat.* Species within the same area occupy different habitats. For example, there are two closely related species of violet in Denmark, *Viola arvensis* and *V. tricolor.* One grows in chalky soil, and the other prefers acid soil. Although these species can occasionally form hybrids, the situation is rare, because they are seldom close enough to breed.

2. *Seasonal or temporal differences in mating patterns.* Although such differences rarely operate as the only isolating mechanism, sometimes closely related species flower or mate at different times, and therefore their gametes never have a chance to meet (see Fig. 17.11).

3. *Behavioral differences in courtship.* The mating behavior of one animal species fails to attract members of another. The main purpose of courtship behavior is to ensure that members of the same species will recognize and mate with each other. In two species of fruit flies, *Drosophila melanogaster* and *D. simulans,* males perform intricate courtship movements to attract females. Laboratory tests have shown that the dance of the *D. melanogaster* male fails to attract the females of *D. simulans.*

4. *Structural differences in reproductive organs.* Fertilization between some species of plants is impossible, because the pollen tube of one species does not grow well in the style of the other, or because the pollen tube cannot grow long enough to reach the ovary. In animals, a basic structural incompatibility of the copulative organs may make mating impossible.

5. *Inhibition of fertilization.* Even when mating is successful, fertilization sometimes does not occur, because the sperm of one species is not attracted to or cannot penetrate the egg of the other.

When postzygotic inhibiting mechanisms are operative, mating and fertilization occur, but the development of viable fertile hybrids is blocked.

1. *Hybrid inviability or weakness.* Some hybrids die during the early stages of embryological development. It is during this time that the genetic material from the two parents must begin to operate as a unit. If the disparity between parental contributions is too great, death will result. With somewhat less disparity, the hybrid survives but is weaker than average members of either parental species and may quickly be eliminated by natural selection.

2. *Hybrid infertility.* Hybrids may be viable, but they are not able to produce offspring, either because their reproductive organs did not develop properly or because gamete formation is disrupted by abnormal gene or chromosome segregation in meiosis. Commonly, the hybrid of the sex that is determined by dissimilar sex chromosomes (usually the male with X and Y chromosomes) tends to be less viable or less fertile than a hybrid of the other sex. Hybrids between hoofed mammals, such as the zebra and the horse, characteristically have abnormal gonads. The most familiar hybrid of this type is the mule, a cross between a horse and a donkey.

3. *Hybrid breakdown.* Although the first generation of hybrids may be viable and fertile, first-generation hybrids mated with each other or with a member of a parental species produce a next generation that is extremely weak in comparison with the parental species or with the first-generation hybrid. For example, the fruit flies *D. pseudoobscura* and *D. persimilis* can be mated, and they can produce male and female hybrids. Although the males are sterile, the females may be mated with their parental species, but the offspring are extremely feeble.

It is obvious that the list of mechanisms that can prevent interspecies breeding is very long. If breeding behavior is known, therefore, it is not difficult to determine whether two populations comprise two species or one. It is also obvious that the creation of new species by hybridization is not a common occurrence.

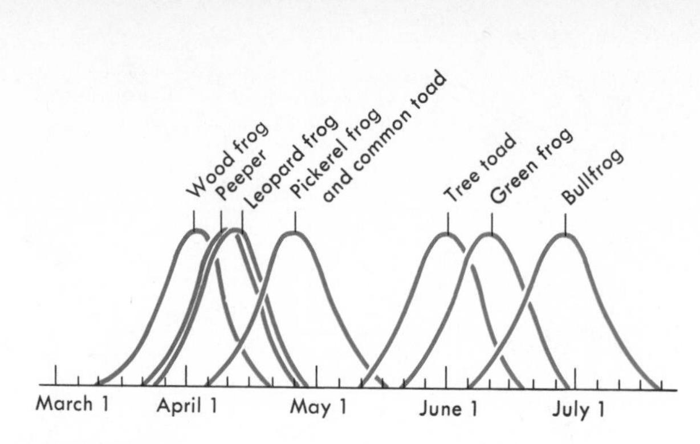

FIGURE 17.11

Reproductive isolating mechanisms. The periods of most active mating are distinct for each species of frog. These seasonal differences prevent interbreeding. Where the mating seasons overlap, however, breeding sites will differ. Both seasonal and habitat differences in mating activity function as genetic isolating mechanisms.

FIGURE 17.12

Speciation. Two species of one genus are (A) the cougar (Felis concolor) *and (B) the jaguar* (Felis onca). *The cougar—also known as the mountain lion or puma—weighs about 175 pounds and ranges in wilderness areas from Canada through South America. A male jaguar, the mightiest of the American cats, may weigh as much as 250 pounds and can break the neck of a full-grown ox. Once known as far north as California, jaguars are now found only in remote regions of South America. At some point in their evolution, the cougar and the jaguar became genetically isolated, and although their territories may overlap, they do not interbreed under normal circumstances.*

HYBRIDIZATION

Despite all the obstacles created by reproductive isolating mechanisms, sometimes speciation can occur by hybridization. Hybridization is a very rapid but rare form of speciation as compared with the division-isolation-modification process. In animals, hybridization is particularly rare, because males will not mate with females of another species unless there are no females of their own species around.

Although new hybrids may have special adaptive traits because they combine the adaptive qualities of both parents, most viable fertile hybrids do not survive, because they are not well adapted to the environment where they grow up. Because a hybrid has qualities that are different from those of either parent, it will generally be less well adapted than its parents to their respective habitats. If parent and hybrid must compete for the same habitat, the hybrid is usually eliminated. However, if a new environment is available, the hybrid may be able to occupy it successfully. It follows logically that hybridization has been a successful form of speciation mainly during periods of great environmental change.

Many successful plant hybrids have been formed when hybridization is followed by chromosome duplication, a process called **hybrid polyploidy.** The duplication solves the problem of infertility due to incompatible chromosome sets, because it ensures that normal meiosis and thus normal gametogenesis can occur. A normal organism contains two sets of similar chromosomes, A and A, but a hybrid contains two dissimilar sets, A and B. During meiosis in hybrids, chromosome incompatibility disrupts normal synapsis and division, and abnormal gametes are produced. If the hybrid chromosomes are duplicated (AA and BB), the similar sets can pair with each other during meiosis. Meiotic division can occur without mishap, and normal gametes will be generated.

Hybrid polyploid individuals are called **allopolyploids.** An example of a plant allopolyploid is the rado-cabbage, a cross between a radish and a cabbage. Both parental species have nine pairs of chromosomes, so the rado-cabbage contains 18 pairs. The plant produces normal gametes and is fertile.

In summary, evolution is a process of modification and speciation. Environmental selection pressures are constantly altering gene pool frequencies. At certain times such factors as disruptive selection, isolation, genetic drift, and hybridization act together with natural selection to give rise to new species. In this way the myriad of highly specialized species we know today have been generated from a relatively few simple forms of prehistoric life.

Summary

1. The biological diversity on the earth today is the result of evolution, a slow but continuous process in which species have undergone change and modification during the course of their descent from pre-existing forms.

2. Although the idea of evolution dates back to ancient times, Darwin and Wallace were the first to understand and scientifically demonstrate the mechanisms of evolutionary change. Darwin concluded that the organisms best adapted to their specific physical environment and best able to compete with other organisms for the limited necessities of life are the fittest to survive. The agent of *natural selection* is the environment, which selects only the fittest individuals to reproduce and perpetuate the species.

3. Since the time of Darwin, the theory of evolution has undergone modifications based on the progressively accumulating knowledge of genetics and biochemistry. In the light of modern genetics, Lamarck's theory of inherited acquired characteristics has proven untenable. The environment can produce phenotypic change, but such modifications are not heritable. Only traits encoded in gamete genotypes can be inherited and become the basis of evolutionary change. Thus natural selection does not produce favorable variations; it can only preserve them if and when they occur.

4. New genotype variations can arise during the production of parental gametes by crossing-over or linkage of genes and during recombination when parental gametes are united. Mutations and changes in the structure and/or number of chromosomes also result in genetic variation. Mutations that affect a number of genes are almost always nonadaptive since they disrupt the delicate balance of existing functions within the organism. However, small mutations acting in concert with normal genes can be adaptive to the organism. Not all mutations are immediately expressed in the next generation.

5. Variation in a population depends on the genetic constitution of its gene pool. Natural selection shapes evolution by determining which genes remain in the gene pool. The frequency of idealized phenotypic expression is algebraically expressed by the Hardy-Weinberg law.

6. When a population becomes isolated by geographical or environmental means, gene flow to or from other populations is prevented. Isolation eventually results in the generation of new species through gene pool modification. A quicker method of speciation is hybridization, the interbreeding between two established species. Hybridization is usually unsuccessful, because prezygotic and postzygotic isolating mechanisms generally prevent mating or negatively affect the viability and fertility of hybrids.

7. Most hybrids are inviable or infertile because their chromosome sets are incompatible. The formation of allopolyploids, in which chromosome number multiplies, allows greater viability, normal gamete formation, and thus successful speciation. Allopolyploidy occurs primarily in plants.

REVIEW AND STUDY QUESTIONS

1. Darwin and Wallace jointly presented papers on their conclusions on evolution by natural selection. Although reached independently, the positions were essentially the same. Yet we almost universally call it "Darwin's theory." Why?

2. What was the basic idea in Lamarck's theory of evolution? In what sense did that theory represent an important step forward?

3. Why did Lyell's *Principles of Geology* have such a great influence on Darwin's thinking?

4. Malthus's essay on population directed Darwin to an important breakthrough. What was the idea in the essay, and what did it lead to?

5. Is a gene pool analogous to a phenotype or to a genotype? Explain.

6. What is the Hardy-Weinberg law? Try to express its meaning in language that would be understandable to someone who has studied no biology at all.

7. What is selection pressure? What is its critical importance in Darwin's concept of evolution?

8. Give one illustration of each of the following: cryptic coloration; aposematic coloration; Batesian mimicry.

REFERENCES

Hamilton, T. H. 1967. *Process and Pattern in Evolution.* Macmillan, New York.

King, K., Jr., and P. E. Hare. 1972. "Amino Acid Composition of Planktonic *Foraminifera:* A Paleobiological Approach." *Science* 175:1461–1463.

Maier, R. A., and B. M. Maier. 1970. *Comparative Animal Behavior.* Brooks/Cole, Belmont, Calif.

Moody, P. A. 1970. *Introduction to Evolution,* 3rd edition. Harper & Row, New York.

Savage, J. M. 1969. *Evolution,* 2nd edition. Holt, Rinehart and Winston, New York

Srb, A. M., R. D. Owen, and R. S. Edgar. 1965. *General Genetics,* 2nd edition. Freeman, San Francisco.

Stebbins, G. L. 1971. *Processes of Organic Evolution,* 2nd edition. Prentice-Hall, Englewood Cliffs, N.J.

SUGGESTIONS FOR FURTHER READING

Clarke, A. C. 1968. *2001: A Space Odyssey.* New American Library, New York.

A fictional and mystical exploration of man's place in the cosmic order.

Darwin, C. 1959. *The Origin of Species.* Variorium edition. University of Pennsylvania Press, Philadelphia.

This edition of the classic document in which Darwin sets forth his theory of organic evolution includes all of Darwin's modifications.

Darwin, C. 1962. *The Voyage of the Beagle.* Doubleday, Garden City, N.Y.

Compilation of the observations by Darwin, which provides useful insight into his research in developing the theory of evolution.

Greene, J. C. 1959. *The Death of Adam: Evolution and Its Impact on Western Thought.* Iowa State University Press, Ames, Iowa (paperback, New American Library, 1961).

A thoughtful analysis of the social and philosophical impact that evolutionary theory has had on Western intellectual life and thought.

Kurten, B. 1968. *The Age of Dinosaurs.* McGraw-Hill, New York.

An exciting description of fossil carnivores, late Tertiary and Quarternary stratigraphy, and other aspects of paleobiogeography by a well-known paleontologist.

Moorehead, A. 1969. *Darwin and the Beagle.* Harper & Row, New York.

A thoroughly enjoyable reconstruction of Darwin's voyage by a popular biographer and historian.

Pfeiffer, J. E. 1972. *The Emergence of Man,* 2nd edition. Harper & Row, New York.

A popular but sensible survey of the present state of our knowledge of human origins. Includes lively asides and suggests future lines of research.

Simpson, G. C. 1953. *Life of the Past.* Yale University Press, New Haven, Conn.

A simple discussion of how the fossil record is read and of the recorded forms of life.

Stokes, W. L., and S. Judson. 1967. *Introduction to Geology.* Prentice-Hall, Englewood Cliffs, N.J.

Straightforward and clearly illustrated introduction to both physical and historical geology.

DIVERSITY OF LIFE: INTRODUCTION

chapter eighteen

Although this photograph shows only a tiny representation of the 4000-species phylum of echinoderms, it illustrates strikingly a principal theme of evolution. Members of this group vary greatly in form, and yet they clearly show that they share certain fundamental characteristics, such as radial symmetry, which reflect their common evolutionary ancestry.

In most of the material considered so far in this text, there has been a stress on the universality of living systems, on the attributes that all organisms share. Since there are underlying similarities in the biochemistry of cells, the cell theory, and genetics, we have been able to study physiological and behavioral processes in general terms within large groups of organisms. But if we find it astonishing that every life form shares certain basic characteristics, we must be overwhelmed by the variety and diversity of these life forms. If all *types* of organisms—referred to as species—currently living on the earth were catalogued, their number would exceed several million.

How would such a catalogue be arranged to give us easy access to its collection of valuable information? And aside from confirming the enormous diversity of life, how might it provide us with insights into the nature of the possible relationships between the species? Did these millions of species arise separately and individually, or are they somehow related?

The science of **taxonomy,** which traces its history to the beginnings of scientific thought, is the biologist's method for attempting to answer these questions. The taxonomist names and identifies species and classifies them into a hierarchical arrangement that reflects their interrelationships. Despite suspicions in certain quarters, the purpose of taxonomy is not just to confuse the undergraduate!

Taxonomy has provided us with the most probable explanation for the diversity of life: Species differ because of their adaptations to diverse environmental conditions, but they are all related by degrees through a fundamental and ancient pattern of organization. In short, species are not separate, unchanging creations. Rather, they have evolved from one another over immensely long periods of time. The most important idea behind taxonomy today is that all species in any taxonomic grouping are related by common ancestry.

History and formation of the classification system

Classifying and ordering the things of this world appears to be a basic tendency of human beings. Plants may be thought of as poisonous or nonpoisonous, as vegetables or weeds. Animals can be viewed as harmful or beneficial, wild or domestic. We place books in a library or goods in a department store into some sort of orderly arrangement. Often these collections or assemblies of objects are classified into a **hierarchical** arrangement, that is, a graded or ranked series of categories.

A stamp collector might choose to arrange a collection in terms of the colors of the stamps or their geometric shapes. This might work well if all other stamp collectors agreed to the same classification system. But the color of the stamp or its shape has less to do with the nature of stamps than does the country of issue. Most stamp collectors therefore arrange stamps first by the country of origin, then by the use of the stamp within each country, and finally by the year of issue within each category of use.

This method, proceeding from the most general category containing the largest number of stamps to the most specific category containing the smallest number of stamps, is an example of both hierarchical and natural classification.

Taxonomists have been classifying living things since the time of the ancient Greeks. The first recorded attempts at systematically classifying animals and plants (two of the general groupings of organisms that we still refer to as kingdoms) were made by Aristotle (384–322 B.C.), who compiled data on the animal life of his experience, and Theophrastus (370–285 B.C.), student of Plato and Aristotle, whose *Enquiry into Plant's* and *On the Causes of Plants* have been preserved.

With the decline of the classical world of Greece and Rome, the Dark Ages descended on Europe and lasted for several hundred years. Progress in scientific thought ceased, not to be revived until about the fourteenth century, the beginning of that general renewal of cultural and intellectual interest known as the Renaissance. Then an eighteenth-century naturalist named Carl von Linné (1701–1778) but better known by his Latinized name, Carolus Linnaeus (Fig. 18.1), published *Systema Naturae* (1735). That book of classification and several others that he published later organized and systematized much of the knowledge about living things that had accumulated. Linnaeus based his hierarchical and natural classification system on morphological, anatomical, and physiological attributes of organisms.

FIGURE 18.1
Linnaeus in his laboratory, as seen by an eighteenth-century artist.

Linnaeus was influenced at least as much by the metaphysics and Christian dogma of his day as by his scientific curiosity. He held that species were fixed entities existing as they had since the time of biblical creation. Linnaeus believed that the 4235 species of animals he classified had existed from the beginning of life on Earth and that they would never change in the future.

Individual differences within the organisms belonging to a particular species—for example, breeds of dogs or cats—were dismissed as unimportant. Linnaeus wrote of a philosophical, intangible archetype that contained the metaphysical essence—the universal dog-ness, horse-ness, human-ness—of each species.

Today's taxonomist owes much to Linnaeus. Even though the foundation of our classification system differs greatly from that of Linnaeus, we have retained his form. We still use his basic classification unit, the species, and we continue to group species into the increasingly general categories Linnaeus established: genus, family, order, class, phylum, and kingdom. Each category contains a larger number of organisms than the category below it in the hierarchy.

However, taxonomists today use the Linnaean system in an expanded form. The prefixes sub- and super- may be attached to any group name. The category Division, used in botanical listings, is equivalent to the zoological Phylum. Additional groupings—branch, grade, series, cohort, tribe—are used when needed. These levels were added to the classification scheme as more species were discovered and differences between them could not be accommodated by the Linnaean system. Table 18.1 shows the classification of the opalescent sea slug and of the human organism according to the modern system.

The philosophic and scientific underpinnings of taxonomy changed radically with the publication of Darwin's *On the Origin of Species*. The ideas of fixity of species and separate creation of each species were doomed. Taxonomy has become thoroughly and permanently associated with evolutionary theory.

Even though scientists agree that the basic kind of organism is the species, agreeing on a clear and exact definition of "species" is difficult. Generally, a particular species is thought to include those organisms that interbreed and produce fertile offspring. But many unicellular organisms show no form of sexual reproduction. Considering a species as formed of individuals with common ancestry that goes back to the first evolutionary branch is correct—but vague and almost impossible to prove.

Perhaps because the term species is so difficult to define, a new idea has gained recent acceptance. It is the statistical concept of the **population** or organisms as the basic systematic unit of classification. In a population, no one ideal or abstract archetype defines a species. Instead, the species is defined by the total of the physiological, behavioral, morphological, biochemical, and genetic characteristics of the members of the population.

TABLE 18.1
Classification of the opalescent sea slug and the human organism according to the contemporary system

Kingdom	Animalia	Animalia
Grade	Coelomata	Coelomata
Series	Proterostoma	Deuterostoma
Phylum	Mollusca	Chordata
Subphylum	—	Vertebrata
Class	Gastropoda	Mammalia
Subclass	—	Eutheria
Order	Opistobranchia	Primates
Suborder	Nudibranchia	Anthropoidea
Family	Phanerobranchiatae	Hominidae
Genus	*Hermissenda*	*Homo*
Species	*crassicornis*	*sapiens*
(Common name)	(opalescent sea slug)	(human being)

Nomenclature

Organizing the species systematically and naming them have now become so complicated that international commissions and congresses meet regularly to oversee the operation of the system.

One of the important considerations of any classification system is its usefulness as a means of communication between the individuals using the system. The method of classification must be clear and understandable. Linnaeus chose Latin, the language of scholars, as the language of classification. Many Latin words were incorporated directly, and words from other languages and names were "Latinized," that is, given Latin endings. Latin remains the language of taxonomists in naming the levels of the classification system.

We also owe to Linnaeus the system we still use for naming all living things: **binomial nomenclature,** or two-word names. The first word in the scientific name is the genus, and the second is the species. (For certain organisms a third word indicates the subspecies.)

But why a scientific name—long, difficult to remember, and written in an unfamiliar language—when a common name seems adequate and is certainly more convenient? As you have already discovered, common names in English have appeared throughout this book. Nonetheless, there are times and places in which common names would cause confusion because they are not precise enough for the requirements of the scientific community, particularly in scientific publications. One species may have several common names. The ordinary blue jay, *Cyanocitta christata,* has also been called corn thief, nest robber, blue coat, jay, and common jay. Conversely, the same common name might be applied to more than one species. Common names are colloquial, confusing, and inexact. They require translating into other languages. Consider, for example, a European white water lily, with 15 English, 44 French, 105 German, and 81 Dutch common names.

The scientific name of an organism may incorporate a physical attribute of that organism: *Quercus rubra,* the red oak. The name may refer to the habitat of the organism or the geographic region where it is found: *Lynx canadensis,* the Canadian lynx. The new organism can be named for a person: *Pasteurella,* a bacterial genus named for Louis Pasteur. Sometimes biologists seem to be imitating Gertrude Stein's "a rose is a rose is a rose" when they identify an organism as *Rattus rattus rattus,* which amounts to "a rat is a rat is a rat."

Techniques for determining relationships

For Linnaeus and his colleagues, the criteria for determining relationships were morphological, anatomical, and physiological. At that time descriptions of plants and animals were based on specimens collected on expeditions to various parts of the world and brought to the home laboratory for study. Organisms were usually classified without regard to behavior or habitat.

The tradition of collections continues today but serves another purpose. In the United States alone, major zoological museums house more than 90 million preserved animals. Collections of preserved plants maintained in herbaria number some 24 million specimens. As related sciences advanced, taxonomists made use of new developments to improve their own science. Taxonomy is now an interdisciplinary science in which the biochemistry, embryology, and genetics of an organism, and even its behavior, habitat, and geographic distribution must all be taken into account.

There has always been some arbitrariness in taxonomy, since it is a science of human judgment. In a sense, it is an artificial pursuit, because scientists impose classification on an organism. Classification does not alter an organism or change its nature; rather, it helps to clarify that nature and the relationship of the organism to other living things.

However, many organisms do not fit neatly into any particular category. The classification of one-celled organisms, many of which show both plant and animal characteristics, has long been disputed. Two organisms can be so similar that determining whether they are the same or different species can rest on one small bit of evidence.

The placement of several Australian mammals in the classification scheme provides a good example of the role of judgment in taxonomy (see Fig. 18.2). There are similarities between certain Australian mammals—the Tasmanian wolf, the flying phalanger, and the wombat—and some North American mammals—the wolf, the flying squirrel, and the ground hog. The animals of Australia occupy biological niches similar to those of their North American counterparts, and their behavior and morphology are similar. However, there are skeletal differences. Furthermore, the wolf, flying squirrel, and ground hog are **placental** mammals, whose young undergo complete intrauterine development. The Australian counterparts are **marsupials,** whose young are born in an immature state and continue development in the external pouch of the mother.

Since marsupials and placentals have mammary glands and nurse their young, both groups are members of the class Mammalia, but they are placed in separate

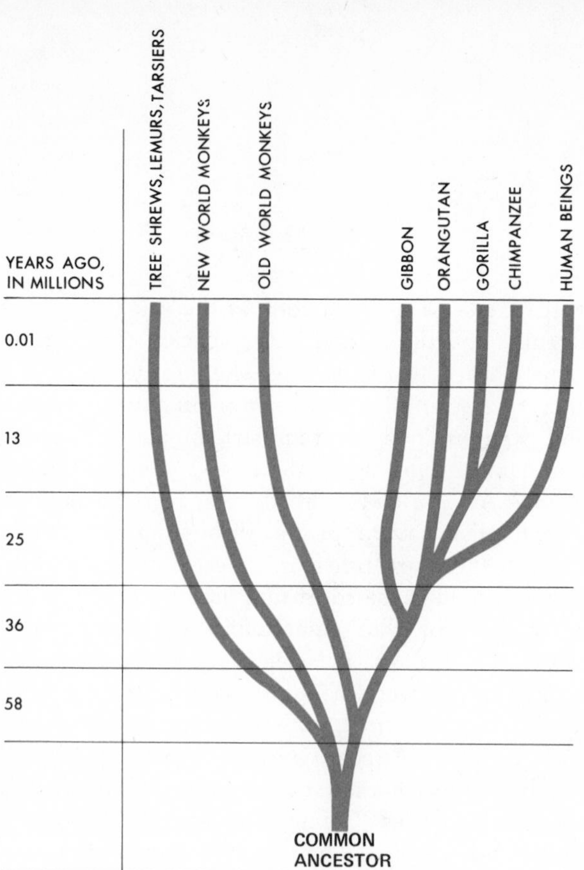

FIGURE 18.3

Phylogeny of the major primates, expressed as a tree diagram. All the primates known to us are thought to have a common ancestor, perhaps similar to today's tree shrew. (Adapted from J. J. W. Baker and G. E. Allen, The Study of Biology, *3rd edition, Addison-Wesley, 1977)*

FIGURE 18.2

Despite their similar appearance, the North American ground hog (A), a placental mammal, and the Australian wombat (B), a marsupial, belong to different orders within the Class Mammalia. Placental mammals and marsupials developed quite independently from a common ancestor, and marsupials are characterized by an abdominal pouch in which they rear their prematurely born young.

subclasses. The Tasmanian wolf, flying phalanger, and wombat have been placed with the kangaroo, another marsupial, in the subclass Metatheria. The North American placental mammals are grouped in the subclass Eutheria.

Often problems posed by classification are more subtle. Several species of crickets differ only in their calls. In theory they might be capable of mating and producing fertile offspring; yet they do not, because the female of each species responds only to the call of the male of her species. A behavioral characteristic thus can determine classification as a separate species.

An important part of taxonomy is the attempt to determine evolutionary relationships and therefore the evolutionary history of living organisms. Since the fossil record is incomplete for most organisms, the evolutionary history, or phylogeny—expressed graphically as a

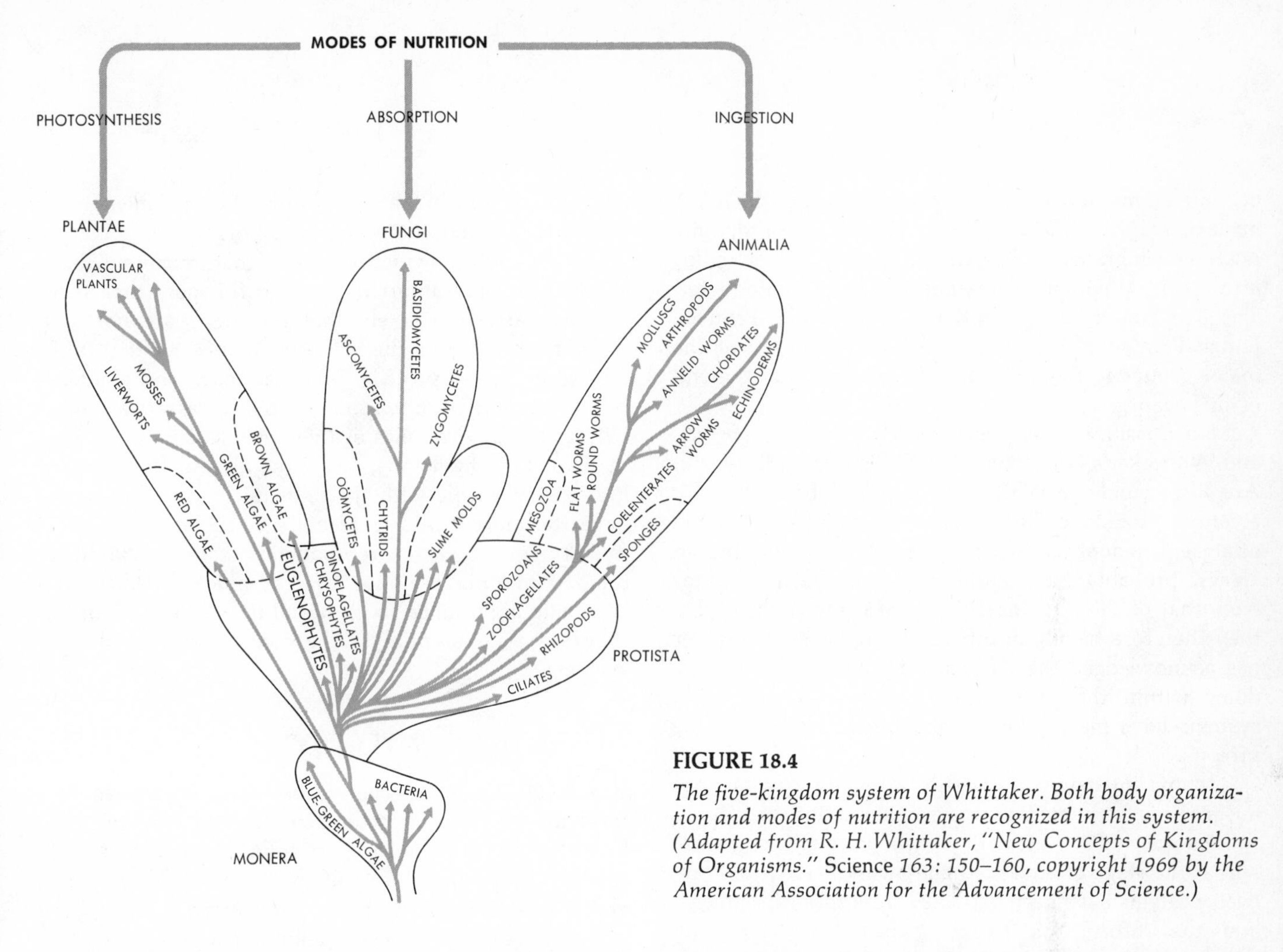

FIGURE 18.4

The five-kingdom system of Whittaker. Both body organization and modes of nutrition are recognized in this system. (Adapted from R. H. Whittaker, "New Concepts of Kingdoms of Organisms." Science *163: 150–160, copyright 1969 by the American Association for the Advancement of Science.)*

tree diagram, as shown in Fig. 18.3—can be only approximate.

Recently developed scientific techniques are also used to determine probable evolutionary relationships. Proteins in the egg whites of penguins, ratites (nonflying birds related to the ostrich), and ducks have been compared. Results of the work have shown more similarity between penguin and duck proteins than between penguin and ratite proteins. Penguins are therefore thought to be more closely related to ducks than to ratites.

The science of taxonomy itself continues to evolve. Some taxonomists still cling to the two-kingdom system of Linnaeus, which placed all living things of the world into either of two kingdoms, Plantae and Animalia. Yet many organisms—bacteria, blue-green algae, fungi, protozoa—do not fit the definitions for plants or animals alone. In 1866, little more than a century after Linnaeus published *Systema Naturae*, E. H. Haeckel, a German biologist and philosopher, proposed a third kingdom, Protista, to contain mostly unicellular organisms.

Advancements in biochemistry and the development of the electron microscope uncovered the fact that bacteria and blue-green algae differ importantly from other unicellular organisms in that they are procaryotes; that is, they have no separate nucleus enclosed in a nuclear membrane. In 1956, an American systematic biologist, H. F. Copeland, placed these special organisms in a separate fourth kingdom, Monera.

Of the many classification schemes suggested in recent years, the one that has now gained widest acceptance is the five-kingdom system proposed by R. H. Whittaker in 1969 (Fig. 18.4). Whittaker continues to recognize the profound difference between the procaryotic monerans and eucaryotic organisms, in which the nuclear material (DNA) is surrounded by a definite nuclear membrane. He has placed fungi and slime molds into a fifth kingdom for reasons we shall examine later. The five kingdoms of Whittaker are Monera, Protista, Fungi, Plantae, and Animalia. A brief summary of the major groups in these five kingdoms appears at the end of this chapter.

No classification system devised to date is perfect, and Whittaker's is no exception. Difficulties still remain. Are slime molds, in Whittaker's words, "aberrant fungi, eccentric protists, or very peculiar animals?" Two phyla of algae, Rhodophyta (red algae) and Phaeophyta (brown algae), probably have a different unicellular ancestor from that of the phylum Chlorophyta (green algae). Do they therefore belong in different kingdoms? Whittaker has acknowledged the difference by means of subkingdoms within the plant kingdom. Other classification systems have placed the red and brown algae with the protists.

Many unanswered questions remain for taxonomists. It is estimated that a million living and fossil species are undiscovered. Phylogenetic relationships are still at the stage of advanced guesswork. Recently, however, proteins have been extracted from fossil specimens, and the information to be gained from analyzing their properties will be extremely valuable. Furthermore, mathematical theory coupled with computer methods capable of storing and handling large amounts of taxonomic information are becoming more sophisticated. Indeed, some scientists now call themselves numerical taxonomists.

But the ultimate tool for taxonomy, the complete sequencing of the bases of the DNA of organisms, has yet to be done for any organism. No doubt taxonomy for some time to come will retain its arbitrary nature and remain a science of judgment.

Kingdom Monera

Schizomycophyta (bacteria) and Cyanophyta (blue-green algae) are the two phyla included in Kingdom Monera. They are unicellular or colonial microorganisms, all of which are procaryotes, lacking membrane-bound nuclei and complex cellular organelles. Recently discovered fossil evidence shows that monerans have existed since Precambrian times, or for more than three billion years. It is likely that monerans are not very different today from the procaryotic ancestor that probably gave rise to protists, fungi, animals, and plants.

Yet primitive organisms are not necessarily simple. Within the tiny confines of their cells (0.1 to 60 microns in diameter), monerans carry out all the functions of life. Barely visible with the light microscope, monerans have the most diverse metabolic machinery seen in all five kingdoms. Their reproduction rate is extremely rapid. Not only are they of tremendous medical and economic importance (both harmful and beneficial), but they are also essential to the continuation of all ecosystems.

PHYLUM SCHIZOMYCOPHYTA

The 15,000 or so species of bacteria are divided into three major groups according to their cell structure. The bacilli are rodlike, the cocci are spherical, and the spirilla are spiral-shaped. Bacteria of the same species may also group together. For example, cocci arrange themselves in pairs, beadlike chains, irregular clusters, and other combinations. Several basic structures that all bacteria have in common are shown diagrammatically in Fig. 18.5. The cell membrane encloses the cytoplasm, which contains granules including ribosomes and polyribosomes. Filamentous DNA is located in a nuclear region, but there is no membrane-bound nucleus.

The bacterial cell is surrounded and protected by a cell wall, which differs from the cellulose cell wall of higher plants in that it is made up of amino acids, sugars, lipids, and carbohydrates. Many bacteria possess thin, whiplike appendages, **flagella,** made of protein subunits. Flagella enable the organism to move. Similar but smaller appendages, **pilli,** may occur. They are important to reproductive processes or they enable the organism to stick to inert surfaces. Certain bacteria secrete a gelatinous slimy material that completely encases the cell. This capsule is composed of polysaccharides, polypeptides, or a combination of the two. Species

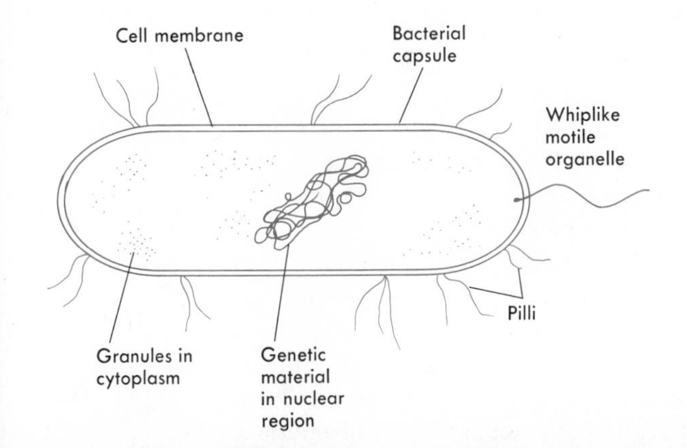

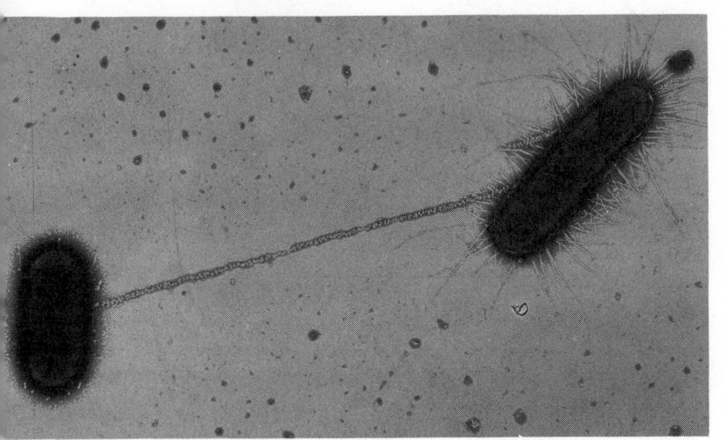

FIGURE 18.5

Drawing and electron micrograph of bacteria. The drawing includes many of the basic structures that bacteria have in common. In the micrograph, a slender tube between a pair of mating Escherichia coli *allows genetic material to pass from one organism to the other.*

of *Pneumococcus* are classified partly by differences in the chemical constituents of their capsules.

Bacteria reproduce primarily by binary fission, which is analogous to but less complex than the mitotic division of eucaryotes. Bacterial DNA is thin and fibrillar and is probably a single long strand. During cell reproduction the DNA duplicates, and then the strands separate. A ring of new cell wall forms on the inner surface of the original cell wall, about halfway between the ends, and then grows inward. Finally the cells separate. These events follow in order, but one stage need not be completed before the next begins.

A second method of asexual reproduction found in bacteria is budding (see Fig. 12.1). An outgrowth, or bud, of cytoplasm appears on a bacterial cell. Then genetic material moves into the bud and cellular separation occurs. Conjugation (see Fig. 12.2), a form of reproduction that is somewhat like the sexual reproduction of eucaryotes, may occur with certain genera of bacteria, particularly *Escherichia, Salmonella, Shigella,* and *Pseudomonas.* First, two bacteria pair off. Pilli are thought to aid in this pairing or to be actually involved in the second stage of the process, in which a fragment of DNA is transferred from one cell to the other. Genetic recombination is followed by the separation of the cells. The result is the appearance of a new phenotype of the same species. This is an advantageous process because it results in an increase in genetic variation and a higher probability of eliminating undesirable genes.

At times, certain bacteria engage in spore formation. Small **endospores** in a dormant state can survive the environmental extremes of high heat, cold, or low humidity. Though not a reproductive process, endospore formation enables the organism to survive adverse conditions. Whereas the vegetative cells of most bacterial cells are killed after five to ten minutes of exposure to 55–65°C moist heat, the sporulated cells of *Clostridium botulinum,* the organism responsible for botulism poisoning in improperly canned food, can survive more than five hours of exposure to 100°C moist heat.

Bacterial reproduction is exceedingly rapid. *Escherichia coli,* a species found normally in the human intestinal tract, may divide as often as every 17 minutes. If bacterial growth continued unchecked, the number of organisms produced would literally overwhelm the planet. However, growth and reproduction are limited by the death of older cells, exhaustion of nutrients, the presence of such harmful substances as metabolic wastes or antibiotics, or the natural life span of the bacterial mass.

The diverse nutritional requirements of bacteria provide another method of classification. Certain bacteria are autotrophs and therefore can synthesize their own organic compounds from simple inorganic molecules. In this respect they resemble plants. However, the energy required to accomplish the synthesis of the amino acids, sugars, and lipids need not come from light.

Bacteria that do use light, the photosynthetic autotrophs, contain in granules rather than in plastids a type of chlorophyll that differs from the chlorophylls *a*, *b*, and *c* of higher plants. Photosynthetic bacteria fix carbon from CO_2 but, unlike plants, do not use water for photosynthesis and never produce oxygen as its product. The hydrogen used to reduce CO_2 to the level of a carbohydrate can be derived from a compound like hydrogen sulfide (H_2S). The product is then elemental sulfur rather than oxygen.

Chemosynthetic bacteria also use CO_2, but the energy for their reaction comes from the oxidation of simple chemical compounds, not sunlight. For example, some soil bacteria oxidize ammonia (NH_3) to nitrate (NO_3^-) and use the energy gained from the reaction to reduce CO_2. Other chemoautotrophs may oxidize hydrogen sulfide, elemental sulfur, hydrogen, or ferrous iron in order to fix carbon from carbon dioxide.

Bacteria differ in their oxygen requirements and in their tolerance to the presence of oxygen. This characteristic provides another means of classification. Bacteria that cannot exist in the absence of oxygen are called **obligate aerobes.** Many species, like the chemosynthetic autotrophs, cannot survive in the presence of oxygen, so they are known as **obligate anaerobes.** Bacteria that are more flexible in their oxygen requirements and can tolerate its presence or absence are the **facultative anaerobes.**

Most bacterial species are heterotrophs. They cannot synthesize their own amino acids or sugars and must therefore obtain them from another source. If that source is another living organism, the bacteria are parasitic. Saprophytic bacteria obtain their nutrients from dead and decaying organic matter.

Bacteria are literally everywhere and play diverse ecological roles. One gram of soil may contain twenty million bacteria. Together with the fungi, bacteria are principally responsible for the decay of organic matter, whether on the forest floor, in the compost heap, in spoiled food, or in the sewage-treatment plant. Nitrogen-fixing bacteria found in nodules on the roots of legumes (peas, beans, clover) fertilize the soil by converting molecular nitrogen to soluble nitrates, which can then be absorbed by plants through their roots. One can correctly think of bacteria as the great recycling agents of the earth.

Industry puts bacteria to work in a number of ways. The dairy industry uses bacterial cultures to produce sour cream, buttermilk, yogurt, and a variety of cheeses. After a yeast has fermented sugar in apple juice to alcohol, bacteria are used to turn the alcohol to cider vinegar.

Fortunately, most species of bacteria are not harmful, but many are. They are the pathogens, agents that produce disease. Pathogenic bacteria gain entrance to—that is, infect—the host, where they cause adverse physiological or anatomical changes. Pathogenic bacteria infect many kinds of other organisms. They cause galls, leaf spot, and leaf wilt in plants. Animals contract anthrax, bubonic plague, brucellosis, and other diseases from bacteria. In humans, bacteria can cause tetanus, plague, strep throat, syphilis, gonorrhea, and many other diseases.

PHYLUM CYANOPHYTA

Procaryotic photosynthetic organisms, the blue-green algae constitute the second phylum of the Kingdom Monera. Although called "blue-green," the various species of these unicellular or colonial algae show many colors—red, yellow, green, purple. They contain the chlorophyll *a* found in higher plants, as well as other pigments. Their classification at present is somewhat unsatisfactory. Many more biochemical and genetic studies need to be made before the classification of this phylum can approach the completeness of bacterial classification.

Cyanophyta differ from bacteria in that the cell is surrounded by a cellulose wall, and the tendency to colonize is much greater. Colonial aggregates of identical cells may be shaped like disks, plates, or branched filaments. Like the bacteria, the cell or colony of cells may be surrounded by capsular material.

Blue-green algae live in fresh water, salt water, and soils. Salt-water species are constituents of **plankton,** that collection of small organisms that drifts near the ocean surface and forms the basis of all aquatic food chains. Occasionally, when blue-green algae reproduce very rapidly, they make the water itself appear colored and are said to **bloom.** One red species is responsible for the periodic coloring of the Red Sea. Cyanophyta can tolerate environmental extremes and are found in salt lakes, dry climates, the frigid Antarctic, or the 85°C waters of hot springs in Yellowstone National Park.

Nutritionally the blue-green algae have simple requirements. Most are autotrophs and manufacture their nutrients through photosynthesis. Unlike bacteria, blue-green algae produce molecular oxygen as a by-product of photosynthesis. Many authorities believe that the development of an oxygen-rich atmosphere in the middle Precambrian period was due to the photosynthetic activity of these microscopic organisms. However, the main product of their photosynthesis is not the starch of higher plants but glycogen, similar to the substance stored in the livers of animals. A few blue-green algae are heterotrophs.

Reproduction of blue-green algae is totally asexual. They are evolutionarily primitive and believed to be at approximately at the same level of development as were their Precambrian ancestors.

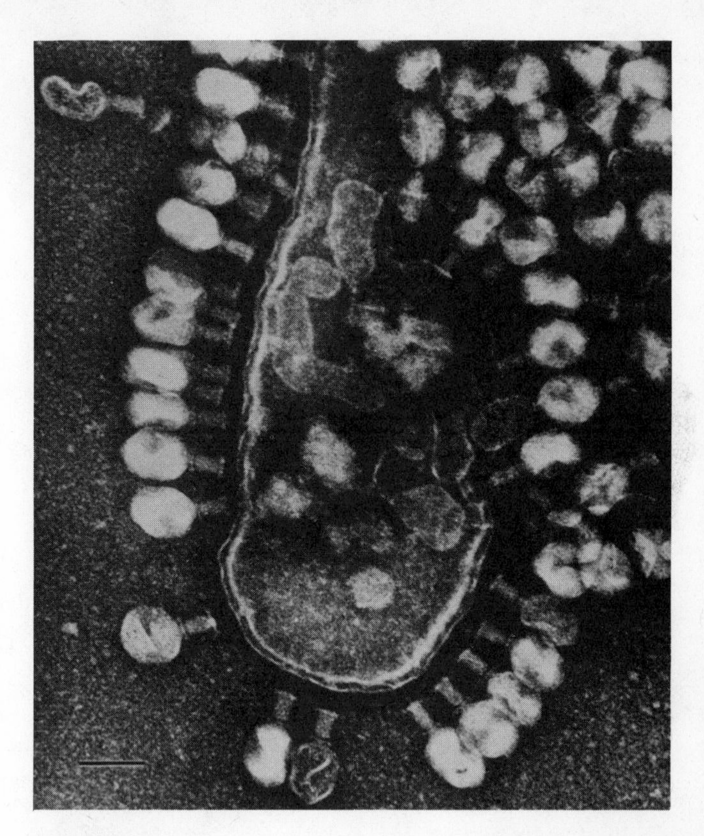

FIGURE 18.6

Viruses are considered intermediate between living and nonliving structures. They can be isolated and crystallized—a characteristic of nonliving chemicals—yet they consist of nucleic acid with a protein coat and are capable of taking over a host cell to reproduce more virus particles. The electron micrograph shows bacteriophages attacking an E. coli *cell. Note the empty heads of viruses that have injected their DNA into the cell.*

THE VIRUSES

The position of viruses relative to living organisms has long been disputed by biologists. The basic question about viruses has concerned whether they are living or not. Because of their chemical simplicity, they are usually discussed when considering the Kingdom Monera.

Viruses are not cells, but they are capable of parasitizing living cells. They are made up of the basic genetic stuff of living things: DNA or RNA, surrounded by a protein coat. Yet they are not alive, because they cannot reproduce themselves outside the host cell. They have no enzymes of their own and no way of producing ATP. Most viruses can be crystallized—a characteristic they share with chemicals.

Virus classification in and by itself has also been troublesome. Should a virus be classified on the basis of the host it invades or the disease it causes? Either system is unwieldy. A more satisfactory classification system is one based on the chemical composition and the structure of viruses. The most general groupings are the Ribovira (containing RNA) and the Deoxyvira (containing DNA). These categories are subdivided on the basis of the shape or symmetry of the virus particle: helical, cubic, or with head and tail.

A virus invades, or infects, the host cell by attaching itself to specific receptor sites on the host cell. Bacterial viruses, for example, have been found to attach to the cell envelope, the pilli, or the flagella. Once the virus is secured to the cell, it must penetrate the cell wall or cell membrane. The whole virus particle generally penetrates animal cells, but only the nucleic acids are sent into the cells of bacteria. Once inside the host cell, the viral nucleic acid takes over the metabolism and genetic

Viruses

The viruses represent life in its simplest known form, and may be said to stand at the boundary between living and nonliving matter. They have a definite structure and the ability, under specific conditions, to replicate that structure. But a virus particle, or virion, cannot perform the functions of life except by parasitizing the living cell of a host. The virion consists of a molecule of nucleic acid, comprising between 10 and 100 genes, enclosed in a jacket of protein which both protects the nucleic acid and provides a mechanism for infecting the host. It is the host that provides the virus with the mechanism and materials for carrying out the instructions encoded in its nucleic acid.

Three hypotheses have been offered in explanation of the extremely simple nature of viruses. One hypothesis is that the viruses represent fragments of genetic material detached from their original substrates in cellular organisms; this might explain their highly specific parasitism. A second is that modern viruses are a modified version of free-living acellular ancestors, which have become parasitic only after the disappearance of the primordial "organic soup." The third hypothesis is that viruses arose from cellular ancestors through the gradual loss of extranuclear function, as a result of extreme specialization for parasitism. Interestingly, viral infection is limited to a relatively few phyla. Most bacteria, angiosperms, vertebrates, and arthropods serve as viral hosts, but not protozoa, the lower plants, or most other animal groups.

The T4 bacteriophage (A and B) has been intensively studied and will serve to illustrate the mechanism of viruses as a group, although it is an unusually complex viral structure. The nucleic acid (DNA) is contained in the head, or capsid. The phage infects a bacterium by attaching itself to a specific site on the bacterium's cell membrane by means of the tail fibers and end plate. The tail sheath contracts to force the DNA from the head through a hollow core and into the host cell (C-1). This event causes the bacterial DNA to be degraded, and the bacterium's metabolism now comes under the control of the viral DNA. Thus the bacterium's synthetic machinery is used to make enzymes (C-2), which replicate viral DNA and viral structural proteins (C-3). As viral DNA forms into new viral heads, the component protein parts come together around them, forming complete new virions (C-4). This process continues for about 25 minutes, until the lysis, or dissolution, of the host cell releases about 200 new virus particles, each capable of infecting a new host (C-5).

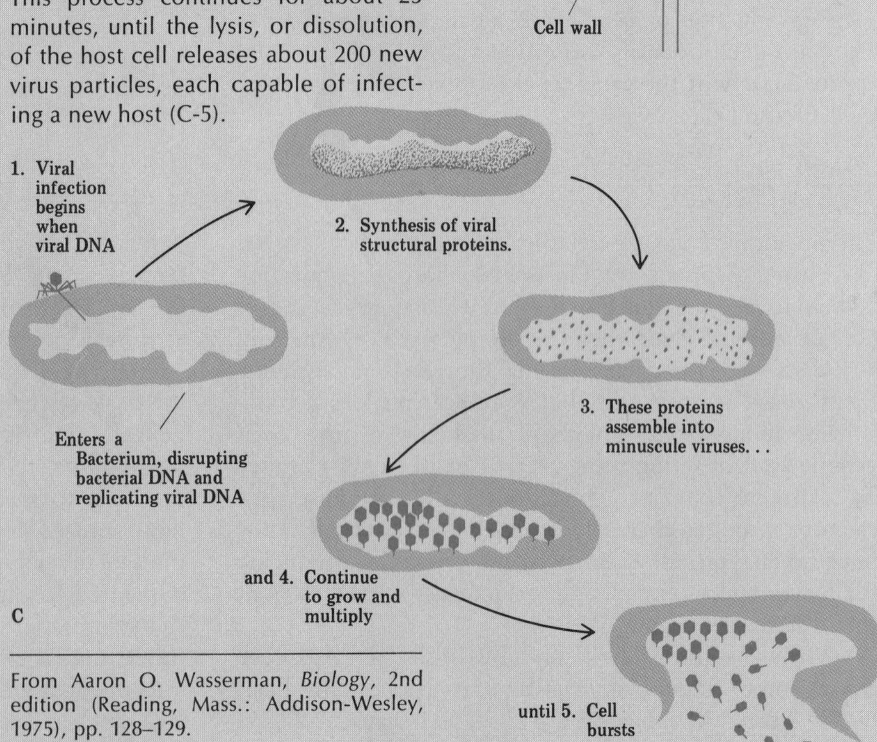

From Aaron O. Wasserman, *Biology*, 2nd edition (Reading, Mass.: Addison-Wesley, 1975), pp. 128–129.

machinery of the host cell to produce hundreds, even thousands, of new virus particles. One infected cell of the tobacco plant may produce 10,000 new tobacco mosaic virus particles. The host cell or the entire organism may die as a result of viral infection.

Viruses infect all types of organisms. They cause mosaic diseases or leaf curl in plants. Animals, including humans, are variously subject to infection by viruses that cause rabies, psittacosis (parrot fever), influenza, the common cold, poliomyelitis, measles, mumps, chicken pox—and many other diseases. Even bacteria are attacked by particular types of viruses called bacteriophages ("bacteria eaters").

Advances in immunology have made it possible to eliminate some viral diseases (e.g., small pox) and to control others. Some viruses are known to transform normal cells in experimental animals to cancer cells, but the link between viruses and human cancer remains unproven.

The evolutionary position of viruses is also the subject of sometimes heated debate. Did they arise independently from other forms of life? Are they the long-sought link between inorganic matter and living organisms? Current thought suggests that viruses are latecomers rather than an early life-related form. They may represent the "ultimate" adaptation to a parasitic form —only a protein envelope and nuclear material. Such a conclusion seems logical since other parasitic forms, whose ancestry is more readily traced, show a progressive loss of structures throughout their evolutionary history until all that remains are those structures needed for invasion and reproduction. Viruses also resemble higher forms of life in that they undergo mutation and genetic recombination. Perhaps the question of viral evolution will never be answered.

Kingdom Protista

The Kingdom Protista can be thought of as a transitional kingdom, in an evolutionary sense, between the primitive procaryotes and the more structurally sophisticated members of the higher kingdoms. Some might view Protista as a "catch-all" kingdom into which taxonomists have placed many unusual organisms that do not belong in the other kingdoms or that possess characteristics of more than one kingdom. Neither the idea for this kingdom nor its name is new, but the organisms classed as protists have varied considerably in those classification systems that have included such a kingdom.

Under Whittaker's five-kingdom system of classification, protists are defined as eucaryotic unicellular or colonial organisms whose cytoplasm contains complex organelles, such as plastids and mitochondria. The nuclear material is surrounded by a nuclear membrane, and cell division usually proceeds by true mitosis. This kingdom exhibits several methods of reproduction: binary cell division, conjugation, and true sexual reproduction.

Protist cells may be as small as those of some bacteria (0.2 microns in diameter) or large enough to be seen with the naked eye (3–4 centimeters), but the average size is 100–300 microns. Many protists contain one or a few long flagella or a great number of cilia. Unlike bacterial flagella, the motile structures of protists are quite complex. They are composed of eleven protein filaments, nine of which form a cylinder containing the other two (9 + 2 arrangement).

Protists obtain nutrients in various ways—ingestion, absorption, photosynthesis, or a combination of methods. They are capable of reacting to conditions in the environment, such as the presence of food or an obstacle.

PHYLUM EUGLENOPHYTA

Typical of the phylum Euglenophyta, a widely distributed group of organisms that show characteristics of more than one kingdom, are members of the genus *Euglena.* These organisms do not have cell walls, and they have one or more whiplike flagella and a photoreceptive pigment or eyespot—animal-like characteristics. Yet *Euglena* also have chlorophyll *a* and *c* in chloroplasts in the cytoplasm, and they perform photosynthesis—plantlike characteristics. Finally, they may also obtain nutrients by absorption—a moneran characteristic.

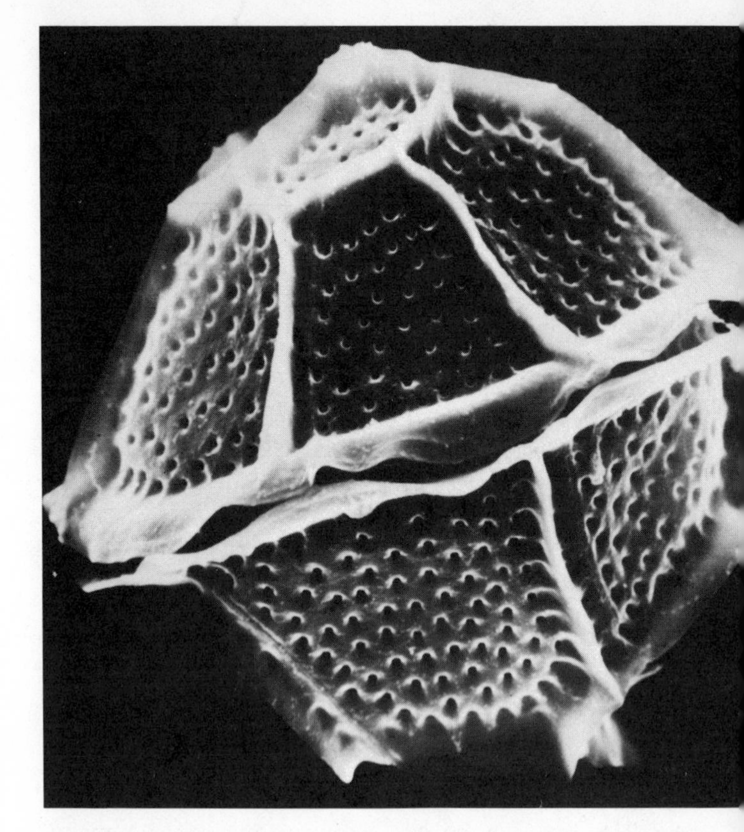

FIGURE 18.7
Freshwater Gonyaulax, *a dinoflagellate closely related to the marine organisms responsible for the "red tides" in both eastern and western coastal waters.*

Reproduction occurs by longitudinal cell division, and no sexual reproduction has been observed. Wastes and water are excreted by means of a contractile vacuole, which when full migrates to the cell membrane and expels its material. It is not surprising that *Euglena* has been classified as a plant by botanists and as an animal by zoologists.

PHYLA CHRYSOPHYTA, PYRROPHYTA

Chrysophyta (diatoms and golden algae) and Pyrrophyta (fire algae) are photosynthetic eucaryotes. They are "armored" organisms, so-called because they have rigid cell walls. The walls of the chrysophytes contain silica. Those of the pyrrophytes are made of cellulose. The coloring of both comes from carotene-related pigments.

Chrysophytes, including diatoms, are usually marine organisms, constituents of plankton. Their beautiful and delicate silica shells in a variety of shapes constitute one means of distinguishing the species. The earth that forms as shells of dead organisms accumulate on the ocean floor is mined and used for insulating, filtering, and abrasive materials. Chrysophytes may be flagellate or nonmotile. Reproduction proceeds asexually or sexually.

Pyrrophytes, many of which are called dinoflagellates, are also components of plankton. Their cellulose cell walls are usually indented by a transverse furrow, around which one of their two flagella is wound. Many are luminescent. Dinoflagellates may undergo periodic blooms. The red-colored *Gonyaulax* are responsible for the "red tides" that have appeared in recent years in both eastern and western coastal waters. (Fig. 18.7). They secrete a poison that kills fish. Molluscs, especially clams and mussels, feed on the dinoflagellates and accumulate the poison, which is toxic to humans.

Knowledge of the reproductive process of dinoflagellates is incomplete, but most investigators believe that no sexual reproduction occurs. The process of cell division closely resembles mitosis, but there are differences. It lies somewhere between that of eucaryotes and procaryotes since the spindle-fiber system does not develop. If the water temperature drops, dinoflagellates may form resistant cysts or spores, which regenerate as vegetative cells when environmental conditions again become favorable.

THREE PHYLA WITH ANIMAL TRAITS

Some protists with animal characteristics are placed in three phyla on the basis of the way they move. In the past these small unicellular eucaryotes were collectively called Protozoa and were often treated as the simplest members of the animal kingdom. In general they ingest food, are quite motile, and can respond to environmental conditions.

Sarcodina The genus *Amoeba* is the most familiar member of the phylum Sarcodina. Though definitely a eucaryote, the amoeba, a blob of protoplasm without definite external structure, has little in the way of distinct organelles (Fig. 18.8). Amoeba generally reproduce by binary fission. They move in a peculiar manner. They

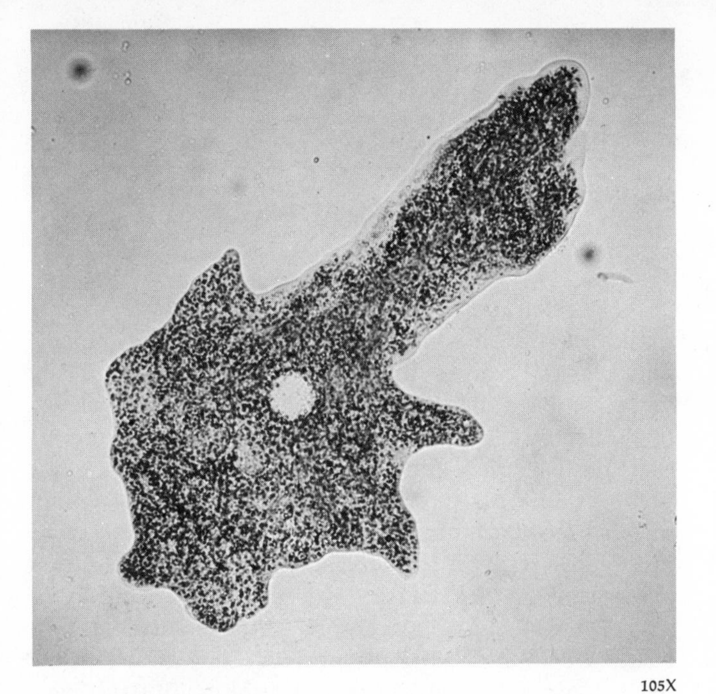

105X

FIGURE 18.8

Amoeba proteus, *a protist of the phylum Sarcodina. The entire organism is a flowing mass of almost transparent cytoplasm. Its nucleus is visible in this micrograph, but it contains little else that can be identified as distinct organelles.*

125X

FIGURE 18.9

A diverse assembly of armored sarcodines. Most striking in the photograph are the delicate skeletal structures.

extend bits of themselves, pseudopodia (false feet), and seem to pull themselves along in what is known as **amoeboid** motion. These protoplasmic projections are also used for engulfing food. The structure of the pseudopodia is one way in which species of amoeba are classified. One harmful species of amoeba causes dysentery in humans.

Related to the amoeba are the armored sarcodines (Fig. 18.9), including marine types, Foraminifera, with calcium carbonate shells through which pseudopodia protrude, and the fresh-water Heliozoa and marine Radiolaria, which have glassy-spiked spherical skeletons. The shapes of these carbonate or silicate shells aid in identification and classification. Shells of dead sarcodines, collected on the ocean floor, have contributed to the fossil record. England's white cliffs of Dover consist almost entirely of foraminiferan shells.

Sporozoa Sporozoa are parasitic protists that infect all types of organisms from other protists to humans. They take their name from the fact that several species are transmitted to the host in spore form, consisting of a single cell or a group of cells encased in a resistant membrane. They are nonmotile as adults, and they absorb rather than ingest their food. The complicated life cycle of sporozoans involves sexual and asexual phases.

Sporozoans of genus *Plasmodium* cause various forms of malaria. *Plasmodia* occupy two hosts at different stages of their life cycle. While they are parasites of the *Anopheles* mosquito, *Plasmodia* undergo a sexual phase of reproduction. After they are transferred to man in the bite of the mosquito, the reproductive process is asexual. The parasitic cycle is completed when an uninfected mosquito bites a person who has active malaria.

Ciliophora The complex unicellular protists in the phylum Ciliophora are most readily distinguished by their large numbers of cilia. Further classification within the phylum is based partly on the type and distribution of cilia. Ciliates are found in a variety of shapes.

The most frequently studied of the ciliates is the slipper-shaped *Paramecium.* Found in abundance in fresh water, *Paramecia* can be easily propagated, and they are large enough to be studied under the light microscope. Within the cell are several complex organelles whose functions are analogous to the organs of higher animals. The *Paramecium* has an oral groove, mouth, food vacuole, contractile vacuole, and an anal pore (see Fig. 8.8). It's cilia may be striated (similar to animal muscle cells), and they are attached to interconnected basal bodies, so that ciliary motion is synchronized and coordinated.

Ciliates possess two types of nuclei, the diploid micronucleus involved in reproduction and the polyploid macronucleus associated with cell metabolism. Reproduction usually proceeds by binary fission. However, genetic material may be exchanged by conjugation, in which two organisms temporarily fuse. During this time, the macronuclei disappear, and the micronuclei divide and fuse before the organisms separate.

The separation of monerans from protists represents a significant evolutionary boundary. Like fungi, animals, and plants, protists are eucaryotic. Protist cells are more complex than moneran cells, which are believed to be ancestral to protists. One recent theory suggests that protists were formed by the gradual symbiotic union of different types of procaryotic organisms within one cell membrane for the mutual benefit of both types of organisms. The chloroplast and the mitochondrion are organelles that may have originated in that way.

Within the phyla of protists, the parasitic sporozoans are thought to have evolved "downward" from more complex organisms that lost the ability to ingest food and the ability to move in the adult state. Evolutionists also believe that other forms of life—plants, animals, and fungi—had their origins in unicellular organisms very much like the protists.

Five-kingdom classification system

Several phylogenetic systems of classification are in use today, and modifications are constantly being made. The system presented here is adapted from R. H. Whittaker (*Science* 163: 150–160). The Whittaker five-kingdom system has gained widespread acceptance since its proposal in 1969. The following list is intentionally abbreviated, including only the major groupings and a few that, though not "major" in terms of size, are mentioned in this text because of their evolutionary importance. No attempt has been made to present groupings below the class level, with two exceptions: angiosperms, the plant class we know best, and mammals, the class to which we belong.

KINGDOM MONERA
- Phylum Schizomycophyta: bacteria
- Phylum Cyanophyta: blue-green algae

KINGDOM PROTISTA
- Phylum Euglenophyta: euglenoids
- Phylum Chrysophyta: diatoms, golden algae
- Phylum Pyrrophyta: fire algae, dinoflagellates
- Phylum Sarcodina: protozoans with pseudopodia
- Phylum Sporozoa: parasitic protozoans
- Phylum Zoomastigina: flagellates
- Phylum Ciliophora: ciliates

KINGDOM PLANTAE
- Subkingdom Rhodophycophyta
  - Division Rhodophyta: red algae
- Subkingdom Phaeophycophyta
  - Division Phaeophyta: brown algae
- Subkingdom Euchlorophyta

Branch Chlorophycophyta
- Division Chlorophyta: green algae
- Division Charophyta: stoneworts

Branch Metaphyta
- Division Bryophyta: liverworts, hornworts, mosses
  - Class Hepaticeae: liverworts
  - Class Antherocerotae: hornworts
  - Class Musci: mosses

Division Tracheophyta: vascular plants
Subdivision Psilopsida: psilopsids
Subdivision Lycopsida: club mosses
Subdivision Sphenopsida: horsetails
Subdivision Pteropsida: ferns
Subdivision Spermopsida: seed plants
Class Pteridospermae: seed ferns
Class Cycadae: cycads
Class Ginkgoae: ginkgos
Class Coniferae: cone-bearers
Class Angiospermae: flowering plants
Subclass Monocotyledonae: plants with single seed leaf
Subclass Dicotyledonae: plants with two seed leaves

KINGDOM FUNGI

Division Myxomycetes: slime molds
Division Zygomycetes: bread molds
Division Ascomycetes: sac fungi
Division Basidiomycetes: club fungi
Lichens: fungi in symbiosis with algae

KINGDOM ANIMALIA

Phylum Porifera: sponges
Phylum Cnidaria: coelenterates
Class Hydrozoa: hydra, hydroids
Class Scyphozoa: marine jellyfish
Class Anthozoa: sea anemones, corals
Phylum Platyhelminthes: flatworms
Class Turbellaria: free-living flatworms
Class Trematoda: flukes
Class Cestoda: tapeworms
Phylum Aschelminthes: cavity worms
Class Nematoda: roundworms
Phylum Acanthocephala: spiny-headed worms
Phylum Entoprocta: marine pseudocoelomates
Phylum Rotifera: microscopic wormlike animals
Phylum Bryozoa: moss animals
Phylum Brachiopoda: lamp shells
Phylum Echiurida: cylindrical marine worms
Phylum Sipunculida: peanut worms
Phylum Annelida: segmented worms
Class Polychaeta: tube worms
Class Oligochaeta: soil, fresh-water, and marine worms
Class Hirudinea: leeches
Phylum Mollusca: unsegmented animals with head, mantle, and foot
Class Monoplacophora: mostly extinct
Class Amphineura: chitons
Class Pelecypoda: bivalves
Class Scaphopoda: tooth shells
Class Gastropoda: snails, limpets, slugs
Class Cephalopoda: octopus, squid, Nautilus
Phylum Arthropoda: arthropods
Class Crustacea: crayfish, shrimp, crabs
Class Arachnida: spiders, ticks, scorpions
Class Insecta: bees, beetles, butterflies, grasshoppers
Class Chilopoda: centipedes
Class Diplopoda: millipedes
Phylum Echinodermata: radially symmetrical marine animals
Class Crinoidea: sea lilies, feather stars
Class Asteroidea: starfish
Class Echinoidea: sea urchins, sand dollars
Class Holothuroidea: sea cucumbers
Phylum Hemichordata: wormlike marine animals
Phylum Chordata: chordates
Subphylum Urochordata: tunicates
Subphylum Cephalochordata: lancelets
Subphylum Vertebrata: vertebrates
Class Agnatha: jawless fish
Class Chondrichthyes: cartilaginous fish
Class Osteichthyes: bony fish
Class Amphibia: frogs, toads, salamanders
Class Reptilia: turtles, snakes, crocodiles, dinosaurs
Class Aves: birds
Class Mammalia: mammals
Subclass Prototheria: egg-laying mammals
Subclass Metatheria: marsupials
Subclass Eutheria: placentals
Order Insectivora: insect eaters (moles, shrews)

Order Edentata: toothless mammals (anteaters, sloths)
Order Rodentia: rodents (rats, mice, squirrels)
Order Artiodactyla: even-toed ungulates (cattle, deer)
Order Perissodactyla: odd-toed ungulates (horses)
Order Proboscidea: elephants
Order Lagomorpha: rabbits, hares
Order Carnivora: meat eaters (cats, dogs, bears, seals)
Order Cetacea: whales, dolphins, porpoises
Order Chiroptera: bats
Order Primates: monkeys, apes, humans

Summary

1. Taxonomy, the science concerned with locating, identifying, and classifying the living things on Earth, began with the ancient Greeks. A Swedish naturalist, Linnaeus, developed in the eighteenth century the binomial nomenclature still in use today.

2. As new scientific theories and techniques developed, taxonomists adopted them to improve and refine the classification system and to determine evolutionary relationships. Taxonomy is now an interdisciplinary pursuit that uses information from biochemistry, genetics, and ecology in addition to morphology, anatomy, and physiology to determine species, the basic kinds of organisms. Increasingly, mathematical theory and computer science have contributed to taxonomy. Taxonomy was and will remain a science of judgment.

3. The most significant difference separating Kingdom Monera (bacteria and blue-green algae) from the other four kingdoms is the organization of the nuclear material. The DNA of procaryotic moneran cells is located in a diffuse area, the nuclear region, and is believed to be a single strand. Other organisms, the eucaryotes, carry their DNA in a distinct nucleus surrounded by a nuclear membrane. Monerans are probably quite similar to their Precambrian ancestors.

4. The Kingdom Protista has been called a "catch-all" classification. It includes eucaryotes that do not clearly belong in the three other eucaryotic kingdoms. With their nuclei, vacuoles, and plastids, they show more complicated intracellular structures than do monerans.

5. Protists may have evolved from the monerans, perhaps by a symbiotic process in which different types of moneran cells became associated within a single cell membrane.

REVIEW AND STUDY QUESTIONS

1. Explain what is meant by the statement "Taxonomy is a science of human judgment."

2. What contribution(s) of Linnaeus do we still consider useful?

3. How would you define "species"?

4. What are the advantages of the accepted system of binomial nomenclature? Are there any disadvantages?

5. What is the most important distinction between the Monera and Protista? What important characteristic do they have in common?

6. Before they were placed with blue-green algae in a separate kingdom, bacteria were usually included in the plant kingdom. In what ways are bacteria similar to plants, and in what ways are they different?

7. What is the principal reason that the classification of viruses is a continuing problem? How would you classify them?

8. Why is Kingdom Protista often called a "catch-all?

REFERENCES

Lwoff, A., and M. Tournier. 1966. "Classification of Viruses." *Annual Review of Microbiology* 20: 45.

Margulis, L. 1974. "The Classification and Evolution of Prokaryotes and Eukaryotes." In R. C. King, ed., *Handbook of Genetics*, Vol. 1. Plenum Press, New York.

Pigott, G. H., and N. G. Carr. 1972. "Homology between Nucleic Acids and Blue-Green Algae and Chloroplasts of *Euglena gracilis*." *Science* 175: 1259–1261.

Rizzo, P. J., and L. D. Nooden. 1972. "Chromosomal Proteins in the Dinoflagellate Alga *Gyrodinium connii*." *Science* 176: 796–797.

Simpson, G. G. 1961. *Principles of Animal Taxonomy*. Columbia University Press, New York.

Whitehouse, H. L. K. 1969. *Towards an Understanding of the Mechanism of Heredity*. St. Martin's Press, New York.

Whittaker, R. H. 1969. "New Concepts of Kingdoms of Organisms." *Science* 163: 151–160.

SUGGESTIONS FOR FURTHER READING

Blackwelder, R. E. 1967. *Taxonomy: A Text and Reference Book*. Wiley, New York.

A zoology reference book in taxonomy containing rules of zoological nomenclature.

Margulis, L. 1970. *Origin of Eukaryotic Cells*. Yale University Press, New Haven, Conn.

Deals with the accumulation and interpretation of data concerning the evolution of eukaryotes from prokaryotes.

Steinbeck, J., and E. F. Ricketts. 1941. *The Sea of Cortez: A Leisurely Journal of Travel and Research*. Appel, New York.

As its subtitle suggests, an enjoyable account of a collecting expedition taken by the novelist John Steinbeck with his friend, biologist Ed Ricketts, in the Gulf of California.

Updike, J. 1972. "Under the Microscope." In *Museums and Women*. Knopf, New York.

A short story about a biologist, included in a collection by the noted novelist and short story writer.

DIVERSITY OF LIFE: PLANTS AND FUNGI

chapter nineteen

Bald cypress trees growing in the Florida marshes are commonly draped with bromeliads, better known as Spanish moss. Each plant is uniquely adapted to its environment. The moss draws its nutrients from air and rain, not from the plants to which it usually attaches itself. The cypress sends out roots and "knees" above the water level to carry air to the waterlogged roots below.

In all systems of classification of living things, whether the systems were formal or informal, the plants have held a firm position as a separate group. Early taxonomists acknowledged just two kingdoms, those of the plants and the animals. Such an essentially simple division served well so long as the taxonomist was concerned with classifying mainly the familiar organisms. Just as horses and cows or bears and lions can be easily placed in the animal group, so can lilies and asters or lilac bushes and maple trees be identified as plants. The problems arose when taxonomists had to begin to classify organisms that were unquestionably living but that were much less easily recognized as plants and animals. Some of these organisms were placed in the two familiar kingdoms, and others were not. In this chapter we shall examine both familiar and unfamiliar organisms that are now considered plants, as well as that group of organisms, the fungi, that have now been assigned to a kingdom of their own.

Kingdom Plantae

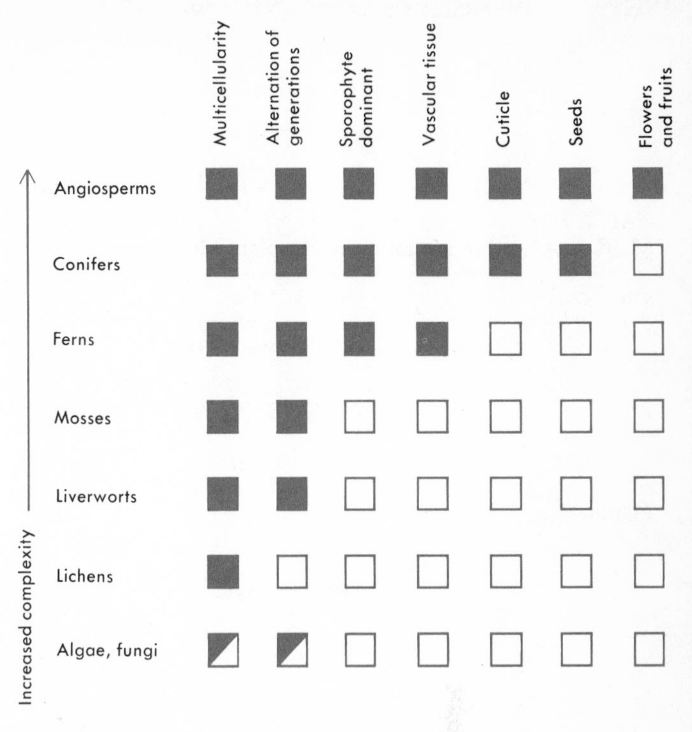

FIGURE 19.1

Selected structural advances in the plant kingdom, showing increased complexity by the accumulation of adaptive features. Shaded squares indicate that a characteristic is present, blank squares that a feature is absent. Plants have become increasingly specialized to survive on land, through water conduction systems and special reproductive features.

Most of the 500,000 species in Kingdom Plantae are easily recognized as green plants: grasses, trees, shrubs, garden flowers, seaweeds, garden vegetables. Important features of plants are summarized in Fig. 19.1.

In general, the plant kingdom includes multicellular eucaryotic organisms that are believed to have evolved from photosynthetic unicellular forms of life, and that exhibit advanced tissue differentiation. (However, we do place in this kingdom some unicellular green algae that do not exactly fit the definition of plant.) Most plants are photosynthetic autotrophs whose cellulose-walled cells contain chlorophyll *a*, *b*, or *c* in plastids. With the exception of some green algae, plants are nonmotile and are fixed to a substrate, such as soil. The vast majority of plants live in a terrestrial (land) environment rather than the aquatic (water) environment of monerans and protists. Their structural diversity has enabled plants to occupy desert, mountain, temperate, tropical, and arctic habitats.

Plants have complex life cycles, generally involving both sexual and asexual reproduction. This combination of sexual and asexual reproduction within one plant species results in an "alternation of generations." Figure 19.2 reviews a generalized life cycle. No single plant species will exhibit all modes of reproduction, and the importance of the phases varies with the particular species.

The reason the term "alternation of generations" is used to describe the life cycle of plants is that a diploid **sporophyte** generation in plants alternates with a haploid **gametophyte** generation. The best way to understand this is to follow a particular example. In the sporophyte generation, the plant has a diploid number of chromosomes—that is, one set of chromosomes from its male parent and one set from its female parent. The plant may exist independently (as in ferns and seed-bearing plants), it may be dependent on the gametophyte generation (as in mosses), or it may be restricted to a single diploid cell (as in unicellular green algae). Sporophytes reproduce by producing spores—hence the term

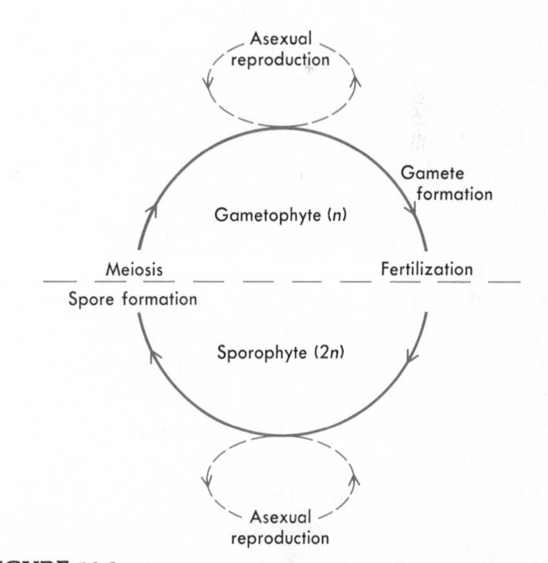

FIGURE 19.2

Generalized life cycle of a plant. Asexual reproduction may or may not occur, depending on the species. In some plants the gametophyte generation is dominant, in others the sporophyte.

TABLE 19.1
Main classification groupings of the plant kingdom

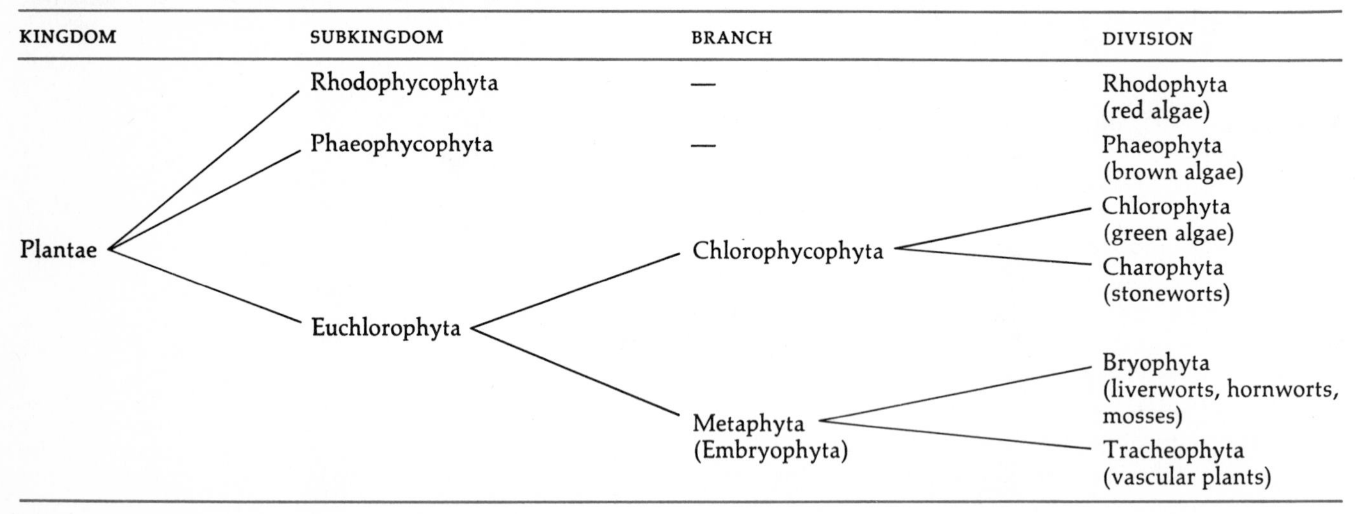

KINGDOM	SUBKINGDOM	BRANCH	DIVISION
Plantae	Rhodophycophyta	—	Rhodophyta (red algae)
	Phaeophycophyta	—	Phaeophyta (brown algae)
	Euchlorophyta	Chlorophycophyta	Chlorophyta (green algae)
			Charophyta (stoneworts)
		Metaphyta (Embryophyta)	Bryophyta (liverworts, hornworts, mosses)
			Tracheophyta (vascular plants)

"sporophyte." In all plants, spores are produced by the process of meiosis and are consequently haploid.

The next stage in a plant life cycle, the gametophyte, is created when a spore begins to divide mitotically to produce a multicellular body. This body, since it is derived from *mitotic* division of a *haploid* cell, is also haploid and is called the gametophyte. The gametophyte produces male and female gametes by mitosis in special organs. When male and female gametes unite, a diploid zygote is formed. In unicellular green algae, this zygote is the only diploid structure in the life cycle, but in multicellular plants, mitotic divisions of the zygote result in a fairly complex and diploid sporophyte.

The kind of alternation of generations just described is universal among green plants, and the life cycles of all plants can be studied as variations on that theme. Important points to bear in mind are that plant meiosis results in haploid spores rather than gametes (as in animals), and that gametes of plants are produced mitotically rather than meiotically (as in animals).

Plant classification is determined by several factors. Plants have been divided by many taxonomists on the basis of tissue differentiation and structural complexity into three main groupings (considered subkingdoms by some, grades by others): Rhodophycophyta, Phaeophycophyta, and Euchlorophyta, which are further divided into branches and divisions, as shown in Table 19.1. Except for the Tracheophyta, all are structurally primitive. They lack specialized tissues and do not differentiate into roots, stems, and leaves. A plant thus lacking in special differentiated parts is referred to as a **thallus.** Metaphyta (also called Embryophyta) are generally more complex structurally. They have developed a multicellular embryo sporophyte from multicellular sex organs.

Botanical divisions (equivalent to the phyla of the other kingdoms) are further distinguished by the types of chlorophyll found, the chemical composition of cell walls, the type of substance manufactured and stored as food, and the reproductive structures.

THE ALGAE

Although divisions Chlorophyta and Charophyta are members of subkingdom Euchlorophyta, we will discuss them first because of their structural simplicity and their tendency to form multicellular colonies. Some classification systems group the green, red, and brown algae into one subkingdom, the Thallophyta. However, they differ in several important ways: photosynthetic and accessory pigments, cell wall composition, food stored. The placement in separate subkingdoms recognizes these facts, as well as the probability that they do not share a recent common evolutionary ancestor.

DIVISION CHLOROPHYTA: GREEN ALGAE

Deposits at Bitter Springs in Australia seem to contain evidence that Precambrian green algae existed as long as a billion years ago. Evidence that they lived in Cambrian (600 million years ago) and Ordovician (500 million years ago) times is more certain. Higher plants are thought to have evolved from organisms much like the simpler chlorophytes found today.

Division Chlorophyta contains photosynthetic organisms that are unicellular, multicellular, or multinucleate. The gametophyte generation predominates. Unicellular or colonial forms found in this division show characteristics of protists and plants and might have been classified in either kingdom. They are placed with other green algae because they appear to share the same evolutionary line as higher plants, and they contain chlorophyll *a* and *b*. Other taxonomic schemes, however, do place chlorophytes and other eucaryotic algae in Kingdom Protista.

Green algae are mostly fresh-water organisms found in lakes, streams, puddles, and even snow banks. They show little tissue differentiation although there are exceptions. The thallus of *Ulva*, the sea lettuce, for example, may be composed of a thin leaflike portion and a rootlike holdfast (Fig. 19.3). Individual cells may show certain features not common to all.

The cell wall is usually two-layered with an inner cellulose layer and an outer gelatinous layer. There are usually flagella at some stage of the life cycle. Flagellated green algae often also have contractile vacuoles. Chloroplasts, one or several in number, contain chlorophyll *a* and *b* and carotenes. The form and location of chloroplasts within the cell provide aid in the classification of green algae. Starch is formed as a result of photosynthesis and stored in the chloroplasts or in leucoplasts.

There are several ways that evolutionary lines may diverge from a primitive motile unicellular organism. Examples of these possibilities are found in living green algal species. Flagella may be lost and movement may become amoeboid. This development is thought to lead to protozoan lines. The green alga *Pleurococcus*, for example, has no flagella.

FIGURE 19.3
Ulva lactuca. *Some green algae of the genus* Ulva *are commonly known as sea lettuce because the thallus has a thin leaflike portion that suggests lettuce.*

Organization of cells into a colony without loss of the motility of most of the individual cells is a second possible line of evolution to higher plants, but one with limited promise. Several green algae, of which *Volvox* is an example, show this type of organization. A development that meets with only moderate success occurs with the loss of both motility and the capacity for vegetative cell division while nuclear division is retained. *Oedogonium*, a filamentous green algae with a chitinous outer cell wall, shows this tendency.

The most successful divergence and the one that probably led to higher plants appears to be the loss of

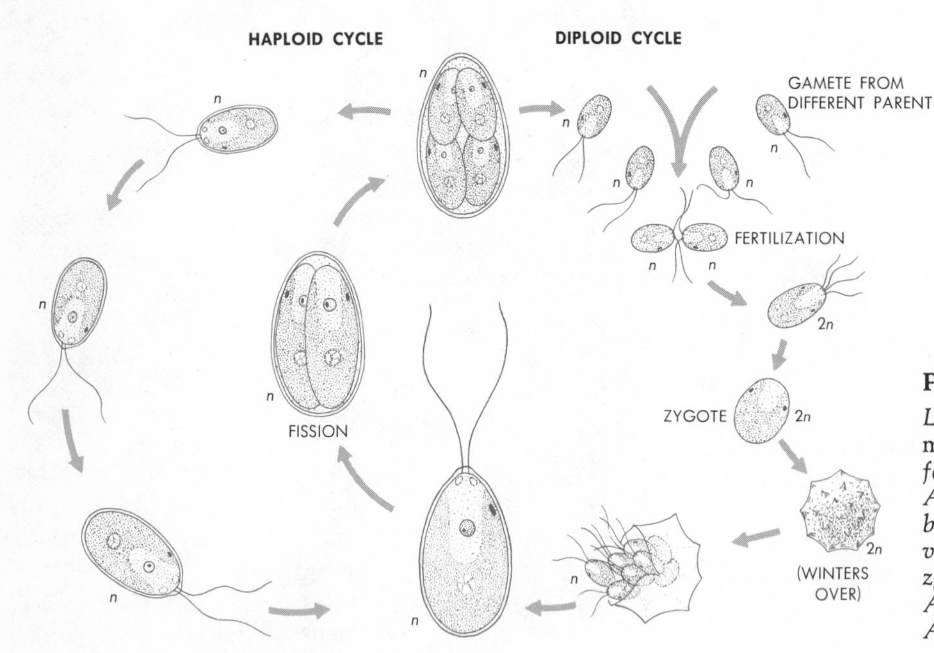

FIGURE 19.4

Life cycle of the green alga Chlamydomonas. *This organism is commonly found in pools, lakes, and damp soil. Asexual reproduction in the form of binary fission may occur, or two individuals may fuse to produce a diploid zygote. (From J. J. W. Baker and G. E. Allen,* The Study of Biology, *3rd edition, Addison-Wesley, 1977)*

cellular motility, except for gametes, with retention of the capacity for vegetative cell division (mitosis). The genera *Ulva* and *Ulothrix*, structurally the most complex of the green algae, illustrate this tendency.

The reproductive cycles of green algae are interesting and show a progression toward complexity. The life cycle of the unicellular *Chlamydomonas* is shown in Fig. 19.4. For most of its life cycle, *Chlamydomonas* exists in the haploid state. The cells divide mitotically, and in one season many generations of haploid cells are produced. Sexual reproduction occurs when some of these cells function as gametes and fuse. When two gametes fuse to form a diploid zygote, it exists only briefly. Meiosis occurs almost immediately and is followed by mitotic divisions to produce from 2 to 16 haploid cells within a single cell wall. The cell wall then breaks apart to release new free-swimming cells.

Table 19.2 shows a progression toward multicellularity formed by different genera of green algae. The highest stage of complexity of a flagellated colony is reached by *Volvox*, which has a definite cellular organization and involved reproductive processes (Fig. 19.5). *Volvox* forms hollow spherical colonies, one cell layer thick, of 500 to several thousand flagellated haploid member cells. In an immature colony, all the cells appear identical, much like *Chlamydomonas* cells. As time passes, some cells (gonidia) lose their flagella, enlarge toward the center of the colony, and divide mitotically until they contain a large number of haploid cells, which may be released to form a new colony.

TABLE 19.2
Representative genera of Chlorophyta showing increasing complexity of cellular arrangements

GENUS	NUMBER OF CELLS	ARRANGEMENT
Chlamydomonas	1	None
Pascheriella	3 to 4	Lateral attachment
Gonium	4, 8, or 16	Platelike in a single plane, one cell thick
Pandorina	4, 8, 16, or 32	Oblong or spherical
Volvox	500 to several thousand	Hollow sphere, one cell thick
Ulothrix	Multicellular	Single-filament or occasionally branched
Ulva	Multicellular	Leaflike, two cell layers thick

However, these enlarged cells can form two different types of gametes. The eggs of *Volvox* cannot move, but the sperm can. The enlarged cells that produce the two different types of gametes are called **antheridia** (male) or **oogonia** (female).

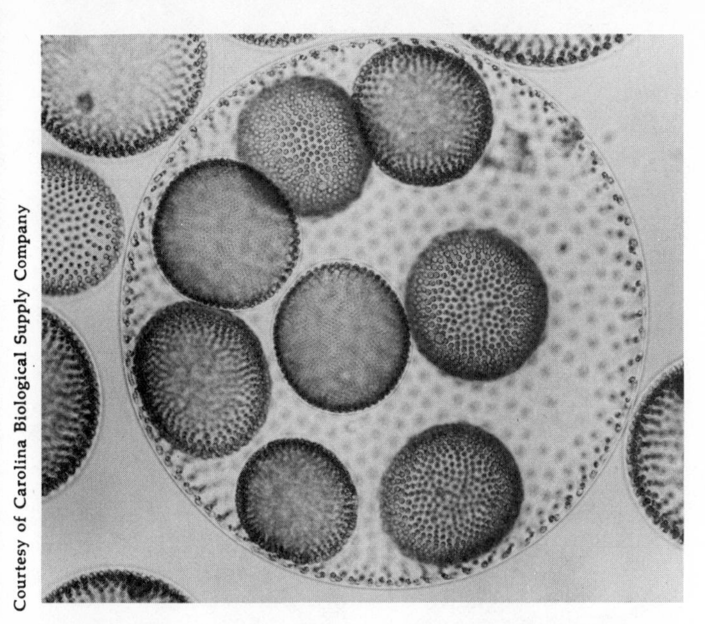

Courtesy of Carolina Biological Supply Company

FIGURE 19.5

Volvox aureus, *a colony of algae. The structure of the colony resembles the blastular stage of development of a multicellular organism. Within the* Volvox *shown here, one can see eight daughter colonies.*

Sperm released from the antheridia migrate through water to the oögonia, where fertilization occurs. The diploid fertilized egg, the zygote, remains dormant after the parent colony dies (usually in autumn). A new haploid colony originates (usually in spring) with a meiotic division to produce haploid zoospores.

Spirogyra, named for its spiral chloroplast, is a filamentous genus of Chlorophyta that does not produce flagellated gametes. Asexual reproduction occurs by simple cell division of the haploid filament cells or by fragmentation of the filament. Sexual reproduction involves a type of conjugation (see Fig. 12.3). Cells from two different filaments produce bulges in their cytoplasms and then fuse. Nuclear material from one cell moves into the other to form the zygote. Fertilization is followed by a resting stage. After a time, meiosis produces four haploid daughter cells, three of which disintegrate. The remaining cell divides mitotically to produce a new haploid filament.

Ulothrix, a filamentous genus that produces **zoospores** (independently motile spores), shows additional structural differentiation with the development of its basal cell into a holdfast that secures the organism to the substrate. Often the basal cell loses its chlorophyll.

Asexual reproduction occurs by simple cell division at the free end, resulting in elongation of the filament. Filaments may also be fragmented. Haploid zoospores, 2 to 32 in number, each with four flagella, may develop within a single specialized cell, the **sporangium,** of the filament. Once released, each zoospore can generate a new filament by losing its flagella and attaching to a substrate. A fourth method of asexual reproduction involves the formation of **aplanospores,** nonmotile spores with a cell wall. They are similar to bacterial endospores, except that mitotic divisions can occur within an aplanospore.

Sexual reproduction is heterothallic; that is, gametes from two different filaments, or thalli, fuse to form the diploid zygote. The gametes are **isogamous** (morphologically alike) but chemically different. Meiosis occurs within the zygote, and the zoospores that are released generate a new haploid filament.

Further structural complexity is evident in the marine genus *Ulva,* whose thallus consists of a leaflike structure of two cell layers and rootlike structures, rhizoids, composed of several filamentous holdfasts. Asexual reproduction in *Ulva* is minimal. Haploid gametes are released near the edges of the leaflike structure. These may produce a new haploid thallus or fuse into a flagellated zygote. The zygote divides *mitotically* to form a diploid thallus that is very similar in appearance to the haploid thallus.

Photosynthetic bacteria are thought to be the predecessors of green algae. The most primitive of the green algae probably was similar to *Chlamydomonas.* The types of organisms represented by *Volvox* (spherical hollow colony), *Ulothrix* (filamentous) and *Ulva* (two-layered leaflike) probably evolved from the *Chlamydomonas* type. The higher forms of Euchlorophyta were probably derived from green algal ancestors.

DIVISION CHAROPHYTA: STONEWORTS

The stoneworts are thought to have diverged from the green algae before the Devonian period. They have many branches and a rhizoid holdfast. Often their cellulose cell walls become calcified. No form of asexual reproduction has been observed. Gametangia, structures that produce gametes, form on the haploid thallus and are shielded by sterile jackets. Distinct antheridia and oögonia are produced on the same thallus.

SUBKINGDOM RHODOPHYCOPHYTA

DIVISION RHODOPHYTA: RED ALGAE

Most of the 3500 species of red algae are multicellular marine forms with a complex filamentous and branched thallus. Some are multinucleate, having as many as 4000 nuclei within a single cell wall. A few unicellular species exist. Some species are found with coral animals in coral reefs. Many fossil forms have been found dating to the Ordovician and Jurassic periods.

Neither vegetative cells nor gametes bear flagella, and there is no centriole. Within the cytoplasm is a large central vacuole. One or more chloroplasts contain chlorophyll *a,* and in many species the chloroplasts contain chlorophyll *d,* which is unique to red algae. Pigments found in the red algae include phycobilins, which give them their red coloration. Phycobilins act as accessory photosynthetic pigments that absorb the higher-energy blue wavelengths of light and transfer this radiation to the chlorophyll pigments. Because blue wavelengths of light penetrate water to greater depths, red algae can survive in deeper water than can most other marine plants.

Food is stored in granules as floridean starch, which is similar to portions of the starch of higher plants. A sugar composed of one molecule of galactose and one molecule of glycerol is also produced. Red algae are the only plants that produce galactose in great quantities.

Gametes of rhodophytes differ from other algae in that they have no flagella. The haploid gametophyte is the predominant form. Life cycles are quite complex and often not well understood. Asexual reproduction occurs through production of amoeboid nonflagellated spores.

The evolutionary position of red algae is uncertain. They share certain features with the blue-green algae, but red algae are eucaryotes, and it is improbable that the eucaryotic cell evolved more than once. The question remains unresolved.

FIGURE 19.6
The red alga Laurencia poitei, *a multicellular marine seaweed. Red algae have special pigments that enable them to absorb light and carry on photosynthesis at great depths, where other algae cannot survive.*

Red algae are sometimes used for food, particularly in Japan, where they are cultivated in tidal waters. From red algae industry derives stabilizers, suspending agents, and moisture retainers for use in a variety of commercial products, including ice cream, chocolate milk, salad dressing, cheese, and marshmallows. Agar produced from red algae is used as a gelatin and as a constituent of culture media for growing bacteria.

SUBKINGDOM PHAEOPHYCOPHYTA

DIVISION PHAEOPHYTA: BROWN ALGAE

Ranging in size from simple microscopic filamentous organisms to large plants with well-developed tissue regions and distinct morphological parts, the brown algae are known as "seaweeds" throughout the world. No unicellular species have been identified. Usually marine, they grow in the intertidal zone or are completely submerged. The brown color comes from the presence of a pigment that masks the green color of chlorophyll *a* and *c*.

Because it contains different kinds of sugars, the cellulose cell wall differs from that of other plants. The outer layer is composed of algin, a substance that is unique to phaeophytes. Food is usually stored as laminarin (a polysaccharide) but never as starch.

Structurally the brown algae are the most highly differentiated of the algae. They have true tissues. Rootlike rhizoids anchor the organism to the substrate, and stemlike structures—stalks or stipes—connect the leaflike blades to the rhizoids.

Reproduction may be asexual or sexual. Genera of phaeophytes show a trend toward dominance of the sporophyte generation over the gametophyte. Both generations are multicellular and exist independently of each other. Phaeophytes are classified largely by variations in their reproductive cycles. In the genus *Ectocarpus,* the sporophyte and gametophyte organisms are so similar that microscopic examination is required to distinguish between them. By contrast, *Laminaria* (kelp) has a microscopic filamentous gametophyte, but its sporophyte is very large, often reaching a length of 50 meters (Fig. 19.7). Spores are produced by meiotic divisions in sporangia on the blade of the thallus. Complete dominance of the sporophyte over the gametophyte is achieved in the genus *Fucus,* where there is *no* fully developed gametophyte organism. Separate sporangia on the sporophyte plant produce small flagellate spores or larger nonmotile spores. Instead of dividing to produce separate gametophytes, these spores function as gametes and fuse to form a zygote. The zygote develops normally into a diploid sporophyte. This anomalous life cycle, which is rather like an animal life cycle, is unique to *Fucus* and its close relatives.

Because of the presence of chlorophyll *c*, a high proportion of carotenes, and similarities in flagella, brown algae are thought to be closely related to the golden algae, which are here included among the protists. Since the brown algae have flagellated gametes, they probably had a flagellated ancestor, but little more is known because the fossil record of brown algae is poor.

FIGURE 19.7

Laminaria, *commonly known as kelp. One of the most familiar of the brown algae, kelp is shown in Oregon coastal water at low tide.*

TABLE 19.3
Comparison of the major characteristics of typical eucaroytic algae

CHARACTERISTIC	CHLOROPHYTA	RHODOPHYTA	PHAEOPHYTA
Habitat	Mostly fresh-water	Mostly marine	Mostly marine
Cellular organization	Unicellular, colonial, multicellular, multinucleate	Branched filaments, some multinucleate	Multicellular; definite tissue regions
Cell Wall Composition			
Inner	Cellulose	Carbohydrate polymers, some cellulose	Cellulose with pentose sugars
Outer	Pectic	Pectic with sulfated galactose	Algin
Photosynthetic pigments	Chlorophyll *a* and *b*	Chlorophyll *a* and *d*	Chlorophyll *a* and *c*
Other pigments	Carotenes, xanthophylls	Carotenes, xanthophylls, phycobilins	Carotenes, xanthophylls
Food stored	Starch or oils	Floridean starch, floridoside	Laminarin
Dominant generation	Gametophyte or shared equally	Gametophyte	Gametophyte, sporophyte, or shared equally
Motility	Flagellated gametes; some flagellated vegetative cells	No flagella observed	Flagellated gametes

Brown algae are sometimes harvested for use as fertilizer. Kelp and other species are important food sources in Japan.

Major characteristics of all algae are summarized in Table 19.3.

SUBKINGDOM EUCHLOROPHYTA: BRANCH METAPHYTA

Plants classified as metaphytes (also called embryophytes) consistently illustrate a well-developed alternation of generations. In mosses and liverworts, the gametophyte is the dominant stage, and the sporophyte is dependent or parasitic on the gametophyte. In the transition to plants that have a conducting system (tracheophytes, or vascular plants), the dominant stages are reversed, and the sporophyte is more evident and better developed. In the more primitive vascular plants (e.g., ferns and horsetails), early development of the sporophyte is dependent on the gametophyte, but once the sporophyte begins photosynthetic activities, the gametophyte degenerates and dies. With the advent of seed plants, the gametophyte becomes quite dependent on the sporophyte. It also becomes so reduced in size that it cannot be seen except with a microscope.

The cell wall of metaphytes is characteristically composed of cellulose, and the cells contain chlorophyll *a* and *b*, as well as other pigments. Food is stored as starch. Sporophytes have stomata and guard cells (not found in lower plants), which facilitate gas and moisture exchange. Vascular tissue enables water, nutrients, and organic molecules to travel throughout the plant. Both vascular tissue and a modification of the sex organs contributed toward freeing the plants from a completely aquatic environment. One result was that land colonization was successful. The bryophytes, however, lack true vascular tissue. They have only a few anatomical remnants that suggest conductive tissue.

DIVISION BRYOPHYTA: LIVERWORTS, HORNWORTS, AND MOSSES

Organisms of this 25,000-species division reproduce by an alternation of generations. Fertilization must occur in an aquatic medium, which may be simply a film of dew or droplets of rain. For this reason and because they have no well-developed vascular system, bryophytes are quite small, and they occupy moist habitats, usually on or near the ground.

Bryophytes, which have been found in the fossil record since Cambrian times, spend the greater part of their life cycle as gametophytes (see Fig. 12.6). The gametophyte is always photosynthetic, and it is larger than the sporophyte. Gametes are produced by specialized multicellular structures on the gametophyte. These structures are the **archegonium,** which produces a single nonmotile egg, and the **antheridium,** which produces hundreds of biflagellate sperm. After fertilization, the diploid sporophyte, usually nonphotosynthetic, grows out of the old archegonium and remains a parasite of the gametophyte. Haploid spores produced by meiotic division within the sporophyte are released and produce the new gametophyte.

Bryophytes are most successful in moist climates. They are very hardy plants, however, and can survive in a dormant stage during dry or cold periods. They are found at high altitudes, in arctic regions, and even in arid localities. The most familiar of the bryophytes are the mosses, but liverworts and hornworts are also in this division. Although the life cycles of all three types are similar, the organisms are readily distinguished by the appearance of the gametophyte and sporophyte generations.

The hornworts are inconspicuous plants but worldwide in distribution. They are characterized by flattened gametophytes that lack any specialization into complex organs. Rhizoids anchor the underside of the thallus to the substrate. Antheridia and archegonia are also produced on the lower surface of the thallus. After fertilization of the egg by a sperm, a zygote is produced in the archegonium. The zygote divides to generate the sporophyte, which in hornworts is reduced to a spindle-

FIGURE 19.8

The liverwort Marchantia. *The bryophytes are the least complex multicellular organisms to exhibit a dimorphic life cycle—an alternation of generations between the sporophyte and the gametophyte. The single flat-topped stalk at the top of the photograph is the antheridial structure on a male plant. The fluted tops of the archegonial receptacles on a female plant are visible just below and to the left, but their stalks are hidden.*

shaped structure only a few millimeters long. Most of the tissue of the mature sporophyte is given over to the production of spores, and the sporophyte is nutritionally dependent on its parent gametophyte.

Some liverworts have a simple, flattened thallus like that of a hornwort. Others have gametophytes that are "leafy" in appearance or otherwise more complex. *Marchantia* is an example of a liverwort with a particularly elaborate gametophyte (Fig. 19.8). In this genus, the antheridia and archegonia are carried on umbrella-like stalks. As in the hornworts, the sporophytes of liverworts are reduced to spore-producing capsules. The sporophytes of liverworts generally have short stalks.

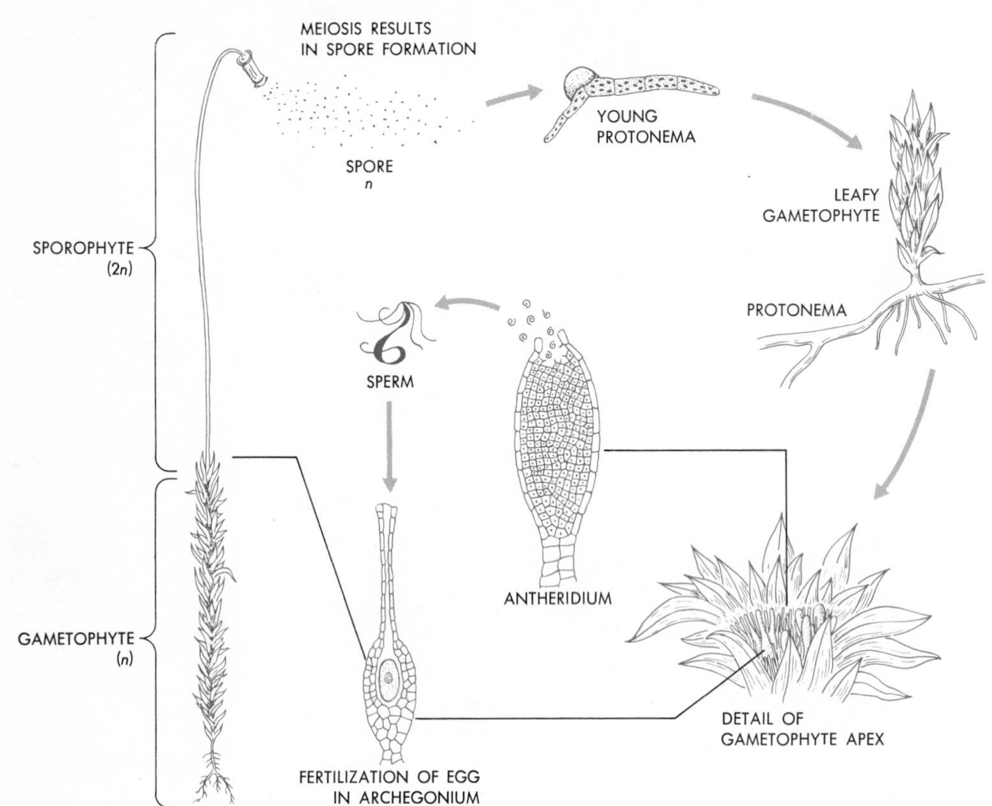

FIGURE 19.9

Life cycle of a moss. (From J. J. W. Baker and G. E. Allen, The Study of Biology, *3rd edition, Addison-Wesley, 1977)*

The conspicuous part of a moss life cycle is, of course, the leafy gametophyte (Fig. 19.9). Hidden among the leaves are the antheridia and archegonia. As the sporophyte of a moss grows up out of the archegonium, it differentiates into a capsule (which will produce the spores), a stalk, and a "foot," which absorbs water and nutrients from the parent gametophyte. Meiosis in the capsule produces haploid spores, which can be dispersed substantial distances by the wind. Upon dividing, these spores first develop into a multicellular filamentous structure, which resembles a filamentous green alga and from which the conspicuous green gametophyte develops.

Some scientists have believed that bryophytes form an evolutionary link between green algae and true vascular plants. More recently, others have suggested that bryophytes may be an offshoot of a more complex ancestral vascular plant in which the gametophyte and sporophyte were independent. Evidence for this hypothesis is the presence of stomata and meristematic tissue on some hornwort sporophytes.

DIVISION TRACHEOPHYTA: VASCULAR PLANTS

The key to success in a land environment was the evolution of the vascular system. Some classification schemes treat the major groups of vascular plants as separate divisions. Others (including Whittaker, whose system we are following in this text) consider them subdivisions of division Tracheophyta. That tracheophytes have led a primarily terrestrial existence is the result of several adaptations.

The development of xylem and phloem enables the plant to transport water and nutrients over longer distances. Consequently, the plant does not need to be very close to or immersed in an aquatic environment. However, a plant that is to grow large without the buoyant support of water must become rigid. Tracheophytes have developed woody stems, stalks, or trunks, all of which allow the photosynthetic tissue to be held high in the air. Lignin in the xylem cells of many vascular plants gives them rigidity. Turgor in cells also increases rigidity.

Algae, bryophytes, and some tracheophytes depend on water as the medium of fertilization. As plants grow larger, however, their reproductive organs are farther away from the aquatic environment and are increasingly exposed to air. As we shall see, tracheophytes have developed special means of protecting their gametes from dehydration and of ensuring fertilization of eggs in a nonaquatic environment.

Tracheophytes show the most diverse tissue differentiation of the plant kingdom. They usually have extensive root and leaf systems. Reproductive organs and the structures associated with them are often ingenious. The sporophyte generation dominates the gametophyte in all subdivisions. The gametophyte may or may not be physiologically independent of the sporophyte.

FIGURE 19.10

Club moss, a member of the genus Lycopodium, *better known as princess or ground pine. Visible in this photograph are conelike strobili, consisting of sporophylls grouped together.*

Subdivision Psilopsida A very small division, the Psilopsida include only three living species. Some authorities believe that living psilopsids (e.g., *Psilotum nudum*) are descended with only minor modifications from the earliest vascular plants, the Rhyniopsida of Silurian and Devonian age (425 million years ago).

Psilopsids are simple vascular plants whose dominant sporophyte is small, a foot or less in length, and rootless, having an underground extension of the stem, the rhizome, for support. *Psilotum* has small scalelike leaves, but plants of the genus *Tmesipteris* have well-developed leaves. Sporangia are borne on the ends of short branches attached to the main stems. The gametophyte plants are tiny, subterranean, and without chlorophyll. Archegonia and antheridia form on the same gametophyte organism. Fertilization proceeds, as with bryophytes, with multiflagellated sperm swimming through water toward the eggs. The zygote divides mitotically and develops into a new sporophyte, which is briefly dependent on the gametophyte.

Subdivision Lycopsida The next step toward increasing complexity in vascular plants is illustrated by the club mosses. Lycopsids have true roots, and their leaves are differentiated and attached to the stem at a node. Growth at the stem proceeds through mitotic divisions of an apical meristem. Sporangia are found on the upper surface of specialized leaves, sporophylls, rather than at the tips of branches as in the psilopsids. Sporophylls are often grouped together in a conelike structure called a **strobilus.**

Lycopsids also made their first appearance in Silurian times. Although only four or five genera have survived to the present, about 1200 species still exist. Consequently, they are considered more successful than psilopsids. Today they are low-growing viny plants that may stretch along the ground for several meters. The genus *Lycopodium*, ground pine, is a familiar ground cover in North American forests (Fig. 19.10). But in the Carboniferous period, lycopsids were the dominant plants, and they grew to heights of 30 to 40 meters. Later their remains contributed to coal formations.

Reproductively, lycopsids may differ from psilopsids. The sporophyte of *Lycopodium*, like that of a psilopsid, produces spores that will develop into a gametophyte capable of producing *both* antheridia and archegonia. However, sporophytes of other genera (e.g., *Selaginella*) produce two different types of spores, large and small. The small ones, microspores, develop into male gametophytes that will produce only antheridia. The larger megaspores will become female gametophytes that will produce only archegonia.

This type of life cycle, in which unisexual gametophytes are produced from different types of spores, is called **heterospory.** Heterosporous plants include some lycopsids, a few ferns, and all seed-bearing plants. In addition to the fact that the gametophytes are unisexual, another remarkable feature of heterosporous plants is that the gametophytes develop and reach maturity within the cell wall of the parent microspore or megaspore. The gametophytes are consequently very small and do not live as independent photosynthetic plants. The alternative condition, **homospory,** is the one we have encountered in the bryophytes. That is, only one kind of spore is produced, and it gives rise to bisexual gametophytes. Apart from bryophytes, homosporous plants include *Lycopodium*, the living sphenopsids, and most ferns.

When eggs have matured, parts of the archegonia disintegrate and become slimy, providing the aqueous pathway for sperm to reach egg. The young sporophyte is at first dependent on the gametophyte.

Subdivision Sphenopsida Included among the sphenopsids are fossil forms that are believed to have evolved from rhyniopsid or psilopsid ancestors. Today only the genus *Equisetum* with about 25 species survives. Sphenopsids, called horsetails and scouring rushes, are found only near bogs and streams, but where they do exist they may be quite abundant. The largest species grows to a height of about one meter.

The only evolutionary advance of any importance exhibited by the sphenopsids is that some fossil species had a cambium layer on their stems. Like lycopsids, sphenopsids have true roots, stems, and leaves. Gametophytes and sporophytes are independent organisms. The genus *Equisetum* is strictly homosporous, but some fossil sphenopsids seem to have been heterosporous.

As in all vascular plants, the sporophyte is the dominant conspicuous phase of the life cycle. *Equisetum* sporophytes consist of above-ground photosynthetic shoots and underground shoots called rhizomes. Both kinds of shoots are conspicuously jointed, with whorls of leaves arising at the joints. The sporangia are carried in umbrella-shaped structures, sporangiophores, which are aggregated into strobili. The gametophyte is tiny and photosynthetic, and it produces eggs and sperm on the same organism.

Subdivision Pteropsida Ferns can be separated from all other nonseeded vascular plants by their large leaves with complex vascular systems. The origins of ferns are obscure, but they probably arose along with other vascular plant groups in the Devonian period. In the Carboniferous period, together with giant lycopsids and sphenopsids, ferns were important components of vegetation and contributed to the formation of coal. Today approximately 10,000 living species are found throughout the world, particularly in shady moist habitats. Their size range is very large. Tropical tree ferns may grow to a height of 20 meters, yet the smallest fern may rise only a few millimeters above the ground.

The dominant leafy sporophyte is structurally complex. In ferns of temperate regions, the shoot is an underground rhizome, and the leaves are the only above-ground structures. (Tree ferns of tropical forests are the only ferns that have conspicuous above-ground stems.) In the spring, new leaves develop from the apical meristem of the rhizome, grow up through the soil, and are seen with their tips coiled into "fiddleheads." The mature leaf is large and complex with a branching vein system. It consists of the leafy blade and the stalklike petiole, which connects the blade to the stem.

Sporangia are located on the undersurface of the leaf, often in a mass (sorus), sometimes on specialized nongreen leaves, or even in seedlike structures (sporo-

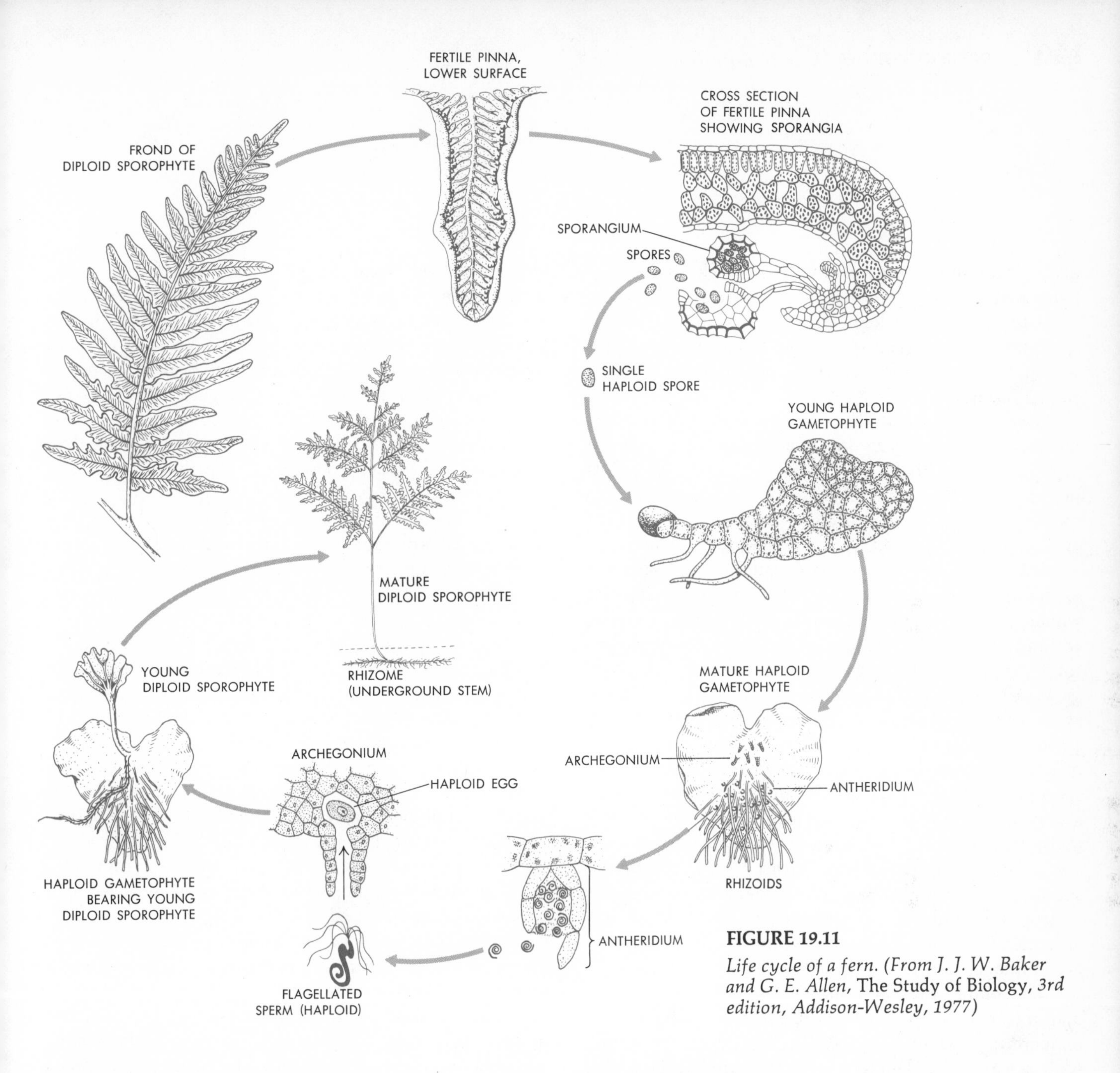

FIGURE 19.11

Life cycle of a fern. (From J. J. W. Baker and G. E. Allen, The Study of Biology, *3rd edition, Addison-Wesley, 1977)*

carps). Most ferns are homosporous and form a distinct independent gametophyte (Fig. 19.11). The small, heart-shaped gametophyte has anchoring rhizoids growing from its lower surface. Both antheridia and archegonia develop on the lower surface of the gametophyte. The multiflagellated sperm swim through water to reach the archegonia, where fertilization takes place. The sporophyte embryo is at first dependent on the gametophyte, but once it has begun to mature, the gametophyte will die.

Subdivision Spermopsida The spermopsids, or seed plants, are the most diverse and familiar members of the plant kingdom. Almost exclusively land plants, they have successfully adapted to nearly every conceivable habitat that can support life. Believed to have ancestral forms related to ferns, the first seed plants probably originated in the Carboniferous period and were well established by the Permian.

Again, the sporophyte generation is the more conspicuous. Now only the immature male gametophyte leads a partially independent existence. The female gametophyte is completely dependent on the sporophyte. Since fertilization no longer takes place in an aqueous environment, seed plants have developed ways of protecting their gametophytes and gametes from environmental hazards.

Seeds are anatomically very complex structures. The condition of heterospory was a necessary development in the evolution of the seed. Megaspores of seed plants are produced by megasporangia, which, together with additional envelopes of protective tissue, form structures called ovules. Within the ovules, meiosis in a single cell produces four megaspores, but only one survives to create a megagametophyte. The megagametophyte remains small for its entire life and reaches maturity within the ovule. Fertilization of the megagametophyte's single egg produces a zygote, and the young sporophyte (embryo) develops for a while before ceasing growth and becoming dormant. A seed, then, results from the fertilization of an ovule, and it contains tissue of the megasporangium, megagametophyte, and embryo (the sporophyte of the next generation). The microsporangium produces microspores meiotically. They divide mitotically to produce partially developed gametophytes, the pollen grains, which are released from the microsporangium, or pollen sac. The gametophytes exist for a time independently of the sporophyte. With a few exceptions, the male gametes within the pollen grains are not flagellated.

The outer walls of pollen grains are very durable and are not destroyed by strong acids or bases, even at temperatures as high as 300°C. The walls are highly sculptured, and pollen grains can usually be identified to the level of genus or species by the details of sculpturing. Pollen grains accumulate in the sediment of lakes and bogs. Subsequent analysis of the sediment gives an indication of the flora present in previous centuries and thus some indication of climatic conditions in the past.

After fertilization, the ripened ovule—the ovary, or seed—is protected by the hardened tissue that becomes the seed coat. The embryo contained inside the seed remains in an arrested stage of development until germination. A food reserve may or may not be included within the seed. Seeds can be extremely tough. In colder climates they must be able to withstand repeated freezing and thawing before they germinate, usually in the spring following their development. In arid climates they resist drying out, sometimes for years, before climatic conditions become right for germination to occur. Seeds can pass undigested through the intestinal tracts of animals and birds. Seeds that were more than 1000 years old when found in a Manchurian lake bed germinated!

Seed plants have also developed extensive root systems, which are capable of extracting minute amounts of water and minerals from the ground and which serve to anchor them to it. Roots may also serve as food reservoirs.

The development of rigid woody stems is characteristic of many spermopsids. Although many flowering plants are **herbaceous** (having little or no woody tissue), they are generally believed to be derived from woody ancestors. A woody stem is formed by the activity of a cylindrical meristem, the vascular cambium, located between the wood and the bark. It forms xylem inwardly and phloem outwardly. Wood is therefore xylem. The bark is formed of phloem and of cork, which arises from another cylindrical meristem, the cork cambium. Cork, the conspicuous part of the bark of most trees, is harvested in commercial quantities from the oak *Quercus suber*. The tremendous strength of wood is due to the presence of lignin, which cross-links the cellulose strands in the xylem cell walls.

Leaves, varied in arrangement, are the principal site of photosynthesis. Chlorophyll *a* and *b* are the photosynthetic pigments, but leaves also contain carotenes and xanthophylls. Food is stored as starch within plastids in the leaves or in the roots.

Spermopsids are divided into two broad categories, the **gymnosperms** (naked-seeded) and the **angiosperms** (enclosed-seeded). Although most classification systems recognize significant enough differences among the various gymnosperms to consider them separate classes, gymnosperms as a group do share certain attributes, and it is convenient to discuss them in general terms. The female gametophyte is always multicellular and contains archegonia. The ovules are carried on scales or leaves and are not otherwise protected. In angiosperms, the megagametophyte is reduced to a few nuclei—usually eight—no archegonia are present, and the ovules are protected by modified leaves called carpels.

Gymnosperms. The classes of seed plants in the gymnosperm group that we will consider here are Pteridospermae (seed ferns), Cycadae (cycads), Ginkgoae (ginkgos), and Coniferae (conifers). All except the conifers are known either exclusively as fossils or principally as fossils, with a few relic living genera.

Although seed ferns were numerous during the Carboniferous period, none survive today. With their fernlike leaves, they presented an interesting puzzle to paleontologists and were for a long time considered to be ferns, pteropsids, rather than seed plants. The clue to their correct classification came when fossils were discovered with the seeds actually attached to the margins of the leaves. These seed ferns were among the dominant plants of the Carboniferous period. They may be considered a form showing development between true ferns and more advanced gymnosperms.

The cycads, with nine surviving genera, are found in tropical and subtropical regions. They resemble palm trees, which are angiosperms, however. The stem is topped by a crown of leaves that is newly formed each year. All cycads are dioecious; that is, the plants bear either mega- or microsporangia, not both. In some cycads the sporophylls are aggregated into conspicuous cones.

Closely related to the cycads are the fossil cycadeoids that were common in the Mesozoic period. The chief difference seems to be in the arrangement of the sporophylls, in which male and female are found closely associated on the same plant. For a time, biologists thought this arrangement was the forerunner of the flower, but when studies uncovered too few anatomical similarities, the idea was dropped.

Only one living species of the ginkgos survives to present times. Called the maidenhair tree because its leaves resemble those of the maidenhair fern, *Ginkgo biloba* is a native of China. Male and female sporophylls

FIGURE 19.12

The bristlecone pine, the oldest known living organism. Some of these trees are estimated to be more than 4000 years old.

are found on separate trees. Female sporangia develop at the end of short stalks, and male cones are similar to those of conifers. Ginkgos and cycads are the only spermopsids that have flagellated sperm. Ginkgo seeds are foul-smelling, but the tree is attractive.

The **conifers** (cone bearers), the most successful surviving gymnosperms, make up a conspicuous portion of the world's flora. Even so, there are only about 500 living species—not nearly so many as one might expect. They are especially successful in the cooler regions of the northern hemisphere, where vast forests of conifers may be made up of a single species. Familiar conifers, most of which are trees, are hemlock, pine, sequoia, redwood, cedar, juniper, larch, and spruce. The oldest and largest living things are conifers. Some bristlecone pines (Fig. 19.12) in the mountains of California and Nevada are more than 4000 years old. Sequoia and redwoods, which grow to heights exceeding 300 feet, are nearly as old. Conifer fossils have come from as far back as the late Carboniferous period.

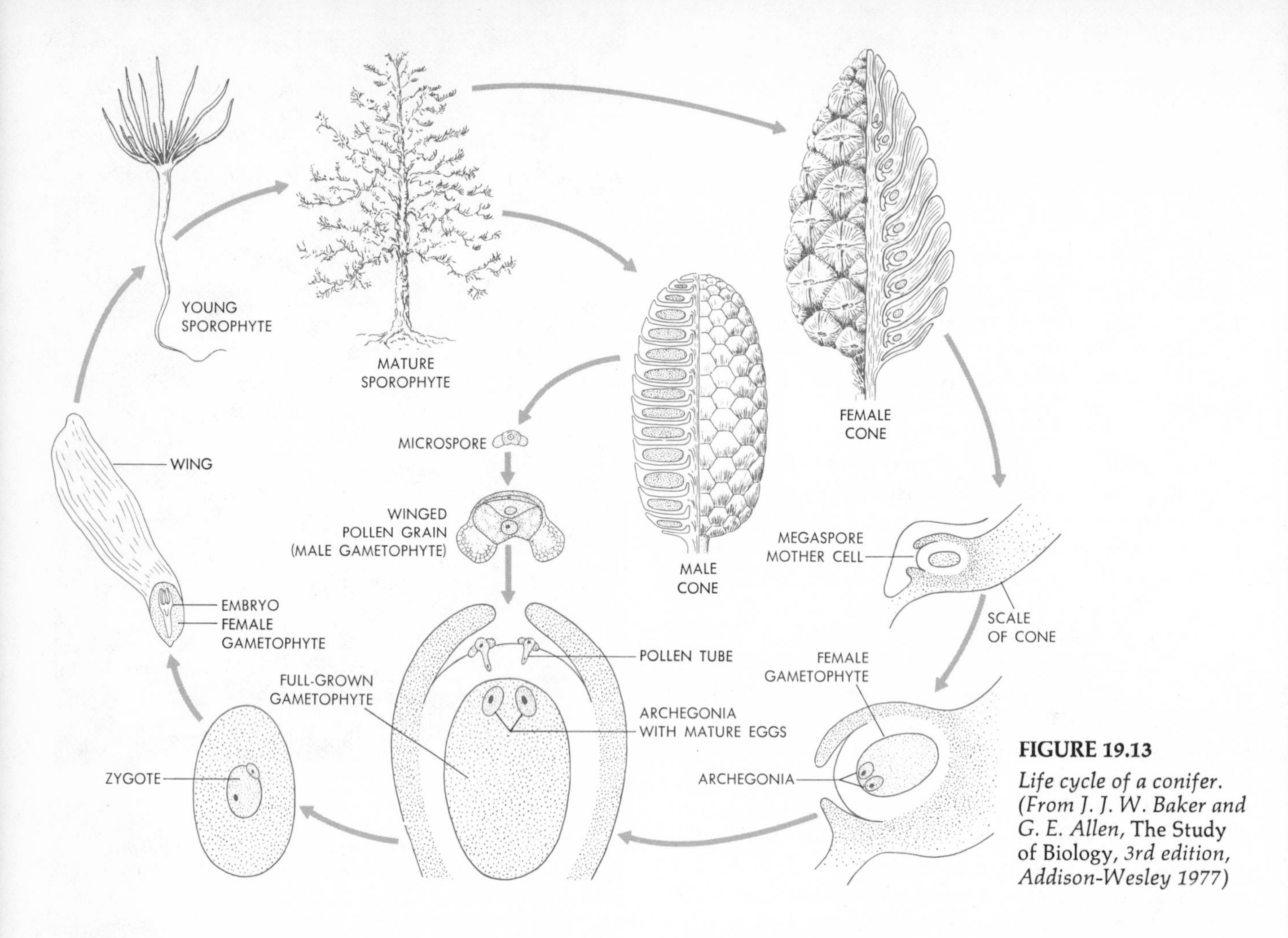

FIGURE 19.13
Life cycle of a conifer. (From J. J. W. Baker and G. E. Allen, The Study of Biology, *3rd edition, Addison-Wesley 1977)*

Leaves of conifers are usually small, simple, and needlelike or scaly. Most conifers are evergreens, keeping their leaves throughout the winter. A few, such as the larch, are deciduous, shedding their leaves in autumn.

Male and female cones are usually found on the same tree. Male cones, which form on short branches, begin development in one growing season and mature the following spring (Fig. 19.13). They are generally small—one to ten centimeters in length. Pollen, often winged, is produced in enormous quantities and spread entirely by the wind. Female cones are larger than the male. After pollination occurs, the seeds may remain in the female cone for varying amounts of time, sometimes for years. Cones of some species of pine may not release seeds until after a forest fire.

In economic terms, conifers are the most important of the gymnosperms. Their wood is used for lumber, plywood, chipboard, matchsticks, and toothpicks. Paper is made from the wood pulp, and turpentine and rosin are made from the sticky sap. Conifers are widely used for landscaping. The Indians of the Pacific Northwest, from Oregon to southern Alaska, made extensive use of the versatile cedar. They hollowed out the trunk for dugout canoes, carved it into totem poles, masks, or food vessels, and split it into house boards. The roots provided them with tough cord.

Angiosperms. It is the development of the flower that sets angiosperms apart from all other plants. Flowering plants include broad-leaved deciduous trees and shrubs, grains and other grasses, garden flowers and vegetables, most weeds, and several poisonous plants.

As with gymnosperms, the sporophyte is the conspicuous generation, with roots, stems, and leaves. But the life span of angiosperms is much more variable. Annuals, which include many flowers and vegetables, live a single year. Biennials live two years and do not produce flowers until the second year. Those that live longer are perennials.

TABLE 19.4
Comparison of monocots and dicots

	DICOTS	MONOCOTS
Cotyledons	2	1
Cambium	usually	occasionally
Secondary growth	usually	occasionally
Petiole	usually	occasionally
Vein patterns	net	parallel
Vascular bundles	circular or fused	scattered
Flower parts	multiples of 4 or 5	multiples of 3
Examples	oaks, roses, beans	palms, lilies, corn

Angiosperms are divided into two groups, depending on the type of seed they produce. Within the seed coat are several structures: the embryo, or immature sporophyte; an optional endosperm, or food storage tissue; and either of two types of leaflike structures, the cotyledons. Monocotyledons (or monocots) have only one cotyledon; dicotyledons (dicots) have two. Examples of monocots are corn, grasses, onions, and lilies. Some of the more numerous dicots are peas, beans, pansies, fruit trees, and sunflowers. Other characteristics of monocots and dicots are given in Table 19.4.

The reproductive organs of the plant are in the flower. Male organs (stamens) and female (pistils) are both found on a "complete" flower. However, some flowers are "imperfect." The stamens of corn, for ex-

FIGURE 19.14
Reproductive mechanisms of Hibiscus, *a flowering plant. The photograph clearly shows the male stamens and female pistils found on a complete flower. A pollen grain produced by the stamen of the hibiscus is shown in the scanning electron micrograph.*

Beidler/Omikron 1300X

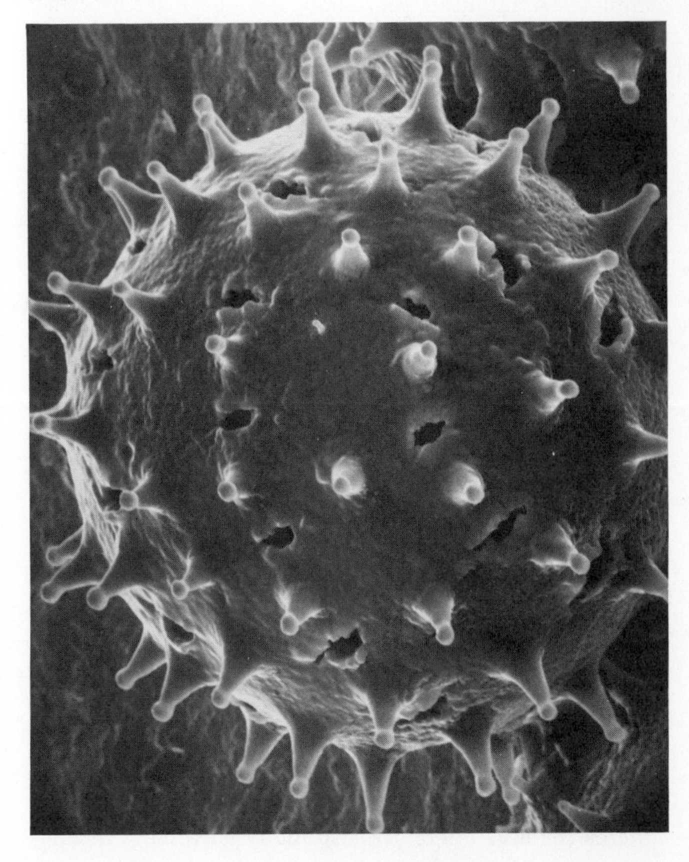

ample, are produced in the tassels (the flowers), but the pistils are found at the bases of leaves. Willows and cottonwoods, among others, bear male and female flowers on separate plants.

Pollen produced by the stamen is transported to the pistil in several ways. Plants with complete flowers may self-pollinate, but they are often prevented from doing so. Stamens and pistils on a given flower may mature at different rates, for example, or the reproductive organs may differ in size. (If the pistil is taller than the stamen, self-pollination is highly improbable.) Pollination may also be effected by the wind. Most cross pollinations between different individual flowers are the work of insects, birds, or bats. Such flowers are usually large and showy, and they often produce nectar that attracts the animal pollinator.

Our knowledge of angiosperm ancestry is very incomplete. A hundred years ago, Charles Darwin called the origin of these plants an "abominable mystery," and the subject remains controversial today. Most evidence on early angiosperm plants comes from study of pollen grains, because harder structures, such as pollen grains, became fossilized to a greater extent than did the softer plant structures like flowers and leaves. Most authorities agree that the earliest definite angiosperm pollen is from the beginning of the Cretaceous period (136 million years ago) and that by the end of the Cretaceous (65 million years ago) a complex angiosperm flora had evolved, containing many groups that are alive today.

The main groups of insects we know today were also fully developed by the late Cretaceous period. Insects, birds, and flowers seem made for one another. Since bees are capable of distinguishing colors at the blue end of the spectrum, they are attracted to white, blue, and yellow flowers. Flowers pollinated at night by moths are often white. Hummingbirds, which can see red as a color, are found especially in the tropics, where there are many red-flowering species of plants.

The anatomy of a flower often ensures that the body of its pollinator will be dusted by pollen, which will in turn be transferred to the stigma of the next flower visited. Since stigmas are often sticky, pollen adheres readily. There are also many surprising mechanisms involved in pollination. For example, the flowers of some orchids look very much like female wasps of a certain species. Male wasps, mistakenly attracted to the flowers, pollinate them.

After the pollen grain has reached the stigma at the tip of the pistil, it germinates. A pollen tube, the male gametophyte, grows through the pistil until it reaches the embryo sac, the female gametophyte (Fig. 19.15). The nonflagellated male gametes contained in the pollen tube reach the female gametophyte, where one male gamete fuses with part of the embryo sac to form the endosperm of the seed. A second male gamete fertilizes the haploid egg cell to form the diploid zygote.

Once fertilization and seed formation have occurred, the seeds are released from the parent plant and dispersed. Some plants have not developed any special mechanisms for seed dispersal, but many have. Some seeds, such as the maple's, are winged; others, the dandelion and milkweed among them, have feathery parasols attached. Seeds of both these types are scattered by the wind. Squirrels carry off and bury the seeds of nut trees—oak, walnut, hickory, pecan, filbert—but they do not consume all they gather. The seeds contained in fruits and berries are often eaten by animals and birds and subsequently excreted undigested. Seeds of burdock, cocklebur, and carrot are barbed, and when caught in the fur of passing animals, they are carried to other locations. The seeds of a number of plants—coconut, water lily, and eelgrass, for example—are carried long distances by water. As fruits of the Scotch broom dry, they act like coiled springs. When touched or moved by the wind, they explode and scatter the enclosed seeds.

The economic importance of flowering plants is very great. Though less used for lumber and paper products than the soft woods of gymnosperms, angiosperm wood, or hard wood, is turned into furniture, flooring, sports equipment, etc. Flowering plants are most important as providers of food for terrestrial animal life. They are the basis of nearly every land-based food chain.

From flowering plants we obtain such staples as cereals, rice, beans, peas, corn, potatoes, and many, many others. Flowering plants directly provide most of

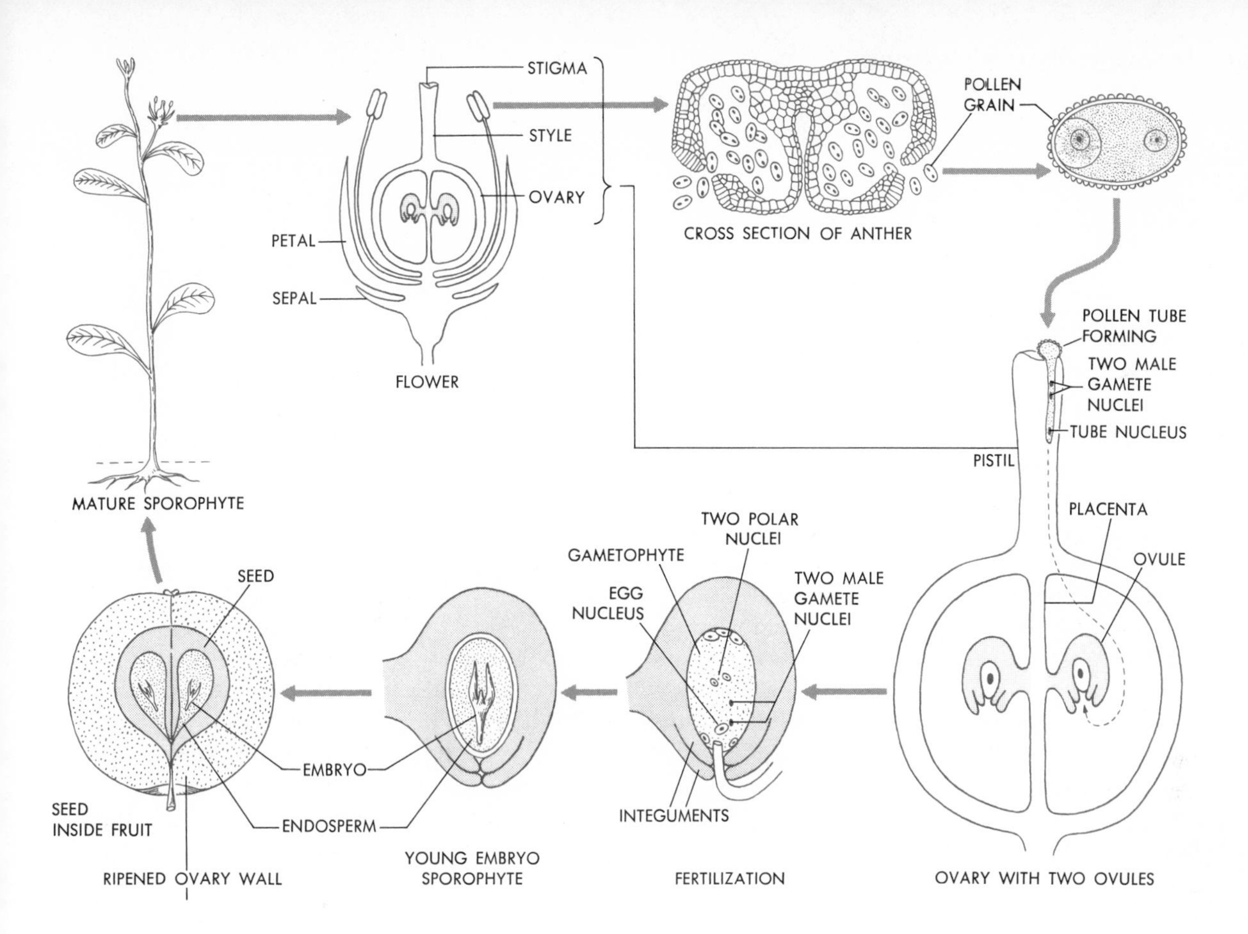

FIGURE 19.15

Life cycle of an angiosperm, or flowering plant. (From J. J. W. Baker and G. E. Allen, The Study of Biology, *3rd edition, Addison-Wesley, 1977)*

the protein for the people of the world. Fruits and vegetables are important suppliers of vitamins and minerals. Sucrose, table sugar, is obtained from sugar cane and beets. Some medicines—digitalis for treating heart disease and quinine for malaria—are still derived from flowering plants, as are herbs and spices, coffee, tea, and tobacco. Although synthetic fabrics are now dominant in the industries that make clothing and rope, we still harvest cotton, flax, and manila hemp.

Kingdom Fungi

There are more than 40,000 species of fungi and probably an equal number of undiscovered species. Nevertheless, there has long been confusion about their classification. Until recently, fungi were usually considered as nonphotosynthetic members of the plant kingdom. However, they are different in other respects as well, and placing them in a kingdom of their own, as Whittaker has done, seems to be the most satisfactory way to handle their classification.

Fungi do not contain chlorophyll or vascular tissue. As heterotrophs, either saprophytic or parasitic, they obtain nourishment from the environment by absorption. They secrete digestive enzymes into the substrate on which they grow and then absorb the nutrients produced. Consequently, they are important decomposers of dead matter in our biosphere. Since this nutritional strategy is similar to that used by some decomposer bacteria, it has been suggested that fungi evolved from primitive heterotrophic bacteria.

Fungi may be unicellular, multicellular, or multinucleate eucaryotes. They include the familiar mushrooms, molds, and yeasts, as well as the less familiar slime molds, rusts, and smuts. But there is much more to many fungi than meets the eye. A mushroom is only the fruiting body of a fungus. A larger portion of the fungus is actually hidden in the substrate, that is, underground.

An individual fungus organism may be microscopic, or it may weigh a few pounds. Except for slime molds and yeasts, most of the vegetative bodies of fungi form slender filaments, or **hyphae,** in the substrate. Collectively, these hyphae make up the **mycelium.**

DIVISION MYXOMYCETES: SLIME MOLDS

The 500 species of slime molds are found on dead vegetation of the forest floor. Some biologists have considered them protozoans, and they are thought to have evolved from flagellated algae.

In the life cycle of a true slime mold, a diploid vegetative multinucleate mass, the plasmodium (Fig. 19.16), is surrounded by a thin membrane. It moves in amoiboid fashion, absorbing or ingesting food so long as conditions remain favorable. When conditions worsen, it may go into a resting state. The diploid plasmodium at other times produces stalked sporangia. Meiosis is followed by the emergence of haploid cells that are amoiboid, have two flagella, and may divide mitotically. After they lose their flagella, nuclear fusion produces diploid zygotes. The zygotes divide mitotically and produce a new plasmodium.

FIGURE 19.16

Plasmodium of a slime mold. The plasmodium consists of strands of semisolid and liquid protoplasm. Its movement is amoeboid, first in one direction, then another. It ingests bacteria, wood particles, and other organic matter as it moves.

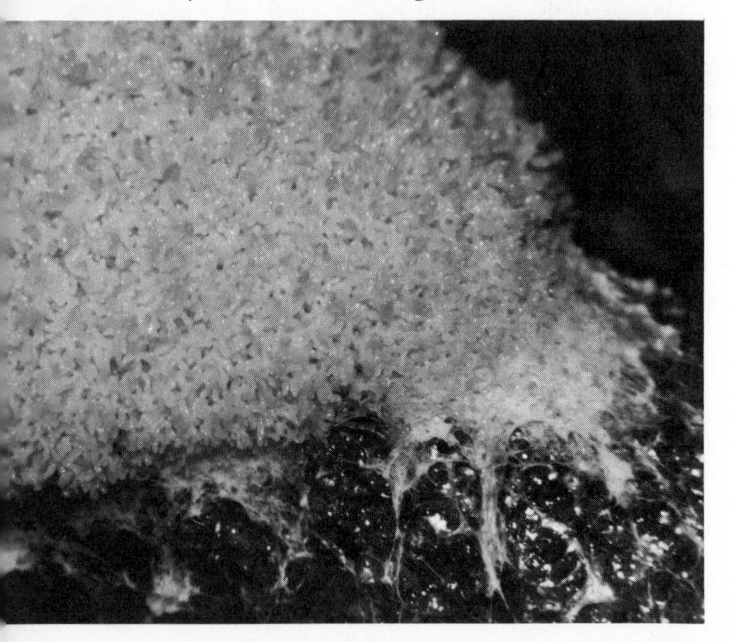

DIVISION ZYGOMYCETES

The ordinary bread mold, *Rhizopus stolonifer*, is one of the most frequently encountered and thoroughly studied of the zygomycetes. The cell wall is composed of chitin, which performs the same function as cellulose by giving the cell rigidity.

Hyphae making up the multinucleate mycelium are haploid. The vegetative mycelium (rhizoids) stays in the substrate but does send vegetatively reproducing hyphae into the air. Nonflagellated haploid asexual spores develop in sporangia. When released, these spores form new haploid organisms in the substrate. Sexual reproduction occurs when two hyphae from different mycelia conjugate to form a diploid zygote. Meiosis within the zygote produces new haploid spores, and they are released.

DIVISION ASCOMYCETES

Like zygomycetes, the ascomycetes, including the yeasts and powdery mildews, have nonflagellated spores and filamentous cells with chitinous walls. However, frequent septa divide the cells into multinucleate sections. They are called ascomycetes, or sac fungi, because sexual spores are produced in a specialized saclike type of hyphae, called an **ascus,** found in the reproductive mycelium.

A fruiting body (ascocarp) becomes differentiated from the vegetative mycelium. It may be several centimeters long and visible above the substrate. Certain haploid hyphae in the fruiting body become differentiated into sexually reproductive structures, antheridia and ascogonia (similar in function to archegonia). A tube grows from one ascogonium to an antheridium.

Nuclei migrate through the tube to the ascogonium, where they pair but do not fuse. As hyphae containing paired nuclei elongate, mitotic divisions occur. But the haploid nuclei remain paired. Septa form, and the section at the end of a hypha contains a single pair of nuclei, which fuse and produce a diploid nucleus. Meiosis is followed by mitotic divisions until typically eight haploid nuclei, the ascospores, are contained in this terminal section, which is now the saclike ascus. The ascospores are released into the environment, where they germinate into new vegetative hyphae.

The life cycle of the yeasts is not so complicated. Generally unicellular, they sometimes reproduce asexually by simple cell division but more frequently by budding. Sexual reproduction occurs when two cells fuse to form a diploid nucleus, which then undergoes meiosis to produce four nuclei. The zygote enlarges and forms an ascus.

A number of organisms are classified as ascomycetes. Ergot attacks rye and other grains. Powdery mildews are parasites of flowering plants. When yeasts ferment the sugars and carbohydrates in bread dough, one of the products, carbon dioxide, makes the dough rise. In wine, beer, and ale making, alcohol is the product of interest in the fermentation process. Truffles and morels are among the most sought-after of the edible fungi.

DIVISION BASIDIOMYCETES

The best-known fungi are the basidiomycetes. The 15,000 species include mushrooms, toadstools, rusts, stinkhorns, bracket fungi, puffballs, coral fungi, jelly fungi, and smuts. Most basidiomycetes are saprophytes, but the rusts and smuts are plant parasites. Among the mushrooms are many tasty edible varieties, but there are also some that are poisonous in the extreme.

Basidiomycetes are found throughout the world in terrestrial habitats. The mycelium of a basidiomycete is an extensive, long-living, underground hyphal mass. Club-shaped basidia are formed within the fruiting body, the basidiocarp—for example, that visible portion of the organism that we call a mushroom or puffball.

FIGURE 19.17
Amanita muscaria. *Like other club fungi,* Amanita muscaria, *a poisonous mushroom, has only a small section growing aboveground. The solid-looking tissue mass, the fruiting body, is actually composed of tightly packed filamentous hyphae differentiated into a stalk and cap. A more extensive mass of hyphae penetrates the soil.*

In functional terms, the basidia are analogous to the asci of the ascomycetes. Within the basidia, nuclear fusion of two nuclei is followed by meiosis. The resultant basidiospores, usually four in number, escape into the air as a fine powder or remain fairly close to the parent plant. Basidiospores in mushrooms are formed on the gills, in bracket fungi in pores, and in puffballs in the center of the spherical portion.

LICHENS

Lichens are composed of two organisms, a blue-green or green alga and a fungus (usually an ascomycete), which exist together in a symbiotic relationship. Until the middle of the nineteenth century, lichens were thought to be a separate type of organism. A widely supported hypothesis at present is that the first lichen association began with the parasitism of an alga by a fungus. The fungus obtains nutrients from the photosynthetic alga. Just what benefits, if any, the alga may gain from the association is uncertain, but they probably include water and minerals.

Lichens are often found on rocks and tree trunks. The growth, which may be several inches across, is usually thin, may be leaflike or scaly, and is often quite colorful. Lichens are important agents in the breakdown of rocks into soil. They occur even at high altitudes and in arctic regions. Lichens called reindeer moss or Icelandic moss provide food for reindeer, caribou, and cattle in northern regions. Dyes, including litmus, are obtained from lichens. Some contemporary Canadian Eskimos gather lichens to make dyes for wool.

FIGURE 19.18
Lichen of the genus Cladonia. *The lichens, a specialized group of organisms that are half fungus and half alga, are generally included with the fungi. More than 17,000 species have been found growing on frozen soil, trees, rocks, and even metal. They are slow but efficient soil formers, and they have recently received attention as possible indicators of atmospheric pollution, because pollution upsets the delicate balance between the fungal and algal partnership.*

Summary

1. The plant kingdom includes multicellular eucaryotic organisms that are believed to have evolved from photosynthetic forms and that exhibit advanced tissue differentiation. Most plants are photosynthetic autotrophs.

2. An important characteristic that differentiates plants from animals is a life cycle involving an alternation of generations between sexual and asexual reproduction.

3. The plant kingdom has three main groupings or subkingdoms. All except part of one subkingdom (the tracheophytes of subkingdom Euchlorophyta) are structurally primitive.

4. Classification of the algae is the subject of continuing debate. Placement of most of the known species in separate subkingdoms within the same kingdom recognizes both their similarities and their differences.

5. Most of the familiar plants are included in Branch Metaphyta, which has two main divisions, the bryophytes (liverworts, hornworts, and mosses) and the tracheophytes (vascular plants). Of the several subdivisions of tracheophytes, the one containing the largest number of familiar plants is subdivision Spermopsida (seed plants), which is further divided into gymnosperms—the naked-seed plants, including the conifers—and angiosperms—the enclosed-seed flowering plants.

6. Fungi have only recently been classified in a kingdom of their own. They were formerly considered nonphotosynthetic plants. The principal characteristic that distinguishes fungi from other organisms is that they absorb their nutrients from the environment. They are major decomposers of dead organic matter.

REVIEW AND STUDY QUESTIONS

1. Name at least five characteristics found in most plants.

2. Trace a plant of your choice through its complete life cycle.

3. Why do green algae seem to be especially difficult to classify?

4. Explain why red algae can survive in deeper water than can most other marine plants.

5. What particularly distinguishes tracheophytes from other plants? Select four tracheophytes that in your opinion demonstrate the great diversity within that division.

6. Why was the evolution of the seed so important?

7. What are the principal differences between plants and fungi?

REFERENCES

Ahmidjian, V. 1967. *The Lichen Symbiosis.* Blaisdell, Waltham, Mass.

Alexopoulos, C. I. 1962. *Introductory Mycology,* 2nd edition. Wiley, New York.

Bold, H. C. *Morphology of Plants,* 3rd edition. Harper & Row, New York.

Margulis, L. 1971. "The Origin of Plant and Animal Cells." *American Scientist* 59: 230–235.

SUGGESTIONS FOR FURTHER READING

Benson, L. 1957. *Plant Classification.* Heath, Boston.
A standard text on taxonomy and classification.

Chamberlain, C. J. 1935. *Gymnosperms, Structure and Evolution.* University of Chicago Press, Chicago.
A classic study of ginkgos, cycads, pines, and relatives.

Dawson, E. Y. 1966. *Marine Botany.* Holt, Rinehart and Winston, New York.
Textbook on marine algae and ecology.

Lawrence, G. H. M. 1951. *Taxonomy of Vascular Plants.* Macmillan, New York.
Standard text of taxonomy and classification of vascular plants.

Solbrig, O. T. 1970. *Principles and Methods of Plant Biosystematics.*
Standard reference in modern plant taxonomy.

Wagner, P. 1974. (June). "Wines, Grape Vines and Climate." *Scientific American* Offprint no. 1298. Freeman, San Francisco.
Explores the reasons for the great differences in wines and grape varieties.

DIVERSITY OF LIFE: ANIMALS

chapter twenty

An example of the diversity of animal life is this quarreling pair, a pelican and a sea lion, two creatures that live in the sea. A swift swimmer and strong flier, the pelican can dive into the water and scoop up fish in the large elastic pouch attached to its bill. The sea lion's streamlined body is protected by a coat of dense, oily fur. A layer of blubber insulates body heat and adds to the animal's buoyancy. Its limbs have evolved into powerful webbed flippers, which can carry it many fathoms beneath the sea.

A general term for multicellular animals is metazoans. The only animal or animal-like organisms treated so far have been the protistans, single-celled organisms that were probably the ancestors of the metazoans. The metazoans are more than an aggregation of cells. They are a complex of cells that behaves as a unit capable of living an independent existence. There is a division of labor among the cells. If separated, the cells cannot function as independent organisms and, at most, will survive only for a limited time. As always, there are exceptions, including those animals, such as the sponge, that are capable of regeneration. Through evolution, the trend has been toward larger size and specialization of specific organs and organ systems, particularly those that are involved with the acquisition of food and sexual reproduction. With increased specialization, larger body masses are generally required, and in turn, additional organ systems are required to maintain metabolic functions at optimum levels.

Although the development of organ systems and related mechanisms to ensure sexual reproduction is of major importance, the metazoans are heterotrophic organisms and directly dependent on producers for their food. Consequently, much of their evolutionary development of organ systems is directly related to the process of food procurement.

	Multicellularity	Central digestive system	Bilateral symmetry	Tube-within-a-tube body plan	Coelom	Segmentation	Closed circulatory system	Jointed appendages	Indeterminate cleavage	Endoskeleton	Deuterostomia	Notochord, tubular nervous system
Chordata	■	■	■	■	■	■	■	■	■	■	■	■
Echinodermata	■	■	□	■	■	□	□	□	■	■	■	□
Arthropoda	■	■	■	■	■	■	■	■	□	□	□	□
Annelida	■	■	■	■	■	■	□	□	□	□	□	□
Mollusca	■	■	■	■	■	□	□	□	□	□	□	□
Aschelminthes	■	■	■	■	□	□	□	□	□	□	□	□
Platyhelminthes	■	■	■	□	□	□	□	□	□	□	□	□
Cnidaria	■	■	□	□	□	□	□	□	□	□	□	□
Porifera	■	□	□	□	□	□	□	□	□	□	□	□

Increased complexity

FIGURE 20.1

Increased complexity in the animal kingdom, as indicated by the presence of certain structural characteristics. Shaded squares mean that a feature is present in the phylum, blank squares that the feature is absent.

Kingdom Animalia

Most animals are motile during some part of their life cycles. Though a sponge cannot move from place to place, its body can contract a little, and its pores open and close. Animal cells are eucaryotic, and they are enclosed by only a flexible membrane. Their tissues are surrounded by salty fluid, either coming from a marine environment or produced by the cells themselves. In general, animals have well-developed nervous and muscular systems, which enable them to respond rapidly to environmental stimuli. The alternation of generations between haplophase and diplophase that is so common in plants is rare in animals.

Many of the animal characteristics summarized above can also be found in other kingdoms. Protists are highly motile; marine algae tolerate salt water; the Venus's-flytrap, a plant, moves to capture insects. But it is the sum total of an organism's characteristics that determines its classification. As with the classification of members of the other four kingdoms, judgment plays an important role in animal classification.

Whittaker divides the animal kingdom into three subkingdoms on the basis of the presence or absence of tissues: Agnotozoa (no tissue differentiation); Parazoa (cell differentiation but limited tissue differentiation); Eumetazoa (extensive tissue differentiation).

Eumetazoans are further divided by body symmetry: radial or bilateral. Animals with bilateral symmetry form their body cavities, or coeloms, in different ways. Animals that have no coelom, such as the flatworms of phylum Platyhelminthes, are termed **acoelomates.** If the body cavity develops from the embryonic blastocoel and is not lined with mesoderm, the animal is considered **pseudocoelomate.** Phyla Aschelminthes, Acanthocephala, and Entoprocta are in this group.

The true **coelomate** has a body cavity lined with mesoderm. If the coelom is formed within the mesoderm, it is a schizocoel. Schizocoels include members of the phyla Bryozoa, Brachiopoda, Echiurida, Sipunculida, Annelida, and Arthropoda. In the remaining phyla—Echinodermata, Hemichordata, Chordata—the coelom is an enterocoel, arising from outpocketing of the embryonic gut.

A further embryonic distinction separates the animal kingdom into two large groups. If the blastopore develops into the mouth, as it does with most animals, the organism is called a **protostome** (from the Greek for first mouth). But in animals with enterocoels, the blastopore becomes the anus, and a new opening becomes the mouth. They are called **deuterostome** (second mouth).

The number of animal phyla is generally placed around 20, with some authorities recognizing as many as 26. Only the most representative and evolutionarily interesting phyla will be discussed here.

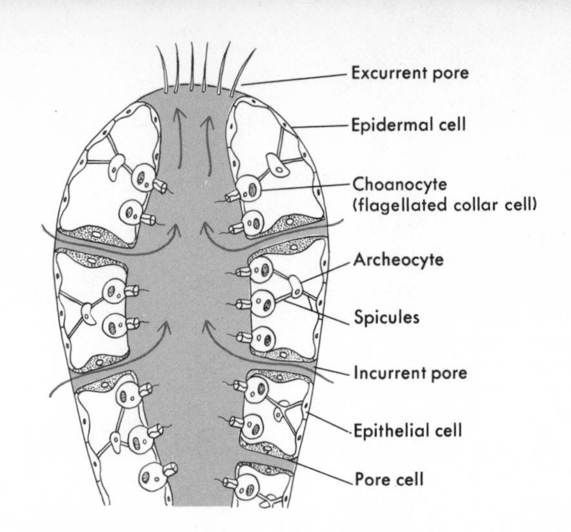

FIGURE 20.2

Body plan of a sponge. Porifera, or sponges, are multicellular animals with no organs of any kind and vaguely defined tissues. In order to gain food and oxygen, a sponge filters vast quantities of water. The beating flagella of the collar cells create water currents that are drawn into the central cavity through numerous pores in the wall of the sponge. Microscopic food particles suspended in these currents adhere to the collar cells and are engulfed into the cell body. Digested food is then diffused to other cells. The water currents, exhaled through an opening at the end of the body, carry away carbon dioxide and nitrogenous wastes.

PHYLUM PORIFERA: SPONGES

Because of their appearance, sponges were once thought to be plants and were not definitively placed in the animal kingdom until 1857. The colors of poriferans ("pore bearers") range from grey to red, blue, yellow, or black. They may grow to be several feet across and live for years.

More than 5000 species of sponges have been classified. They were the first animals to have developed multicellularity, a major advance in the course of animal evolution. Most all of them are marine, except for a few freshwater forms. They are very simple in structure and primitive in their tissue development (see Fig. 20.2). In this simplicity of structure, they are similar to colonial flagellated protists. Water, along with particles of food, enters through a number of openings (ostia) into a central cavity (spongeocoel) and leaves through a single terminal opening (osculum). There is an outer epithelial layer of cells and an inner epithelium which includes flagellated collar cells (choanocytes). The choanocytes move water in through the ostia by their flagellar beating and may trap and ingest some of the food particles or pass the food on to cells found in a middle region (mesoglea). In the mesoglea are various cells that may move about in an amoeboid fashion or form conspicuous needlelike structures called spicules. The spicules are made of either calcium carbonate or siliceous materials and provide the basic support of the sponge.

Some sponges, especially those harvested for use as bath sponges, are supported by a protein network that is similar in composition to hair, nails, and feathers. Classification of sponges rests largely on skeletal composition and arrangement.

Sponges reproduce sexually and asexually. In sexual reproduction, both sperm and eggs are formed from cells on the organism. Both types of gametes may be formed on the same organism (hermaphroditic (or **monoecious**) or on different organisms **(dioecious).** In either case, the sperm is transferred to the egg by other cells within the sponge or carried by water currents. Sponges also reproduce asexually by budding.

Sponges are remarkable in their powers of regeneration. If a portion of the organism is cut away, the sponge will grow back. Fragments of only a few cells will regroup to form a new sponge. Sponges are attached (sessile) to the substrate and often utilized by

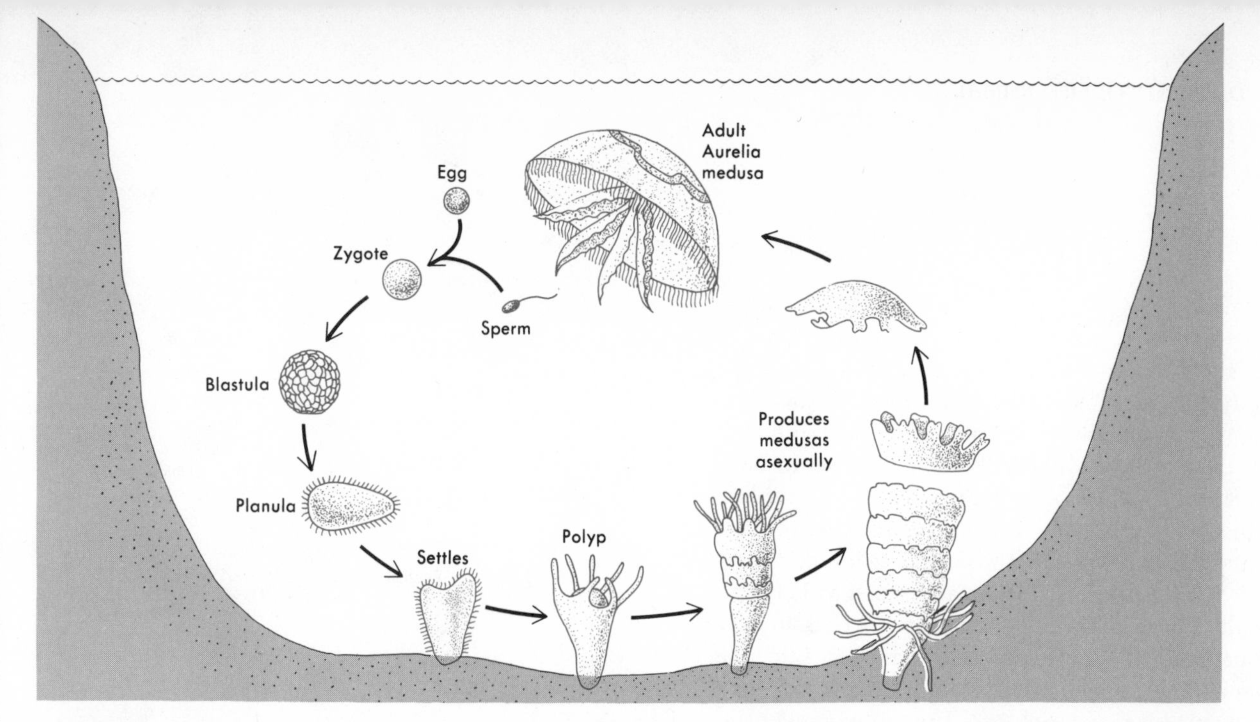

FIGURE 20.3

Life cycle of a jellyfish. Like other coelenterates, the jellyfish Aurelia *includes in its life cycle a medusa phase, when reproduction is sexual, and a polyp phase with asexual reproduction. Adult medusae are either male or female. Released sperm pass into the mouth of the female medusa and fertilize eggs in the coelenteron. The resulting zygotes develop into small ciliated larvae, the planulae. These larvae swim free of the parent and settle in the substrate. There they develop into polyps, which are eventually transformed into a series of transverse stacked sections. These structures separate, swim free, and develop into eight-lobed adult medusae.*

other animals for camouflage. Ordinarily they are not too destructive, and few organisms prey on them because of their spicules and poisonous excretions.

PHYLUM CNIDARIA: COELENTERATES

Commonly referred to as coelenterates, these organisms, about 10,000 species, live mostly in salt water. They include the familiar jellyfish (which is not a real fish since it lacks a backbone), sea anemones, Portuguese men-of-war, corals, and hydras. Fossil coelenterates date from the Cambrian period, and large fossil reefs of stony corals have been found in the United States and Europe. Aristotle considered organisms of this group to be a life form intermediate between plants and animals. We should not be surprised to discover that the sea "anemone" looks like a beautiful underwater flower. Linnaeus classified coelenterates with the starfish, and they were not placed in a separate phylum until 1888.

Coelenterates exhibit radial symmetry and the development of distinct tissue layers, such as an outer epidermis and an inner gastrodermis, which lines a simple or chambered internal cavity (coelenteron). It is this cavity that marks the major evolutionary advance of these organisms. A gelatinous layer, mesoglea, lies between the two tissues.

External to and surrounding the cavity are numerous tentacles that contain stinging cells, nematocysts. Coelenterates are mostly carnivorous. Prey—insect larvae, small crustaceans, or small fish—are stung and moved into the cavity, where digestive enzymes are secreted by the gastrodermis. Digestion is extracellular in this case. Some species engulf bits of undigested food, amoeboid fashion, and digest them intracellularly. The sting of some coelenterates is very painful to humans. Coelenterates have muscle fibers, connective tissue, and nerve cells (including eye spots).

In addition to certain morphological characteristics, the three classes of coelenterates show variations in their polymorphic life cycles, which may include several long developmental stages. Generally there are two stages to the coelenterate life cycle: the polyp, or sessile, stage; and the medusa, or free-swimming, stage (Fig. 20.3).

In the class Hydrozoa, both polyp and medusa stages are equally evident. The medusa stage is the

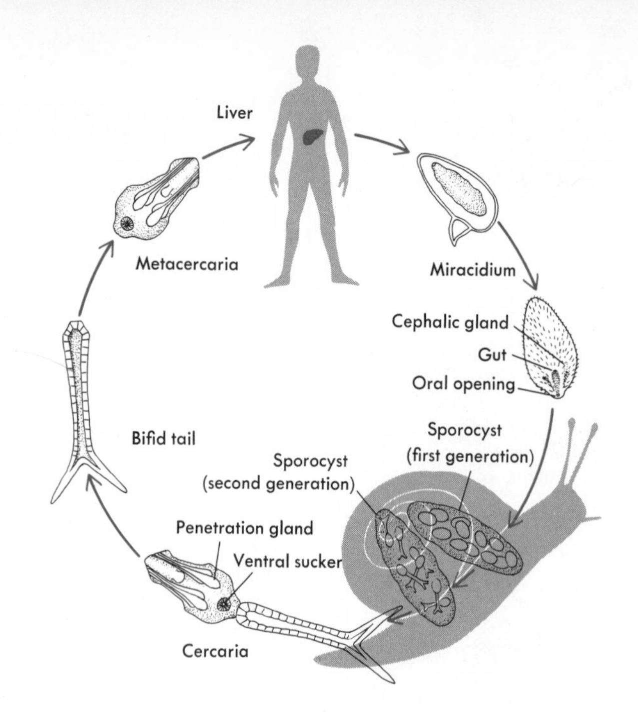

FIGURE 20.4

Life cycle of a flatworm. The phylum Platyhelminthes, consisting of the flatworms, includes a number of species that have become specially adapted to a parasitic way of life. The blood liver fluke, Schistosoma mansoni, *has a life cycle involving two different hosts. The adult worm lays eggs in the tiny blood vessels of the human intestinal wall, and the hatched eggs are discharged in the feces. Where human excrement is used as fertilizer, the eggs find their way into water in rice fields, irrigation ditches, or rivers. There they hatch into* miracidia, *which then penetrate the tissues of certain snails and repeatedly reproduce asexually as they change from one form to another through the life cycle. The new individuals burrow through the skin of a human being and into the vascular system. Passing through the heart and lungs, they eventually mature in the intestines, lay eggs, and thus initiate a new cycle.*

dominant stage in class Scyphozoa (which includes the marine jellyfish, such as *Aurelia*), whereas in the class Anthozoa (sea anemones and colonial corals) the medusae are absent and only polyps are present. Both forms have the basic morphology, with one opening, the mouth, leading to the coelenteron. Many of the polyps have an exoskeleton that may be of chitinous material or a hard calcareous substance such as is found in the corals. These animals are dioecious or occasionally monoecious, and they release their gametes into the water. Depending on the species, the fertilized egg may form other medusae or polyps. The polyps may reproduce asexually by budding, often forming medusae, and in some groups, the medusae carry on asexual reproduction.

Coelenterates have some powers of regeneration. A hydra cut in half will grow into two small hydra. Some sea anemones reproduce asexually by dividing themselves transversely or laterally.

PHYLUM PLATYHELMINTHES: FLATWORMS

Scattered among three classes are 15,000 described species of flatworms. Although they are mostly parasites, there are free-living marine and fresh-water specimens.

The flatworms show certain evolutionary trends for the first time. They have a definite **bilateral symmetry**—both sides of the animal are similar. They also have a front end containing sensory organs (brain and eye spots) and a rear end, as well as dorsal and ventral (top and bottom) sides. They have a distinct mesodermal layer instead of mesoglea; obvious muscle tissue; a complex excretory system that complements a more complex organ system, with its greater metabolic requirement; and a centralized nervous system instead of the typical network type found in the previous phyla.

The platyhelminthes are divided into three classes. The free-living marine and fresh-water organisms are grouped in class Turbellaria. In clean, rapidly moving fresh-water streams and under rocks or in dark places, planaria can be found gliding along the rock surface. Planaria are commonly used in laboratories to illustrate such phenomena as regeneration and learning behavior.

Flukes (class Trematoda) and tapeworms (Cestoda) are both ecto- and endoparasites of vertebrates and have great economic importance. Many of the endoparasites have two hosts (see Fig. 20.4). It appears that a mollusc, particularly a snail, is often one of the hosts, and quite often a vertebrate is the other. The parasites lodge in vital organs and can do tremendous damage, sometimes leading to death.

Platyhelminthes are hermaphroditic, containing male and female sex organs within the same organism. Fertilization may occur internally when two organisms mate and each releases sperm into the other's genital atrium (cross-fertilization). Self-fertilization is also possible. Planaria reproduce asexually by transverse fission when one organism splits in two behind the pharynx. The two fragments then mature. Planaria are also able to regenerate when cut into small pieces, each of which will produce a new organism.

PHYLA ASCHELMINTHES, ACANTHOCEPHALA, ENTOPROCTA

Of these three phyla, it is the Aschelminthes (cavity worms) that typify the group of organisms referred to as pseudocoelomates. The other phyla are small, either endoparasites in the digestive tracts of vertebrates, as are the Acanthocephala, or sessile and mostly marine entoprocts. More than 12,000 cavity worms have been named; they inhabit fresh and salt waters or land and may be free-living, commensal, or parasitic.

In the previous phyla, or acoelomates, the space between the alimentary canal and the epidermis was either absent or filled with mesoderm and mesenchyme. In these three phyla, there is actually a space between the alimentary canal and epidermis without the presence of any structure or body fluid. In the embryological formation of the space (pseudocoel), the blastocoel remains in part, so that a chamber exists. The alimentary canal has an anterior opening (mouth) and a posterior opening (anus). The alimentary canal in these organisms is generally straight, with some swollen regions associated with different functions, such as temporary storage or digestion.

Although muscles that function primarily for movement are found just underneath the epidermis, the digestive tract is nonmuscular. No circulatory or respiratory system is necessary since most of these organisms are small and can exchange gases by diffusion. They have a well-developed excretory system that rids them of their metabolic wastes and in some instances maintains their internal osmotic pressure.

Most species of aschelminthes are members of the class Nematoda (roundworms). These are usually small,

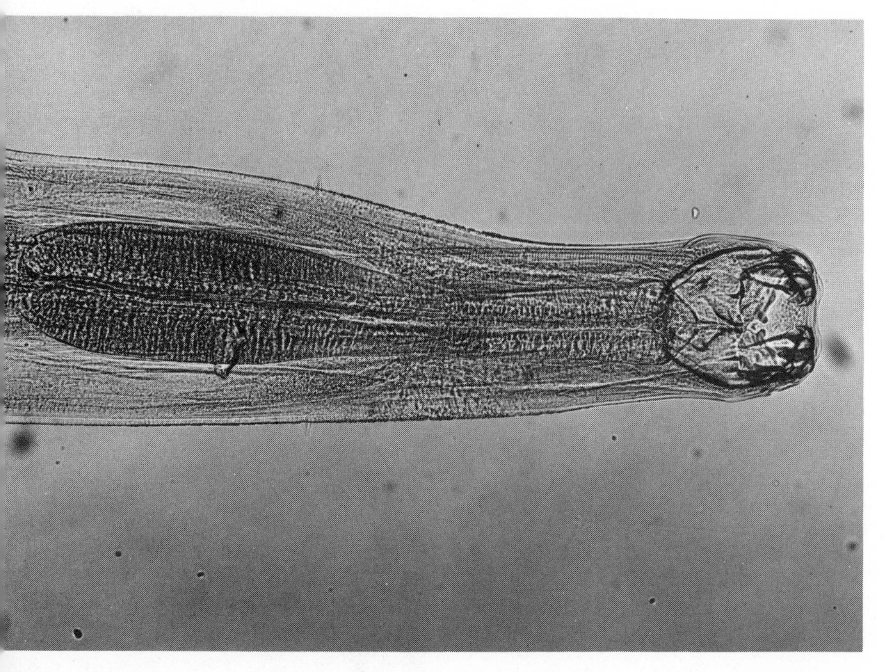

FIGURE 20.5

The hookworm Ancylostoma, *showing anterior sucking mouth, pharynx, ventral and dorsal teeth, and sawteeth. This small animal parasite enters the body of the host through the skin and attaches itself by its sucking mouth to the lining of the small intestine.*

80X

but some have been measured to lengths of more than seven meters. They exist everywhere, in almost any kind of habitat. The more widely known members are endoparasites. The hookworm, *Necator*, was once a serious problem in the southern United States, particularly for those who walked barefoot. *Trichinella*, the worm that is responsible for trichinosis, may pass into the human body when the individual eats poorly cooked pork. In both the pig and man, the larvae ultimately migrate to muscles, where they encyst and may remain for up to 20 years, producing a painful crippling.

The fossil record of the pseudocoelomates is limited since, like most acoelomates, these fleshy organisms lack any kind of skeleton that can be preserved.

PHYLA BRYOZOA, BRACHIOPODA, ECHIURIDA, SIPUNCULIDA

Four phyla have been classified in a group known collectively as the Tentaculata. These aquatics all have bilateral symmetry. They also have free-living trochophore larvae, which are characteristic of the annelid-mollusc branch of evolutionary development. Trochophore larvae are tiny, ciliated, often pear-shaped larvae, which have a ring or wheel of cilia running around the body behind the mouth.

The bryozoans, moss animals, are often mistaken for seaweeds, but they are actually small colonies of animals that attach to underwater marine objects, including ship bottoms. They have been in existence since at least the Ordovician period, when their calcified skeletons began to contribute to the formation of lime-bearing rocks.

Brachiopods, which have two shells, resemble clams. They have left an extensive fossil record and were more common in the past (30,000 fossil species recorded) than in the present (250 living species). They are marine, and a few survive to depths of 16,000 feet.

The echiurids and sipunculids are wormlike bottom feeders, living in burrows on mudflats but capable of some movement. Since they do not contain a limy or chitinous skeleton, the fossil record is sparse. Echiurids have nervous, excretory, and circulatory systems like those of the Annelids. A male echiurid is much smaller than the female in whose body it lives as a parasite.

PHYLUM MOLLUSCA

In this diverse group of about 45,000 living species and about as many fossil species are animals with soft bodies, usually surrounded by a hard exoskeleton. Included in the molluscs are clams, mussels, oysters, snails, slugs, squid, octopus, chitons, and others. Many molluscs inhabit salt water—some to depths of 35,000 feet—and brackish estuaries, as well as fresh water, and a few species even live on land. They have at least the capacity of moving about, although many live more or less permanently in burrows (clams) or attached to rocks (chitons, mussels). Some crawl along slowly (snails, slugs), and others swim freely (squid, octopus). The molluscs are one of the most successful animal groups with respect to structure, function, and adaptability.

The molluscs are coelomate unsegmented animals, characterized by the presence of a calcareous secreting organ (mantle) and a large muscular structure (foot), which is located posterior to the mouth and functions in locomotion. They possess a well-developed digestive system that, in many species, includes a rasping structure (radula) located near the mouth. Gas exchange is carried out through gill structures or, in some species, by a portion of the mantle that is modified into a lung. Blood is pumped to the various organs by a dorsal heart. A pair of excretory organs may also function as a pathway for discharged gametes. Some species have a well-developed nervous system made up of usually three ganglionic masses that control muscles and sensory organs. The animals are dioecious, and fertilization may occur either internally or externally. The zygote develops into the ciliated trochophore larva. Larval types are important clues in determining the course of animal evolution.

There are six classes of molluscs: Monoplacophora, Amphineura, Scaphopoda, Gastropoda, Pelecypoda, and Cephalopoda. Species of Monoplacophora were thought to be completely extinct, but several years ago a living genus, *Neopilina*, was discovered. It differs from other

Diversity of Living Forms

1
Diatoms are golden algae belonging to the phylum Chrysophyta. They are the principal photosynthetic food source for animal life in the ocean and contribute to the earth's atmospheric oxygen.

2
The flame angelfish (phylum Chordata) and the sea anemone (phylum Cnidaria) are two colorful life forms inhabiting the shallow waters of tropical oceanic reefs.

3
The Octopus, a timid sea animal, is a cephalopod mollusc. It has a round saclike body, and its eight flexible arms surround a parrotlike beak with which it tears apart small prey. A discharge of thick, inky liquid covers emergency escapes.

4
The common starfish, of class Asteroidea in phylum Echinodermata, generally preys on molluscs by extruding its stomach through an opening in the shell and digesting its way into the soft tissues. The hermit crab, of class Crustacea in phylum Arthropoda, adopts mollusc shell homes for protection. When threatened, it presents a formidable barricade of pincers.

1

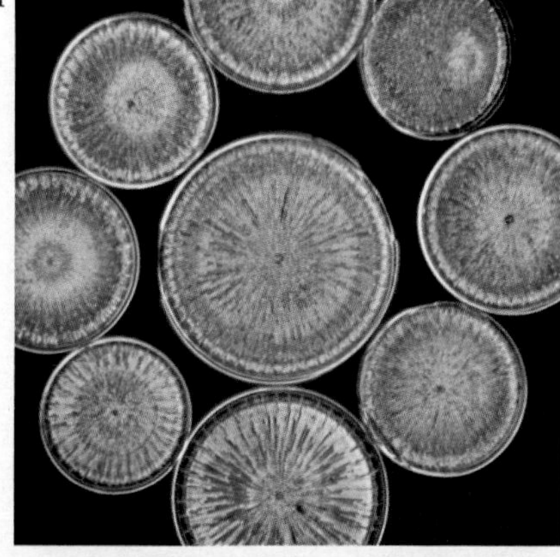

2

3

4

5
In Central and South America dwell several species of small, colorful frogs whose skin glands secrete a highly toxic protein used by Indians of the region on their arrow tips. (Vertebrata, Amphibia)

6
The Indian rock python (Vertebrata, Reptilia) often attains a length of nearly 20 feet and a weight of some 200 pounds.

7
Several beautiful tropical mantids (Arthropoda, Insecta) closely resemble the flowers on which they reside. These mantids have voracious appetites and spend most of their time lying in wait for insects drawn to the colorful blossoms. In fact, they are even cannibalistic; a female often devours her mate after copulation.

8
Liquid excreted from the spider's spinnerets hardens on exposure to air and forms a thread of protein. Although only a millionth of an inch thick, this remarkable substance has considerable strength and elasticity. (Arthropoda, Arachnida)

6

5

7

8

9

10

11

12

13

9
Ripe thistleseed is the caviar in the diet of the goldfinch (Vertebrata, Aves), and in the spring thistledown and wild grasses are woven into the finch's nest. In return, the birds scatter the winged seeds on the wind, which eventually carries some to fertile soil.

10
The black-chinned hummingbird (Vertebrata, Aves) acts as a pollinating agent as it extracts nectar from the cactus flower. Its proportionately large wing muscles permit it to fly up, down, sideways, and backwards or to hover in a haze of rapidly beating wings.

11
The male frigate bird (Vertebrata, Aves) attracts a mate by inflating his scarlet throat pouch. Tirelessly, the female brings him twigs while he roosts on the empty nest. Each mated pair must protect their home and young from the cannibalistic, stick-stealing members of the frigate bird colony.

12
A comparison of mushrooms and maple leaves illustrates the great diversity of plant life. The saprophytic mushrooms, most familiar of the fungi, are no larger than a single leaf from the large, photosynthetic angiosperm.

13
The legume pods of Florida's rosary pea vine contain one of the most toxic agents known. A single brilliant red and black seed can kill a human child.

14
The 30-foot killer whale (Vertebrata, Mammalia) is actually a carnivorous dolphin that hunts in packs, using its remarkable sonar system to track down prey: fish, penguins, seals, and even the large baleen whales. Intelligent, aggressive, and loyal to their comrades, killer whales are surprisingly gentle in captivity.

15
No other mammal can match the bizarre facial colors of the male mandrill (Vertebrata, Mammalia). Troops of these baboons live mainly in the forests of West Central Africa. The powerful males are armed with saber-like canine teeth and make a dangerous adversary, even for a leopard.

16
The largest tigers (Vertebrata, Mammalia) live in Siberia and western China; males often approach 12 feet in length and weigh over 500 pounds. These beautiful, endangered animals are among the limited number of big cats that enjoy water and a cooling swim.

14

15

16

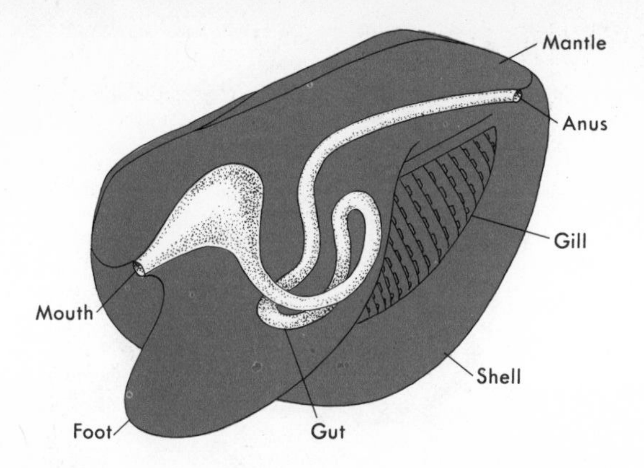

FIGURE 20.6

The freshwater clam, an example of one type of sedentary, filter-feeding mollusc. A bilaterally symmetrical, "headless" bivalve, the clam is propelled by a muscular foot that protrudes between the two valves of the shell. In common with other molluscs, the bivalves are characterized by soft bodies covered by a mantle, which secretes a shell. Other molluscan forms include the slow-moving gastropods, like the snail, and the tentacled cephalopods, like the squid, whose tubelike foot can expel water to provide jet propulsion.

molluscs in that it has gills, nephridia, and muscles in a segmented arrangement.

The marine chitons are the most familiar of the Amphineura. Found on rocks, they are protected by overlapping dorsal plates. They may grow to several centimeters in length and are a popular food item in certain parts of the world.

Scaphododa, the tooth shells, are marine molluscs that dwell primarily in mud. The shells of *Dentalium* were used as money by the Indians of the North American Pacific coast.

Snails, limpets, nudibranches, and slugs are in the class Gastropoda. Most of them have only one shell and are called univalves, although some of the nudibranches and slugs have no shell. A radula enables these organisms to feed on many different plants or, in some cases, to be the predators of other molluscs. The aquatic gastropods utilize gills for respiration, but lunglike organs are found in the land forms. In numbers of species and individuals, the gastropods are the largest of the molluscan classes.

They are found in diverse marine and terrestrial habitats from the depths of the oceans to the flanks of the Himalayan mountains. Many edible species are considered gourmet treats. The French revere their escargots (snails) as the Californians do abalone. Shells of gastropods were "wampum" for many native American tribes. Some gastropods are harmful. The larvae of certain parasitic platyhelminthes mature in snails. Slugs in areas with considerable rainfall, such as the coastal Pacific Northwest, grow to be several inches long. They can devastate seedlings in a garden overnight.

The members of the class Pelecypoda are called bivalves and include clams, scallops, and oysters. Most of these are marine with a worldwide distribution, although a few inhabit fresh waters (Fig. 20.6). The bodies of these organisms are completely enclosed in a mantle that secretes a surrounding shell, whose halves are held tightly together by powerful muscles. Annual growth rings may be quite prominent on their shells.

Bivalves move about by the use of the foot, which may be extended between the valves, or by forcing water out of a tubular structure—the excurrent siphon—in a jet-propulsion action. The movement of cilia located on the gills directs water containing microscopic food particles into the incurrent siphon, from which the water passes over the gills and out the excurrent siphon. The food becomes entangled in mucus, which the gills secrete, and is then passed to the mouth by ciliary action. No radula is present in these organisms. Most pelecypods are sedentary, dependent on water currents to bring plankton and food particles to them. Hence they are often found in regions where organic content is high. Pelecypods are dioecious, but otherwise males and females look alike. Fertilization usually occurs outside the adult in water.

Economically, pelecypods are the most important of the molluscs. They have been a source of food for coastal people for thousands of years. For archeologists, deposits of discarded shells left by primitive tribes have been important indicators of early human habitation. Today clams, scallops, and oysters are harvested by the thousands of tons. Oyster farming is practiced, and the oyster, *Pinctada*, yields the pearls used in jewelry.

The most specialized molluscs are Cephalopoda, or the octopus, squid, and *Nautilus*. Their shells can be greatly reduced (squids), absent (octopus), or very well developed *(Nautilus)*. The presence of the shell of the *Nautilus* may be an example of a vestigial characteristic. The foot has become modified into tentacles that surround the mouth. The mantle contains a series of cavities, which fill up with water when certain muscles relax.

FIGURE 20.7

The land snail Helix, *a slow-moving mollusc that travels on a broad, undulating ventral foot. A sticky solution produced by the snail acts as a protective roadway. It is so effective that snails have been known to crawl across razor blades without injury.*

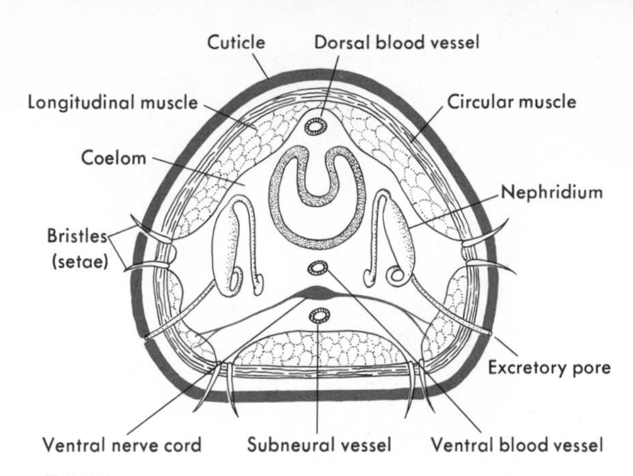

FIGURE 20.8

The body plan of annelids, represented here by the earthworm. Annelids are distinguished from the coelenterates and the flatworms by the presence of a cavity, or coelom. In this three-layered plan, which first evolved in the molluscs, the gut is a straight tube that extends within the coelom from mouth to anus. It is suspended within the body wall by the mesodermal mesenteries. The coelom plays an especially important part in nutrition. Food and waste products diffuse into the coelomic fluid and through it reach their ultimate destination.

When these muscles contract, water is forcibly ejected through a funnel-like structure. The animal may be propelled in any of several directions, depending on the funnel position. The organism accomplishes differential movement by varying the force of ejection, enabling itself to move abruptly and swiftly. Cephalopods are thus able to escape enemies and be successful predators. They have well-developed eyes, which are structurally very similar to vertebrate eyes. The origin of the eyes, however, is different—an excellent example of convergent evolution. All the cephalopods except the *Nautilus* possess an ink gland that secretes an opaque substance, thereby duping a predator into thinking its prey is as large as the ink cloud it perceives.

Male and female reproductive organs are borne on separate individuals. Eggs are fertilized internally and laid externally. There is no larval stage of development; the young that hatch from the eggs look like their parents.

The fossil record of the molluscs is quite extensive, dating back some 600 million years. The most primitive molluscs were aquatic, originating in the sea. Although there is no agreement on their evolution, many authorities suggest that the molluscs may have developed from an annelid ancestor. Their partial segmentation and the trochophore larvae support this suggestion. The recently discovered fossil of the segmented *Neopilina* also affirms the link between the annelids and the molluscs.

PHYLUM ANNELIDA

Organisms of all animal phyla examined so far have unsegmented bodies. The annelids have segmented bodies. The segments are repeated from anterior to posterior, and certain organ systems are repeated from one segment to the next. It is this feature of segmentation that occurs, in varying degrees, in all the higher forms of animal life. A mouth and anus are present, and chambers along the length of the digestive tract perform specific functions (Fig. 20.8). The exterior of the body is covered by a protective layer, the cuticle.

The fluid-containing segments may be compressed by contractions of the musculature, thereby elongating the organism, or expanded so that the organism is shortened (see Fig. 10.3). Most segments have hairlike structures (setae) that, when coordinated with the muscular contractions, make the animal move either forward or backward. These organisms have a well-developed circulatory system. In the anterior end, surrounding the esophagus, are five pairs of muscular chambers, which have been called "hearts." However, they do not pump blood in the earthworm; rather, they regulate the blood pressure. The actual pumping is performed by the dorsal blood vessel. All the vessels are interconnected in what is thus a closed circulatory system. Also in the anterior end are nerve ganglia that act like a brain to control various nervous responses. Various ganglia are located

FIGURE 20.9

Amphitrite, *a polychaete worm. The polychaetes are primarily segmented marine worms. Some swim freely; others burrow under rocks or other materials on the sea floor. The distinct head is marked by sensory appendages, and several setae sprout from each ringlike segment.*

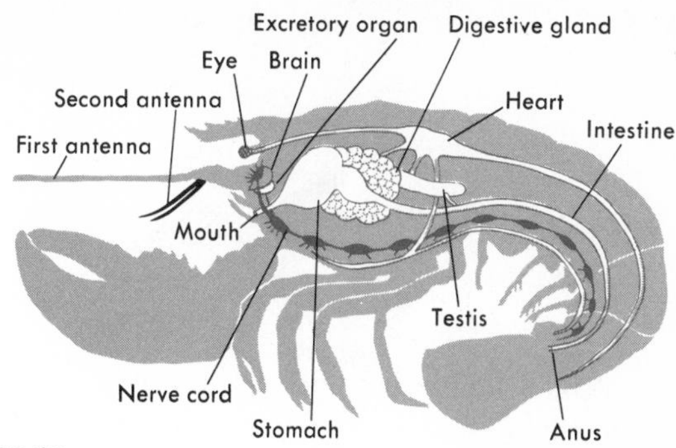

FIGURE 20.10

Principal body structures of an arthropod. The arthropods are distinguished from other invertebrates by a jointed, chitinous exoskeleton, jointed legs, and a complex musculature and nervous system. This basic body plan has been modified in numerous ways to produce forms as diverse as the butterfly and the lobster.

along the ventral nerve cord. Smaller nerve cords extend from these to various organs in the body. An excretory organ system is present in each segment and helps to eliminate most of the metabolic wastes from the fluid-filled segments and from the capillaries that surround some of the nephridia—the tubular structures that function like kidneys.

The members of the three classes of annelids are quite common. The brightly colored Polychaeta are mostly marine (Fig. 20.9). Although most of them are small, some of them are more than three meters long. Sexual reproduction is common, and fertilization occurs in the sea. In many worms, reproduction occurs only during a few days or nights during the year. The Oligochaeta, such as the earthworms, may be aquatic but are mostly terrestrial. The sex organs of these worms are in specific segments, and although the organisms are hermaphrodites, reciprocal copulation generally occurs. The third class (Hirudinea) consists of the leeches. These are ectoparasites that fasten to vertebrates and feed on their blood. The leeches are hermaphrodites and practice reciprocal copulation. Whereas the polychaetes develop a trochophore larva—with a band of cilia located anterior to the mouth—the earthworms and leeches lack such a larval stage.

Annelids and molluscs are thought to have a common evolutionary ancestor. Some members of these phyla develop at first into trochophore larvae. All exhibit bilateral symmetry and have a complete digestive system, even in the larval stage. Annelids also show some similarity to echiurids in that the latter have similar excretory, circulatory, and nervous systems and exhibit some segmentation in the larval stage.

PHYLUM ARTHROPODA

Among the invertebrates, the most successful group is the arthropods. More than a million species are classified, and possibly a million or more are still unclassified. Included in the group are crabs, lobsters, spiders, insects, millipedes, centipedes, and many others. Arthropods live in every conceivable habitat, from mountaintops to ocean floors, from polar regions to the equator, in the air or water, on the ground or beneath it. They range in size from microscopic to lengths of up to three meters. Arthropod fossils, mostly related to crustaceans and arachnids, have been found in Cambrian and Ordovician rocks.

Like many other animals, arthropods develop from three germ layers and exhibit bilateral symmetry. Their bodies are segmented, but they differ from annelids in significant ways. They do not pass through a ciliated trochophore larval stage; they are not hermaphroditic; sexes are separate; they have compound eyes.

Arthropod appendages are jointed—a significant step in the evolution of higher animals—and their bodies are segmented and covered by a chitinous exoskeleton (Fig. 20.10). Generally a distinct head, thorax, and abdomen are evident, with concurrent localization of particular organ systems in certain body regions and loss of others. The organisms have a reduced coelom, a modified circulatory system, and a well-developed nervous system with ganglia in the head region and pairs of ganglia at each segment along the ventral nerve cord. Also present among the arthropods is a variety of respiratory mechanisms, which include direct diffusion, gills, and the presence of tracheae or specialized branched tubes that directly contact the various tissues. Certain species of arthropods have appendages present on all the segments, but in many other species the appendages are either modified or lacking.

Traditionally, arthropods are divided into several classes, many of which are encountered in everyday life. In this chapter, we will discuss class Crustacea (crayfish, shrimp, barnacles, crabs), class Insecta (grasshoppers, beetles, bugs, bees, butterflies), class Arachnida (spiders, ticks, scorpions), and classes Chilopoda and Diplopoda (centipedes and millipedes). We will first examine briefly the subphylum Onychophora, sometimes considered a class of arthropods.

Onychophorans, small wormlike species of the genus *Peripatus*, are unusual in that they share certain annelid characteristics. Their nephridia are segmented and their reproductive ducts contain cilia. Many experts feel that *Peripatus* may be a descendant of certain arthropods that evolved from the annelids, forming an invertebrate "missing link."

Most crustaceans occupy marine habitats. The crustacean body is composed of segments that subdivide the three main body regions (head, thorax, abdomen) and that contain paired appendages. The crayfish, for example, has a total of 19 segments, five in the head, eight in the thorax, and six in the abdomen. The head and thorax are not greatly differentiated and are referred to as a cephalothorax. Crustaceans obtain oxygen by gills, which are shielded from the environment by a carapace, a hard shell. The nervous system is quite well developed. There is a "brain" located in the head, and several nerves are connected to the eyes and antennae. Other ganglia and nerves along the length of the body connect with a ventral nerve cord. Most species are dioecious, and often both sexes are distinguishable (dimorphic). The female carries the eggs on her body until they hatch. The young larvae may be miniatures of their parents, or they may appear quite different and undergo several stages of development (metamorphosis).

Young crustaceans in particular can regenerate lost appendages. They feed primarily on other small animals and on dead animal matter as scavengers. Some eat plants or parasitize fish. Many crustaceans—crayfish, lobsters, crabs, shrimp—are prized as food by humans.

Insects have only three pairs of legs although they have many more than three segments. Some of their appendages are modified into mouth parts for chewing or directing food into the mouth. Their distinct head region is quite complex with a well-developed brain, highly sensitive sense organs for sight and touch, chemical receptors, and hormonal glands that control development and molting. All these characteristics have contributed to the ability of insects to become one of the dominant groups of organisms on the face of the earth. Insects have developed complicated behaviors, including the complex social relationships found among such groups as the ants and bees. Their major limitation has been the presence of an exoskeleton, because it has to be shed periodically, and a new one must then be grown to replace it. Molting is an inefficient method of growth, and it renders the organisms particularly vulnerable to predators while it is going on. Along the sides of the body are paired spiracles, openings into which air passes. These are connected to cuticle-lined tracheae that spread throughout the insect body.

Insects are classified into three major groups, depending on the type of metamorphosis the young undergo. Some—springtails, for example—have no metamorphic stage of development, and the young look like miniature adults. Others—dragonflies, roaches, grasshoppers, termites, bugs—produce a nymph that resembles the adult somewhat, but that must go through developmental changes as it grows.

In the third group—butterflies, moths, bees, beetles, flies—the young are considered larvae. The gypsy moth, *Porthetria dispar*, illustrates the typical stages of the insect life cycle. The moth deposits its eggs in a mass. From the eggs hatch voracious caterpillars, or larvae, which grow to 6 cm. Then each caterpillar spins a cocoon in a secluded place and, suspended from a branch, pupates for a period of several weeks. After this period there emerges a sexually mature adult moth that is capable of mating and depositing its eggs on the surface of

FIGURE 20.11
The green darner dragonfly, Anax junius. *This beautiful water insect eats and mates in flight. As it flies, it holds its six spiny legs together to form a basket in which to capture insects. It preys on many insects harmful or annoying to man.*

a branch or any other suitable object. Whereas the larvae are most often destructive organisms, the adult assists in sexual reproduction in many flowering plants by carrying pollen. It is of evolutionary significance to note that the appearance of the insect is often adapted to feed on the flowers of a particular plant, and it is believed that as both species evolved, concurrent modifications developed with specific methods of pollination.

Insects play diverse roles in the living community. Some are the chief pollinators of flowering plants. From bees we obtain honey and beeswax; from the silkworm, silk. Scavenger insects or their larvae feed on dung and dead animal bodies. Many animals and birds are dependent on insects for food.

Unfortunately, some insects destroy fruit and vegetable crops or damage conifers. Termites help to decompose dead trees in the forest, but they also destroy the wood in buidings they invade. Insects that bite humans are a nuisance at best, and some of them carry such diseases as yellow fever, malaria, and bubonic plague.

Ticks, spiders, mites, and scorpions are included in the class Arachnida. Arachnids are mostly terrestrial and therefore lack gills, but each has a cephalothorax, an abdomen, and four pairs of legs. Arachnids have a type of lung in the abdomen composed of several plates. Air circulates around these plates, and the blood is oxygenated. Spiders trap insects by spinning silken webs. Mites often parasitize plants and other animals. Ticks, which transmit Rocky Mountain spotted fever, obtain nourishment from the host—reptile, bird or mammal—that they bite. Horseshoe and king "crabs" are actually arachnids, not crustaceans.

Centipedes and millipedes have apparent similarities in that they are both terrestrial and distinctly segmented. Yet the millipedes have two pairs of legs per segment, are herbivores, move slowly, and live in damp, shaded habitats. The centipedes have only one pair of legs per segment, are carnivores, move rapidly, are often poisonous, and live in a variety of habitats.

PHYLA ECHINODERMATA, HEMICHORDATA

Two phyla, together with the chordates, are thought to belong to a second evolutionary branch that diverged millennia ago from the branch that includes molluscs, annelids, and arthropods. One important characteristic of all three phyla is the formation of the anus from the blastopore in the embryo. Beginning with the platyhelminthes, all of the earlier phyla, the protostomes, formed the mouth from the blastopore. The phyla forming the mouth from a later opening are called the deuterostomes, a term that means second mouth.

The name Echinodermata means spiny skin, and the phylum includes such organisms as sea stars, sand dollars, sea urchins, brittle stars, sea cucumbers, and sea lilies. They are all marine, found mostly in muddy bottoms and sandy or rocky areas, from fairly deep waters to shallow areas. Generally sluggish animals, they range in size from 2 cm to 1 m. More than 5000 species have been classified.

Echinoderms are unique among the invertebrates, and their morphological peculiarities tend to separate them from all the other species. They are most closely related to the chordates. They are enterocoels whose coelom forms as pouches in the gut of the developing embryo. The blastopore becomes the anus in the adult. Early cleavages in the fertilized egg are similar. The skeleton is internal (endoskeleton) and develops from mesoderm. Hemichordate and echinoderm larvae are similar. Although they exhibit the primitive characteris-

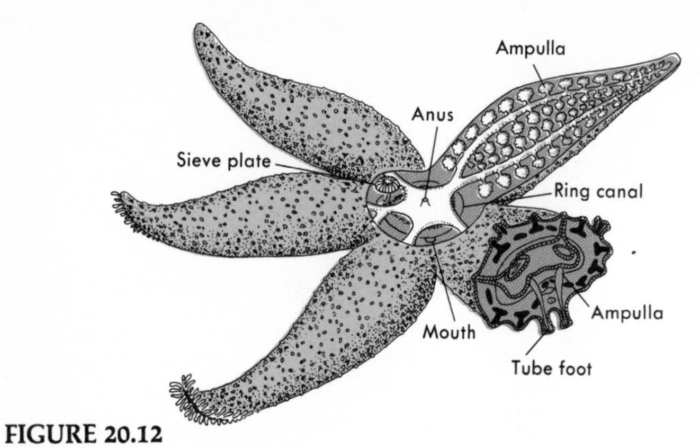

FIGURE 20.12

The unique water-vascular system of echinoderms, illustrated here by the starfish. This system serves as a means of locomotion and a food-capturing device. Water enters through the sieve plate and proceeds by ciliary action down the radial canal to the bulbous ampulla. When the ampulla contracts, water is forced into the tube foot, which is thus stiffened and elongated. If the foot touches and attaches to an object by its suckerlike tip, the longitudinal muscles of the foot will contract. Water is forced back into the ampulla, shortening the foot and drawing the body forward. By exerting a constant pull through steady contraction of the tube foot, a starfish is also able to force open clam shells.

tic of radial symmetry, their larvae are bilaterally symmetrical. In addition to developmental differences, the bipinnaria larvae are unlike the annelid trochophore larvae in appearance, because the cilia surround the mouth instead of encircling the entire organism.

The adults lack segmentation, with no distinguishable head, thorax, or abdomen. They have simple respiratory, reproductive, circulatory, and nervous systems and no specialized excretory system. Of evolutionary importance is their development of an endoskeleton and a water-vascular system, the latter found in no other group of organisms (see Fig. 20.12). The calcareous endoskeleton consists of fused plates. In some instances the plates are fused together to form a solid covering. The sexes are separate, and the gametes are released into the surrounding water for external fertilization. The zygote often develops into a free-swimming ciliated larva. After a brief planktonic existence, the larva is transformed into an adult by metamorphosis. The adults are some of the most beautiful marine organisms found, often brightly colored and delicately shaped. Although most of the echinoderms are mobile, certain ones, such as the sea lilies, are sessile—attached by their base to a substrate. Even with an abundant fossil record dating back some 600 million years, it is not known from what group of organisms the echinoderms might have evolved.

A small phylum of about 100 species, the Hemichordates are marine animals. The most familiar are the acorn worms and tongue worms, which range in size from about 25mm to 3cm. Hemichordates are perhaps most significant from an evolutionary point of view. It is thought that they may be closely related to a common ancestor of both the echinoderms and the chordates. The hemichordate larva is similar to the echinoderm larva. In all three phyla, the anus develops from the blastopore. Like embryonic chordates, the hemichordates have paired gill slits and a dorsal nerve cord.

PHYLUM CHORDATA

This is considered the most recently developed and most advanced group of organisms in the animal kingdom. The phylum may be subdivided into three subphyla: Urochordata (tunicates), Cephalochordata (lancelets), and Vertebrata (vertebrates). Humans and such animals as birds, snakes, lizards, frogs, salamanders, and fish all have backbones and are called vertebrates. The other two subphyla are less known but nevertheless are of evolutionary importance. All three groups are segmented (not too conspicuous in the vertebrates), have a notochord (gelatinous supporting rod above the gut), a dorsal hollow nerve cord, and paired pharyngeal gill slits on the side of the pharynx. They also exhibit bilateral symmetry.

The tunicates and lancelets are small marine animals with no vertebral column and no distinct cranium or brain. Most of the tunicates are sessile, probably a primitive condition. There is a free-swimming stage from which it is thought the vertebrates may have evolved, thus eliminating the sessile stage. The lancelets, which are nonsessile, are thought to have evolved from sessile forms.

The free-swimming tunicate larva looks somewhat like a frog tadpole. But when it attaches to a rock, shell, or other immobile object, the notochord and nerve cord become smaller and are absorbed by the body. A protective covering, the tunic, then encloses the body. One individual may be contained in its own tunic, or several individuals may form a colony and share a single tunic. Reproduction is either sexual or asexual by budding.

A frequently studied lancelet, *Amphioxus*, resembles a fish in its external appearance. It contains a notochord and segmented muscular regions along its body. The digestive tract is straight, and the circulatory system with its dorsal aorta is similar to that of vertebrates, but it lacks a heart. Reproduction is sexual and the larvae are ciliated.

FIGURE 20.13
A Gila monster feasting on the eggs of a desert quail. This large poisonous lizard is native to the deserts of the Southwest. It grows to a length of about 18 inches. The venom flows along grooves at the base of the animal's jaw into wounds made by the teeth.

Vertebrates Although the vertebrates are the most complex of the animals, they are not so abundant or diverse as insects. As already noted, the vertebrates do not have a conspicuous segmentation except in embryonic development and in their vertebrae. The notochord is wholly or partially replaced by the cushions of tissue between the vertebrae. The vertebrae arch over the nerve cord to form a tunnel-like protection, and at the anterior end, a protective case of cartilage or bone develops around the brain.

Earliest fossil vertebrates probably lacked jaws and were covered by scales. They were most likely aquatic having an internal skeleton with the notochord encased by vertebrae.

Class Agnatha includes the oldest and most primitive vertebrates, which are represented today by lampreys and hagfish, both parasites of fish. These jawless fish have a circular sucking disk to attach to a fish. A tooth-like projection and its tongue serve to rasp an opening in the fish so that the blood and body fluids may be sucked out. These parasites have damaged many fish, destroying much of the fishing industry in certain areas, such as the Great Lakes. The notochord persists in the adult, and the skeleton is cartilaginous. Agnatha have a two-chambered heart, dorsal and ventral aortas, and several aortic arches in the gill region. Reproduction is totally sexual, with fertilization of eggs occurring externally. The young of the lamprey hatch as tiny larvae, and they do not reach maturity for three to seven years. Hagfish do not go through a larval stage of development.

The members of class Chondrichthyes are cartilaginous fish and have no true bone structure. These include mostly marine animals like sharks, skates, and rays. A cartilaginous skeletal system is considered to be very primitive and a precursor to bone. The cartilaginous fish were the first vertebrates to have jaws and paired appendages, fins. The brain and nervous system are more complex than the lamprey's, but as with the lamprey, there are no lungs or swim bladder. The heart is two-chambered, and aortic arches are present. All blood flowing through the heart is venous. The notochord is present in the adult. The chondrichthyans probably originated as fresh-water animals, but they migrated to the oceans very early in their history.

Fertilization is external, and most give birth to live young that do not go through a larval stage. They have very tough skin made of pointed sharp scales, which are similar in structure to their teeth. It is the similarity of their teeth to the teeth of higher vertebrates that suggests that teeth have evolved from scales. Fossil teeth date the chondrichthyans back some 450 million years.

The true fish of the class Osteichthyes are the primary vertebrates that live in an aquatic environment. They include fresh-water and salt-water species, such as trout, salmon, sturgeon, swordfish, eels, and seahorses. Salmon and eels spend a portion of their life cycle in salt water and another period in fresh water. The 20,000 species of fish may be found at all latitudes, at all

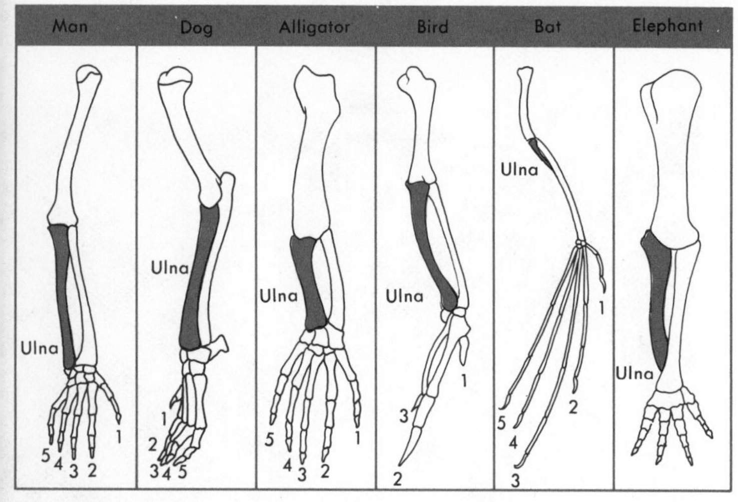

FIGURE 20.14

Comparison of the forelimbs and hindlimbs of several vertebrates. Although the homologous forelimbs are used for quite different functions, their patterns are similar, varying only in emphasis of specific parts of certain bones. These modifications reflect adaptations to a variety of needs. In the dog, for example, the forelimbs are extended by elongating and raising the wrist and carpal bones, thus providing an appendage well adapted for running. The large and heavy bones of the elephant's forelimb, in comparison, are adapted to provide maximum support for a heavy body. The relatively general ized structure of the human arm and hand, with its unique opposable thumb, have provided man with a maneuverable appendage that can be used to manipulate his surroundings.

depths, and in all waters, including those that are heavily polluted.

They have a true bony skeleton, with a brain case of bone. It is thought that the first fish had lungs as well as gills, and a few species of lungfish still survive. The lung in most fish has been modified to a swim bladder. This organ controls the buoyancy of the fish as the amount of air contained within the bladder is changed. The notochord is often reduced in the adult.

All these fish most likely originated in fresh water, and even those species now found exclusively in salt water remained in fresh water much longer than the cartilaginous fish did.

In the class Amphibia, most of the animals spend a portion of their life cycle on land and a portion in the water, with the exception of a few aquatic salamanders. Terrestrial members like the frogs and toads return to the water to breed, and fertilization is external. The larvae of the amphibians are fishlike with gills but are generally transformed into adults that look very little like the larvae. The aquatic salamanders are the exception and change very little as the adult develops. The adult frogs and toads lose their gills and have lungs which function in respiration.

Several important adaptations allow most adult amphibians to spend large amounts of time on land. The appendages in the adult are well developed for land travel, although amphibians have retained their swimming capabilities. Their skin can tolerate exposure to air for relatively long periods. The circulatory system in the larval stage, which is much like that of fish, has a two-chambered heart and several aortic arches. In the adult, to accommodate circulation to lungs, the heart is three-chambered with two auricles and one ventricle. Only one aortic arch persists, and both venous and arterial blood flows through the heart. Amphibians probably evolved from the very primitive lobe-finned fish. These fish, known only from fossils before a live one (*Latimeria*) was actually captured in 1939, had functional lungs, nostrils that opened into the mouth, and fins that were structured in such a way that the fish probably could crawl from one stream to another during times of drought.

The class Reptilia includes the first truly terrestrial animals—lizards, snakes, alligators, crocodiles, and turtles. Reptiles are found mostly in tropical and subtropical regions, but they also inhabit desert, alpine, and cold temperate regions. In areas with harsher environmental conditions, many reptiles hibernate. Like frogs and fishes, they cannot regulate their internal temperature. However, they are better suited to the terrestrial environment than are amphibians.

Besides the evolutionary development of limbs on which they could move about on land (see Fig. 20.14), reptiles have a better developed lung system and a circulatory system with a heart that has specialized compartments which pump the blood to specific regions. Their skin is dry and scaly, often with an underlying armor of bony dermal plates that protect them against drying out. Some well-defined mechanisms of breeding behavior have evolved. For example, in some reptile species the eggs develop outside the female. She often deposits them in a sunny place under the ground, where the warmth of the sun may speed the hatching process. The development of the "land egg"—that is, an egg that does not need to be laid in watery surroundings—probably also played a considerable role in freeing reptiles from the water. There is no larval stage in the life cycle of reptiles. The young develop ultimately into sexually mature adults. The reptiles have an excellent fossil record, which indicates that more than 200 million years ago, they were the dominant animals on the earth.

Some dinosaurs were herbivores, some were carnivores, and they varied considerably in size. Some had extremely small and primitive brains although the brains of some carnivorous species were quite large. Pterodactyls were flying dinosaurs, and ichthyosaurs inhabited the sea. Dinosaur fossils have been found frequently in the western United States, and a section of fossil-rich land in Colorado has been set aside as Dinosaur National Monument.

Recently there has been renewed interest in the ancestry and evolution of reptiles. Dinosaurs were once thought to be closely related only to reptiles and at an evolutionary dead end. By studying the fossil record, scientists have now come to believe that, in contrast to the reptiles, dinosaurs were able to regulate their internal temperature and that two different groups of these creatures were ancestral to birds and mammals.

Nearly 9000 species of birds, the feathered flying vertebrates of class Aves, are recognized. They occupy diverse habitats. Except at nesting time, the albatross lives entirely on the sea. Many birds prefer meadows. Others frequent brushy areas or inhabit trees. Birds are found from the Antarctic continent to the tropics, and from sea level to elevations of 20,000 feet. They may be very tiny or taller than a person. For example, one hummingbird weighs only 1/10 ounce, but the ostrich grows to seven feet and may weigh as much as 300 pounds.

Physical adaptations have helped birds succeed in vastly different habitats. Feet and legs are specialized for wading (herons), swimming (geese), grasping (eagles and hawks), perching (finches), climbing (woodpeckers). Beaks are used for cracking seeds (grosbeaks), probing into sand (snipes), catching insects (flycatchers), tearing prey (eagles). Wings are mostly used for flying, but some birds (penguins and grebes) use them for swimming. The brilliant feathers of some birds (parrots and cockatoos) contrast with the more muted plumage of others. Their songs aid in establishing territory, warning of danger, and guiding the young.

FIGURE 20.15

The red-tailed hawk, a large, heavy-bodied bird of prey. Its curved beak, sharp talons, and powerful feet are adaptations for efficiently killing small birds and animals for food. Occasional raids on poultry are more than offset by its destruction of many harmful rodent pests.

Birds are built for flight. The body is streamlined, the wing an airfoil, the tail a rudder. Without adding much weight, feathers greatly increase the surface area of the wings, which are powered by large chest muscles. Bones are hollow and light in weight, and the sense organs of sight and hearing are highly developed.

Birds exhibit several evolutionary advances over other lower contemporary vertebrates. They have a high metabolic rate and can maintain a steady body temperature. Their bodies are insulated by feathers. Birds have four-chambered hearts, the most efficient type, in which blood from pulmonary and systemic circulations does not mix.

The fossil record is not extensive. The oldest bird fossils, *Archaeopteryx* and *Archaeornis* dating from the Jurassic period, were found in Germany. Some evidence for the evolution of birds from certain dinosaurs has come from Russia, where a small thecodont was found to have had scales that were adapted to provide insulation (ancestral feathers).

Mammals, like birds, are found throughout the world. Seals, whales, walruses, and manatees spend much or all of their lives in the oceans from the Arctic to temperate regions. Beavers and otters occupy freshwater habitats. Hoofed animals—antelope, giraffes, buffalo—are found in grasslands. Some rodents spend much of their time in burrows in the soil or in trees. Not surprisingly, carnivores—wolves, foxes, lions, jaguars—are found wherever there are herbivores. Bats are the only mammals capable of true flight.

The smallest mammals are mice and shrews, which may weigh less than an ounce. The largest are the elephants and whales. The blue whale, which can weigh over 100 tons, is the largest known animal ever to have lived on Earth.

In general, mammal bodies are wholly or partially covered with an outer insulating material, hair, and a subcutaneous layer of fat. The heart is four-chambered, and oxygenated blood is separated from deoxygenated blood. Mammals are able to regulate their internal temperature and maintain a high metabolic rate. Mammalian teeth are complex and differentiated for various purposes—biting, grinding, tearing. The lower jaw is a single bone.

Fertilization is internal, and most mammalian young develop within a uterus. After birth, the young obtain nourishment through the mammary glands of the female. These developments ensure a higher survival rate for the young. Behaviors, sense organs, and voices are all more highly developed in mammals than in other chordates. Many mammals are capable of showing emotion and changing their behaviors.

Mammals are divided into three subclasses based on the way the young develop. Protheria, such as the duck-billed platypus of Australia, lay eggs but suckle their young. Metatheria are the marsupials—wombats, kangaroos, Tasmanian wolves—which bear their young in a relatively undeveloped state and nourish them within an external pouch. The young of Eutheria, the placental mammals, have a much more complete intrauterine development.

The more familiar orders of the Eutheria are: Insectivora (moles, shrews), Primates (monkeys, humans), Chiroptera (bats), Edentata (sloths), Rodentia (mice, rats), Lagomorpha (hares, rabbits), Carnivora (dogs, cats), Cetacea (whales, dolphins), Proboscidea (elephants), Perissodactyla (odd-toed hooves—horses), and Artiodactyla (even-toed hooves—cows, deer).

We humans have always owed much to other mammals. In both past and present, they have provided food and clothing. They work for us, and some have even contributed to medical research.

Summary

1. Animals are characterized by their ability to move, well-developed muscular and nervous systems, quick responsiveness to changes in the environment, and the absence of cell walls.

2. The members of the animal kingdom have evolved toward larger size and greater specialization of organ systems. The coelom is first found in the Mollusca, segmentation in the Annelida, jointed appendages in the Arthropoda, an endoskeleton in the Echinodermata, and a notochord in the Chordata.

3. Evolutionary adaptations have enabled many animals to live on land. Higher animals have lungs, jointed appendages, a skin tolerant to air, internal regulation of body temperature, support of the body by a bony endoskeleton, and development of young internally or externally in a waterproof egg.

4. Birds and mammals have a high metabolic rate and are capable of regulating their internal body temperature. Their success has been in part due to the following features: four-chambered heart, the separation of venous and arterial blood, two circulations (pulmonary and systemic), and body insulation (fur, feathers, hair, or fat).

REVIEW AND STUDY QUESTIONS

1. Give a *very general* description that could apply to most organisms found in the animal kingdom.

2. Until fairly recent times, sponges were thought to be plants rather than animals. What characteristics of sponges suggest to you reasons for this misclassification?

3. Why is "jellyfish" an incorrect name for that familiar animal? What points of similarity characterize the phylum to which the jellyfish belongs?

4. Why has bilateral symmetry become a characteristic of evolutionary importance in animals but not in plants?

5. Most of the molluscs that are familiar to us are found in two classes. Name the two classes and give three examples of each.

6. Name as many arthropods as you can, grouping them by class.

7. What two groups of animals are chordates but not vertebrates?

8. Name as many familiar vertebrates as you can.

REFERENCES

Bakker, R. T. 1975 (April). "Dinosaur Renaissance." *Scientific American* Offprint no. 916. Freeman, San Francisco.

Gierer, A. 1974 (December). "Hydra as a Model for the Development of Biological Form." *Scientific American* Offprint no. 1309. Freeman, San Francisco.

Leggett, W. C. 1973 (March). "The Migrations of the Shad." *Scientific American* Offprint no. 1268. Freeman, San Francisco.

Margulis, L. 1971. "The Origin of Plant and Animal Cells." *American Scientist* 59: 230–235.

Warren, J. W. 1974 (November). "The Physiology of the Giraffe." *Scientific American* Offprint no. 1307. Freeman, San Francisco.

SUGGESTIONS FOR FURTHER READING

Blackwelder, R. E. 1967. *Taxonomy: A Text and Reference Book.* Wiley, New York.
A zoology reference book in taxonomy containing rules of zoological nomenclature.

Buchsbaum, R. M., and L. J. Milne. 1960. *The Lower Animals: Living Invertebrates of the World.* Doubleday, Garden City, N.Y.
In clear, popular writing the authors describe each of the invertebrate classifications.

Burt, W. H., and R. P. Grossenheider. 1964. *A Field Guide to the Mammals.* Houghton Mifflin, Boston.
A clear, well-illustrated, and easy-to-use field guide to the mammals of North America. Emphasis is on the diagnostic points that differentiate species and races.

Hyman, L. H. 1940–1967. *The Invertebrates.* McGraw-Hill, New York.
A classic, authoritative reference on the invertebrates and their evolution.

Mayr, E. 1969. *Principles of Systematic Zoology.* McGraw-Hill, New York.
This is a standard textbook for zoological taxonomy.

Romer, A. S. 1970. *The Vertebrate Body.* 4th edition. Saunders, Philadelphia.
Covers comparative vertebrate morphology and evolution.

Schmalhausen, I. I. 1968. *The Origin of Terrestrial Vertebrates.* Academic Press, New York.
Reference work. Develops an evolutionary sequence of land vertebrates.

Updike, J. 1972. "During the Jurassic." In *Museums and Women.* Knopf, New York.
A short story by a noted writer, of interest to students of classification.

PRINCIPLES OF ECOLOGY

chapter twenty-one

The marsh pond, an important ecosystem, is well supplied with essential abiotic components, such as gases, solar radiation, and moisture. Among its principal producers are cattails (genus Typha*), rooted in the shallow water, and floating green plants, mainly algae. Its consumers range from microscopic organisms to large fish and birds.*

Ecology—a term derived from the Greek word *oikos*, meaning "house"—is the study of the relationships between organisms and their environments or, broadly speaking, their houses. The object of study is not the individual organism, however, but the community—a group of individuals of various species living and interacting with one another in a given area. A community, together with its nonliving physical environment, constitutes an ecological functional unit called an **ecosystem.** All the ecosystems of the earth in turn constitute the **biosphere.** Returning to our definition, we can more precisely explain ecology as the study of ecosystems.

Abiotic components of the ecosystem

By definition, an ecosystem has two components—the biotic, consisting of living things, and the abiotic, consisting of things that are *not* alive. The nonliving components include habitat, gases, solar radiation, temperature, moisture, and inorganic and organic nutrients. A **habitat** is the physical environment in which organisms live. There are two major habitat types, terrestrial and aquatic. Although the atmosphere is a medium for organism transportation and dispersal, it is not considered a major habitat, because it does not serve as an actual locale of residence. Ecosystems contain a great variety of habitats, and they contain all the essential abiotic elements that organisms require.

TERRESTRIAL ENVIRONMENT

The major life-supporting element of the terrestrial environment, soil, is composed of mineral matter interspersed with varying quantities of organic substances, air, and water. Soil generally consists of three layers. At the surface is the topsoil, which contains particulate mineral matter as well as appreciable quantities of organic substances that give the layer a darkish hue. Beneath the topsoil is the subsoil, a layer of mineral matter comparatively devoid of organic substances. The bedrock layer—composed of rocks—is the deepest of the three.

The topsoil is of key importance to most terrestrial habitats. The combined presence in the topsoil of minerals, organic compounds, and relatively large quantities of air and water allows this layer to support many forms of plant and animal life. In fact, maximum root extension for most plants goes no deeper than the topsoil or upper subsoil layers.

AQUATIC ENVIRONMENT

Approximately 70 percent of the earth's surface consists of aquatic habitats, of which there are three major types —fresh-water, salt-water, and estuarine. Fresh-water habitats include lakes, ponds, streams, rivers, swamps, and springs. Salt-water habitats, which make up most of the planet's aquatic environment, include the seas and the oceans. Intermediate between the fresh-water and the salt-water habitats is the estuary—a river mouth where salt waters and fresh waters mix.

Essentially the three habitats differ from each other by their degrees of salinity, but each contains the dissolved gases, minerals, and organic substances necessary to support life. The salt-water habitat is the most chemically stable of the three because its pH, temperature, and salt, carbon dioxide, and oxygen concentrations fluctuate the least. However, the seas are often physically variable as the result of waves, currents, and tides that are caused by the gravitational effects of the moon and the sun. Fresh waters, on the other hand, are physically more equable, yet undergo greater fluctuations in salinity, temperature, pH, and concentrations of dissolved gases.

GASES

Almost all organisms require certain gases in order to live. Green plants need CO_2 to carry on photosynthesis, all respiring organisms require oxygen for oxidative phosphorylation, and certain organisms utilize gaseous nitrogen and sulfur compounds as part of their metabolic activities. Essential gases are dissolved in water and present in pockets in the soil, but the major source of gases is, of course, the atmosphere. The atmosphere is a mixture of gases: 78 percent nitrogen, 21 percent oxygen, and 0.03 percent carbon dioxide. Other gases, such as hydrogen, helium, and methane, are present in trace amounts. Near the earth's surface, water vapor composes up to 4 percent of atmospheric volume. Besides its role as a gas reservoir, the atmosphere serves other functions essential to ecosystems. It provides a means of transportation for both plants and animals, and in conjunction with solar radiation, it is the medium in which weather is generated.

SOLAR RADIATION, TEMPERATURE, MOISTURE

Solar radiation is an essential abiotic component, because directly or indirectly all organisms derive from it the energy necessary for life. Plants, which directly

FIGURE 21.1

The ptarmigan's white winter plumage blends with the snow, and its speckled summer coat merges into a background of lichen and rock. This color change results from a series of endocrinal and hormonal activities triggered by the amount of light that penetrates the bird's eyes. Thus intensity of sunlight and length of day directly affect the ptarmigan's seasonal molt, a protective adaptation that helps the bird avoid predators.

utilize solar energy, vary in their requirements for sunlight, and their ability to live in a certain habitat is determined in part by the intensity of sunlight and the length of day in that environment. Thus sunlight is a determinant of the distribution of plants and of animals that depend on plants for food.

In addition, solar radiation is responsible directly or indirectly for the earth's climate and weather, which in turn determine two other abiotic factors, temperature and moisture. Solar radiation varies according to latitude and thus creates the three major climatic zones—tropical, temperate, and polar. The generation of weather involves the interaction of hotter and cooler portions of the atmosphere with warmer and cooler areas of the earth's surface and oceans.

Although the process is too complex to be considered here, the role of solar radiation can be sketched out briefly. Some solar radiation is absorbed by the atmosphere, thereby warming it and generating winds. Even more solar energy is absorbed by the earth's land and oceans. Some of this energy is radiated back into the atmosphere as heat, warming the atmosphere and generating air movement. Another portion causes the evaporation of water, which rises into the atmosphere as water vapor. In conjunction with other factors, atmospheric water vapor forms clouds and precipitation. Winds and precipitation are responsible for the distribution of life-giving moisture over the surface of the globe. Solar radiation absorbed by the oceans is a determinant of ocean currents, which also have an important role in generating weather. Finally, because of the heat-absorbing properties of water, the oceans, in conjunction with solar energy, serve as a heat buffer, absorbing solar radiation as heat and preventing large temperature fluctuations.

INORGANIC AND ORGANIC NUTRIENTS

All forms of life require certain biogenic salts that contain minerals essential to vital life functions, such as protein formation, enzyme activity, photosynthesis, etc. Salts are absorbed by plants from the water in their habitat and acquired by animals from water or food.

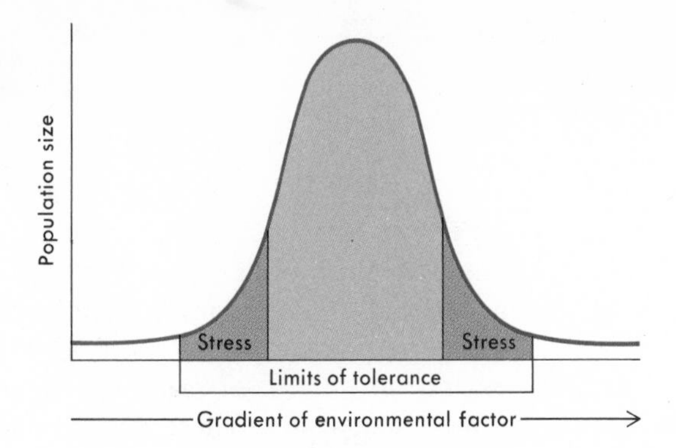

FIGURE 21.2

Limits of tolerance. A normal (bell-shaped) distribution curve is typical of natural populations. The greatest concentration of individuals appears where environmental conditions are optimal. Concentrations fall off at the extremes.

Minerals required in large amounts—such as nitrogen, phosphorus, sulfur, calcium, magnesium, potassium, and the iron salts—are called **macronutrients.** Those required in trace amounts—such as manganese, copper, and cobalt—are called **micronutrients,** or trace elements.

In addition, living organisms require organic nutrients. Recall that autotrophs get both inorganic and organic substances from their environment and they make their own nutrients. Heterotrophs get their organic nutrients by eating autotrophs.

All organisms depend on their abiotic environment for survival, and they will survive only if the various abiotic factors are maintained within critical limits. Any factor within the ecosystem that tends to deter growth is called a **limiting factor.**

The effect of limiting factors is expressed in what is known as the *Liebig-Blackman law of the minimum,* which states that the growth of a population is dependent on and limited by the amount of an essential abiotic factor that is present. This law is well illustrated by the fact that crop yields sometimes depend on a seemingly insignificant but relatively scarce abiotic factor, such as boron, that is needed in only small amounts. More usually, however, crop growth is limited by the amounts of the easily exhausted macronutrients—nitrates, phosphates, and potassium. Consequently, most inorganic fertilizers used in modern farming consist primarily of these three macronutrients. Yields are seldom limited by such factors as carbon dioxide and water, since these are usually abundant.

In addition to lower limits, there are upper limits of tolerance, and thus too much of something can be a limiting factor as well as too little. This is illustrated by the fact that small amounts of arsenic in the human diet have a tonic effect, but large doses are fatal. V. E. Shelford formulated a concept of biological tolerance in 1913. His *law of tolerance* states that for each ecological parameter to which an organism responds, there is a minimum critical value and a maximum critical value, which together are known as the **limits of tolerance** (Fig. 21.2). Each species, of course, has its own limits of tolerance for each abiotic factor. Since these tolerances are so relative, two prefixes—*steno-,* meaning narrow range, and *eury-,* meaning wide range—are employed to characterize them. An organism tolerant of a narrow temperature range would be stenothermal, and one tolerant of a wide range, eurythermal. Within the limits of tolerance, whether narrow or wide, there is also an **optimal range** at which the organism functions or grows best.

The subject of limits and tolerances has recently been one of the favorite problems studied by systems ecologists. Inasmuch as changes in quantities of simple inorganic and organic materials can be modeled, computer simulations can often predict the cycles of blooms of lakes and other such changes in population sizes. This is one of the principal ways in which ecology has changed from a qualitative, descriptive form of natural history to a quantitative, predictive science.

Abiotic-biotic interactions: ecosystem balance

An ecosystem is not a static unit of abiotic and biotic components. Instead it is a changing, self-modifying system in which abiotic and biotic elements affect one another, thereby ensuring the continuance (or in some cases the collapse or modification) of the ecosystem.

Thus the study of ecology does not treat organisms as discrete and isolated entities but as integral functioning parts of a larger ecological machine. Ecologists use the term **ecological niche** to designate the position or status of an organism within its community and ecosystem resulting from the organism's structural adaptations, physiological responses, and specific behavior. Whereas the habitat is the place where an organism lives, the niche is the function it performs. In anthropomorphic terms, we can consider the habitat the address of the organism and the niche its role or job. In drawing such an analogy, however, we must be careful to remember that organisms do not purposely choose to fulfill certain roles. Rather, they have come to their niches by a long evolutionary history of adaptation, in which they have survived by carrying out a limited set of functions within the community.

A niche is defined by such factors as the organism's position in various food chains; its predators and prey; its limits of tolerance for abiotic chemicals, sunlight, and moisture; the time and place it breeds; its diseases; etc. Because the niche is determined by many many factors, it would probably be next to impossible to fully detail a single organism's role in the ecosystem. However, a niche is defined by an organism's reactions to and effects on both abiotic and biotic factors in the ecosystem. When we examine competition as an interspecific interaction later in this chapter, we will consider an important part of the definition of niche, namely, that only one species can occupy a niche within a given habitat area.

BIOGEOCHEMICAL CYCLES

Organic compounds are recycled again and again through an ecosystem by way of the food chains. Substances that make up producers are ingested and assimilated by consumers. The matter composing both producers and consumers is later broken down by decomposers to a form that can be used again by new generations of producers. (We shall consider these three categories of organisms in more detail when we examine the biotic components of the ecosystem later in this chapter.) Similarly, inorganic elements are also recycled by a series of interactions between biotic and abiotic ecosystem components. These interactions are known as inorganic cycles or **biogeochemical cycles,** and they are dependent on organisms that, in occupying certain niches, perform activities that perpetuate the cycles. Five major biogeochemical cycles are those of carbon, oxygen, phosphorous, nitrogen, and water. These cycles ensure the availability and reuse of inorganic compounds necessary for life.

Carbon cycle From the carbon dioxide in the atmosphere and in water, life-sustaining organic compounds, such as carbohydrates, proteins, and lipids, are formed through photosynthesis. These compounds are broken down either by the plants themselves or by the animals that eat the plants or other animals. At each step CO_2 is released into the air or water through energy-producing respiration (Fig. 21.3). Not all of the environment's carbon is recycled in this manner, however. Some organic materials are excreted as waste products that, together with dead plants and animals, are decomposed by microorganisms. Bacteria and fungi play a key role in the carbon cycle by returning additional CO_2 to the atmosphere and water through respiration.

Completion of the cycle along these carbon pathways can take a few minutes, or it can take years. Furthermore, there is another important path taken by carbon, one that is of very long duration. Some dead animals and plants that escape decomposition by microorganisms are deposited between rock layers or on lake bottoms and river beds and compressed by tremendous geological force over thousands of years. This process produces carbon-containing rocks, such as limestone and diamond, and the earth's fossil fuels—coal, oil, and gas. This carbon remains trapped until the rocks are weathered by wind and water or the fuels are burned. It then returns to the atmosphere again in the form of carbon dioxide. Because we are currently burning fossil fuels at an unprecedentedly rapid rate, the increase in atmospheric carbon dioxide may have profound effects on our climate.

Oxygen cycle The biogeochemical cycle of oxygen is closely linked with the carbon cycle, and it also involves the processes of photosynthesis and cellular respiration

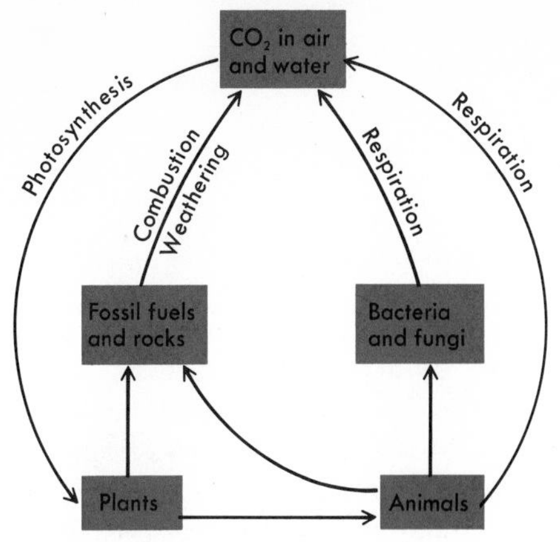

FIGURE 21.3
Some major pathways in the carbon cycle.

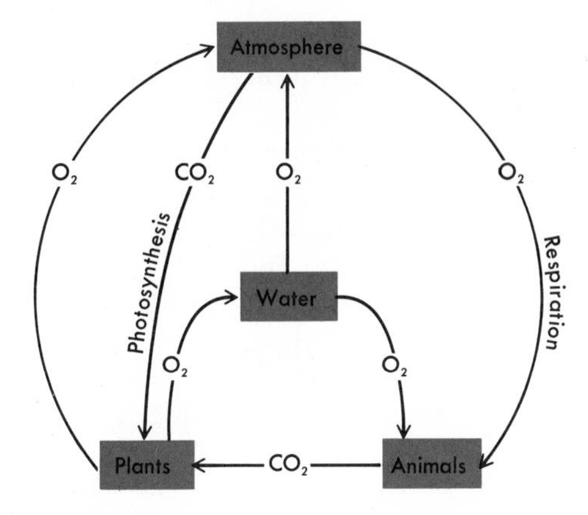

FIGURE 21.4
The movement of oxygen through the oxygen cycle.

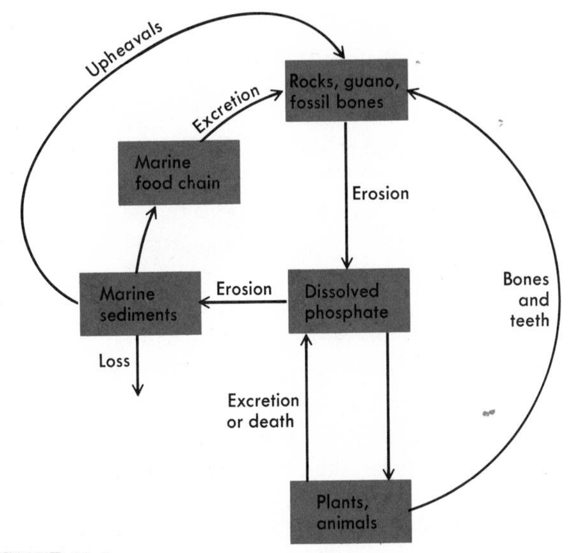

FIGURE 21.5
Principal pathways in the phosphorus cycle.

(Fig. 21.4). Photosynthetic plants use the carbon dioxide and water in the environment to produce organic compounds and molecular oxygen, O_2. Oxygen in this form makes up about 20 percent of the earth's atmosphere. There is also a considerable amount of oxygen in the ocean's waters. Terrestrial and aquatic organisms use the molecular oxygen in respiration, giving off carbon dioxide and water, which are again available to photosynthetic green plants. Thus the carbon and oxygen cycles—photosynthesis and respiration—are complementary.

Phosphorus cycle Phosphorus is necessary for life as an essential component of phospholipids in cell membranes, nucleic acids, and high-energy compounds, such as ATP. The major sources of phosphorus are phosphate-containing rocks formed in previous geologic eras. Phosphates are leached from rocks or eroded in particulate form by the action of water. Phosphates may be deposited in the soil or carried into streams, lakes, and the ocean, where they are precipitated as salts of calcium, aluminum, and iron. They then sink to the sediment at the bottom. Phosphates are also derived from the decaying bodies of plants and animals. Soil or shallow-water phosphates are absorbed by plants. Phosphates from shallow marine sediment are returned to the land by means of a fish-bird food chain (Fig. 21.5). Sediment-eating fish are consumed by sea birds, whose phosphorus-rich excrement, guano, is dropped on the land. Phosphates in ocean sediment can also be recycled with the uplifting of the ocean floor, but in general, phosphates in deep ocean sediments are not recycled and consequently are lost to the biosphere. Because geologic uplifting is rare and the fish-bird reclamation process is relatively insignificant, more phosphates are lost in the ocean than are recycled. In the future a lack of available phosphate may be the critical limiting factor in the world's ability to feed its enormous human population.

Nitrogen cycle The most complex of the biogeochemical cycles is that of nitrogen. Although 78 percent of the atmosphere is nitrogen gas, this atmospheric nitrogen

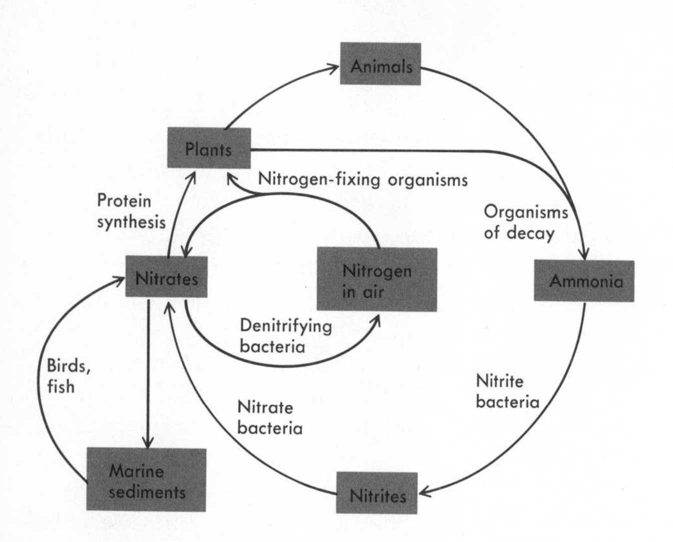

FIGURE 21.6

The nitrogen cycle. Although four-fifths of the gas in the atmosphere is nitrogen, it is not directly available to animals or most plants. The nitrogen cycle in the biosystem depends on nitrogen-using bacteria that are able to fix atmospheric nitrogen or nitrogen from soil and water, making it available to consumers in the form of nitrates. Nitrogen is returned to the soil in wastes and through decay.

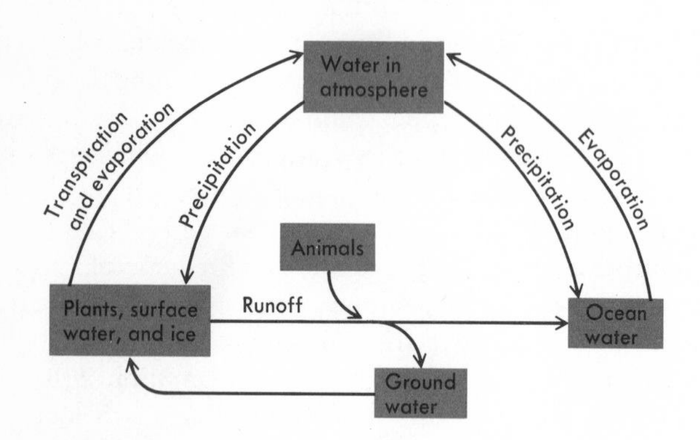

FIGURE 21.7

The water cycle. Water is present on the earth in a limited amount. What water there is becomes available for life functions through the work of the water cycle. The cycle is driven by solar radiation, which causes water to evaporate from streams, lakes, oceans, and living organisms. Precipitated out of the atmosphere as rain or snow, water becomes available to the biosphere. As surface or groundwater it is exposed to evaporation, and so the cycle is repeated.

can be used only by a few blue-green algae, some free-living bacteria, and certain species of bacteria that are symbiotically associated with the roots of legumes. The latter species of bacteria incorporate gaseous nitrogen into chemical compounds, a process called **nitrogen fixation.** The nitrogenous compounds (primarily the amino acids) can then be assimilated by plants. Consumers acquire nitrogenous compounds by eating plants. These compounds are eventually returned to the environment with the decomposition of dead plants and animals (Fig. 21.6).

When nitrogenous compounds are returned to the environment, they undergo a process of decomposition called **ammonification,** in which the nitrogen is released as ammonium ions through the action of microorganisms. This ammonia may be reassimilated by plants, may evaporate, or may be oxidized by microorganisms into nitrate or nitrite—a process called **nitrification.** Some of these nitrates and nitrites are assimilated by plants, but some may stay in the soil or undergo **denitrification**—a microbial process in which the nitrates are transformed into nitrogen or nitrous oxide, gases that reenter the atmosphere.

Although some nitrogen is lost from the cycle through deposition into the oceans, the loss is compensated for by gaseous nitrogen released during volcanic activity.

Water cycle All materials of life require water as a transport system. Water itself is constantly recycled through the biosphere in a pattern that is familiar to us (Fig. 21.7). Water vapor in the atmosphere forms clouds through condensation. When weather conditions are right, this water falls to the ground and into the oceans as rain or snow (precipitation). About 97 percent of the earth's water is found in the oceans, and this water returns to the atmosphere through evaporation caused by the heat of the sun. When ocean water evaporates, its salts are left behind; therefore the rainwater that returns to earth is fresh water.

The water that falls on the ground may follow several different paths. Some filters through the topsoil and merges into streams, rivers, and lakes. Water that follows this path is called runoff. The water that seeps below the topsoil and forms underground streams is ground water. Plants pick up water in the soil through their roots and use it in photosynthesis. Some of this water evaporates off the leaf surfaces in the process called **transpiration.** When animals eat the plants, they excrete water in urine that becomes part of the runoff. Animals also exhale water vapor as a product of respiration. Fresh water in lakes and streams, accounting for only one percent of the earth's water, either joins the oceans or evaporates into the atmosphere to begin the cycle anew.

Transported in this cycle of condensation, precipitation, and evaporation is a flowing laboratory of chemical reaction. Rivers carry materials from the soil to the sea. Minerals in the atmosphere fall to the ground with the rain. Plants, animals, and the soil exchange nutrients through water transport. All the biogeochemical cycles are constantly interacting and recycling to nurture life on earth.

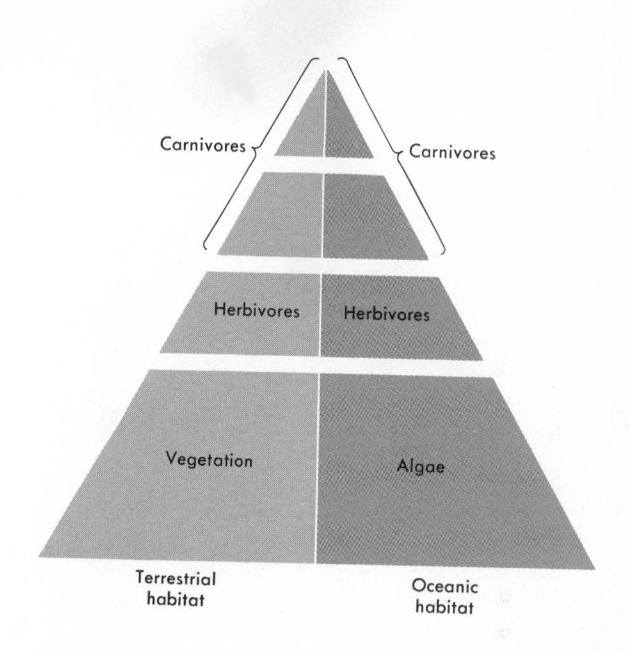

FIGURE 21.8

Food pyramid. As shown here graphically, there is a structural similarity between terrestrial and oceanic (or aquatic) habitats. In each habitat the bottom of the pyramid consists of producers: lichens, mosses, grasses, trees, etc., in the terrestrial habitat; algae in the oceanic habitat. Higher levels consist of primary, secondary, and tertiary consumers. The plant life is the food of the herbivores, which in turn support the carnivores and ultimately the decomposers (not shown).

Biotic components of the ecosystem

An ecosystem consists of three biotic elements—**producers, consumers,** and **decomposers.** The producers are autotrophic organisms, chiefly green plants, that utilize radiant energy to manufacture food from simple inorganic substances. In terrestrial ecosystems, the major autotrophs are the flowering plants; in the oceans, they are the microscopic phytoplankton, especially the diatoms. Consumers, on the other hand, are heterotrophic organisms (primarily animals), and they are either directly or indirectly dependent on autotrophs for food. Consumers may be herbivores, organisms that feed on plant life, or carnivores, organisms that consume only animals and thus depend on producers indirectly. Consumers that feed on both animal and plant materials are called omnivores. Like consumers, decomposers—bacteria and fungi—are heterotrophic organisms. They feed on dead protoplasm, breaking down its complex organic components into simple compounds that can be used by producers. Thus the continuous functioning of an ecosystem hinges on the activity of decomposers in recycling organic compounds.

A good example of an ecosystem and its components is the typical pond (Fig. 21.9). The necessary abiotic substances in the pond are water, carbon dioxide, oxygen, nitrogen, calcium, and organic compounds, such as amino acids, humus, and vitamins. The three biotic components—producers, consumers, and decomposers—are each represented by several species. There are two types of producers: large plants, either rooted or floating, which generally grow in shallow water only; and phytoplankton—minute floating plants, usually algae, that are dispersed throughout the pond as deep as light will penetrate. Among the consumers are insect larvae, crustacea, and fish. Decomposers consist of aquatic bacteria

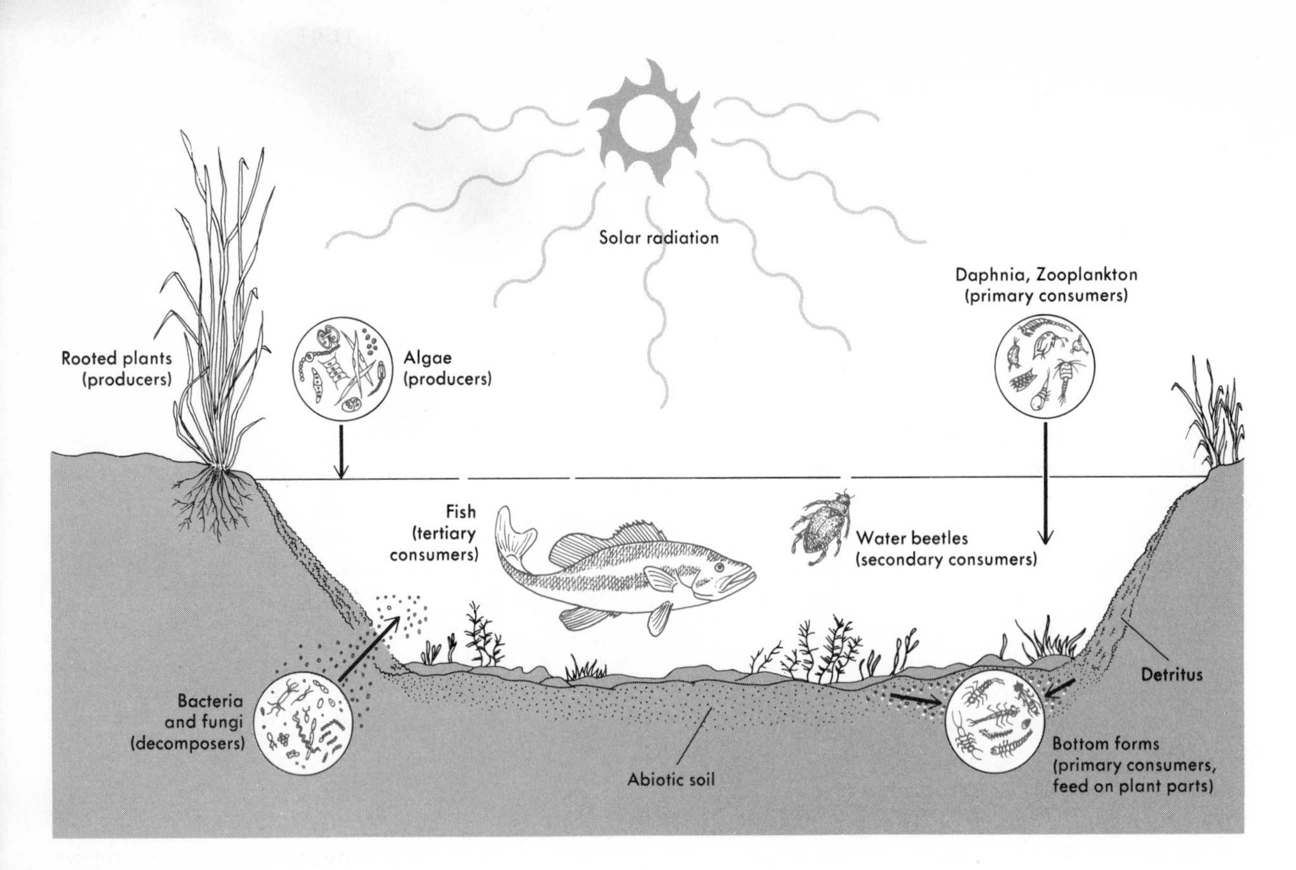

FIGURE 21.9

The freshwater pond, a small but complete ecosystem. Energy that drives the system comes from solar radiation. It directly supports producers such as algae and rooted plants. Zooplankton such as daphnia are the primary consumers, and they in turn are the prey of larger animals such as the predacious diving beetle and the scavenging water beetle. Tertiary consumers such as fish may in turn be eaten by larger animals. Decomposers break down organic wastes and return materials to the soil and water.

and fungi, species that are especially abundant along the mud at the bottom of the pond, where dead plant and animal materials accumulate. They decompose the complex materials in dead organisms, rendering them simple enough for use by the producers for their own growth, thus completing the cycle.

FOOD CHAINS

Organic nutrients are transferred from producers to consumers and decomposers as organisms eat and in turn are eaten by other organisms. This succession is called a **food chain** (Fig. 21.10). Like the food pyramid (see Fig. 21.8), the food chain is divided into **trophic levels.** On each trophic level are all the organisms whose food reaches them after the same number of steps. In other words, the producers—the green plants—constitute the first trophic level. The herbivores (primary consumers), those organisms that consume the plants, constitute the second trophic level. On the third trophic level are the carnivores (secondary consumers), which consume the herbivores. There may be higher trophic levels as well. Carnivores that consume third-level carnivores constitute the fourth trophic level (tertiary consumers), and so on. All chains begin at the producer level, and all end at the decomposer level—that level occupied by the bacteria and fungi responsible for the decay of dead tissues and cells. From level to level, energy is lost. Between each two successive trophic levels, the loss in conversion efficiency is from 10 to 20 percent.

In most ecosystems, many different food chains overlap and are interrelated with one another to form what is called a community **food web.** In short, the populations within the food web are linked by the various food chains through which organic nutrients are transferred. Should there be an excessive interruption of nutrient flow within one food chain, other chains could conceivably be seriously affected. In the more

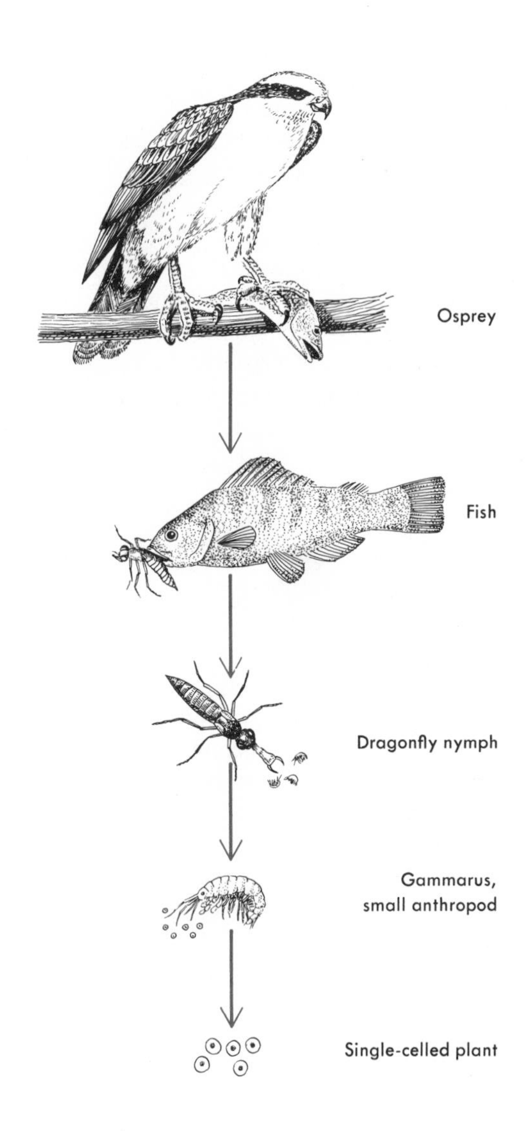

FIGURE 21.10

A typical food chain. Because a large proportion of energy is lost from one trophic level to the next, each organism must eat large numbers of its prey to satisfy its nutrient requirements. One result is the biological concentration of such nonexcretable materials as pesticides. With each step along the food chain, the concentration of the material increases with respect to body size. The osprey, at the end of the chain, may carry so great a concentration as to threaten its survival, because pesticides in its body interfere with egg-shell formation.

complex food webs, however, alternative food sources may be available.

ENERGY FLOW

Just as organic nutrients are circulated through a community along food chains, so energy, an inherent component of food, is also transferred to community members along the food chains. This energy transfer is governed by the first and second laws of thermodynamics, which we encountered in Chapter 5. Recall that according to the first law, energy may be converted from one form to another but is never created or destroyed. Although energy cannot be destroyed, the second law of thermodynamics states that it can be dissipated and lost as heat. Thus energy is transferred from lower to higher trophic levels, but much energy is lost from the ecosystem during conversions between trophic levels. Not all nutrients that are eaten are used for energy or assimilated as protoplasm. Some energy may be lost from the ecosystem through evaporation, and much is lost through the process of respiration, in which food energy is utilized but with a large attendant release of heat. This large loss of energy makes energy transfer different from nutrient transfer. Organic compounds are recycled through the ecosystem, but energy cannot be reused. Thus energy flow is a linear process that ends at the decomposer level. To maintain an ecosystem, there must be a constant input of energy to replenish supplies reduced by loss. This input comes mainly from solar radiation. Plants use the energy in solar radiation to manufacture food, and thus plants form the base of the energy flow process. The herbivores, by consuming plants and by being consumed by carnivores and omnivores, are the pivotal link in the process by which energy is transferred from producers to consumers at higher trophic levels.

ECOLOGICAL PYRAMIDS

Relationships between the various trophic levels of a community have been expressed visually by means of bar-type graphs showing the producer level on the bottom and successive trophic levels stacked on top. Graphs have been made to show energy, numbers of individuals, and biomass, and because all of these have a triangular shape, they have been termed ecological pyramids (see Fig. 21.11). The **pyramid of energy** depicts the amount of energy flow to each successive trophic level in a community. Because energy is always lost in transfer, each successive trophic level receives less total energy than the level below. Thus the producer level always forms a wide pyramid base, and successive levels narrow to an apex at the highest trophic level.

The **pyramid of numbers** depicts the relative numbers of individuals present at each trophic level in the community. The pyramid may be upright or inverted depending on the size of the producers in the community. Larger producers, such as the oak tree, will be fewer in number than the smaller organisms, such as phytoplankton and grasses. If the producer element consists of larger and hence fewer individuals, the pyramid appears inverted.

The **pyramid of biomass** depicts the relative distribution of living material among the trophic levels of a community. (The amount of biomass is based on total dry weight or caloric value.) The biomass pyramid may be inverted if the turnover rate of the producers is much more rapid than that of the consumers, as in many aquatic ecosystems.

POPULATIONS

The biotic component of ecosystems consists of many plant and animal populations living and interacting together. A population is a collective group of organisms of the same species that occupies a particular space and has various characteristics that are unique to the individuals as a group but are not applicable to them separately. Some of these characteristics are growth, biotic potential, patterns of increase, and density. For each population, these traits may be determined by abiotic factors in the environment, interactions between population members, and interactions with other populations.

Growth The dynamic feature most fundamental to a species population is growth—the capacity for the increase of individual numbers. The rate of growth of a population is expressed as the number of individuals by which the population increases in a given amount of time. The rate of growth is a function of two factors—natality (the birth rate, or the production of new individuals per unit of time), and mortality (the death rate, or the death of individuals per unit of time). Both rates may be influenced by the supply of abiotic and biotic nutrient requirements, as well as by biotic interactions, such as competition and overcrowding.

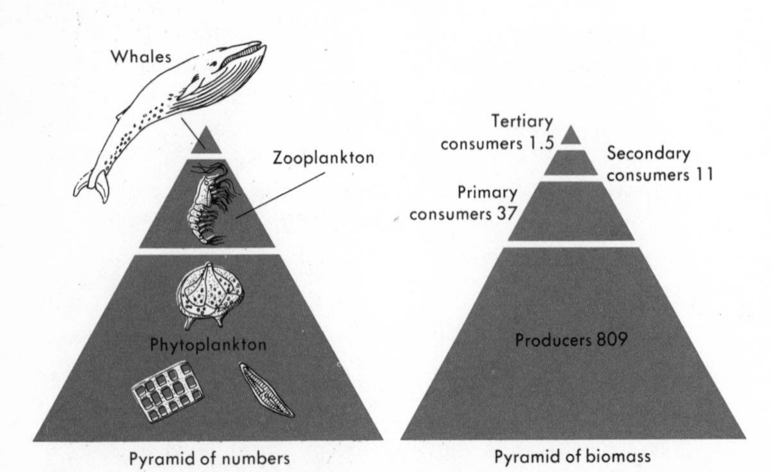

FIGURE 21.11

Pyramids of numbers and biomass. This particular oceanic food pyramid is unusual in that the largest organisms feed directly on the smallest. Nevertheless, the principle is the same in any pyramid of numbers. Millions of planktonic organisms are required to support a single whale. On the pyramid of biomass, which represents the aquatic ecosystem of Silver Springs, Florida, the figures are in grams of dry biomass per square meter.

Biotic potential The reproductive potential—or biotic potential—of a population is the population's theoretical rate of growth if reproduction occurs at maximum rate under optimal conditions in an idealized environment. Under natural conditions, however, the biotic potential is limited by **environmental resistance**—the action of one or more factors of the environment that has the effect of decreasing natality, increasing mortality, or both. Again, influential factors may be abiotic and/or biotic.

Patterns of increase All populations have characteristic growth patterns, or population growth curves. These consist of two basic types: the **J-shaped growth curve** and the **S-shaped** or **sigmoid growth curve.** These growth forms are determined by the carrying capacity of the environment, the upper limit of population density that can be sustained by an environment under a given set of conditions. Such a limit is based on environmental resistance, usually taking the form of curtailment of food supply or living space. When a carrying-capacity limit is imposed suddenly, the growth curve is J-shaped —the biotic potential falls away abruptly. This often happens when the population increases beyond its carrying capacity for a given food supply or abiotic requirement. The supply is temporarily exhausted, and the population size decreases rapidly as the result of an increased mortality due to starvation. When environmental resistance is stable or increases gradually, the population growth is S-shaped, or sigmoid. In this pattern, growth is slow at first. Thereafter, increase is rapid for a time but soon slows gradually as increased environmental resistance is met. Eventually, a somewhat stable equilibrium level is reached and maintained.

Density The density of a population is the total number of individuals inhabiting a particular area of the habitat for a given period of time. Population density tends to fluctuate above and below the particular carrying-capacity level. These fluctuations are influenced by environmental factors, called **density-independent** and **density-dependent** factors. Density-independent factors are extrinsic conditions, such as drought, flood, and temperature (in short, all physical abiotic factors of the environment). Density-dependent factors are conditions intrinsic to the population, such as overcrowding, competition, predation, and parasitism.

Biotic-biotic interactions

All organisms under natural conditions interrelate with members of their own species, as well as with organisms from the populations of other species. Generally, individuals interact in carrying out their vital functions of nutrition, growth, and reproduction. These relationships may be **intraspecific interactions,** relationships with members of the same species, or **interspecific interactions,** relationships with members of other species.

INTRASPECIFIC INTERACTIONS

The relationships of organisms within the *same* population involve several aspects of interaction. These are distribution of population, isolation and territoriality, and social rank.

Population distribution The pattern of distribution within a population is usually simply the clumping together of individuals in varying degrees of association. Individuals aggregate in a clumped arrangement for a number of reasons. Often they are responding to differences in the local habitat, where certain areas are more suitable for living than others. They may also be responding to daily and seasonal fluctuations in weather, or the action may be the result of reproductive requirements and social attractions. There are shortcomings to this type of aggregation, such as increasing competition for food and living space, but they are offset by a single primary benefit, the increased survivability of the group. Compared with isolated organisms, individuals in groups often experience a lower mortality rate during unfavorable periods or during attacks by enemies.

Two other broad patterns of distribution are random and uniform. Random distribution of individuals within a population is rarely seen in nature. It usually occurs where conditions throughout the habitat are the same, so that individuals are not prompted to aggregate. Where competition is severe, uniform distribution may occur as each individual claims an area to support itself. Since needs are similar, spatial distribution tends to be uniform.

Isolation and territoriality Among the vertebrates and higher invertebrates, individuals in pairs, family groups, or cliques often confine their activities to a definite area called the **home range.** A section of the home range that is actively defended by an individual or a pair is called a **territory.** Territoriality serves three main functions—to protect a nest or den, to provide a cache for food supplies, and to preempt an area for courtship displays. Competition and actual directed antagonism result in the **isolation** of individuals who lose out.

Social rank Many individuals of the same species may live together in a society where intraspecific relationships are partially determined by a social hierarchy. These forms of social aggregation are seen among various insect species, as well as vertebrates. One type of social hierarchy is the *pecking order,* so called because the phenomenon was first quantitatively studied in chickens (see Fig. 21.12). The pecking-order hierarchy ranges from a dominant individual to one that is completely subordinate, with a gradation of social rank in between. Among chickens, for example, the highest chicken in the hierarchy pecks all the others in the group, but they will not peck it. The second chicken pecks all chickens of lower rank and is not pecked itself, except by the first chicken. Successive chickens peck subordinate chickens and are pecked by superior chickens.

Among primates, such as the baboon, there are hierarchical societies headed by **dominance cliques** rather than by dominant individuals. Sometimes social position is partly inherited. For example, the top female baboons form a clique, and their offspring enjoy a higher social status than the offspring of mothers of lower social rank.

FIGURE 21.12
A flock of hens in line crossing a farm field. The alpha (dominant) hen, which holds the top position on the pecking order, leads the way. The social structure of many species of animals includes such a dominance hierarchy, a form of social organization that limits aggression and stabilizes social interactions.

Among some insects, such as ants, termites, social wasps, and certain species of bees, a very sophisticated social order has evolved. It is based on **castes,** which are social classes comprising individuals with specialized structures that enable them to perform specialized functions. Certain individuals, for example, are especially suited for reproduction. They belong to the primary reproductive caste, whose sole function is to establish the colony and to reproduce the individuals that will make up the other castes. Members of the worker caste are generally sterile females, and their only functions are to gather or produce the food for the entire colony and to enlarge or maintain existing living chambers. A third group of individuals constitutes the soldier caste. Members are typically equipped with large jaws and often sting or secrete repellent substances. Their function is to guard the entrances to the colony and to protect the other castes. Some insect groups have a secondary reproductive caste. Members are often wingless, and they serve as a reproductive contingency body, becoming reproductively active if members of the primary reproductive caste are injured or die.

INTERSPECIFIC INTERACTIONS

The relationships of organisms with members of other species range from complete mutual independence to total mutual dependency. Some types of interspecific interactions are neutralism, commensalism, mutualism, predation, parasitism, and competition.

Neutralism In this situation, individuals of different species function in often very close association but do not affect one another adversely or beneficially. One reason for this situation is that the organisms differ considerably in their breeding behavior, food procurement, and other vital processes.

Commensalism In this interaction, one organism benefits from the association while the other neither benefits nor, under normal conditions, is adversely affected. This type of relationship is illustrated by the interaction of the remora fish with certain species of shark. The remora has a large sucking disc on its anterior end by which it attaches itself to the outer surface of the shark. The remora is thus transported to the shark's feeding areas, where it detaches itself, swims around to feed on the scraps left over from the shark's meal, and reattaches itself to be transported elsewhere. Although the remora benefits immensely from the association, the shark is unaffected.

Mutualism In mutualism (also called **obligatory mutualism**) organisms are dependent on each other for survival. An example of this type of relationship is the lichen—an association of an alga with a fungus. The alga cells are protected by the fungus and derive moisture from it. The fungus obtains the carbon compounds photosynthesized by the alga, as well as growth factors (vitamins) and certain nitrogenous compounds. For another example of mutualism, see Fig. 21.13.

FIGURE 21.13
Mutualism in insects. A symbiotic relationship that is beneficial to both species is referred to as mutualism. Some species of ant feed on the honeydew produced by aphids, their "ant cows," which they herd in and out of winter quarters and protect from predators.

FIGURE 21.14

Predation. The Alaskan brown bear, officially the largest living carnivore, is actually omnivorous, as are all bears. It feeds on plants, insects, and animals, such as the dog salmon, which it fishes from Alaskan streams. Predator and prey exercise reciprocal control over each other's populations. The sick and unfit are eliminated by the process of natural selection, and a balance is maintained between supply of and demand for food.

Predation In the predator-prey relationship, members of one population (predators) kill and feed on members of another (prey). The predator is dependent on its prey for nutrients and therefore for survival. Although a few predators are plants (e.g., the Venus's-flytrap), most are animals (see Fig. 21.14).

Parasitism Like predation, parasitism consists of the use by one population of resources derived from another, while it is also dependent on the other for survival. The basic difference between a predator and a parasite is that a predator kills and consumes its prey, whereas a parasite passes much of its life on or in the body of its living host, from which it derives nutrients in a manner that is damaging but not fatal to the host. Among the internal parasites that may infest the human body are blood and liver flukes, hookworms, and tapeworms. Particularly in tropical areas, large numbers of people suffer from serious debilitating diseases caused by these parasites. Two of them, malaria and schistosomiasis, affect an estimated one-third of the world's human population.

Competition Whereas neutralism is an interaction between two different populations in which they do not affect one another, competition is an interaction whereby each population is in mortal struggle with the other for the necessities of life (such as nutrients and living space). The competition occurs when two or more species of organisms have similar requirements and are trying to occupy the same niche. Gause's **competitive exclusion** principle states that two species cannot occupy the same niche within a given habitat. One will either be excluded or modify its requirements.

The phenomenon of competition can be illustrated experimentally with two species of the flour beetle *Tribolium*—insects that can complete their entire life history in a relatively simple habitat, such as a jar of flour or wheat bran. In the experiment, the species *T. castaneum* and *T. confusum* are placed in the same jar of flour, and the conditions of humidity and temperature are experimentally altered. *T. castaneum* prevails under hot and moist conditions, but *T. confusum* wins out when conditions are cool and dry. When these variables are intermediate, sometimes one species survives, sometimes the other. In this way, the experiment shows that the ability to compete is related to the organism's physiological makeup, as well as the conditions within its environment—that is, not only *where* it lives but *how*.

COMMUNITY CHANGE

The interactions of animals and plants with their natural surroundings is often characterized by states of transition and flux. The term for this type of transition is **ecological succession,** the orderly sequence of different communities over a period of time in some particular area (Fig. 21.15). The two general forms of transition are primary succession and secondary succession.

Primary succession When a biotic component is the first to become established on a particular substrate on which no living matter has previously lived, this group of plants and animals is called the **pioneer community.** A sequence of communities that began as a pioneer community on virgin soil is called **primary succession.** As the pioneer community becomes firmly entrenched, it affects the abiotic components of the environment, so that they are gradually altered. Consequently, the environment is likely to be modified in a way that renders it more favorable to the entrenchment of new and different species, and thus a new type of community develops on the site of the old pioneer community. As succession proceeds, a single site supports a series of communities, each of which is in a state of transition. The successive communities collectively are called a **sere;** any one community is a **seral stage.** The types of substrate suitable for the establishment of pioneer communities are bare mineral soil, exposed rock surfaces, sand dunes, volcanic ash, new islands emerging in streams or along coastal shores, and even new dams sunken vessels, and pier supports.

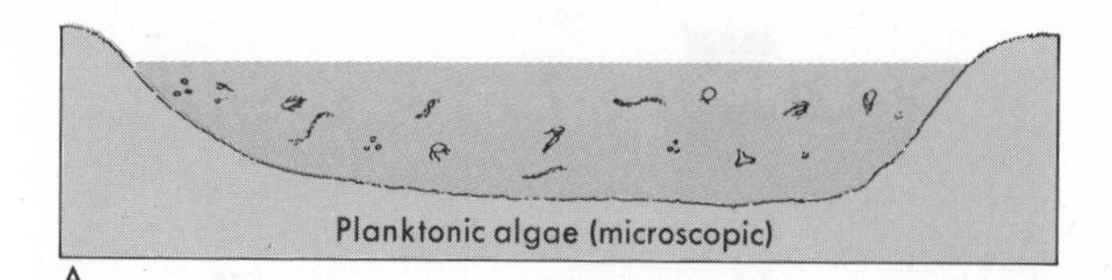

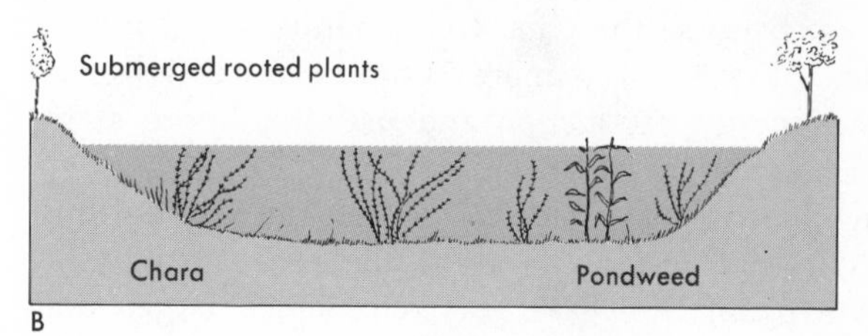

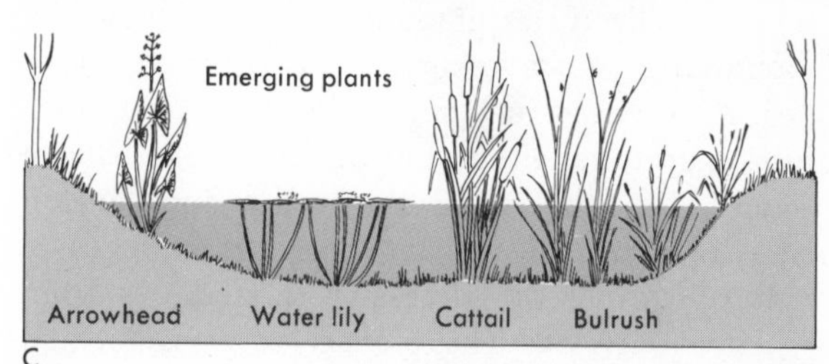

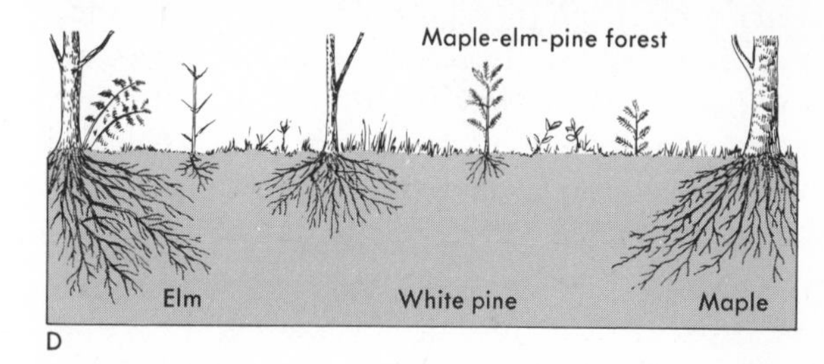

FIGURE 21.15

Ecological succession. (A) The bottom of a newly formed small pond becomes covered with silt. Rooted plants (B) prepare the underwater soil to receive and nourish emerging plants, such as the cattail and water lily (C). With further deposition of aquatic debris over a long period of time, the pond becomes shallower. Terrestrial plants gradually encroach from the margins, until the pond habitat is replaced, or succeeded by meadow or forest (D).

Secondary succession In primary succession the substrate has never supported life, but in secondary succession the land has been the site of previous growth. In the latter type of sequence, some kind of occurrence —such as a fire, a drastic alteration in climate, or man's intervention—caused the previous inhabitants to be eliminated. The secondary succession begins as new organisms reoccupy the land. The early seral stage of a secondary succession differs considerably from the communities that would have existed on this site in a primary succession.

Whether the transition sequence is one of primary succession or secondary succession, the final or stable community characteristic of a specific climate and type of soil is called a **climax community.** The climax community is self-perpetuating and in a state of equilibrium with the physical habitat. The basis for this equilibrium is that there is no net annual accumulation of organic matter. The annual production and import of organic materials is balanced by the annual consumption by the community of this material and its export.

Trends In all forms of ecological succession, certain trends are observed. First, the kinds of plant and animal species change continuously. Those species that are important in the pioneer stages are not likely to be so in the climax. Initially the changes are rapid; later they are more gradual. Second, as succession continues, the total biomass as well as the nonliving organic matter increases in sequential seres. Third, the food chains and food webs become more complex, so that the energy flow within the community increases. Finally, as the succession progresses, the amount of food energy that is utilized gradually comes to equal the amount that is produced. Because the climax community has a greater diversity of species, a more complex food web, and balanced energy production and use, it is more stable than the earlier seral stages.

An example of ecological succession is the transformation that has occurred in the area surrounding Lake Michigan. This lake was once much larger than it is at present. The water gradually retreated from its original boundaries, uncovering successive stretches of sand dunes. As soon as each successive dune dried out, land plants began to take root, but because the substrate was so sandy, the early seral stages did not develop quickly. As a result, the sequence of seral stages is clearly visible in order on successive sand dunes from the lake's edge outward. The pioneer stage is located at the water's edge, and its vegetation consists of cotton wood and beach grass. Increasingly older stages are evident as one proceeds away from the shore. In order, they are jack pine forest, black oak dry forest, oak and oak-hickory moist forest, and finally the climax community of beech-maple forest.

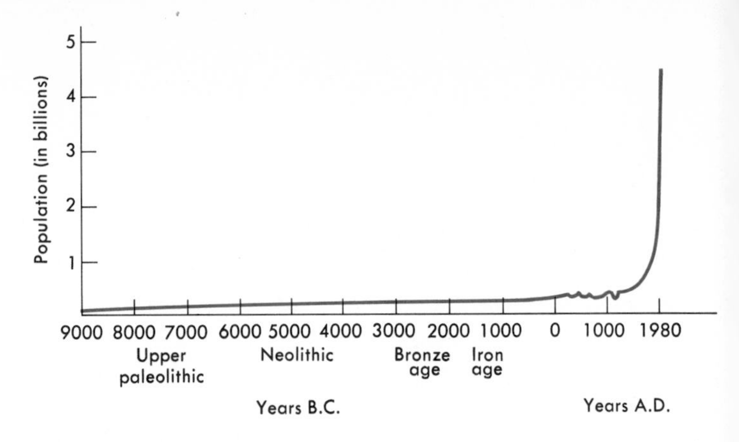

FIGURE 21.16

Profile of human population growth. In early years the human population remained fairly constant, growing with time like any other natural population. The sharp dip shortly after the 1000 A.D. mark reflects the effect of the Black Plague. Since the start of the industrial age the pattern has changed, however. Growth has become exponential, and in early 1976 we passed the four billion mark. Today overpopulation may threaten to become the most serious problem of the future.

Environmental problems

No reasonably alert human resident of the world in the latter half of the twentieth century can be unaware of the importance of the environment to our continuing existence or of the magnitude of the problems involving the environment. To discuss such problems in more than a superficial way, however, is beyond the scope of the present book. Those who wish to pursue this field of interest in greater depth will find some helpful suggestions for further reading at the end of this chapter.

MAN AND HIS HABITAT

As the human population has grown and the world economy has expanded to include more agriculture and more industry, man has had an ever-increasing effect on the ecosystems of which he is a part. He has destroyed old and introduced new producers, broken food chains and altered food webs, modified environmental abiotic factors, and changed environmental carrying capacities. Because ecosystem balance is very complex and, for the most part, not very well understood, the only categorical statement that can be made about man's role in environmental change is that his full impact is unknown. Nevertheless, the growing concern over environmental disruption is based on an ever-increasing understanding of the drastic effects caused by overpopulation and various forms of pollution.

Most of the environmental problems facing us today are a direct result of man's population increase (see Fig. 21.16). The world's overall population growth rate is now approximately 2 percent each year. This growth rate results from both the reduction of infant mortality and the extension of life span. The effect of a burgeoning world population and increased technological skill is an increased demand for newer and better products whose production tends to increase the pollution of land, air, and water, both marine and fresh.

AIR POLLUTION

Since the beginning of the Industrial Revolution, industrial processes have been emitting enormous quantities of contaminants in gaseous and particulate form. Inorganic gaseous pollutants include carbon monoxide, sulfur dioxide, carbon dioxide, hydrogen sulfide, hydrogen chloride, nitrogen dioxide, ammonia, and ozone. Organic gaseous pollutants include alcohols and hydrocarbons. Among the particulate emissions are smoke and dust.

The problems caused by these emissions are compounded by their reactions with one another, catalyzed by sunlight. Smog, a combination of smoke and fog, is the best known example of such reactions in the extreme. Smog is produced by stagnant air, sunlight, and reactions involving nitrogen dioxide, ozone, and hydrocarbons.

Effects of air pollution Air pollution affects abiotic and biotic components of the biosphere, including climate and temperature, energy flow, biogeochemical cycles, and individual ecosystems.

There are two theories of how climate and temperature may be affected by air pollution. Atmospheric carbon dioxide tends to absorb and trap heat radiated from the earth. Since industrial emissions have increased the concentration of carbon dioxide in the past 30 to 40 years from 15 to 25 percent, there is speculation that this could possibly raise the temperature of the earth and melt the polar ice, a phenomenon called the greenhouse effect. Temperature is also influenced by atmospheric particulate matter, which tends to reflect incoming solar radiation, thereby reducing the amount reaching the earth. This has a cooling effect and could possibly influence global temperatures and weather. Neither heating nor cooling trends have been observed, and there is currently a running argument between proponents of the two theories.

Atmospheric pollution can affect energy flow in an ecosystem by interfering with all three levels of the food chain—producers, consumers, and decomposers. Green plant producers may be killed or damaged by gaseous pollutants that enter through their stomata. There is such a reduction in photosynthetic activity as a result that fewer nutrients are available to consumers. Air pollution can also affect the consumers directly. For example, smog is known to impair the fertility of female mice. Furthermore, human mortality rates in Los Angeles from 1962 to 1965 showed a direct correlation with the concentration of atmospheric carbon monoxide. Air pollutants also affect decomposers, since gases such as ozone, sulfur dioxide, and nitrogen dioxide are lethal to most microorganisms. In fact, ozone and sulfur dioxide are used industrially as disinfectants.

The nitrogen and sulfur biogeochemical cycles are affected by air pollutants because the contaminants damage essential microorganisms or overload a part of a cycle. Carbon monoxide impairs the ability of nitrogen-fixing bacteria to incorporate atmospheric nitrogen, and sulfur dioxide tends to decrease soil acidity, making the soil less favorable for the growth of nitrifying bacteria. Sulfate is normally present in the atmosphere and incorporated by certain organisms into organic compounds. However, the quantities of sulfate generated by the burning of fossil fuels cannot be utilized naturally, and the excess remains in the atmosphere.

Finally, air pollution has been known to destroy entire ecosystems. In the early part of the twentieth century, several thousand acres of high forest in the vicinity of Ducktown, Tennessee, were turned into a desert as the result of the release of huge quantities of sulfur dioxide from a smelter in the area. Observations of similar effects from the activity of smelters indicate quite clearly that atmospheric pollution can set back the orderly processes of ecological succession.

MARINE POLLUTION

The most damaging kind of marine pollution has been oil spills from oil tankers, the normal operation of watercraft, and the leaking of oil from offshore wells.

This oil affects marine plants and animals as well as certain biogeochemical cycles. When swimming and diving birds, for example, are covered with oil, their feathers become matted—a condition that reduces their buoyancy and prevents them from flying (see Fig. 21.17). Moreover, the feathers no longer insulate the birds, and they quickly die from their exposure to the cold water. The effect of oil pollution on biogeochemical cycles is illustrated by the destruction of the murre —the most abundant species of bird in certain Arctic regions. These birds normally feed on bottom fish, and

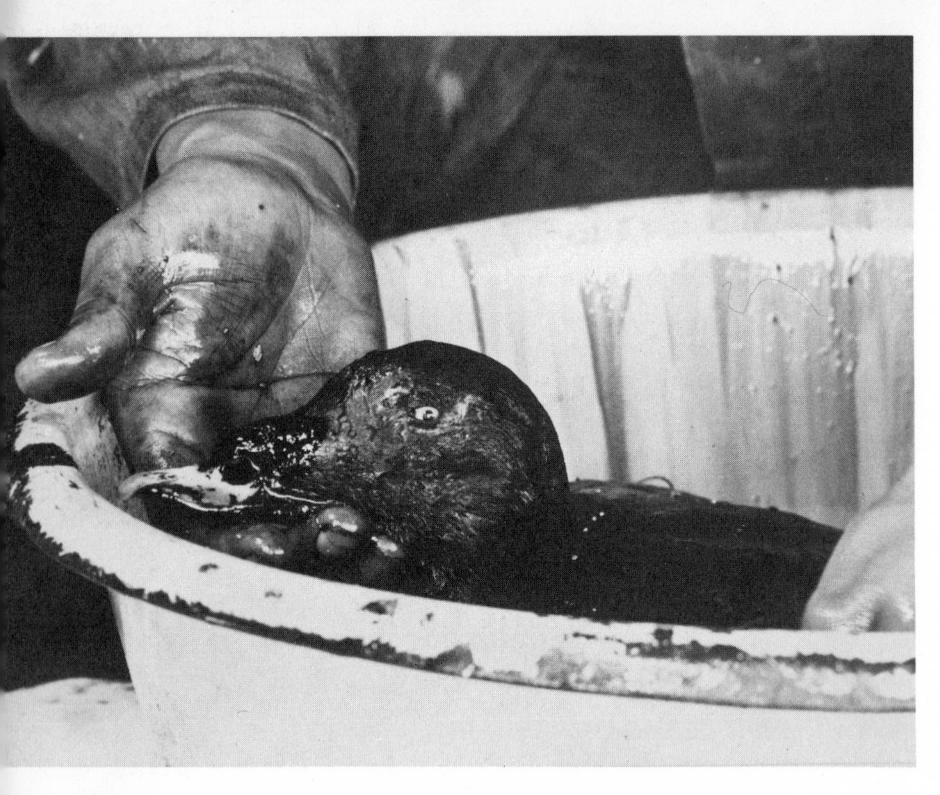

FIGURE 21.17

Cleaning birds after an oilspill in San Francisco Bay. This scoter is one of many birds that have become victims of advancing technology and the pressures to increase offshore drilling. Oil-soaked plumage makes flight and therefore feeding impossible, and many birds drown or die from starvation.

their excretions release to the surface those minerals, such as nitrate and phosphate, that ordinarily would remain on the bottom. This fertilizing effect serves to maintain the fertility of surface Arctic waters. Yet tankers in the Newfoundland region, by flushing their oil wastes into the sea, kill approximately 1000 of these birds each day. Thus a key component of the biogeochemical cycles of the Arctic is upset—and the ultimate consequences of this interference are as yet unknown.

FRESH-WATER POLLUTION

The pollution of fresh-water streams, rivers, and lakes is caused by many forms of domestic and industrial wastes and consists of both chemical and physical components.

Chemical pollutants consist of such organic chemicals as proteins, fats, and carbohydrates, as well as synthetic detergents and such inorganic constituents as acids and alkalies, salts of heavy metals, and soluble salts. Physical pollutants are such particles as the coloring matter of organic dyes, which are suspended in water and make it cloudy.

Organic pollutants damage the fresh-water habitat by adding nutrients that provide a favorable medium for bacterial growth. Such growth requires oxygen and results in a diminution in the amount of oxygen dissolved in the water. This bacterial demand for oxygen is called the **biological oxygen demand (BOD).** It is used as an index of pollution, especially that from organic sources. Organic pollutants have contributed to the almost total depletion of dissolved oxygen from 1400 square miles of the bottom waters of the central basin of Lake Erie. Oxygen scarcity has caused the decline of certain of the lake's fish species.

The acid and alkali wastes cause damage by drastically altering the pH in fresh-water habitats, which in general lack buffer capacity. Salts of the heavy metals, such as lead, zinc, nickel, copper, and cadmium, are damaging to all living things. Soluble salts, such as phosphates, may do damage by altering the ecological balance imposed by a limiting factor. Phosphorus was once a limiting factor in certain fresh-water systems. With the introduction of phosphate-containing detergents, the amount of phosphorus in these fresh-water habitats has accumulated to the point that it is no longer a limiting factor. The result has been an abundance-depletion cycle of algae growth. First, the algae grow rapidly and increase the oxygen content of the water through their increased photosynthetic activity. As a result there is an excess of dead algae, since greater numbers of algae result in greater numbers that die. There is intense bacterial growth, fed by the abundance of algae, and ultimately a severe depletion of oxygen, which is used by the bacteria.

Physical pollutants may affect an ecosystem in several ways. They may alter the intensity of sunlight that passes through the water, inhibiting plant growth. They may kill fish by coating their gills. They may also silt a stream bed, smothering fish eggs.

Thermal pollution The primary cause of thermal pollution—the heating of natural waterways to temperatures above the normal—is the release into them of water that has been heated while being used in industrial processes. Increases in water temperature alter the solubility of oxygen. The warmer the water is, the less oxygen it may hold. Thus less oxygen is dissolved into the water from the atmosphere, and less oxygen produced photosynthetically is retained. This can severely alter the biotic as well as the abiotic elements of the ecosystem.

In addition, a rise in water temperature may disturb the behavior of certain organisms. Fish, for example, often migrate and spawn in response to temperature cues, and their eggs may not hatch if the water is too warm. Water temperature also affects the lifespans of organisms. The water flea *Daphnia* lives 108 days at a water temperature of 46°F, but only 29 days at 82°F. Although some acclimatization to temperature increases is possible when the change is gradual, sudden temperature changes are particularly lethal.

THE FUTURE

Man has already succeeded in placing the environment in which he lives and on which he depends in danger of irreparable damage. Although the technology exists that can correct the situation, economic and political priorities so far have overridden pressure to preserve and restore the environment. Some such priorities are difficult to dismiss. In areas of the world where there are millions of people dying of malnutrition and millions

FIGURE 21.18

The awesome mushroom cloud of nuclear explosion. Does it hold the future in jeopardy?

more living in primitive, poor, and totally unmechanized societies, it is impossible to make a case for protecting the environment at the expense of industrial and agricultural development. However, if the biosphere is ultimately to be preserved, there will have to be a grassroots change in personal values. The health of the environment must be held more important than a car, unlimited air conditioning, or the choice of 10 different cereals at breakfast.

The ultimate problem—so inconceivable that most of us refuse even to think about it—may be the destructive power of the nuclear explosion. All man's attempts to create weapons, from the slingshot to the atomic bomb, have arisen from the recognition of an inherent shortcoming. Almost without parallel in the kingdom he shares with other animals, man has neither fang nor claw. Unfortunately he has succeeded too well in overcoming that lack. Today the United States and the Soviet Union together possess nuclear explosive capability at a level that is equivalent to about 15 tons of TNT per man, woman, and child. If we should ever lose control of our tempers and start a no-win nuclear war, all other ecological problems will fade to relative insignificance.

Summary

1. Ecology is the study of ecosystems and the relationships between organisms and their environments. Ecosystems consist of abiotic and biotic elements. The nonliving components are habitat, gases, solar radiation, temperature, moisture, and inorganic and organic nutrients.

2. The life-supporting agent of terrestrial habitats is soil, whose three layers are topsoil, subsoil, and bedrock. Aquatic environments consist of three major types of habitat—freshwater, saltwater, and estuary. The atmosphere is not a habitat but serves as a source of gases and a medium for organism transportation and the generation of weather conditions. Solar radiation is the source of energy for living things and a determinant of climate and weather.

3. The extent of biological growth is determined by abiotic limiting factors. The Liebig-Blackman law of the minimum states that the growth of a population is dependent on and limited by the essential abiotic factor that is present in an amount close to the organism's minimum requirement. Shelford's law of tolerance states that for each ecological parameter there is a minimum and a maximum critical value.

4. The three biotic components of an ecosystem are producers, consumers, and decomposers. Organic nutrients are transferred from producers to consumers and decomposers along food chains. Nutrients are continuously recycled through the community.

5. Community energy flow is a one-way process. Energy is transferred along the food chains from producers to consumers and finally to decomposers. Relationships between the various trophic levels of a community may be expressed by means of graphs called ecological pyramids, which include the pyramid of energy, the pyramid of numbers, and the pyramid of biomass.

6. An ecosystem is perpetuated through a dynamic balance between its biotic and abiotic components. Each organism's niche or role in the community is determined by its relationship with abiotic and biotic elements. Likewise the density of community populations is limited by abiotic and biotic factors. Finally, the essential inorganic compounds are made available to living things through interactions between the environment and community organisms.

7. Intraspecific interactions between organisms affect population distribution, isolation and territoriality, and social rank. Some forms of social rank are the pecking order, dominance clique, and caste. Important types of interspecific interactions are neutralism, commensalism, mutualism, predation, parasitism, and competition.

8. Communities are not static units. In time they change and replace one another in a process called ecological succession. A pioneer community is replaced by a series of seral stages that modify the soil in such a way that it eventually favors the growth of a stable climax community. Primary succession occurs on virgin soil, whereas secondary succession occurs on land where previous communities have lived.

9. Most environmental problems today are a direct or indirect result of human population increases. The effects of a burgeoning world population and increased technological skill are the pollution of air, marine water, and fresh water, as well as the thermal pollution of rivers and streams.

10. Air pollution, caused by the emission of gaseous and particulate contaminants, affects climate and temperature, energy flow, biogeochemical cycles, and individual ecosystems. Marine pollution, caused primarily by oil spills, affects marine biota and certain biogeochemical cycles. Fresh-water pollution, caused by the dumping of domestic and industrial wastes, affects oxygen supply and abiotic limiting factors. Thermal pollution, caused by the release of heated water into natural waterways, affects oxygen supply and reproductive cycles.

REVIEW AND STUDY QUESTIONS

1. What is the meaning of the root word from which both "economy" and "ecology" are derived? Considering the meaning of the two English words, give your explanation for deriving them from the same Greek root.

2. State the Liebig-Blackman law of the minimum, and give a specific example that illustrates it.

3. When a bird is sighted in unfamiliar surroundings, the news spreads rapidly, and bird lovers travel long distances in the hope of seeing it. State the biological principle that makes such a sighting an unusual occurrence.

4. What is an "ecological niche"?

5. Starting with carbon dioxide in the air, indicate the sequential steps in the carbon cycle.

6. Compare neutralism, commensalism, mutualism, and parasitism.

7. Explain ecological succession by tracing the history of an area from pioneer community to climax community.

8. In your local area what are the ecological problems of greatest concern?

REFERENCES

Alexander, M. 1961. *Introduction to Soil Microbiology.* Wiley, New York.

Angino, E. E., L. M. Magnuson, T. C. Waugh, O. K. Galle, and J. Bredfeldt. 1970. "Arsenic in Detergents: Possible Danger and Pollution Hazard." *Science* 168: 389–390.

Babich, H., and G. Stotzky. 1972. "Ecologic Ramifications of Air Pollution." In *International Conference on Transportation and the Environment.* Society of Automotive Engineers, New York.

Boughey, A. S. 1973. *Ecology of Populations,* 2nd edition. Macmillan, New York.

Brady, N. C. 1974. *The Nature and Properties of Soils,* 8th edition. Macmillan, New York.

Brandt, C. S., and W. W. Heck. 1968. "Effects of Air Pollutants on Vegetation." In A. C. Stern, ed., *Air Pollution,* Vol. 1. Academic Press, New York.

Brodine, V., ed. 1973. *Air Pollution.* Harcourt Brace Jovanovich, New York.

Delwiche, C. C. 1970 (September). "The Nitrogen Cycle." *Scientific American* Offprint no. 1194. Freeman, San Francisco.

Greve, P. A. 1971. "Chemical Wastes in the Sea: New Forms of Marine Pollution." *Science* 173: 1021–1022.

Harte, J., and R. H. Socolow. 1971. *The Patient Earth.* Holt, Rinehart and Winston, New York.

Hunt, G. S. 1965. "The Direct Effects on Some Plants and Animals of Pollution in the Great Lakes." *BioScience* 15: 181–186.

Knight, C. B. 1965. *Basic Concepts of Ecology.* Macmillan, New York.

Niering, W. A. 1968. "The Effects of Pesticides." *BioScience* 18: 869–875.

Odum, E. P. 1971. *Fundamentals of Ecology,* 3rd edition. Saunders, Philadelphia.

Oort, A. H. 1970 (September). "The Energy Cycle of the Earth." *Scientific American* Offprint no. 1189. Freeman, San Francisco.

Orians, G. H., and E. W. Pfeiffer. 1970. "Ecological Effects of the War in Vietnam." *Science* 168: 544–554.

Penman, H. L. 1970 (September). "The Water Cycle." *Scientific American* Offprint no. 1191. Freeman, San Francisco.

Sargent, F., II. 1967. "Adaptive Strategy for Air Pollution." *BioScience* 17: 691–697.

Southwick, C. 1972. *Ecology and the Quality of Our Environment.* Van Nostrand, Princeton.

Stokinger, H. E., and D. L. Coffin. 1968. "Biologic Effects of Air Pollutants." In A. C. Stern, ed., *Air Pollution,* Vol. I. Academic Press, New York.

Tebbens, B. D. 1968. "Gaseous Pollutants in the Air." In A. C. Stern, ed., *Air Pollution,* Vol. I. Academic Press, New York.

Turk, A., J. Turk, and J. Wittes. 1972. *Ecology, Pollution, Environment.* Saunders, Philadelphia.

Wagner, R. H. 1974. *Environment and Man,* 2nd edition. Norton, New York.

Westing, A. H. 1969. "Plants and Salt in the Roadside Environment." *Phytopathology* 59: 1174–1181.

SUGGESTIONS FOR FURTHER READING

Cailliet, G. M., P. Y. Setzer, and M. S. Love. 1971. *Everyman's Guide to Ecological Living.* Macmillan, New York.

An attempt to answer the average citizen's quest for things he can do to help solve the ecological crisis in the United States. Sponsored by the Santa Barbara Underseas Foundation, whose first concern involved in the 1969 oil spill, the book contains chapters on what to do with packaging materials, wood and paper products, and solid waste disposal, for example.

Cairns, J. 1975 (November). "The Cancer Problem." *Scientific American* Offprint no. 1330. Freeman, San Francisco.

Examination of the fact that almost all cancers seem to be due directly or indirectly to environmental factors. Identification of those factors and their elimination is consequently the most promising approach to cancer control.

Ehrlich, P. R., and A. H. Ehrlich. 1972. *Population Resources Environment.* Freeman, San Francisco.

A major work on the ecological problems resulting from the population explosion, with emphasis on the underdeveloped areas of the United States and the world at large and the problems that are caused by the poverty of such overpopulated areas.

Fallis, A. M., ed. 1971. *Ecology and Physiology of Parasites.* University of Toronto Press, Toronto.

A special report, based on a 1970 symposium, that deals with selected species and their evolution. Well illustrated. Includes discussions that raise interesting and conflicting viewpoints.

Heyerdahl, T. 1971. "Atlantic Ocean Pollution and Biota Observed by the 'Ra' Expedition." *Biological Conservation* 3: 164–167.

An alarming description of the oceanic pollution observed on the two "Ra" expeditions, as well as predictions regarding potential ocean pollution.

Idyll, C. P. 1973 (June). "The Anchovy Crisis." *Scientific American* Offprint no. 1273. Freeman, San Francisco.

A discussion of the ecological damage to the world's largest fishery, which is in the Peru Current.

Leinwand, G., ed. 1969. *Air and Water Pollution.* Washington Square Press, New York.

A dramatic and readable introduction to the problems of air and water pollution, including specific sources and solutions. The readings in this book were chosen for their informativeness and ability to compel the reader to contribute to solving the problem.

Spilhaus, A. 1972. "Ecolibrium." *Science* 175 (4023): 711–715.

A prescription for a balanced relationship between organisms and their environment. The author contends that what is needed is careful planning of people's needs, both basic and secondary (leisure activities and communication, for example).

Woodwell, G. W., P. P. Craig, H. A. Johnson. 1971. "DDT in the Biosphere: Where Does It Go?" *Science* 174: 1101–1110.

Contamination by DDT of the soil, atmosphere, and oceans, examined in terms of its implications for life.

ZONES OF LIFE
chapter twenty-two

A moose wades to the shore of a beaver pond cupped in a mountain valley—a setting typical of a temperate mountain biome. The natural world is divided into several such biological communities, each characterized by its own distinctive plant and animal life. However, as industrial civilization extends its control of natural resources, such scenes become increasingly rare. Natural wilderness in the United States has become confined to a few protected or inaccessible outposts.

We have examined life on many levels: the vital chemical reactions that occur within the cells of living organisms; the structures and interactions of cells that compose organisms; the interactions between organisms that make up populations; and relationships between populations that constitute an ecological community. In this chapter we will look at yet another level of biological organization, the way in which communities are distributed within the biosphere.

Communities are groups of organisms that inhabit the same environment and interact with one another. Terrestrial communities that have similar climate and vegetation tend to be grouped together on the surface of the globe, and collectively they make up a **biome.** Examples of terrestrial biomes include the tundra, the desert, and the tropical rain forest.

Far bigger than the animal society or the ecological community, the biome is one of the largest functional units studied by ecologists. A terrestrial biome usually covers large areas of land and includes many varieties of plants and animals. Despite this diversity, the biome is distinguished by the predominance of certain plant species that are characteristic of the biome and are referred to as dominant species.

Biomes, then, are composed of similar communities, and as you will recall, communities are not static; rather, they are slowly changing in the process known as **succession.** In a sere, a series of successive communities, only the final climax community is relatively stable. Within the biome, many communities are in the climax state. Species characteristic of climax communities are the dominant species of the biome, and they give the biome its characteristic appearance. The diversity within the biome is attributable to two facts: (1) besides climax communities, the biome contains many seral communities progressing toward climax; and (2) in many areas man has disrupted normal succession and caused the growth of different kinds of plants.

Environmental conditions

Biomes are relatively stable units. Nevertheless, such forces as wind, fire, disease, and man can be very disruptive. The kinds of organisms present in any area are there because they have become adapted to that area's particular set of environmental conditions. Any disturbance may modify the environment so that the existing species may be somewhat unfit for the new conditions. If so, new species, better adapted to the new conditions, are likely to occupy the area.

Biomes that have relatively few dominant species are particularly vulnerable to disruption. For example, in a forest biome of Canada, vast areas are dominated by a single tree species, the Canadian spruce. A tree disease caused by the spruce budworm may cause complete collapse of this biome. If the biome were more diversified, another plant species might fill the niche vacated by the spruce.

The character and extent of community succession in a given area depends on climate and soil conditions. The climax community is achieved only if environmental factors permit a full sequence of seral stages. Throughout a single biome the climate and soil conditions are relatively uniform, and they favor the growth of a certain kind of climax community. Thus conditions in the central plains of the United States favor the development of grassland climax communities, whereas conditions in the plains in Canada favor the growth of evergreen forest climax communities. Environmental factors that determine the character of biomes are temperature, precipitation, wind, and soil.

TEMPERATURE

Temperature distribution on the earth is determined by the length of time that areas are exposed to solar radiation and by the angle of this exposure. These factors are determined by the earth's daily rotation and the tilt in the earth's axis. Daily rotation accounts for the existence of night and day and related temperature fluctuations. Seasonal temperature variations are the result of the tilt in the earth's axis, which seasonally changes the angle at which sunlight reaches the surface of the globe. The effect of tilt is felt differently at different latitudes. There is little seasonal temperature variation near the equator but a considerable amount of it at the poles. Of course, temperature in general decreases from the equator to the poles.

PRECIPITATION AND WIND

The amount of precipitation is critical to any biome, for it provides the water necessary for life. The distribution of precipitation (rain, snow, hail, etc.) is related to the flow of air currents, size of land masses, and the occurrence of geologic features, such as mountains and valleys. Areas on the coast or near large landlocked bodies of water receive the most precipitation, whereas interior areas, particularly those behind high mountains, tend to receive the least. A phenomenon known as **rain shadow** develops when moisture-laden winds blowing from the ocean drop their collected moisture on the ocean side of coastal slopes, producing a drier region on the inland-facing slopes. However, not all the precipitation in an area can be used by biological systems. The amount of moisture potentially available to plants and animals is reduced by evaporation and runoff of water flowing into streams, rivers, lakes, and oceans.

Winds act with precipitation and temperature to determine moisture levels. Wind direction influences the movement of storms and thus the distribution of precipitation. In addition, wind velocity and temperature determine evaporation rate, a factor which has great effect on the kinds of plants that can survive in an area. In sum, the interplay of temperature, precipitation, and wind determines the climate of a geographic region. Each of the major biomes has a characteristic range for each of these three factors.

SOILS

All organisms grow in or on some type of medium from which they derive physical support, nutrients, and water. The major type of growth medium for most higher plants is soil. A soil is composed of mineral particles, organic matter, air, and water in various proportions. Soils are formed by a combination of climatic and geological processes and the action of living organisms.

Climate contributes to soil formation by causing rocks to be broken down into smaller particles. For example, highly variable temperatures cause alternate freezing and thawing of the earth, which in turn crumbles rocks. High winds may shift stones so that they grind against and pulverize each other. Precipitation erodes or weathers stones, breaking them up into tiny particles. Certain organisms, such as lichens and mosses, also play a role in soil formation. Growing on or between rocks, they break the rocks as they grow.

The mineral content of a soil is determined by the geological history of the region, and rocks are classified by their geological origins. Sedimentary rocks are formed on the bottoms of bodies of water. Igneous rocks are formed by volcanic action. Glacial rocks are those left after a large ice sheet has receded. Each type of rock yields a different type of soil. The organic component of soil is derived from the decay of dead organisms. Soil formation occurs most rapidly in the tropics, where there is an abundance of water to erode rocks and to prompt the various soil-forming plants to grow. By contrast, soil formation occurs very slowly, if at all, in the polar regions.

BIOME DISTRIBUTION

Although soil conditions are important, the type of vegetation that grows in a biome is determined primarily by climate. Temperature appears to be a particularly critical factor. Evidence of this fact is seen in the patterns of biome distribution (Fig. 22.1).

FIGURE 22.1
Distribution of biomes around the world.

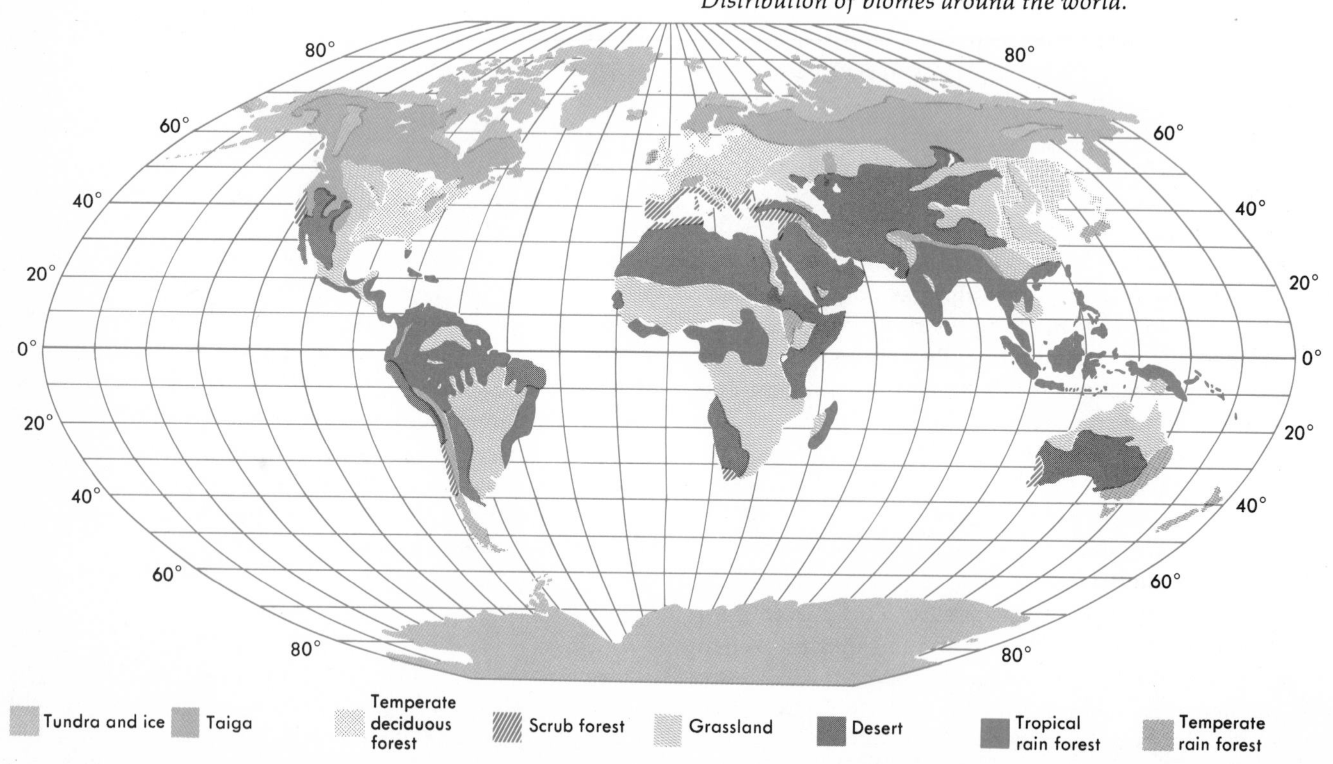

In the northern hemisphere, at least, the lines dividing the biomes tend to run parallel to the lines of latitude, and the biomes form a series of bands extending down from the North Pole. This phenomenon is most evident in the Old World, but latitudinal zonation is also seen in eastern North and South America and across the northern parts of North America. In western North and South America, mountain ranges running north to south tend to break up the latitudinal pattern of distribution. Yet on these high mountain ranges there are vegetation zones distributed altitudinally that are similar to biomes. The highest and coldest zone is very similar to the most northerly and coldest biome. The next highest zone is similar to the next coldest biome, and so forth.

FIGURE 22.2
Tundra. The musk ox is one of the few large animals able to thrive under the harsh conditions of the tundra. In the short summer, weeks of sunshine encourage the growth of hardy lichens, grasses, and dwarfed vegetation, but winter ushers in a long, icy period of night. In recent years, this relatively fragile ecosystem has been endangered by the discovery of oil in Alaska. The ecological consequences of the pipeline are not yet fully known.

North American biomes

A terrestrial biome is among the most easily recognizable units studied in ecology because it has a rather uniform appearance and covers a large area. It is dominated by a type of vegetation unique to the area and to the area's climate, but it is also a mosaic of many species of plants and animals, both living and dead. We will now briefly consider each of the major North American biomes and the effect of man in changing their native conditions.

TUNDRA

Arctic tundra forms a latitudinal band across the continent south of the ice and snow-covered polar regions. Tundra is a Siberian term that means a treeless marshy plain. The land is flat or gently rolling and reticulated (given the appearance of a net) by ponds, small lakes, and bogs. The climate is severe, with long, hard, dry winters and short, cool summers. Annual precipitation is low, averaging only about 10 inches. With such low precipitation one might wonder why the tundra is so wet. The reason is that a few inches below the soil surface is a layer of permanently frozen subsoil called *permafrost*, which prevents surface water drainage. Bad drainage, coupled with a low evaporation rate and flat terrain, turns the area into a marshy plain in summer. In winter, of course, it is frozen solid.

Only a few kinds of vegetation can survive in the tundra environment. The dominant plants are grasses, sedges, mosses, and lichens, although in some places there are a few dwarfed woody species. When warm summer temperatures melt the ice and snow on the soil surface, plants rapidly renew growth from subsurface parts that have withstood the long periods of freezing temperatures. The actual net growth of plant species in the tundra from year to year is very slight.

The variety of animal life in the tundra is also limited. The mammals fall into two groups in terms of size and behavior: large migratory species and relatively small nonmigratory species. The larger species, herd animals such as the caribou, musk ox, and reindeer, migrate from one suitable feeding ground to the next, followed by such large predators as the arctic wolves. The nonmigrants, mainly rodents, occupy a defined territory and are preyed on by the lynx and arctic fox. During the summer months, large numbers of migratory waterfowl inhabit the tundra. Black flies, gnats, and mosquitoes are also abundant.

Because the growing season is so short and there are so few plant and animal species, the tundra has a very fragile ecological balance, which is easily disturbed. Already man has upset this biome by disrupting predator control. By killing the predators of the large migratory herbivores, he has allowed caribou to overpopulate and overgraze some areas.

The greatest potential threat to the tundra, however, is the tapping of recently discovered oil fields in Alaska. A pipeline carrying heated oil will create a permanent physical barrier, especially disruptive to migratory species. In addition, in summer the use of wheel or track vehicles for transportation and to lay pipelines can destroy the tundra's fragile plant life and soil structure. Because of the slow growth rate, tundra plants require a long period of time to recover from disturbances. A vehicle running through the mushy soil leaves a ditch-like track that persists for many years. Considerable erosion occurs in these tracks because there are few plant roots to bind the soil.

BOREAL CONIFEROUS FOREST

The boreal or northern coniferous forest is commonly called the **taiga.** In North America, this biome extends in a wide band across the northern United States and stretches into Canada and Alaska. It covers most of the land area between the 45th and 57th north latitudes.

Climate in the taiga ranges from cool to cold. Average temperature ranges from a low of −30°F in winter to a high of 70°F in summer. Precipitation is not very heavy, although it amounts to more here than in the tundra. Most of the rain falls during the summer, when the range of precipitation extends from the 10 to 20 inches in the west to as high as 50 inches in the east near the Great Lakes. Snowfall in winter may accumulate to depths of 17 feet or more, a factor that explains the selective value of the taiga's spire-shaped trees. That is, the snow tends to slide off the trees rather than break their branches with its tremendous weight.

The dominant vegetation of taiga climax communities consists of several species of cone-bearing evergreens with needle-shaped leaves, including spruces, firs, and hemlocks. On sites where local conditions prevent the climax condition, birches, poplars, and some species of pine dominate.

The dominant animals include large migratory herbivores: woodland caribou, moose, and deer. Most of the herbivores in this biome are browsers, eating only twigs and leaves from trees and shrubs.

As in the tundra, the major predator is the wolf, but the lynx, wolverine, black bear, and grizzly bear are also present. Many bird species live in the taiga, especially during the summer months. Migratory shore birds and waterfowl flock to the lakes, ponds, and bogs that are abundant in the taiga. Insects like the spruce budworm, spruce bark beetle, and tent caterpillar have an unusually great impact on ecosystem balance here because they are capable of wiping out the vast areas of taiga occupied by only a single species of tree. Man too has endangered the climax condition of the taiga. In the latter half of the nineteenth century and the early part of the twentieth, lumbering was a major industry over much of this region. Thinking that the timber supply was limitless, lumbermen destroyed large areas of virgin forest. Furthermore, the piles of branches they left everywhere dried and became a tremendous fire hazard. Forest fires raged over the area between the 1870s and early 1900s. Because of this wholesale destruction, birches and poplars now cover areas where the coniferous forests once grew in Michigan, Wisconsin, and the neighboring Canadian provinces.

TEMPERATE DECIDUOUS FOREST

A large part of North America, including all of the eastern United States, lies in the biome known as temperate deciduous forest. As the label "temperate" implies, most climatic factors in the biome are moderate. The northern and southern boundaries of the biome are determined by temperature—cold is the limiting factor in the north and heat in the south. The biome extends westward until lack of available moisture becomes a limiting factor.

FIGURE 22.3

Temperate deciduous forest. Abounding with diverse life forms, the temperate forest supports a complex web of food chains. This whitetail doe and her fawns occupy a position near the center of a chain. They will graze on forest vegetation and will be preyed on, in turn, by large carnivores, such as the bobcat, cougar, and bear.

Precipitation tends to fall uniformly throughout the year, and frost occurs during the winter months.

Unlike the taiga, this biome varies in appearance from winter to summer. As indicated in the label "deciduous" in its title, its trees and shrubs lose their leaves during a portion of the year. In winter when the leaves are gone, the biome appears very bare and open. In summer, however, the leaves on the tall trees are so dense that very little sunlight reaches the forest floor. This seasonal change in leaf cover and sunlight presents some survival problems for the biome's animals and low vegetation. Many animal species, especially birds, migrate south in the fall when the protective leaf cover disappears. Some plant species, such as dogwood and *Trillium,* adapt by flowering early in the spring before the canopy trees have leafed out, blocking out the sun.

Some of the dominant types of vegetation of the deciduous forest biome are oak, hickory, maple, holly, walnut, and beech. Disturbed areas, where cutting, burning, or agriculture have taken their toll, are dominated frequently by such species as the willow, cottonwood, and the fast-disappearing elm. In autumn at the end of the growing season, these trees transform the biome into a blaze of color by turning many shades of red, yellow, and brown.

The animals of the deciduous forest include deer, a few black bears, bobcats, foxes, squirrels, chipmunks, and raccoons. Tree-dwelling birds are very common, especially in the summer months, and bluejays, owls, and woodpeckers are year-round residents. Reptiles and amphibians are much more common in this biome than in the taiga.

The arrival of European man has had a huge impact on this biome. In the eastern United States, practically all the original climax forest has been cut down. With the creation of more partially wooded and open land, the deer population has increased greatly in some areas.

Forested areas have sometimes been affected by man unintentionally. For example, the native chestnut was once a common species. However, some Chinese chestnuts planted in New York City around the turn of the century were infected with a strain of fungus not native to this country. American chestnuts were not adapted to resist the fungus, and as a result, they were decimated by the disease. The Appalachian and Blue Ridge mountain forests were particularly affected because about 80 percent of their tree population was once chestnut.

TEMPERATE GRASSLAND

The grassland or prairie biome in North America covers much of the central continent, stretching westward from the temperate deciduous forest to the edge of the Rocky Mountains, and from the Canadian provinces south to the Texas Gulf Coast. The grasslands also extend in a southwesterly direction to the edge of the arid southwestern deserts.

The topography, or appearance, of the land ranges from flat to gently rolling. Prairie soil is rich, and the grasses undulate rhythmically in the almost constant

FIGURE 22.4

Temperate grassland. Much of the central United States once abounded with tall grasses and grassland game, such as the bison and the pronghorn antelope. Now the area is largely agricultural, and corn and wheat have replaced the native grasses.

wind. The grassland biome provides a very clear example of how climate determines vegetation. Although the soil is fertile, the annual rainfall pattern, the hot summer, and the constant wind combine to make the prairie uninhabitable for the deciduous forest, which otherwise might occupy the area. Annual precipitation is irregular, occurring mostly in the spring and fall. In summer the wind and the heat cause the evaporation of almost all available water, and the area becomes too dry for trees.

Irregular precipitation and a wide annual range of temperature from above 100°F in summer to below zero in winter make the climate in the grassland biome highly variable. Grasses are particularly well adapted to live in this type of climate. They do not require large amounts of water, and even though their surface growth is destroyed annually by frost, their dense fibrous root systems can survive the severe winters and scorching summers.

Grass-root systems have contributed to the development of the dark organic layer of surface soil so characteristic of the prairies. Rich in minerals, this organic layer is about three feet thick in the eastern portions of the grassland and has been responsible for the great success of agriculture in the midwestern United States.

Most of the plants in the grassland are low-growing herbaceous species. Besides grasses, they include goldenrod, aster, and sunflower.

Many grassland animals live near to or in the ground, where grasses provide protective cover. Some

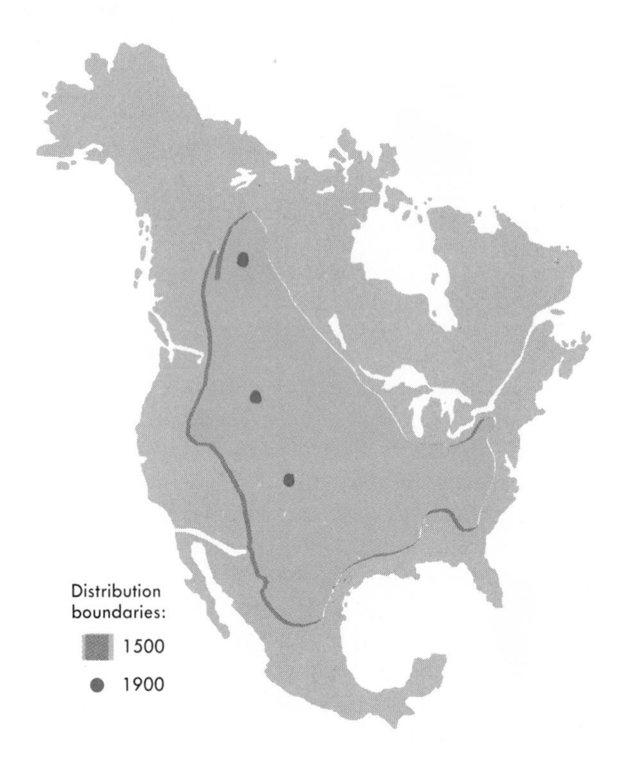

FIGURE 22.5

One result of man's alteration of a major grasslands area—the dwindling of the North American bison herds. Bison once roamed practically the entire continent but were killed off by settlers as they and the railroads moved westward.

animals, such as rabbits, prairie dogs, and ground squirrels, live in burrows. Many species of song birds build nests on the ground, and the multitude of insect species inhabit the vegetation itself. The largest prairie species, the pronghorn antelope and the bison, rely on their speed, strength, and fellow herd members for protection from enemies. The major predators are coyotes, foxes, badgers, hawks, and snakes.

Almost all of the eastern half of the native grassland has been replaced by cultivated crops. Man has utilized nearly every suitable acre for agricultural purposes and converted the rest into pasture land for cattle and sheep. In the western half of the prairie, man has all but eliminated the large native herbivores, such as the bison, and replaced them with domestic grazing animals (Fig. 22.5). As a result, many of these areas have been overgrazed, and the original grasses have been replaced by cacti, sagebrush, and junipers.

FIGURE 22.6

Desert. Although the desert biome may appear to be a lifeless place, it is in fact the habitat of many insects, reptiles, and mammals. Dominated by a few plant species, such as sagebrush and the saguaro cactus, it appears devoid of animal life, but actually, most of the animals that occupy the desert have become nocturnal in order to avoid the heat of day.

DESERT

In North America the desert biome occurs at low to medium elevations from southern Oregon and Idaho in the United States to the states of Chihuahua and Sonora in Mexico. One climatic determinant of this biome is very low annual precipitation, averaging less than 10 inches, because the desert areas in the United States are in a rain shadow, as described earlier in this chapter.

North American deserts are characterized by driving winds, great seasonal changes in temperature, and bright sunny days. Desert soils are generally coarse (gravel and sand), and many areas are covered with closely packed stones ("desert pavement"). Desert rainfall occurs mostly as very hard thundershowers that usually last only a few minutes. The dry hard-packed soil surface allows little of this water to enter, and most of it runs off to basins in lower regions. Runoff water entering these basins is usually high in minerals and small soil particles, such as clay and silt. As a result of the high evaporation rate in the desert region, many of the basins dry up, and permanent desert lakes like the Great Salt Lake in Utah have very high salt concentrations.

In the desert regions of Oregon, Idaho, Nevada, and Utah, the dominant vegetation consists of such shrubs as the sagebrush and saltbush. They are the only plants that can withstand the dry conditions and cold winters of the region. As a result of both the latitudinal position of the desert and altitudes of 4000 to 8000 feet, frost may occur in any month of the year. There are relatively few ground animals, of which the major species are the coyote, badger, kit fox, kangaroo rat, ground squirrel, and jackrabbit. Bird species include such predators as the golden and bald eagles, hawks, prairie falcons, and owls.

In the hotter deserts of New Mexico, Arizona, Texas, and Mexico, the dominant plants are the creosote bush, yucca, smaller cacti, and century plant. A unique bird, the roadrunner, and a small desert pig, the peccary, are found in certain areas. Other animals include coyotes, jackrabbits, ground squirrels, kangaroo rats, lizards, and snakes.

The deserts have suffered relatively little at the hands of man, because the aridity renders most areas unsuitable for human habitation. Nonetheless, when careful management of water reserves and irrigation make the crucial abiotic factor of water available, the desert can sometimes be transformed into lush farmlands, as has been done in parts of the Middle East. This result may have its own costs, however, since bad irrigation techniques tend to cause severe salination of rivers. For example, by the time the Rio Grande becomes the Texas-Mexico border, its water is unfit for either drinking or irrigation.

MONTANE ZONATION

The types of biomes discussed so far are distributed in latitudinal bands on the continents in the northern portion of the globe, and these biomes occur roughly in sequence from tundra in the north to desert in the south. This same pattern of flora and fauna distribution occurs in smaller scale on the slopes of mountains, where distinct bands called **montane zones** are distributed altitudinally (see Fig. 22.7). From mountaintop to bottom, the montane zones are alpine tundra (above timberline), coniferous forest, coniferous woodland, grassland, and semiarid desert. With small variations each zone has flora and fauna similar to that in a larger biome. The main exception is the coniferous woodland zone, which contains a significant proportion of coniferous trees, unlike the corresponding deciduous biome.

The reason for the similarity between the montane zones and the biomes is that the climates of each montane zone and its corresponding biome are very much alike. The importance of climate in determining vegetation is illustrated by the fact that as one goes south, where warmer temperatures and less precipitation occur,

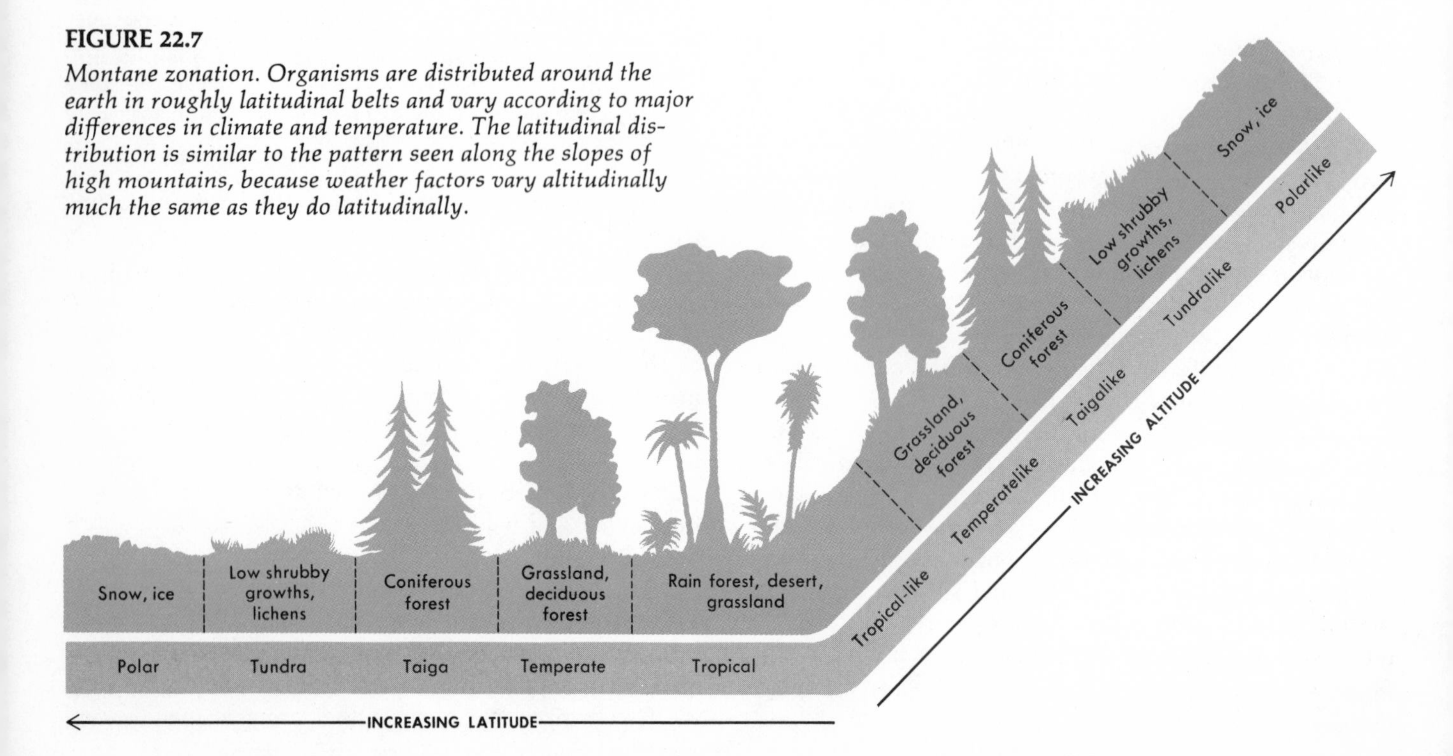

FIGURE 22.7

Montane zonation. Organisms are distributed around the earth in roughly latitudinal belts and vary according to major differences in climate and temperature. The latitudinal distribution is similar to the pattern seen along the slopes of high mountains, because weather factors vary altitudinally much the same as they do latitudinally.

montane zones are found at higher and higher altitudes. For example, alpine tundra in southern Arizona occurs at altitudes greater than 12,000 feet, but the tundra on the north slopes of mountains in northern Colorado is found as low as 9000 feet. The montane zones also occur at lower elevations on northern than on southern slopes because temperatures on the slopes facing north away from the sun are colder and conditions are more moist.

In North America the complete series of montane zones can be observed in the mountain ranges of the west from Canada to Mexico. There the mountains are high and conditions at the base are warm and dry, creating large climatic differences and the full range of montane zones.

FIGURE 22.8

Subtropics. The alligator is a dominant carnivore in the aquatic subtropical biome of the New World. The Florida Everglades constitute a precariously balanced habitat situated on the gently sloping margin of the continent. Since the alligator has become a protected species, its population has made a remarkable recovery.

SUBTROPICAL FOREST

The subtropical biome in the United States is confined mostly to south Florida and the Florida Keys. The climate is warm throughout the year, and frosts are rare. Precipitation falls mainly from February to November. Rainfall averages 50 inches during this period, whereas from November to February it averages only 10 inches.

Much of this biome is less than 25 feet above sea level. Where moisture is plentiful there are extensive areas of meadowlike, freshwater marshland, such as the Everglades, where common plants include saw grass and the bald cypress.

Where salty marine water encroaches on the marshland, conditions favor the growth of mangrove swamps. Here the dominant vegetation is stratified, generally occurring on two levels. The upper level consists of large trees, including the gumbo limbo, mahogany, royal palm, and long-leaf pine. The lower level consists of small trees and shrubs, including the palmettos and the redbay. Lower-level species occur on upland areas and on the hammocks, highland areas in the marshes.

The more spectacular animal species of the subtropical biome are the black bear, the mountain lion, and the diamondback rattlesnake. More common animals include the alligator, the garfish, softshell turtles, and a profusion of shore and song birds.

In attempting to control water flow, especially in the Everglades region of south Florida, man has grossly altered the structure of the subtropical biome. Because this biome is very diverse, it can withstand such natural phenomena as fire and hurricanes, but only if the area receives large amounts of fresh water annually. Recently the Army Corps of Engineers' flood control projects have diverted to the east and west fresh water that otherwise would have flowed from central Florida south into the Everglades. The loss of fresh water allows sea water to flow further inland and results in the replacement of the saw grass and cypress marshes with mangrove swamps. In addition, because there is now so little fresh water in the dry season, huge fires frequently burn out of control over large areas.

Aquatic environments

The North American continent contains a multitude of aquatic habitats as well as its terrestrial ones. Fresh-water ponds, lakes, and streams punctuate the landscape, and marine environments border the continent. Just as terrestrial life is distributed in large homologous biomes, so fresh-water and marine organisms tend to live in specific underwater zones that constitute aquatic biomes. The zones differ in such environmental factors as availability of sunlight and bottom strata, water pressure, salinity, temperature, etc., and most organisms are adapted to occupy one zone.

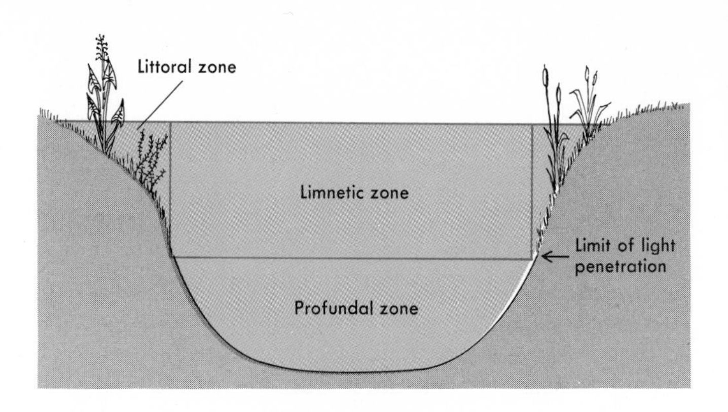

FIGURE 22.9

The three zones of fresh still-water habitats. Most of the visible plant life consists of the rooted cattails, bullrushes, water lilies, etc., of the littoral zone, but the microscopic organisms of the limnetic zone can exceed the rooted plants in food production. The limit of light penetration is an important line of demarcation, because without sunlight the profundal zone cannot support photosynthesizing producers.

FRESH-WATER ENVIRONMENTS

Fresh-water environments may have either still or running water, but we shall focus on still water to show the kinds of zonation that occur in aquatic biomes.

Still-water habitats, such as ponds and lakes, are generally divided into three zones, as shown in Fig. 22.9. The **littoral zone** consists of bottom and shallow-water areas where sunlight penetrates. Deeper-water areas are divided into the remaining two zones, the **limnetic zone,** consisting of the water areas only to the depth of light penetration, and the **profundal zone,** consisting of the bottom and the water below the depth of light penetration. Sunlight penetration is important because it determines whether and to what extent a zone can support photosynthesizing producers.

Littoral zone

The littoral zone contains the greatest number of different species. Its producers include both rooted and free-floating plants. Of the rooted variety, cattails and bullrushes abound in the shallowest water. Water lilies, with surface leaves and deep water roots, grow in deeper water, and totally submerged species like waterweed and muskgrass inhabit the deepest water. Free-floating species are microscopic phytoplankton, including algae, and diatoms. Blue-green algae play an important role in the nitrogen biogeochemical cycle because they use molecular nitrogen to form nitrates.

Both algae and diatoms are an important source of food for consumers. However, if water nutrient levels rise either naturally or as the result of pollution, these organisms can proliferate quickly. Then nitrates produced by the excess algae make the water smell and taste bad, and high diatom concentrations, called algal bloom, make the water appear a sickly green color.

Littoral zone consumers come from almost every phylum of the animal kingdom. They include snails, rotifers, flatworms, crustaceans, dragonflies and other insect nymphs, and, of course, fish. Some are herbivorous, and others are carnivorous.

Limnetic zone

Most of the inhabitants of the limnetic zone are microscopic. Consequently, to the naked eye the area seems devoid of life. Limnetic producers are diatoms and algae. Despite their small size, they can exceed rooted plants in food production per unit area. One characteristic of limnetic zone producers is great yearly fluctuations in population density. In response to the availability of nutrients, phytoplankton tend to bloom in the early spring and fall, but their population drops in summer and winter.

Limnetic zone consumers consist primarily of zooplankton, or microscopic animals. Some are herbivorous filter feeders, straining bacteria and phytoplankton from the water. Others are carnivorous. Fish do occupy the limnetic zone, but they generally feed near the bottom of the littoral zone.

Profundal zone

Because light does not penetrate here, the inhabitants are dependent on the other zones for food and thus are all consumers. Bacteria and fungi predominate. These organisms perform a vital service for the inhabitants of other zones because they break down and recycle essential nutrients. These compounds are carried to higher zones by water currents.

FIGURE 22.10
Marine habitat. The marine biome supports a tremendous abundance of individual organisms. In the rather small area shown here, there are snails, mussels, a sea anemone, and an array of ocean plant life.

There is actually considerable diversity and variety among standing bodies of water. Because ponds are generally smaller in surface area and shallower than lakes, the littoral zone in a pond is relatively larger, and the limnetic and profundal zones are smaller or completely absent. Bodies of water that form only during a season of high precipitation (or high flooding and runoff) are also called ponds. Such ponds are frequented by organisms that can survive the absence of water in a dormant stage.

In addition, the following special categories of lakes have been described by scientists.

Bog lakes are very acid, and their margins are filled with peat, which is partially decomposed vegetation that has become compressed and carbonized. When extracted from such bogs and dried, peat can be used for fuel.

Deep, ancient lakes were formed during the Mesozoic age, the age of reptiles. Such lakes frequently contain many species of animals and plants found nowhere else on earth. The deepest lake in the world, Lake Baikal in southern Siberia, descends more than a mile.

Desert salt lakes occur in very dry climates where precipitation is low and water comes primarily by drainage. Because the climate is very dry, water loss by evaporation exceeds water gain by precipitation. The salts dissolved in the water are retained in the lake, and their concentration increases. Organisms that live in such lakes must be adapted to high salt concentrations.

Volcanic lakes are associated only with active volcanos, and they have extreme chemical conditions. They may be acid or alkaline.

Artificial lakes have been made recently by human engineering activity, often the damming up of a river or stream. They may be characterized by fluctuating water levels and a high concentration of suspended materials that give the lake a dark and cloudy or muddy appearance (high turbidity).

MARINE ENVIRONMENTS

Approximately 70 percent of the earth's surface is covered with water, and most of it is sea or ocean water. The important physical factors affecting life in the oceans are currents, tides, depth, temperature, light penetration, salinity, and pressure. These factors vary greatly throughout the earth's marine environment, and as a result, there are many different types of marine habitats, including salt marshes, tidal beaches, estuaries, and open ocean. Conditions in tidal and surface layers of marine habitats are particularly favorable, and the greatest number of plant and animal species inhabit these areas. At depths below 300 feet, relatively little life is present.

Like lakes, oceans can be divided into different areas, and we will briefly examine zonation in the sea. Marine habitats are divided into shallow-water **neritic habitats** on the continental shelf and **oceanic habitats** beyond the continental shelf. In both habitats, bottom and open-water areas to the depth of light penetration constitute the **euphotic zone,** where most oceanic life occurs.

The ocean, of course, extends much deeper than the level of light penetration. Ecologists have used the rather poetic adjectives bathyal, abyssal and hadal to describe the successively deeper zones of the ocean. Some organisms occupy or penetrate these areas to various depths, but compared with surface areas, these zones have relatively few organisms.

Marine organisms may dwell on the bottom or in open water, and the ocean floor is replete with fixed, or **sessile,** organisms that cling to the bottom. Although the forms of life that exist in the ocean are far too extensive for consideration here, it is important to note that the major producers of the ocean are the microscopic phytoplankton that occupy the euphotic zone. Phytoplankton not only form the base of most marine food chains but also generate much of the earth's atmospheric oxygen.

Biogeography

Biogeography is a science that studies the past history of plants and animals in relation to their distribution over the earth's surface. The natural communities seen today on a particular continent did not exist in the same form millions of years ago. Furthermore, the world's landmasses did not always have the shape and location they have today.

There is a considerable amount of both geological and paleontological evidence that landmasses now separated must at one time have been joined. Rocks of similar structure, chemical content, and age have been found in eastern South America and western Africa, and the coastlines of those two landmasses have long been recognized as remarkably complementary. Plant and animal fossils found in widely separated parts of today's world, such as Antarctica, South Africa, and China, are clearly related. Most scientists are now convinced that there was once only one large continent on earth, and that over hundreds of millions of years it divided in successive stages, ultimately to become the world we know today. At various times, many landmasses have risen from and sunk back into the seas, and continents have not only separated from each other but have continued to move after separating, a phenomenon known as **continental drift.** Established plants and animals moved with the drifting continents and thus left related fossil remains in widely scattered areas. Animals also moved across land bridges that later disappeared, cutting off those earlier paths of movement. The concept of continental drift has provided some answers to questions posed by paleontologists in the realm of biogeography.

During the nineteenth century many scientific expeditions and voyages were made for the purpose of studying the geography of plants and animals. Alfred Wallace, who with Darwin first proposed the theory of natural selection as the basis for evolution, was the first to publish a book of biogeography, and this work is considered the beginning of the science of biogeography. In his book Wallace divided the world into several biogeographical regions on the basis that each area appeared to have been isolated from other regions during significant periods of evolutionary development. The boundaries he proposed, shown in Fig. 22.11, are for the most part still used. These biogeographic regions are the Australian, Oriental, Palearctic, Ethiopian, Nearctic, and Neotropical.

AUSTRALIAN REGION

In addition to Australia itself, the Australian region includes New Zealand, Tasmania, and the eastern part of Indonesia. This most isolated of the biogeographical regions is separated from the Oriental region by an indefinite boundary referred to as Wallace's Line, which runs roughly northeast to southwest between the Philippines and Moluccas, Borneo and the Celebes, and Bali and Lombok.

The Australian region, which is referred to as an island continent, has been separated from Eurasia for many millions of years. The fauna and flora here are both distinct and unusual. Most groups apparently arrived by island hopping from the Oriental region. The predominant mammals are the marsupials, animals in which the embryo is transferred at an early stage of development from the female's uterus to a pouch on the abdomen to complete development. Marsupials were apparently the only mammalian group to enter the region before a land bridge to Eurasia was cut off early in the Age of Mammals, 50 million years ago. The marsupials

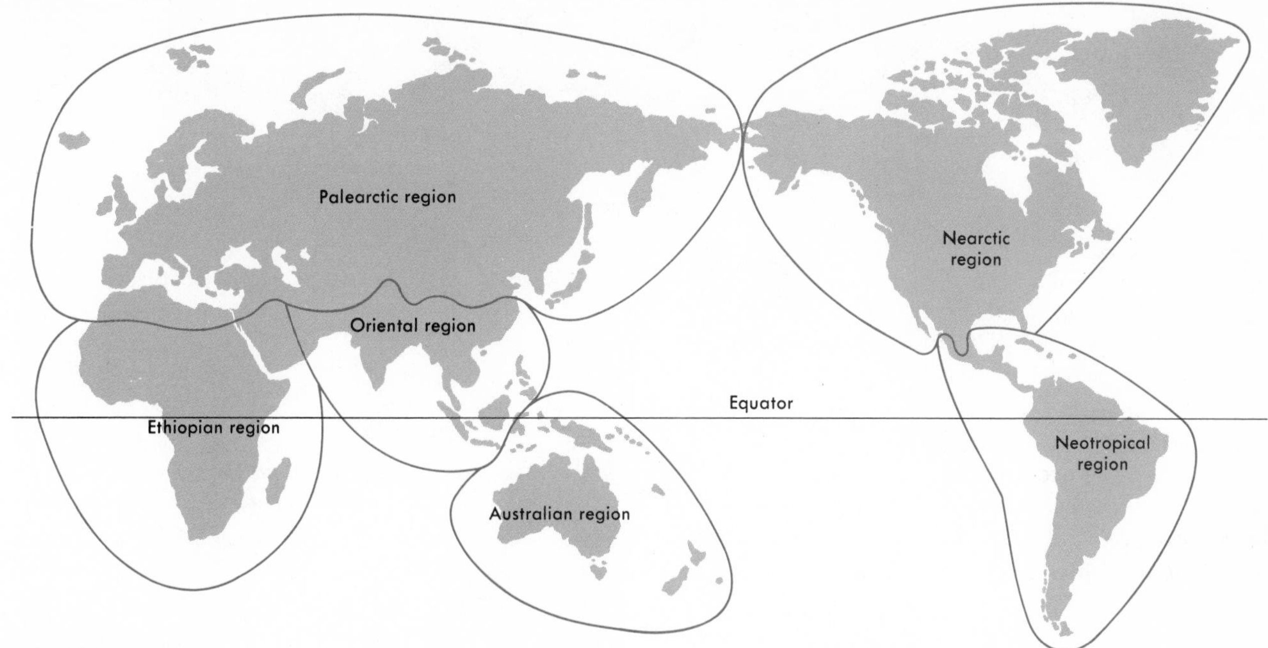

FIGURE 22.11

The six biogeographical regions of the world as proposed by Alfred Wallace. Each region is categorized by the types and distribution of organisms living there. The Siberian and Central American land bridges between the Nearctic region and the Palearctic and Neotropical regions have greatly influenced the course of evolution.

underwent a rapid period of adaptive radiation shortly after their arrival, a process which occurred in several millions of years. Marsupials today fill the niches that placental mammals fill in other regions of the world. Frequently the marsupials even look like their placental ecological equivalents.

In very recent times man introduced mammals such as the rabbit and deer into the Australian region. These species multiplied rapidly, disturbed many habitats, and may be causing the extinction of several unique marsupial herbivores.

ORIENTAL REGION

The Oriental region lies southeast of the Himalayan Mountains and extends from India through Indochina and up to southern China. It also includes the western part of the Malay Archipelago. It is largely a tropical region with a few endemic species, that is, species occurring in one biogeographical area only. They include the tree shrew and the tarsier. The area contains many primates, large cats like the tiger, pachyderms like the Indian elephant and rhinoceros, and reptiles like the crocodile. These species are related to some in the Ethiopian region, because 25 million years ago there was a moist tropical land bridge in Arabia that connected India and Africa. Across it in both directions many ancestral groups of present-day plants and animals migrated. Some groups, such as bears, tapirs, and deer, could not cross the land bridge, because for them the moist tropical landmass was a barrier. Later, when the climate of this Arabian area dried, no species could migrate across it, and a barrier was effected between the two regions.

ETHIOPIAN REGION

Africa south of the Sahara Desert and Atlas Mountains, as well as part of Arabia, are included in the Ethiopian region. Like the Neotropical region in the New World, this region has a wide variety of climates and includes everything from alpine tundra to arid deserts. Much of this region is covered by large tropical forests, savannas (grassland with widely spaced trees throughout), and grasslands. The region ranks a close second to the Neotropical region in the number of endemic species and includes the greatest variety of vertebrates.

Notable similarities between this region and the Oriental region are the African elephant, rhinoceros, and such primates as the chimpanzee and gorilla. The region also contains large numbers of reptiles and amphibians that are related to those in the Oriental region.

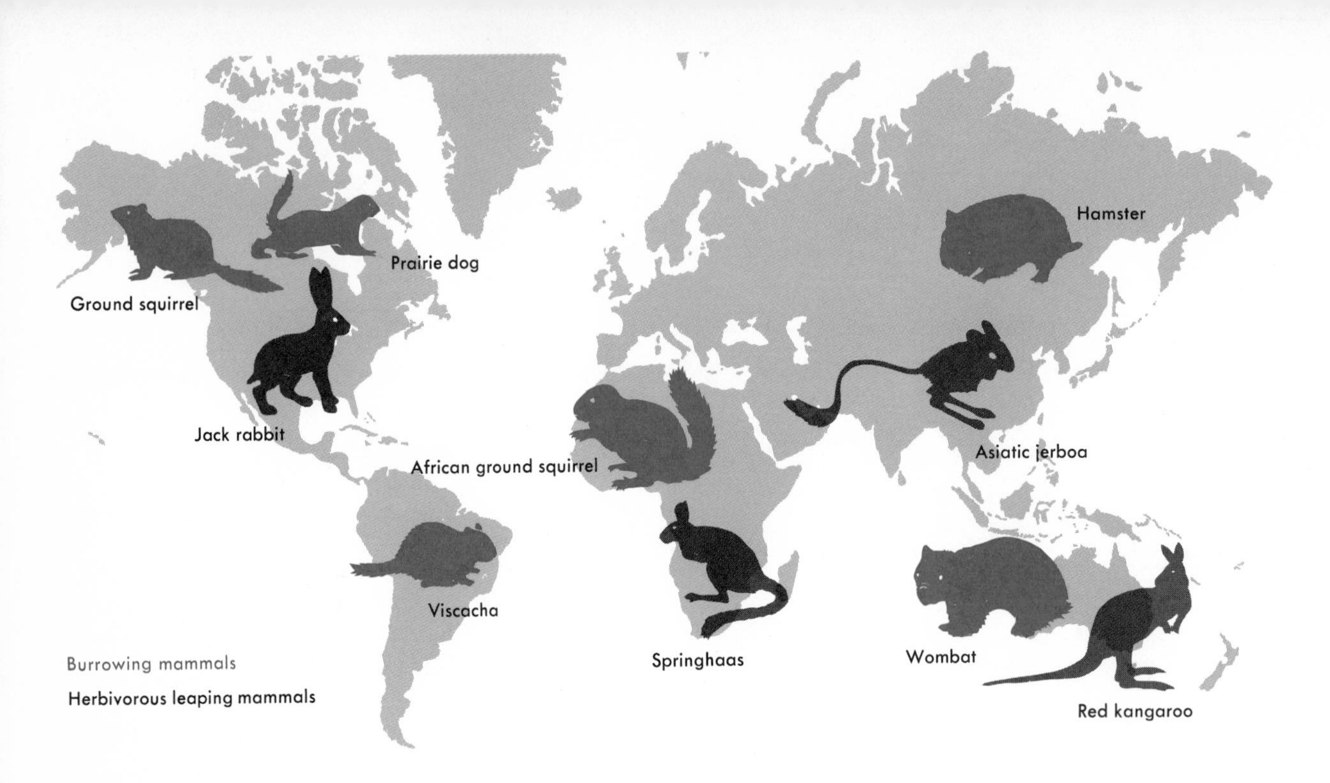

FIGURE 22.12

Common environments have produced ecological equivalents among two groups of mammals—the one a burrowing animal that feeds aboveground, and the other a leaping herbivorous type—in the grasslands biome of each of the major biogeographical regions.

PALEARCTIC REGION

North of the Oriental and Ethiopian regions lies the large Palearctic region, which includes parts of Asia north of the Himalayan Mountains, all of Europe, northern Arabia, and coastal North Africa. The Palearctic region has few endemic species because much of the fauna and flora came from the Old World tropics and New World temperate regions.

The Palearctic region shows the greatest affinities with the Nearctic region. The two areas were once connected by a land bridge across the Bering Sea. Because of this bridge, Eurasia and North America have similar types of plants and animals. Their biomes are almost exactly alike, and the stag, reindeer, and elk of Eurasia are strikingly similar to the elk, caribou, and moose of North America.

NEARCTIC REGION

All of North America south to the Tropic of Cancer in Mexico lies in the Nearctic region, which also includes Greenland. The vegetation types are similar to those in the Palearctic, but the animals, in some respects, are quite different. The Nearctic has many more endemic species and a greater diversity of reptiles and amphibians. Notable endemics are the kangaroo rat, pocket gopher, and pronghorn antelope. Much of the fauna and flora of the region came from the temperate Old World and tropical New World.

NEOTROPICAL REGION

Lying south of the Nearctic region and including southern Mexico, Central America, all of South America, and the West Indies, the Neotropical region is joined with the Nearctic by the isthmus of Panama. This land bridge was open during the early part of the Age of Mammals, and several groups of placentals and marsupials entered the Neotropical region at that time from the north. Shortly thereafter, the isthmus sank into the sea and the region was isolated. Later, about 15 million years ago, the isthmus rose again and the land bridge was reopened, but a very diverse and distinct fauna and flora had developed during the period of isolation. After the reopening, many animal groups from the Nearctic migrated to the south, but few came north. Most of the present-day fauna in the Neotropical region came from the Nearctic after the land bridge had been reopened. Nevertheless, the Neotropical region is richest in endemic species of all biogeographical areas. Endemics include several families of bats, many species of rodents and guinea pigs, the New World monkeys, sloths, true anteaters, and armadillos.

Summary

1. A biome is a large ecological unit composed of many similar communities, the majority of which are in the climax state. Climax community plant species dominate the biome and give the biome its rather uniform appearance.

2. The character of a biome is determined largely by its climate and soil. Only the plants and animals best adapted to the biome's environmental conditions will live there. Biomes in the northern hemisphere tend to be distributed according to climate in sequential latitudinal bands extending southward from the polar regions.

3. The major North American biomes are tundra, taiga, temperate deciduous forest, temperate grassland, desert, and subtropical forest. On the slopes of the high western mountains are montane zones, biomelike areas of characteristic vegetation distributed altitudinally. In order of decreasing altitude, the zones are alpine tundra, coniferous forest, coniferous woodland, grassland, and semiarid desert. Man has had a significant role in modifying the earth's biomes.

4. Aquatic environments are classed as either freshwater or marine. Ecologists have found that aquatic flora and fauna are distributed in underwater zones according to organisms' physiological requirements for sunlight, bottom strata, temperature, water pressure, and salinity. In lakes and ponds there are three zones—littoral, limnetic, and profundal. In the ocean, most life is found in the euphotic zone of the neritic and oceanic habitats.

5. Biogeography is the study of the evolutionary history of plants and animals through the examination of their past and present distribution on the globe. Alfred Wallace, the father of biogeography, divided the earth into six regions on the basis that fossil and anatomical evidence indicated that each region had been isolated from all others during significant periods of evolutionary history, and thus each had a distinct evolutionary history of its own. Wallace's biogeographical regions are identified as Australian, Oriental, Palearctic, Ethiopian, Nearctic, and Neotropical.

REVIEW AND STUDY QUESTIONS

1. Distinguish between community and biome.

2. Explain how each of the following environmental factors affects the character and extent of community succession: temperature, precipitation, wind, soil.

3. Explain montane zonation and relate it to general biome distribution.

4. Compare tundra and taiga in terms of vegetation and animal life. Is there an exact dividing line between them? If so, what is it?

5. How do extensive lumbering activities affect ecosystem balance?

6. What biome do you live in? What are its characteristics?

7. What evidence of continental drift does a standard map of the world offer? What other observations have supporters of that theory proposed as further evidence?

REFERENCES

Billings, W. D. 1966. *Plants, Man, and the Ecosystem.* Wadsworth, Belmont, Calif.

Cain, S. A. 1971. *Foundations of Plant Geography.* Hafner, New York.

Keast, A. 1971. "Continental Drift and the Evaluation of the Biota on Southern Continents." *Quarterly Review of Biology* 46: 335–378.

Kormondy, E. J. 1969. *Concepts of Ecology.* Prentice-Hall, Englewood Cliffs, N.J.

Low, R. M. 1971. "Interspecific Territoriality in a Pomacentrid Reef Fish, *Pomacentrus flavicanda* Whitley." *Ecology* 52: 648–654.

Odum, E. P. 1971. *Fundamentals of Ecology,* 3rd edition. Saunders, Philadelphia.

Shelford, V. E. 1963. *The Ecology of North America.* University of Illinois Press, Urbana.

Smith, R. L. 1974. *Ecology and Field Biology,* 2nd edition. Harper & Row, New York.

SUGGESTIONS FOR FURTHER READING

Bishop, J. A., and L. M. Cook. 1975 (January). "Moths, Melanism, and Clean Air." *Scientific American* Offprint no. 1314. Freeman, San Francisco.

A return look at species of moths that darkened in nineteenth-century England. As the air has improved in quality, the moths have noticeably reverted to lighter shades.

Carson, R. 1961. *The Sea Around Us,* revised edition. Oxford, New York.

A classic on the oceans and sea life.

Denison, W. C. 1973 (June). "Life in Tall Trees." *Scientific American* Offprint no. 1274. Freeman, San Francisco.

Examination of the high forest canopy, where entire communities of other plants and animals dwell, helping to provide the trees with nitrogen.

Fritts, H. C. 1972 (May). "Tree Rings and Climate." *Scientific American* Offprint no. 1250. Freeman, San Francisco.

The ring pattern of trees as a source of information not only of the trees' age but of climates of the past.

Horn, H. S. 1975 (May). "Forest Succession." *Scientific American* Offprint no. 1321. Freeman, San Francisco.

Use of a New Jersey wooded area as the basis of a predictive model for the succession of trees in a mixed forest.

Ponnamperuma, C., and H. P. Klein. 1970. "The Coming Search for Life on Mars." *The Quarterly Review of Biology* 45: 235–258.

Reviews of experiments on the origin of life and the case for extraterrestrial life. The author conjectures as to whether life on other planets will be similar to ours. An excellent bibliography on exobiology is included.

Richards, P. W. 1973 (December). "The Tropical Rain Forest." *Scientific American* Offprint no. 1286. Freeman, San Francisco.

The current plight of the evergreen tropical forest, one of the oldest ecosystems on Earth and a treasury of genetic diversity.

Shepard, F. P., and H. R. Wanless. 1971. *Our Changing Coastlines.* McGraw-Hill, New York.

A morphological description of the coastlines of North America. Informally written, this work contains useful information, although some of its conclusions are inaccurate because they are based on aerial photographs.

Wiggins, I. L., and D. M. Porter. 1971. *Flora of the Galápagos Islands.* Stanford University Press, Stanford, Calif.

The site of Darwin's discovery of the adaptive radiation of birds, tortoises, and lizards, the Galápagos Islands have been relatively unexplored by botanists. This book is a thorough examination and classification of all the flora on the islands. A scholarly work that should also appeal to the general public.

GLOSSARY

Abscisic acid (ab-siz′ik) Plant hormone that tends to slow activity of plants by inducing leaf abscission and dormancy in buds and seeds, weakening actions of active hormones, and enabling plant to survive adverse conditions.

Abscission (ab-sizh′un) Dropping of leaves.

Accelerator center (ak-sel′e-rā-tor) One of two control centers in the brain that transmits impulses along the sympathetic nerves to the pacemaker of the heart, causing heartbeat rate to increase.

Acid A compound that releases hydrogen (H+) ions when dissolved in water.

ACTH (adrenocorticotropic hormone) (a-drē′no-kor-ti-kō-trō′pik) Hormone secreted by the anterior pituitary that induces the adrenal cortex to produce corticoids.

Activation energy The minimum energy required to initiate a chemical reaction.

Active transport A process, occurring at the cell membrane, in which a cell expends energy to move a particle or ion through the membrane, often against a concentration gradient.

Adaptation The process through which a species becomes increasingly able to survive and breed in its particular environment.

Adaptive radiation The development of several highly adapted species from a single less-specialized species.

ADP (adenosine diphosphate) (a-den′ō-sin dī′fos-fāt) The compound that when bonded to a phosphate becomes high-energy-containing ATP.

Adrenal cortex (a-drē′nal kor′teks) The outer layer of the adrenal gland; secretes corticoids, which regulate the body's response to stress and the liver's release of glucose.

Adrenal medulla (me-dul′a) The inner portion of the adrenal glands, located over kidneys; secretes hormones (epinephrine and norepinephrine), which speed certain body functions in response to stress.

Aerobic (ār-ō′bik) Requiring oxygen.

Albumin (al-bū′men) A protein in blood plasma.

Aldosterone (al-dos′te-rōn) One of the mineralocorticoid hormones secreted by the adrenal cortex. It promotes the retention of sodium and water in the bloodstream.

Allele (a-lēl′) One of two or more alternative genes that control the same characteristic and occupy the same place on similar chromosomes.

Allopatric (a-lō-pat′rik) Inhabiting different areas.

Allopolyploid (a-lō-pol′i-ploid) Hybrid organism whose chromosome sets have been duplicated.

Allosteric regulation (a-lō-ster′ik) Modification of enzyme activity caused by a change in the shape of the enzyme.

Alpha helix (al′fa hē′liks) A spiral molecular configuration, common in proteins, which is formed and maintained by hydrogen bonds.

Alveoli (alvē′o-lī) Air sacs where gas exchange occurs in lungs.

Amino acids (a-mē′no) The building blocks of proteins, consisting of an organic acid, an amino group ($-NH_2$), and one of about 20 characteristic side groups.

Ammonification (a-mon-i-fi-kā′shun) A natural process of decomposition in which nitrogenous components (primarily the amino acids) are acted on by microorganisms to release nitrogen in the form of an ammonium ion.

Anaerobic (an-ār-ō′bik) Requiring no oxygen.

Anaphase (an′a-fāz) The third phase of mitosis, in which the chromatids separate and migrate toward the centrioles at opposite ends of the cell. In meiosis, first anaphase distributes one set of homologous chromosomes to each new cell; second anaphase generates the haploid set of chromosomes in each cell.

Aneuploidy (an′yū-ploi-dē) The loss or gain of one or more chromosomes from a complete set.

Angiosperm (an′ji-ō-sperm) A flowering plant.

Angstrom unit (A) (ang′strom) One ten-millionth of a millimeter.

Anterior pituitary gland (pi-tū′i-tā-rē) Master gland of the body secretes growth hormone, gonadotropins, thyroid-stimulating hormone, adrenocorticotropic hormone, and prolactin. In large part under the control of the hypothalamus.

Anther (an′ther) The male spore-forming body on stamens of flowers. It produces pollen grains.

Antheridium (an-the-rid′i-um) The structure in nonvascular plants that produces male gamates.

Antibody (an′tē-bo-dē) Substance in blood serum characterized by specific reactivity to an antigen.

Anticodon (an-ti-kō′don) Adapter portion of the tRNA molecule, which consists of a nucleotide triplet complementary to that of a specific codon on mRNA.

Antigen (an′ti-jen) Substance that will stimulate production of antibodies and react with them.

Aorta (ā-ōr′ta) Main artery of body, which accepts blood pumped by left ventricle.

Aortic valve Heart valve between the ventricles and the aorta.

Apical dominance (ā'pi-kal) Elongation of the main stem together with inhibition of lateral bud growth in response to auxins produced by the apical bud.

Aplanospore (a-plan'ō-spōr) A nonmotile asexual spore, common in true fungi.

Aposematic coloration (a-poz-i-mat'ik) Warning coloration on organisms that are distasteful or have strong defenses.

Archegonium (ahr-ki-gō'ni-um) The female sex organ of liveworts, mosses, and ferns.

Arterioles (ar-tēr'i-ōlz) Small arteries that regulate blood flow.

Artery (ar'te-rē) Thick-walled muscular tube, which conducts blood away from the heart.

Astral fibrils (as'tral fib'rilz) Fibers radiating from centrioles.

Atom The smallest structural unit of an element that retains the chemical and physical properties of that element.

Atomic number The number of protons in the nucleus of an atom of a given element.

Atomic weight The relative weight of an atom, using carbon (^{12}C) as the standard.

ATP (adenosine triphosphate) (a-den'ō-sin trī'fos-fāt) The molecule that acts as an energy carrier in the cell. The energy is stored in a high-energy bond between the second and third phosphates.

Atrium (at'ri-um) One of the two thin-walled upper chambers of the heart, which receives incoming blood.

Autonomic nervous system (ot-ō-nom'ik) The system of neurons in vertebrates that is not under voluntary control. It innervates the visceral organs, the heart, glands, and smooth muscle.

Autosomes (ot'ō-sōmz) All chromosomes except the sex-determining chromosomes.

Autotroph (ot'ō-trof) An organism that synthesizes its own food from the simple inorganic compounds it absorbs from its environment.

Auxins (oks'inz) Plant growth hormones, manufactured and secreted by meristematic tissue.

Avoidance reaction Movement or series of movements away from a noxious stimulus.

Axon (aks'on) A long unbranched or slightly branched nerve fiber that carries impulses away from the cell body of the neuron.

Baroreceptor (ba-rō-ri-sep'tor) Pressure sensitive nerve ending in carotid artery, which forms part of the feedback loop that slows heartbeat rate.

Basal body (kinetosome) (bā'sal; ki-net'o-sōm) An organelle, derived from the division of the centrioles, from which grow cilia or flagella.

Base A compound that releases hydroxyl ion (OH−) when dissolved in water.

Batesian mimicry (bāt'zi-an mim'ik-rē) Mimicry of an aposematic animal; markings on a palatable unprotected species.

Behavior Action(s) that alters the relationship between the organism and its external environment.

Binary fission (mitosis) (bī'nā-rē fish'on; mī-tō'sis) An even division of cytoplasm and nucleus into two daughter cells.

Binomial nomenclature (bī-nō'mi-al nō'men-clā-chur) The system of applying scientific names, usually in Latin, to species. Each species is given a generic and a specific name, such as *Drosophila melanogaster*, the common fruit fly.

Biogeography The study of the history of the dispersal of plants and animals throughout the world.

Biological magnification The accumulation of pesticides (especially fat-soluble ones) in the bodies of organisms in increasing concentration at each successive level of the food chain.

Biological oxygen demand (BOD) An index of water pollution by organic material that can be related to the bacterial population's degree of oxygen use.

Biome (bī'ōm) A complex of all communities with similar animals and vegetation and characterized by specific climate and geographical area.

Biosphere (bī'ō-sfer) All the area above, on, and below the earth's surface where life exists, including air, earth, and water.

Bladder Sac that stores urine extracted from the blood by the kidneys.

Blastocoel (blas'tō-sēl) The cavity within the blastula.

Blastula (blas'chu-la) An animal embryo after cleavage and before gastrulation. Typically a single layer of cells forming a hollow sphere.

Blood A liquid connective tissue consisting of erythrocytes, leucocytes, and platelets suspended in a liquid matrix, the plasma.

Blood pressure Force exerted against artery walls by blood being pumped from the heart.

Bolus (bō'lus) A moist mass of food formed in the mouth for swallowing.

Bone A connective tissue consisting of living cells embedded in a matrix of calcium salts and collagen.

Bowman's capsule (bō-manz cap'sūl) Cuplike body where wastes, salts, and fluids are filtered from the blood by the kidney.

Bronchi (bron'kē) Two branching tubes that connect the trachea with the bronchioles and air sacs of the lungs.

Bronchioles (bron'kē-ōlz) Finely branching tubes leading to alveoli in lungs.

Bud In plants, the end of a stem or branch that contains the meristematically derived leaf and stem primordia.

Budding A form of asexual reproduction in which repeated mitotic division of a given region produces a new organism, which then breaks off from the parent organism.

Buffer A chemical substance that resists rapid or extreme changes in pH by releasing or binding hydrogen ions according to hydrogen ion concentration.

Calorie (kal'o-rē) The amount of heat required to raise the temperature of 1 gram of water 1° Centigrade.

Calvin cycle The dark reactions of photosynthesis, in which sugar is formed by the reduction of carbon dioxide.

Cambium (kam'bi-um) A layer of meristematic tissue in the stems and roots that gives rise to xylem, phloem, and parenchyma.

cAMP (cyclic adenosine monophosphate) (sik'lik a-den'o-sin mon-ō-fos'fāt) Nucleotide that may serve as a messenger within cells, then stimulating cellular activites.

Capillaries (kap'il-ār-ēz) Tiny blood vessels where gas and nutrients can diffuse into and out of the bloodstream.

Carbohydrates (kar-bō-hī'drātz) Group of compounds of carbon, hydrogen, and oxygen; includes sugars, starches, and celluloses.

Cardiac (kar'di-ak) Of or pertaining to the heart.

Cardioinhibitory center (kar'di-ō-in-hib'i-tō-rē) One of two heart control centers in the brain that transmit impulses by the vagus nerve to the pacemaker of the heart; slows the rate of heartbeat.

Carnivore (kar'ni-vōr) A heterotroph that eats only flesh.

Carotenoids (ka-roh'ten-oidz) Yellow and orange pigments contained in some chromoplasts.

Carotid artery (ka-roh'tid) One of two large blood vessels that supply the head and neck regions.

Carrier molecules Molecules (perhaps proteins in membrane) that bond to molecules outside the cell and actively transport them across the membrane into the cell.

Cartilage (kar'ti-lej) The flexible yet stiff connective tissue that constitutes the embryonic skeleton and is retained in certain parts of the adult body.

Castes (kastz) A type of social hierarchy in certain species of insects in which social order is based on the possession by certain individuals of specialized structure and function.

Catalyst (kat'a-list) A substance that increases the rate at which other substances react but is itself not altered by the reaction.

Catastrophism (kat-as-trōf'izm) The eighteenth-century geological theory that the earth was molded and reshaped by a series of cataclysmic events.

Cell The smallest unit of living tissue that performs all the functions essential to life: energy use, molecule synthesis, growth, replication, internal environment regulation, and reaction to external environmental changes.

Cell membrane The living partition, made of phospholipid and protein, which forms the outermost edge of the cell and acts as a barrier between the cell and its environment.

Cell plate A membrane that forms a wall between two daughter cells in plant cytokinesis.

Cellulose (sel'ū-lōs) A polysaccharide produced by plants as a supporting material. Because it cannot be hydrolyzed easily, man and most other animals cannot digest it.

Central nervous system (CNS) In vertebrates, the brain and spinal cord, containing the cell bodies and most of the neurons, and exerting a great deal of control over the rest of the nervous system.

Centrioles (sen'tri-ōlz) A pair of cytoplasmic organelles, each composed of two cylindrical bodies perpendicular to each other. Centrioles are active during cell division.

Centromere (kinetochore) (sen'tro-mēr; kin-et'a-kōr) Central body in early mitosis on chromosomes, where chromatids are joined and where spindle fibers become attached.

Cerebellum (se-re-bel'um) An expansion of the dorsal side of the brain near its hind end. It is a coordinating center for proprioceptive stimuli and complex muscular movements.

Cerebrum (se-rē'brum) The expansion of the dorsal and lateral sides of the brain at its front end; the chief center for conditioned, or learned, responses; also the seat of learning and memory.

Chemoreceptors (kē-mō-ri-sep'torz) Sensory cells that respond to specific chemical stimuli, e.g., taste or smell.

Chemotaxis (kē-mō-tak'sis) Behavioral response to chemical or physical stimuli.

Chiasma (ki-az'ma) Intersection of parts of two chromatids during meiosis.

Chlorophyll (klō'ro-fil) Green plant pigment that functions as receptor of light energy in photosynthesis.

Chloroplast (klō'ro-plast) A plastid containing chlorophyll.

Cholesterol (ko-les'te-rol) A kind of lipid found in blood plasma, which serves as a precursor molecule that may be converted by enzymes to important hormones.

Chondrocytes (kon'dro-sītz) Connective cells in cartilage that secrete the matrix fibers.

Chromatid (krō'ma-tid) One of two identical fibers constituting a newly replicated chromosome.

Chromatin (krō'ma-tin) Body of nuclear material, containing nucleic acids, which is present during interphase.

Chromosome (krō-mo-sōm) A rod-shaped or filamentous structure found in the nucleus, which is composed of DNA and protein and within which the genes are located.

Chyme (kīm) Loose, watery mass constituting partially digested contents of the stomach and intestines.

Cilium (sil'i-um) A short, fine, protoplasmic organelle on the surface of a cell, which beats rhythmically to move a unicellular organism or to move particles over the surface of the cell.

Circadian (ser-kā'di-an) Occurring at or close to 20-hour intervals.

Classical conditioning The process whereby an organism is taught to make a new automatic reflex response to a formerly neutral stimulus.

Cleavage (klē'vej) The successive cell divisions of the developing egg that transform it into a multicellular blastula.

Climax community The final, stable stage of a series of communities. Typically contains certain plant and animal species in specific climates and areas of the world.

Codominance (kō-dom'i-nans) The full expression of both genes of an allelic pair.

Codon (kō'don) Unit of DNA code composed of three nucleotide bases. Each triplet represents a specific amino acid.

Coenzyme (kō-en'zīm) Substance associated with an enzyme that plays an essential role in the catalytic activity of the enzyme.

Coleoptile (kō-li-op'til) The first leaf of a monocot seedling.

Collagen (kol'a-jen) A fibrous protein that forms part of the basic structural material of bone matrix of many different kinds of connective tissues.

Collecting tubule (tū'byūl) Tubule that channels urine from the nephrons into the pelvis and the kidney.

Collenchyma (kol-en'ki-ma) Supportive tissue in plants, usually consisting of long, thick-walled cells.

Colloid (kol'oid) A liquid mixture in which particles larger than molecules of dissolved solute but smaller than large particles of a suspension are evenly distributed throughout the liquid and do not settle out when colloid is allowed to stand. It may have two forms: sol and gel.

Colon (kō'lon) The large intestine, where water secreted into the alimentary canal during digestion is reabsorbed into the body.

Commensalism (ko-men'sal-izm) An interspecific interaction in which one organism benefits from the association while the other neither benefits nor is adversely affected.

Community All organisms that occupy the same habitat and interact with one another.

Companion cell A thin-walled plant cell containing a nucleus and other organelles, which is thought to play a role in the metabolic activities of associated sieve elements in angiosperms.

Compound Two or more atoms of different elements that, joined together, have different chemical and physical properties than the component elements.

Concentration gradient (grā'di-ent) The difference in concentration between two areas.

Condensation reaction The linkage of two monomers through the release of a water molecule at the point of linkage.

Conditioned reflex A learned, automatic response to a given stimulus.

Conditioning The process of developing new automatic responses to certain stimuli.

Conjugated protein (kon'ja-gā-ted) A protein that is attached to another class of organic compound, such as carbohydrates, lipids, nucleic acids, or metal atoms.

Conjugation (kon-ja-gā'shun) The temporary union of two unicellular organisms, in which there is a transfer of genetic material.

Connective cells Cells that make up the animal tissues that integrate and hold together the body parts in a single entity —for instance, bone, cartilage, or blood.

Continental drift The separation and movement of land masses over hundreds of millions of years.

Contraception Prevention of impregnation of ovum by sperm.

Contractile vacuole (kon-trak'til vak'yū-ōl) Membrane-enclosed sac in the cytoplasm of some unicellular organisms, which accumulates water and wastes and then contracts, expelling them from the cell.

Copulation (kop-ū-lā'shun) The introduction of sperm cells from the male animal into the reproductive tract of the female.

Coronary system (ko'ro-nā-rē) Blood vessels serving the heart muscle.

Corpora cavernosa (kor'po-ra ka-ver-nō'sa) Two bodies of spongy erectile tissue in the penis that become engorged with blood during sexual excitement.

Corpus luteum (kor'pus lū'ti-um) The follicle that becomes filled with a fatty yellow substance after ovulation and then produces the steroid hormone progesterone.

Cortex (kor'teks) In the egg cell, a gelatinous, colloidal surface layer about three microns thick, located beneath the surface of the egg. It is believed to have organizing properties and is activated by the sperm during fertilization.

Cortisone (kor'ti-sōn) One of a group of hormones (corticoids) secreted by the adrenal cortex.

Cotyledon (kot-l-ē'don) Part of a plant embryo that stores and digests food; a seed leaf.

Countercurrent flow Mechanism that accounts for efficient gas exchange in gills. The flow of water is in the opposite direction to the flow of blood in the capillaries.

Courtship A diverse group of elaborate instinctive behavioral patterns in vertebrate animals to ensure sexual excitement resulting in the synchronous release of gametes.

Covalent bond (kō-vā'lent) A chemical bond that results from the sharing of electrons.

Creatinine (krē-at'i-nēn) A by-product of muscle metabolism.

Crenation (kre-nā'shun) The shrinkage of animal cells due to water loss.

Cretinism (krēt'in-izm) Condition caused by insufficient level of thyroid hormone during childhood; characterized by mental retardation and dwarfism.

Cristae (kris'tē) The projections formed inside a mitochondrion by the folds of the inner membrane. They are the sites where respiration occurs within the mitochondrion.

Crossing-over The exchange of genetic material between chromatids during synapsis.

Cryptic coloration (krip'tik) Camouflage.

Cuticle (kyū-ti-kl) A waxy coating secreted by epidermal cells of the leaf, which prevents excessive water loss and wards off invasion by bacteria and viruses.

Cutin (kyū-tin) Waxy substance that coats the upper and lower epidermis of a leaf.

Cyclosis (sī-klō'sis) Internal streaming of cell cytoplasm.

Cystine (sis'tēn) An amino acid which is really two amino acids bound together by two sulfur atoms; can form peptide bonds with four other amino acids.

Cytochromes (sī'tō-krōmz) Pigment molecules that carry electrons in the breakdown of sugar during respiration and in the synthesis of sugar in photosynthesis.

Cytokinesis (sī-tō-ki-nē'sis) Division of cytoplasm.

Cytokinins (sī-tō-kī'ninz) Plant hormones that have effects oposite to those of auxins. They stimulate transverse growth and lateral branching and have an anti-aging effect.

Cytopharynx (sī-tō-far'inks) Mouthlike groove in single-celled organism like the paramecium.

Cytoplasm (sī-tō-plazm) Protoplasm located outside the nucleus and inside the cell membrane.

Deamination (dē-am-i-nā'shun) The removal of the amino group from an amino acid.

Decomposers (dē-kom-pōz'erz) Lower organisms (bacteria, fungi) in an ecosystem that convert dead organic material into plant nutrients.

Deletion (di-lē'shun) Loss of a segment of chromosome.

Denaturation (dē-nā-che-rā'shun) Loss of shape of protein by the destruction of hydrogen bonds as a result of excessive heat or acidity.

Dendrite (den'drīt) The relatively short, usually highly branched portion of the neuron, which carries impulses toward the cell body.

Denitrification (dē-nī-tri-fi-kā'shun) The microbial process whereby the nitrates are transformed back into nitrogen or to ammonia oxide, which reenter the atmosphere as gases.

Desquamation (des-kwa-mā'shun) Flaking off or shedding of superficial cells, as in stomach lining, mucous membranes, etc.

Determination The developmental process by which the prospective fate of a cell or group of cells becomes progressively narrowed until a specific cell type results.

Deuterostome (dū'ter-o-stōm) Any animal whose embryonic blastopore becomes the anal opening, with later formation of mouth from another opening.

Development The progressive production of the phenotypic characteristics of an organism through the processes of cellular differentiation, morphogenesis, and growth.

Diabetes insipidus (dī-a-bēt'ēs in-sip'i-dus) A disease in which kidneys do not reabsorb water normally, so urine is highly diluted; can be caused by tumor in posterior pituitary gland.

Diabetes mellitus (dī-a-bēt'ēs mel'i-tus) Disease caused by inadequate insulin secretion in which sugar is improperly metabolized; high blood sugar, weakness, frequent urination, and sometimes death are the result.

Diaphragm (dī'a-fram) Muscle at base of chest cavity, which creates the pressure differentials that cause air to move into and out of the lungs.

Diastole (dī-as'to-lē) Relaxation period after contraction of the heart.

Dicot (dī'kot) Plant whose seed has two cotyledons and which usually has a taproot system.

Differentiation A developmental process by which an unspecialized cell or tissue that has wide potential becomes progressively specialized, both functionally and structurally.

Diffusion (dī-fū'shun) The movement of molecules from an area where their concentration is high to one where it is lower.

Digestion The process in which food is broken down by enzymes in the digestive system into molecules which may be absorbed and utilized by cells.

Diploid (dip'loid) Containing two identical sets of chromosomes, a characteristic of somatic cells.

Directional selection Selection that acts to eliminate genes at one end of the gene pool spectrum while favoring genes at the other end.

Disruptive selection Selection that tends to divide the gene pool by favoring two or more adaptive types of genes.

DNA (Deoxyribose nucleic acid) (de-ok-si-rī'bōs nū-klē'ik) A nucleic acid that carries the chemical codes that determine the heredity of the cell.

Dorsal lip (dor'sal) The inductive tissue above the blastopore at the gastrula stage.

Drive Strong motivation that arises from a basic physiological need.

Duodenum (dū-o-dēn'um) The first portion of the small intestine, which receives chyme from the stomach and in which enyzmes and bile salts secreted by the pancreas and liver carry on intestinal digestion.

Echolocation (ek'ō-lō-kā'shun) The process by which an animal (e.g., a bat) can orient itself to its surroundings by emitting high-frequency sounds and interpreting their reflections.

Ecological niche (ē-kō-loj'i-kal nich) Designates the position or status of an organism within its community and ecosystem, resulting from the organism's structural adaptations, physiological responses, and specific behavior.

Ecological succession The orderly sequential transition of different communities over a period of time in some particular area.

Ecology (ē-kol'o-jē) The study of the relationships of organisms with each other and with their environment.

Ecosystem (ē-kō-sis'tem) Includes all the populations residing in a given area together with the nonliving physical (abiotic) environment.

Ectoderm (ek'tō-derm) The outermost layer of cells in the gastrula. It gives rise to the nervous system, skin, and sense organs.

Ectotherm (ek'tō-therm) An animal whose body heat comes in part from the environment. Also called poikilotherm.

Electron (e-lek'tron) A negatively charged particle that orbits the nucleus of an atom.

Element A substance that cannot be broken down by chemical reactions into substances of simpler composition.

Embryo sac (em'bri-ō) The female gametophyte of a seed plant.

Endergonic reaction (en-der-gon'ik) A chemical reaction in which energy is absorbed.

Endorcrine system (en'do-krin) The hormone-secreting organs of the body.

Endoderm (en'dō-derm) The inner layer of cells in the gastrula. It forms the primitive gut.

Endometrium (en-dō-mē'tri-um) The glandular, highly vascular lining of the uterus, which, in animals exhibiting a menstrual cycle, undergoes a cyclic hormone-induced proliferation and is then shed if fertilization does not occur.

Endoplasmic reticulum (en-dō-plaz'mik re-tik'yū-lum) An organelle located in the cytoplasm and composed of elaborate foldings of the unit membrane. Smooth ER is simply membrane; rough ER is lined with ribosomes and is the site of protein synthesis.

Endoskeleton An internal skeleton found in vertebrates, composed of bone and/or cartilage and moved by sets of opposing muscles.

Endosperm Tissue that provides nourishment for the developing plant embryo in a seed.

Endospore (en'dō-spōr) A spore formed within a parent cell.

Endotherm (en'dō-therm) An animal that can maintain a constant body temperature from within itself. Also called homeotherm.

Enzymatic hydrolysis (en-zī-mat′ik hī-drol′i-sis) The process accomplished by enzymes in which a large molecule is broken into its smaller component parts when water molecules are released from certain linkage points.

Enzyme (en′zīm) An organic catalyst.

Epicotyl (ep′i-kot-l) In plants, the beginning of the embryonic bud and leaves.

Epidermis (ep-i-der′mis) Protective tissue formed by the outermost layer of plant or animal cells.

Epiglottis (ep-i-glot′is) A flap that when closed prevents swallowed food from entering the trachea.

Epinephrine (adrenalin) (ep-i-nef′rin; a-dren′a-lin) Hormone secreted by the adrenal medulla, which stimulates certain body functions in stress response. Causes increased heart rate, blood pressure, flow of blood to brain, respiratory rate, etc.

Epistatic (ep-i-sta′tik) Characterizing a gene that is capable of suppressing the effect of another gene.

Epithelial cells (ep-i-thē′li-al) Animal cells that perform a protective and supportive function and are found on every surface of the body, both internal and external.

Erythrocyte (e-rith′rō-sīt) The red blood cell, which contains hemoglobin, the protein responsible for O_2 delivery and CO_2 removal.

Erythropoietin (e-rith-rō-poi′e-tin) Hormone formed in the blood by action of an enzyme produced in the kidney on a serum factor produced in the liver. The hormone stimulates the production of red blood cells and bone marrow, particularly during conditions of oxygen deficiency.

Esophagus (e-sof′a-gus) The tube connecting the pharynx to the stomach.

Estrogen (es′tro-jen) Female sex hormone secreted by the ovaries and adrenal cortex; affects growth and muscle tone in smooth muscles, as well as female secondary sex characteristics and normal reproductive functions.

Estuary (es′chu-ăr-ē) A river mouth where salt water and fresh water mix as the result of tidal action.

Ethology (ē-thoh′lo-jē) The study of animal behavior in a natural environment.

Ethylene (eth′i-lēn) A normal plant metabolite that acts as a hormone, inducing flowering, dropping of leaves, fruit ripening, and lateral growth.

Eucaryotic cell (yū-ka′ri-o-tik) Cell containing a nucleus and distinct organelles.

Euphotic zone (yū-foh′tik) In marine environments, that portion of both the neritic and oceanic habitats into which light penetrates.

Euploidy (yū′ploi-dē) The loss or gain of a complete set of chromosomes.

Exergonic reaction (eks-er-gon′ik) A chemical reaction in which energy is given off.

Exoskeleton A hinged external skeleton common to such invertebrates as arthropods, crustaceans, and molluscs.

Expressivity (eks-pres-iv′i-tē) The extent to which a trait governed by a gene manifests itself in an individual.

Extracellular digestion The breakdown of food that occurs outside individual cells.

Fallopian tube (fa-lōp′i-an) One of two oviducts that have an open end next to the ovary. The other end empties into the uterus.

Fat One of the four classes of lipids, composed of glycerol and fatty acids.

Feces (fē′sēz) The waste products of digestion, which include indigestible vegetable matter, dead intestinal bacteria, desquamated cells from the stomach mucosa, water, and bile pigment.

Feedback A regulatory process in which the output returns to the original system as input—for instance, when the concentration of an end product of a reaction serves to inhibit or stimulate the rate.

Fermentation The anaerobic breakdown of glucose into either lactic acid or ethyl alcohol and CO_2, with the attendant formation of ATP.

Fertilization The merging of sperm genetic material with ovum genetic material to produce a diploid zygote.

Fibrin (fī′brin) The insoluble fibrous protein, formed from fibrinogen, which creates a meshwork to catch blood cells during blood clotting at the site of injury.

Fibrinogen (fī-brin′o-jen) A protein in blood plasma that functions in the blood-clotting mechanism.

Fibrous root system (fī′brus) Several roots of roughly similar size that branch off into smaller side roots. Typical of monocots.

Fission (fish′un) A form of asexual reproduction in which a unicellular organism divides to produce two organisms.

Fixed action patterns Innate automatic behavior that involves a whole organism; formerly called instincts.

Flagellum (fla-jel′um) A long, whiplike organelle protruding from the posterior part of a unicellular organism, used in movement of the organism.

Flame cells Cells that perform an excretory function in planaria.

Florigen (flōr′i-jen) Plant hormone that induces or promotes flowering.

Fluorescence (flū-or-es′ens) Unused potential energy released as light when an electron falls back to its original level after excitation to a higher level in photosynthesis.

Follicle (fol′i-kl) One of many fluid-filled sacs in the ovary containing the developing egg cell; manufactures the steroid hormone estrogen.

Follicle-stimulating hormone (FSH) Released from the anterior pituitary under stimulus from the hypothalamus. It stimulates the development of the Graafian follicle.

Food chain The means by which food energy is transferred from organisms that produce their own food (autotrophs) to organisms that must acquire food from sources outside themselves (heterotrophs).

Food vacuole (vak′yū-ōl) A cellular structure found in organisms that carry on intracellular digestion. Food particles are encased within it, where they are broken down by lysosomes.

Food web Several interlocking food chains within a community or over several communities.

Gamete (ga-mēt′) A reproductive cell containing a haploid number of chromosomes.

Gametogenesis (ga-mēt-o-jen′e-sis) The specialization of haploid cells to form mature gametes.

Gametophyte generation (ga-mēt′o-fīt) The haploid stage of a plant life cycle during which gametes are produced by mitosis.

Gas absorption The process of absorbing gaseous compounds from air or water. Autotrophs absorb CO_2 and O_2; heterotrophs absorb O_2.

Gastrula (gas′tru-la) An embryo in the two-layered stage consisting of ectoderm and endoderm. In higher animals a third layer, the mesoderm, arises between the first two.

Gause's Principle of Competitive Exclusion Two populations with the same ecological niche cannot coexist; one will either be excluded or, to survive, will modify its requirements.

Gene (jēn) Unit of heredity.

Gene flow The transfer of genes from gene pool to gene pool by sexual reproduction.

Gene pool The sum of all the genes held by members of a breeding population.

Genetic drift (je-net′ik) Fluctuation in gene frequencies caused by chance acting in small populations.

Genitalia (je-ni-tāl′i-a) The internal and external organs of reproduction in both sexes.

Genome (jē′nōm) The sum of all the different genes in an organism.

Genotype (jē-no-tīp) The genetic makeup of an organism.

Geotropism (jē-ot′tro-pizm) The orientation of plant parts to the pull of gravity. Roots tend to point downward, demonstrating a positive geotropism, and stems to point upward, demonstrating a negative geotropism.

Germination The resumption of growth by a spore or seed after a quiescent period, characterized by an increased metabolic and enzymatic activity and increased H_2O absorption.

Gibberellins (ji-ber-el′inz) Plant growth hormones that promote cell elongation and help to reactivate metabolism in dormant plants.

Gill (gil) Specialized evaginated organ that accomplishes gas exchange in water-dwelling animals and is also involved in osmoregulation.

Globulin (glob′yū-lin) A protein in blood plasma, which may function as an antibody that protects the body from invading organisms.

Glomerulus (glo-mer′a-lus) Network of capillaries contiguous with the Bowman's capsule in a kidney nephron.

Glucagon (glū′ka-gon) Hormone secreted by the islets of Langerhans in the pancreas that stimulates the breakdown of glycogen to glucose in the liver.

Glucocorticoids (glū-kō-kor′ti-koidz) A group of hormones, secreted by the adrenal cortex, which regulate carbohydrate metabolism (derivation of glucose from protein, glycogen storage in the liver), help body adapt to stress, and minimize inflammations caused by infection or allergies.

Glucogenesis (glū-kō-gen′i-sis) Formation of glucose from carbohydrates.

Glucose (glū′kōs) A six-carbon monosaccharide ($C_6H_{12}O_6$) used by organisms to store food energy in an easily accessible form.

Glycemia (glī-sēm′i-a) Blood sugar level.

Glycemic regulation (glī-sēm′ik) Regulation of blood sugar level.

Glycogen (glī′ko-jen) A polysaccharide (carbohydrate) storage unit for monosaccharide glucose molecules in most animals and fungi. Hydrolysis converts the glycogen back into glucose.

Glycolysis (glī-koh′li-sis) The enzymatic breakdown of sugars into simpler compounds with release of energy.

Golgi apparatus (gōl′jē) An organelle of the cytoplasm consisting of foldings of unit membrane. The Golgi apparatus appears to be a collecting and processing center for lipids and proteins altering these substances to suit the needs of the cell. It may also be involved with cellular secretion.

Gonadotropins (go-nad-ō-trō′pinz) A group of hormones that regulate certain activities of the gonads.

Gonads (gō′nadz) The essential reproductive organs of multicellular animals in which gametes are formed. They also produce the male and female hormones—testosterone, progesterone, and estrogen.

Graafian follicle (grahf′i-an fol′i-kl) A mature ovarian follicle, containing an ovum.

Grana (grā′na) A grouping of photosynthetic membranes in a chloroplast.

Grooming Cleaning an animal's coat or plumage to get rid of parasites and dirt.

Guard cells Kidney-shaped cells that surround the stomata and regulate the width of stomata openings.

Gymnosperm (jim′nō-sperm) A vascular plant that produces naked seeds.

Habitat (hab′i-tat) The specific place where a particular organism lives.

Habituation (ha-bich-yū-ā′shun) The phenomenon in which an organism becomes accustomed to a certain stimulus, so that the stimulus ceases to evoke any response.

Haemocytes (hē′mo-sītz) Cells in the blood of lower animals (e.g., insects) that help to seal wounds and defend organisms from invading parasites.

Haploid (hap′loid) Containing one set of chromosomes, half the chromosomes typically found in somatic cells.

Hematopoiesis (hē-mat-a-poi-ē′sis) The formation of red blood cells in the red marrow of the bone. Also called hemopoiesis.

Hemocyanin (hē-mo-sī′a-nin) Protein that carries oxygen in the blood of certain invertebrates.

Hemoglobin (hē′mo-glō-bin) The protein in erythrocytes that carries O_2 to cells and helps to remove their waste CO_2.

Hemolysis (hē-mol′i-sis) The destruction of red blood cells with release of hemoglobin.

Hepatic portal system (he-pat′ik por′tal) Series of blood capillaries and vessels that carries nutrients from the capillary bed of the intestines to the capillary bed of the liver.

Herbaceous (er-bāsh′us) Relating to a plant without woody tissue.

Herbivore (er′bi-vōr) An animal that eats only plants.

Hermaphroditic (her-ma-frō-dit′ik) Possessing both male and female gamete-producing organs.

Heterogamous (het-er-oh′ga-mus) Producing unlike gametes.

Heterospory (het′er-ō-spōr-ē) Production of more than one kind of spore.

Heterotroph (het′er-ō-trōf) An organism that cannot manufacture its own food and must obtain nourishment by consuming other organisms or particulate organic matter.

Heterozygote superiority (het-er-ō-zī′gōt) Phenomenon in which heterozygotes are more fit than homozygous individuals.

Heterozygous (het-er-ō-zī′gus) Having two different alleles for a given characteristic.

Homeostasis (hō-mē-ō-stā′sis) The maintenance of optimal internal living conditions through continuous adjustment of life processes.

Homeotherm (hō′mē-ō-therm) An animal that maintains a constant body temperature. Also called endotherm.

Homing instinct The ability of an organism to return to an area from which it has been removed.

Homolog (hōm′o-log) One of a pair of chromosomes having nearly identical patterns of gene arrangement.

Homospory (hō′mō-spōr-ē) Production of only one kind of spore.

Homozygous (hō-mō-zī′gus) Having two like genes for a given characteristic.

Hormone (hōr′mōn) A substance produced in minute quantities in one part of the body and transported to another region where it produces its effects.

Human chorionic gonadotropin (HCG) (kō-ri-on′ik go-nad-ō-trō′pin) Hormone secreted by the human placenta in early pregnancy that stimulates the ovaries to maintain high output of female hormones.

Hybrid polyploidy (hī′brid pol′i-ploi-dē) Hybridization followed by duplication of chromosome sets.

Hybridization Generation of a new species by the interbreeding of two distinct species.

Hydrogen bond Weak electrostatic attraction between hydrogen atoms in molecules.

Hydrolysis (hī-drol′i-sis) The decomposition of a chemical substance by the introduction of a molecule of water; the reverse of condensation.

Hydrostatic skeleton (hī-drō-stak′ik) A mass of body fluid, as in the earthworm, that functions as a structural unit manipulated by muscles.

Hydroxl ion (hī-drok′sl) The OH- ion.

Hypertonic (hī-per-ton′ik) Containing more nonpenetrating particles on one side of a semipermeable membrane than on the other side and thus exerting more osmotic pull.

Hypocotyl (hī′pō-kot-l) In plants, the part of the embryonic axis that will develop into part of the primary root or stem.

Hypothalamus (hī-pō-thal′a-mus) The floor and sides of the brain beneath the cerebral hemispheres. It is the primary neurosecretory center of the brain and regulates hormonal concentrations throughout the body. Master control for the autonomic nervous system.

Hypotonic (hī-pō-ton′ik) Containing fewer nonpenetrating particles on one side of a semipermeable membrane than on the other side and thus exerting less osmotic pull.

Imprinting A special form of learning that occurs at a particular phase in early life and results in the development of a strong, stable response to certain stimuli.

Incomplete dominance Heterozygous genotype results in a phenotype that is a blending of allele characteristics.

Induction (1) A developmental process by which undifferentiated cells are chemically or physically induced by their microenvironment to become differentiated along a specific pathway. (2) The stimulation of structural gene activity by the presence of certain substances in the substrate.

Insight learning The ability to solve problems by applying past learning to a new situation.

Instinct *See* **Fixed action pattern.**

Insulin (in′su-lin) Hormone secreted by the islets of Langerhans in the pancreas. Regulates the metabolism of carbohydrates and stimulates the liver to store glucose in the form of glycogen.

Interkinesis (in-ter-ki-nē′sis) Period between the first and second divisions of meiosis. It is similar to interphase in mitosis, but no genetic replication occurs.

Intermediate lobe of the pituitary (pi-tū′i-tā-rē) Sandwiched between the anterior and posterior lobes, this section secretes MSH, which stimulates the dispersal of pigment granules within pigment-containing skin cells in amphibians and reptiles.

Internode The area between two successive nodes of a plant stem, from which cell elongation results in stem growth.

Interphase Period between mitotic divisions in which chromosomes are duplicated.

Intracellular digestion The breakdown of food particles that occurs within a cell, a method generally employed by unicellular organisms like the amoeba. Food is surrounded and drawn into cells in a food vacuole and then broken down by enzymes secreted into the vacuole by the cell.

Invagination (in-vaj-i-nā′shun) An infolding that encloses a hollow space or pocketlike cavity within a solid structure.

Inversion Reversal of the sequence of genes on a chromosome.

Ion (ī′on) Electrically charged atom or molecular group.

Ionic bond (ī-on′ik) A chemical bond that results from an electrostatic attraction between oppositely charged atoms or molecular groups.

Islets of Langerhans (ī′letz lahng′er-hahnts) Cells located in the pancreas that secrete hormones (insulin, glucagon) that regulate the blood sugar level by controlling the amount of glucose stored as glycogen.

Isogamous (ī-sohg′a-mus) Producing undifferentiated gametes.

Isomer (ī′so-mer) A compound that has the same molecular formula as one or more other compounds but that differs from the others in physical structure.

Isotonic (ī-so-ton′ik) Containing equal amounts of nonpenetrating particles on either side of a semipermeable membrane.

Isotopes (ī′so-tōpz) Substances whose atomic nuclei contain the same number of protons but a different number of neutrons.

Kidney Excretory organ that filters out the blood metabolic wastes and excess fluids and helps to regulate blood pH and salt concentration.

Kinesthetic sense (kin-es-thet′ik) A proprioceptive system monitoring posture, movement, and maintenance of equilibrium.

Kinetic energy (ki-net′ik) The energy of motion.

Kinetin (kī-ne-tin) The most important of the cytokinins.

Kinetochore (ki-net′o-kōr) Location on chromosomes where chromatids are joined and spindle fibers attach.

Krebs cycle (krebs sī′kl) The oxidation of pyruvic acid with the attendant formation of CO_2 and large amounts of ATP.

Lamina (lam′i-na) The blade of a leaf or the leaf itself.

Lateral line system A receptor system in fish consisting of specialized hair cells inside a canal-like depression running longitudinally down the side of the body.

Law of mass action The rate of a chemical reaction is directly proportional to the molecular concentrations of the reacting compounds.

Laws of thermodynamics (ther-mō-dī-nam′iks) (1) In a closed system, the total amount of energy remains constant;

energy can be neither created nor destroyed. (2) The amount of *usable* energy is always decreasing.

Lenticel (len'ti-sel) Area of loosely packed parenchyma cells where gas exchange occurs in bark-covered stems.

Leucocyte (lū'ko-sīt) White blood cell, which helps to protect the body from infecting agents such as bacteria.

Lignin (lig'nin) A waterproof material secreted by xylem cells.

Limiting factor Any abiotic factor within the ecosystem that tends to deter growth of a population.

Limnetic zone (lim-net'ik) In freshwater environments, the areas beyond the littoral zone to the depth of light penetration.

Linkage A condition in which several genes, located on the same chromosome, tend to be inherited as a group.

Lipase (lī'pās) An enzyme that breaks down lipids by enzymatic hydrolysis.

Lipids (li'pidz) A group of organic compounds, insoluble in water, that includes fats, phospholipids, waxes, and steroids.

Lipogenesis (lī-pō-jen'e-sis) The formation of fat from glucose.

Littoral zone (lit'or-al) The bottom and shallow-water areas of freshwater environments.

Liver Largest organ of the body. It performs homeostatic and secretory functions: secretes bile, regulates glucose level in blood, removes poisons and dead or injured erythrocytes from blood, deaminates excess amino acids.

Loop of Henle (hen'lē) Long loop of nephron tubule.

Lumen (lū'men) Inner channel in tracheidlike cell of xylem tissue, which provides plant support.

Lungs Invaginated internal membranous sacs where gas exchange is accomplished in higher animals.

Luteinizing hormone (LH) (lū'ten-īz-ing) Released from the anterior pituitary by stimulus from the hypothalamus. Its presence in the ovary results in ovulation and stimulation of the corpus luteum.

Lymph (limf) Clear tissue fluid resembling blood plasma. Consists of fluids that filter through the walls of the blood capillaries.

Lymph nodes Masses of tissue that filter out undesirable components of tissue fluid and produce lymphocytes.

Lymph vessels Tubes that conduct animal tissue fluids to blood.

Lymphatic system (lim-fat'ik) Series of capillaries and vessels through which lymph is carried.

Lysosome (lī-so-sōm) A saclike organelle containing digestive enzymes.

Malpighian tubules (mal-pig'i-an tū'byūlz) Excretory structures in insects.

Mammary glands (mam'a-rē) Milk-producing glands found only in mammals.

Marsupials (mar-sū'pi-alz) Group of mammals with brief intrauterine development. They deliver their young from the uterus while still relatively young to a pouchlike unfolding of the mother's skin for further development.

Medulla (me-dul'a) (1) The medulla oblongata, a portion of the hindbrain in vertebrates, connected with the spinal cord. (2) The inner part of such organs as the kidneys or endocrine glands.

Meiosis (mī-ō'sis) Nuclear division, characteristic of gamete formation, in which the number of chromosomes in the daughter cells is half the number in the original cell.

Menstruation A periodic shedding of the endometrium of the uterus in some mammals.

Meristem cells (mer'i-stem) Undifferentiated cells in a plant, which are responsible for growth of new tissues. Found in stems, leaves, roots, and flowers.

Mesentery (mez'en-ter-ē) A thin sheet of mesoderm connecting the enteron with the body wall and containing neural and vascular elements.

Mesoderm (mez'ō-derm) The cell layer between the ectoderm and the endoderm in the embryo of all animals above the level of the coelenterates. It gives rise to the skeleton, muscle, circulatory, and excretory systems, and to most of the reproductive system.

Messenger RNA (mRNA) RNA made from the DNA template that carries the code of protein synthesis from nucleus to cytoplasm, where synthesis occurs.

Metabolism (me-tab'o-lizm) The sum of physical and chemical processes by which living organisms are produced and maintained, and the transformation by which energy is made available to organisms.

Metamorphosis (met-a-mōr'fo-sis) A developmental process common to insects and amphibians characterized by an abrupt, hormonally induced and regulated transformation from a larval to an adult form.

Metaphase (met'a-fāz) The phase of mitosis in which chromosomes line up at the cell equator and become attached to spindle fibers, and the kinetochores divide. In the first metaphase of meiosis the kinetochores do not divide, forming haploid sets of chromosomes; in second metaphase the kinetochores divide to create single-stranded chromosomes.

Micron (μ) (mī′kron) Unit of length equal to one thousandth of a millimeter.

Micronutrients (mī-krō-nū′tri-entz) Elements that are necessary for life but required by organisms only in minute quantities.

Migration The seasonal or time-related movement of animals to and from a specific area.

Mineralocorticoids (min-e-ral-o-kōr′ti-koids) A group of hormones secreted by the adrenal cortex that maintain proper electrolytic balance in the blood.

Mitochondrion (mī-tō-kon′dri-on) A cytoplasmic organelle, which is the site of aerobic respiration.

Mitosis (mī-tō′sis) Nuclear division, characteristic of somatic cells, in which the number of chromosomes in the daughter cells is the same as that in the original cell.

Mitral valve (mī′tral) Heart valve between the left atrium and ventricle.

Molecule (mol′e-kyūl) Two or more atoms of the same or different elements joined together.

Molt A developmental process consisting of the hormonally induced shedding of an exoskeleton as the organism grows to maturity.

Monocot (mon′o-kot) A plant whose seed contains one cotyledon and that characteristically has a fibrous root system.

Monocyte (mon′o-sīt) A leucocyte without cytoplasmic granules.

Monomer (mon′o-mer) A basic chemical unit, which when linked together with other similar units results in long repetitive chain compounds called polymers.

Monosaccharide (mon-ō-sak′a-rīd) A class of six-carbon sugars having the molecular formula $C_6H_{12}O_6$.

Monotremes (mon′o-trēmz) Certain primitive mammals that still lay eggs.

Morphogenesis (mō-fō-jen′e-sis) The development of the characteristic form and structure of a cell or an organism.

Motor neuron (nū′ron) A neuron whose cell body is located in the ventral portion of the gray matter of the spinal cord; transmits excitation directly to an effector.

Motor unit Collectively refers to the muscle fibers and the motor neuron that supplies them.

Mullerian mimicry (mul-ir′i-an mim′ik-rē) Mutual coloration imitation by two aposematic organisms.

Muscle A tissue specialized for the production of movement by a contraction or shortening of specifically organized arrays of contractile proteins in the cells making up its fibers. Muscle tissue is found in sheets or bundles.

Mutation (myū-tā′shun) Random change in genetic material, which may occur on chromosomal or molecular level.

Mutation pressure The effect of mutation in changing gene pool frequencies.

Mutational load The sum of the unfavorable mutations in a gene pool not yet eliminated by natural selection.

Mutualism (myū′tyū-al-izm) An interspecific interaction in which each organism is dependent on the other for survival.

Myelin (mī′e-lin) A fatty material that surrounds the axons of neurons in the central nervous system of vertebrates.

Myocardium (mī-ō-kar′di-um) Muscular wall of the heart.

Myofibrils (mī-ō-fī-brilz) The longitudinal protein contractile filaments in muscle cells, composed of the proteins myosin and actin.

Myometrium (mī-ō-mē-tri-um) The muscular layer of the uterus.

Myxedema (mik-se-dē′ma) Condition caused by insufficient level of thyroid hormone in the adult; characterized by weight gain, coarsened hair, thickened and furrowed tongue, puffy features, and general slowdown of physical and mental processes.

NADP (nicotinamide adenine dinucleotide phosphate) One of the most important of the substances that act as electron carriers in photosynthesis.

Nastic movement A response of a plant to a stimulus (e.g., light, heat) in which the movement is oriented relative to the plant body, not to the stimulus.

Natural selection The struggle for survival that eliminates the less fit and allows only the strongest, best-adapted organisms to survive and breed.

Negative pressure Vacuum or low air pressure.

Nephridium (nef-rid′i-um) Excretory structure in simple land animals like the earthworm, consisting of a coiled tubule embedded in capillaries.

Nephron (nef′ron) The basic filtering unit in the kidney, consisting of a glomerulus, Bowman's capsule, two convoluted tubule portions, and the loop of Henle.

Nephrostome (nef′ro-stōm) The opening of a nephridium through which tissue fluid enters.

Neritic habitat (ne-rit′ik) In marine environments, a shallow-water area on the continental shelf.

Nerve A group or bundle of neuron fibers.

Nerve net A network of radiating nerve cells without any differentiation between axons and dendrites and lacking any integrative or coordinative ability.

Neuron (nū'ron) A nerve cell, including the cell body, its dentritic projections, and axons.

Neurulation (nū-ru-lā'shun) The formation of the nervous system in vertebrates. It is a process by which surface ectoderm is induced by the underlying mesoderm to fold up, form a tube, invaginate beneath the surface, and then evaginate to form the CNS as development continues.

Neutralism (nū'tral-izm) An interspecific interaction in which neither population is affected because of considerable differences in the life processes of each population.

Neutron (nū'tron) A heavy particle with no charge, which may be found in the atomic nucleus of certain elements.

Neutrophil (nū'tro-fil) A leucocyte with cytoplasmic granules.

Niche (nich) The biological role played by a species in its community, which includes where it lives, what it eats, what preys on it, etc.

Nitrification (nī-tri-fi-kā'shun) Process in which ammonium ions are oxidized by microorganisms into nitrate or nitrite, forms that can be assimilated by plants.

Nitrogen fixation (nī'tro-jen fik-sā'shun) Conversion of atmospheric nitrogen into organic nitrogen compounds available to green plants. This is a process that can be carried out only by certain soil bacteria.

Nonadaptive traits Characteristics with no apparent adaptive value.

Norepinephrine (nōr-ep-i-nef'rin) Hormone secreted by the adrenal medulla, which stimulates certain body functions in response to stress. Causes increased heart rate, blood pressure, flow of blood to the brain, respiratory rate, etc.

Nucleic acids (nū-klē'ik) A class of molecules that includes ribonucleic acid (RNA) and deoxyribonucleic acid (DNA) and that consists of long polymers of ribo- and deoxyribonucleotides. They are found primarily in the cell nucleus and play a role in the transmission of hereditary characteristics, the control of cellular activities, and protein synthesis.

Nucleolus (nū-klē'o-lus) A dense ovoid body within the nucleus, composed largely of RNA and thought to be the site of RNA synthesis.

Nucleotide (nū'klē-o-tīd) Consisting of a nitrogen-containing base (a purine or pyrimidine), a sugar (ribose or deoxyribose) and a phosphate, this class of molecule is involved in the transfer of energy within the cell and the formation of nucleic acids.

Nucleus (nū'klē-us) (1) Structure in all eucaryotic cells containing the chromatin. (2) The central body in an atom, which may be composed of protons and neutrons and about which electrons orbit.

Nutrition The processes by which an organism obtains and assimilates the substances necessary for life.

Oceanic habitat In marine environments, an area beyond the continental shelf.

Ommatidium (om-a-tid'i-um) A functionally independent photoreceptor unit of a compound eye.

Omnivore (om'ni-vōr) A heterotroph that eats both plant and flesh.

Oogenesis (ō-a-jen'e-sis) Formation of the ova (egg cells) in the ovary.

Operator gene Gene that acts to switch on or off the activity of a structural gene.

Operon (op'er-on) A group of structural genes and the operator gene that controls them.

Organ A grouping of specialized tissues working in concert.

Organelle (or-gan-el') One of a number of microscopic intracellular structures, such as mitochondria or chloroplasts, which are specialized to perform specific roles in cell functioning.

Organic compounds (or-gan'ik) Substances that are composed primarily of carbon, hydrogen, oxygen, and nitrogen and that in nature are produced only by living things.

Osmosis (oz-mō'sis) The diffusion of water molecules through a semipermeable membrane. The water molecules move from an area of higher concentration to an area of lower concentration of water.

Osteoblast (os'tē-ō-blast) A connective cell in bone that secretes collagen fibers and matures into an osteocyte.

Ovary (ō'va-rē) Female reproductive organ that produces hormones and ova.

Oviduct (ō'vi-dukt) The pathway (fallopian tube) in which the ovum travels on its way to the uterus.

Ovulation (ov-yū-lā'shun) A cyclic, hormonally induced release of a mature egg into the oviduct.

Ovum (ō'vum) The unfertilized egg.

Oxidation The loss by an atom of one or more electrons to another atom that is thereby reduced. Upon oxidation, the atom has an increased positive valence.

Oxygen debt The accumulation of lactic acid in muscles, resulting from the fact that during intense physical activity the body resorts to glycolysis to supplement the energy obtained from aerobic respiration.

Oxytocin (ok-sē-tōs'in) Hormone secreted by posterior pituitary, which stimulates contractions of the uterus and mammary glands.

Pacemaker Tissue in the muscles of the heart that controls heartbeat rate.

Pacinian corpuscles (pa-sin′i-an kōr′pus-lz) Pressure receptors, onionlike in shape, located in the deeper layers of the skin and in many internal organs.

Palisade mesophyll (pal′i-sād mez′o-fil) Column-shaped cells, closely packed, that form the upper layer of mesophyll cells in a leaf.

Pangenesis (pan-jen′e-sis) An early theory of inheritance.

Parasitism (par′a-sit-izm) A form of symbiosis in which one organism, the parasite, obtains shelter and/or nourishment at the expense of another, the host.

Parasympathetic nervous system (par-a-sim-pa-thet′ik) A subdivision of the autonomic nervous system, regulating the homeostatic mechanisms by stimulating digestion and inhibiting other functions.

Parathyroid glands (par-a-thī′roid) Glands embedded in the rear surface of the thyroid gland. They secrete a hormone (parathyroid hormone) that tends to raise the level of calcium in the blood.

Parathyroid hormone (PTH) Hormone secreted by the parathyroid glands, which stimulates bone osteoclasts to release calcium from bone and which stimulates the kidneys to reabsorb excreted calcium.

Parenchyma (pa-ren′ki-ma) Thin-walled, relatively undifferentiated cells that make up the soft parts of plants.

Parthenogenesis (par-then-o-jen′e-sis) Development of an organism from an unfertilized egg.

Pathogen (path′o-jen) A disease-causing organism.

Pecking order The hierarchy of dominance in poultry flocks, in which each individual according to its rank can peck subordinate fowl and in turn must submit without retaliation to pecking by however many fowl are superior to it.

Penetrance (pen′e-trans) The ability of a gene to express itself in conjunction with other genetic factors.

Penis (pē′nis) The male organ of the external genitalia that swells with blood during sexual excitement.

Peptide bond (pep′tīd) The bond formed between carbon and nitrogen atoms that links amino acids together into polypeptides.

Peristalsis (per-i-stol′sis) Wavelike muscle contractions in the walls of esophagus, stomach, and small and large intestines, which propel food along the alimentary canal.

Petiole (pet′i-ōl) The stem of the leaf.

PGAL (phosphoglyceraldehyde) (fos-fō-glis-er-al′de-hīd) The simple carbohydrate that is produced during the dark phase of photosynthesis.

pH A symbol used to express the acidity of a compound. A pH of 7 is neutral; higher values are basic (alkaline), and lower values are acidic.

Phagocytosis (fag-ō-sī-tō′sis) The process by which large particles are enveloped in a vacuole and drawn into the cell.

Pharynx (far′inks) The part of the digestive tract connecting the oral cavity to the esophagus. In vertebrates it is also part of the respiratory passage.

Phenotype (fē′nō-tīp) The somatic manifestation of a genotype.

Pheromones (fer′o-mōnz) Substances similar to hormones that are secreted in trace amounts by certain organisms and that serve as a means of identification and communication between individuals, usually of the same species.

Phloem (flō′em) Specialized plant tissue through which plant nutrients are carried up and down the plant.

Phospholipid (fos-fō-lip′id) Kind of lipid found in blood plasma and in cellular membranes.

Phosphorylation (fos-for-i-lā′shun) The addition of phosphate to a molecule.

Photolysis (fō-tol′i-sis) Decomposition resulting from the action of light.

Photon (fō′ton) Unit of light energy.

Photoperiodism (fō-tō-pēr′i-od-izm) The phenomenon of an organism's response to light periods of variable duration. Usually reproductive in function.

Photosynthesis (fō-tō-sin′the-sis) The process in which light energy is captured by chlorophyll and used to form glucose and oxygen from carbon dioxide and water.

Phototaxis fō-tō-tak′sis) Behavioral response to light.

Phototropism (fō-toh′tro-pizm) Movement of plant leaves and stems toward a light source.

Phytochrome (fī′to-krōm) Protein-pigment complex in plant leaves, which is light sensitive and plays a role in enzyme formation for protein synthesis and plant flowering.

Pineal body (pineal gland) (pī′nē-al) Structure located in the center of brain, which during evolutionary history may once have been a light-sensitive regulator of body cycles; releases the hormone melatonin.

Pinocytosis (pin-o-sī-tō′sis) The process conducted by a cell in which liquid or small particles are enveloped in a vacuole and brought into the cell.

Pioneer community The first biotic component to become established on a particular substrate on which no living matter has previously lived.

Pistil (pis′tl) The female spore-forming organ of a flower.

Placenta (pla-sen′ta) A temporary reproductive organ of most mammals, by which the embryo is attached to the wall of the uterus and through which homeostatic exchanges take place between mother and embryo.

Plankton (plank′ton) Floating or weakly swimming small plant and animal life of a body of water.

Plaque (plak) Fatty deposits in blood vessels.

Plasma (plaz′ma) Amber fluid in which are suspended the solid blood components—red cells, white cells, and platelets.

Plasmolysis (plaz-moh′li-sīs) The shrinkage in plant cells of the cytoplasmic contents due to water loss.

Plastid (plas′tid) A relatively large organelle found in plant cells, which functions in photosynthesis or nutrient storage.

Platelet (plāt′let) Nonnucleated, disc-shaped element of blood that functions in the clotting of blood.

Pneumatophore (nū-mat′o-fōr) Specialized tube for gas absorption that extends from a root to above the surface of the soil.

Poikilotherm (poi-kē′lō-therm) An animal whose body temperature varies with temperature of the environment. Also called ectotherm.

Point mutation Random change in genetic material on the gene level. Such changes include addition, deletion, or substitution of nucleotide(s) in a DNA or RNA molecule.

Polar bodies Small cells produced during meiotic divisions of oogenesis in ovaries.

Polar charges Opposite and equal electrostatic charges on different sides or ends of a molecule.

Pollen (pol′en) The male spore of seed plants; gives rise to the male gametophyte.

Pollen tube The male gametophyte of seed plants.

Polypeptide (pol-i-pep′tīd) A chain of amino acids linked together by peptide bonds.

Population In ecology, any group of individuals of one species.

Portal system Vascular pathways with capillaries at both ends; for instance, the hepatic portal system from intestines to liver.

Positive pressure High air pressure.

Posterior pituitary (pō-ster′i-or pi-tū′i-tā-rē) Posterior lobe of the pituitary gland located at the base of the brain; secretes hormones (vasopressin, oxytocin) that regulate water reabsorption in the kidneys and contraction of smooth muscles in uterus and mammary glands.

Postzygotic isolating mechanisms (post-zī-gōt′ik) Mechanisms that reproductively isolate species by damaging hybrid viability and fertility.

Precapillary sphincters (pre-cap′i-lā-rē sfink′ters) Muscles in walls of arterioles that help to regulate blood pressure and blood distribution within the body.

Predation (pre-dā′shun) An interspecific relationship in which one population eats members of another and is thus dependent on its victims for survival.

Prezygotic isolating mechanisms (prē-zī-gōt′ik) Mechanisms that reproductively isolate species by preventing mating and fertilization.

Procaryotic cell (prō-ka′ri-o-tik) A cell, such as a bacterium or blue-green alga, which contains no organelles or nucleus.

Profundal zone (prō-fun′dal) In freshwater environments, the bottom and the water areas below the depth of light penetration.

Progesterone (prō-jes′ti-rōn) A steroid hormone, produced in the corpus luteum, that inhibits the sloughing of the endometrium and thus plays an important role in maintaining pregnancy.

Prophase (prō′fāz) The first phase of mitosis, in which chromosomes become visible and spindle formation occurs. In meiosis, the first prophase includes synapsis or the pairing of homologous chromosomes. In the second prophase, no synapsis can occur because the cells are haploid.

Proprioceptors (prō-pri-ō-sep′tōrz) Specialized sensory cells that receive stimuli from internal organs and systems and transmit them to the brain so that they can be monitored and homeostasis maintained.

Proteases (prō-tē-ās′ez) Enzymes that break down proteins by enzymatic hydrolysis.

Protein (prō′tēn) Molecule formed by linkage and complex folding of polypeptide chains.

Proton (prō′ton) A positively charged particle located in the nucleus of an atom.

Protostome (prō′tō-stōm) Any coelomate animal whose embryonic blastopore becomes its mouth.

Pseudopodium (sū-dō-pō′di-um) "False foot," used for locomotion in organisms like the amoeba.

Pulmonary artery (pul′mon-ār-ē ar′te-rē) Artery that takes unoxygenated blood to the lungs.

Pulmonary circulatory system (pul′mon-ār-ē ser′kū-la-to-rē) The blood vessels going to, from, and within the lungs.

Pulmonary vein (pul′mon-ār-ē vān) Vein that channels oxygenated blood from lungs.

Pyloric sphincter (pī-lor′ik sfink′ter) A muscle at the bottom of the stomach that acts as a valve regulating the passage of chyme from the stomach to the small intestine.

Pyramid of biomass (bī'ō-mas) Measures the amount of living material present in any ecological system, based on total dry weight or caloric value in different trophic levels.

Pyramid of energy Shows energy relationships between various trophic levels of a particular food chain. Plants are at the bottom of the pyramid and thus represent the utilization of the greatest amount of energy; herbivores are next, then carnivores and omnivores.

Pyramid of numbers Shows the number of organisms at each trophic level of an ecosystem, to suggest the structure of the system.

Pyruvic acid (pī-rū'vik) An organic acid that occupies a key position in the metabolism of carbohydrates.

Radicle (rad'i-kl) In plants, the embryonic part of the seed-plant embryo that will develop into the primary root.

Receptor (rē-sep'tor) A structure specialized to transduce some particular kind of stimulus into a decipherable electrical impulse.

Recombination The union of genes from two parents as a result of sexual reproduction and the gene combinations formed by this process.

Reduction The gain by an atom of one or more electrons from another atom that is thereby oxidized. The reduced atom has a decreased positive valence.

Reflex A response to a specific stimulus. A reflex may be simple or complex, unconditioned or conditioned.

Reflex arc A series of neurons transmitting an excitation from a receptor through the central nervous system to an effector. The simplest arc is monosynaptic, since only two neurons are involved—a sensory and a motor neuron. Arcs increase in complexity with the number and type of interneurons between the receptor and motor neurons.

Refractory period (re-frak'to-rē) Period of approximately one millisecond during which a neuron is incapable of responding to another stimulus regardless of its strength.

Regeneration The replacement of lost or injured tissue, permitted by the ability of some cells to dedifferentiate and develop in a new way.

Regulator gene Gene that regulates the activity of a structural gene according to needs of organism.

Releaser External stimulus that elicits an instinctive response.

Releasing factors Hormones secreted by the hypothalamus that stimulate or inhibit the production of specific pituitary hormones.

Renin (rē'nin) Hormone secreted by the kidneys that stimulates the adrenal cortex's synthesis and secretion of aldosterone and affects blood pressure.

Repression Repression of structural-gene activity in the presence of a certain concentration of a compound in the cytoplasm.

Respiration The processes by which a cell releases the energy in glucose, producing ATP. These processes include aerobic respiration (glycolysis, Krebs cycle, and electron and hydrogen transport) and anaerobic respiration, or fermentation. Also refers to the intake of O_2 and release of CO_2 in breathing.

Reverse mutation The change of a mutant gene back to its original form.

Rhizoids (rī'zoidz) Simple rootlike structures specialized to absorb water and nutrients from soil.

Rhizome (rī'zōm) Underground horizontal stem.

Rhodopsin (rō-dop'sin) A visual pigment contained in the rod cells of the human retina.

Ribosomes (rī'bo-sōmz) Organelles composed largely of RNA found in association with the endoplasmic reticulum or free in the cytoplasm. Ribosomes play an important part in the synthesis of proteins from amino acids.

RNA (ribonucleic acid) (rī-bō-nū-klē'ik) A nucleic acid that is important in the synthesis of proteins (*see* **Messenger RNA** *and* **Transfer RNA**).

Root The part of a vascular plant that usually responds to geotropic and hydrotropic stimuli and thus grows downward, anchoring the plant and absorbing water and nutrients.

Root hairs Tiny extensions of root epidermal cells, which increase root surface area, thereby increasing the root's efficiency in absorbing water.

Runner An above-ground horizontal stem by means of which some plants (e.g., strawberry) reproduce asexually.

Salmonella (sal-mon-el'a) A bacterium that, when consumed in contaminated foods, may cause diarrhea, nausea, and vomiting.

Saprophytes (sap'rō-fītz) Organisms that sustain themselves on nonliving organic matter.

Sarcolemma (sar-kō-lem'a) A thin membrane surrounding a muscle fiber.

Sclerenchyma (skle-ren'ki-ma) Fibers that make up part of angiosperm xylem tissue and provide support.

Secretin (si-krē'tin) A hormone produced by cells in the duodenum that stimulates the pancreas to secrete digestive hormones.

Seed An embryo sporophyte surrounded by the endosperm and enveloping layers derived from the ovule of the parent sporophyte.

Selection pressure Environmental force that changes gene pool frequencies by eliminating nonadaptive traits and their genes.

Seminiferous tubules (sim-i-nif′er-us tū-byūls) Highly convoluted tubes in the testis within which spermatogenesis takes place.

Septum (sep′tum) Partition of muscle that divides the right and left sides of the heart.

Seral stage (sir′al) One of a series of transient communities that lead to a climax community and collectively form a sere.

Sere (sēr) A succession of communities, each characterized by specific flora and fauna, which lead up to and include a climax community.

Sessile (ses′il) Attached to the ocean bottom.

Sex-influenced gene Gene that may be carried on any chromosome that is recessive in one sex and dominant in the other.

Sex-limited genes Genes that may be carried on any chromosome but that are normally expressed in only one sex.

Sex-linked genes Genes located on the X chromosome.

Sex steroids (stir′oidz) A group of hormones secreted by the gonads and adrenal cortex, including male (testosterone), and female (estrogen, progesterone), plus other hormones that have the effects of a weaker form of testosterone.

Sexual reproduction The union of male and female gametes to form a zygote. It usually involves two partners, but self-fertilization is practiced by some lower organisms.

Sickle-cell anemia (sik′l-sel a-nē′mi-a). An inherited blood disease. Red blood cells contain faulty hemogobin, which has reduced oxygen-carrying capacity.

Sieve plates (siv plātz) Groupings of pores in the end walls of sieve tubes.

Sieve tube Long, thin-walled cells active in food transport in gymnosperm and angiosperm phloem tissue.

Simple reflex Innate automatic response involving stimulus to a single reflex arc.

Sinus (sī′nus) Indistinct channels in the circulatory system where blood may pool.

Small intestine (in-tes′tin) The long, narrow, muscular tube between the stomach and the colon in which digestion is completed and nutrients are absorbed.

Smog Type of air pollution consisting of smoke and fog, the particulate portion often reacting due to the catalytic action of sunlight.

Solution A liquid mixture in which ions or molecules of the dissolved substance (solute) are distributed evenly among the molecules of the dissolving substance (solvent).

Somatic mutation (so-mat′ik) Mutation in a body as opposed to a germ cell.

Speciation (spē-shē-ā′shun) The generation of new species.

Species (spē′shēz) All similar organisms that occupy the same or contiguous regions, live at the same geologic time, and reproduce sexually with each other but with no other types or organisms.

Sperm The highly motile male gamete in animals.

Spermatogenesis (sper-mat-o-jen′e-sis) The process by which spermatogonia develop into mature sperm.

Spindle Organelle formed of fibrils between two centrioles.

Spiracles (spir′a-klz) External pores through which gas enters and leaves the internal tracheae in insects.

Spongy mesophyll (mez′ō-fil) Loosely packed, irregularly shaped cells that form the lower layer of mesophyll cells in a leaf.

Sporangium (spō-ran′ji-um) A specialized structure in which spores are produced.

Spore An asexually produced, single-cell reproductive unit protected from adverse environmental conditions by an external capsule that enables the organism to remain dormant for long periods of time.

Sporophyte generation (spōr′ō-fīt) The stage of plant reproduction in which spores are produced by meiosis.

Sporulation (spōr-yū-lā′shun) Spore formation, a form of asexual reproduction.

Stabilizing selection Selection that generally maintains constant gene pool frequencies.

Steroids (stir′oidz) One of the four classes of lipids.

Stoma (stō′ma) An opening in the epidermis of a leaf through which gases pass between the mesophyll and external environment.

Stomach The muscular sac that stores food, reduces food to a liquid with hydrochloric acid, and starts the digestion of protein with the secretion of pepsin.

Striated muscle (strī′ā-ted) Voluntary skeletal or involuntary cardiac muscle in vertebrates.

Structural gene Gene that contains the blueprint for a specific kind of molecule.

Subspecies (sub′spē-shēz) Race; group of individuals of one species that have distinctive characteristics but can interbreed with other members of the species.

Substrate (sub′strāt) Molecule involved in chemical reaction on which an enzyme acts to speed reaction rate.

Succession The slow, orderly, ecological progression of changes in the types of vegetation in the development of any particular area.

Sucrose (sū′krōs) Disaccharide formed of fructose and glucose.

Surface tension Cohesion of water molecules at water's surface caused by hydrogen bonding between molecules.

Suspension A liquid mixture in which solid particles, aggregates of molecules stuck together, are distributed throughout the liquid. These particles will settle to the bottom of the mixture if the suspension is allowed to stand undisturbed.

Swim bladder An air sac found in bony fish that enables them to control their buoyancy.

Symbiosis (sim-bi-ō′sis) A close living association of two organisms of different species.

Sympathetic nervous system A subdivision of the autonomic nervous system that is stimulated during times of stress. It has an inhibitory effect on digestion but a stimulatory effect on other functions.

Sympatric (sim-pat′rik) Pertaining to two or more populations inhabiting the same area.

Synapse (sin′aps) The region of contact between two successive neurons.

Synapsis (si-nap′sis) The intertwining of the chromatids of paired chromosomes during prophase 1 of meiosis.

Syngamy (sin′ga-mē) The joining of the two haploid chromosome complements in fertilization.

Systemic circulatory system (sis-tem′ik ser′kū-la-tō-rē) The blood vessels that carry blood throughout the body (except to the lungs).

Systole (sis′to-lē) Contraction of the heart.

Taiga (tī′ga) Boreal coniferous forest biome characterized by cool climate and evergreen vegetation.

Taproot system A single large root from which may grow many lateral roots; typical of dicots.

Taxis (tak′sis) Innate behavioral movement directed toward or away from the direction from which a stimulus is received.

Telophase (tel′o-fāz) The last stage of mitosis, in which chromosomes elongate, the spindle disappears, the nuclear membranes form, and cytokinesis occurs. Also the final stage in both division sequences of meiosis. In telophase 1, the chromosome number is haploid, but the chromosomes are double-stranded. In telophase 2, the chromosomes are single-stranded.

Temperature A measure of heat energy, the kinetic energy of molecules.

Territory A definite area that is defended against intruders.

Testis (tes′tis) The male gonad, which produces sperm and the male hormone testosterone.

Testosterone (te-stos′te-rōn) Male sex hormone secreted by the testes and adrenal cortex; stimulates development of male secondary sex characteristics and normal reproductive functions; also affects functioning of skeletal muscle, adolescent growth, and sex drive.

Thalamus (thal′a-mus) A part of the vertebrate forebrain just posterior to the cerebrum; an important intermediate and integrative center between all other parts of the nervous system and the cerebrum.

Thallus (thal′us) A simple plant body not differentiated into root, stem, or leaves.

Thermotaxis (ther-mō-tak′sis) Behavioral response to temperature.

Thigmotropism (thig-moh′trō-pizm) Sensitivity of a plant to touch.

Thoracic duct (tho-ras′ik) Duct through which contents of lymph system are emptied into the superior vena cava.

Threshold The quantitative value that must be exceeded if a cell is to generate an action potential in response to a stimulus.

Thromboplastin (throm-bō-plas′tin) A lipid-protein complex in platelets that, when released into blood plasma, triggers the blood-clotting process.

Thrombus (throm′bus) Blood clot.

Thyrocalcitonin (thī-rō-kal-si-tō′nin) Hormone secreted by the thyroid gland, which inhibits the movement of calcium from bone to blood.

Thyroid gland (thī′roid) Gland located in the lower front portion of the neck, which secretes hormones (thyroid hormone, thyrocalcitonin) that regulate metabolism, maturation, and the movement of calcium from bones to the bloodstream.

Thyroid-stimulating hormone (TSH) Hormone secreted by the anterior lobe of the pituitary, which stimulates the production of thyroid hormone by the thyroid gland.

Tissue Sheets or aggregates of specialized cells that are similar in structure and function.

Topsoil The surface layer of the terrestrial environment containing organic matter, water, mineral matter, and air in proportions varying with locality.

Trachea (trā′kē-a) The cartilaginous tube through which air passes to reach the lungs; the windpipe.

Tracheae (trā′kē-ē) Branching air tubes in which gas exchange occurs in insects.

Tracheids (trā′kē-idz) Water-conducting tubes, which are part of xylem tissue.

Tracheophytes (trā′kē-o-fītz) Plants characterized by the presence of specialized conducting tissues, xylem and phloem, in their roots, stems, and leaves; the vascular plants.

Transfer RNA (tRNA) Short-stranded RNA, which bonds with amino acids and carries them to the ribosomes for protein synthesis.

Translocation Attachment of chromosome segment to nonhomologous chromosome.

Transpiration Loss of water molecules through evaporation from leaves in plants.

Tropism (trō′pizm) The movement of a plant in response to the nature and direction of a specific environmental stimulus.

Troposphere (trōp′o-sfēr) The layer of the atmosphere that reaches an elevation of approximately 10 kilometers above sea level.

Tuber (tū′ber) The enlarged end portion of a rhizome.

Tundra (tun′dra) Biome characterized by flat marshy land, mossy low vegetation, and arctic or mountain climate.

Turgor (ter′ger) The taut, pressurized condition of a plant cell due to absorption of water to the limit of distensibility of the cell wall. A homeostatic mechanism, since it offers some protection to a plant cell in a hypotonic environment.

Turgor pressure Force created by absorbed water molecules pressing outward from the cell interior.

Ultrafiltration A renal process by which the fluid component of the blood, with its salts, wastes, and other chemical constituents, but not the larger protein components, are filtered into the nephron by the hydrostatic pressure in the glomerulus.

Uniformitarianism (yū-ni-form-i-tā-ri-an-izm) The geological theory, first proposed in the eighteenth century that the Earth was formed and molded through a gradual day-by-day process according to natural laws.

Urea (yū-rē′a) Nitrogen-containing compound, an end product of nitrogen metabolism.

Ureter (yū′re-ta) Tube leading from the kidney to the bladder.

Urethra (yū-rēth′ra) Tube carrying urine from the bladder to the exterior in mammals.

Uric acid (yū′rik) Nitrogen-containing compound, the end product of nucleic acid metabolism.

Uterus (yūt′e-rus) A pear-shaped enlargement of the lower end of the oviduct. It is divided anatomically into a body, or corpus, and a neck, or cervix. The lining of the corpus undergoes hormonally induced cyclic changes.

Vacuole (vak′yu-wōl) A sac found in the cytoplasm bounded by the unit membrane and filled only with an aqueous solution or suspension.

Vagina (va-jīn′a) An elastic, glandular, and convoluted canal between the cervix of the uterus and the external genitalia.

Valence (vā′lens) The capacity of an element to gain or lose electrons through bonding chemically with other elements.

Vasculature (vas′kū-lā-cher) The arteries, veins, and capillaries of the circulatory system as a group.

Vasopressin (antidiuretic hormone, ADH) (vā-sō-pres′in; an-ti-dī-ya-ret′ik) Hormone secreted by the posterior pituitary, which regulates water reabsorption in the kidneys.

Vegetative propagation (vej′e-tā-tiv pro-pa-gā-shun) An asexual method of reproducing a complete new plant from a part of another plant of the same species.

Vein (vān) Support and transport tissue in the mesophyll of a leaf composed of xylem and phloem cells. Also, a blood vessel that transports blood toward the heart.

Venae cavae (vē-nē cā-vē) The two great veins that empty into the right atrium of the heart.

Ventricle (ven′tri-kl) One of the two lower chambers of the heart, which pump blood away from the heart.

Vessel elements Long, open-ended tubes connected end-to-end to form water-conducting pipelines that are found in xylem tissue.

Villi (vil′ī) Fingerlike projections that increase the absorptive surface of the small intestine to facilitate the absorption of nutrients.

Vitamin (vī′ta-min) An organic compound essential in the diet in small amounts.

Wax One of the four classes of lipids.

Xylem (zī′lem) Plant tissue composed of tracheids, vessel elements, and fiber tracheids, specialized to conduct water.

Yolk The stored food substances in the egg cell.

Zygote (zī′gōt) The diploid cell formed by the union of male and female gametes.

PHOTOGRAPH ACKNOWLEDGMENTS

(numbers refer to figure numbers)

Chapter 1 opening photograph, page 1, Marshall Henrichs; 1.2, Klaus Schwarz, Veterans Administration Hospital, Long Beach; 1.3, Walter Dawn/Photo Researchers, Inc.; page 30, Jerome Gross, Harvard Medical School; page 46, Gregory Antipa, Wayne State University; 4.1, The Bettmann Archive; 4.2A, Armed Forces Institute of Pathology (neg. no. 66-1836-1); 4.2B, Unitron Instrument Co., Newton Highlands, Mass.; 4.3, Stuart J. Coward, University of Georgia; 4.7, Samuel F. Conti, University of Kentucky; 4.9, Omikron; 4.10, George E. Palade, Yale University School of Medicine; 4.11A, Omikron; 4.11C, Don W. Fawcett and Susumu Ito, Harvard Medical School; page 76, R. B. Hoit/Photo Researchers, Inc.; 5.6, Lewis K. Shumway, Washington State University; page 108, Karen Tweedy-Holmes/Animals Animals © 1976; 6.1, Karl Maslowski/Photo Researchers, Inc.; 6.2, Jeremy Pickett-Heaps, University of Colorado, Boulder; 6.5, Hugh Spencer/Photo Researchers, Inc.; 6.6, Lovett Williams, Jr./Photo Researchers, Inc.; 6.7, Riker Laboratories, Northridge, Calif.; 6.8, Len Lee Rue III/Tom Stack & Associates; 6.10, Courtesy of the American Museum of Natural History; page 136, Russ Kinne/Photo Researchers, Inc.; 7.2, Runk/ Schoenberger—Grant Heilman, Inc.; 7.6, Gary Breckon, University of Wisconsin; 7.7A, Grant Heilman; 7.7B, Courtesy of Carolina Biological Supply Company; 7.12, Des Bartlett/Photo Researchers, Inc.; 7.18, Peck-Sun Lin, Tufts University School of Medicine; 7.20, Alex Ferenczy, Jewish General Hospital, Montreal; page 168, H. Armstrong Roberts; 8.7, George Rodger/Magnum Photos; 8.19, Len Lee Rue III/Tom Stack & Associates; page 194, Richard Trump/Omikron; 9.11, Stephen Dalton/Photo Researchers, Inc.; page 214, E. R. Lewis, T. E. Everhart, Y. Y. Zeevi, University of California, Berkeley; 10.1, Runk/Schoenberger—Grant Heilman, Inc.; 10.2, Peck-Sun Lin, Tufts University School of Medicine; 10.13, Courtesy of the American Museum of Natural History; 10.14, M. W. F. Tweedie/Photo Researchers, Inc.; 10.16, Heather Angel/Biofotos; page 244, Toni Angermayer/Photo Researchers, Inc.; 11.4, Ylla/Photo Researchers, Inc.; 11.5, Baron Hugo von Lawick, © National Geographic Society; 11.6, Tom Stack; 11.9, Frank Stevens/Photo Researchers, Inc.; 11.10, Charles Summers, Jr./Tom Stack & Associates; 11.11A & B, Allan Cruickshank/Photo Researchers, Inc.; 11.11C, A. A. Francesconi/Photo Researchers, Inc.; 11.11D, Karl Maslowski/Photo Researchers, Inc.; 11.12, Ylla/Photo Researchers, Inc.; 11.13, J. J. Languepin/Photo Researchers, Inc.; 11.14, Karl Maslowski/Photo Researchers, Inc.; page 268, Robert Mitchell/Tom Stack & Associates; 12.1A, Omikron; 12.1B, Philip Feinberg/Omikron; 12.2, Courtesy of Carolina Biological Supply Company; 12.11, Lynwood M. Chace/Photo Researchers, Inc.; 12.12, Len Lee Rue III/Photo Researchers, Inc.; 12.17A, From E. Anderson, *J. Cell Biol.* 37:514–539 (1968), by courtesy of the author and permission of the Rockefeller University Press; 12.17B, Don Fawcett, Harvard Medical School; page 290, William T. Jackson, Dartmouth College; 13.1, Peck-Sun Lin, Tufts University School of Medicine; 13.5, Courtesy of Ward's Natural Science Establishment, Inc.; 13.7, Courtesy of Carolina Biological Supply Company; page 304, Des Bartlett/Photo Researchers, Inc.; 14.1, The Bettmann Archive; 14.7, Lynn McLaren/Photo Researchers, Inc.; 14.8, Peck-Sun Lin, Tufts University School of Medicine; 14.9, Geoffrey Gove; 14.11, The Bettmann Archive; page 334, Authenticated News International; 15.1, Wide World Photos; 15.3, The Bettmann Archive; 15.4, R. E. Franklin and R. Gosling, *Nature* 171: 740 (1953); 15.5, Wide World Photos; 15.10, B. A. Hamkalo, University of California, Irvine, in *Int. Rev. Cytol.* 33:1 (1972); page 360, Robert Mitchell/Tom Stack & Associates; 16.4, Carnegie Institution, Dept. of Embryology, Davis Division; 16.8, Courtesy of Carolina Biological Supply Company; 16.15, Richard H. Falk and Ernest M. Gifford, Jr., University of California, Davis; 17.1, The Bettmann Archive; 17.2, Courtesy of the American Museum of Natural History; 17.3, J. Cooke/Animals © 1976; 17.7, Runk/Schoenberger—Grant Heilman, Inc.; 17.9A, Robert Ellison/Photo Researchers, Inc.; 17.9B, Jack Dermid/Photo Researchers, Inc.; 17.12A, Len Lee Rue III/Tom Stack & Associates; 17.12B, Karl Weidmann/Photo Researchers, Inc.; page 410, Courtesy of the American Museum of Natural History; 18.1, The Bettmann Archive; 18.2A, Len Lee Rue III/Photo Researchers, Inc.; 18.2B, New York Zoological Society Photo; 18.5, Dr. Charles C. Brinton, Jr., and Judith Carnahan, University of Pittsburgh; 18.6, Lee D. Simon, Rutgers University; 18.7, Gregory Antipa, Wayne State University; 18.8, Runk/Schoenberger—Grant Heilman, Inc.; 18.9, Runk/Schoenberger—Grant Heilman, Inc.; page 430, Jack Dermid/Photo Researchers, Inc.; 19.3, Grant Heilman; 19.5, Courtesy of Carolina Biological Supply Company; 19.6, Runk/Schoenberger—Grant Heilman, Inc.: 19.7, F. B. Grunzweig/Photo Researchers, Inc.; 19.8, Grant Heilman; 19.10, Grant Heilman; 19.12, Georg Gerster/Photo Researchers, Inc.; 19.14A, E. R. Degginger; 19.14B, Beidler/Omikron; 19.16, Robert W. Mitchell/Tom Stack

& Associates; 19.17, Grant Heilman; 19.18, Hugh Spencer/Photo Researchers, Inc.; page 456, Carl Roessler, The Sea Library; 20.5, Runk/Schoenberger—Grant Heilman, Inc.; 20.7, Runk/Schoenberger—Grant Heilman, Inc.; 20.9, Runk/Schoenberger—Grant Heilman, Inc.; 20.11, Grant Heilman; 20.13, Tom Brakefield/Tom Stack & Associates; 20.15, Charles Summers, Jr./Tom Stack & Associates; page 476, Hal Harrison/Grant Heilman, Inc.; 21.1A, Len Lee Rue III/Tom Stack & Associates; 21.1B, Charles Summers, Jr./Tom Stack & Associates; 21.12, Joe Munroe/Photo Researchers, Inc.; 21.13, Jerome Wexler/Photo Researchers, Inc.; 21.14, Joseph Branney/Tom Stack & Associates; 21.17, Georg Gerster/Photo Researchers, Inc.; 21.18, Wide World Photos; page 502, Jen and Des Bartlett/Photo Researchers, Inc.; 22.2, Jerry L. Hout/Photo Researchers, Inc.; 22.3, Len Lee Rue III/Photo Researchers, Inc.; 22.4, Allan Cruickshank/Photo Researchers, Inc.; 22.6, Russ Kinne/Photo Researchers, Inc.; 22.8, Allan Cruickshank/Photo Researchers, Inc.; 22.10, Runk/Schoenberger—Grant Heilman, Inc.

COLOR PORTFOLIO

Diatoms, Bruce Coleman, Inc.; *Octopus,* Grant Heilman; *Starfish & crab,* Runk/Schoenberger—Grant Heilman, Inc.; *Flower mantis,* M. P. L. Fogden/Bruce Coleman, Inc.; *Arrow frog,* Alan Blank/Bruce Coleman, Inc.; *Goldfinch,* Thase Daniel/Bruce Coleman, Inc.; *Tiger,* Bruce Coleman, Inc.; *Hummingbird,* Bob and Clara Calhoun/Bruce Coleman, Inc.; *Rosary pea,* James H. Carmichael, Jr./Bruce Coleman, Inc.; *Rock python,* E. R. Degginger; *Angelfish,* E. R. Degginger; *Mushrooms,* E. R. Degginger; *Mandrill,* Bob Putzbach/Tom Stack & Associates; *Frigate,* Lloyd McCarthy/Tom Stack & Associates; *Killer whale,* William Eastman III/Tom Stack & Associates; *Spider,* Larry C. Moon/Tom Stack & Associates.

B&W PORTFOLIO 1

1.1, Jean Paul Revel, California Institute of Technology; 1.2, Barbara J. Panessa, New York University Medical Center; 1.3, Courtesy of Dr. Harry Ohanian, NCI, and Coates and Welter Instrument Corporation; 1.4, Howard J. Arnott, University of Texas, Arlington; 1.5, Albert L. Jones, Veterans Administration Hospital, San Francisco; 1.6, William A. Jensen, University of California, Berkeley; 1.7, From K. R. Porter and M. A. Bonneville, *Fine Structure of Cells and Tissues,* 4th edition (Lea and Febiger, 1973); 1.8, E. B. Small.

B&W PORTFOLIO 2

2.1, Coates and Welter Instrument Corporation; 2.2, Peck-Sun Lin, Tufts University School of Medicine; 2.3, © Carroll H. Weiss, RBP, 1973, Camera M. D. Studios, Inc.; 2.4, Graziadai/Omikron; 2.5, Susumu Ito, Harvard Medical School; 2.6, Susumu Ito, Harvard Medical School; 2.7, Courtesy Carl Zeiss, Inc., New York; 2.8, Coates and Welter Instrument Corporation.

B&W PORTFOLIO 3

3.1, Mia J. Tegner, Scripps Institution of Oceanography; 3.2, William T. Jackson, Dartmouth College; 3.3, E. B. Small, G. A. Antipa, and D. S. Marszalek; 3.4, Frederick C. Steward, *Growth and Organization of Plants* (Addison-Wesley, 1968).

INDEX

Numerals in italics indicate references to illustrative material.